SI UNITS USED IN MECHA

Quantity	Unit	
(*Base Units*)		
Length	meter*	m
Mass	kilogram	kg
Time	second	s
(*Derived Units*)		
Acceleration, linear	meter/second^2	m/s^2
Acceleration, angular	radian/second^2	rad/s^2
Area	meter^2	m^2
Density	kilogram/meter^3	kg/m^3
Force	newton	N (= kg·m/s^2)
Frequency	hertz	Hz (= 1/s)
Impulse, linear	newton-second	N·s
Impulse, angular	newton-meter-second	N·m·s
Moment of force	newton-meter	N·m
Moment of inertia, area	meter^4	m^4
Moment of inertia, mass	kilogram-meter^2	kg·m^2
Momentum, linear	kilogram-meter/second	kg·m/s (= N·s)
Momentum, angular	kilogram-meter^2/second	kg·m^2/s (= N·m·s)
Power	watt	W (= J/s = N·m/s)
Pressure, stress	pascal	Pa (= N/m^2)
Product of inertia, area	meter^4	m^4
Product of inertia, mass	kilogram-meter^2	kg·m^2
Spring constant	newton/meter	N/m
Velocity, linear	meter/second	m/s
Velocity, angular	radian/second	rad/s
Volume	meter^3	m^3
Work, energy	joule	J (= N·m)
(*Supplementary and Other Acceptable Units*)		
Distance (navigation)	nautical mile	(= 1.852 km)
Mass	ton (metric)	t (= 1000 kg)
Plane angle	degrees (decimal)	°
Plane angle	radian	—
Speed	knot	(1.852 km/h)
Time	day	d
Time	hour	h
Time	minute	min

*Also spelled *metre*.

Reprinted, by permission, from J. L. Merian "Conversion Factors" and "SI Units used in Mechanics," in *Engineering Mechanics, Volume 1 Statics* Front Matter, Wiley, New York, 1978

APPROXIMATE NUMERICAL VALUES USEFUL IN STRESS ANALYSIS

Atmospheric pressure = 14.7 psi = 100 kPa
A stress of 10 ksi is 70 MPa
Elastic modulus of aluminum = $10(10)^6$ psi = 70 GPa
Elastic modulus of steel = $30(10)^6$ psi = 210 GPa
Weight density of concrete = 150 lb/ft^3 = 24 kN/m^3
Weight density of water = 62 lb/ft^3 = 10 kN/m^3
Mass density of water = 1000 kg/m^3

Library of Congress Cataloging in Publication Data:

Cook, Robert Davis.
Concepts and applications of finite element analysis.

Includes bibliographical references and index.
1. Structures, Theory of. 2. Finite element
method. I. Title.
TA646.C66 1981 624.1'71 80-26255
ISBN 0-471-03050-3

Printed in the United States of America

10 9 8 7 6 5 4 3 2 1

Concepts and Applications of Finite Element Analysis

Second Edition

Robert D. Cook

Department of Engineering Mechanics

University of Wisconsin - Madison

John Wiley & Sons

New York · Chichester · Brisbane · Toronto

Preface

The finite element method is a generally applicable method for getting numerical solutions. Problems of stress analysis, heat transfer, fluid flow, electric fields, and others have been solved by finite elements. This book emphasizes stress analysis and structural mechanics. Other areas are treated in a way that is easy for stress analysts to understand. The formulation and computation procedures of finite elements are much the same in all areas of application.

This text is introductory and is inclined toward practical application. Theory is presented as needed. The book contains enough material for a two-semester course sequence. By studying this book, an engineer can learn to use finite elements with confidence and effectiveness, and those who want to do advanced work will have a sound physical understanding from which to proceed.

The background for the book is as follows. Undergraduate courses in statics, dynamics, and mechanics of materials must be mastered. Mathematically, little is needed but the ability to differentiate and integrate sines, cosines, and polynomials. I presume knowledge of matrix operations such as multiplication, transposition, differentiation, and the meaning of an inverse (the calculation procedure for inversion is not essential). Students should be competent in Fortran programming, to the extent of being able to use subroutines, COMMON blocks, and storage such as disc files. Other areas—theory of elasticity, plates and shells, energy methods, numerical analysis—are desirable but not essential. Fortunately, we seldom call on these areas and, when we do, it is usually for their elementary concepts.

The following list more fully describes the content and orientation of the book.

- It is more concerned with analysis of continua than with special methods for framed structures. However, truss and beam elements are often used as simple but useful vehicles for explanations.
- Linear static analysis is emphasized.
- Elements discussed in detail are based on assumed displacement fields, are not restricted to a special shape, and have no "excessive" nodal continuities. Of the multitude of elements available, only a few are discussed with emphasis on the isoparametric type.
- To show unambiguously the steps of certain processes, Fortran coding is given when it is simple, helpful, and not very long. This coding may not be the most recent or most efficient. Complete programs are available from other sources.
- Topics are emphasized that seem useful and well enough established to be enduring. Less durable topics, such as explanations of currently popular computer programs, are omitted.

I have tried to avoid second-edition prolixity. First-edition topics not appropriate to the level and orientation of the book have been omitted except for citations of the literature. New topics have been added only if clearly suitable (see Chapters 14, 17, and 18). There are more numerical examples. Less obvious are rearrangement of sections and improvement in wording. Also incorporated are useful thoughts culled over the years from the literature and from the comments of students and colleagues. Many homework problems have been added; answers are given in the back of the book. The problems illustrate principles and procedures and foster insight. Most require neither a computer nor extensive numerical calculations.

The practical success of the finite element method depends on a reliable computer program. Students in a first course say that a programming project is an excellent learning device. They prefer to write and test a simple but complete program instead of coding subroutines. Sample elements and situations on which to base a programming project are listed here. Each can invoke Gauss quadrature for element stiffness formation if so desired. For students with interests outside structural mechanics, nonstructural problems of similar extent are possible.

1. Standard beam element, two d.o.f. (degrees of freedom) per node.

2. Same as item 1, but add internal d.o.f.

3. Shear beam on an elastic foundation, one d.o.f. per node. (See Fig. 9.5.2. For the beam, consider only w and energy stored by γ_{xy}.)

4. Beam as a degenerate isoparametric plate element (Figs. 9.5.1 or 9.5.2).

5. Tapered bar element, two nodal and one nodeless d.o.f.

6. Plane disc of constant or variable thickness, with annular elements and torsional loads only.

7. Same as item 6, but add nodeless d.o.f.

8. Same as item 6, but use radial loads only.

9. Same as item 6, but add an elastic foundation and allow only transverse shear stiffness and lateral load.

10. Same as item 5, but consider torsional action only.

11. Plane frame, three d.o.f. per node.

12. Same as item 11, but use the element of part 4.

13. Rectangular elements, one d.o.f. per corner, for the harmonic equation (soap film, seepage flow, and so on).

14. Application of Eq. 4.10.4 to a tapered bar with a distributed axial load.

A programming project can also be assigned in a second course. A possible project involves Fourier series components of loading and superposition of the separate solutions. Examples include plane stress problems that involve circular regions with annular elements and asymmetric loads, and finite strip analysis of simply supported plates. A project that involves natural frequencies of vibration or dynamic response is also good. Alternatively, if a general purpose program with interactive graphics is available, students will profit by using it to solve a practical problem.

I am grateful to the students whose questions led to better explanations and improved homework problems. The substance of the book comes mostly from published papers. Their authors have my appreciation. In language matters, I have become inordinately sensitive to the words "that" and "which", but thank the Wiley editors for teaching me the distinction. Rules given by Strunk and White in *The Elements of Style,* especially Rule 13, helped tighten the prose. Of the six typists who worked on the manuscript, Pat Klitzke did quality work quickly.

Robert D. Cook

1 Intro, pre exercises

2 From CAD to model... Formulation, Method selection

3 Major Concepts of the model. Ele formulation, Ele types

4 Discretization Error Ctrl: Convergence

5 MESH. Techniques, compatibility, common pbls

6 Modeling Process: steps, techniques, exercises

7 TYPES: Thermal
nonlinear
Modal
Buckling
Dynamic

8 Design optimization Topological
Structural

9 In the Design Process
Geom differences {CAD v FEA}
Software Integration (C v F)

Chapters from Paul Kurowski's book

Contents

Notation

This is a list of principal symbols. Locally used notation and modifications (as by addition of a subscript) are defined where used. Similarly, a symbol that has different meanings in different contexts is defined where it is used. Matrices are denoted by boldface type.

MATHEMATICAL SYMBOLS

$[\ \ ]$	A rectangular or square matrix.
$\lceil\ \ \rfloor$	A diagonal matrix.
$\lfloor\ \ \rfloor$	A row vector.
$\{\ \ \}$	A column vector. *Note*. $\begin{Bmatrix} u \\ v \end{Bmatrix} \equiv \{u\ v\}$.
$[\ \ ]^{-1}$	Matrix inverse.
$[\ \ ]^T$	Matrix transpose (also applies to row and column vectors).
$[\ \ ]^{-T}$	Inverse transpose; $[\ \ ]^{-T} \equiv ([\ \ ]^{-1})^T \equiv ([\ \ ]^T)^{-1}$.
	Note. The foregoing brackets and braces may be omitted from submatrices and from the separate matrices of a matrix product that is bracketed.
$\cdot$	Time differentiation; for example, $\dot{u} = du/dt$, $\ddot{u} = d^2u/dt^2$.
,	Partial differentiation if the following subscript(s) is literal; for example, $w_{,x} = \partial w/\partial x$, $w_{,xy} = \partial^2 w/\partial x \partial y$.
$\bar{\ }$	Amplitude; for example, $u = \bar{u} \sin \omega t$. (Other meanings are numerous.)
$\left\{\dfrac{\partial \mathbf{\Pi}_p}{\partial \mathbf{a}}\right\}$	Represents $\left\{\dfrac{\partial \mathbf{\Pi}_p}{\partial a_1}\ \dfrac{\partial \mathbf{\Pi}_p}{\partial a_2}\ \cdots\ \dfrac{\partial \mathbf{\Pi}_p}{\partial a_n}\right\}$, where $\mathbf{\Pi}_p$ is a scalar function of parameters $a_1, a_2, \ldots, a_n$.

LATIN SYMBOLS

A	Area.
$[\mathbf{A}]$	Relates $\{\mathbf{d}\}$ to $\{\mathbf{a}\}$; $\{\mathbf{d}\} = [\mathbf{A}]\{\mathbf{a}\}$.
$\{\mathbf{a}\}$	Generalized coordinates.
B	Semibandwidth of a matrix.
$[\mathbf{B}]$, $[\mathbf{B}_a]$	The "strain-displacement" matrix (Section 4.3).
C_n	Continuity of degree n (Section 4.2).
$C(\mathbf{K})$	Condition number of $[\mathbf{K}]$ (Section 15.6).
$[\mathbf{C}]$	Damping matrix or constraint matrix.
D	Displacement.
d.o.f.	Degree (or degrees) of freedom.
$\{\mathbf{D}\}$	Nodal d.o.f. of a structure (global d.o.f.).

$\{\mathbf{d}\}$	Nodal d.o.f. of an element.
$[\mathfrak{D}]$	Flexural rigidity matrix of a plate.
e	Elongation.
E, E_s	Elastic modulus (E), secant modulus (E_s) (Fig. 13.10.1).
$[\mathbf{E}]$	Matrix of elastic stiffnesses (Section 1.5).
$\{\mathbf{F}\}$	Body forces per unit volume.
$\{\mathbf{f}\}$	A displacement field; $\{\mathbf{f}\} = \{u\ v\ w\}$ in 3-space.
G	Shear modulus.
$[\mathbf{H}]$	Used in the way that $[\mathbf{Q}]$ is used.
I	Moment of inertia of a beam.
$[\mathbf{I}]$	The unit matrix (also called the identity matrix).
J	Determinant of $[\mathbf{J}]$, known as the Jacobian.
$[\mathbf{J}]$	The Jacobian matrix.
k	Spring stiffness. Thermal conductivity (Chapter 17).
$[\mathbf{K}]$	Structure (global) stiffness matrix.
$[\mathbf{k}]$	Element stiffness matrix (conductivity matrix in Chapter 17).
$[\mathbf{K}_\sigma]$	Structure (global) stress stiffness matrix.
$[\mathbf{k}_\sigma]$	Element stress stiffness matrix.
L, ℓ	Length.
ℓ, m, n	Direction cosines.
$[\mathbf{M}]$	Structure (global) mass matrix.
$[\mathbf{m}]$	Element mass matrix.
M, N	Bending moment (M), membrane force (N).
M BAND	Same as B. (First used in Chapter 2.)
N, NEQ	Number of equations. (First used in Chapter 2.)
N DOF	Number of d.o.f. per node. (First used in Chapter 2.)
NUMEL	Number of elements in a structure. (First used in Chapter 2.)
NUMNP	Number of nodes in a structure. (First used in Chapter 2.)
$[\mathbf{N}], \lfloor \mathbf{N} \rfloor$	Matrix of shape functions; $\{\mathbf{f}\} = [\mathbf{N}]\{\mathbf{d}\}$.
O	Order; for example, $O(h^2)$ = a term of order h^2.
$[\mathbf{O}], \{\mathbf{O}\}$	Null matrix, null vector.
p_i, q_i	Concentrated forces on node i (Chapters 1 and 2).
P	Force.
$\{\mathbf{P}\}$	Vector of externally applied loads on structure nodes.
$[\mathbf{Q}], \{\mathbf{Q}\}$	Matrix of various uses. Defined where used.
q	Lateral load (surface or line).
R	Residual (Section 15.9; Chapter 18).
$\{\mathbf{R}\}$	Total load on structure nodes; $\{\mathbf{R}\} = \{\mathbf{P}\} + \Sigma\,\{\mathbf{r}\}$.
$\{\mathbf{r}\}$	Forces applied by element to nodes (Eq. 4.3.5).
$\{\bar{\mathbf{r}}\}$	$\{\bar{\mathbf{r}}\} = -\{\mathbf{r}\}$ (Section 2.4).
S	Surface.
s, t	Coordinate directions, usually Cartesian.
T	Temperature
t	Thickness. Time.

[**T**]	Transformation matrix.
U, U_0	Strain energy, strain energy per unit volume.
u, v, w	Displacement components.
V	Volume.
x, y, z	Cartesian coordinates.
x', y', z'	Local Cartesian coordinates.

GREEK SYMBOLS

α	Coefficient of thermal expansion.
α, β, γ	Area coordinates (Section 7.9).
β	Angle, relaxation factor, foundation modulus, and the like.
$[\mathbf{\Gamma}]$	The Jacobian inverse; $[\mathbf{\Gamma}] = [\mathbf{J}]^{-1}$.
Δ	Small change operator; for example, Δt is a time increment.
δ	Virtual operator; for example, δu is a virtual displacement.
$\{\boldsymbol{\epsilon}\}$, $\{\boldsymbol{\epsilon}_0\}$	Strains, initial strains (Section 1.6).
θ	Angle. Circumferential coordinate.
$\{\boldsymbol{\kappa}\}$	Vector of curvatures (as in plate bending).
λ	An eigenvalue. A Lagrange multiplier.
ν	Poisson's ratio.
ξ, η, ζ	Isoparametric coordinates (Chapter 5).
$\mathbf{\Pi}$	A functional. ($\mathbf{\Pi}_p$ = total potential energy).
π	3.1415926536 . . .
ρ	Mass density.
$\{\boldsymbol{\sigma}\}$, $\{\boldsymbol{\sigma}_0\}$	Stresses, initial stresses (Section 1.6).
ϕ	A dependent variable. Meridian angle of a shell (Chapter 10).
$\{\mathbf{\Phi}\}$	Vector of surface tractions (Section 1.3).
ω	Circular frequency in radians per second.

GRAPHIC SYMBOLS

Force or displacement vector.

Moment or rotation vector (by the right-hand rule).

Spring or elastic support.

Roller support (resists positive or negative normal force).

Pinned support (resists all forces but does not resist moment).

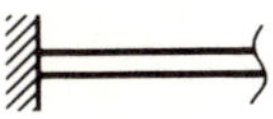

Fixed support (resists all forces and moments).

CHAPTER 1

INTRODUCTION

1.1 THE FINITE ELEMENT METHOD

The finite element method is a numerical procedure for solving a continuum mechanics problem with an accuracy acceptable to engineers.

Imagine that stresses and displacements of the structure in Fig. 1.1.1*a* must be found. Numerical answers are not found in any book. Classical methods describe the problem with partial differential equations but yield no answers because the geometry and loading are too complicated. In practice, *most* problems are too complicated for a closed-form mathematical solution. A numerical solution is required, and the most versatile method that provides it is the finite element method.

Figure 1.1.1.*b* shows a finite element model. The quadrilateral and triangular regions are *finite elements*. Black dots are *nodes* where elements are connected to one another. A *mesh* is an arrangement of nodes and elements. This particular mesh shows triangular and quadrilateral elements, some with side nodes and some with only corner nodes. (Information about mesh layout is in Section 7.13 and Appendix B.)

In a sense, finite elements are pieces of the actual structure. But we cannot convert Fig. 1.1.1*a* to Fig. 1.1.1*b* by simply making saw cuts until only wisps of material at the nodes hold the pieces together. Such a structure is greatly weakened. Also, the pieces would have strain concentrations at the nodes and would tend to overlap or separate along the saw cuts. Clearly, the actual structure does not act this way, so a finite element must deform in restricted ways. For example, if element edges are constrained to remain straight, as in Fig. 1.1.1*c*, adjacent elements will neither overlap nor separate.

To formulate an element, we must find the nodal forces that produce the various element deformation modes. We can find these forces by elementary theory for a "natural" finite element such as a beam or a bar. But for elements that the analyst defines by drawing lines on a continuum, such as those in Figs. 1.1.1*b* and 1.1.1*c*, new procedures are needed (Chapters 3, 4, 5, 8, 9, 10, 17, and 18).

The finite element method is not restricted to problems of structural mechanics (Fig. 1.1.2). Figure 1.1.2 also suggests how the smoothly varying ϕ surface can be modeled by elements of various types. If modeled by triangles, the ϕ surface is approximated by flat facets. The four- and eight-node elements display warped and curved surfaces, respectively, and better approximate the actual function. The approximation improves as more elements are used.

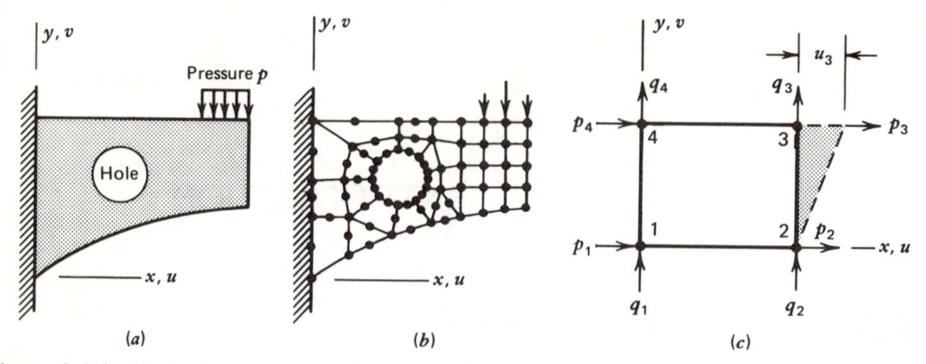

Figure 1.1.1. (*a*) A plane structure of arbitrary shape. (*b*) A possible finite element model of the structure. (*c*) A plane rectangular element, showing nodal forces p_i and q_i. The dashed line shows the deformation mode associated with x-direction displacement of node 3.

Within any triangular element in Fig. 1.1.2, ϕ is a linear function of x and y. The element's elevation and inclination are defined by the three nodal values of ϕ. No two elements need have the same elevation or slope. This sketch shows the essence of the finite element method: *piecewise approximation of a function ϕ, by means of polynomials, each defined over a small region (element) and expressed in terms of nodal values of the function.*

Two additional structural problems help convey the versatility of the method. The rocket nozzle in Fig. 1.1.3 is a solid of revolution. Each element is a toroidal ring of triangular cross section. We seek the stresses produced by temperature gradient and

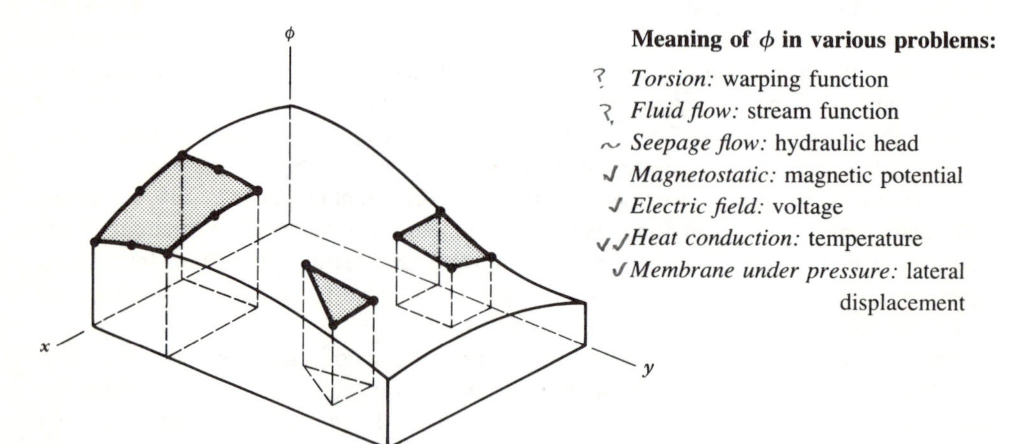

Figure 1.1.2. A continuous function $\phi = \phi(x, y)$ and typical elements that might be used to approximate it.

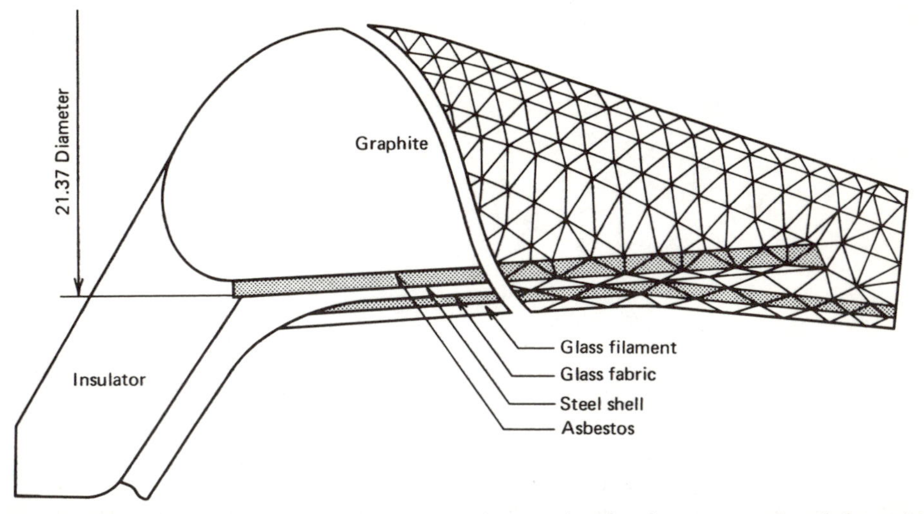

Figure 1.1.3. Cross section of a multimaterial rocket nozzle, showing construction (left portion) and possible finite element mesh (right portion). This problem was solved in the early days of finite element technology [8.1].

internal pressure. Figure 1.1.4 shows three ways of modeling an arch dam by isoparametric elements (discussed in Chapter 5). We might ask for stresses under static load or for the dynamic response to earthquake load.

Additional problems met in industrial work and solved by finite element methods appear in Figs. 1.1.5 to 1.1.9 and Appendix B.

Versatility is an outstanding feature of finite elements. The method can be applied to various problems. The region under analysis can have arbitrary shape, loads, and boundary

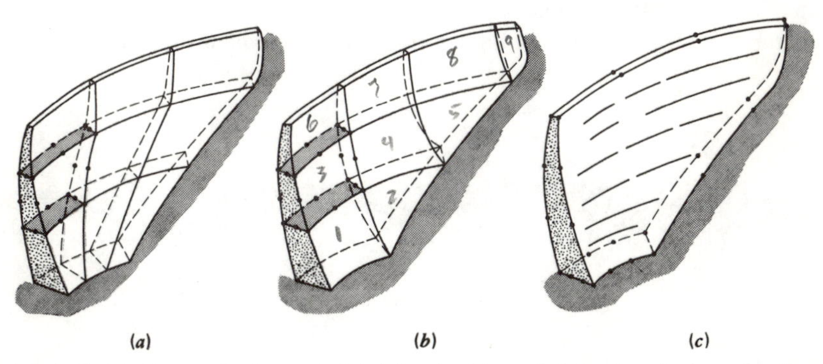

Figure 1.1.4. Half of an arch dam, modeled by (*a*, *b*) quadrilateral and triangular ''quadratic'' elements, and (*c*) a single ''cubic'' element [1.1]. Nodes of a typical element are shown by dots.

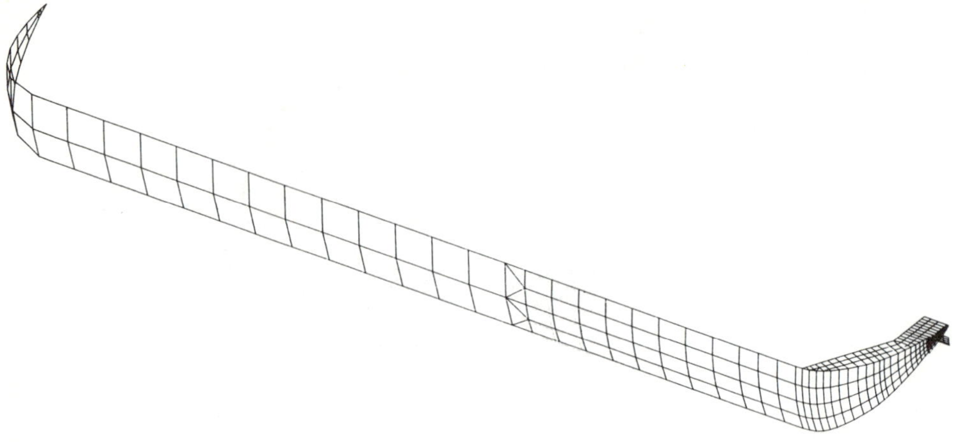

Figure 1.1.5. A special-purpose pressure vessel with reinforcements in one head. Symmetry is exploited by modeling a symmetric segment and applying appropriate boundary conditions to the cut surfaces. (*Courtesy of A. O. Smith Corp., Data Systems Division, Milwaukee, Wisconsin.*)

conditions. The mesh can mix elements of different types, shapes, and physical properties. This great versatility often is contained within a single computer program: user-prepared input data control the selection of problem type, geometry, boundary conditions, element selection, and so on.

Another feature of finite elements is the close physical resemblance between the mesh and the actual structure. The mesh is not a mathematical abstraction that is hard to visualize.

The finite element method also has disadvantages. A specific numerical result is found for a specific problem: there is no closed-form expression that permits analytical study of the effects of changing various parameters. A computer and a reliable program are essential. Experience and good engineering sense are needed to construct a good mesh. Many input data are required, and voluminous output data must be sorted and understood. However, these drawbacks are not unique to the finite element method.

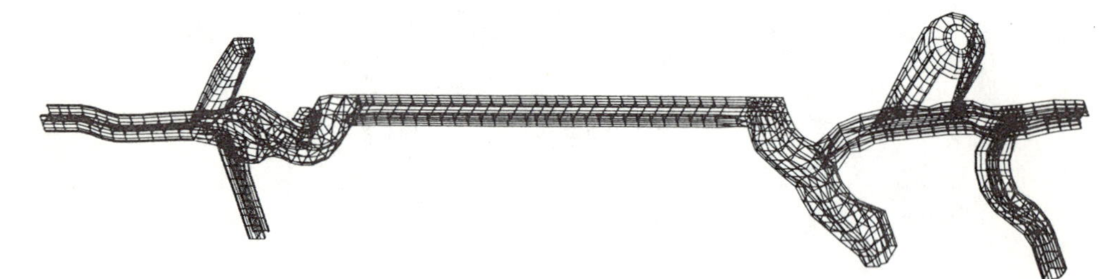

Figure 1.1.6. A detailed model of half of an automobile frame, used to find deformations, stresses, natural frequencies, and mode shapes. (*Courtesy of A. O. Smith Corp., Data Systems Division, Milwaukee, Wisconsin.*)

A summary of finite element history. Beginning in 1906 and sporadically thereafter, researchers suggested the "lattice analogy" to solve continuum problems [1.2–1.4].[1] Here the continuum is approximated by a regular mesh of elastic bars. The method seeks to capitalize on well-known methods for analysis of framed structures. In a 1941 mathematics lecture, published in 1943, Courant suggested piecewise polynomial interpolation over triangular subregions as a way to get approximate numerical solutions [1.5]. He recognized this approach as a Rayleigh-Ritz solution of a variational problem. This is the finite element method as we know it today. Courant's work was forgotten until engineers had independently developed it.

None of the preceding work was practical at the time because there were no computers to do the calculations. By 1953 engineers were writing stiffness equations in matrix notation and solving the equations with digital computers [1.6]. The classic paper by Turner, Clough, Martin, and Topp appeared in 1956 [1.7]. With this paper and others [1.8], explosive development of finite element methods in engineering began. The name "finite element" was coined in 1960 [1.9]. By 1963 [4.1] the method was recognized as rigorously sound, and it became a respectable area of study for academicians. As late

[1] Numbers in brackets indicate references listed at the back of the book.

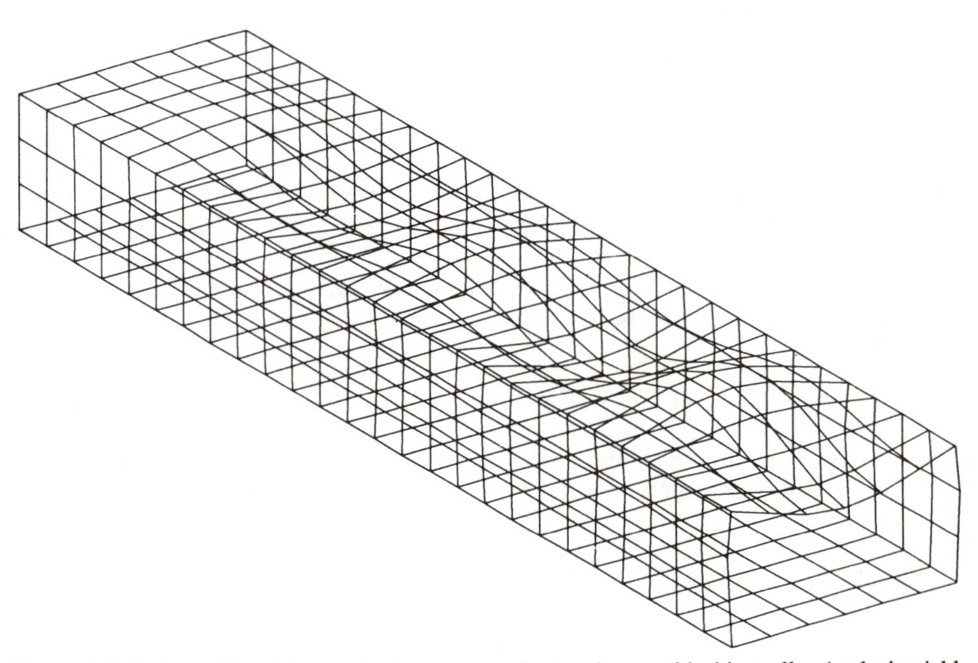

Figure 1.1.7. A local buckling mode in a rectangular box beam with thin walls. Analysis yields the buckling load as well as the buckling mode. (*Courtesy of A. O. Smith Corp., Data Systems Division, Milwaukee, Wisconsin.*)

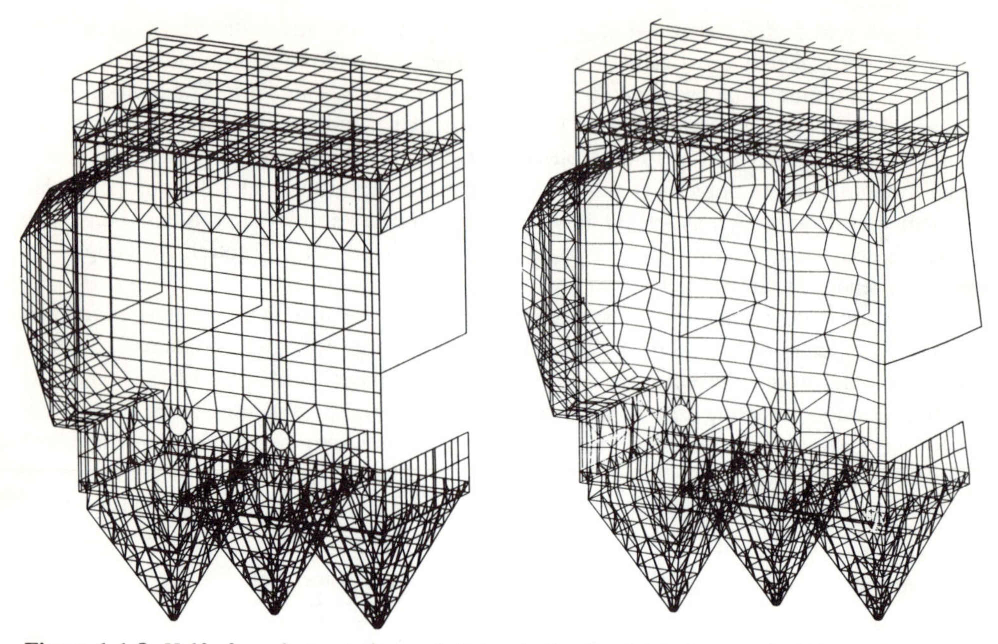

Figure 1.1.8. Half of an electrostatic precipitator, built of welded plates and stiffeners. The other half is not needed because symmetry is exploited. The right-hand figure shows an exaggeration of deformations produced by external pressure. (*Courtesy of A. O. Smith Corp., Data Systems Division, Milwaukee, Wisconsin.*)

as 1967, engineers and mathematicians worked with finite elements in apparent ignorance of one another [1.10]. (Today the two camps are aware of one another, but mathematicians are rarely interested in engineering problems, and engineers are rarely able to understand mathematicians.) Ten papers about finite elements were published in 1961, 134 in 1966, and 844 in 1971 [5.10]. By 1976, two decades after engineering applications began, the cumulative total of publications about finite elements exceeded 7000 [1.19].

1.2 SOME IMPORTANT MATRICES AND EQUATIONS

For now, we define a degree of freedom (d.o.f.) as the displacement or rotation of a node. Then, for an element with n d.o.f., we can write the equations

$$
\begin{aligned}
k_{11} d_1 + k_{12} d_2 + \cdots + k_{1n} d_n &= \bar{r}_1 \\
k_{21} d_1 + k_{22} d_2 + \cdots + k_{2n} d_n &= \bar{r}_2 \\
\vdots \qquad\qquad\qquad\quad \vdots\quad &\quad\;\; \vdots \\
k_{n1} d_1 + k_{n2} d_2 + \cdots + k_{nn} d_n &= \bar{r}_n
\end{aligned}
\tag{1.2.1}
$$

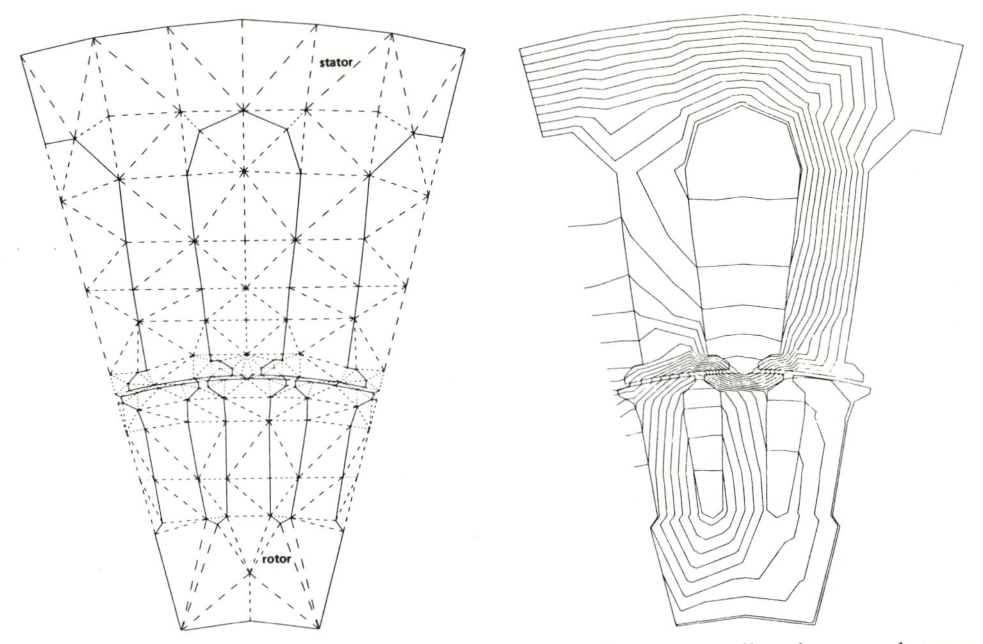

Figure 1.1.9. An induction motor. Elements model the solid parts as well as the spaces between them. Symmetry is exploited by modeling only a half-pole. The computed magnetic flux contours for zero rotor speed are shown by the right-hand figure. (*Courtesy of A. O. Smith Corp., Data Systems Division, Milwaukee, Wisconsin.*)

where d_i is the ith d.o.f. and $\bar{r}_i$ is the corresponding force or moment applied to the element. The k_{ij} are stiffness coefficients. If gathered into matrix form, Eqs. 1.2.1 are

$$[\mathbf{k}]\,\{\mathbf{d}\} = \{\bar{\mathbf{r}}\} \tag{1.2.2}$$

where $[\mathbf{k}]$ is the *element stiffness matrix*, $\{\mathbf{d}\}$ is the *element nodal displacement vector*, and $\{\bar{\mathbf{r}}\}$ is the *vector of element nodal loads*. (We are describing the *stiffness* or *displacement* method, in which displacements are the primary unknowns to be computed. Stress is a secondary variable, computed from displacements. The displacement method is the most popular form of the finite element method in structural mechanics.)

To describe the meaning of $[\mathbf{k}]$, we consider Fig. 1.2.1. Equation 1.2.2 becomes

$$\underset{4\times 4}{[\mathbf{k}]}\,\{w_1\ \theta_1\ w_2\ \theta_2\} = \{\bar{\mathbf{r}}\} \tag{1.2.3}$$

If all d.o.f. are zero but the jth and if $d_j = 1$, then we see that $\{\bar{\mathbf{r}}\} = \{\mathbf{k}_{ij}\}$, the jth column of $[\mathbf{k}]$. In words, the jth column of $[\mathbf{k}]$ is the vector of forces (and perhaps moments) that

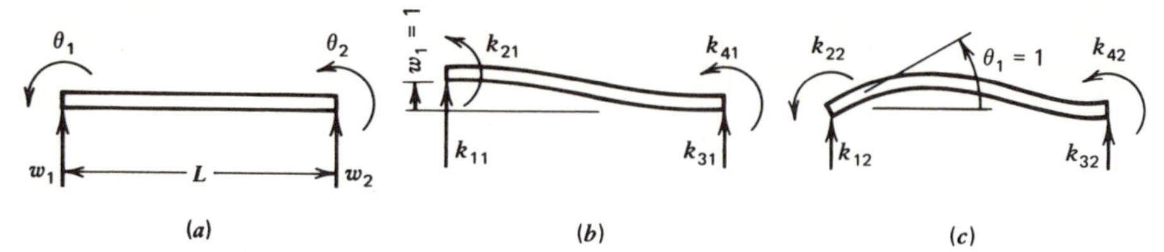

Figure 1.2.1. (*a*) The d.o.f. $\{\mathbf{d}\}$ of a standard beam element. (*b*) The deformation mode $\{\mathbf{d}\} = \{1 \ 0 \ 0 \ 0\}$ and required forces k_{i1}. (*c*) The deformation mode $\{\mathbf{d}\} = \{0 \ 1 \ 0 \ 0\}$ and required forces k_{i2}.

must be applied to the element to make $d_j = 1$ and preserve static equilibrium while the other d's are zero. This statement applies to *any* element stiffness matrix. The first two of the four possible beam modes $d_j = 1$ are shown in Figs. 1.2.1*b* and 1.2.1*c*. The "forces" k_{ij} are shown in the assumed positive sense, that is, in the same directions as the d.o.f. in Fig. 1.2.1*a*. Clearly, k_{31} and k_{32} must be negative numbers. For this simple element, beam theory yields the k_{ij} in terms of L and the bending stiffness EI.

Element nodes are assigned numbers in Fig. 1.1.1*c*. We could have used letters. Either serves as dummy labels that have no significance after elements have been put together to form a structure.

The interpretation given to columns of $[\mathbf{k}]$ also applies on the structure level. In Fig. 1.2.2*a*, by assigning a unit displacement to every node in turn, each time writing the necessary forces as a column in a 4-by-4 matrix, we find

$$\begin{bmatrix} k_1 & (-k_1 + 0) & (0 + 0) & 0 \\ -k_1 & (k_1 + k_2) & (-k_2 + 0) & 0 \\ 0 & (0 + -k_2) & (k_2 + k_3) & -k_3 \\ 0 & (0 + 0) & (0 + -k_3) & k_3 \end{bmatrix} \begin{Bmatrix} u_1 \\ u_2 \\ u_3 \\ u_4 \end{Bmatrix} = \begin{Bmatrix} -P \\ 0 \\ 0 \\ 0 \end{Bmatrix} \tag{1.2.4}$$

Each dashed line in Eq. 1.2.4 encloses an element stiffness matrix, as can be seen by considering $u_i = 1$ and then $u_{i+1} = 1$ in Fig. 1.2.2*b*. The arrangement of terms in Eq. 1.2.4 suggests—and we will see it clearly in Chapter 2—that the structure matrix can be built by adding element matrices in an overlapping fashion. Node 4 in Fig. 1.2.2*a* is fixed against motion; that is, $u_4 = 0$. Thus the equations that relate active d.o.f. are

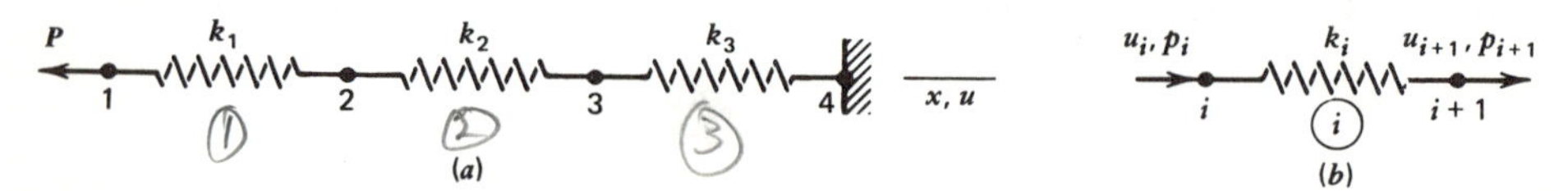

Figure 1.2.2. (*a*) A structure that has three active d.o.f. (u_1, u_2, u_3). Its "finite elements" are three linear springs of stiffness k_1, k_2, and k_3. (*b*) Nodal d.o.f. and forces of a typical element i.

$$\begin{bmatrix} k_1 & -k_1 & 0 \\ -k_1 & k_1 + k_2 & -k_2 \\ 0 & -k_2 & k_2 + k_3 \end{bmatrix} \begin{Bmatrix} u_1 \\ u_2 \\ u_3 \end{Bmatrix} = \begin{Bmatrix} -P \\ 0 \\ 0 \end{Bmatrix} \tag{1.2.5}$$

$$\text{or} \quad [\mathbf{K}]\{\mathbf{D}\} = \{\mathbf{R}\} \tag{1.2.6}$$

Matrices in Eq. 1.2.6 are named with the same terms used for matrices in Eq. 1.2.2, except that "structure" or "global" replaces "element."

Equations 1.2.6 can be solved for $\{\mathbf{D}\}$. Hence, because $\{\mathbf{d}\}$ for each element is contained in $\{\mathbf{D}\}$, deformations of all elements are known. From deformations we compute stresses. Then the solution is complete. This process is generally applicable: it is not restricted to spring or bar elements.

How large a problem can be solved? What is "large" depends on the user, the money available, and the computer resources. Today a problem with more than 5000 d.o.f. might be considered large. Problems with more than a million d.o.f. have been successfully solved.

1.3 THEORY OF ELASTICITY

Elasticity theory [1.20] helps us understand the finite element method. We summarize concepts in this section, using Cartesian coordinates.

Figure 1.3.1 shows a differential (not finite!) element. Body forces F_x and F_y have dimensions of force per unit volume and can arise from gravity, acceleration, a magnetic

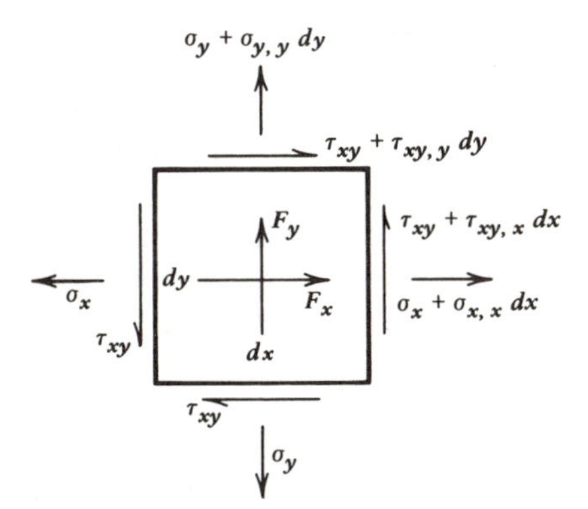

Differential equations of equilibrium:

$$\sigma_{x,x} + \tau_{xy,y} + F_x = 0$$

$$\tau_{xy,x} + \sigma_{y,y} + F_y = 0$$

Figure 1.3.1. Stresses and body forces per unit volume that act on a plane differential element of constant thickness t. The comma notation indicates partial differentiation. For example, $\sigma_{x,x} = \partial\sigma_x/\partial x$.

field, and so on. They are positive when they act in the positive coordinate directions. On each unit of volume ($dV = t\,dx\,dy$), F_x and F_y produce *total* forces $F_x\,dV$ and $F_y\,dV$. In general, stresses are functions of the coordinates. Thus, for example, $\sigma_{x,x}$ is the rate of change of σ_x with respect to x, and $\sigma_{x,x}\,dx$ is the change of σ_x over a distance dx. Equilibrium of forces in the x direction requires that

$$-\sigma_x t\,dy - \tau_{xy} t\,dx + (\sigma_x + \sigma_{x,x}\,dx)t\,dy + (\tau_{xy} + \tau_{xy,y}\,dy)t\,dx + F_x t\,dx\,dy = 0 \qquad (1.3.1)$$

There is a corresponding y-direction equilibrium equation. Both are shown in their simplified form in Fig. 1.3.1. In *three dimensions,* the body force vector is

$$\{\mathbf{F}\} = \{F_x \quad F_y \quad F_z\} \qquad (1.3.2)$$

and the differential equations of equilibrium are

$$\begin{aligned} \sigma_{x,x} + \tau_{xy,y} + \tau_{zx,z} + F_x &= 0 \\ \tau_{xy,x} + \sigma_{y,y} + \tau_{yz,z} + F_y &= 0 \\ \tau_{zx,x} + \tau_{yz,y} + \sigma_{z,z} + F_z &= 0 \end{aligned} \qquad (1.3.3)$$

When an elastic body is deformed, no cracks appear in stretching, no kinks appear in bending, and no part overlaps another. Stated more elegantly, this is the *compatibility condition:* the displacement field is continuous and single-valued. As shown in Section 1.4, in plane problems the strains ϵ_x, ϵ_y, and γ_{xy} are derived from only two displacement field quantities, $u = u(x, y)$ and $v = v(x, y)$. Therefore, if compatibility prevails, the three strains are not independent. The relation between them is called the "compatibility equation." We will not have to use it because we will start with displacements, not strains. Thus the compatibility condition is satisfied automatically.

Boundary conditions apply to stress and displacement, and are part of the problem definition. For example, in Fig. 1.1.1a, the left edge does not move ($u = v = 0$ on $x = 0$); where the pressure acts, $\sigma_y = -p$ and $\tau_{xy} = 0$; on the right edge, $\sigma_x = \tau_{xy} = 0$; and so on. Where displacements are prescribed, stresses are unknown and assume values implicitly dictated by the solution. Similarly, where stresses are prescribed, displacements are unknown.

Stress boundary conditions on an arbitrarily oriented edge lead to the following equations. Surface tractions Φ_x and Φ_y in Fig. 1.3.2 have stress units and act on the *total* (not the projected) boundary area $t\,ds$. Equilibrium equations in x and y directions reduce to the equations in Fig. 1.3.2. Here ℓ and m are direction cosines of the boundary normal. In *three dimensions* the surface traction vector is

$$\{\Phi\} = \{\Phi_x \quad \Phi_y \quad \Phi_z\} \tag{1.3.4}$$

and the equations that relate surface tractions to internal stresses at the boundary are

$$\begin{aligned} \Phi_x &= \ell\sigma_x + m\tau_{xy} + n\tau_{zx} \\ \Phi_y &= \ell\tau_{xy} + m\sigma_y + n\tau_{yz} \\ \Phi_z &= \ell\tau_{zx} + m\tau_{yz} + n\sigma_z \end{aligned} \tag{1.3.5}$$

where ℓ, m, and n are direction cosines of the boundary normal.

If a stress or displacement field satisfies equilibrium, compatibility, and boundary conditions, then a solution has been found. If the load versus response behavior is linear, then the solution is unique. How do these observations relate to finite element analysis? If elements are based on displacement fields, so that nodal displacements are the primary unknowns, then the compatibility condition is satisfied within elements. Suitably chosen fields also provide comptability *between* elements and satisfy displacement boundary conditions. Equilibrium equations and boundary conditions on stress are satisfied only approximately. The approximation improves as more d.o.f. are used and, barring computational difficulties, the exact solution is achieved in the limit of an infinitely refined mesh.

Other finite element methods are based on fields other than displacement. A stress-field model would satisfy equilibrium equations *a priori*, and mesh refinement would yield a better approximation of compatibility conditions.

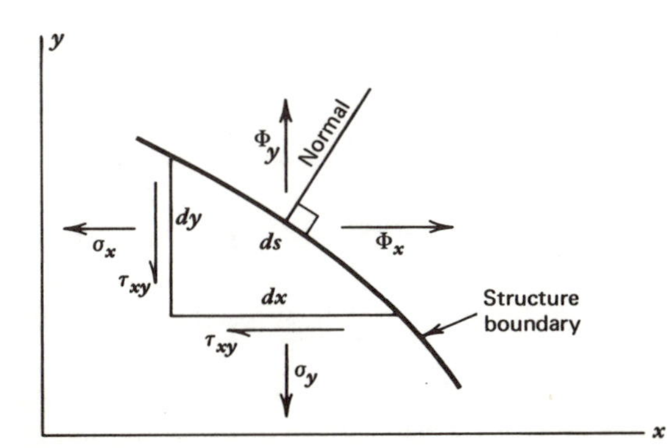

Surface tractions:

$$\Phi_x = \ell\sigma_x + m\tau_{xy}$$
$$\Phi_y = \ell\tau_{xy} + m\sigma_y$$

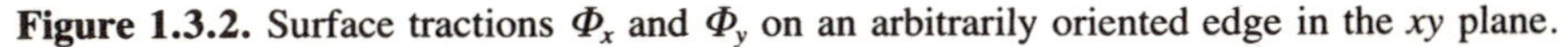

Figure 1.3.2. Surface tractions Φ_x and Φ_y on an arbitrarily oriented edge in the xy plane.

1.4 STRAIN-DISPLACEMENT RELATIONS

The relation between displacement and strain is essential in element formulation. Here we consider the two- and three-dimensional cases in Cartesian coordinates [1.20]. Expressions for plates, solids of revolution, and so on are given where they are used.

In Fig. 1.4.1, a general strain field converts configuration 012 to configuration 0′1′2′. Like stresses in Fig. 1.3.1, displacements u and v are functions of the coordinates. We assume that increments such as $u_{,x}\,dx$ are small in comparison with u and v. By definition, normal strain is the ratio of change in length to original length, so

$$\epsilon_x = \frac{L_{0'2'} - L_{02}}{L_{02}} = \frac{[dx + (u + u_{,x}\,dx) - u] - dx}{dx} = u_{,x} \tag{1.4.1}$$

A similar analysis yields the y-direction normal strain as

$$\epsilon_y = v_{,y} \tag{1.4.2}$$

Shear strain in the "engineering definition" is defined as the amount of change in a right angle. Because displacement increments are small, $\beta_1 \approx \tan\beta_1$ and $\beta_2 \approx \tan\beta_2$, so the engineering shear strain is

$$\gamma_{xy} = \beta_1 + \beta_2 = \frac{(u + u_{,y}\,dy) - u}{dy} + \frac{(v + v_{,x}\,dx) - v}{dx} = u_{,y} + v_{,x} \tag{1.4.3}$$

The foregoing *strain-displacement relations* can be stated in matrix-operator form. Here they are, along with their counterparts in three dimensions.

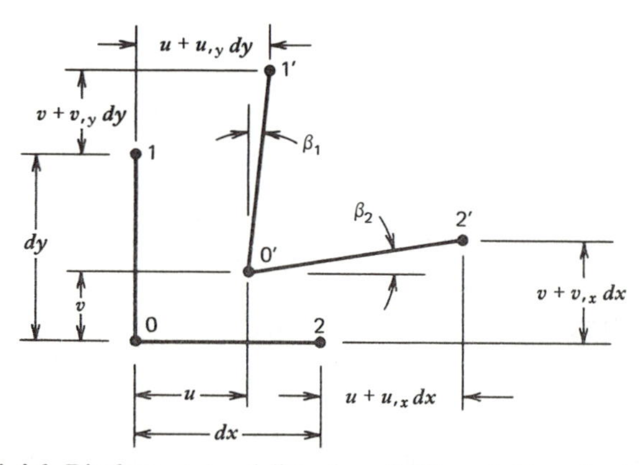

Figure 1.4.1. Displacement and distortion of differential lengths dx and dy.

$$\begin{Bmatrix} \epsilon_x \\ \epsilon_y \\ \gamma_{xy} \end{Bmatrix} = \begin{bmatrix} \frac{\partial}{\partial x} & 0 \\ 0 & \frac{\partial}{\partial y} \\ \frac{\partial}{\partial y} & \frac{\partial}{\partial x} \end{bmatrix} \begin{Bmatrix} u \\ v \end{Bmatrix}, \qquad \begin{Bmatrix} \epsilon_x \\ \epsilon_y \\ \epsilon_z \\ \gamma_{xy} \\ \gamma_{yz} \\ \gamma_{zx} \end{Bmatrix} = \begin{bmatrix} \frac{\partial}{\partial x} & 0 & 0 \\ 0 & \frac{\partial}{\partial y} & 0 \\ 0 & 0 & \frac{\partial}{\partial z} \\ \frac{\partial}{\partial y} & \frac{\partial}{\partial x} & 0 \\ 0 & \frac{\partial}{\partial z} & \frac{\partial}{\partial y} \\ \frac{\partial}{\partial z} & 0 & \frac{\partial}{\partial x} \end{bmatrix} \begin{Bmatrix} u \\ v \\ w \end{Bmatrix} \tag{1.4.4}$$

1.5 STRESS-STRAIN RELATIONS

The stress-strain relation is abbreviated as

$$\{\boldsymbol{\epsilon}\} = [\mathbf{C}]\{\boldsymbol{\sigma}\} \quad \text{or as} \quad \{\boldsymbol{\sigma}\} = [\mathbf{E}]\{\boldsymbol{\epsilon}\} \tag{1.5.1}$$

where [**C**] is a matrix of material compliances, [**E**] is a matrix of material stiffnesses, and $[\mathbf{E}] = [\mathbf{C}]^{-1}$. Written out, the stress vector $\{\boldsymbol{\sigma}\}$ and the stain vector $\{\boldsymbol{\epsilon}\}$ are, respectively,

$$\{\boldsymbol{\sigma}\} = \{\sigma_x \; \sigma_y \; \sigma_z \; \tau_{xy} \; \tau_{yz} \; \tau_{zx}\}, \qquad \{\boldsymbol{\epsilon}\} = \{\epsilon_x \; \epsilon_y \; \epsilon_z \; \gamma_{xy} \; \gamma_{yz} \; \gamma_{zx}\} \tag{1.5.2}$$

Initial stresses and strains can also exist. These are considered in Section 1.6. Equations 1.5.1 express Hooke's law. For real solids the "law" is an approximation limited to small strains.

Matrix [**E**] is symmetric (and therefore so is [**C**]). This is easily shown if we grant in advance the validity of Eq. 3.4.4. By differentiation,

$$\frac{\partial^2 U_0}{\partial \epsilon_x \partial \epsilon_y} = \frac{\partial^2 U_0}{\partial \epsilon_y \partial \epsilon_x}, \qquad \text{so } E_{12} = E_{21} \tag{1.5.3}$$

and so on. Thus, in the most general case, [**E**] (and [**C**]) contains 21 independent coefficients.

There are nine independent coefficients in [E] if the material is orthotropic. If axes x, y, and z coincide with principal directions of the material, we can write

$$\begin{aligned}
\epsilon_x &= +\frac{1}{E_x}\sigma_x - \frac{\nu_{yx}}{E_y}\sigma_y - \frac{\nu_{zx}}{E_z}\sigma_z, \qquad \gamma_{xy} = \frac{\tau_{xy}}{G_{xy}} \\
\epsilon_y &= -\frac{\nu_{xy}}{E_x}\sigma_x + \frac{1}{E_y}\sigma_y - \frac{\nu_{zy}}{E_z}\sigma_z, \qquad \gamma_{yz} = \frac{\tau_{yz}}{G_{yz}} \\
\epsilon_z &= -\frac{\nu_{xz}}{E_x}\sigma_x - \frac{\nu_{yz}}{E_y}\sigma_y + \frac{1}{E_z}\sigma_z, \qquad \gamma_{zx} = \frac{\tau_{zx}}{G_{zx}}
\end{aligned} \tag{1.5.4}$$

Each Poisson ratio ν_{ij} characterizes strain in the j direction produced by stress in the i direction ($i, j = x, y, z$). For example, uniaxial stress σ_z produces strains $\epsilon_z = \sigma_z/E_z$, $\epsilon_x = -\nu_{zx}\epsilon_z$, and $\epsilon_y = -\nu_{zy}\epsilon_z$. Because of the symmetry requirement,

$$E_x\nu_{yx} = E_y\nu_{xy}, \qquad E_y\nu_{zy} = E_z\nu_{yz}, \qquad E_z\nu_{xz} = E_x\nu_{zx} \tag{1.5.5}$$

From Eqs. 1.5.4 we can write [C]; then $[\mathbf{E}] = [\mathbf{C}]^{-1}$. If x, y, and z are not principal directions, [E] must be transformed (Eq. 6.2.13).

An isotropic material has only two independent elastic coefficients, E and ν. Thus nonzero terms in the upper triangle of [E] are

$$\begin{aligned}
E_{11} = E_{22} = E_{33} = (1-\nu)c, \qquad & E_{12} = E_{13} = E_{23} = \nu c \\
& E_{44} = E_{55} = E_{66} = G
\end{aligned} \tag{1.5.6}$$

where

$$c = \frac{E}{(1+\nu)(1-2\nu)} \qquad \text{and} \qquad G = \frac{E}{2(1+\nu)}$$

Consider next the two-dimensional problem of a body that lies in the xy plane. Then, by definition, $\tau_{yz} = \tau_{zx} = \gamma_{yz} = \gamma_{zx} = 0$. If the problem is plane stress, then $\sigma_z = 0$ and ϵ_z need not enter the solution process. Thus, for an isotropic material, Eqs. 1.5.4 and $[\mathbf{E}] = [\mathbf{C}]^{-1}$ become

$$\begin{aligned}
\epsilon_x &= \frac{1}{E}\sigma_x - \frac{\nu}{E}\sigma_y \\
\epsilon_y &= -\frac{\nu}{E}\sigma_x + \frac{1}{E}\sigma_y \qquad [\mathbf{E}] = \frac{E}{1-\nu^2}\begin{bmatrix} 1 & \nu & 0 \\ \nu & 1 & 0 \\ 0 & 0 & \frac{1-\nu}{2} \end{bmatrix} \\
\gamma_{xy} &= \frac{\tau_{xy}}{G} = \frac{2(1+\nu)}{E}\tau_{xy}
\end{aligned} \tag{1.5.7}$$

If the problem is plane strain, then $\epsilon_z = 0$ and σ_z need not enter the solution process. By discarding rows 3, 5, and 6 from the 6-by-6 [E], we find, for an isotropic material,

$$[\mathbf{E}] = \frac{E}{(1+\nu)(1-2\nu)} \begin{bmatrix} 1-\nu & \nu & 0 \\ \nu & 1-\nu & 0 \\ 0 & 0 & \dfrac{1-2\nu}{2} \end{bmatrix} \tag{1.5.8}$$

For plane stress or plane strain conditions to prevail in the xy plane, the xy plane must be a plane of elastic symmetry. Thus, if the material is orthotropic, the z-axis must be a principal material direction. If, in addition, the x- and y-axes are principal material directions, $E_{13} = E_{31} = E_{23} = E_{32} = 0$.

Material property matrices for solids of revolution and for plates appear in Sections 8.2 and 9.1, respectively.

Poisson's ratio ν is little affected by temperature. Modulus E is affected more: for stainless steel E decreases about 20% when the temperature rises from 0°C to 450°C. Barring plastic flow, elastic properties are almost independent of stress. For example, an increase in hydrostatic pressure from 0 MPa to 350 MPa increases the moduli of steel and aluminum 0.8% and 2.6%, respectively [1.11].

When strain rates are high, as in wave propagation, modulus E is higher than its static value. The difference is negligible for common metals but appreciable for rubberlike materials.

Capabilities of the finite element method far exceed the knowledge of material behavior on which an analysis must be based. If test data are lacking, as is often the case with anisotropic materials, we can only estimate the elastic constants. Even when the constants are known, anisotropy has an adverse effect on the accuracy of finite element solutions [6.8].

1.6 TEMPERATURE EFFECTS. INITIAL STRESS AND STRAIN

With additional stress and strain vectors, Eq. 1.5.1 becomes

$$\{\boldsymbol{\sigma}\} = [\mathbf{E}]\big(\{\boldsymbol{\epsilon}\} - \{\boldsymbol{\epsilon}_0\}\big) + \{\boldsymbol{\sigma}_0\} \tag{1.6.1}$$

where $\{\boldsymbol{\epsilon}_0\}$ and $\{\boldsymbol{\sigma}_0\}$ are vectors of initial strains and initial stresses, respectively. As examples, $\{\boldsymbol{\epsilon}_0\}$ might describe moisture-induced swelling and $\{\boldsymbol{\sigma}_0\}$ might describe stresses produced by heating. Alternatively, both effects can be placed in $\{\boldsymbol{\epsilon}_0\}$, or $\{\boldsymbol{\epsilon}_0\}$ and $\{\boldsymbol{\sigma}_0\}$ can be viewed as different ways to express the same thing. For example, free expansion of an orthotropic material with principal axes xyz produces the strains

$$\{\boldsymbol{\epsilon}_0\} = \{\alpha_x T \quad \alpha_y T \quad \alpha_z T \quad 0 \quad 0 \quad 0\} \tag{1.6.2}$$

where T is temperature above an arbitrary reference temperature and the α's are coefficients of thermal expansion. Now we can use Eq. 1.6.2 and $\{\boldsymbol{\sigma}_0\} = 0$ in Eq. 1.6.1. The alternative method is to use $\{\boldsymbol{\epsilon}_0\} = 0$ and $\{\boldsymbol{\sigma}_0\} = -[\mathbf{E}]\{\boldsymbol{\epsilon}_0\}$ in Eq. 1.6.1. Both methods yield the same stresses, $\{\boldsymbol{\sigma}\} = -[\mathbf{E}]\{\boldsymbol{\epsilon}_0\}$, when mechanical strains $\{\boldsymbol{\epsilon}\}$ are prohibited. The negative sign means that heating produces compressive stress.

If boundary conditions permit unrestrained expansion and contraction, the material is homogeneous, and the temperature field $T = T(x, y, z)$ is a linear function of x, y, and z, then all stresses are zero.

If $\{\boldsymbol{\epsilon}_0\}$ and $\{\boldsymbol{\sigma}_0\}$ are known with respect to principal material axes but are needed with respect to global axes, the transformations described in Section 6.2 can be used.

In the special case of isotropy we have, in three dimensions,

$$\{\boldsymbol{\sigma}_0\} = -\frac{E\alpha T}{1 - 2\nu}\{1 \quad 1 \quad 1 \quad 0 \quad 0 \quad 0\} \tag{1.6.3}$$

and, in plane stress,

$$\{\boldsymbol{\epsilon}_0\} = \{\alpha T \quad \alpha T \quad 0\}, \qquad \{\boldsymbol{\sigma}_0\} = -\frac{E\alpha T}{1 - \nu}\{1 \quad 1 \quad 0\} \tag{1.6.4}$$

and, finally, in plane strain,

$$\{\boldsymbol{\epsilon}_0\} = (1 + \nu)\{\alpha T \quad \alpha T \quad 0\}, \qquad \{\boldsymbol{\sigma}_0\} = -\frac{E\alpha T}{1 - 2\nu}\{1 \quad 1 \quad 0\} \tag{1.6.5}$$

Temperature-dependent moduli can be accommodated by using the [E] appropriate to the temperature that prevails. A temperature-dependent expansion coefficient is more troublesome. By definition, $\alpha = \partial\epsilon/\partial T$ when deformation is unrestrained, so $\epsilon = \alpha T$ only if α is independent of T. Otherwise,

$$\epsilon = \int_0^T \alpha \, dT = \bar{\alpha} T, \qquad \text{where} \qquad \bar{\alpha} = \frac{1}{T}\int_0^T \alpha \, dT \tag{1.6.6}$$

Here $\bar{\alpha}$ is an effective expansion coefficient, valid over the temperature range at hand.

The thermal *conductivity* coefficient k, used in Chapter 17, is no more sensitive to stress than is E (Section 1.5) [1.12]. However, k is sensitive to temperature change, and $\partial k/\partial T$ may be positive or negative.

Some have taken the viewpoint that if E depends on T, then α depends on stress [1.13]. For uniaxial stress, the argument is

$$\frac{\partial \alpha}{\partial \sigma} = \frac{\partial}{\partial \sigma}\left(\frac{\partial \epsilon}{\partial T}\right) = \frac{\partial}{\partial T}\left(\frac{\partial \epsilon}{\partial \sigma}\right) = \frac{\partial}{\partial T}\left(\frac{1}{E}\right) = -\frac{1}{E^2}\frac{\partial E}{\partial T} \tag{1.6.7}$$

Consider the heating of a bar that already carries stress σ. Strain ϵ can be calculated in two ways.

$$\epsilon = \frac{\sigma}{E} + \left(\alpha + \frac{\partial \alpha}{\partial \sigma}\sigma\right)T, \qquad \text{or} \qquad \epsilon = \frac{\sigma}{E + \dfrac{\partial E}{\partial T}T} + \alpha T \tag{1.6.8}$$

The former equation takes the viewpoint that α depends on σ but E is independent of T. The latter equation takes the viewpoint that E depends on T but α is independent of σ. The two equations are alternate ways of saying the same thing: with Eq. 1.6.7, both yield the same ϵ. The latter equation is preferable because it does not require an iterative procedure to find thermally induced stresses.

1.7 FINDING ADDITIONAL INFORMATION

Information about numerical structural mechanics is found in textbooks, journals, proceedings of conferences and symposia, and reports sponsored or published by government agencies. Indexes, abstracts, and digests are available [1.14–1.17]. There are voluminous bibliographies with author and key word indexes [1.18, 1.19]. Doctoral dissertations are indexed in *Dissertation Abstracts International*. To pursue ongoing developments in a particular area, the *Science Citation Index* can be helpful: listed under an indexed paper are subsequently published papers that have used the indexed paper as a reference. Literature searches, done by computer and of either broad or highly specific coverage, are offered by the National Technical Information Service, NASA, and the Defense Documentation Center.

Few industrial users find it worthwhile to develop their own software. It is more cost effective to use existing programs and modify them if necessary. Descriptions of the capability and availability of software are found in the indexes [1.14–1.17], in the *SMCP Symposium* (see References in the back of the book), in the *Structural Mechanics Software Series* (N. Perrone and W. Pilkey, eds., University Press of Virginia), and in other sources found in a library. Listings of programs and subroutines appear in some journals, government reports, and doctoral dissertations.

There are regular meetings of user's groups such as the annual *NASTRAN User's Colloquium*. They publish proceedings that are often useful. Program manuals, such as the *NASTRAN Theoretical Manual* [15.27], can be bought without buying the whole program and can be educational.

There are consulting firms that sell services ranging from the processing of a data set prepared by a client to doing the complete modeling and analysis job. Such a firm usually has more than one general-purpose analysis program to choose from.

1.8 THE COMPUTER SIREN

Computers are so fascinating that it is easy to trust them too much. There are many uncertainties in analysis. We are unsure about material properties, construction errors, and the stiffness of connections and fasteners. Loads are often of uncertain magnitude and distribution and, in analysis, we use only a few of the many possible load cases. Perhaps the structure will be used for a purpose not intended. These uncertainties make us view computed results with some reserve and question the need for "exact" analysis. However, an analysis should not be so crude or so casual that it adds to the confusion.

In analysis, we build a mathematical model and pose a question about it, such as "What are the deflections?" If the program works, it will tell us about deflections *of the mathematical model*, not about deflections of the structure.

Furthermore, the analysis will say nothing about buckling unless we ask, even if buckling is the actual mode of failure. We cannot presume that a structure is obliged to behave as a computer says it should, no matter how expensive the program, how elegant the graphic display, or how many digits are printed out. Computation assists engineering judgment but must not replace it.

Powerful programs (NASTRAN, ANSYS, MARC, and others) cannot be used without training and cannot be trusted when used by people who have heard only introductory lectures about finite elements. Programs work properly on test problems supplied with the program, but bugs may appear when other analyses are attempted. Accordingly, an analysis budget must allow time for checking and revision.

PROBLEMS

1.1 (a) Draw the two beam deformation modes not shown in Fig. 1.2.1. Determine the algebraic signs of the 16 k_{ij}. Judging by algebraic signs, what property does $[\mathbf{k}]$ display?
(b) Use elementary beam theory to derive the 4-by-4 stiffness matrix of a uniform beam of length L and bending stiffness EI (see Eq. 4.4.5 for the answer).

1.2 The sketch shows a typical bar of an elastic plane frame. Without numerical calculations:
(a) Find the algebraic signs of the k_{ij} in columns 1, 3, and 5 of $[\mathbf{k}]$.
(b) Find the algebraic signs of the k_{ij} in columns 2, 4, and 6 of $[\mathbf{k}]$.
(c) If $\beta = 0$, which k_{ij} are zero? Use standard small-displacement approximations.

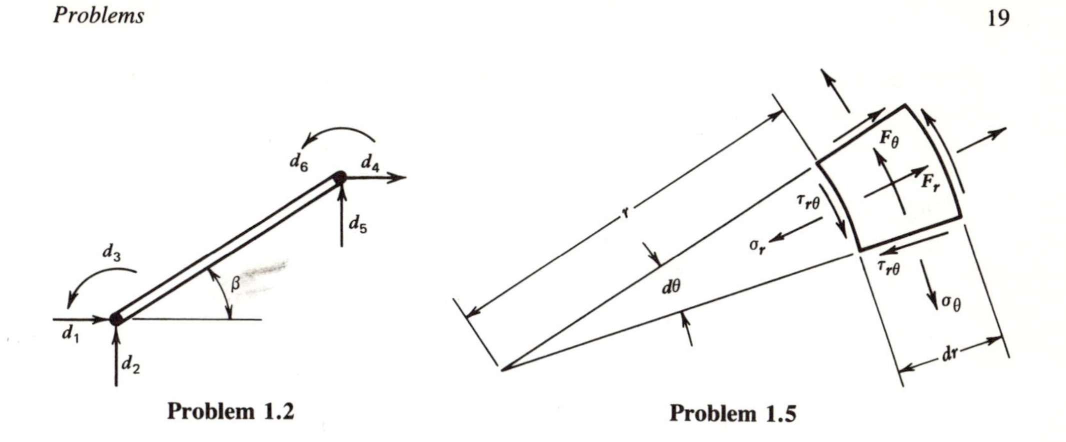

Problem 1.2 **Problem 1.5**

1.3 Find u_1, u_2, and u_3 in Fig. 1.2.2*a* by elementary (not matrix) methods. Show that these values satisfy Eq. 1.2.5.

1.4 Write the equilibrium equations of Fig. 1.3.1 and Eq. 1.3.3 in a matrix-operator form such as that used in Eq. 1.4.4.

1.5 (a) The sketch shows a plane differential element similar to that in Fig. 1.3.1 but in polar coordinates. Derive the differential equations of equilibrium in polar coordinates.
(b) Similarly, use cylindrical coordinates to derive equations analogous to Eqs. 1.3.3.

1.6 (a) Derive the equations in Fig. 1.3.2. Show that the special cases of a horizontal boundary and a vertical boundary yield the expected results.
(b) Derive Eqs. 1.3.5.
(c) What change must be made in the equations in Fig. 1.3.2 if the coordinates are polar instead of Cartesian?

1.7 Let stresses in a plane problem be given by

$$\sigma_x = a_1 + a_2x + a_3y, \qquad \sigma_y = a_4 + a_5x + a_6y, \qquad \tau_{xy} = a_7 + a_8x + a_9y$$

where each a_i is a constant. Let all body forces vanish. For the following special cases, state whether equilibrium is satisfied or what relation among the a_i is needed to satisfy equilibrium.
(a) All $a_i = 0$ except a_1, a_4, and a_7.
(b) $a_3 = a_5 = a_8 = a_9 = 0$.
(c) $a_2 = a_6 = a_8 = a_9 = 0$.
(d) All of the a_i are nonzero.

1.8 Section 1.3 mentions a compatibility equation for plane problems. Derive this equation, using information in Section 1.4.

1.9 (a) Write **[C]** from the three strain equations in Eq. 1.5.7. Hence, find **[E]** = $[\mathbf{C}]^{-1}$.
(b) Similarly, for Eq. 1.5.8, write **[C]** and find **[E]** = $[\mathbf{C}]^{-1}$.

1.10 Combine Eqs. 1.5.7, the strain-displacement relations, and the equilibrium equations with $F_x = F_y = 0$, and show that

$$u_{,xx} + u_{,yy} = (1 + \nu)(u_{,yy} - v_{,xy})/2$$

This equation and its companion are known as the equilibrium equations expressed in terms of displacements.

1.11 Let displacements in a plane stress problem be given by

$$u = a_1 + a_2x + a_3y + a_4x^2 + a_5xy + a_6y^2$$
$$v = a_7 + a_8x + a_9y + a_{10}x^2 + a_{11}xy + a_{12}y^2$$

where each a_i is a constant. Let all body forces vanish. For the following special cases, state whether equilibrium is satisfied or what relation among the a_i is needed to satisfy equilibrium. The medium is isotropic.

(a) $a_4 = a_5 = a_6 = a_{10} = a_{11} = a_{12} = 0$.
(b) $a_4 = a_6 = a_{10} = a_{12} = 0$.
(c) $a_6 = a_{10} = 0$.
(d) $a_4 = a_{12} = 0$.
(e) All of the a_i are nonzero.
(f) Are any of the cases (a) to (e) unsuitable for a finite element formulation because they fail to satisfy equilibrium? Explain.

1.12 Verify Eqs. 1.6.3, 1.6.4, and 1.6.5.

1.13 (a) Verify that Eqs. 1.6.8 predict the same strain.
(b) Explain why the first of Eqs. 1.6.8 requires an iterative solution while the second does not.

CHAPTER **2**

THE STIFFNESS METHOD AND THE PLANE TRUSS

2.1 INTRODUCTION

In this chapter the stiffness method introduced in Section 1.2 is explained in detail. We consider how to assemble elements to form a structure stiffness matrix, how to impose boundary conditions, and how to solve structure equations. The plane truss is used to illustrate these concepts, which apply with equal force to all finite element structures.

The bars of a truss are its finite elements. A truss is a "natural" finite element structure, since no conceptual division into elements is needed.

Truss analysis by the stiffness method is practical and simple. It is easy to derive stiffness matrices by direct physical argument. This is not true with most finite element structures.

In this chapter, each bar is assumed to be uniform, linearly elastic, pin-connected at its ends, and axially loaded. Displacements shown in sketches are *greatly exaggerated*. Actual displacements are assumed to be small enough that if θ is the angle of rotation of any bar, $\sin \theta = \theta$ and $\cos \theta = 1$. We consider only static problems. Within these restrictions, the analysis is exact, not approximate.

2.2 STRUCTURE STIFFNESS EQUATIONS

Consider, for example, the three-bar truss of Fig. 2.2.1. Nodes and elements (bars) are numbered arbitrarily. For element i, where $i = 1, 2, 3$, let A_i = cross-sectional area, E_i = elastic modulus, and L_i = length. From elementary mechanics of materials, axial force F_i and change in length e_i have the relation

$$e_i = \frac{F_i L_i}{A_i E_i} \tag{2.2.1}$$

Stiffness is defined as the ratio of force to displacement and is given the symbol k. Thus, for bar i,

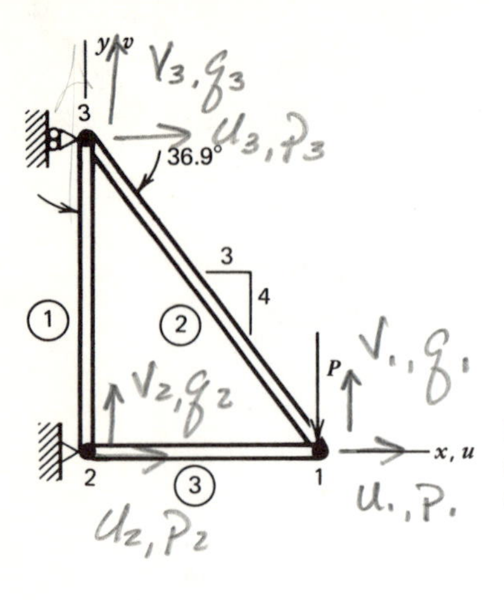

Figure 2.2.1. A three-element plane truss. Degrees of freedom u_1, v_1, and v_3 are active, while d.o.f. u_2, v_2, and u_3 are restrained.

$$k_i = \frac{F_i}{e_i} = \frac{A_i E_i}{L_i} \tag{2.2.2}$$

We illustrate structure stiffness concepts by treating the structure as a whole. Supports at nodes 2 and 3 are temporarily removed. Let each node be displaced a small amount, first in the x direction and then in the y direction, while all other nodes are held at zero displacement. In each of these six cases we calculate forces that must be applied to the nodes to maintain static equilibrium of the deformed truss. The first two free-body diagrams are shown in Fig. 2.2.2.

As an example of the computation, consider the forces in Fig. 2.2.2*a*. Because u_1 is small, its component along bar 2 is $e_2 = 0.6u_1$. This deformation produces the axial force $F_2 = k_2 e_2$, whose horizontal and vertical components have the respective magnitudes $0.6F_2 = 0.36k_2u_1$ and $0.8F_2 = 0.48k_2u_1$. Bar 3 has elongation $e_3 = u_1$ and contributes forces $F_3 = k_3e_3 = k_3u_1$.

Let $\{\mathbf{Q}_1\}$ represent the vector of forces in Fig. 2.2.2*a* for *unit* displacement, $u_1 = 1$. Then, for $u_1 \neq 1$, the forces are

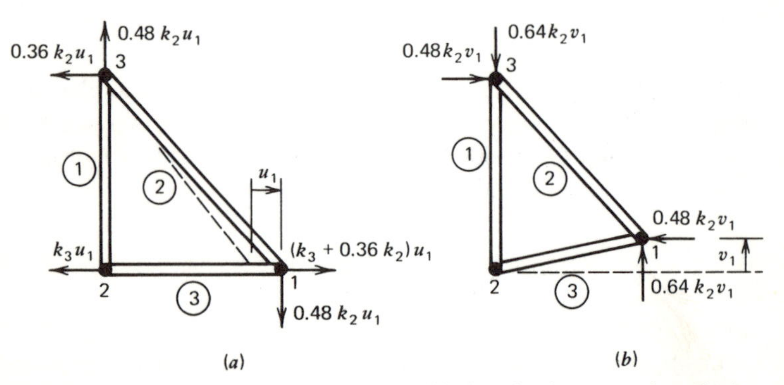

Figure 2.2.2. Forces required to maintain static equilibrium in the respective displacement states $\{\mathbf{D}\} = \{u_1\ 0\ 0\ 0\ 0\ 0\}$ and $\{\mathbf{D}\} = \{0\ v_1\ 0\ 0\ 0\ 0\}$.

$$\{\mathbf{Q}_1\}\, u_1 = \{k_3 + 0.36k_2 \quad -0.48k_2 \quad -k_3 \quad 0 \quad -0.36k_2 \quad 0.48k_2\}u_1 \tag{2.2.3}$$

Similarly, forces in Fig. 2.2.2*b* are

$$\{\mathbf{Q}_2\}\, v_1 = \{-0.48k_2 \quad 0.64k_2 \quad 0 \quad 0 \quad 0.48k_2 \quad -0.64k_2\}\, v_1 \tag{2.2.4}$$

Let $\{\mathbf{Q}_3\}$, $\{\mathbf{Q}_4\}$, $\{\mathbf{Q}_5\}$, and $\{\mathbf{Q}_6\}$ represent force vectors for the remaining four unit displacement states $u_2 = 1$, $v_2 = 1$, $u_3 = 1$, and $v_3 = 1$. Now, if nodal displacements appear *simultaneously,* in arbitrary combination, nodal forces are the sum of the six cases.

$$[\mathbf{Q}_1\ \mathbf{Q}_2\ \mathbf{Q}_3\ \mathbf{Q}_4\ \mathbf{Q}_5\ \mathbf{Q}_6]\begin{Bmatrix} u_1 \\ v_1 \\ u_2 \\ v_2 \\ u_3 \\ v_3 \end{Bmatrix} = \begin{Bmatrix} p_1 \\ q_1 \\ p_2 \\ q_2 \\ p_3 \\ q_3 \end{Bmatrix} \tag{2.2.5}$$

where nodal forces that correspond to u_i and v_i are p_i and q_i. A force and its corresponding displacement are considered positive when they act in the positive coordinate direction. Written out, Eq. 2.2.5 is

$$\begin{bmatrix} k_3 + 0.36k_2 & -0.48k_2 & -k_3 & 0 & -0.36k_2 & 0.48k_2 \\ -0.48k_2 & 0.64k_2 & 0 & 0 & 0.48k_2 & -0.64k_2 \\ -k_3 & 0 & k_3 & 0 & 0 & 0 \\ 0 & 0 & 0 & k_1 & 0 & -k_1 \\ -0.36k_2 & 0.48k_2 & 0 & 0 & 0.36k_2 & -0.48k_2 \\ 0.48k_2 & -0.64k_2 & 0 & -k_1 & -0.48k_2 & k_1 + 0.64k_2 \end{bmatrix}\begin{Bmatrix} u_1 \\ v_1 \\ u_2 \\ v_2 \\ u_3 \\ v_3 \end{Bmatrix} = \begin{Bmatrix} p_1 \\ q_1 \\ p_2 \\ q_2 \\ p_3 \\ q_3 \end{Bmatrix} \tag{2.26}$$

Or, in our standard abbreviation, these structure stiffness equations are

$$[\mathbf{K}]\,\{\mathbf{D}\} = \{\mathbf{R}\} \tag{2.2.7}$$

where $[\mathbf{K}]$ is the *structure stiffness matrix*. As shown by the development, Eqs. 2.2.7 are equilibrium equations.

The j*th column of* $[\mathbf{K}]$ *is the vector of nodal forces that must be applied to the nodes to maintain static equilibrium when the* j*th d.o.f. has unit displacement and all other d.o.f. have zero displacement*. For a frame, "forces" include forces and moments and "displacements" include displacements and rotations. By this procedure—displacing one node at a time—we can generate the stiffness matrix of any truss or frame, regardless of the number of elements or the degree of static indeterminancy. There are always as many equations as there are nodal d.o.f. If the structure is not a truss or a frame, stiffness coefficients must be formulated indirectly, as in Chapters 4 to 10.

2.3 THE NATURE OF [K]. SOLUTION FOR UNKNOWNS

Each diagonal coefficient K_{ii} is positive. If K_{ii} were zero, displacement D_i would generate no resisting force R_i, which implies that the structure is unstable. If K_{ii} were negative, the force and its corresponding displacement would be oppositely directed, which is physically unreasonable (imagine here that $D_i > 0$ while other d.o.f. in $\{\mathbf{D}\}$ are zero).

Each column of $[\mathbf{K}]$ sums to zero because each column represents a set of nodal forces that satisfies static equilibrium. *Caution.* If $\{\mathbf{D}\}$ contains linear *and rotational* d.o.f., as for a beam or a frame, columns do *not* sum to zero (Problem 2.6).

The stiffness matrix is symmetric. This is true of any structure that has a linear force-displacement relationship, as may be shown by the Betti-Maxwell reciprocal theorem[1] (Problem 2.4).

The stiffness matrix in Eq. 2.2.6 is singular and cannot be inverted to solve for displacements. Its rank is six, but its order only three. The physical reason is that the truss can have rigid-body motion. Any number of rigid-body motions are possible, but there are only three *linearly independent* rigid-body motions for a plane structure. For our example truss, consider the rigid-body motions

$$\{\mathbf{D}\}_1 = \{\bar{u} \quad 0 \quad \bar{u} \quad 0 \quad \bar{u} \quad 0\} \qquad \{\mathbf{D}\}_2 = \{0 \quad \bar{v} \quad 0 \quad \bar{v} \quad 0 \quad \bar{v}\}$$
$$\{\mathbf{D}\}_3 = \left\{\bar{u} \quad \frac{\bar{v}}{2} \quad \bar{u} \quad \frac{\bar{v}}{2} \quad \bar{u} \quad \frac{\bar{v}}{2}\right\} \qquad \{\mathbf{D}\}_4 = \{4\theta \quad 3\theta \quad 4\theta \quad 0 \quad 0 \quad 0\} \tag{2.3.1}$$

where θ is a small angle of rotation. (The reader should sketch these motions.) Of the first three, any one can be found by a linear combination of the other two, so they are linearly dependent. Subject to the small-rotation restriction, *any* rigid-body motion $\{\mathbf{D}\}$ yields zero forces; that is, $[\mathbf{K}]\{\mathbf{D}\} = 0$.

We suppress rigid-body motion by imposing displacement boundary conditions. Figure 2.2.1 shows that conditions appropriate to this problem are

$$u_2 = v_2 = u_3 = 0 \tag{2.3.2}$$

and that the force boundary conditions are

$$p_1 = 0 \qquad q_1 = -P \qquad q_3 = 0 \tag{2.3.3}$$

[1]See a text on analysis of framed structures or theory of elasticity.

The remaining three forces and three displacements are as yet unknown. Note that when a D_i is known, the corresponding R_i is unknown, and vice versa.

In practice, matrix rearrangement and inversion operations are avoided, as shown in subsequent sections. But here we use these operations to display a formal solution for the unknowns. Let subscripts c and x designate known and unknown quantities, respectively. After rearrangement of coefficients, the structure equations can be partitioned.

$$\begin{bmatrix} \mathbf{K}_{11}\mathbf{K}_{12} \\ \mathbf{K}_{21}\mathbf{K}_{22} \end{bmatrix} \begin{Bmatrix} \mathbf{D}_x \\ \mathbf{D}_c \end{Bmatrix} = \begin{Bmatrix} \mathbf{R}_c \\ \mathbf{R}_x \end{Bmatrix} \tag{2.3.4}$$

Accordingly,

$$\begin{aligned} [\mathbf{K}_{11}]\{\mathbf{D}_x\} + [\mathbf{K}_{12}]\{\mathbf{D}_c\} &= \{\mathbf{R}_c\} \\ [\mathbf{K}_{21}]\{\mathbf{D}_x\} + [\mathbf{K}_{22}]\{\mathbf{D}_c\} &= \{\mathbf{R}_x\} \end{aligned} \tag{2.3.5}$$

We find displacements $\{\mathbf{D}_x\}$ from the first of Eqs. 2.3.5.

$$\{\mathbf{D}_x\} = [\mathbf{K}_{11}]^{-1}\big(\{\mathbf{R}_c\} - [\mathbf{K}_{12}]\,\{\mathbf{D}_c\}\big) \tag{2.3.6}$$

Last, we find $\{\mathbf{R}_x\}$ from the second of Eqs. 2.3.5. In the present example $\{\mathbf{D}_c\} = 0$, but this need not be so in all problems.

2.4 ELEMENT STIFFNESS EQUATIONS

In Section 2.2 we generated the structure stiffness matrix [**K**] by a direct attack on the entire structure. This approach clarifies the physical meaning of [**K**], but it is not a systematic method that is easy to program. Alternatively, we can generate [**K**] as the sum of element stiffness matrices [**k**]. This approach *is* easy to automate. We now derive the necessary [**k**] matrix for a plane truss (bar) member.

Let the element of Fig. 2.4.1 have constant cross-sectional area A and elastic modulus E. Everything needed to generate [**k**] can be found from A, E, and the four nodal coordinates x_i, y_i, x_j, and y_j. First, we compute

$$\begin{gathered} L = [(x_j - x_i)^2 + (y_j - y_i)^2]^{1/2} \\ s = \sin\beta = \frac{y_j - y_i}{L}, \qquad c = \cos\beta = \frac{x_j - x_i}{L} \end{gathered} \tag{2.4.1}$$

Next, as in Section 2.2, we generate columns of [**k**] by activating each d.o.f. in turn while keeping the others zero. The first of these four cases is shown in Fig. 2.4.2. Axial shortening cu_i produces an axial compressive force $F = (AE/L)cu_i$, whose x and y

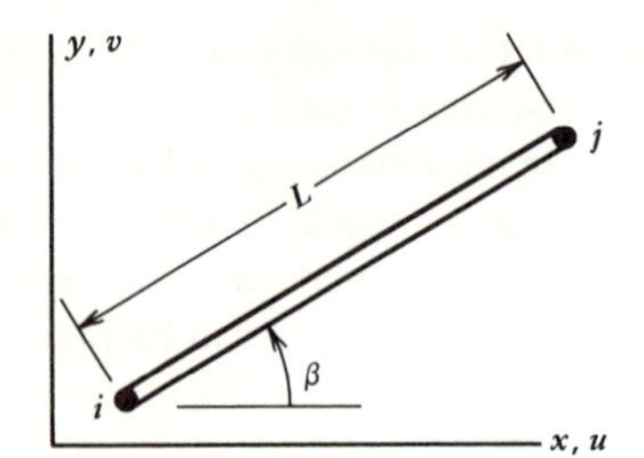

Figure 2.4.1. Truss element, arbitrarily oriented in the xy plane.

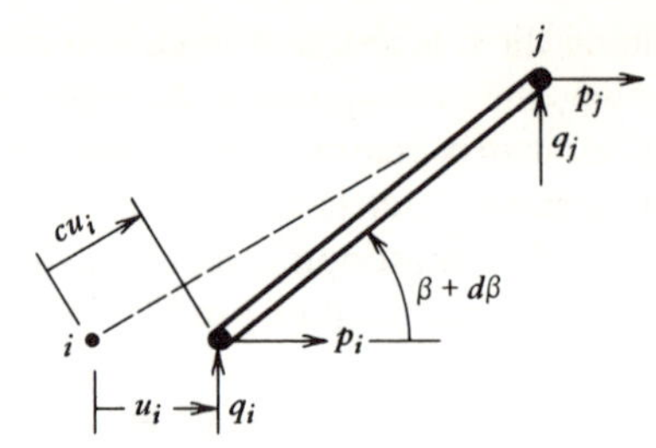

Figure 2.4.2. Truss element after imposed displacement $u_i \neq 0$, $v_i = u_j = v_j = 0$.

components are $p_i = -p_j = Fc$ and $q_i = -q_j = Fs$. These components provide static equilibrium. Thus

$$\frac{AE}{L}\begin{Bmatrix} c^2 \\ cs \\ -c^2 \\ -cs \end{Bmatrix} u_i = \begin{Bmatrix} p_i \\ q_i \\ p_j \\ q_j \end{Bmatrix} \tag{2.4.2}$$

Similar results are given by the remaining displacements, v_i, u_j, and v_j. For arbitrary values of nodal d.o.f. we superpose results, just as in Eq. 2.2.5, and find

$$\frac{AE}{L}\begin{bmatrix} c^2 & cs & -c^2 & -cs \\ cs & s^2 & -cs & -s^2 \\ -c^2 & -cs & c^2 & cs \\ -cs & -s^2 & cs & s^2 \end{bmatrix} \begin{Bmatrix} u_i \\ v_i \\ u_j \\ v_j \end{Bmatrix} = \begin{Bmatrix} p_i \\ q_i \\ p_j \\ q_j \end{Bmatrix} \tag{2.4.3}$$

where the square matrix, including the factor AE/L, is the element stiffness matrix $[\mathbf{k}]$. We abbreviate Eq. 2.4.3 as

$$[\mathbf{k}]\{\mathbf{d}\} = \{\bar{\mathbf{r}}\} \tag{2.4.4}$$

The jth column of $[\mathbf{k}]$ is the array of forces that must act on the element to provide static equilibrium when $d_j = 1$ and all other nodal d.o.f. are zero.

It will be necessary to distinguish between forces applied by the nodes and forces applied by the element. Forces $\{\bar{\mathbf{r}}\}$ are forces applied *by* the nodes *to* the element.

Later we illustrate several concepts by using truss elements that lie on the x-axis and are permitted only x-direction displacements. For this special case, Eq. 2.4.3 reduces to

$$\frac{AE}{L}\begin{bmatrix} 1 & -1 \\ -1 & 1 \end{bmatrix} \begin{Bmatrix} u_i \\ u_j \end{Bmatrix} = \begin{Bmatrix} p_i \\ p_j \end{Bmatrix} \tag{2.4.5}$$

2.5 ASSEMBLY OF ELEMENT STIFFNESS MATRICES

In this section and the next we use structures with a small number of d.o.f. to illustrate the assembly process. We will see that element node labels, such as i and j in Fig. 2.4.1, are used only in generating element matrices. In the assembly process it is the *structure* node labels, such as 1, 2, and 3 in Fig. 2.2.1, that determine the location in $[\mathbf{K}]$ to which each coefficient in $[\mathbf{k}]$ is assigned. This is true of any finite element, regardless of its type, size, shape, or number of nodes.

We can visualize the construction of a structure as follows. Structure nodes are positioned in space and assigned numbers, such as 1, 2, and 3 in Fig. 2.2.1. Elements are at first unassembled, but each element node is tagged with a structure node number to show where it is to be placed. One by one, elements are attached to the appropriate structure nodes. With each addition, the structure gains stiffness.

Symbolically, this process is that of starting with a null structure stiffness matrix, then adding the stiffness matrix of each element to it. Indeed, we can directly add element matrices if they all operate on identical displacement vectors and all have as many rows and columns as $[\mathbf{K}]$. This is called "expansion to structure size." (Expansion may seem cumbersome. It is—but in Section 2.6 we will simplify it.)

Example

Consider the truss of Fig. 2.2.1. First, we write the element matrices for each bar as follows. Coefficients k_1, k_2, and k_3 represent the AE/L factors of the respective bars.

Bar 1: $i = 2, j = 3, \beta = 90°, c = 0, s = 1.$

$$[\mathbf{k}]_1\{\mathbf{d}\}_1 = k_1 \begin{bmatrix} 0 & 0 & 0 & 0 \\ 0 & 1 & 0 & -1 \\ 0 & 0 & 0 & 0 \\ 0 & -1 & 0 & 1 \end{bmatrix} \begin{Bmatrix} u_2 \\ v_2 \\ u_3 \\ v_3 \end{Bmatrix} \tag{2.5.1}$$

Bar 2: $i = 1, j = 3, \beta = 126.9°, c = -0.6, s = 0.8.$

$$[\mathbf{k}]_2\{\mathbf{d}\}_2 = k_2 \begin{bmatrix} 0.36 & -0.48 & -0.36 & 0.48 \\ -0.48 & 0.64 & 0.48 & -0.64 \\ -0.36 & 0.48 & 0.36 & -0.48 \\ 0.48 & -0.64 & -0.48 & 0.64 \end{bmatrix} \begin{Bmatrix} u_1 \\ v_1 \\ u_3 \\ v_3 \end{Bmatrix} \tag{2.5.2}$$

Bar 3: $i = 1, j = 2, \beta = 180°, c = -1, s = 0.$

$$[\mathbf{k}]_3\{\mathbf{d}\}_3 = k_3 \begin{bmatrix} 1 & 0 & -1 & 0 \\ 0 & 0 & 0 & 0 \\ -1 & 0 & 1 & 0 \\ 0 & 0 & 0 & 0 \end{bmatrix} \begin{Bmatrix} u_1 \\ v_1 \\ u_2 \\ v_2 \end{Bmatrix} \tag{2.5.3}$$

Each [**k**] must be expanded to size 6 by 6. We do this by adding two rows and two columns of zeros—at the start in Eq. 2.5.1, in the middle in Eq. 2.5.2, and at the end in Eq. 2.5.3. Thus each displacement vector becomes identical.

$$\{\mathbf{d}\}_1 = \{\mathbf{d}\}_2 = \{\mathbf{d}\}_3 = \{\mathbf{D}\} = \{u_1 \; v_1 \; u_2 \; v_2 \; u_3 \; v_3\} \tag{2.5.4}$$

Adding rows and columns of zeros can be physically justified as follows. Consider, for example, element 2. Element 2 and node 2 are not connected, so there can be no {**D**} that makes element 2 produce forces at node 2. This accounts for the two rows of zeros. Also, no displacement of node 2 can make element 2 apply a force to any node. This accounts for the two columns of zeros.

The reader can check that the expanded [**k**]'s do indeed sum to the [**K**] in Eq. 2.2.6. Summing [**k**]'s to get [**K**] is sometimes called the *direct stiffness method.*

Example

Consider Fig. 2.5.1. Element matrices are

$$[\mathbf{k}]_1\{\mathbf{d}\}_1 = \begin{bmatrix} a_1 & a_2 & a_3 \\ a_4 & a_5 & a_6 \\ a_7 & a_8 & a_9 \end{bmatrix} \begin{Bmatrix} \phi_i \\ \phi_j \\ \phi_k \end{Bmatrix}, \qquad [\mathbf{k}]_2\{\mathbf{d}\}_2 = \begin{bmatrix} b_1 & b_2 & b_3 \\ b_4 & b_5 & b_6 \\ b_7 & b_8 & b_9 \end{bmatrix} \begin{Bmatrix} \phi_i \\ \phi_j \\ \phi_k \end{Bmatrix} \tag{2.5.5}$$

Here it does not matter how the a's and b's are calculated. Also, we ignore symmetry of the [**k**]'s to show more clearly what happens to the a's and b's. Nodal loads in element 1, in element and structure node numberings, respectively, are

$$\begin{aligned} p_i &= a_1\phi_i + a_2\phi_j + a_3\phi_k, & p_1 &= a_1\phi_1 + a_2\phi_4 + a_3\phi_2 \\ p_j &= a_4\phi_i + a_5\phi_j + a_6\phi_k, & p_4 &= a_4\phi_1 + a_5\phi_4 + a_6\phi_2 \\ p_k &= a_7\phi_i + a_8\phi_j + a_9\phi_k, & p_2 &= a_7\phi_1 + a_8\phi_4 + a_9\phi_2 \end{aligned} \tag{2.5.6}$$

The equation $p_3 = 0$ is added to the second group of equations; then the coefficients are reordered in these equations to suit the format

$$\{p_1 \; p_2 \; p_3 \; p_4\} = [\mathbf{k}]_{1 \text{ expanded}} \{\phi_1 \; \phi_2 \; \phi_3 \; \phi_4\} \tag{2.5.7}$$

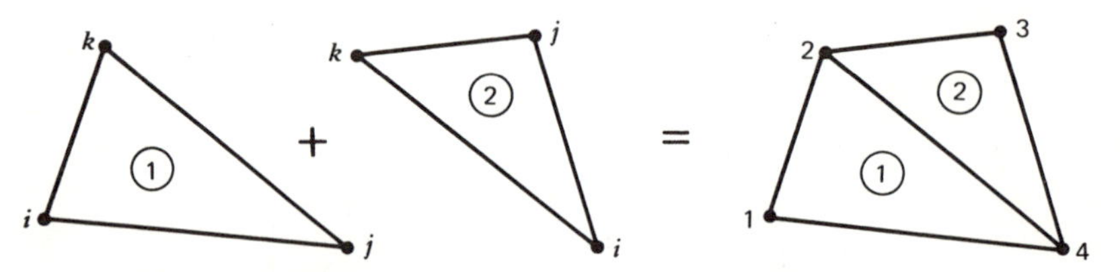

Figure 2.5.1. A four-node, four-d.o.f. structure built of two triangular elements like the triangle in Figure 1.1.2. Each node has a single d.o.f. ϕ_i. Nodes are numbered arbitrarily.

Element 2 is treated similarly. Now, because the two matrices $[\mathbf{k}]_1$ and $[\mathbf{k}]_2$ have the same size and operate on the same vector $\{\mathbf{D}\} = \{\phi_1\ \phi_2\ \phi_3\ \phi_4\}$, they can be directly added to yield the structure matrix $[\mathbf{K}]$. Thus

$$[\mathbf{K}] = \begin{bmatrix} a_1 & a_3 & 0 & a_2 \\ a_7 & a_9 & 0 & a_8 \\ 0 & 0 & 0 & 0 \\ a_4 & a_6 & 0 & a_5 \end{bmatrix} + \begin{bmatrix} 0 & 0 & 0 & 0 \\ 0 & b_9 & b_8 & b_7 \\ 0 & b_6 & b_5 & b_4 \\ 0 & b_3 & b_2 & b_1 \end{bmatrix} \tag{2.5.8}$$

Coefficients below the diagonal of a $[\mathbf{k}]$ matrix may appear above the diagonal in $[\mathbf{K}]$. This happens in the present example (but not in the truss example) because expansion *and rearrangement* are needed to make $\{\mathbf{d}\}_1$ and $\{\mathbf{d}\}_2$ the same as $\{\mathbf{D}\}$.

If element node labels are switched but structure node labels are maintained, coefficients in $[\mathbf{k}]$ are reordered but find their way to the same positions in $[\mathbf{K}]$ as before.

2.6 ASSEMBLY OF EQUILIBRIUM EQUATIONS

We continue with the structure of Fig. 2.5.1 before returning to the truss and making general statements. Consider, for example, the load p_4 required at node 4 to maintain equilibrium under arbitrary values of the nodal d.o.f. Each element that contains node 4 contributes to p_4. With the aid of Eqs. 2.5.5, we write

$$p_4 = (a_4\phi_i + a_5\phi_j + a_6\phi_k)_1 + (b_1\phi_i + b_2\phi_j + b_3\phi_k)_2 \tag{2.6.1}$$

Now switch from element node labels to structure node labels on the ϕ's and gather terms.

$$\begin{aligned} p_4 &= a_4\phi_1 + a_5\phi_4 + a_6\phi_2 + b_1\phi_4 + b_2\phi_3 + b_3\phi_2 \\ p_4 &= a_4\phi_1 + (a_6 + b_3)\phi_2 + b_2\phi_3 + (a_5 + b_1)\phi_4 \\ p_4 &= K_{41}\phi_1 + K_{42}\phi_2 + K_{43}\phi_3 + K_{44}\phi_4 \end{aligned} \tag{2.6.2}$$

Equation 2.6.2 produces the K_{ij} in the last row of Eq. 2.5.8. Note that the location in $[\mathbf{K}]$ to which a k_{ij} is assigned depends on the *structure* node labels. An *element* node label is an *arbitrary* letter or number. For assembly, all we must know is to what structure node it corresponds.

This example shows that expansion to structure size is a conceptual device that need not actually be carried out. Instead, we use an algorithm that assigns coefficients of $[\mathbf{k}]$ to their proper locations in $[\mathbf{K}]$. Fortran code is shown in Fig. 2.6.1. Arrays S and SE contain $[\mathbf{K}]$ and $[\mathbf{k}]$, respectively. Structure node labels that correspond to element node labels i, j, and k have already been stored in rows 1, 2, and 3 of array NOD, which has as many columns as there are elements. Subroutine ELEMNT, not shown, returns $[\mathbf{k}]$ in array SE. Information in SE is repeatedly created and destroyed as SE is used for each element in turn. It is assumed that array S has previously been cleared to zero.

Before writing equilibrium equations for the truss, we introduce two element load vectors. They are loads applied *to* nodes *by* elements. If gravity acts in the $-y$ direction of Fig. 2.4.1, a uniform bar of weight W applies forces $\{\mathbf{r}_W\}$ to the nodes.

$$\{\mathbf{r}_W\} = \frac{W}{2}\{0 \quad -1 \quad 0 \quad -1\} \tag{2.6.3}$$

If a fully restrained member is heated T° it sustains an axial compressive force of αEAT, where α is the coefficient of thermal expansion. Therefore forces $\{\mathbf{r}_T\}$ are applied to the nodes.

$$\{\mathbf{r}_T\} = \alpha EAT\{-c \quad -s \quad c \quad s\} \tag{2.6.4}$$

where again $c = \cos\beta$ and $s = \sin\beta$ (Fig. 2.4.1). The same forces $\{\mathbf{r}_T\}$ arise from force fitting a bar αLT units too long.

Forces $[\mathbf{k}]\{\mathbf{d}\}$ are produced by nodal displacements and are defined as forces applied *by* nodes *to* an element. Equal and opposite forces are applied *to* nodes *by* the element. Accordingly, the net forces $\{\mathbf{r}_{\text{net}}\}$ applied *to* the nodes are

$$\{\mathbf{r}_{\text{net}}\} = -[\mathbf{k}]\{\mathbf{d}\} + \{\mathbf{r}_W\} + \{\mathbf{r}_T\} \tag{2.6.5}$$

Physically, nodes can be viewed as pins that connect the bars.

Structural equilibrium equations are written by isolating each node as a free-body diagram and stating that loads applied to it sum to zero. Loads consist of externally applied forces $\{\mathbf{P}\}$ and forces $\{\mathbf{r}_{\text{net}}\}$ from each bar. The set of equilibrium equations, two for each node, is

$$\{\mathbf{P}\} + \sum_{n=1}^{\text{numel}} \{\mathbf{r}_{\text{net}}\}_n = 0 \tag{2.6.6}$$

where numel = number of elements. Equations 2.6.5 and 2.6.6 yield

$$[\mathbf{K}]\{\mathbf{D}\} = \{\mathbf{R}\} \tag{2.6.7}$$

where

$$[\mathbf{K}] = \sum_{n=1}^{\text{numel}} [\mathbf{k}]_n, \qquad \{\mathbf{R}\} = \{\mathbf{P}\} + \sum_{n=1}^{\text{numel}} (\{\mathbf{r}_W\}_n + \{\mathbf{r}_T\}_n)$$

Summations imply expansion of element matrices to "structure size," with possible reordering of coefficients, as discussed in Section 2.5, so that $\{\mathbf{d}\}_n$ of each element n becomes identical to $\{\mathbf{D}\}$. Equation 2.6.7 is a set of nodal equilibrium equations that implies assembly of element matrices.

Figure 2.6.2 shows Fortran code for assembly of the plane truss. There are two d.o.f. per node, not one as in Fig. 2.6.1, and the code looks slightly more complicated. Subroutine ELEMNT returns $[\mathbf{k}]$ in array SE and element loads $\{\mathbf{r}_W\} + \{\mathbf{r}_T\}$ in array RE. Element loads are added into an initially null structure load vector R with the same expansion to structure size and possible reordering that is used for rows of $[\mathbf{k}]$. External loads $\{\mathbf{P}\}$, if any, must be added into $\{\mathbf{R}\}$ *after completion* of the assembly process in Fig. 2.6.2.

Figure 2.6.2 generates a singular $[\mathbf{K}]$. Displacement boundary conditions are needed to enable a solution (Section 2.9).

Other arrangements of the assembly process are possible. One is to generate matrices for each element in turn and write them on a mass storage unit as soon as they are generated. Primary storage that contained data used in element formulation is then free to contain structure matrices $[\mathbf{K}]$ and $\{\mathbf{R}\}$, which are built by recalling element matrices and assembling them. Another arrangement is to work through the structure node by node. Each element that contains the node is formulated, and its contribution to equilibrium equations of the node is assembled into the structure. Thus each element is formulated as many times as it has nodes. The relative efficiency of the two methods depends on how complicated it is to generate $[\mathbf{k}]$ and on the relative cost of CPU time and mass storage.

A compromise assembly arrangement is to consider nodes in numerical order, as in the second arrangement, but to add *all* coefficients of element matrices into $[\mathbf{K}]$ and $\{\mathbf{R}\}$ as soon as they are formulated. Mass storage is not used. Unnecessary generation of element matrices is avoided by flagging each element number after it is used.

```
      DO 500 N=1,NUMEL
      CALL ELEMNT
      KK(1) = NOD(1,N)
      KK(2) = NOD(2,N)
      KK(3) = NOD(3,N)
      DO 400 I=1,3
      K = KK(I)
      DO 300 J=1,3
      L = KK(J)
      S(K,L)=S(K,L)+SE(I,J)
  300 CONTINUE
  400 CONTINUE
  500 CONTINUE
```

Figure 2.6.1 Assembly of 3-by-3 element stiffness matrices (as for the elements of Fig. 2.5.1). NUMEL = number of elements.

```
      DO 500 N=1,NUMEL
      CALL ELEMNT
      KK(2) = 2*NOD(1,N)
      KK(4) = 2*NOD(2,N)
      KK(1) = KK(2) - 1
      KK(3) = KK(4) - 1
      DO 400 I=1,4
      K = KK(I)
      R(K) = R(K) + RE(I)
      DO 300 J=1,4
      L = KK(J)
      S(K,L)=S(K,L)+SE(I,J)
  300 CONTINUE
  400 CONTINUE
  500 CONTINUE
```

Figure 2.6.2 Assembly of stiffness matrices and element load vectors for a plane truss. NUMEL = number of elements.

2.7 NODE NUMBERING THAT YIELDS A BANDED MATRIX

A matrix is "banded" if all nonzero coefficients cluster about the diagonal. Banding is a simple way to exploit matrix sparsity because zeros outside the band need be neither stored nor processed.

The *number* of nonzero coefficients in [**K**] is independent of how nodes are numbered. But changing the node numbers changes their *arrangement*. The truss of Fig. 2.7.1 yields the stiffness matrix of Fig. 2.7.2*a*. While zeros appear inside the band because some bars are horizontal or vertical, *only* zeros appear outside the band. The assembly process dictates that in the stiffness matrix of *any* structure, not only the plane truss, the entry in column j to the right of the diagonal in row i is zero unless the ith and jth d.o.f. are both in the same element.

The *semibandwidth,* given the symbol B, is six in Fig. 2.7.2*a* (B is defined more clearly later). The total bandwidth includes coefficients to the left of the diagonal and is therefore $2B - 1$. A small bandwidth is usually achieved by placing consecutive node numbers across the shorter dimension of a structure.

The entire information content of a symmetric matrix resides in the NB coefficients of the semiband. In practice, matrix order N may greatly exceed B, so there is obvious merit in storing NB coefficients rather than all N^2 matrix coefficients (see Appendix B). Also, we will see in Section 2.11 that equation-solving expense decreases by a factor of about $3B^2/N^2$ if we operate on only NB coefficients instead of on all coefficients in the upper triangle.

A simple storage format for the semiband is shown in Fig. 2.7.2*b*. Rows are shifted left: 1 space for row 2, 2 spaces for row 3, and $i - 1$ spaces for row i. Thus all diagonal coefficients of the matrix appear in the first column of the array. To program the assembly of [**K**] in band form, we change the innermost loop of Fig. 2.6.2 to the form shown in Fig. 2.7.3. The IF statement avoids using coefficients of the element matrix that would fall below the diagonal of the structure matrix.

Semibandwidth B is defined as follows. Find the column number j of the last nonzero entry in row i of [**K**] and compute $b_i = 1 + (j - i)$. Do this for all rows and identify the largest b_i as B. In a computer program we find B by scanning the elements. Let L be the number of unrestrained d.o.f. per node ($L = 2$ for the plane truss). Also, let J and K be the smallest and largest node numbers in the element, however many nodes it

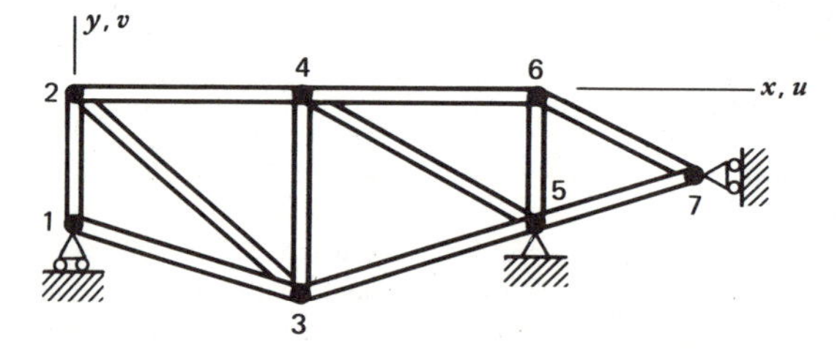

Figure 2.7.1 Plane truss with node numbering that yields a banded matrix.

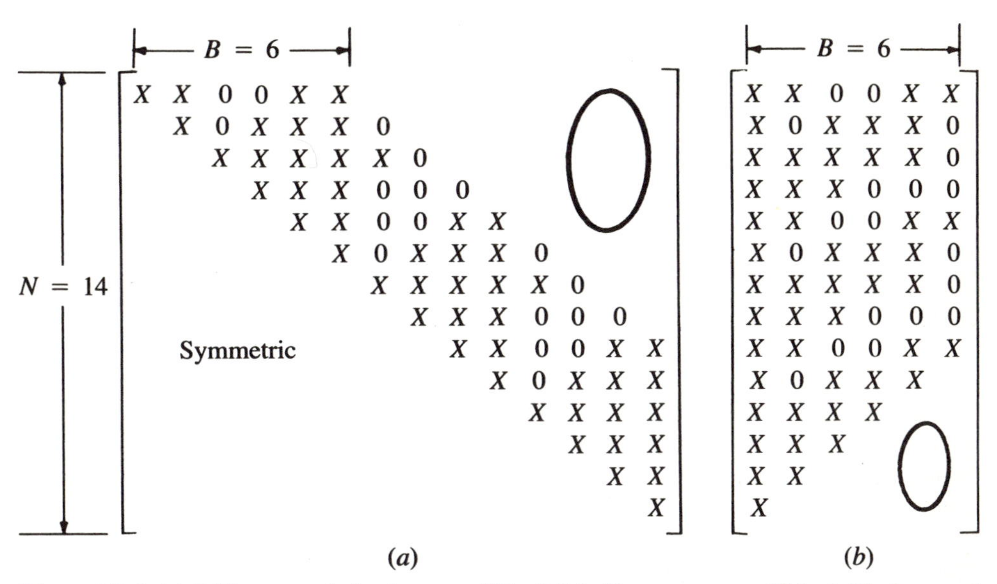

Figure 2.7.2. (*a*) Stiffness matrix for the truss of Fig. 2.7.1. X = nonzero coefficient. Displacement boundary conditions have not yet been imposed. (*b*) Band form storage of the same stiffness matrix.

has. When expanded to structure size, an element matrix spans columns $i = 1 + (J - 1)L$ to $j = LK$. Included in this span are $b = 1 + (j - i) = L(K - J + 1)$ coefficients. The largest b for any element is B.

The terms "skyline" and "profile" are in common use [2.11]. The *skyline* or *envelope* drapes itself over the first nonzero coefficient in each column (Fig. 2.7.4). The *profile* P is the number of coefficients between the diagonal and the skyline, including the diagonal. Thus $P = 12$ in Fig. 2.7.4*a* and $P = 9$ in Fig. 2.7.4*b*. (Note that $B = 4$ for both.) Symbolically, if p_j is the number of rows in column j from the diagonal to the skyline, $P = \Sigma p_j$ where j runs from 1 to N. Efficient equation solvers have a solution time proportional to Σp_j^2.

Band storage format is simple but may not be the most effective sparse-matrix format. Instead, we might store the successive columns under the skyline in a one-dimensional array whose length depends on the profile. For Fig. 2.7.4*b*, this option takes less space than band storage. Another reason to prefer the second arrangement in Fig. 2.7.4 is that equation solving creates "fills": zeros under the skyline become nonzeros. There are three such fills in Fig. 2.7.4*a* but none in Fig. 2.7.4*b*. Storage space must be reserved for fills, and fills must be processed in subsequent operations. We return to these matters in Sections 2.12 and 2.13. Information is abundant; for example, see Refs. 2.1 to 2.3.

```
    DO 300 J=1,4
    IF (KK(J) .LT. K)  GO TO 300
    L = KK(J) - K + 1
    S(K,L) = S(K,L) + SE(I,J)
300 CONTINUE
```

Figure 2.7.3. Altered form of the innermost loop of Fig. 2.6.2 to achieve the storage format of Fig. 2.7.2*b*.

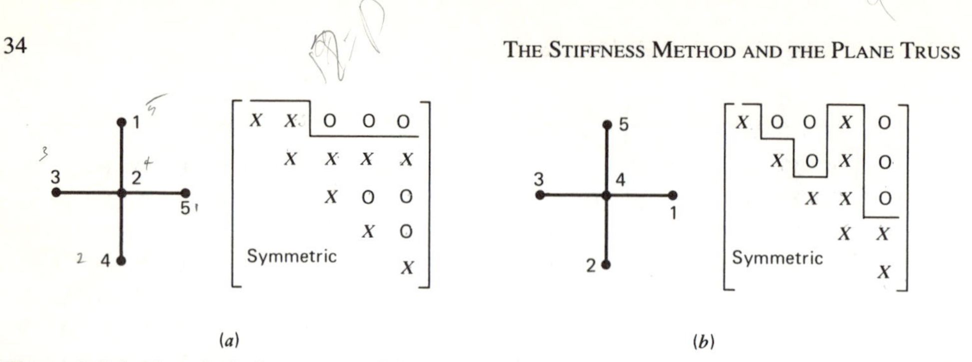

Figure 2.7.4. Hypothetical structure with one d.o.f. per node. Two different numberings and their corresponding stiffness matrices are shown. The solid line is the skyline.

2.8 AUTOMATIC RENUMBERING OF NODES

Imagine that a node must be added to the left of the truss of Fig. 2.7.1. Two new bars will connect it to existing nodes 1 and 2. The entire structure must be renumbered if low bandwidth is to be preserved. It would be far more convenient if the new node could be given the next available number, 8, and the computer program could do the rest. That is, the computer should accept arbitrary node numbers, adopt new numbers for efficient internal operations, and produce results in the user's original numbering system. It can be done. In this section one of many algorithms is outlined.

Renumbering methods exploit graph theory and its terminology. Nodes are *adjacent* if they are attached to the same element. The *degree* of a node is the number of other nodes adjacent to it. The reverse Cuthill-McKee method can now be outlined [2.3, 2.4]. It is simple but effective, and it is the basis of still better schemes.

Pick a starting node, usually one of low degree, and number it 1. Node 1 is in Level 1. Nodes adjacent to node 1 are then numbered 2, 3, and so on, in order of increasing degree. These are nodes in Level 2, and unnumbered nodes adjacent to those in Level 2 are in Level 3. Nodes in Level 3 are then numbered, again in order of increasing degree. Subsequent levels are treated similarly. Ties in degree are broken arbitrarily. When all n nodes have been numbered, the numbering is *reversed* by replacing number i by $n - i + 1$, for $i = 1, 2, \ldots, n$. Reversal does not change the bandwidth, but it usually reduces the profile and never increases it. The entire renumbering may be repeated with a different starting node to see if there is any improvement.

Figure 2.7.4*a* is a Cuthill-McKee numbering, and Fig. 2.7.4*b* is its reversed form. Figure 2.8.1 is a structure that would be difficult to number effectively by hand. The (unreversed) Cuthill-McKee numbering, not shown, has a profile of 86.

Renumbering schemes rarely attempt to minimize anything, nor do they even guarantee improvement. But the goal of reducing bandwidth—or profile, or fills—is usually achieved. The goal selected merely permits us to choose the best among different numberings. No single renumbering strategy is best for all problems. The Cuthill-McKee procedure considers only node numbers; it ignores the number of d.o.f. per node. Greater sophistication may not be worthwhile. Schemes that renumber elements instead of nodes are used for wavefront equation solvers (Section 2.13).

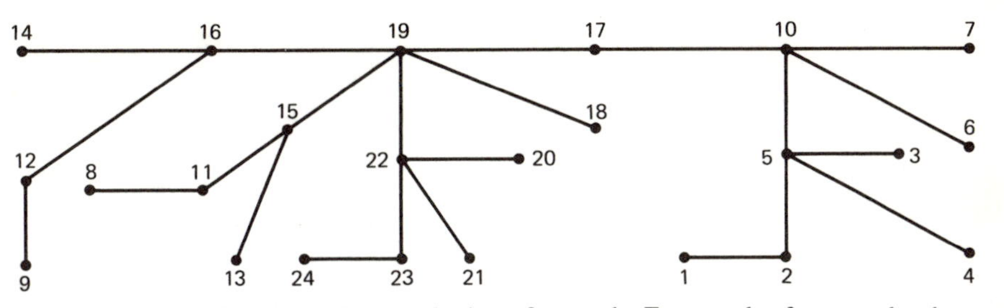

Figure 2.8.1. Reverse Cuthill-McKee numbering of a graph. For one d.o.f. per node, the semi-bandwidth is 8 and the profile is 63.

A table of nodal connectivity is needed for node renumbering. There can be appreciable cost in generating this table and doing the renumbering. But automatic renumbering is usually cost effective, especially if the same numbering is used in repeated solutions, as is the case in nonlinear problems. Renumbering is always worthwhile from the viewpoint of user convenience.

At present, Ref. 2.5 may be the best node renumbering algorithm. Results are as good as those of reverse Cuthill-McKee and are obtained at roughly one-tenth the cost [2.5, 2.11].

2.9 DISPLACEMENT BOUNDARY CONDITIONS

We now are able to generate the structure stiffness equations. But the stiffness matrix is singular: a unique solution is possible only after rigid-body motion is prevented.

The partitioning method of Eq. 2.3.4 enlarges bandwidth and profile and is not suited to computer programming. A method is used that effectively generates the first set of Eqs. 2.3.5 by not writing the second set [2.6]. Thus we generate equations whose displacement vector $\{\mathbf{D}\}$ contains only unknowns and whose load vector $\{\mathbf{R}\}$ contains only knowns. For the time being we assume that *all prescribed d.o.f. are fixed;* that is, $\{\mathbf{D}_c\} = 0$ in Eq. 2.3.5.

The treatment uses an array ID that lists the equation number corresponding to each d.o.f. of the *unsupported* structure. Array ID has as many rows as the maximum number of d.o.f. per node and as many columns as there are nodes.

Consider, for example, the truss of Fig. 2.7.1. Imagine that it is to be analyzed with a general-purpose program that allows six d.o.f. per node (three displacements and three rotations). Therefore ID has six rows and seven columns. We start with ID null and, by input data, insert a 1 corresponding to each d.o.f. to be excluded from the analysis (see Fig. 2.9.1*a*). The last four rows of 1's state that all z-direction translations and all nodal rotations are to be excluded.[2] Entries ID(2,1) = ID(1,5) = ID(2,5) = ID(1,7) = 1 mean

[2]Rotation and z-direction motion of structure nodes must be prohibited because the truss elements do not resist these motions. If retained, these d.o.f. would be associated with zero stiffness, and $[\mathbf{K}]$ would be singular. Elements can still rotate in the xy plane because the suppressed rotations are not among their nodal d.o.f.

$$[\mathbf{ID}] = \begin{bmatrix} 0&0&0&0&1&0&1 \\ 1&0&0&0&1&0&0 \\ 1&1&1&1&1&1&1 \\ 1&1&1&1&1&1&1 \\ 1&1&1&1&1&1&1 \\ 1&1&1&1&1&1&1 \end{bmatrix} \qquad [\mathbf{ID}] = \begin{bmatrix} 1&2&4&6&0&8&0 \\ 0&3&5&7&0&9&10 \\ 0&0&0&0&0&0&0 \\ 0&0&0&0&0&0&0 \\ 0&0&0&0&0&0&0 \\ 0&0&0&0&0&0&0 \end{bmatrix}$$

(*a*) (*b*)

Figure 2.9.1. Array ID for the truss of Fig. 2.7.1. (*a*) After input data supply 1's for d.o.f. to be suppressed. (*b*) After the program of Fig. 2.9.2 converts it to a table of equation numbers.

that $v_1 = u_5 = v_5 = u_7 = 0$, so these d.o.f. are also to be excluded. Next, we number the 0's in ID by working down the successive columns and replace the original 1's by 0's (see Fig. 2.9.1*b*). The result is interpreted as stating that u_1 is the first d.o.f. in $\{\mathbf{D}\}$, u_2 is the second, v_2 is the third, and so on, until v_7 is the tenth and last. Matrix $[\mathbf{K}]$ will be 10 by 10. The conversion of Fig. 2.9.1*a* to Fig. 2.9.1*b* is accomplished by Fig. 2.9.2, which also yields NEQ = 10.

Assembly of structural equations requires a slight modification of Figs. 2.6.2 and 2.7.3 (see Fig. 2.9.3). Array KK is filled with structural equation numbers for each bar in turn: for example, bar 1-3 of Fig. 2.7.1 yields KK entries 1, 0, 4, and 5. The first IF statement discards rows associated with suppressed d.o.f., such as the preceding KK(2) = 0. The second IF statement discards columns of $[\mathbf{K}]$ associated with suppressed d.o.f. and also discards k_{ij} that would fall below the diagonal of $[\mathbf{K}]$.

For stress computation, described in Section 2.10, the d.o.f. of each element are needed. That is, for element i, we must extract $\{\mathbf{d}\}_i$ from the solution vector $\{\mathbf{D}\}$. Vector $\{\mathbf{d}\}_i$ may contain zeros that correspond to suppressed d.o.f. excluded from $\{\mathbf{D}\}$. See Fig. 2.9.4.

Next, we relax an initial restriction and permit *prescribed, nonzero d.o.f.* Consider the truss of Fig. 2.2.1. We can closely approximate its fixed support conditions by Fig. 2.9.5*a* if we make bars 4, 5, and 6 very stiff, say 10^6 times as stiff as the others. All six d.o.f. of the original truss are retained in $\{\mathbf{D}\}$. Now imagine, for example, that u_2 is not to be zero as in Fig. 2.2.1, but the prescribed value $\bar{u}_2$. We need only apply a force

```
      NEQ = 0
      DO 62 N=1,NUMNP
      DO 60 J=1,NDOF
C --- TRANSFER IF D.O.F. IS FIXED. OTHERWISE INCREMENT NEQ.
      IF (ID(J,N) .GT. 0)  GO TO 58
      NEQ = NEQ + 1
      ID(J,N) = NEQ
      GO TO 60
   58 ID(J,N) = 0
   60 CONTINUE
   62 CONTINUE
```

Figure 2.9.2. Fortran statements that generate a table of equation numbers, such as Fig. 2.9.1*b*. NUMNP = number of structure of nodes. NDOF = number of d.o.f. allowed per node. NEQ = number of equations to be solved for displacements.

```
      DO 500 N=1,NUMEL
      CALL ELEMNT
      I = NOD(1,N)
      J = NOD(2,N)
      KK(1) = ID(1,I)
      KK(2) = ID(2,I)
      KK(3) = ID(1,J)
      KK(4) = ID(2,J)
      DO 400 I=1,4
      IF (KK(I) .LE. 0)  GO TO 400
      K = KK(I)
      R(K) = R(K) + RE(I)
      DO 300 J=1,4
      IF (KK(J) .LT. K)  GO TO 300
      L = KK(J) - K + 1
      S(K,L) = S(K,L) + SE(I,J)
  300 CONTINUE
  400 CONTINUE
  500 CONTINUE
```

Figure 2.9.3. Assembly of active stiffness equations in the banded format described in Section 2.7.

to the right at node 2 large enough to stretch bar 5 the prescribed amount $\bar{u}_2$. In physical terms, we add a big spring to the three-bar truss, then apply a big force to stretch it. The comparatively flimsy truss is simply dragged along. In matrix terms, we add a large stiffness $\bar{K}$ to the third diagonal coefficient in $[\mathbf{K}]$ and replace the third coefficient in $\{\mathbf{R}\}$ by the large force $\bar{K}\bar{u}_2$. *Caution*. Relatively stiff members can be connected to supports but should not be placed *within* an elastic structure, since numerical difficulty is then likely (Section 15.5). Also, stiff support members should be aligned with directions of nodal d.o.f. If skew, as in Fig. 2.9.5*b*, they generate large diagonal *and* off-diagonal coefficients in $[\mathbf{K}]$ and again make numerical difficulty likely. The latter problem can be avoided by adopting new d.o.f., normal and tangential at the skew support (Section 6.4).

Another way to prescribe d.o.f. is shown in Fig. 2.9.6. The matrices are *arbitrary* and do not represent any particular structure. The condition $u = \bar{u}$ is imposed by substituting $u = \bar{u}$ into all equations, placing all knowns on the right side, nulling the first row and column, and writing the trivial statement $u = \bar{u}$ as the first equation. The symmetry and band structure of the matrix are preserved. Solution automatically gives $u = \bar{u}$, where $\bar{u}$ may or may not be zero. The procedure of Fig. 2.9.6 can be repeated for other d.o.f. If *all* d.o.f. are prescribed, the coefficient matrix becomes a unit matrix.

```
      M = 0
      DO 220 K=1,NNEL
C --- N IS THE STRUCTURE NUMBER OF NODE K OF THE NTH ELEMENT.
      N = NOD(K,NTH)
      DO 200 L=1,NDOF
      M = M + 1
      DE(M) = 0.
      J = ID(L,N)
C --- J IS ZERO ONLY IF THE D.O.F. IS FIXED.
      IF (J .GT. 0)  DE(M) = D(J)
  200 CONTINUE
  220 CONTINUE
```

Figure 2.9.4. Fortran code to extract the displacement vector DE of the *N*th element from the structure solution vector D. NNEL = number of nodes per element. NDOF = number of degrees of freedom per node allowed by the program (NDOF = 6 in Fig. 2.9.1*b*).

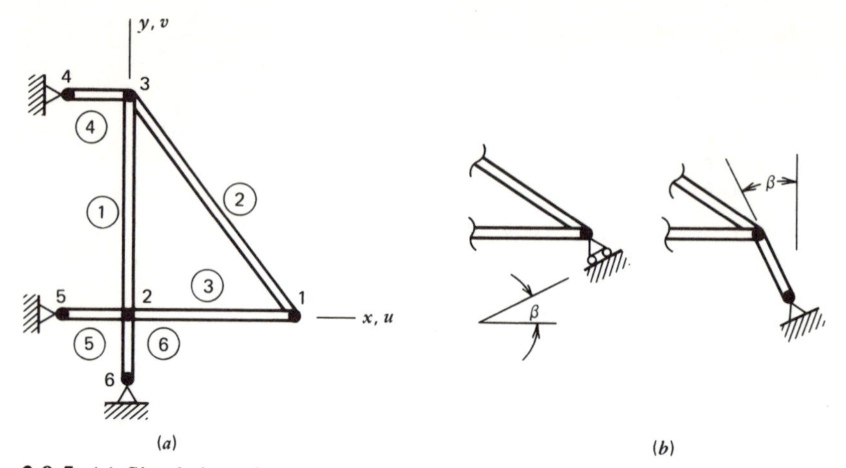

Figure 2.9.5. (*a*) Simulation of supports in Fig. 2.2.1 by adding three stiff bars. (*b*) Simulation of a skew support (see caution in the text).

The method of Fig. 2.9.6 is exact, not approximate. Also, like the method of Fig. 2.9.3, it can be applied to the separate element matrices prior to assembly. This is desirable because compact storage formats for the structure matrix can make it awkward to operate on.

2.10 SUPPORT REACTIONS. STRESS COMPUTATION

Procedures in Section 2.9 discard information that would yield forces associated with prescribed d.o.f. (the second of Eqs. 2.3.5). If these reactions are required, how can they be found? One method is to write the necessary rows of [**K**] in a mass storage file before they are discarded or destroyed. After solving for displacements, recall the rows and multiply each by the displacement vector.

$$\sum_j k_{ij}D_j = R_i, \qquad \text{or} \qquad \sum_m \left(\sum_j k_{ij}\, d_j \right) = R_i \tag{2.10.1}$$

The first expression sums the j nonzero entries in row i. The second is a similar sum, taken over the m elements that contain the ith d.o.f. In the second expression, the coefficients k_{ij} could be recomputed instead of stored and then recalled.

$$\begin{bmatrix} K_1 & K_2 & K_3 \\ K_4 & K_5 & K_6 \\ K_7 & K_8 & K_9 \end{bmatrix} \begin{Bmatrix} u \\ v \\ w \end{Bmatrix} = \begin{Bmatrix} a \\ b \\ c \end{Bmatrix}, \qquad \begin{bmatrix} 1 & 0 & 0 \\ 0 & K_5 & K_6 \\ 0 & K_8 & K_9 \end{bmatrix} \begin{Bmatrix} u \\ v \\ w \end{Bmatrix} \begin{Bmatrix} \bar{u} \\ b - K_4\bar{u} \\ c - K_7\bar{u} \end{Bmatrix}$$

Figure 2.9.6. Alternative method of treating prescribed d.o.f.

Another method is used with the auxiliary support members of Fig. 2.9.5*a*. If a nonzero displacement is *not* prescribed—that is, if a rigid support is approximated—the (very small) displacement of the support member is multiplied by its (very large) stiffness to find the force it exerts. If a nonzero displacement *is* prescribed, the following argument applies. Two springs are in parallel: a small structure stiffness K and a large support stiffness $\bar{K}$. The load assigned is $\bar{K}\bar{D}$, where $\bar{D}$ is the prescribed d.o.f. Thus

$$(K + \bar{K})\,D = \bar{K}\bar{D}, \qquad \text{and} \qquad R = KD = \bar{K}(\bar{D} - D) \tag{2.10.2}$$

where R is the required reaction. Factor $\bar{D} - D$ should contain (say) three significant digits or more, so $\bar{K}$ must not be so large that it obliterates K. If $\bar{D} = 0$, Eq. 2.10.2 reduces to the first case of this paragraph.

Stress in a plane truss element is computed as follows. In Fig. 2.4.2, the elongation of the bar is

$$e = (u_j - u_i)\cos\beta + (v_j - v_i)\sin\beta \tag{2.10.3}$$

which can be computed after nodal u's and v's are known. Stress associated with e must be superposed on any initial stress that may be present. Consider, for example, a bar with thermal expansion coefficient α. If heated $T°$ before nodal displacements are permitted, it carries an axial compressive stress $E\alpha T$. Therefore the net stress is

$$\sigma = E\epsilon - E\alpha T = E\left(\frac{e}{L} - \alpha T\right) \tag{2.10.4}$$

If the bar is free to expand, $\sigma = 0$ and $e = \alpha LT$.

2.11 GAUSS ELIMINATION SOLUTION OF EQUATIONS

The application of Gauss elimination to stiffness equations $[\mathbf{K}]\{\mathbf{D}\} = \{\mathbf{R}\}$, where the coefficient matrix $[\mathbf{K}]$ is *symmetric,* is described. The first equation is symbolically solved for D_1, then substituted into the subsequent equations. The second equation is similarly treated, then the third, and so on. This forward-reduction process alters $\{\mathbf{R}\}$ and changes $[\mathbf{K}]$ to upper triangular form, with 1's on the diagonal. Finally, unknowns are found by back-substitution, so that the numerical value of D_1 is found last.

The *i*th equation is used to eliminate the *i*th unknown: all pivots are diagonal coefficients K_{ii}. No search is made for the largest pivot in each elimination. This simple approach is acceptable because pivots are not small unless the structure is nearly unstable or badly modeled (see Chapter 15).

Example

Consider three equations, so that indexes i and j on the K_{ij} run from 1 to 3. Elimination of D_1 is accomplished as follows. Divide the first equation by K_{11}. Multiply it by K_{21} and subtract it from the second equation, then multiply it by K_{31} and subtract it from the third equation. Thus

$$\begin{bmatrix} 1 & K_{12}/K_{11} & K_{13}/K_{11} \\ 0 & K_{22} - K_{21}(K_{12}/K_{11}) & K_{23} - K_{21}(K_{13}/K_{11}) \\ 0 & K_{32} - K_{31}(K_{12}/K_{11}) & K_{33} - K_{31}(K_{13}/K_{11}) \end{bmatrix} \begin{Bmatrix} D_1 \\ D_2 \\ D_3 \end{Bmatrix} = \begin{Bmatrix} R_1/K_{11} \\ R_2 - K_{21}(R_1/K_{11}) \\ R_3 - K_{31}(R_1/K_{11}) \end{Bmatrix} \qquad (2.11.1)$$

Because of symmetry, $K_{ij} = K_{ji}$. The reduced matrix remains symmetric, and computations can be restricted to the upper triangle. The next step of forward reduction is to process the second equation similarly as if it were the first equation of a new 2-by-2 system. Completion of forward reduction yields the 3-by-3 system $[\mathbf{K}']\{\mathbf{D}\} = \{\mathbf{R}'\}$, where $[\mathbf{K}']$ is upper triangular and $K'_{ii} = 1$. Finally, by back-substitution,

$$D_3 = R'_3, \qquad D_2 = R'_2 - K'_{23}D_3, \qquad D_1 = R'_1 - K'_{12}D_2 - K'_{13}D_3 \qquad (2.11.2)$$

A numerical form of this example appears in Fig. 2.11.1. The first two reductions are shown, and the third requires only the division of row 3 in Fig. 2.11.1*d* by $\frac{1}{3}$. Then, by back-substitution,

$$D_3 = 24, \qquad D_2 = 4 - (-\tfrac{2}{3})24 = 20, \qquad D_1 = 2 - (-\tfrac{1}{2})20 = 12 \qquad (2.11.3)$$

We see that the semibandwidth $B = 2$ is preserved throughout reduction and that each reduction involves only B rows and up to B coefficients per row.

In physical terms, each elimination frees a d.o.f. to move. In Fig. 2.11.1, elimination of D_1 frees D_1, so that bars 1 and 2 act as two springs in series with a net stiffness of $\frac{1}{2}$. This stiffness combines with the unit stiffness of bar 3 to yield the stiffness of $\frac{3}{2}$ in Fig. 2.11.1*c*. Elimination of D_2 allows three bars to act in series. They have a net stiffness of $\frac{1}{3}$, as seen in Fig. 2.11.1*d*. In general, each elimination releases a constraint, so the d.o.f. can move as dictated by loads and the elastic properties of the rest of the structure. This argument indicates that if the structure is stable, diagonal coefficients continually decrease but remain positive at all stages of reduction.

Sometimes we want to re-solve the equations using a new load vector but the same stiffness matrix. Reduction of $[\mathbf{K}]$ is by far the most costly part of the solution, so we want to restrain the previously reduced coefficient matrix and do only right-side reduction and back-substitution. We see from Eq. 2.11.1 that the first reduction of the right side requires only K_{11} and the reduced coefficients in the first row of the coefficient matrix. A similar argument applies to subsequent reductions. Accordingly, $\{\mathbf{R}\}$ can be processed from the reduced $[\mathbf{K}]$ if diagonal coefficients are not converted to unity.

Now Fig. 2.11.2 can be understood. This equation solver [2.7] is simple and reasonably good. It is not optimal; indeed, it is as rustic a routine as should be used. But it is far better than a matrix-inversion solution or ignoring the sparsity of the coefficient matrix.

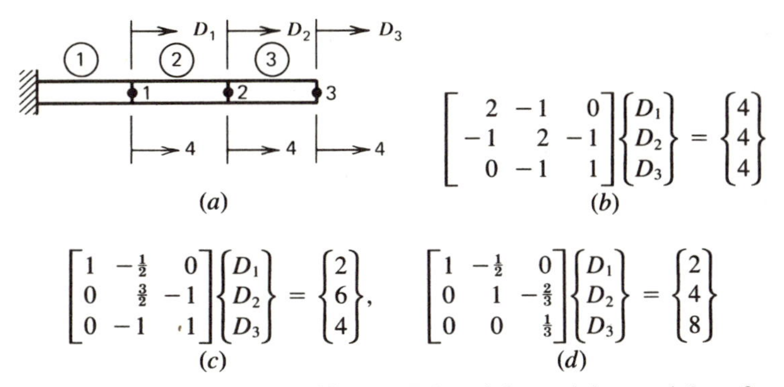

Figure 2.11.1. (*a*) Three-bar truss with stiffness $AE/L = 1$ for each bar, and three four-unit loads. (*b*) The corresponding structural equations when the left end is fixed. (*c, d*) The structural equations after one and two reductions, respectively.

For a single load vector and $N >> B$ we find, by counting the number of multiplications in the innermost loop, that solution time in Fig. 2.11.2 is proportional to $NB^2/2$. This compares with $N^3/6$ for a full but symmetric matrix. Note also that the inverse of a band matrix is full, not banded.

Here is a brief explanation of Fig. 2.11.2. The DO 790 and DO 830 loops start one of the NEQ reductions. The DO 780 and DO 820 loops substitute an equation into MBAND subsequent equations. The DO 750 loop modifies the affected coefficients in a single equation. The "proper" value S(I,1) = 1.0 for each reduced diagonal coefficient is used implicitly during back-substitution. The IF statements avoid "do-nothing" computations when S(N,L) happens to be zero. The IF statements also exploit the triangular block of zeros (see Fig. 2.7.2*b*) to avoid row subscripts larger than NEQ. Array S is cleared to zero before stiffness coefficients are added into it. Therefore IF comparisons between nonintegers are meaningful: we look for coefficients that are *exactly* zero. During reduction, index K and L need run only to MBAND because only coefficients in the "active triangle" are changed (see Fig. 2.12.1).

2.12 GAUSS AND CHOLESKI SOLUTION ALGORITHMS

Choleski decomposition is introduced because we use it in subsequent discussions and because it leads to better equation solvers than that of Fig. 2.11.2.

A symmetric, positive-definite matrix $[\mathbf{K}]$ can be expressed as

$$[\mathbf{K}] = [\mathbf{U}]^T[\mathbf{U}] \tag{2.12.1}$$

where $[\mathbf{U}]$ is an upper triangular matrix. This is the Choleski decomposition of $[\mathbf{K}]$. The U_{ij} can be induced by writing out the product in Eq. 2.12.1. They are

$$
\begin{aligned}
U_{ij} &= 0, \qquad \text{for} \qquad i > j \qquad \text{(upper triangular form)} \\
U_{11} &= \sqrt{K_{11}}, \qquad U_{1j} = K_{1j}/U_{11} \\
U_{ii} &= \left(K_{ii} - \sum_{k=1}^{i-1} U_{ki}^2 \right)^{1/2}, \qquad \text{for} \qquad i > 1 \\
U_{ij} &= \left(K_{ij} - \sum_{k=1}^{i-1} U_{ki} U_{kj} \right) \Big/ U_{ii}, \qquad \text{for} \qquad i > 1 \qquad \text{and} \qquad j > i
\end{aligned}
\tag{2.12.2}
$$

```
C --- TREAT THE CASE OF ONE OR MORE INDEPENDENT EQUATIONS.
      IF (MBAND .GT. 1)  GO TO 690
      DO 680 N=1,NEQ
  680 R(N) = R(N)/S(N,1)
      RETURN
  690 GO TO (700,800), IFLAG
C --- FORWARD REDUCTION OF THE COEFFICIENT MATRIX.
  700 DO 790 N=1,NEQ
      DO 780 L=2,MBAND
      IF (S(N,L) .EQ. 0.)  GO TO 780
      I = N + L - 1
      C = S(N,L)/S(N,1)
      J = 0
      DO 750 K=L,MBAND
      J = J + 1
  750 S(I,J) = S(I,J) - C*S(N,K)
      S(N,L) = C
  780 CONTINUE
  790 CONTINUE
C --- FORWARD REDUCTION OF THE VECTOR OF CONSTANTS.
  800 DO 830 N=1,NEQ
      DO 820 L=2,MBAND
      IF (S(N,L) .EQ. 0.)  GO TO 820
      I = N + L - 1
      R(I) = R(I) - S(N,L)*R(N)
  820 CONTINUE
  830 R(N) = R(N)/S(N,1)
C --- SOLVE FOR UNKNOWNS BY BACK-SUBSTITUTION.
      DO 860 M=2,NEQ
      N = NEQ + 1 - M
      DO 850 L=2,MBAND
      IF (S(N,L) .EQ. 0.)  GO TO 850
      K = N + L - 1
      R(N) = R(N) - S(N,L)*R(K)
  850 CONTINUE
  860 CONTINUE
      RETURN
      END
```

Figure 2.11.2. Gauss elimination equation solver for a symmetric and nonsingular coefficient matrix stored in the band form of Fig. 2.7.2*b*. NEQ = number of equations, MBAND = semi-bandwidth, IFLAG = option switch, R = constant vector, and S = coefficient matrix, stored in an NEQ-by-MBAND array. The solution vector replaces the constant vector. Pivots are diagonal coefficients, taken in order. Except for the special case MBAND = 1, start at statement 700 for the first constant vector or at statement 800 if another constant vector is to be processed. Section 15.8 describes a recommended improvement.

For accuracy, sums should be accumulated before subtraction.

Consider the application of Eq. 2.12.1 to a matrix [**K**] of semibandwidth $B = 2$. In row 1 we find $U_{1j} = 0$ for $j > 2$. Hence, dropping products that involve $U_{ij} = 0$, we find

$$\begin{aligned} U_{12}^2 + U_{22}^2 &= K_{22}, & U_{23}^2 + U_{33}^2 &= K_{33}, \\ U_{22}U_{23} &= K_{23}, & U_{33}U_{34} &= K_{34}, \\ U_{22}U_{2j} &= K_{2j} = 0 \quad \text{for} \quad j > 3, & U_{33}U_{3j} &= K_{3j} = 0 \quad \text{for} \quad j > 4 \end{aligned} \tag{2.12.3}$$

and so on. Nonzero U_{ij} are computed in this order: U_{11}, then U_{12} and U_{22}, then U_{23} and U_{33}, then U_{34} and U_{44}, and so forth (this grouping of terms serves a subsequent argument). All U_{ij} outside the band of [**K**] are zero. When a U_{ij} is computed, the corresponding K_{ij} is no longer needed. Accordingly, like Gauss elimination, Choleski decomposition can be programmed to exploit bandedness and to overwrite the stored band of [**K**] with the result of forward reduction [2.2, 2.8].

The determinant of an n-by-n matrix [**K**] is

$$\det[\mathbf{K}] = (U_{11}U_{22} \cdots U_{nn})^2 = \left(\prod_{i=1}^{n} U_{ii} \right)^2 \tag{2.12.4}$$

To evaluate det[**K**] by Gauss elimination, we can use Fig. 2.11.2. After completing the loop on statement 790,

$$\det[\mathbf{K}] = \mathsf{S(1,1)} * \mathsf{S(2,1)} * \cdots * \mathsf{S(NEQ,1)} \tag{2.12.5}$$

which is the product of the diagonal coefficients at the time they act as pivots.

Equation solving by the Choleski algorithm proceeds as follows.

$$\begin{aligned} &\text{Given:} \quad [\mathbf{K}]\{\mathbf{D}\} = \{\mathbf{R}\} \quad \text{or} \quad [\mathbf{U}]^T[\mathbf{U}]\{\mathbf{D}\} = \{\mathbf{R}\} \\ &\text{Define:} \quad [\mathbf{U}]\{\mathbf{D}\} = \{\mathbf{Y}\} \\ &\text{Solve:} \quad [\mathbf{U}]^T\{\mathbf{Y}\} = \{\mathbf{R}\} \quad \text{for} \quad \{\mathbf{Y}\} \quad \text{by forward-substitution} \\ &\text{Solve:} \quad [\mathbf{U}]\{\mathbf{D}\} = \{\mathbf{Y}\} \quad \text{for} \quad \{\mathbf{D}\} \quad \text{by back-substitution} \end{aligned} \tag{2.12.6}$$

Square-root operations in Eq. 2.12.2 can be avoided [2.9, 2.10], but root extraction is a minor expense if there are many equations. Then the Gauss and Choleski algorithms have practically the same efficiency *if the matrix is full*. They differ when applied to sparse matrices because they exploit sparsity in different ways, as we now describe.

Study of Section 2.11 shows that Gauss elimination is *row* oriented when applied to a banded matrix. A step of reduction *completes* modification of the upper row in the active triangle of Fig. 2.12.1*a*. The next step *begins* modification of another column. If all NB coefficients cannot fit in primary memory an obvious strategy is to place only the active triangle in primary memory and shift one row out and one column in after each step of reduction.

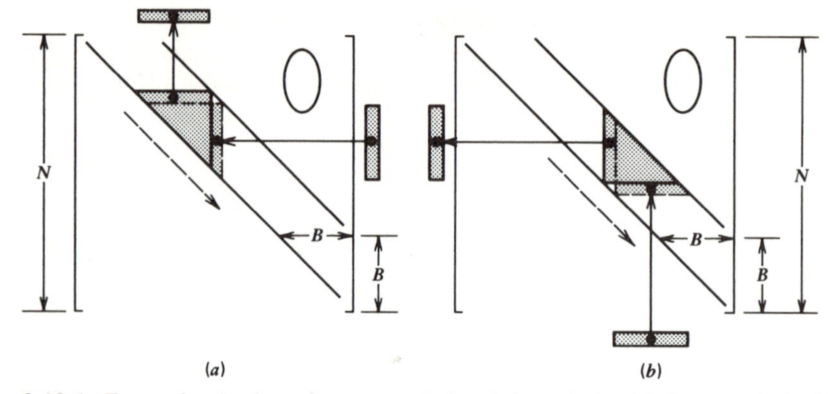

Figure 2.12.1. Forward reduction of a symmetric banded matrix by (*a*) Gauss and (*b*) Choleski algorithms. The "active triangle" represents terms affected by one of the N reduction steps. The triangle shifts down one step after each reduction.

Analogous remarks apply to Choleski decomposition, which is *column* oriented (see the sentence that follows Eq. 2.12.3). A step of reduction *completes* modification of the first column in the active triangle, and the next step *begins* modification of another row (Fig. 2.12.1*b*). Note also from the last of Eqs. 2.12.2 that each U_{ij} is computed as the scalar product of columns above row i. This feature, not present in Gauss elimination, is significant for parallel or pipeline computers, which have special prowess in vector operations [2.33].

The relative merits of Gauss and Choleski algorithms are suggested by Fig. 2.12.2. The efficiency of a Gauss elimination solver improves if the semibandwidth B (MBAND in Fig. 2.11.2) is not a constant maximum value but takes the value B_i appropriate to each row i. Note, however, that B_i pertains to the row width *when the row is eliminated*. Thus, in Fig. 2.12.2*b*, we must use $B_2 = 3$, not $B_2 = 1$. The Gauss solver recognizes that $K_{13} = 0$ and does not attempt the useless substitution of row 1 into row 3. But when it substitutes row 1 into row 2 it does not recognize that K_{23} is not changed. An IF test to recognize this must be in the innermost loop, where it is likely to use more time than it saves.

Choleski decomposition (Fig. 2.12.2*c*) skips the useless processing of a zero seen in Fig. 2.12.2*b*. We need not begin to accumulate the products $U_{ki}U_{kj}$ of Eqs. 2.12.2 until

$$\begin{bmatrix} X & X & & X & \\ X & X & & & \\ & & X & X & \\ X & & X & X & X \\ & & & X & X \end{bmatrix} \qquad \begin{bmatrix} X & X & & X & \\ & X & 0 & F & \\ & & X & X & \\ & & & X & X \\ & & & & X \end{bmatrix} \begin{matrix} B_1 = 4 \\ B_2 = 3 \\ B_3 = 2 \\ B_4 = 2 \\ B_5 = 1 \end{matrix} \qquad \begin{bmatrix} X & X & & X & \\ & X & & F & \\ & & X & X & \\ & & & X & X \\ & & & & X \end{bmatrix}$$

(*a*) (*b*) (*c*)

Figure 2.12.2. (*a*) X's denote nonzero terms in a symmetric matrix. (*b*) The matrix after Gauss reduction. F = fill, a created nonzero term. 0 = a zero that is processed but remains zero. (*c*) The matrix after Choleski decomposition.

k is large enough to reach the top of the lower of the two columns (thus, for columns 3 and 4, $U_{33}U_{34}$ is the first product).

An ''active column'' consists of all coefficients from the skyline to the diagonal in that column. Fills in Choleski (and Gauss) algorithms are confined to the active columns. Storage of the active columns in Fig. 2.12.2 requires 10 spaces. Storage of all coefficients in rows within the variable band of Fig. 2.12.2*b* requires 12 spaces. Rows (or columns) can be stored one after another in a one-dimensional array.

For most finite element structures, [**K**] is arranged so that active-column storage takes less space then variable-band row storage. This favors the column-oriented Choleski method over Gauss elimination. Also, the Choleski method is better than the variable-band Gauss method at skipping multiplications by zero.

It is possible to assemble [**U**] directly from corresponding element factors. Thus decomposition of [**K**] is avoided [2.12].

A *disadvantage* of the Choleski method is that indefinite matrices require special treatment to avoid the square root of a negative number (''mixed'' elements generate indefinite matrices).

2.13 REMARKS ABOUT EQUATION SOLVERS

Preceding sections suggest different options in equation-solving. There are many additional algorithms, some so closely related that terminology is confused. There are also many ways to store sparse matrices, preserve their sparsity, and skip multiplications by 1's and 0's. The number of permutations is enormous, and there are hundreds of papers on the subject [2.3]. Equation solving is indispensable in most areas of applied mathematics. It may account for roughly one-quarter of the ''number-crunching'' cost in a static analysis and up to three-quarters of the cost in a nonlinear or dynamic analysis. Here we note some items of particular value in structural mechanics.

There can be no ''best'' equation solver. Different classes of problems, and specific problems within a class, yield differently organized matrices. An approach that is good in one situation may be poor in another [2.13, 2,14]. Unless there are at least several hundred d.o.f., it usually makes little difference whether the algorithm is clever or dull.

The *wavefront* or *frontal* method is an arrangement of Gauss elimination in which assembly of equations alternates with their solution [2.1-2.3, 2.10, 2.13-2.15, 2.25, 2.31]. Equations associated with element 1 only are reduced first. Then the adjacent element, number 2, makes its contribution to [**K**]. If additional equations are now fully summed—that is, if some d.o.f. are common only to elements 1 and 2—these equations are reduced. The next reduction awaits further contributions to the partially formed stiffness equations. The assembly-solution process can be viewed as a ''wave'' that sweeps over the structure. Effective element numbering increases efficiency because processing proceeds in element-number order. However, an assembly-solution alternation can also be programmed in node-number order.

A frontal algorithm requires little main storage but demands appreciable bookkeeping and attention to data transfers between primary and mass storage. Data transfer is important: the efficiency of an equation solver depends less on the mathematical algorithm than on the effectiveness of data handling [2.32].

A complicated equation solver may require almost 1000 lines to list. Programming must be meticulous, and perhaps adapted to a particular computer, if expected gains in efficiency are to be realized. It helps to exploit the sparsity of {**R**}, especially if there are many load cases. Speed is increased if [**K**] is stored one dimensionally. But space must be reserved for fills, and extra bookkeeping is needed to attach the proper row and column numbers to the stored coefficients. It is possible to reserve exactly the space needed for fills and avoid all multiplications by zero. Such a routine [2.3, 2.16, 2.17] is particularly efficient when applied to network problems, and it is competitive with other good solvers when applied to structural problems.

Listings of many equation solvers have been published. A sampling is Refs. 2.1, 2.2, and 2.18 to 2.25. These are *direct* solvers, which require a fixed number of steps for a given matrix topology, regardless of the numerical values of the coefficients.

2.14 INDIRECT EQUATION SOLVERS

Indirect methods require an indefinite number of operations to solve equations. They terminate when a convergence test is satisfied. They are less popular than direct methods but have attractive features.

Gauss-Seidel iteration is simple and was popular in the early days of finite elements. Consider, for example, its application to a 3-by-3 system $[\mathbf{K}]\{\mathbf{D}\} = \{\mathbf{R}\}$. Let D_i be initial estimates of the unknowns (such as all $D_i = 0$), and let D_i' be the newly computed values.

$$\begin{aligned} D_1' &= (R_1 \phantom{- K_{21}D_1'} - K_{12}D_2 - K_{13}D_3)/K_{11} \\ D_2' &= (R_2 - K_{21}D_1' \phantom{- K_{12}D_2} - K_{23}D_3)/K_{22} \\ D_3' &= (R_3 - K_{31}D_1' - K_{32}D_2' \phantom{- K_{13}D_3})/K_{33} \end{aligned} \tag{2.14.1}$$

Note that new values are used as soon as they are available. To symbolize the next sweep, we change D_i to D_i' and D_i' to D_i'' in Eq. 2.14.1. Convergence can be speeded by introducing an overrelaxation factor β. We write

$$D_i' = \beta D_i' + (1 - \beta)D_i = D_i + \beta(D_i' - D_i) \tag{2.14.2}$$

where the first "=" means "is replaced by," as in Fortran. Written out for N equations, Eq. 2.14.2 becomes

$$D_i' = D_i + \frac{\beta}{K_{ii}}\left(R_i - \sum_{j=1}^{i-1} K_{ij}D_j' - \sum_{j=i}^{N} K_{ij}D_j\right) \tag{2.14.3}$$

If $\beta > 1$ the method is called SOR, for *successive overrelaxation* [2.26]. If $\beta = 1$ we again have Gauss-Seidel iteration. The method converges if $0 < \beta < 2$ and $[\mathbf{K}]$ is positive definite, as it is for a stable structure. The optimum β is problem dependent, but it is often about 1.6. If the matrix is sparse, nonzero contributions to the sums in Eq. 2.14.3 number much less than N.

A physical interpretation of Eq. 2.14.3 is that the quantity in parentheses is a force unbalance between the applied load R_i and the elastic resistance of the structure. Iteration seeks D_i that reduce the unbalance to zero.

The partitioning seen in Eq. 2.3.6 is implied in Eq. 2.14.3. That is, $\{\mathbf{D}\}$ contains only unknowns and $\{\mathbf{R}\}$ contains only knowns. In Eq. 2.14.3 this means that the R_i contain the "$K_{ij}D_j$ effect" of prescribed d.o.f. D_j (see Fig. 2.9.6*b*). Equations that correspond to prescribed d.o.f. are skipped.

Energy minimization is another indirect method. The total potential energy of a stable system is a relative minimum. Mathematical programming can be used to seek the D_i that yield a minimum [2.27-2.29, 2.34].

Advantages of indirect methods include [2.1-2.3]:

1. They are easier to program than direct methods.
2. They can start from an approximate solution, if one is known.
3. Few iterations are needed to find a low-accuracy solution.
4. No storage need be reserved for fills. There are none.

Disadvantages of indirect methods include [2.1-2.3]:

1. Since there is no reduced $[\mathbf{K}]$, multiple load vectors cannot be processed rapidly.
2. Convergence may be slow, especially if the equations tend to be ill conditioned.
3. It is hard to estimate solution time in advance.
4. Speed does not increase if the matrix is symmetric (symmetry halves the cost of a direct method).

At present, direct solvers are more popular. Indirect solvers are competitive in the analysis of three-dimensional solids where, because of large bandwidth, direct methods generate many fills but indirect methods converge rapidly [2.27, 2.30]. Indirect solvers can be useful in nonlinear problems and in reanalysis during design, where one solution vector is a good approximation of the next.

2.15 PUTTING IT ALL TOGETHER

Figure 2.15.1 shows how topics in this chapter can be combined into a program for static analysis. The figure outlines an austere program, without pre- or postprocessors, analysis options, and other bells and whistles that are common in user-oriented programs. Even so, alternatives come to mind. If element data and structural arrays can fit simultaneously into primary storage, the file-writing of Step 3 can be omitted. Similarly, Step 8 is omitted

1. Input number of nodes and elements, nodal coordinates, node numbers for each element, material properties, temperature change, loads, and boundary conditions.
2. Set up array ID (Fig. 2.9.2) to designate fixed d.o.f. and equation numbers.
3. For each element: generate [**k**], {$\mathbf{r}_T$}, and {$\mathbf{r}_W$} of Sections 2.4 and 2.6; fill array KK of Fig. 2.9.3. Write all this information on a mass storage device.
4. Clear primary storage to make space for [**K**] and {**R**}. Recall the information of Step 3, element by element, and assemble structure arrays (Fig. 2.9.3). Calculate bandwidth by scanning KK arrays. Add external loads {**P**} into {**R**}.
5. Solve equations (Fig. 2.11.2).
6. For each element: extract {**d**} from {**D**} (Fig. 2.9.4) and compute stresses.
7. Print displacements and stresses.
8. Repeat Steps 5 through 7 for next load vector, if any. In the equation solver, skip decomposition of the coefficient matrix.

Figure 2.15.1. Flow of a simple finite element program.

if all load cases are processed in a single pass through the equation solver (columns of an array [**R**] contain the separate load vectors). We could insert a node-renumbering algorithm between Steps 1 and 2 and add much other finery elsewhere. Reference 2.1 lists a complete program of about 1000 lines, much like the one outlined in Fig. 2.15.1.

Example

We use Fig. 2.15.2 to illustrate the steps in Fig. 2.15.1.

Step 1

Read NUMNP = 4 and NUMEL = 3. Read nodal coordinates and numerical values of A, E, T, and α for each bar (here we carry symbols instead for clarity). Read NOD(1,1) = 1 and NOD(2,1) = 2 for bar 1, NOD(1,2) = 2 and NOD(2,2) = 3 for bar 2, and NOD(1,3) = 3 and NOD(2,3) = 4 for bar 3 (see notation in Fig. 2.6.2). Fill initial ID array as shown in the following step.

Step 2

Assume that the program permits three d.o.f. per node, as for space truss analysis. The initial and converted ID array is shown next.

$$\underset{\text{initial}}{[\mathbf{ID}]} = \begin{bmatrix} 1 & 0 & 0 & 1 \\ 1 & 1 & 1 & 1 \\ 1 & 1 & 1 & 1 \end{bmatrix}, \qquad \underset{\text{converted}}{[\mathbf{ID}]} = \begin{bmatrix} 0 & 1 & 2 & 0 \\ 0 & 0 & 0 & 0 \\ 0 & 0 & 0 & 0 \end{bmatrix}$$

Step 3

$$\begin{matrix}[\mathbf{k}] \\ \text{each bar}\end{matrix} = \frac{AE}{L}\begin{bmatrix} 1\;0\;0 & -1\;0\;0 \\ 0\;0\;0 & 0\;0\;0 \\ 0\;0\;0 & 0\;0\;0 \\ -1\;0\;0 & 1\;0\;0 \\ 0\;0\;0 & 0\;0\;0 \\ 0\;0\;0 & 0\;0\;0 \end{bmatrix}$$

array KK contains:
Bar 1: 0 0 0 1 0 0
Bar 2: 1 0 0 2 0 0
Bar 3: 2 0 0 0 0 0

$$\{\mathbf{r}_T\}_1 = \{\mathbf{r}_T\}_2 = AE\alpha T\,\{-1\;0\;0\;1\;0\;0\}, \qquad \{\mathbf{r}_T\}_3 = 0$$

Step 4
Let dots indicate positions that receive no contribution from an element.

$$[\mathbf{K}] = \frac{AE}{L}\begin{bmatrix} 1 & \cdot \\ \cdot & \cdot \end{bmatrix} + \frac{AE}{L}\begin{bmatrix} 1 & -1 \\ -1 & 1 \end{bmatrix} + \frac{AE}{L}\begin{bmatrix} \cdot & \cdot \\ \cdot & 1 \end{bmatrix} = \frac{AE}{L}\begin{bmatrix} 2 & -1 \\ -1 & 2 \end{bmatrix}$$

$$\{\mathbf{R}\} = AE\alpha T\{1\;\;\cdot\} + AE\alpha T\{-1\;\;1\} + AE\alpha T\{0\;\;\cdot\} + \{-F\;\;0\}$$

$$\{\mathbf{R}\} = \{-F\;\;AE\alpha T\}, \qquad \text{MBAND} = 2 \qquad \text{(see Problem 2.22)}$$

Step 5

$$D_1 = u_2 = -\frac{2LF}{3AE} + \frac{\alpha LT}{3}, \qquad D_2 = u_3 = -\frac{LF}{3AE} + \frac{2\alpha LT}{3}$$

Step 6

$$\{\mathbf{d}\}_1 = \{0\;\;u_2\}, \qquad \sigma_1 = E\left(\frac{u_2 - 0}{L} - \alpha T\right) = -\frac{2F}{3A} - \frac{2E\alpha T}{3}$$

$$\{\mathbf{d}\}_2 = \{u_2\;\;u_3\}, \qquad \sigma_2 = E\left(\frac{u_3 - u_2}{L} - \alpha T\right) = \frac{F}{3A} - \frac{2E\alpha T}{3}$$

$$\{\mathbf{d}\}_3 = \{u_3\;\;0\}, \qquad \sigma_3 = E\left(\frac{0 - u_3}{L} - 0\right) = \frac{F}{3A} - \frac{2E\alpha T}{3}$$

We see that the given temperature field effectively loads node 3 only. Thermal stress is constant over the entire length, as expected from Fig. 2.15.1*b*.

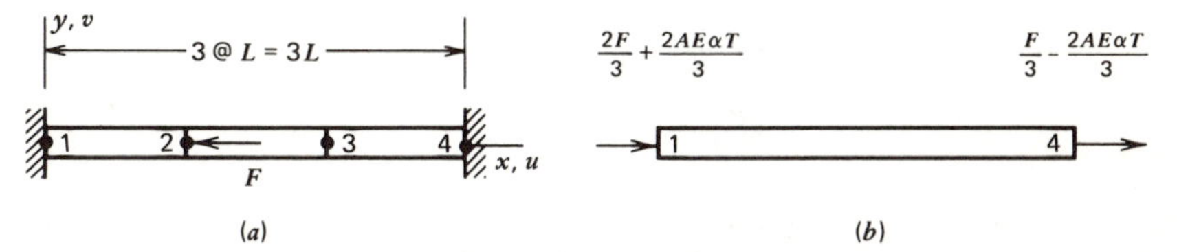

Figure 2.15.2. (*a*) Example problem. The bar is divided into three identical elements. Elements 1 and 2 *only* are heated $T°$. (*b*) Support reactions predicted by elementary mechanics of materials.

2.16 SOME RELATED NONSTRUCTURAL PROBLEMS

Equations that describe current flow in an electrical network resemble stiffness equations that describe a truss. Networks have only one d.o.f. per node. Here we outline a network analysis method.

Ohm's law applied to Fig. 2.16.1*a* yields

$$I_i = \frac{1}{r}(V_i - V_j) \qquad \text{or} \qquad \begin{Bmatrix} I_i \\ I_j \end{Bmatrix} = \frac{1}{r}\begin{bmatrix} 1 & -1 \\ -1 & 1 \end{bmatrix}\begin{Bmatrix} V_i \\ V_j \end{Bmatrix} \tag{2.16.1}$$
$$I_j = \frac{1}{r}(V_j - V_i)$$

where V_i and V_j are nodal voltages. Continuity demands that $I_i = -I_j$. For node 1 of Fig. 2.16.1*b* we note that (current flow in) = (current flow out) and write

$$\begin{aligned} \bar{I}_1 &= (I_1)_{\text{element 1}} + (I_1)_{\text{element 2}} + (I_1)_{\text{element 3}} \\ \bar{I}_1 &= \frac{1}{r_1}(V_1 - V_2) + \frac{1}{r_2}(V_1 - V_3) + \frac{1}{r_3}(V_1 - V_4) \\ \bar{I}_1 &= \left(\frac{1}{r_1} + \frac{1}{r_2} + \frac{1}{r_3}\right)V_1 - \frac{1}{r_1}V_2 - \frac{1}{r_2}V_3 - \frac{1}{r_3}V_4 \end{aligned} \tag{2.16.2}$$

Clearly, Eqs. 2.16.2 resemble Eqs. 2.6.1 and 2.6.2. The method is the same; symbols differ. The last of Eqs. 2.16.2 is the first row of the structure equations. It can also be obtained from the element matrices

$$\begin{Bmatrix} I_1 \\ I_2 \end{Bmatrix} = \frac{1}{r_1}\begin{bmatrix} 1 & -1 \\ -1 & 1 \end{bmatrix}\begin{Bmatrix} V_1 \\ V_2 \end{Bmatrix}, \quad \begin{Bmatrix} I_1 \\ I_3 \end{Bmatrix} = \frac{1}{r_2}\begin{bmatrix} 1 & -1 \\ -1 & 1 \end{bmatrix}\begin{Bmatrix} V_1 \\ V_3 \end{Bmatrix}, \text{ etc.} \tag{2.16.3}$$

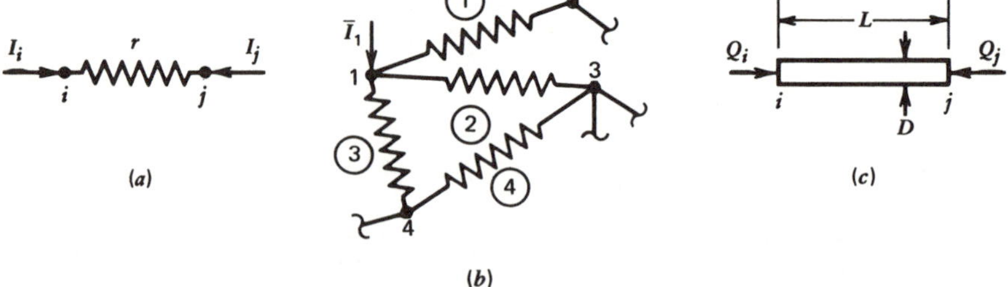

Figure 2.16.1. (*a*) Element of resistance r, with current flows I_i and I_j *into* nodes i and j, respectively. (*b*) Portion of a network of resistors, with external current input $\bar{I}_i$ into node 1. (*c*) Fluid flow rates Q_i and Q_j into a pipe of inside diameter D and length L.

by expanding them to structure size and adding them.

For flow in a pipe (Fig. 2.16.1*c*) the equations resemble Eqs. 2.16.1.

$$Q_i = c\left(\frac{P_i - P_j}{L}\right)^e \quad \text{and} \quad Q_j = c\left(\frac{P_j - P_i}{L}\right)^e \tag{2.16.4}$$

where P_i and P_j are nodal pressures. Coefficient c depends on D, Reynolds number, and pipe roughness. Unfortunately, it is only for laminar flow that $e = 1$, so that Q is directly proportional to the pressure gradient. For turbulent flow, $e \approx 0.5$, and the network equations are nonlinear.

PROBLEMS

2.1 For the truss of Fig. 2.2.1, sketch the remaining four free-body diagrams not shown in Fig. 2.2.2 and write the four equations analogous to Eqs. 2.2.3 and 2.2.4.

2.2 Consider the truss of Fig. 2.2.1. Write rigid-body motion vectors for the following cases and show that each produces zero forces $\{\mathbf{R}\}$. Are the three cases linearly independent?
   (a) Translation in the direction of bar 2.
   (b) Rotation through a small angle about node 1 (the point $x = L_3$, $y = 0$).
   (c) Rotation through a small angle about the point $x = L_3$, $y = L_1$.

2.3 Sketch the rigid-body motion $\{\mathbf{D}\} = \{-3\ 3\ 0\ 0\ -4\ -4\}$ for the truss of Fig. 2.2.1. Why is the product $[\mathbf{K}]\{\mathbf{D}\}$ not zero?

2.4 The Betti-Maxwell reciprocal theorem states that if two sets of loads ($\{\mathbf{R}_1\}$ and $\{\mathbf{R}_2\}$) act on a structure, work done by the first set in acting through displacements caused by the second set is equal to work done by the second set in acting through displacements caused by the first set. Symbolically, $\{\mathbf{D}_1\}^T\{\mathbf{R}_2\} = \{\mathbf{D}_2\}^T\{\mathbf{R}_1\}$. Substitute $\{\mathbf{D}_1\} = [\mathbf{K}]^{-1}\{\mathbf{R}_1\}$ and $\{\mathbf{D}_2\} = [\mathbf{K}]^{-1}\{\mathbf{R}_2\}$ and show that $[\mathbf{K}]$ is symmetric.

2.5 Let $k_1 = k_2 = k_3 = k$ in the truss of Fig. 2.2.1. Solve for $\{\mathbf{D}_x\}$ and $\{\mathbf{R}_x\}$ by Eqs. 2.3.5 and 2.3.6. Check the computed $\{\mathbf{R}_x\}$ by static equilibrium equations.

2.6 (a) Must diagonals k_{ii} of an element stiffness matrix be greater than zero? Can they be less than zero?
   (b) A beam element has four d.o.f. (see Fig. 1.2.1.*a*). Physically, why do rows of its stiffness matrix not sum to zero? And why, apart from symmetry, do its columns not sum to zero?

2.7 (a) Imagine that at node j in Fig. 2.4.2, u_j' and p_j' are *parallel* to the bar and v_j' and q_j' are *perpendicular* to the bar. Change nothing at node i. Find the element stiffness matrix that operates on $\{u_i\ v_i\ u_j'\ v_j'\}$.
   (b) Check that $[\mathbf{k}]\{\mathbf{d}\} = 0$ for the following rigid-body motions: x translation, y translation, and a small rotation about node i.

2.8 Write a Fortran subroutine that generates the stiffness matrix and element load vector of a truss element. Let element node coordinates, A, E, α, T, and weight density γ be transmitted to the subroutine (see Eqs. 2.4.3, 2.6.3, and 2.6.4).

2.9 Change the structure node labels in Fig. 2.5.1 from 1, 2, 3, and 4 to 1, 3, 4, and 7, respectively. What is the specific row and column position in a 7-by-7 [**K**] to which each a_i and b_i of Eq. 2.5.5 is assigned?

2.10 Permute the element node labels in Fig. 2.5.1 so that i becomes j, j becomes k, and k becomes i. Reorder the coefficients in Eqs. 2.5.5 to correspond to the new labels and show that Eq. 2.5.8 is again produced.

2.11 Manually apply the instructions of Fig. 2.6.1 to the element matrices of Eqs. 2.5.5. The result should be that of Eq. 2.5.8.

2.12 (a) Write the element matrix of Eq. 2.5.1 in the "a_i form" of Eq. 2.5.5. Manually apply the instructions of Fig. 2.6.2 to place the a_i in their proper positions in [**K**]. Then give the a_i their appropriate values from Eq. 2.5.1. Check your result against Eq. 2.2.6.
(b) Repeat part (a) using the element matrix of Eq. 2.5.2.
(c) Repeat part (a) using the element matrix of Eq. 2.5.3.

2.13 Imagine that coefficients in the plane truss element stiffness matrix are arranged to suit the order of d.o.f. $\{u_i \ u_j \ v_i \ v_j\}$. If the *structure* d.o.f. still have the order $\{u_1 \ v_1 \ u_2 \cdots v_N\}$, revise Fig. 2.6.2 as needed.

2.14 Verify the number and location of X's in Fig. 2.7.2*a*. Use the procedure of Fig. 2.2.2 and take note of symmetry. Without numerical calculation, what is the algebraic sign of each coefficient?

2.15 (a) Number nodes of the unsupported plane truss shown so that [**K**] will have as small a bandwidth as possible.
(b) Consider the two diagonal terms of [**K**] that correspond to node A. In your numbering scheme, what are the row and column numbers of these terms? Which elements make nonzero contributions to the first of these terms? And which to the second?
(c) Consider the K_{ij} to the right of the diagonal coefficient K_{ii}. In your numbering scheme, which of these K_{ij} are nonzero in each of the two rows containing diagonal coefficients that correspond to node A? What are the algebraic signs of these K_{ij}?

2.16 Repeat Problem 2.15 for the truss shown.

2.17 With reference to the truss of Fig. 2.7.1 and its stiffness matrix in Fig. 2.7.2:
(a) Calculate the semibandwidth using the element-by-element approach suggested in Section 2.7.
(b) Find, if possible, a different node numbering that yields the same semibandwidth.
(c) Find a node numbering that yields the largest possible bandwidth.
(d) What is the profile of the matrix in Fig. 2.7.2?

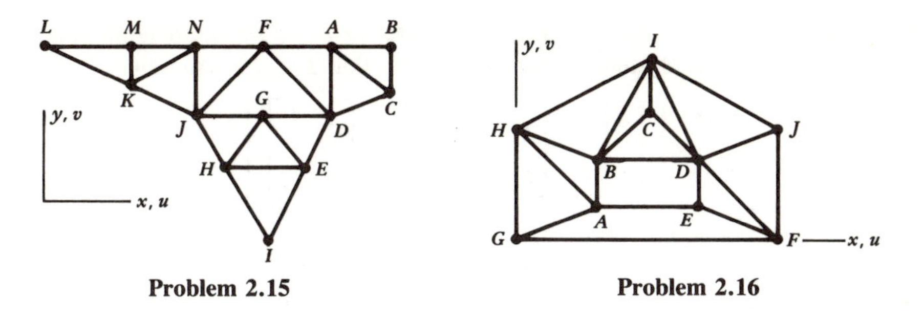

Problem 2.15 **Problem 2.16**

2.18 Section 2.7 outlines an algorithm that calculates the semibandwidth B of a matrix by scanning the elements. Outline a similar algorithm that calculates the profile P.

2.19 Devise reverse Cuthill-McKee numberings for the following structures. Compute semibandwidth and profile from a sketch of $[\mathbf{K}]$ that shows the location of nonzero coefficients. For simplicity, assume one d.o.f. per node.
(a) Figure 2.7.1, starting at node 5.
(b) Figure 2.7.1, starting at node 7.
(c) Figure 2.8.1, starting at node 8.
(d) Figure 2.8.1, starting at node 14.

2.20 Verify that $B = 8$ and $P = 63$ for the graph of Fig. 2.8.1.

2.21 For the following structures, generate an ID array analogous to that in Fig. 2.9.1*a*. Then work through Fig. 2.9.2 by hand to convert it to a form similar to the one in Fig. 2.9.1*b*.
(a) Truss of Fig. 2.2.1.
(b) Truss of Fig. 2.7.1 if node 1 is free and node 3 is fully fixed.
(c) Imagine that Fig. 2.7.1 is a *frame,* so that bars are welded together at the nodes. Each node now has a rotational d.o.f. as well as translational d.o.f.

2.22 If many d.o.f. are suppressed, actual semibandwidth B may be less than that computed by scanning the node numbers of each element. Explain why. Then devise an algorithm that computes the actual semibandwidth B by scanning information in array ID.

2.23 Use the loads and elastic properties already given in Fig. 2.11.1*a* but impose the prescribed displacement $\overline{D}_2 = -6$. Do so by:
(a) The "big spring-big force" method discussed in connection with Fig. 2.9.5*a*. Solve for the D_i and the force at node 2.
(b) The method of Fig. 2.9.6. Solve for the D_i and the force at node 2.

2.24 How might you impose a prescribed *relative* displacement, say a value of $v_4 - v_3$ in Fig. 2.7.1? Give a physical explanation; then state exactly what matrix coefficients must be changed and how. Is there a reason *not* to adopt the method you propose?

2.25 Consider the 2-by-2 system of equations [**K**] $\{u\ v\} = \{a\ b\}$. Imagine that $u = \bar{u}$ is to be enforced. By the two methods of Section 2.9, we write either

$$\begin{bmatrix} K_1 + c & K_2 \\ K_3 & K_4 \end{bmatrix} \begin{Bmatrix} u \\ v \end{Bmatrix} = \begin{Bmatrix} c\bar{u} \\ b \end{Bmatrix} \quad \text{or} \quad \begin{bmatrix} 1 & 0 \\ 0 & K_4 \end{bmatrix} \begin{Bmatrix} u \\ v \end{Bmatrix} = \begin{Bmatrix} \bar{u} \\ b - K_3\bar{u} \end{Bmatrix}$$

where c is a large number.

(a) Compute $\{u\ v\}$ in both cases. Show that the two agree as c becomes large.

(b) Compute the support reaction associated with $\bar{u}$ by the two methods of Section 2.10. Show that they agree as c becomes large.

2.26 If T in Eq. 2.10.4 is only the *average* temperature in a bar, is our truss analysis procedure still valid? Explain.

2.27 Use Eqs. 2.4.5, 2.6.4, and 2.10.4 to show that $\sigma = 0$ for a one-element structure, loaded only by a temperature change T. Let $u_1 = 0$.

2.28 (a) Which of the zeros within the band of Fig. 2.7.2 will be converted to nonzero coefficients by a Gauss elimination solution?

(b) If numerical values replace the X's in Fig. 2.7.2 and the equation solver of Fig. 2.11.2 is applied with NEQ = 14, it will terminate with an error message. What message do you expect and why?

2.29 Using hand calculation, apply the Gauss elimination solver of Fig. 2.11.2 to the equations of Fig. 2.11.1*b*.

2.30 Imagine that several load vectors are stored in the first NUMLOD columns of an array [**R**]. Modify the algorithm of Fig. 2.11.2 so that all load vectors are processed in a single pass through the subroutine. Avoid NUMLOD repetitions of IF statements if you can.

2.31 (a) Convert the equation solver of Fig. 2.11.2 to a form that uses MBAND values that may be different for each row, as suggested in the latter part of Section 2.12.

(b) Devise an algorithm that computes the required MBAND values from the node and element data.

2.32 Find the U_{ij} for a 3-by-3 matrix by writing out the product in Eq. 2.12.1. Show that Eqs. 2.12.2 produce the same result.

2.33 (a) Use the Choleski method to solve the equations of Fig. 2.11.1*b*.

(b) Find the determinant of [**K**] in Fig. 2.11.1*b*. Use Eq. 2.12.4.

(c) Repeat part (b) using Eq. 2.12.5.

2.34 Write a Fortran subroutine that solves equations by the Choleski method. Assume that the coefficient matrix is (a) full, and (b) stored in the band form of Fig. 2.7.2*b*. *Caution*. In some compilers, a DO loop is executed once even if the lower limit exceeds the upper limit.

2.35 Apply Eqs. 2.14.1 and carry out five iterations in the solution of the equations of Fig. 2.11.1*b*. Let the load vector be {0 0 4} instead of the one given. Start with all $D_i = 0$.
(a) Process the D_i in the order given in Eq. 2.14.1.
(b) Process the D_i in *reverse* order.
(c) Combine parts (a) and (b) by alternating directions of sweep.

2.36 Repeat Problem 2.35 using Eq. 2.14.2 with $\beta = 1.6$.

2.37 Connect a bar of stiffness $AE/L = 1$ from the wall to node 3 in Fig. 2.11.1*a*. With the structure thus augmented, repeat Problem 2.35.

2.38 Repeat Problem 2.37 using Eq. 2.14.2 with $\beta = 1.6$.

2.39 Write a subroutine that solves equations by Gauss-Seidel iteration (Eq. 2.14.3). Assume that the coefficient matrix is N by N, nonsymmetric, and full. Incorporate a convergence criterion.

2.40 Using Fig. 2.15.1 as a guide, write a computer program for the analysis of plane trusses.

CHAPTER 3

POTENTIAL ENERGY AND THE RAYLEIGH-RITZ METHOD

3.1 INTRODUCTION

The *total potential energy* of a structure or system is designated Π_p and is expressed as a function of displacements. When Π_p is minimized with respect to the displacements, it yields equilibrium equations. Indeed, for a truss, it yields the same nodal equilibrium equations $[\mathbf{K}]\{\mathbf{D}\} = \{\mathbf{R}\}$ found by direct physical argument in Chapter 2.

Why, then, do we bother with Π_p? Because direct physical argument formulates only the simpler elements, such as bars and beams. In Section 4.3 we derive from Π_p an expression for the element stiffness matrix. The expression does not presuppose a particular element geometry or number of d.o.f. In this chapter we define Π_p and explain how it is used in the *Rayleigh-Ritz method*—a method that reduces a continuum problem to one with a finite number of d.o.f. As we will see, the finite element method can be viewed as a form of the Rayleigh-Ritz method. Other viewpoints include weighted residual methods (Chapter 18).

This chapter is restricted to static structural mechanics problems. It is more mathematical than Chapter 2. We do not demean physical insight, which is responsible for the early rapid development of the finite element method and for its ready appeal to stress analysts. Nevertheless, a theoretical understanding suggests tactics that physical reasoning does not. Theory also opens the way to applications other than stress analysis and to use of expressions other than Π_p as a basis.

3.2 TOTAL POTENTIAL ENERGY

A *system* is a structure and the forces that act on it. The system is *conservative* if, when the system is displaced from any configuration and then led back to it again, the forces do zero net work regardless of the path taken. Thus the current potential energy depends only on the current configuration, not on how the system got there or where it was before. A configuration or a displacement field is *admissible* if it violates neither internal compatibility (Section 1.3) nor essential boundary conditions.

Boundary conditions are of two types: *essential* (or *principal* or *forced*), and *nonessential* (or *natural*). When displacements are the primary unknowns, essential boundary conditions are prescriptions of displacement and nonessential boundary conditions are prescriptions of stress. (Displacement boundary conditions are also called *geometric* or *kinematic*.) For example, a uniformly loaded cantilever beam has as essential conditions the requirements that the fixed end have no displacement and no rotation and as nonessential conditions the requirements that the free end have zero moment and zero shear force. Further remarks about boundary conditions appear in Section 3.8, and additional terminology is noted in Section 17.2.

The *principle of minimum potential energy* states that:

> *Among all admissible configurations of a conservative system, those that satisfy the equations of equilibrium make the potential energy stationary with respect to small variations of displacement. If the stationary condition is a minimum, the equilibrium state is stable.*

This principle is applicable whether or not the force versus displacement relation is linear. A nonlinear system may have more than one equilibrium configuration.

As a simple example, consider the conservative system of Fig. 3.2.1. Admissible configurations are defined by the single d.o.f. D. Valid expressions for Π_p may differ by a constant. Two are

$$\Pi_p = \tfrac{1}{2}kD^2 + P(H - D), \qquad \Pi_p = \tfrac{1}{2}kD^2 - PD \tag{3.2.1}$$

The term $kD^2/2$ is work done, and strain energy stored, in stretching the spring D units. When block A is $H - D$ units distant from stop B, load P can do $P(H - D)$ units of work before hitting the stop. So we write $P(H - D)$ as the potential of P in the first of Eqs. 3.2.1. A similar viewpoint is that when P and D are positive in the same sense, displacement D reduces the capacity of P to do work. The reduction, PD units, appears in the second of Eqs. 3.2.1. The latter equation expresses the work an observer must do, against both elastic forces and loads, to change the configuration from the zero-energy datum $D = 0$.

Potential Π_p is stationary with respect to a displacement increment dD from the static equilibrium configuration D_{eq}. So, from either of Eqs. 3.2.1,

$$\frac{d\Pi_p}{dD} = 0, \qquad kD_{eq} - P = 0, \qquad D_{eq} = \frac{P}{k} \tag{3.2.2}$$

Since H in Eqs. 3.2.1 disappears on differentiation, we see that *the datum for potential of external loading is arbitrary* in that it does not affect the equilibrium configuration (see Fig. 3.2.2.).

Figure 3.2.2. shows that small displacements from D_{eq} do not change Π_p. Symbolically,

$$d\Pi_p = (kD_{eq} - P)\, dD = 0 \tag{3.2.3}$$

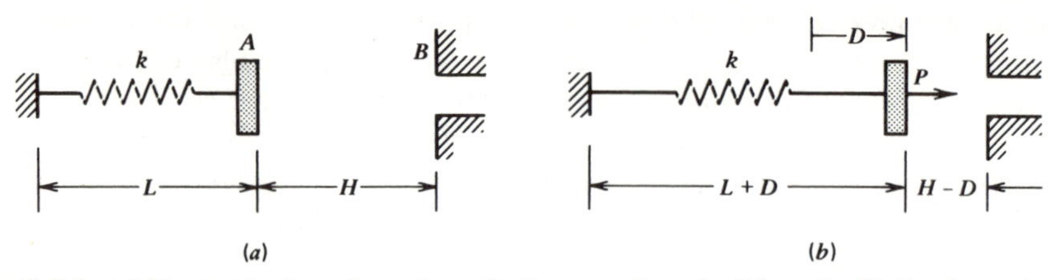

Figure 3.2.1. (*a*) Unstretched configuration of a linear spring of stiffness k. (*b*) Configuration after force P is applied to block A and stretches the spring D units. We assume the equilibrium displacement D_{eq} to be less than H.

which is an expression of the virtual work principle: zero net work is done during an infinitesimal displacement dD from an equilibrium configuration.

In the second of Eqs. 3.2.1 we write $-PD$, not $-PD/2$. This is because P always acts at full intensity and owes its potential to its magnitude and its capacity to displace, not to linear properties of the spring. We could bypass the stationary principle by imagining that D is produced by a gradually increasing load. The load-displacement relation is linear because the spring is linear. So, equating work done to energy stored,

$$\tfrac{1}{2}PD_{eq} = \tfrac{1}{2}kD_{eq}^2, \qquad D_{eq} = \frac{P}{k} \tag{3.2.4}$$

This energy-balance argument is valid but rarely helpful. It yields only one equation, but usually there are several d.o.f. that must be found.

3.3 SEVERAL DEGREES OF FREEDOM. MATRIX MANIPULATIONS

A system has n degrees of freedom if n independent quantities are needed to define its configuration. The n d.o.f. are also called *generalized coordinates*. They may differ in type, even in a single problem. For example, a beam problem uses both linear d.o.f. (translations) and angular d.o.f. (rotations).

Potential Π_p is a function of the generalized coordinates, which we name D_i. Symbolically, $\Pi_p = \Pi_p(D_1, D_2, D_3, \ldots, D_n)$. The n values of D_i define admissible configurations. The total differential is

$$d\Pi_p = \frac{\partial \Pi_p}{\partial D_1} dD_1 + \frac{\partial \Pi_p}{\partial D_2} dD_2 + \cdots + \frac{\partial \Pi_p}{\partial D_n} dD_n = \left\{\frac{\partial \Pi_p}{\partial \mathbf{D}}\right\}^T \{d\mathbf{D}\} \tag{3.3.1}$$

The stationary principle in Section 3.2 states that equilibrium prevails when the D_i define a configuration such that $d\Pi_p = 0$ for *any* admissible departure $\{d\mathbf{D}\}$ from the configuration. This means, for example, that the only nonzero term in $\{d\mathbf{D}\}$ might be dD_1, or

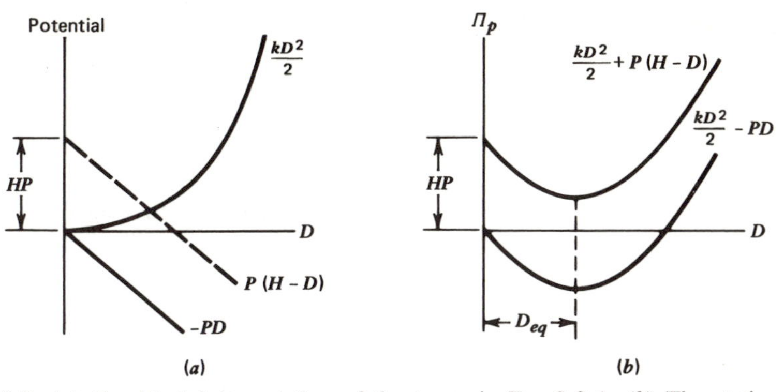

Figure 3.2.2. (*a*) Graphical interpretation of the terms in Eq. 3.2.1. (*b*) The static equilibrium configuration is a relative minimum.

it might be dD_2, and so on—for *any* choice, $d\Pi_p$ must vanish. This is possible only if coefficients of the dD_i vanish separately. Thus

$$\frac{\partial \Pi_p}{\partial D_i} = 0, \qquad i = 1, 2, 3, \cdots, n; \qquad \text{or} \qquad \left\{\frac{\partial \Pi_p}{\partial \mathbf{D}}\right\} = 0 \tag{3.3.2}$$

There are n equations to be solved for the n values of D_i that define the static equilibrium configuration.

Example

The structure of Fig. 3.3.1 illustrates the application of Eqs. 3.3.2. The total potential is

$$\Pi_p = \tfrac{1}{2}k_1 D_1{}^2 + \tfrac{1}{2}k_2 (D_2 - D_1)^2 + \tfrac{1}{2}k_3 (D_3 - D_2)^2 - P_1 D_1 - P_2 D_2 - P_3 D_3 \tag{3.3.3}$$

Equations 3.3.2 and 3.3.3 yield, for $i = 1, 2, 3,$

$$\begin{aligned} k_1 D_1 - k_2(D_2 - D_1) - P_1 &= 0 \\ k_2(D_2 - D_1) - k_3(D_3 - D_2) - P_2 &= 0 \\ k_3(D_3 - D_2) - P_3 &= 0 \end{aligned} \tag{3.3.4}$$

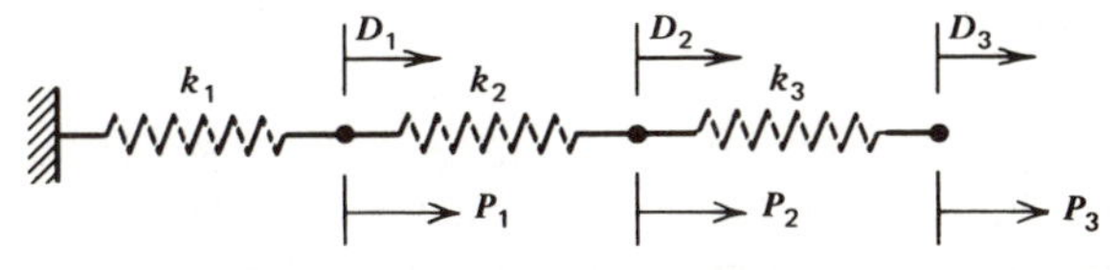

Figure 3.3.1. A three d.o.f. system of three linear springs and three applied loads P_1, P_2, and P_3. D.o.f. D_i are displacements relative to a fixed point, such as the left support. The springs are unstretched when $D_1 = D_2 = D_3 = 0$.

In the matrix form [**K**] {**D**} = {**R**}, Eqs. 3.3.4 are

$$\begin{bmatrix} k_1 + k_2 & -k_2 & 0 \\ -k_2 & k_2 + k_3 & -k_3 \\ 0 & -k_3 & k_3 \end{bmatrix} \begin{Bmatrix} D_1 \\ D_2 \\ D_3 \end{Bmatrix} = \begin{Bmatrix} P_1 \\ P_2 \\ P_3 \end{Bmatrix} \tag{3.3.5}$$

If $k_1 = k_2 = k_3 = 1$ and $P_1 = P_2 = P_3 = 4$, Eqs. 3.3.5 are identical to those in Fig. 2.11.1*b* which were formed by less elegant arguments.

Equation 3.3.3 can be written in the form

$$\Pi_p = \tfrac{1}{2}\{\mathbf{D}\}^T [\mathbf{K}] \{\mathbf{D}\} - \{\mathbf{D}\}^T \{\mathbf{R}\} \tag{3.3.6}$$

This is a *generally applicable* expression for Π_p when [**K**], {**D**}, and {**R**} are defined in the usual way (see Notation).

Example

Further confidence in the present method can be gained from the truss element of Figs. 2.4.1 and 2.4.2. Its potential is

$$\Pi_p = \tfrac{1}{2}ke^2 - p_i u_i - q_i v_i - p_j u_j - q_j v_j \tag{3.3.7}$$

where $k = AE/L$ and elongation e is given by Eq. 2.10.3. With these substitutions, the four equations

$$\left\{ \frac{\partial \Pi_p}{\partial u_i} \quad \frac{\partial \Pi_p}{\partial v_i} \quad \frac{\partial \Pi_p}{\partial u_j} \quad \frac{\partial \Pi_p}{\partial v_j} \right\} = 0 \tag{3.3.8}$$

are found to be the same as Eqs. 2.4.3.

From the foregoing examples we draw the following conclusions. They are true in general.

1. The stiffness matrix of a system with linear load versus displacement characteristics is symmetric; that is, $K_{ij} = K_{ji}$. This is true because each symmetrically located pair of off-diagonal coefficients comes from a single term in Π_p whose form is a constant times D_iD_j.
2. If D_i is a nodal displacement or rotation, each equation $\partial\Pi_p/\partial D_i = 0$ is a nodal equilibrium equation stating that forces or moments applied to node i sum to zero. This is evident in Eqs. 3.3.4 and in the way Eqs. 2.2.6 and 2.4.3 were derived.
3. When displacement d.o.f. D_i are the primary variables, static indeterminancy is of no consequence. For example, if in Fig. 3.3.1 we add springs that connect nodes

2 and 3 directly to the left support, more coefficients appear in [**K**], but the procedure is unchanged and three d.o.f. still suffice.

Formulas for Matrix Manipulation. We close this section with some useful rules that make it easy to get equilibrium equations from a Π_p expression. Define a column vector $\{\mathbf{X}\} = \{x_1\ x_2\ .\ .\ .\ x_n\}$ and a column vector $\{\mathbf{Y}\} = \{y_1\ y_2\ .\ .\ .\ y_m\}$. Let [**A**] be an n-by-n symmetric matrix and [**B**] be an n-by-m rectangular matrix, both independent of the x_i and y_i. Finally, define two scalars, $\phi = \frac{1}{2}\{\mathbf{X}\}^T[\mathbf{A}]\{\mathbf{X}\}$ and $\psi = \{\mathbf{X}\}^T[\mathbf{B}]\{\mathbf{Y}\}$. Then

$$\{\mathbf{X}\}\{\mathbf{X}\}^T, \qquad [\mathbf{B}]^T[\mathbf{B}], \qquad \text{and} \qquad [\mathbf{B}]^T[\mathbf{A}][\mathbf{B}] \qquad \text{are symmetric} \tag{3.3.9}$$

$$\left\{\frac{\partial\phi}{\partial x_1}\ \frac{\partial\phi}{\partial x_2}\ \cdots\ \frac{\partial\phi}{\partial x_n}\right\} = [\mathbf{A}]\{\mathbf{X}\} \tag{3.3.10}$$

$$\left\{\frac{\partial\psi}{\partial x_1}\ \frac{\partial\psi}{\partial x_2}\ \cdots\ \frac{\partial\psi}{\partial x_n}\right\} = [\mathbf{B}]\{\mathbf{Y}\} \tag{3.3.11}$$

$$\text{Since } \psi \text{ is a scalar,} \qquad \psi = \psi^T = \{\mathbf{Y}\}^T[\mathbf{B}]^T\{\mathbf{X}\} \tag{3.3.12}$$

$$\left\{\frac{\partial\psi}{\partial y_1}\ \frac{\partial\psi}{\partial y_2}\ \cdots\ \frac{\partial\psi}{\partial y_m}\right\} = [\mathbf{B}]^T\{\mathbf{X}\} \tag{3.3.13}$$

Note that [**A**] or [**B**] can be a unit matrix. Thus we can write further special forms of these equations.

3.4 EXPRESSIONS FOR TOTAL POTENTIAL ENERGY

First, we derive a Π_p expression for the most general circumstance: all six stresses and all six strains active, with initial stresses and strains. A linear stress-strain relation is presumed. The resulting Π_p expression is used in Section 4.3 to derive a general formula for an element stiffness matrix. Expressions for Π_p in special stress states appear at the end of this section.

Stresses $\{\boldsymbol{\sigma}\}$ and strains $\{\boldsymbol{\epsilon}\}$ are,

$$\{\boldsymbol{\sigma}\} = \{\sigma_x\ \sigma_y\ \sigma_z\ \tau_{xy}\ \tau_{yz}\ \tau_{zx}\}, \qquad \{\boldsymbol{\epsilon}\} = \{\epsilon_x\ \epsilon_y\ \epsilon_z\ \gamma_{xy}\ \gamma_{yz}\ \gamma_{zx}\} \tag{3.4.1}$$

The engineering definition of shear strain is used; for example, $\gamma_{xy} = u_{,y} + v_{,x}$ in rectangular coordinates. The stress-strain relation (Eq. 1.6.1) is

$$\{\boldsymbol{\sigma}\} = [\mathbf{E}]\{\boldsymbol{\epsilon}\} - [\mathbf{E}]\{\boldsymbol{\epsilon}_0\} + \{\boldsymbol{\sigma}_0\} \tag{3.4.2}$$

Let U_0 be strain energy per unit volume. It represents work done (and energy stored) by internal stresses. On a *unit* cube, stress equals force and strain equals displacement. Thus work $\sigma\, d\epsilon$ is done by a force σ as it moves through a displacement $d\epsilon$. Hence, including *all* stresses, infinitesimal straining of a unit volume changes U_0 by the amount

$$dU_0 = \{\boldsymbol{\sigma}\}^T\{d\boldsymbol{\epsilon}\} = \sigma_x\, d\epsilon_x + \sigma_y\, d\epsilon_y + \cdots + \tau_{zx}\, d\gamma_{zx} \tag{3.4.3}$$

Changes in $\{\boldsymbol{\sigma}\}$ produced by changes $\{d\boldsymbol{\epsilon}\}$ are discarded from dU_0 because they produce higher-order terms; for example, $(\sigma_x + d\sigma_x)\, d\epsilon_x \approx \sigma_x\, d\epsilon_x$. From Eq. 3.4.3 we conclude that

$$\frac{\partial U_0}{\partial \epsilon_x} = \sigma_x, \qquad \frac{\partial U_0}{\partial \epsilon_y} = \sigma_y, \qquad \cdots, \qquad \frac{\partial U_0}{\partial \gamma_{zx}} = \tau_{zx} \tag{3.4.4}$$

Expressing Eq. 3.4.4 in matrix form and using Eq. 3.4.2, we find the six equations

$$\left\{\frac{\partial U_0}{\partial \boldsymbol{\epsilon}}\right\} = \{\boldsymbol{\sigma}\} = [\mathbf{E}]\,\{\boldsymbol{\epsilon}\} - [\mathbf{E}]\,\{\boldsymbol{\epsilon}_0\} + \{\boldsymbol{\sigma}_0\} \tag{3.4.5}$$

Integration with respect to the strains yields

$$U_0 = \tfrac{1}{2}\{\boldsymbol{\epsilon}\}^T[\mathbf{E}]\,\{\boldsymbol{\epsilon}\} - \{\boldsymbol{\epsilon}\}^T[\mathbf{E}]\,\{\boldsymbol{\epsilon}_0\} + \{\boldsymbol{\epsilon}\}^T\{\boldsymbol{\sigma}_0\} \tag{3.4.6}$$

Application of Eqs. 3.3.10 and 3.3.11 to Eq. 3.4.6 yields Eq. 3.4.5, which shows that the integration is correct. A constant of integration has been discarded. It is superfluous because it always disappears when Eqs. 3.3.2 are applied.

Displacements of a point in the x, y, and z coordinate directions are

$$\{\mathbf{f}\} = \{u \quad v \quad w\} \tag{3.4.7}$$

where u, v, and w are in general functions of x, y, and z. Body forces $\{\mathbf{F}\}$ and surface tractions $\{\boldsymbol{\Phi}\}$, defined in Section 1.3, lose potential when displacements $\{\mathbf{f}\}$ take place. For example, in a unit volume,

$$\text{Potential change} = -F_x u - F_y v - F_z w \tag{3.4.8}$$

The same positive sense is assumed for F_x and u, for F_y and v, and for F_z and w.

Accordingly, a body of volume V and surface area S has total potential

$$\Pi_p = \int_V U_0\, dV - \int_V \{\mathbf{f}\}^T\{\mathbf{F}\}\, dV - \int_S \{\mathbf{f}\}^T\{\boldsymbol{\Phi}\}\, dS - \{\mathbf{D}\}^T\{\mathbf{P}\} \tag{3.4.9}$$

where U_0 is given by Eq. 3.4.6. In the surface integral, $\{\mathbf{f}\}$ is evaluated on S. Forces $\{\mathbf{P}\}$ are concentrated loads not included in the surface integral, and $\{\mathbf{D}\}$ are their displacements. Written out, $\{\mathbf{D}\}^T\{\mathbf{P}\} = D_1P_1 + D_2P_2 + \cdots$. Usually the P_i and D_i are nodal forces and displacements. They are considered positive in the same sense.

Equation 3.4.9 is not restricted to rectangular Cartesian coordinates. Its derivation requires only that x, y, and z refer to three mutually perpendicular directions at a material point.

For the special cases of plane stress or plane strain, $[\mathbf{E}]$ becomes 3 by 3 and is defined by Eq. 1.5.7 or 1.5.8. Each stress or strain vector contains only three coefficients: all terms with z subscripts are dropped. Also, w, F_z, Φ_z, and the z components of $\{\mathbf{D}\}$ and $\{\mathbf{P}\}$ are discarded.

For *uniaxial stress* σ_x,

$$U_0 = \tfrac{1}{2}\epsilon_x{}^T E \epsilon_x - \epsilon_x{}^T E \epsilon_0 + \epsilon_x{}^T \sigma_0 \tag{3.4.10}$$

where E = modulus of elasticity. Strain ϵ_x is a scalar, so it looks foolish to write $\epsilon_x{}^T$, but the notation is useful in stiffness matrix derivations because it leads to forms that are easy to manipulate by Eqs. 3.3.9 to 3.3.13. Tractions $\{\Phi\}$ can be incorporated in $\{\mathbf{F}\}$ (why?). Then, for uniaxial stress,

$$\Pi_p = \int_V (U_0 - Fu)\, dV - \{\mathbf{D}\}^T\{\mathbf{P}\} \tag{3.4.11}$$

where F is body force per unit volume. Displacements u and D_i and forces F and P_i all have the same positive sense and act parallel to σ_x.

Example

An example makes Eq. 3.4.11 more plausible. Consider Fig. 3.4.1*a*, with $F = 0$ and ϵ_0 and σ_0 constant along the length. Then, $\epsilon_x = D/L$, and

$$\Pi_p = \left(\frac{ED^2}{2L^2} - \frac{E\epsilon_0 D}{L} + \frac{\sigma_0 D}{L}\right) AL - PD \tag{3.4.12}$$

We find D from $d\Pi_p/dD = 0$, then σ from Eq. 3.4.2.

$$D = L\epsilon_0 - \frac{L\sigma_0}{E} + \frac{PL}{AE} \tag{3.4.13}$$

$$\sigma_x = E\frac{D}{L} - E\epsilon_0 + \sigma_0, \qquad \text{hence} \qquad \sigma_x = \frac{P}{A} \tag{3.4.14}$$

These results are expected, since nothing inhibits the tendency of ϵ_0 or σ_0 to elongate or shorten the bar.

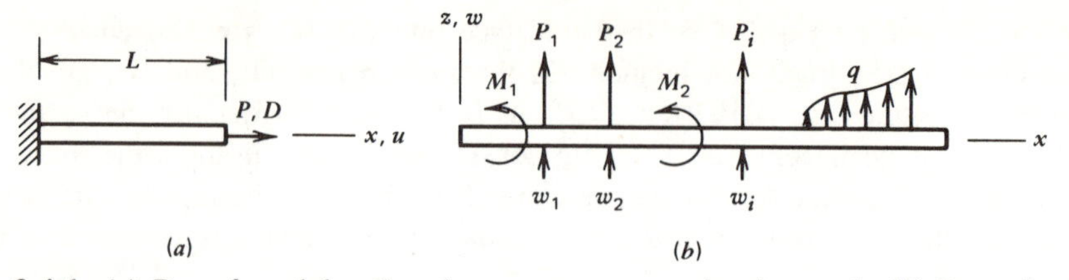

Figure 3.4.1. (*a*) Bar of modulus E and constant cross-sectional area A. (*b*) Beam loaded by distributed load q (force per unit length), concentrated forces P_i and moments M_i.

In *beam bending* (Fig. 3.4.1*b*) each differential element of volume carries uniaxial stress σ_x. From Eq. 9.1.2, the strain at distance z from the centroidal axis of a cross section is $\epsilon_x = -zw_{,xx}$, where $w_{,xx}$ is curvature. We define

$$\epsilon_0 = -z\kappa_0 \qquad \text{and} \qquad \sigma_0 = -m_0 z/I \tag{3.4.15}$$

where κ_0 is initial curvature, positive when concave up like $w_{,xx}$, and m_0 is initial bending moment, positive when it produces positive curvature. The moment of inertia, I, is the integral of $z^2\, dA$ over a cross-sectional area A. We incorporate body force in lateral load q and account for the work of concentrated moments. With these substitutions and integration over dA in the volume element $dV = dA\, dx$, Eq. 3.4.11 yields

$$\Pi_p = \int_L w_{,xx}^T\left(\frac{EI}{2}w_{,xx} - EI\kappa_0 + m_0\right) dx - \int_L qw\, dx - \{\mathbf{w}\}^T\{\mathbf{P}\} - \{\boldsymbol{\theta}\}^T\{\mathbf{M}\} \tag{3.4.16}$$

where $\{\mathbf{w}\} = \{w_1\ w_2 \cdots\}$, $\{\boldsymbol{\theta}\} = \{\theta_1\ \theta_2 \cdots\}$, and $\{\mathbf{M}\} = \{M_1\ M_2 \cdots\}$. In small deflection theory, $\theta = w_{,x}$. Stress in the beam is $\sigma_x = -Ezw_{,xx} - E\epsilon_0 + \sigma_0$. If $\epsilon_0 = \sigma_0 = 0$, the bending moment is $M = EIw_{,xx}$. A demonstration of Eq. 3.4.16, analogous to Eqs. 3.4.12 to 3.4.14, is possible (Problem 3.7).

3.5 THE RAYLEIGH-RITZ METHOD

Structures with discrete members, such as trusses, have a finite number of d.o.f. But continua, such as solids and shells, have infinitely many d.o.f., namely the displacements of every point. Their behavior is described by simultaneous partial differential equations. Excepting a structure of quaint simplicity, there is little chance of solving the differential equations. So we avoid them by using the Rayleigh-Ritz method, which uses interpolation to express the displacement of each point in terms of a finite number of d.o.f. These d.o.f. are found by solving simultaneous algebraic equations. A Rayleigh-Ritz solution is rarely exact, but it becomes more accurate with the use of more d.o.f.

The Rayleigh-Ritz method is a general technique; here we discuss only its application with displacement fields and potential energy expressions.

Consider an elastic solid. We are to find the displacements caused by applied loads. A Rayleigh-Ritz solution begins with an assumed displacement field:

$$
\begin{aligned}
u &= \sum a_i f_i, \quad \text{where} \quad f_i = f_i(x, y, z) \quad \text{and} \quad i = 1, 2, \cdots, \ell \\
v &= \sum b_j g_j, \quad \text{where} \quad g_j = g_j(x, y, z) \quad \text{and} \quad j = 1, 2, \cdots, m \\
w &= \sum c_k h_k, \quad \text{where} \quad h_k = h_k(x, y, z) \quad \text{and} \quad k = 1, 2, \cdots, n
\end{aligned}
\tag{3.5.1}
$$

Each function f_i, g_j, and h_k must be *admissible* (Section 3.2). Usually, but not necessarily, these functions are polynomials. The a_i, b_j, and c_k are generalized coordinates, chosen as follows. Substitute Eqs. 3.5.1 into Eqs. 1.4.4 to find $\{\epsilon\}$, then use Eq. 3.4.9 to find Π_p. Thus Π_p is a function of a_i, b_j, and c_k, just as Π_p is a function of the generalized coordinates D_i in Eq. 3.3.1. The stationary condition is

$$
\frac{\partial \Pi_p}{\partial a_i} = 0 \qquad \frac{\partial \Pi_p}{\partial b_j} = 0 \qquad \frac{\partial \Pi_p}{\partial c_k} = 0 \tag{3.5.2}
$$

where i, j, and k range over the values in Eq. 3.5.1. So we have $\ell + m + n$ linear algebraic equations in as many d.o.f., to be solved for numerical values of the a_i, b_j, and c_k. Thus displacements u, v, and w of Eqs. 3.5.1 are completely determined. Differentiation yields strains, which enter the stress-strain law to yield stresses.

Equations of equilibrium are not satisfied everywhere because the approximate solution has only a finite number of d.o.f. with which to represent the infinitely many d.o.f. of a continuum. Equilibrium errors decrease as more d.o.f. are used.

The numerical solution process of Eqs. 3.5.1 and 3.5.2 can be stated this way in words: first, establish a trial family of admissible solutions; second, apply a criterion to select the best member of the family. Here the criterion is that Π_p be stationary. Other criteria can be used (Chapter 18).

Equations 3.5.2 yield stiffness equations that can be written in the usual form $[\mathbf{K}]\{\mathbf{D}\} = \{\mathbf{R}\}$, where $\{\mathbf{D}\}$ contains the generalized coordinates a_i, b_j, and c_k. Vector $\{\mathbf{R}\}$ contains generalized forces. Not all D_i have units of displacement and not all R_i have units of force, as the following example shows. But each product $R_i D_i$ has units of work.

Example

The elastic bar of Fig. 3.5.1*a* is used to illustrate the Rayleigh-Ritz method. Equation 3.4.11 becomes

$$
\Pi_p = \int_0^\ell \frac{E}{2} u_{,x}^2 A \, dx - \int_0^\ell qu \, dx \tag{3.5.3}
$$

where, in Eq. 3.4.11, F is equivalent to q/A and $dV = A\,dx$. After finding an approximate displacement field $u = u(x)$, the uniaxial stress σ_x will be found from $\sigma_x = E\epsilon_x = Eu_{,x}$.

The simplest *admissible* assumption for this problem is

$$u = a_1 x \tag{3.5.4}$$

It satisfies the essential condition $u = 0$ at $x = 0$. It does not satisfy the nonessential condition $\sigma = 0$ at $x = \ell$, but it is not required to. We substitute Eq. 3.5.4 and $q = cx$ into Eq. 3.5.3 and find the Rayleigh-Ritz solution.

$$\Pi_p = \frac{AE\ell}{2} a_1^2 - \frac{c\ell^3}{3} a_1, \qquad \frac{d\Pi_p}{da_1} = 0 \quad \text{yields} \quad a_1 = \frac{c\ell^2}{3AE}$$

$$\text{Hence} \quad u = \frac{c\ell^2}{3AE} x \quad \text{and} \quad \sigma_x = \frac{c\ell^2}{3A} \tag{3.5.5}$$

The next more capable admissible assumption is

$$u = a_1 x + a_2 x^2 \tag{3.5.6}$$

We substitute Eq. 3.5.6 into Eq. 3.5.3, then write $\partial\Pi_p/\partial a_1 = \partial\Pi_p/\partial a_2 = 0$, and find

$$AE\ell \begin{bmatrix} 1 & \ell \\ \ell & 4\ell^2/3 \end{bmatrix} \begin{Bmatrix} a_1 \\ a_2 \end{Bmatrix} = \frac{c\ell^2}{12} \begin{Bmatrix} 4 \\ 3\ell \end{Bmatrix}, \qquad \begin{Bmatrix} a_1 \\ a_2 \end{Bmatrix} = \frac{c\ell}{12AE} \begin{Bmatrix} 7\ell \\ -3 \end{Bmatrix}$$

$$u = \frac{c\ell}{12AE}(7\ell x - 3x^2), \qquad \sigma_x = \frac{c\ell}{12A}(7\ell - 6x) \tag{3.5.7}$$

Figures 3.5.1*b* and 3.5.1*c* compare exact and approximate results. As might be expected, approximate displacements are better than approximate stresses because of the differentiation in the computation $\sigma_x = Eu_{,x}$. (As a specific example, consider the functions $r = 1 - x^2$ and $s = \cos(\pi x/2)$. In the range $-1 < x < 1$, r looks much like s, but successive derivatives of r and s have less and less resemblance.)

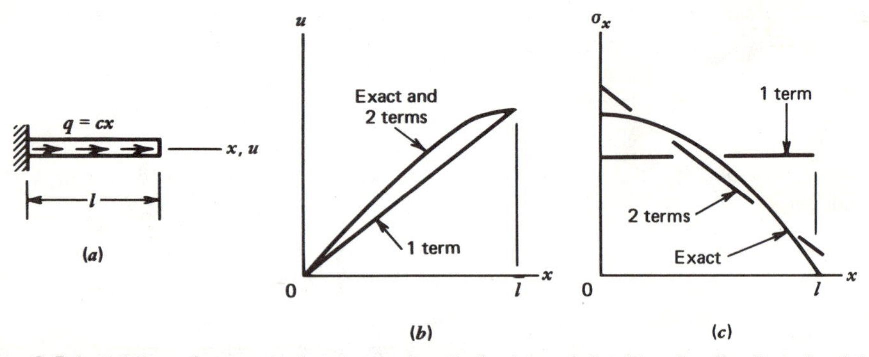

Figure 3.5.1. (*a*) Bar of cross-sectional area A and elastic modulus E under distributed axial load $q = cx$. Constant c has units of force over length squared. (*b*) Exact and approximate displacements. (*c*) Exact and approximate stresses.

In general, the admissible polynomial assumption for the foregoing example is

$$u = a_1 x + a_2 x^2 + a_3 x^3 + a_4 x^4 + \cdots + a_n x^n \tag{3.5.8}$$

A Rayleigh-Ritz solution that uses terms through $a_3 x^3$ is *exact* because the exact shape really is a cubic:

$$u = \frac{c}{6AE}(3\ell^2 x - x^3) \tag{3.5.9}$$

If we use still more terms, the Rayleigh-Ritz solution does not change. We find

$$a_1 = \frac{c\ell^2}{2AE} \qquad a_3 = -\frac{c}{6AE} \qquad a_2 = a_4 = a_5 = a_6 = \cdots = 0 \tag{3.5.10}$$

This example shows that a polynomial *capable* of being nth degree is not *obliged* to be. It can be of lower degree: Eq. 3.5.8 is cubic if a_4 through a_n happen to be zero.

3.6 COMMENTS ON THE RAYLEIGH-RITZ METHOD

Assumed fields must be admissible and should be simple. Fancy functions abound but only polynomials, and occasionally sines and cosines, are simple enough to be practicable. Beyond this, there are no easy answers to important questions: What sort of assumed field is best? How good is the approximate solution?

Let a given problem be solved repeatedly, each time with another term added to the assumed field. We expect the sequence of trial solutions to converge: to the exact Π_p, to the exact displacements, and to the exact stresses. A necessary condition for convergence is that the trial field be *complete*.

Completeness is achieved if the exact displacements, *and* their derivatives that appear in Π_p, can be matched arbitrarily closely if enough terms appear in the trial field. A polynomial series in one or more variables is complete if it is of high enough degree and if no terms are omitted. Fourier series are also complete.

Completeness demands that low-order terms be included. For example, imagine that the bar of Fig. 3.5.1a carries only an axial end load P. If we omit the term $a_1 x$ from Eq. 3.5.8 we omit the very term that contains the exact answer: $u = (P/AE)x$. Completeness is destroyed, and we cannot get the exact answer no matter how large n becomes. (The term $a_1 x$ represents the essential *constant strain capability,* which we say more about in Sections 4.8 and 4.9.)

Derivatives higher than those present in the Π_p expression may not converge toward exact values as more terms are added to a complete trial field. For example, consider beam bending, for which second derivatives of w appear in Π_p (Eq. 3.4.16). A polynomial for w may be complete and yield convergent answers for Π_p, displacement w, slope $w_{,x}$, and curvature $w_{,xx}$. But third derivatives $w_{,xxx}$, which are proportional to transverse shear force, do not converge [3.1].

A Rayleigh-Ritz solution is either exact or it is too stiff. This happens because the structure is permitted to displace into only shapes that can be described by superposing terms of the assumed displacement field. Therefore the correct shape is excluded, unless the assumed field happens to contain it. Effectively, the assumed field imposes constraints that prevent the structure from deforming the way it wants to. Constraints stiffen a structure. In effect, the solution method creates a substitute structure that is stiffer than the real one.

The work W done by loads that gradually increase from zero to $\{\mathbf{P}\}$ is $W = \{\mathbf{D}\}^T \{\mathbf{P}\}/2$ if the structure is linearly elastic. An approximate solution yields a $\{\mathbf{D}\}$ such that W is less than the exact value. This does not mean that *every* d.o.f. in $\{\mathbf{D}\}$ is underestimated. But if the structure carries a single load P, we can say that its computed displacement D is a lower bound to the correct magnitude. (See a further remark in Problem 3.13.)

Strain energy U is equal to W, so the approximate solution underestimates U when loads are prescribed. If displacements are prescribed, U is overestimated because extra force is needed to deform an overly stiff structure. When loads *and* displacements are prescribed, U may be high or low.

Stresses are calculated from displacements, so we expect that a too-stiff structure will underestimate stress magnitudes. However, as seen in Fig. 3.5.1*c*, stresses may be too low in one place but too high in another.

3.7 FINITE ELEMENT FORM OF THE RAYLEIGH-RITZ METHOD

We use the problem of Fig. 3.5.1*a* for illustration. It is sketched again in Fig. 3.7.1*a* where, for convenience, $3L$ replaces ℓ.

Consider a Rayleigh-Ritz solution based on the assumed displacements

$$\begin{aligned} u &= b_1 + b_2 x \quad &\text{for}\quad & 0 \leqq x \leqq L \\ u &= b_3 + b_4 x \quad &\text{for}\quad & L \leqq x \leqq 2L \\ u &= b_5 + b_6 x \quad &\text{for}\quad & 2L \leqq x \leqq 3L \end{aligned} \tag{3.7.1}$$

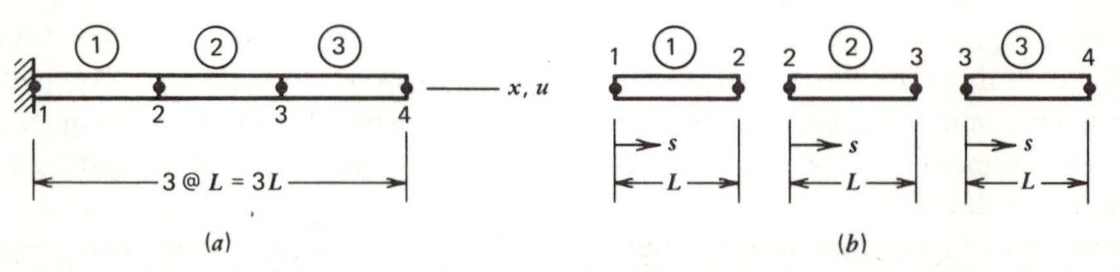

Figure 3.7.1. (*a*) Axially loaded bar of cross-sectional area A and elastic modulus E. (*b*) Finite elements of the bar, with local coordinate s originating at the left end of each.

where the b_i are d.o.f. to be determined. To make Eqs. 3.7.1 admissible, we must set $b_1 = 0$ so that $u = 0$ at $x = 0$. Also, the first and second polynomials must yield the same u at $x = L$ and the second and third the same u at $x = 2L$. These conditions give $b_3 = (b_2 - b_4)L$ and $b_5 = (b_2 + b_4 - 2b_6)L$. Next, b_3 and b_5 are eliminated, so only b_2, b_4, and b_6 remain as independent d.o.f. Potential Π_p follows from Eq. 3.4.11, with one integration spanning $x = 0$ to $x = L$, another $x = L$ to $x = 2L$, and a third $x = 2L$ to $x = 3L$. The three equations $\partial\Pi_p/\partial b_2 = \partial\Pi_p/\partial b_4 = \partial\Pi_p/\partial b_6 = 0$ yield b_2, b_4, and b_6. Thus the displacement field is completely defined, and stresses $\sigma_x = Eu_{,x}$ can be found.

The Rayleigh-Ritz solution outlined here is a finite element solution because it uses a piecewise fit to the displaced shape. The classical Rayleigh-Ritz method uses a single assumed field, often a quite lengthy polynomial, that spans the entire structure. The finite element Rayleigh-Ritz method uses several assumed fields, each a simple polynomial that is defined over only a portion of the structure. This is piecewise polynomial interpolation, and it is the essence of the finite element method.

However, a solution based on Eqs. 3.7.1 has two undesirable properties. First, the assumed displacement fields cannot be used immediately; they must be adjusted to match at $x = L$ and at $x = 2L$. Second, the b's lack an obvious physical meaning. A standard finite element formulation avoids these features by exchanging the b's for nodal d.o.f. This exchange does not alter the computed results, but makes it easy to write a computer program that accepts a description of the structure and its material properties, then generates the algebraic equations, solves them for displacements, and computes stresses. We now illustrate this form of the finite element method.

The separate regions, or elements, are shown in Fig. 3.7.1b. We assume a linear displacement field in each. Thus, in bar 1,

$$u = b_1 + b_2 s, \qquad \text{or} \qquad u = \lfloor 1 \quad s \rfloor \{b_1 \; b_2\} \tag{3.7.2}$$

Now $u = u_1$ at $s = 0$ and $u = u_2$ at $s = L$. We apply these conditions to Eq. 3.7.2 and find

$$\begin{Bmatrix} u_1 \\ u_2 \end{Bmatrix} = [\mathbf{A}] \begin{Bmatrix} b_1 \\ b_2 \end{Bmatrix}, \qquad \text{where} \qquad [\mathbf{A}] = \begin{bmatrix} 1 & 0 \\ 1 & L \end{bmatrix} \tag{3.7.3}$$

Therefore

$$\begin{Bmatrix} b_1 \\ b_2 \end{Bmatrix} = [\mathbf{A}]^{-1} \begin{Bmatrix} u_1 \\ u_2 \end{Bmatrix}, \qquad \text{where} \qquad [\mathbf{A}]^{-1} = \frac{1}{L} \begin{bmatrix} L & 0 \\ -1 & 1 \end{bmatrix} \tag{3.7.4}$$

$$u = \lfloor 1 \quad s \rfloor [\mathbf{A}]^{-1} \{u_1 \; u_2\}, \qquad \text{or} \qquad u = \lfloor \mathbf{N} \rfloor \{\mathbf{d}\} \tag{3.7.5}$$

where the "shape function" matrix $\lfloor \mathbf{N} \rfloor$ and element d.o.f. $\{\mathbf{d}\}$ are

$$\lfloor \mathbf{N} \rfloor = \left\lfloor \frac{L - s}{L} \quad \frac{s}{L} \right\rfloor, \qquad \{\mathbf{d}\} = \{u_1 \; u_2\} \tag{3.7.6}$$

Other elements also have a linear field, so

$$\begin{aligned} &\text{In element 2:} \quad u = \lfloor \mathbf{N} \rfloor \{\mathbf{d}\}, \qquad \{\mathbf{d}\} = \{u_2 \; u_3\} \\ &\text{In element 3:} \quad u = \lfloor \mathbf{N} \rfloor \{\mathbf{d}\}, \qquad \{\mathbf{d}\} = \{u_3 \; u_4\} \end{aligned} \tag{3.7.7}$$

where $\lfloor \mathbf{N} \rfloor$ is again given by Eq. 3.7.6. In each element u has the same *form*—a linear field—but not the same *value* because $\{\mathbf{d}\}$ is generally different for each element. Compatibility between elements is assured because elements share a common d.o.f. where they meet; for example, at node 2, $u = u_2$ in element 1 *and* in element 2.

The preliminary step is now complete: we have written assumed displacement fields with nodal d.o.f. as generalized coordinates. The finite element Rayleigh-Ritz solution can now begin.

Strain ϵ_x is

$$\epsilon_x = u_{,x} = u_{,s}, \qquad \text{since} \qquad dx = ds \qquad \text{and} \qquad \frac{d}{dx} = \frac{d}{ds} \tag{3.7.8}$$

Therefore, in a typical element,

$$\epsilon_x = \lfloor \mathbf{B} \rfloor \{\mathbf{d}\}, \qquad \text{where} \qquad \lfloor \mathbf{B} \rfloor = \frac{d}{ds} \lfloor \mathbf{N} \rfloor = \left\lfloor -\frac{1}{L} \quad \frac{1}{L} \right\rfloor \tag{3.7.9}$$

We substitute into Eq. 3.4.10 and integrate over each element. Strain energy U in a typical element is, with $dV = A\,dx$,

$$U = \frac{1}{2}\{\mathbf{d}\}^T \left[\int_0^L \lfloor \mathbf{B} \rfloor^T EA \lfloor \mathbf{B} \rfloor \, ds \right] \{\mathbf{d}\}, \qquad \text{or}$$

$$U = \frac{1}{2}\{\mathbf{d}\}^T [\mathbf{k}] \{\mathbf{d}\}, \qquad \text{where} \qquad [\mathbf{k}] = \frac{AE}{L} \begin{bmatrix} 1 & -1 \\ -1 & 1 \end{bmatrix} \tag{3.7.10}$$

It should be no great surprise that $[\mathbf{k}]$ is the truss element stiffness matrix of Eq. 2.4.5. (In Fig. 3.7.1 each element has the same length but, in general, elements can have different A, E, and L.)

Let the axial loading consist of only $F = (c/A)x$, where c is a constant. This is the same loading as used in the example in Section 3.5. Therefore, in the respective elements,

$$F = \frac{c}{A}s, \qquad F = \frac{c}{A}(L + s), \qquad F = \frac{c}{A}(2L + s) \tag{3.7.11}$$

The contribution of F in Eq. 3.4.11 can now be evaluated. There are three integrals. Each spans $s = 0$ to $s = L$. The first is

$$\int_V Fu \, dV = \{\mathbf{d}\}^T \int_0^L \lfloor \mathbf{N} \rfloor^T \frac{cs}{A} A \, ds = \frac{cL^2}{6} \{\mathbf{d}\}^T \begin{Bmatrix} 1 \\ 2 \end{Bmatrix} \tag{3.7.12}$$

where u comes from Eq. 3.7.5 but is written as u^T to make subsequent differentiation more obvious.

Total potential Π_p is the sum of the strain energy U in the three elements minus the three integrals over loads F. As in Section 2.5, we expand element matrices to structure size so that each $\{\mathbf{d}\}$ is replaced by the vector of structural d.o.f., $\{\mathbf{D}\} = \{u_1 \; u_2 \; u_3 \; u_4\}$. Thus

$$\Pi_p = \tfrac{1}{2}\{\mathbf{D}\}^T \left(\frac{AE}{L} \begin{bmatrix} 1 & -1 & 0 & 0 \\ -1 & 1 & 0 & 0 \\ 0 & 0 & 0 & 0 \\ 0 & 0 & 0 & 0 \end{bmatrix} + \frac{AE}{L} \begin{bmatrix} 0 & 0 & 0 & 0 \\ 0 & 1 & -1 & 0 \\ 0 & -1 & 1 & 0 \\ 0 & 0 & 0 & 0 \end{bmatrix} \right. \tag{3.7.13}$$

$$\left. + \frac{AE}{L} \begin{bmatrix} 0 & 0 & 0 & 0 \\ 0 & 0 & 0 & 0 \\ 0 & 0 & 1 & -1 \\ 0 & 0 & -1 & 1 \end{bmatrix} \right) \{\mathbf{D}\} - \{\mathbf{D}\}^T \left(\frac{cL^2}{6} \begin{Bmatrix} 1 \\ 2 \\ 0 \\ 0 \end{Bmatrix} + \frac{cL^2}{6} \begin{Bmatrix} 0 \\ 4 \\ 5 \\ 0 \end{Bmatrix} + \frac{cL^2}{6} \begin{Bmatrix} 0 \\ 0 \\ 7 \\ 8 \end{Bmatrix} \right)$$

Equation 3.3.2, $\{\partial \Pi_p / \partial \mathbf{D}\} = 0$, yields the nodal equilibrium equations.

$$\frac{AE}{L} \begin{bmatrix} 1 & -1 & 0 & 0 \\ -1 & 2 & -1 & 0 \\ 0 & -1 & 2 & -1 \\ 0 & 0 & -1 & 1 \end{bmatrix} \begin{Bmatrix} u_1 \\ u_2 \\ u_3 \\ u_4 \end{Bmatrix} = \frac{cL^2}{6} \begin{Bmatrix} 1 \\ 6 \\ 12 \\ 8 \end{Bmatrix} \tag{3.7.14}$$

The essential boundary condition $u_1 = 0$ is imposed by striking out the first equation and the first column of the matrix, as explained in Section 2.9. Solution of the remaining equations yields

$$\{u_2 \quad u_3 \quad u_4\} = \frac{cL^3}{3AE}\{13 \quad 23 \quad 27\} \tag{3.7.15}$$

These nodal u_i happen to be exact. Elsewhere, u is approximate. For example, at $x = 3L/2$, we apply Eq. 3.7.5 to element 2 and find

$$u = \left\lfloor \frac{L - L/2}{L} \quad \frac{L/2}{L} \right\rfloor \{u_2 \quad u_3\} = \frac{6cL^3}{AE} \tag{3.7.16}$$

The exact answer, from Eq. 3.5.9 with $x = \ell/2$ and $\ell = 3L$, is $u = 6.1875cL^3/AE$.
From Eq. 3.7.9, the stress calculation is

$$\sigma_x = E\epsilon_x = E\lfloor \mathbf{B} \rfloor \{\mathbf{d}\} \tag{3.7.17}$$

where $\{\mathbf{d}\}$ takes different values for the different elements. Exact and finite element results are plotted in Fig. 3.7.2. Note, stresses that are discontinuous across nodes are less accurate than displacements.

The foregoing example is similar to the examples of Figs. 2.11.1*a* and 3.3.1. The following comparison applies to this structure. *If loads are discrete* and applied at the nodes, the finite element solution is exact. A classical Rayleigh-Ritz solution is approximate but is improved by adding more terms. *If loads are distributed,* the finite element solution is approximate. It improves as more elements are used, which corresponds to adding more terms to a classical solution. The classical solution may be exact if enough terms are used. *Regardless of loading,* an approximate solution by either approach is too stiff, as noted in Section 3.6. If the solution happens to be exact, adding more terms or more elements does not change the results.

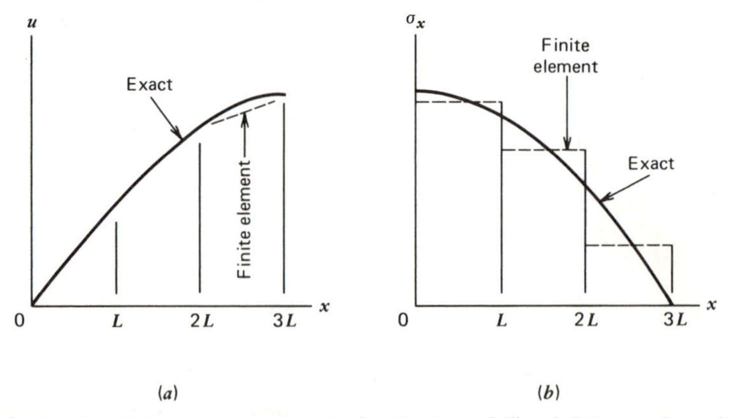

Figure 3.7.2. Exact and finite element results for the bar of Fig. 3.7.1*a*, under axial load $F = (c/A)x$. (*a*) Axial displacement u. (*b*) Axial stress σ_x.

3.8 CONCLUDING REMARKS

An integral expression for Π_p, such as Eq. 3.4.9, is called a *functional*. The term indicates Π_p is not simply a function of displacements and displacement derivatives but depends on their integrated effect. We have converted the functionals to algebraic expressions by saying that displacement fields are governed by a finite number of d.o.f. But this step is not necessary: the principle of stationary potential energy can be applied *directly* to a functional. Then the stationary condition $d\Pi_p = 0$ must be written by applying the *calculus of variations*, and $d\Pi_p = 0$ produces differential equations instead of algebraic equations. Provided the displacement field is admissible, the statement $d\Pi_p = 0$ is found to demand that (1) differential equations of equilibrium (Eqs. 1.3.3) must be satisfied within the body, and (2) boundary equilibrium equations (Eqs. 1.3.5) must be satisfied on portions of the boundary where displacements are not prescribed. Thus we conclude that $d\Pi_p = 0$ demands all components of a valid solution: satisfaction of equilibrium, compatibility, and boundary conditions.

The Rayleigh-Ritz method can be applied to functionals other than Π_p. Instead of displacement, the assumed field may be stress, or a stress function, or temperature, or some other. In each case the functional Π can be tested for correctness by applying the calculus of variations to see if $d\Pi = 0$ yields the appropriate governing differential equations and nonessential boundary conditions.

The field assumed for a Rayleigh-Ritz solution must be admissible, so we must be able to recognize essential boundary conditions. Here is the rule: if $2m$ is the highest-order derivative on the primary field variable in the governing differential equation, then essential boundary conditions involve derivatives of order zero through $m - 1$ on the primary variable. For example, the axially loaded bar of Fig. 3.5.1*a* has u as the primary variable, and the governing differential equation is $AEu_{,xx} + q = 0$. Therefore $2m = 2$, and essential conditions involve only u itself. Nonessential conditions apply to $u_{,x}$, which is proportional to stress. For a beam problem (Fig. 3.4.1*b*), the governing equation is $EIw_{,xxxx} - q = 0$ if EI is constant. Therefore $2m = 4$, and essential conditions are those on displacement w and rotation $w_{,x}$. Nonessential conditions apply to $w_{,xx}$ and to $w_{,xxx}$, which are proportional to moment and shear, respectively.

A Rayleigh-Ritz solution can be interpreted as one that minimizes the squares of the errors in stress [3.7, 3.8]. This interpretation suggests that the finite element method shares features with other least-squares approximations. These include the following, which are in fact observed: (1) approximate stresses tend to oscillate about the correct values as we pass from one element to another; and (2) absolute errors in stress are of the same order for all stress components. This accounts for the larger percentage error where stresses are low and the small stresses calculated at boundaries that should be stress free (see also Section 15.4).

Until now we have demanded that displacement fields be continuous. Elements exist that generate a discontinuous field by overlapping or by separating from one another. At first glance this behavior is alarming, but we will see that it can be acceptable and even desirable. However, the bound is lost: we cannot say for sure that a structure of incompatible elements is too stiff.

Mathematical rigor and further details about energy methods, variational principles, and the calculus of variations appear in several references [3.2–3.6].

PROBLEMS

3.1 Redefine D_2 and D_3 in Fig. 3.3.1 so that D_2 is a displacement relative to D_1 and D_3 is a displacement relative to D_2. For the case $k_1 = k_2 = k_3 = k$ and $P_1 = P_2 = P_3 = P$, solve for the D_i and show that they give the same *absolute* nodal displacements as Eqs. 3.3.3.

3.2 Add springs of stiffness k_4 to k_5 to Fig. 3.3.1, as suggested in point 3 after Eq. 3.3.8. Find the new equation analogous to Eq. 3.3.5.

3.3 Write Π_p for Problem 3.2 if D_2 and D_3 are the relative d.o.f. defined in Problem 3.1.

3.4 (a) Verify that Eqs. 3.3.3 and 3.3.6 are the same.
(b) Verify that Eq. 3.3.8 yields Eqs. 2.4.3.
(c) Use Eq. 3.3.2 to verify the [**k**] found in Problem 2.7.
(d) Verify Eqs. 3.3.10, 3.3.11, and 3.3.13 for the case $m = n = 2$.

3.5 Remove the supports from the truss of Fig. 2.2.1 and write Π_p as a function of the six d.o.f. $u_1, v_1, \cdots, v_3$. Then show that the six equations $\partial\Pi_p/\partial u_i = \partial\Pi_p/\partial v_i = 0$ yield Eqs. 2.2.6.

3.6 Write Eq. 3.4.6 for a state of plane stress in the xy plane. Then consider an isotropic material with $\epsilon_y = -\nu\epsilon_x$ and $\gamma_{xy} = 0$ and see if Eq. 3.4.10 results.

3.7 Consider a cantilever beam of length L that is loaded by moment M_L at its free end $x = L$ and by constant κ_0 and m_0 (Eq. 3.4.15). Let $w = ax^2$. Find a by $d\Pi_p/da = 0$, then evaluate stress and end deflection. Do the results agree with elementary mechanics of materials?

3.8 Consider a cantilever beam, fixed at end $x = 0$. Let end $x = L$ be subjected to the prescribed displacement $w = c$. Is the displacement field

$$w = \frac{cx^2}{L^2} + \sum a_i \sin\frac{i\pi x}{L}$$

admissible (the a_i are generalized coordinates)? If it is not admissible, say why and suggest a field that is.

3.9 (a) Verify the results given in Eq. 3.5.7.
(b) Obtain Eq. 3.5.9 by using three terms of Eq. 3.5.8 in a Rayleigh-Ritz solution.

3.10 Consider the one exact and two approximate solutions for the example problem in Section 3.5. See if the differential equations of equilibrium are satisfied by each of the three solutions.

3.11 A uniformly loaded beam of constant flexural stiffness EI is simply supported at its ends $x = 0$ and $x = L$. Find the deflection and bending moment predicted at $x = L/2$ by a Rayleigh-Ritz solution that has a single d.o.f. Compare exact and approximate results.
(a) Use one term of a polynomial series.
(b) Use one term of a sine series.
(c) Why should you anticipate that part (b) will be better than part (a) if part (a) is the simplest admissible function?

3.12 Section 3.6 considers using the series $u = a_2x^2 + a_3x^3 + \cdots$ for a bar under axial end load P. Find Rayleigh-Ritz solutions for u at $x = L$ based on (*a*) the first term, and (*b*) the first two terms of this series.

3.13 Imagine that the bar of Fig. 3.5.1*a* is not of constant cross section but has area $2A$ for $0 < x < \ell/2$ and area A for $\ell/2 < x < \ell$. Assume $u = (D/\ell)x$, where D is the displacement at $x = \ell$. Set $q = 0$ and find a Rayleigh-Ritz solution for D in the following two cases of thermal loading. Compare exact and approximate results.
(a) Heat the left half only ($\sigma_0 = -E\alpha T$ for $0 < x < \ell/2$).
(b) Heat the right half only ($\sigma_0 = -E\alpha T$ for $\ell/2 < x < \ell$).
Results should show that D may be too large or too small when the assumed field is not exact and enters the load calculation as well as the stiffness calculation.

3.14 The cantilever beam shown has constant stiffness EI and carries two equal forces P. Compute deflection and flexural stress at A, B, and C by the classical Rayleigh-Ritz method. Compare approximate and exact solutions. Use a polynomial for w that contains (*a*) one term, and (*b*) two terms. (*c*) How many terms would be needed to produce exact answers? Explain.

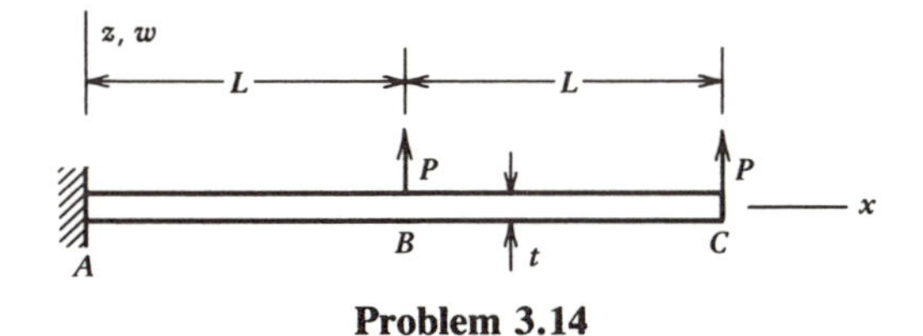

Problem 3.14

3.15 Repeat Problem 3.14 if forces P are replaced by a single clockwise couple M_0 at C.

3.16 Repeat Problem 3.14 if forces P are replaced by a uniformly distributed upward load q over the entire beam.

3.17 Repeat Problem 3.14 if forces P are replaced by a temperature distribution that varies linearly from $+T$ on the lower surface to $-T$ on the upper surface.

3.18 (a) Compute the strain energy in the bar of Fig. 3.5.1*a* by the first integral in Eq. 3.5.3. Use the one-term solution, then the two-term solution, and then the exact solution. Is the trend of answers reasonable?

(b) Similarly, evaluate the second integral of Eq. 3.5.3 for all three solutions. Is the trend reasonable? How do these results correlate with those of part (a), and why?

3.19 Use the solution vector of Eq. 3.7.15 and compute strain energy in the finite element model of Fig. 3.7.1 by applying the first group of terms in Eq. 3.7.13. Also, compute work done by loads by using the last group of terms in Eq. 3.7.13. Compare these results with exact answers.

3.20 Consider a bar of length ℓ, like that in Fig. 3.5.1*a*. Let the loading be $q = c(\ell - x)$, where c is a constant. Generate a finite element solution like that in Section 3.7, using (*a*) one element, (*b*) two elements of length $\ell/2$, and (*c*) three elements of length $\ell/3$.

3.21 (a) Repeat Problem 3.20*a*, but let area A be $A = A_0(3\ell - 2x)/\ell$, where A_0 is the cross-sectional area at the free end. Continue to assume that u is linear in x.

(b) What is the *exact* 2-by-2 stiffness matrix of this tapered element?

3.22 Repeat the finite element solution of Section 3.7 but let the elements have lengths $\ell/6$, $\ell/3$, and $\ell/2$ (reading left to right).

3.23 Consider the beam of Problem 3.14. Take **[k]** from Eq. 4.4.5 and work out a two-element solution analogous to the three-element solution (of a different problem) in Section 3.7. Compare exact and finite element results.

3.24 Find a one-element solution for a cantilever beam that carries a uniformly distributed lateral load q. Compare exact and approximate results for tip deflection, root moment (see Eq. 4.4.4), strain energy stored, and work done by the applied load. See Eq. 4.4.5 for **[k]** and Eq. 4.4.7 for **{r}**.

CHAPTER 

ELEMENTS BASED ON ASSUMED DISPLACEMENT FIELDS

4.1 INTRODUCTION

The most popular elements in structural mechanics are based on the assumption that displacements within an element are adequately described by a simple polynomial. This chapter is devoted to such elements. (Alternatives are discussed in Section 16.1.) We derive formulas for element matrices, explain why some displacement fields work while others do not, and describe how elements are likely to behave.

Engineers find physical intuition more valuable than Rayleigh-Ritz concepts in understanding element behavior. For example, the finite element solution in Fig. 3.7.2 is motivated by linear interpolation: any smooth curve looks like a straight line if we look at a small enough piece of it. On a higher level of sophistication, we can look at a larger piece of a curve and model it with a better function, such as a cubic segment defined by ordinates and slopes at its ends. In general, we claim that a portion of *any* smooth function can be approximated well enough by a polynomial. The quality of the match depends on the size of the portion, the smoothness of the function, and the degree of the polynomial.

An analysis category, such as static stress analysis in two dimensions, can be tackled with any of several kinds of displacement-based elements. It is hard to say which element is best. For now we do not try to decide (see remarks in the latter part of Section 9.2).

4.2 INTERPOLATION

To interpolate is to find the value of a function between known values by operating on the known values with a formula different from the function itself. Thus exact and interpolated curves match at the end points (the nodes) but may differ elsewhere. We will illustrate this behavior by example interpolation formulas. These formulas can be used as assumed displacement fields, and finite elements can be generated from them. In the finite element context, however, interpolation does not imply that nodal values are exact.

Linear interpolation is illustrated in Section 3.7. Equation 3.7.5 shows that the interpolated u has the values $u = u_1$ at $s = 0$ and $u = u_2$ at $s = L$, as it must. But Fig.

3.7.2*a* shows that even if u_1 and u_2 are exact, the interpolated u is approximate for $0 < s < L$ unless the exact function happens to be a straight line.

Next consider cubic interpolation for the function $w = w(x)$. We have four parameters to work with, and can match ordinates *and* slopes at the end points (Fig. 4.2.1*a*). This is called "Hermitian" interpolation. We begin with

$$w = a_1 + a_2 x + a_3 x^2 + a_4 x^3 = \lfloor \mathbf{X} \rfloor \{\mathbf{a}\},$$

where (4.2.1)

$$\lfloor \mathbf{X} \rfloor = \lfloor 1 \ x \ x^2 \ x^3 \rfloor, \qquad \{\mathbf{a}\} = \{a_1 \ a_2 \ a_3 \ a_4\}$$

As in Section 3.7, we want to replace the a's by nodal values, so we apply the end conditions

$$\begin{array}{llllll} w = w_1 & \text{and} & w_{,x} = \theta_1 & \text{at} & x = 0 \\ w = w_2 & \text{and} & w_{,x} = \theta_2 & \text{at} & x = L \end{array} \tag{4.2.2}$$

where, in anticipation of solving small-deflection beam problems, slope $w_{,x}$ is considered equal to rotation θ. Equations 4.2.1 and 4.2.2 yield

$$\{\mathbf{d}\} = \lfloor \mathbf{A} \rfloor \{\mathbf{a}\},$$

where (4.2.3)

$$\{\mathbf{d}\} = \begin{Bmatrix} w_1 \\ \theta_1 \\ w_2 \\ \theta_2 \end{Bmatrix}, \qquad [\mathbf{A}] = \begin{bmatrix} 1 & 0 & 0 & 0 \\ 0 & 1 & 0 & 0 \\ 1 & L & L^2 & L^3 \\ 0 & 1 & 2L & 3L^2 \end{bmatrix}$$

Therefore $\{\mathbf{a}\} = [\mathbf{A}]^{-1} \{\mathbf{d}\}$, and

$$w = \lfloor \mathbf{N} \rfloor \{\mathbf{d}\}, \qquad \text{where} \qquad \lfloor \mathbf{N} \rfloor = \lfloor \mathbf{X} \rfloor [\mathbf{A}]^{-1} \tag{4.2.4}$$

Matrix $\lfloor \mathbf{N} \rfloor$ is called the *shape function matrix*. Here it is 1 by 4 and contains the entries

$$\begin{array}{ll} N_1 = 1 - \dfrac{3x^2}{L^2} + \dfrac{2x^3}{L^3} & N_3 = \dfrac{3x^2}{L^2} - \dfrac{2x^3}{L^3} \\ N_2 = x - \dfrac{2x^2}{L} + \dfrac{x^3}{L^2} & N_4 = -\dfrac{x^2}{L} + \dfrac{x^3}{L^2} \end{array} \tag{4.2.5}$$

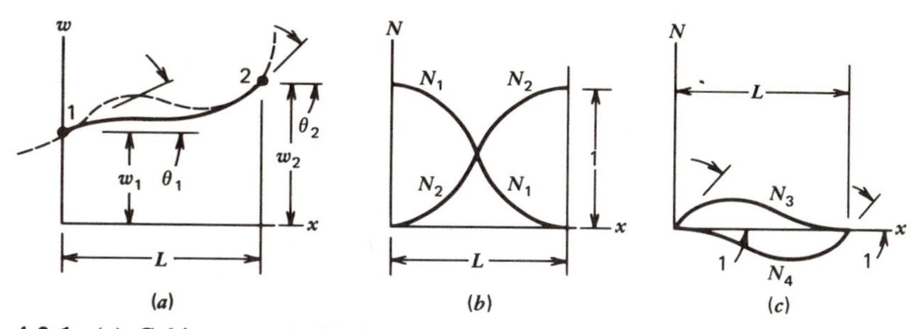

Figure 4.2.1. (*a*) Cubic curve (solid line) fitted to an exact curve (dashed line) by matching ordinates and slopes at points 1 and 2. For a better fit, L should span less of the dashed curve. (*b*) Plots of the shape functions given by Eq. 4.2.5. If displacements are small, a prismatic beam assumes these shapes.

The reader may check that these N_i behave as shown in Fig. 4.2.1; specifically, at $x = 0$ or at $x = L$, $\lfloor \mathbf{N} \rfloor$ and $d\lfloor \mathbf{N} \rfloor/dx$ each contain a single one and three zeros, in locations such that w and $w_{,x}$ match the nodal values in $\{\mathbf{d}\}$.

A shape function $\mathrm{N_i}$ *defines displacements within an element when the* ith *element d.o.f. has unit value and all other element d.o.f. are zero.*

Lagrange's interpolation formula defines a polynomial $u = u(x)$ that passes through n points (Fig. 4.2.2*a*). The polynomial has the form

$$u = a_1 + a_2 x + a_3 x^2 + \dots + a_n x^{n-1} \qquad \text{or} \qquad u = \lfloor \mathbf{N} \rfloor \{u_1 \; u_2 \; u_3 \dots u_n\} \qquad (4.2.6)$$

Lagrange's formula states the N_i directly. It avoids the labor seen in Eqs. 4.2.2 to 4.2.4, where a's are replaced by N's. The N_i are

$$N_1 = \frac{(x - x_2)(x - x_3)(x - x_4) \dots (x - x_n)}{(x_1 - x_2)(x_1 - x_3)(x_1 - x_4) \dots (x_1 - x_n)}$$

$$N_2 = \frac{(x - x_1)(x - x_3)(x - x_4) \dots (x - x_n)}{(x_2 - x_1)(x_2 - x_3)(x_2 - x_4) \dots (x_2 - x_n)}$$

$$\vdots \qquad (4.2.7)$$

$$N_n = \frac{(x - x_1)(x - x_2)(x - x_3) \dots (x - x_{n-1})}{(x_n - x_1)(x_n - x_2)(x_n - x_3) \dots (x_n - x_{n-1})}$$

Each N_i is a polynomial of degree $n - 1$. Note that each $N_i = 1$ and each $N_j = 0$ if $x = x_i$, where i and j are different integers. Thus the curve meets all required ordinates u_i. Between nodes, u depends on *all* the u_i. The simplest special case of Eq. 4.2.6 is $n = 2$, for which

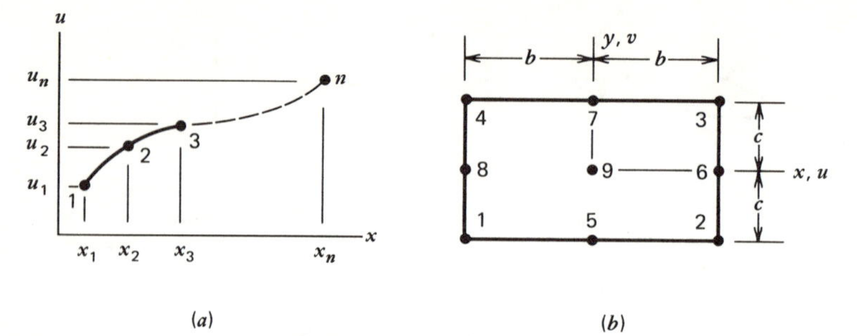

Figure 4.2.2. (*a*) A polynomial of degree $n - 1$ fitted to n points. (*b*) Displacements u and v in a rectangle are to be interpolated from the nine nodal u_i and nine nodal v_i by a quadratic polynomial. This is a quadratic Lagrange element (also called biquadratic, since quadratic interpolation is used in two dimensions).

$$u = \frac{x - x_2}{x_1 - x_2} u_1 + \frac{x - x_1}{x_2 - x_1} u_2 \tag{4.2.8}$$

This is the linear interpolation of Eq. 3.7.6, where $s = x - x_1$ and $L = x_2 - x_1$.

If $x_n - x_1$ in Fig. 4.2.2*a* spans the entire region of interest in a physical problem, Eq. 4.2.6 can be used in a classical Rayleigh-Ritz solution. Finite element alternatives would use lower-order polynomials in smaller spans. The smallest span is $x_{i+1} - x_i$, which could be fitted by linear interpolation, or by the cubic curve of Fig. 4.2.1*a* if derivative d.o.f. are acceptable. A larger span, such as $x_{i+1} - x_{i-1}$, could be fitted with a quadratic polynomial and three d.o.f.

Lagrange interpolation can also be used in two and three dimensions. It leads to plane and solid finite elements with displacement fields that are linear, or quadratic, or higher in the spatial coordinates. Consider, for example, a plane quadratic element (Fig. 4.2.2*b*). Specializing Eq. 4.2.7 to three points $x_1 = -b$, $x_2 = 0$, and $x_3 = b$, and to three points $y_1 = -c$, $y_2 = 0$, and $y_3 = c$, we find

$$\begin{aligned} N_{1x} &= \frac{x(x - b)}{2b^2}, & N_{1y} &= \frac{y(y - c)}{2c^2} \\ N_{2x} &= \frac{(x + b)(x - b)}{-b^2}, & N_{2y} &= \frac{(y + c)(y - c)}{-c^2} \\ N_{3x} &= \frac{(x + b)x}{2b^2}, & N_{3y} &= \frac{(y + c)y}{2c^2} \end{aligned} \tag{4.2.9}$$

Next we apply Eq. 4.2.6 to find u on three x-parallel lines. We name these deflections u_A, u_B, and u_C.

$$\begin{aligned}
&\text{Along line } y = +c, &\quad u_A &= N_{1x}u_4 + N_{2x}u_7 + N_{3x}u_3 \\
&\text{Along line } y = 0, &\quad u_B &= N_{1x}u_8 + N_{2x}u_9 + N_{3x}u_6 \\
&\text{Along line } y = -c, &\quad u_C &= N_{1x}u_1 + N_{2x}u_5 + N_{3x}u_2
\end{aligned} \tag{4.2.10}$$

Then, along any y-parallel line,

$$u = N_{1y}u_C + N_{2y}u_B + N_{3y}u_A \tag{4.2.11}$$

Substitution of Eqs. 4.2.10 into Eq. 4.2.11 yields the desired result. The y-direction displacement is found by replacing all u's by v's. Equation 4.2.11 can be generalized to allow m lines of nodes parallel to the y-axis and n lines of nodes parallel to the x-axis.

$$u = \sum_{i=1}^{m} \sum_{j=1}^{n} N_{ix}N_{jy}u_{ij}, \qquad \text{or} \qquad u = \lfloor \mathbf{N} \rfloor \{\mathbf{u}\} \tag{4.2.12}$$

where u_{ij} and $\{\mathbf{u}\}$ are nodal values of u and $\lfloor \mathbf{N} \rfloor$ contains products of N_{ix} and N_{jy}. A further generalization to three dimensions is possible.

Elements based on Eq. 4.2.12 are called *Lagrange* elements. Some of the many possibilities are shown in Fig. 4.2.3. In Chapter 5 we will see similar elements without internal nodes. And we will see that Lagrange elements need not be rectangular: their sides can be curved if we replace x by ξ and y by η in Eq. 4.2.12 and proceed as we do with the "isoparametric" elements of Section 5.3 (see also Section 7.3).

For future use, we introduce the following symbolism to describe the degree of continuity of a field. A field has C_m continuity if derivatives through order m are continuous. (Some authors write C^m instead of C_m.) Thus $u = u(x)$ is C_0 continuous if u is continuous but $u_{,x}$ is not. It is C_1 continuous if u and $u_{,x}$ are continuous but $u_{,xx}$ is not. In Fig. 3.7.2, the finite element u field is C_0 continuous. In Fig. 4.2.1, w is C_1 continuous across nodes. With all elements of Fig. 4.2.3 the displacements would be C_0 continuous in a mesh. In general, derivatives of order m must be used as nodal d.o.f. if a finite element field is to be C_m continuous.

4.3 FORMULAS FOR ELEMENT MATRICES

The derivation is straightforward and can be verbally summarized as follows [4.1]. Displacements are taken as the primary unknowns, so the appropriate functional is Π_p. We assume a displacement field, defined in piecewise fashion so that displacements within any element are interpolated from nodal d.o.f. of that element. A Rayleigh-Ritz solution is invoked. Thus we generate algebraic equations to be solved for nodal d.o.f. $\{\mathbf{D}\}$. During the argument we identify certain expressions as element stiffness and load matrices.

Displacements $\{\mathbf{f}\} = \{u\ v\ w\}$ in an element are interpolated from element nodal d.o.f. $\{\mathbf{d}\}$ by assumed fields.

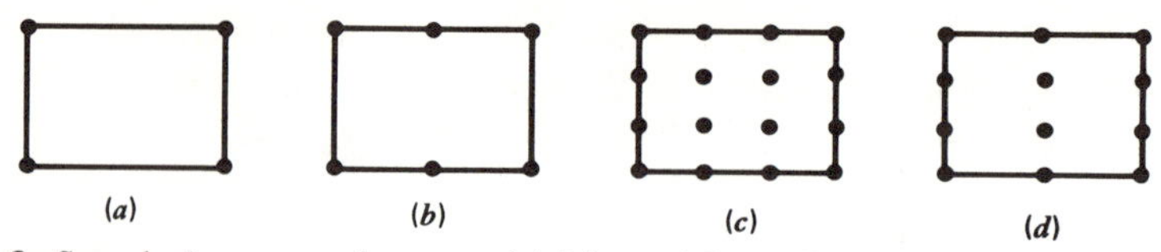

Figure 4.2.3. Sample Lagrange elements. (*a*) Linear (also called bilinear). (*b*) Linear-quadratic. (*c*) Cubic (or bicubic). (*d*) Quadratic-cubic.

$$\{\mathbf{f}\} = [\mathbf{N}]\{\mathbf{d}\} \tag{4.3.1}$$

where [**N**] is the shape function matrix, now written to allow for more than one row.

The so-called *strain-displacement matrix,* [**B**], operates on {**d**} to produce quantities significant in the strain energy expression. These quantities are indeed strains for plane and solid elasticity problems (Eq. 3.4.6). Then $\{\boldsymbol{\epsilon}\} = [\mathbf{B}]\{\mathbf{d}\}$. In bending problems, such as beams and plates, it is customary to express strain energy in terms of curvature (Eq. 3.4.16). Then $\{\boldsymbol{\kappa}\} = [\mathbf{B}]\{\mathbf{d}\}$, where $\{\boldsymbol{\kappa}\}$ is an array of curvatures. In either case, [**B**] is derived from [**N**]. For example, in the respective cases of plane elasticity (Eq. 1.4.4) and beam bending (Eq. 3.4.16),

$$\underset{3\times 1}{\{\boldsymbol{\epsilon}\}} = [\mathbf{B}]\{\mathbf{d}\}, \quad \text{where} \quad [\mathbf{B}] = \begin{bmatrix} \partial/\partial x & 0 \\ 0 & \partial/\partial y \\ \partial/\partial y & \partial/\partial x \end{bmatrix} [\mathbf{N}] \qquad \Bigg| \qquad w_{,xx} = \lfloor \mathbf{B} \rfloor \{\mathbf{d}\}, \quad \text{where} \quad \lfloor \mathbf{B} \rfloor = \frac{d^2}{dx^2} \lfloor \mathbf{N} \rfloor \tag{4.3.2}$$

We take Π_p from Eq. 3.4.9, with the understanding that U_0 may have simpler forms than Eq. 3.4.6, as noted in Section 3.4. Substitution of $\{\mathbf{f}\} = [\mathbf{N}]\{\mathbf{d}\}$ and $\{\boldsymbol{\epsilon}\} = [\mathbf{B}]\{\mathbf{d}\}$ into Eq. 3.4.9 yields

$$\Pi_p = \frac{1}{2} \sum_{1}^{\text{numel}} \{\mathbf{d}\}^T[\mathbf{k}]\{\mathbf{d}\} - \sum_{1}^{\text{numel}} \{\mathbf{d}\}^T\{\mathbf{r}\} - \{\mathbf{D}\}^T\{\mathbf{P}\} \tag{4.3.3}$$

where the summation signs say that we include contributions from all *numel* elements of the structure, and we have defined

$$\boxed{[\mathbf{k}] = \int_V [\mathbf{B}]^T [\mathbf{E}][\mathbf{B}]\, dV} \tag{4.3.4}$$

$$\boxed{\begin{aligned} \{\mathbf{r}\} = & \int_V [\mathbf{B}]^T [\mathbf{E}]\{\boldsymbol{\epsilon}_0\}\, dV - \int_V [\mathbf{B}]^T \{\boldsymbol{\sigma}_0\}\, dV \\ & + \int_V [\mathbf{N}]^T \{\mathbf{F}\}\, dV + \int_S [\mathbf{N}]^T \{\boldsymbol{\Phi}\}\, dS \end{aligned}} \tag{4.3.5}$$

where V denotes the volume of an element and S its surface. In the surface integral, $[\mathbf{N}]$ is evaluated on S. *We identify* $[\mathbf{k}]$ *as the element stiffness matrix and* $\{\mathbf{r}\}$ *as loads applied by an element to its nodes*. These are general expressions. When a specific displacement field is assumed, they yield specific $[\mathbf{k}]$ and $\{\mathbf{r}\}$ matrices. We will say more about these matrices after we finish the derivation.

Every d.o.f. in an element vector $\{\mathbf{d}\}$ also appears in the structure vector $\{\mathbf{D}\}$. Therefore, as in Sections 2.5 and 2.6, $\{\mathbf{D}\}$ replaces $\{\mathbf{d}\}$ if we expand $[\mathbf{k}]$ and $\{\mathbf{r}\}$ to structure size. Thus Eq. 4.3.3 becomes

$$\Pi_p = \tfrac{1}{2}\{\mathbf{D}\}^T[\mathbf{K}]\{\mathbf{D}\} - \{\mathbf{D}\}^T\{\mathbf{R}\} \tag{4.3.6}$$

where

$$[\mathbf{K}] = \sum_{1}^{\text{numel}}[\mathbf{k}], \qquad \{\mathbf{R}\} = \{\mathbf{P}\} + \sum_{1}^{\text{numel}}\{\mathbf{r}\} \tag{4.3.7}$$

Summations indicate the assembly of element matrices, as in Section 2.6. Now Π_p is a function of generalized coordinates $\{\mathbf{D}\}$ so, according to Eq. 3.3.2, static equilibrium prevails when $\{\mathbf{D}\}$ satisfies the equation

$$[\mathbf{K}]\{\mathbf{D}\} = \{\mathbf{R}\} \tag{4.3.8}$$

Differentiation rules in Section 3.3 extract Eq. 4.3.8 from Eq. 4.3.6. Note that Eqs. 4.3.6 and 3.3.6 are the same, as are Eqs. 4.3.8 and 2.6.7.

The element stiffness matrix (Eq. 4.3.4) is symmetric because $[\mathbf{E}]$ is symmetric. Another form of Eq. 4.3.4 is

$$[\mathbf{k}] = \sum_{i=1}^{n}\sum_{j=1}^{n}\int_V \lfloor\mathbf{B}\rfloor_i^T E_{ij} \lfloor\mathbf{B}\rfloor_j \, dV \tag{4.3.9}$$

where $\lfloor\mathbf{B}\rfloor_i$ and $\lfloor\mathbf{B}\rfloor_j$ are the ith and jth rows of $\lfloor\mathbf{B}\rfloor$, E_{ij} is the single coefficient in row i and column j of $[\mathbf{E}]$, and n is the order of $[\mathbf{E}]$. This expression looks cumbersome, but it may suggest ways to skip zeros in computer programming, which is desirable because $[\mathbf{B}]$ is usually sparse [4.2]. Subsequent sections and chapters contain several examples of element stiffness matrix calculation.

In generating $[\mathbf{k}]$ and $\{\mathbf{r}\}$ for specific elements, it is easier to start with matrices of appropriate order than to specialize Eqs. 4.3.4 and 4.3.5 from the three-dimensional case. Consider $[\mathbf{k}]$, for example. From Eq. 4.3.3, the strain energy of a one-element structure is

$$U = \tfrac{1}{2}\{\mathbf{d}\}^T[\mathbf{k}]\{\mathbf{d}\} = \int_V (\text{strain energy density})\, dV \tag{4.3.10}$$

Thus forming [k] is much the same as forming a strain energy expression. In plane stress U is a function of only three strains: ϵ_x, ϵ_y, and γ_{xy}. Accordingly, in Eq. 4.3.4, [E] is 3 by 3 and [B] has three rows. In uniaxial stress U is a function of a single strain, so we use $[\mathbf{E}] = E$ to find [k] for a truss element. In beam bending U can be written in terms of a single curvature, so EI replaces [E] and integration spans length rather than volume (Eq. 4.4.5).

Equation 4.3.5 gives forces applied to nodes by distributions of initial strain, stress, and force. Any of the four integrals may vanish. For example, the surface integral is zero unless the element has an edge on the structure boundary *and* the edge carries a traction. Then we integrate over only that edge. All four integrals vanish if nodal loads {P} make up the entire load vector {R}.

Let $\{\mathbf{p}\} = \{p_x\ p_y\ p_z\}$ be a concentrated force that acts on an element but not at a node. Its contribution to {r} is evaluated from the last integral in Eq. 4.3.5 by saying that a large traction acts on a small area ΔS, so that $\{\mathbf{p}\} = \{\boldsymbol{\Phi}\}\ \Delta S$. Thus, for n concentrated forces $\{\mathbf{p}\}_i$ on an element:

$$\text{Due to loads } \{\mathbf{p}\}, \qquad \{\mathbf{r}\} = \sum_{i=1}^{n} [\mathbf{N}]_i^T \{\mathbf{p}\}_i \tag{4.3.11}$$

where $[\mathbf{N}]_i$ is the value of [N] at the location of $\{\mathbf{p}\}_i$. Similarly, if n concentrated moments $\{\mathbf{m}\}_i = \{m_x\ m_y\ m_z\}$ act at n nonnodal locations on an element:

$$\text{Due to loads } \{\mathbf{m}\}, \qquad \{\mathbf{r}\} = \sum_{i=1}^{n} [\mathbf{N}']_i^T \{\mathbf{m}\}_i \tag{4.3.12}$$

where $[\mathbf{N}']_i$ contains derivatives of [N], evaluated at the location of $\{\mathbf{m}\}_i$. (For the plane beam of Fig. 3.4.1*b*, $[\mathbf{N}'] = d\lfloor\mathbf{N}\rfloor/dx$ and $\{\mathbf{m}\}_i = M_1$ or M_2.) Derivatives are needed for reasons stated in the next paragraph.

Forces {r} can be called *work-equivalent generalized forces*. We say "generalized" because some components of {r} are moments if {d} contains derivatives as nodal d.o.f., as it does for a beam. The "work-equivalent" terminology can be understood from Eq. 4.3.5 by premultiplying {r} by $\{\mathbf{d}\}^T$. Thus, taking for example the surface integral and using Eq. 4.3.1,

$$\{\mathbf{d}\}^T \{\mathbf{r}\} = \int_S \{\mathbf{d}\}^T [\mathbf{N}]^T \{\boldsymbol{\Phi}\}\ dS = \int_S \{\mathbf{f}\}^T \{\boldsymbol{\Phi}\}\ dS \tag{4.3.13}$$

Now consider that $\{\mathbf{d}\} = 0$ except for a single nonzero entry d_i, which we regard as a virtual displacement. Then d_i times its corresponding r_i is virtual work, equal to the work done by distributed tractions $\{\boldsymbol{\Phi}\}$ as they act through the displacement field produced by d_i and the shape function N_i. If d_i is a rotation, r_i is a moment instead of a force. And if $\{\boldsymbol{\Phi}\}$ represents a moment field instead of a traction or pressure field, {f} must represent a rotation field so that the integral of Eq. 4.3.13 yields work.

***Examples of* {r}**

To illustrate the calculation of {r}, we consider edge *DE* in Fig. 4.3.1. If $\{\mathbf{r}_v\}$ denotes the vertical components of {r} at *D* and *E*,

$$\{\mathbf{r}_v\} = \int_0^b \left\lfloor \frac{b-x}{b} \quad \frac{x}{b} \right\rfloor^T \frac{q_2 x}{b}\, dx = \begin{Bmatrix} P_2 \\ 2P_2 \end{Bmatrix} \tag{4.3.14}$$

where $P_2 = q_2 b/6$. At *D*, P_2 combines with P_1 from edge *CD*.

The calculation procedure leads to the following general result. Let x be a coordinate parallel to an edge ij of an element of (constant) thickness t, and let L be the distance from node i to node j. If edge-normal traction Φ varies linearly with x from $\Phi = \Phi_i$ at node i to $\Phi = \Phi_j$ at node j and if the edge-normal displacement is also linear in x, then the equivalent nodal forces are

$$r_i = \frac{Lt}{6}(2\Phi_i + \Phi_j), \qquad r_j = \frac{Lt}{6}(\Phi_i + 2\Phi_j) \tag{4.3.15}$$

where r_i and r_j act normal to the edge. Edge-*parallel* tractions that vary in the same way would be allocated to nodes in the same proportions. These results are independent of the orientation of the edge. The x and y components of r_i and r_j, if needed, can be found by trigonometry.

Forces {r} are called "consistent" when we calculate them from the same shape functions that go into [k] (Eq. 4.3.4). Alternatively, ad hoc "lumping" is possible. For example, multiplying b by the average load intensity in each span of Fig. 4.3.1 gives the same total upward force but a different allocation of nodal loads. The consistent method leads to more accurate displacements and stresses in a course mesh. Consistent and lumped methods become identical as a mesh is repeatedly subdivided (except near the loaded nodes, where errors of the lumped method persist). In Fig. 4.3.1 the load is too complicated, or the mesh too coarse, for either method to be good. Further examples of {r} appear in Sections 4.4 and 5.8.

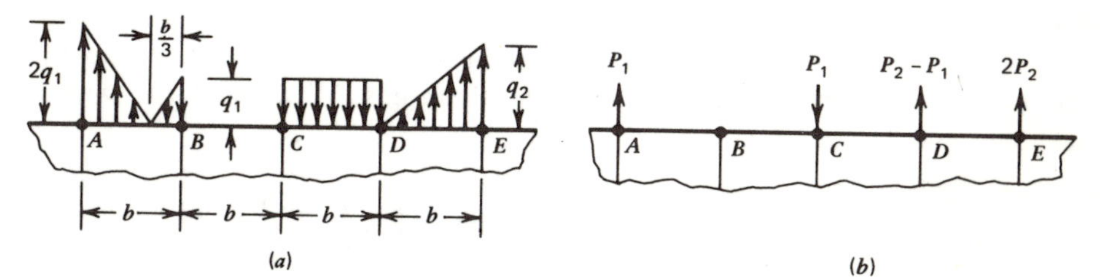

Figure 4.3.1. (*a*) Tractions on the boundary of a plane mesh. (*b*) Equivalent nodal forces if [N] prescribes a linear edge displacement between nodes. $P_1 = q_1 b/2$ and $P_2 = q_2 b/6$, where q_1 and q_2 are forces per unit length.

a-Basis Formulation. Usually, as in Eqs. 4.2.1 to 4.2.4, for example, generalized coordinates $\{\mathbf{a}\}$ are replaced by nodal d.o.f. $\{\mathbf{d}\}$ at the outset. Shape functions N_i then operate directly on the d_i. But the exchange can be postponed until last, as we now show.

Displacements and strains can be written directly in terms of the a_i. In matrix notation

$$\{\mathbf{f}\} = [\mathbf{N}_a]\{\mathbf{a}\} \qquad \text{and} \qquad \{\boldsymbol{\epsilon}\} = [\mathbf{B}_a]\{\mathbf{a}\} \tag{4.3.16}$$

But $\{\mathbf{a}\} = [\mathbf{A}]^{-1}\{\mathbf{d}\}$, so

$$[\mathbf{N}] = [\mathbf{N}_a][\mathbf{A}]^{-1} \qquad \text{and} \qquad [\mathbf{B}] = [\mathbf{B}_a][\mathbf{A}]^{-1} \tag{4.3.17}$$

Substitution of Eqs. 4.3.17 into Eqs. 4.3.4 and 4.3.5 yields the "d-basis" matrices $[\mathbf{k}]$ and $\{\mathbf{r}\}$.

$$[\mathbf{k}] = [\mathbf{A}]^{-T}[\mathbf{k}_a][\mathbf{A}]^{-1}, \qquad \{\mathbf{r}\} = [\mathbf{A}]^{-T}\{\mathbf{r}_a\} \tag{4.3.18}$$

where we have defined the "a-basis" matrices

$$[\mathbf{k}_a] = \int_V [\mathbf{B}_a]^T[\mathbf{E}][\mathbf{B}_a]\, dV, \qquad \{\mathbf{r}_a\} = \int_V [\mathbf{B}_a]^T[\mathbf{E}]\{\boldsymbol{\epsilon}_0\}\, dV - \cdots \tag{4.3.19}$$

Further comments about the a-basis appear in Sections 7.1 and 7.11. An application of Eqs. 4.3.19 appears in Section 10.9.

Matrix $[\mathbf{A}]$ can be singular. If this happens it indicates that the assumed displacement field does not permit nodal d.o.f. d_i to be independent of one another.

4.4 MATRICES FOR TRUSS, BEAM, AND FRAME ELEMENTS

Bar. Consider first the bar of Fig. 4.4.1a. Its stiffness matrix has already been calculated in Section 3.7, where we now recognize Eq. 3.7.10 as an instance of Eq. 4.3.4.

To illustrate the calculation of $\{\mathbf{r}\}$, let the bar be ΔL units too long, so that the initial strain is $\epsilon_0 = \Delta L/L$. Also, let it be heated $T°$ so that (with expansion prohibited) the initial stress is $\sigma_0 = -E\alpha T$. Finally, let there be a body force F, and a concentrated force p at $x = 2L/3$, both in the $+x$ direction. From Eqs. 3.7.6 and 3.7.9, with x the axial coordinate instead of s,

$$\lfloor \mathbf{N} \rfloor = \left\lfloor \frac{L-x}{L} \quad \frac{x}{L} \right\rfloor, \qquad \lfloor \mathbf{B} \rfloor = \left\lfloor -\frac{1}{L} \quad \frac{1}{L} \right\rfloor \tag{4.4.1}$$

Then, from Eqs. 4.3.5 and 4.3.11,

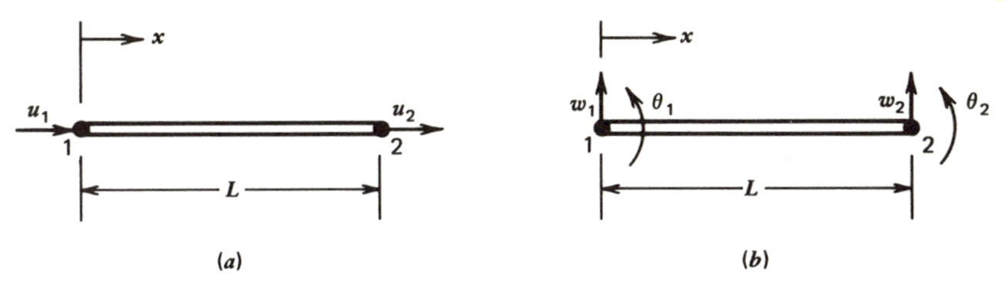

Figure 4.4.1. (*a*) Truss element with cross-sectional area A, elastic modulus E, and d.o.f. $\{\mathbf{d}\} = \{u_1\ u_2\}$. (*b*) Beam element with moment of inertia I, elastic modulus E, and d.o.f. $\{\mathbf{d}\} = \{w_1\ \theta_1\ w_2\ \theta_2\}$.

$$\underset{2\times 1}{\{\mathbf{r}\}} = \int_0^L \lfloor\mathbf{B}\rfloor^T E \frac{\Delta L}{L} A\, dx - \int_0^L \lfloor\mathbf{B}\rfloor^T (-E\alpha T) A\, dx + \int_0^L \lfloor\mathbf{N}\rfloor^T FA\, dx + \begin{bmatrix} \frac{1}{3} & \frac{2}{3} \end{bmatrix}^T p \tag{4.4.2}$$

$$\underset{2\times 1}{\{\mathbf{r}\}} = EA\frac{\Delta L}{L}\begin{Bmatrix} -1 \\ 1 \end{Bmatrix} + EA\alpha T\begin{Bmatrix} -1 \\ 1 \end{Bmatrix} + \frac{FAL}{2}\begin{Bmatrix} 1 \\ 1 \end{Bmatrix} + \frac{p}{3}\begin{Bmatrix} 1 \\ 2 \end{Bmatrix} \tag{4.4.3}$$

where, in $\{\mathbf{r}\}$, forces r_1 and r_2 correspond to d.o.f. u_1 and u_2. Forces $\{\mathbf{r}\}$ are applied *to* the nodes by the element. These results can be written immediately by direct physical argument, but such is not the case for complicated elements. Note that the temperature term can be treated as either $\sigma_0 = -E\alpha T$ or $\epsilon_0 = \alpha T$.

Beam. Consider next the beam element of Fig. 4.4.1*b*. Equations 4.2.4 and 4.2.5 give the shape functions. Then, from Eq. 4.3.2,

$$\lfloor\mathbf{B}\rfloor = \begin{bmatrix} -\frac{6}{L^2} + \frac{12x}{L^3} & -\frac{4}{L} + \frac{6x}{L^2} & \frac{6}{L^2} - \frac{12x}{L^3} & -\frac{2}{L} + \frac{6x}{L^2} \end{bmatrix} \tag{4.4.4}$$

Curvature $w_{,xx}$ is the displacement derivative needed in the strain energy integral. So Eqs. 3.4.16, 4.3.4, and 4.4.4 yield

$$[\mathbf{k}] = \int_0^L \lfloor\mathbf{B}\rfloor^T EI \lfloor\mathbf{B}\rfloor\, dx = \frac{EI}{L^3}\begin{bmatrix} 12 & 6L & -12 & 6L \\ 6L & 4L^2 & -6L & 2L^2 \\ -12 & -6L & 12 & -6L \\ 6L & 2L^2 & -6L & 4L^2 \end{bmatrix} \tag{4.4.5}$$

where $[\mathbf{k}]$ operates on nodal d.o.f. in the order listed in Fig. 4.4.1. If we view the four curves in Figs. 4.2.1*b* and 4.2.1*c* as (greatly exaggerated) beam deflections, the respective columns of $[\mathbf{k}]$ are nodal forces and moments that produce the respective curves.

Imagine that loads on the beam element are q, P, and M, as shown in Fig. 4.4.2. From Eqs. 4.3.5, 4.3.11, and 4.3.12, and with $\lfloor \boldsymbol{\theta} \rfloor = d\lfloor \mathbf{N} \rfloor / dx$,

$$\underset{4\times 1}{\{\mathbf{r}\}} = \int_0^L \lfloor \mathbf{N} \rfloor^T q\, dx + \lfloor \mathbf{N} \rfloor_{L/2}^T P + \lfloor \boldsymbol{\theta} \rfloor_{L/2}^T (-M) \tag{4.4.6}$$

$$\underset{4\times 1}{\{\mathbf{r}\}} = q \begin{Bmatrix} L/2 \\ L^2/12 \\ L/2 \\ -L^2/12 \end{Bmatrix} + P \begin{Bmatrix} 1/2 \\ L/8 \\ 1/2 \\ -L/8 \end{Bmatrix} + M \begin{Bmatrix} 3/2L \\ 1/4 \\ -3/2L \\ 1/4 \end{Bmatrix} \tag{4.4.7}$$

These loads $\{\mathbf{r}\}$, applied *to* the nodes as always, are shown in Fig. 4.4.2.

Frame. A plane frame element can both stretch and bend and so has all six d.o.f. seen in Fig. 4.4.1. To form its 6-by-6 stiffness matrix that operates on $\{\mathbf{d}\} = \{u_1\ w_1\ \theta_1\ u_2\ w_2\ \theta_2\}$, we first expand the 2-by-2 $[\mathbf{k}]$ of the truss element to size 6 by 6. Specifically, we add four rows and four columns of zeros, so that $k_{11} = k_{44} = -k_{14} = -k_{41} = AE/L$, but all other $k_{ij} = 0$. Similarly, we expand the 4-by-4 $[\mathbf{k}]$ of the beam element so that, in size 6 by 6, its first and fourth rows and columns are null. Finally, the two 6-by-6 matrices are added. The result is the $\beta = 0$ case in Fig. 4.4.3.

The foregoing plane truss, beam, and frame elements all lie on the x-axis. They can be arbitrarily oriented by invoking the coordinate transformations of Section 6.3 and 6.5 (see the $\beta \neq 0$ case in Fig. 4.4.3). A *space* truss element has three translational d.o.f. at each node. A *space* frame element has three translational and three rotational d.o.f. at each node and has torsional stiffness and bending stiffness in two directions.

4.5 THE CONSTANT STRAIN TRIANGLE

This element has the six d.o.f. shown in Fig. 4.5.1. It is one of the earliest finite elements [1.7] and is easy to formulate. We will use the element and equations associated with it in subsequent discussions. However, better elements are now available. An assemblage of four constant strain triangles (Fig. 7.2.1*a*) behaves rather like the element of Table 5.12.1.

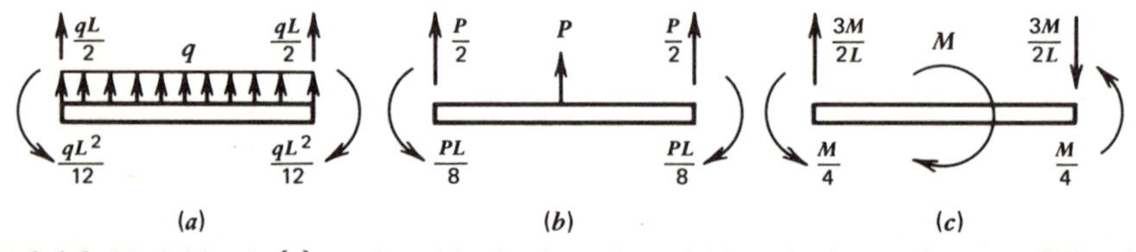

Figure 4.4.2. Nodal loads $\{\mathbf{r}\}$ produced by loads q, P, and M on the beam element of Fig. 4.4.1*b*. Distributed load q is uniform and has units of force divided by length. Force P and moment M act at midspan, $x = L/2$.

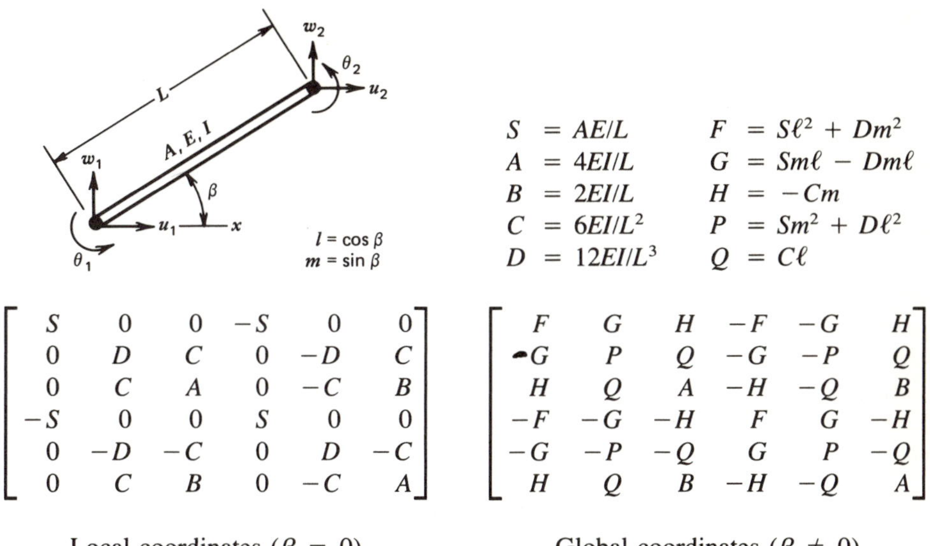

$$
\begin{aligned}
S &= AE/L & F &= S\ell^2 + Dm^2 \\
A &= 4EI/L & G &= Sm\ell - Dm\ell \\
B &= 2EI/L & H &= -Cm \\
C &= 6EI/L^2 & P &= Sm^2 + D\ell^2 \\
D &= 12EI/L^3 & Q &= C\ell
\end{aligned}
$$

$$
\begin{bmatrix}
S & 0 & 0 & -S & 0 & 0 \\
0 & D & C & 0 & -D & C \\
0 & C & A & 0 & -C & B \\
-S & 0 & 0 & S & 0 & 0 \\
0 & -D & -C & 0 & D & -C \\
0 & C & B & 0 & -C & A
\end{bmatrix}
\qquad
\begin{bmatrix}
F & G & H & -F & -G & H \\
G & P & Q & -G & -P & Q \\
H & Q & A & -H & -Q & B \\
-F & -G & -H & F & G & -H \\
-G & -P & -Q & G & P & -Q \\
H & Q & B & -H & -Q & A
\end{bmatrix}
$$

Local coordinates ($\beta = 0$) Global coordinates ($\beta \neq 0$)

Figure 4.4.3. The stiffness matrix of a plane frame element.

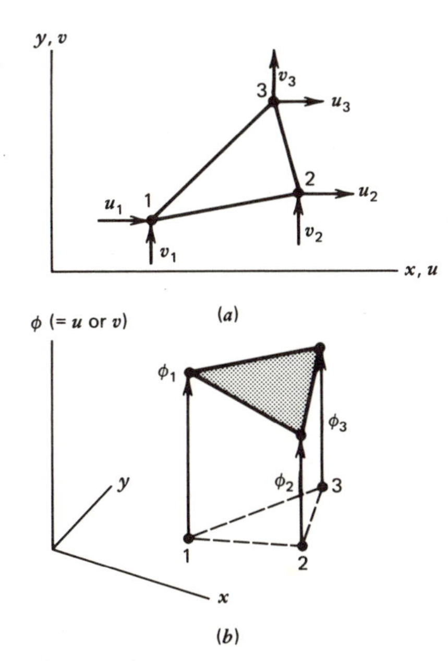

Figure 4.5.1. (*a*) A constant strain triangle and its six nodal d.o.f. (*b*) An illustration of the displacement field with ϕ shown normal to the xy plane for visualization only.

Figure 4.5.2. Possible mesh of constant strain triangles for plane strain analysis of a gravity dam and part of its foundation. The loads are dead weight and water pressure on the left face.

The assumed displacement field is linear in x and y:

$$u = \lfloor \mathbf{X} \rfloor \{a_1 \ a_2 \ a_3\}, \qquad v = \lfloor \mathbf{X} \rfloor \{a_4 \ a_5 \ a_6\} \tag{4.5.1}$$

where $\lfloor \mathbf{X} \rfloor = \lfloor 1 \ x \ y \rfloor$. Thus $\epsilon_x = a_2$, $\epsilon_y = a_6$ and $\gamma_{xy} = a_3 + a_5$ (hence the name "constant strain triangle"). Also, whatever the nodal displacements, sides remain straight. These attributes are not suited to beam problems, where strains must vary through the depth and element edges parallel to the beam axis must curve. *Wherever* bending is significant, many constant strain triangles are needed. Acceptable accuracy with few elements requires that we use different elements, based on more competent displacement fields.

Stiffness matrix formulation follows the now-familiar pattern. Since u and v fields are identical, we need look at only one, say u. We evaluate Eq. 4.5.1 at the three nodes and find

$$\begin{Bmatrix} u_1 \\ u_2 \\ u_3 \end{Bmatrix} = [\mathbf{A}] \begin{Bmatrix} a_1 \\ a_2 \\ a_3 \end{Bmatrix}, \qquad \text{where} \qquad [\mathbf{A}] = \begin{bmatrix} 1 & x_1 & y_1 \\ 1 & x_2 & y_2 \\ 1 & x_3 & y_3 \end{bmatrix} \tag{4.5.2}$$

Hence

$$[\mathbf{A}]^{-1} = \frac{1}{2A} \begin{bmatrix} x_2y_3 - x_3y_2 & x_3y_1 - x_1y_3 & x_1y_2 - x_2y_1 \\ y_2 - y_3 & y_3 - y_1 & y_1 - y_2 \\ x_3 - x_2 & x_1 - x_3 & x_2 - x_1 \end{bmatrix} \tag{4.5.3}$$

where

$$A = \text{area of the triangle} = \tfrac{1}{2}\det[\mathbf{A}] \tag{4.5.4}$$

The determinant becomes negative if nodes 1, 2, and 3 are ordered clockwise around the element, instead of counterclockwise as in Fig. 4.5.1.

In terms of nodal d.o.f. instead of the a_i, Eqs. 4.5.1 become

$$u = \lfloor N_1 \ N_2 \ N_3 \rfloor \{u_1 \ u_2 \ u_3\} \tag{4.5.5}$$

where

$$\lfloor N_1 \ N_2 \ N_3 \rfloor = \lfloor 1 \ x \ y \rfloor \, [\mathbf{A}]^{-1} \tag{4.5.6}$$

So, in one package, the displacement field $\{\mathbf{f}\} = [\mathbf{N}]\{\mathbf{d}\}$ is

$$\begin{Bmatrix} u \\ v \end{Bmatrix} = \begin{bmatrix} N_1 & 0 & N_2 & 0 & N_3 & 0 \\ 0 & N_1 & 0 & N_2 & 0 & N_3 \end{bmatrix} \{u_1\ v_1\ u_2\ v_2\ u_3\ v_3\} \tag{4.5.7}$$

The first of Eqs. 4.3.2 is applied to [N] to produce [B] (see Eq. 7.9.16). The differential volume is $dV = t\,dA$, where t is element thickness. Matrix [B] contains only constants, so if neither t nor [E] varies over the element, Eq. 4.3.4 yields

$$\underset{6\times 6}{[\mathbf{k}]} = \int_A \underset{6\times 3}{[\mathbf{B}]^T} \underset{3\times 3}{[\mathbf{E}]} \underset{3\times 6}{[\mathbf{B}]}\, t\, dA = [\mathbf{B}]^T [\mathbf{E}] [\mathbf{B}]\, tA \tag{4.5.8}$$

Equations 1.5.7 and 1.5.8 give [E] for an isotropic material in plane stress or plane strain.

A linearly varying thickness can be interpolated from nodal thicknesses t_i. Following Eq. 4.5.5, we write

$$t = \lfloor N_1\ \ N_2\ \ N_3 \rfloor \{t_1\ \ t_2\ \ t_3\} \tag{4.5.9}$$

Using t from Eq. 4.5.9 in the integral of Eq. 4.5.8 shows that t in the last of Eqs. 4.5.8 can be interpreted as the average thickness $t = (t_1 + t_2 + t_3)/3$. This result is easily seen by using the area coordinates of Section 7.9.

Figure 4.5.2 shows an application of constant strain triangles. The foundation may have a different modulus than the dam and might be orthotropic, as for layered rock. The mesh shown is rather coarse.

Numerical results appear in Fig. 4.5.3. There are many d.o.f, but results compare unfavorably with those in Tables 5.12.1 and 5.12.2.

Generalization of the constant strain triangle to the analogous solid, a four-node tetrahedron, is straightforward (Problem 4.20).

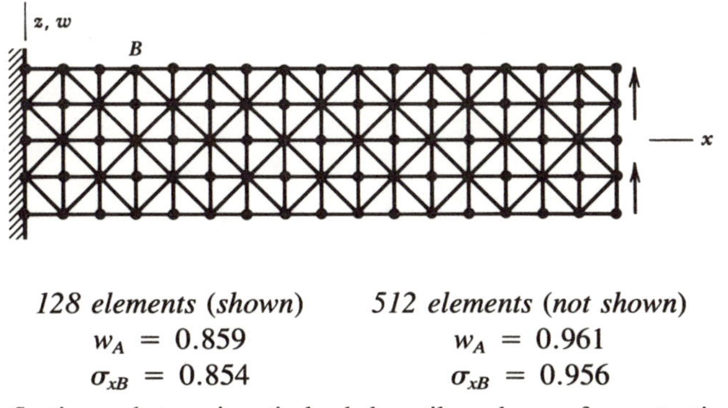

128 elements (shown)	*512 elements (not shown)*
$w_A = 0.859$	$w_A = 0.961$
$\sigma_{xB} = 0.854$	$\sigma_{xB} = 0.956$

Figure 4.5.3. Deflection and stress in a tip-loaded cantilever beam of aspect ratio 4/1 modeled by constant strain triangles [11.13]. Deflection w_A and stress σ_{xB} are each the ratio of computed value to exact value.

4.6 A LINEAR RECTANGULAR ELEMENT

The eight d.o.f. element of Fig. 4.6.1*a* has the assumed displacement field

$$\begin{Bmatrix} u \\ v \end{Bmatrix} = \begin{bmatrix} 1 & x & y & xy & 0 & 0 & 0 & 0 \\ 0 & 0 & 0 & 0 & 1 & x & y & xy \end{bmatrix} \{a_1\ a_2 \cdots a_8\} \qquad (4.6.1)$$

This field, and this element, are sometimes called "bilinear" because coefficients in the matrix come from the product of two linear expressions: $(1 + x)$ times $(1 + y)$. But, like the constant strain triangle, the element is "linear" because sides remain straight when the element deforms. For example, if $y = c$, then u and v are linear in x, so edge 3–4 is always straight. Adjacent elements are therefore compatible with one another.

The procedure of Eqs. 4.5.2 to 4.5.6 exchanges the eight a_i for the eight nodal displacements in Fig. 4.6.1*a*. But the desired shape functions can be found *directly* from Lagrange's interpolation formula, as was done for a more complicated case in Eqs. 4.2.9. We find

$$\{u\ v\} = [\mathbf{N}]\{\mathbf{d}\}, \qquad \text{where} \qquad [\mathbf{N}] = \begin{bmatrix} N_1 & 0 & N_2 & 0 & N_3 & 0 & N_4 & 0 \\ 0 & N_1 & 0 & N_2 & 0 & N_3 & 0 & N_4 \end{bmatrix}$$

$$\text{and} \qquad \{\mathbf{d}\} = \{u_1\ v_1\ u_2\ v_2\ u_3\ v_3\ u_4\ v_4\} \qquad (4.6.2)$$

and the N_i are shown in Fig. 4.6.1*b*. As always, if the coordinates of a node i are inserted, $N_i = 1$ and $N_j = 0$ if $i \neq j$.

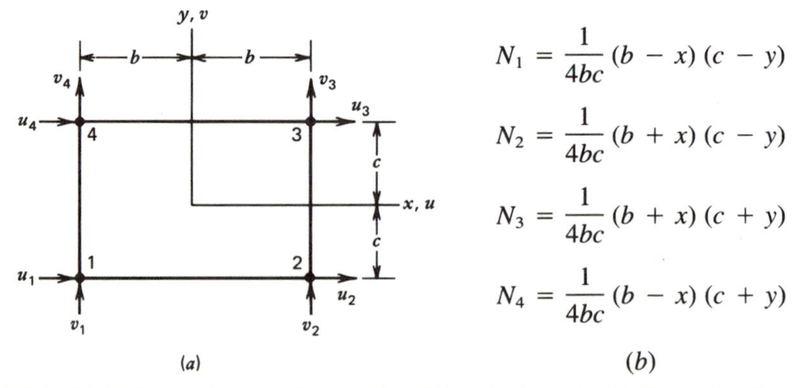

Figure 4.6.1. (*a*) Eight d.o.f. linear (also called bilinear) element. (*b*) Shape functions for this element.

The first of Eqs. 4.3.2 yields

$$\underset{3\times 8}{[\mathbf{B}]} = \frac{1}{4bc}\begin{bmatrix} -(c-y) & 0 & (c-y) & 0 & \text{etc.} \\ 0 & -(b-x) & 0 & -(b+x) & \text{etc.} \\ -(b-x) & -(c-y) & -(b+x) & (c-y) & \text{etc.} \end{bmatrix} \tag{4.6.3}$$

We see that ϵ_x depends on y, ϵ_y depends on x, and γ_{xy} depends on both x and y.

Now we are ready to evaluate the stiffness matrix

$$\underset{8\times 8}{[\mathbf{k}]} = \int_{-c}^{c}\int_{-b}^{b} [\mathbf{B}]^T [\mathbf{E}][\mathbf{B}]\, t\, dx\, dy \tag{4.6.4}$$

where t is the element thickness. If t varies, we can use Eq. 4.5.9 but with four N_i and four t_i. An alternative representation of a tapered body is one with steps, where t is constant over each element but changes from one element to another. A stepped representation is less accurate unless the actual structure is also stepped [4.3]. (Thickness variations affect $[\mathbf{k}]$ and the mass matrix $[\mathbf{m}]$ but not the stress stiffness matrix $[\mathbf{k}_\sigma]$.)

The stiffness matrix of an improved form of this element is given in Table 7.7.3.

4.7 EQUILIBRIUM AND COMPATIBILITY IN THE SOLUTION

An approximate solution does not fulfill all the requirements for equilibrium and compatibility that an exact solution demands. What can we expect? It depends mainly on the kinds of elements we use:

1. *Equilibrium is usually not satisfied within elements*. Stresses, given by Eq. 4.10.1, are functions of the coordinates. They usually do not satisfy the differential equations of equilibrium (Eqs. 1.3.3). But Eqs. 1.3.3 are *always* satisfied by the constant strain triangle. They are satisfied by other elements *if* the element happens to be in a state of constant strain. For example, the rectangular element of Section 4.6 displays constant strain only if $a_4 = a_8 = 0$ in Eq. 4.6.1. Nevertheless, in a varying strain field and for a given number of d.o.f., a mesh of the rectangles usually gives more accurate answers than a mesh of the triangles.

2. *Equilibrium is usually not satisfied between elements*. For example, in Fig. 4.7.1*a* the only stresses are $\sigma_x' = 0$ and $\sigma_x'' \neq 0$. So a differential element that spans the boundary is not in equilibrium. Other types of elements behave in the same way. Similarly, elements that share a node display different states of stress at the node. Computed normal and shear stresses at a free edge are not likely to vanish as theory demands, but they should be small in comparison with the greatest stresses in the mesh.

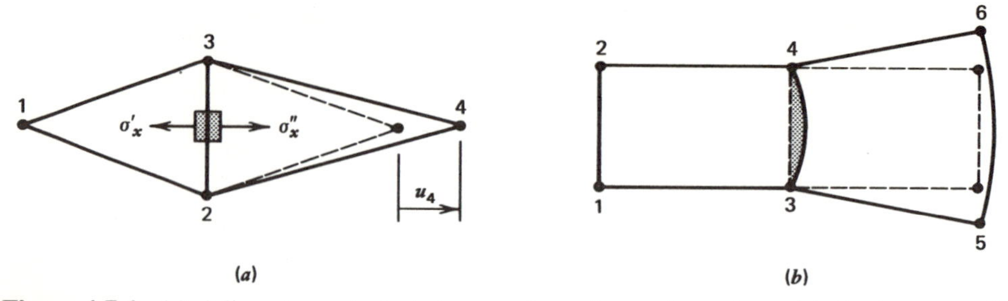

Figure 4.7.1. (*a*) Adjacent constant strain triangles, with u_4 the only nodal displacement. (*b*) Adjacent incompatible elements. All nodes but 5 and 6 have zero displacement.

3. *Equilibrium of nodal forces and moments is satisfied.* The structural equations $\{\mathbf{R}\} - [\mathbf{K}]\{\mathbf{D}\} = 0$ are nodal equilibrium equations. Therefore the solution vector $\{\mathbf{D}\}$ is such that resultant forces and moments applied to each node are zero.
4. *Compatibility is satisfied within elements* if the assumed element displacement field is continuous.
5. *Compatibility may or may not be satisfied along interelement boundaries.* The elements of Sections 4.5 and 4.6 remain straight sided when deformed and therefore provide interelement compatibility. The element of Fig. 4.7.1*b* does not, as shown by the interelement gap. Incompatible elements can be acceptable and even desirable, as explained in Section 4.9.
6. *Compatibility is enforced at nodes,* as is obvious: this is where elements are connected to one another.

An element type that does well in one situation may do badly in another. The ''best'' element works well in a variety of test cases. It is apt to be an element that violates the equilibrium and compatibility requirements of an exact solution. The violations tend toward zero as more and more elements are used, provided the elements are ''proper'' as defined in the next two sections.

4.8 CONVERGENCE REQUIREMENTS. PATCH TEST

''Convergence requirements'' are conditions that guarantee that exact answers will be approached as more and more elements are used to model an arbitrary structure. Practical work schedules may allow only one mesh. Yet we feel more comfortable using elements that would provide convergence if we had enough time and money to use finer and finer meshes.

The first requirement is

1. The displacement field within an element must be continuous.

This requirement is so easily met by polynomial displacement fields that we need not mention it again.

The next requirement leads to much discussion and reinterpretation, which we begin in this section and continue in the next. It is:

2. The element must be able to assume a state of constant strain.

A physical argument can be made for Requirement 2. Imagine that a bar of length L is under distributed axial load, and let b be a small portion of its length. If $L >> b$, the *variation* of strain over b is negligible in comparison with the average strain over length b. Thus, in element 1 of Fig. 4.8.1, as b shrinks, the smooth curve of ϵ_x versus x is modeled arbitrarily closely in stair-step fashion (see also Fig. 3.7.2*b*). Element 2 in Fig. 4.8.1 provides only a saw-tooth pattern: as b shrinks, the teeth get sharper, never model a smooth curve, and convergence is to an incorrect result (see Problem 4.26).

In beams and plates, where bending is involved, Requirement 2 is concerned with constant *curvature,* which produces constant strain in a layer parallel to the midsurface. The general intent of Requirement 2 is that displacement derivatives in the applicable strain energy expression be able to assume constant values.

Requirement 2 is a part of the completeness requirement of Section 3.6. Another part is that still lower terms in the displacement expansion be present, so that rigid-body motion without strain is possible. This is Requirement 3, which we take up in Section 4.9.

The *patch test* [4.4, 4.5] is a simple numerical test of element validity. We describe the test first, then explain its significance.

Assemble a "patch" of elements in such a way that at least one node is completely surrounded by elements. Apply to the boundary nodes either displacements or consistent nodal loads {**r**} that correspond to a state of constant strain (Fig. 4.8.2, for example).

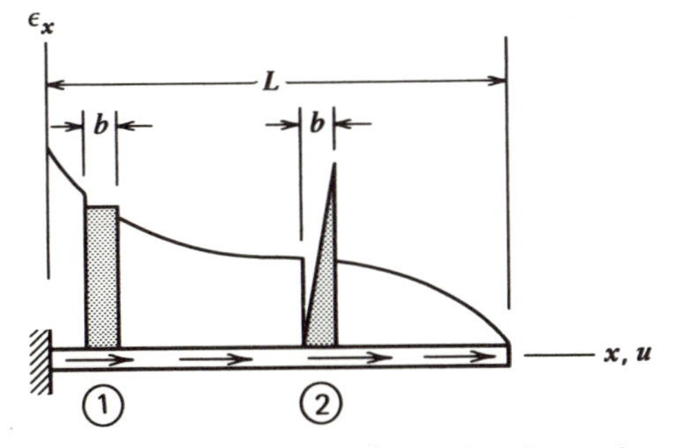

Figure 4.8.1. Strain distribution in a bar under distributed axial load, modeled by elements of length b. Element 1 can represent constant strain, but element 2 can represent only a strain that varies linearly from zero at its left end.

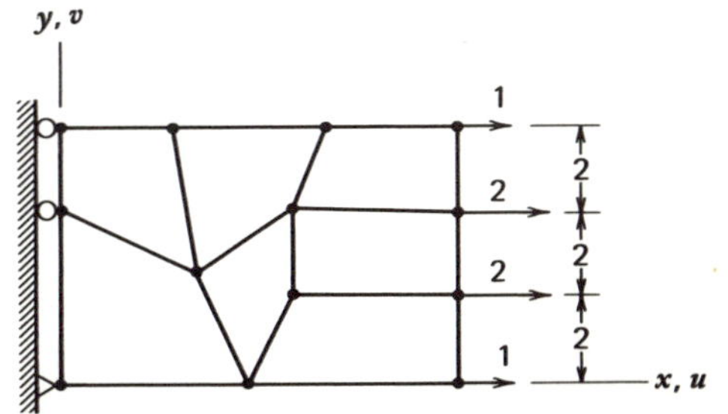

Figure 4.8.2. Possible mesh of plane quadrilateral elements for a patch test. Prescribed nodal loads on the right edge are consistent with constant stress $\sigma_x = 1$, unit thickness, and a linear variation of u between nodes.

Nodes not on the boundary are neither loaded nor restrained. Solve for all nodal d.o.f. that are not prescribed and compute strains (or stresses) in the elements. The patch test is passed if, at every point in every element, computed strains agree with exact values to the limit of computer accuracy.

In an alternate version of the patch test, we compute nodal forces $\{\mathbf{R}\}$ as the product $[\mathbf{K}]\{\mathbf{D}\}$ when *all* D_i are assigned values consistent with a state of constant strain. If the R_i of all nodes inside the boundary of the patch are zero, the patch test is passed.

The significance of the patch test is this. Subject to exceptions noted in the following three paragraphs, an element that passes the patch test provides convergence to exact answers as we indefinitely refine the mesh of an arbitrary structure built of these elements. Thus passing the patch test means that all convergence requirements are satisfied, not just Requirement 2. This is true of both compatible and incompatible elements. The patch test has nothing to say about the *rate* of convergence (see Sections 15.3 and 15.4).

A defective element may satisfy some of the required constant states but not others. For example, a plate bending element may display constant curvatures $w_{,xx}$ and $w_{,yy}$ when asked to but be incapable of displaying constant twist $w_{,xy}$.

Certain quadrilateral elements pass the test if they are rectangles or parallelograms but fail if they are arbitrarily shaped. Certain triangular elements pass in some mesh arrangements but fail in others (see meshes in Fig. 7.13.1).

Strictly, then, a single patch test is not *sufficient* to guarantee convergence to correct results. However, neither is it *necessary:* repeated subdivision of a mesh may cause elements to approach valid shapes or to display constant states that had been missing. Accordingly, a given element should be patch-tested in more than one geometry, mesh layout, and strain state.

A patch test failure may indicate that constant strain modes are lacking in the displacement field. Or, if present, strain *gradient* modes may be activated at the same time. Failure may also be associated with Requirement 3 or 4 of Section 4.9. Nevertheless, it is *possible* for an element that fails to give better answers in a coarse mesh than an element that passes. See also the last paragraph of Section 8.2.

Higher-order patch tests can be contemplated. For example, a patch of plane elements should be able to model pure bending if the u and v fields of each element contain all the quadratic terms (x^2, xy, and y^2). In typical problems, a given number of these elements will be more accurate than an equal number of linear elements.

The patch test can be reinterpreted for application to nonstructural problems, as the discussion of Section 4.9 suggests.

4.9 FURTHER REQUIREMENTS FOR CONVERGENCE AND INVARIANCE

Here we state further requirements for convergence, first in structural mechanics language, then in more general terms at the end of the section.

In the limit of mesh refinement, but not necessarily in larger elements:

3. Rigid-body modes must be present. When nodal d.o.f. {**d**} correspond to rigid-body motion, the element must exhibit zero strain [**B**]{**d**} and zero nodal forces [**k**]{**d**}.
4. Elements must be compatible. There must be no interelement gaps or overlaps. Bending elements, such as beams and plates, must have no sudden change in slope between elements.

Some axisymmetric shell elements satisfy Requirement 3 only as the angle subtended by a meridional arc approaches zero (see Table 10.3.1). But convergence toward exact answers is achieved, and elements of *finite* size work well enough to be useful. However, *good* elements usually satisfy Requirement 3 even in a coarse mesh. Problem 4.27 illustrates the import of Requirement 3.

Elements that violate Requirement 4 in a coarse mesh are called *incompatible* or *nonconforming*. These elements are valid, and convergence is achieved, if the incompatibilities disappear with increasing mesh refinement, and the element approaches a state of constant strain. Section 7.7 discusses an incompatible element based on the element of Section 4.6. See also Problem 4.32 and Section 9.2.

Incompatible elements are often the best kind. Overlaps, gaps, or kinks tend to soften a structure. Softening counters the inherent overstiffness of an assumed-displacement solution (Section 3.6). A good balance of the two effects leads to good answers with a coarse mesh. However, the upper-bound nature of the approximation is lost: there is no guarantee that a mesh of incompatible elements is stiffer than the actual structure. Also, in problems that should be independent of Poisson's ratio ν, incompatible elements may display a dependence on ν [4.6]. (So do the elements of Chapter 5, if they are under-integrated.) The dependence dwindles as the mesh is refined.

For convergence toward exact answers we must satisfy Requirements 1 through 4 only as the mesh is refined. The requirements can be violated in a coarse mesh, yet the elements can be valid and convergent.

The patch test checks the constant strain requirement (Requirement 2). In addition it checks Requirement 4, since constant strain and interelement compatibility must be *simultaneously* displayed by elements of an infinitely fine mesh. Requirement 3 is also tested because some elements of the patch have rigid-body motion superposed on straining modes. Indeed, rigid-body motion can be viewed as a constant strain of zero magnitude.

A desirable but not mandatory requirement is:

5. An element should have no preferred directions. In other words, an element should be geometrically *invariant* (also called geometrically isotropic or spatially isotropic).

In Fig. 4.9.1, if the s-parallel displacements or stresses differ for the two orientations, the element is not invariant. This defect is annoying but does not prevent convergence to correct answers.

Any element can easily be made invariant by formulating its stiffness matrix in local coordinates such as st of Fig. 4.9.1, then transforming to global coordinates xy. Transformation during stress computation may also be required [5.18]. These operations are implicit in the isoparametric elements of Chapter 5.

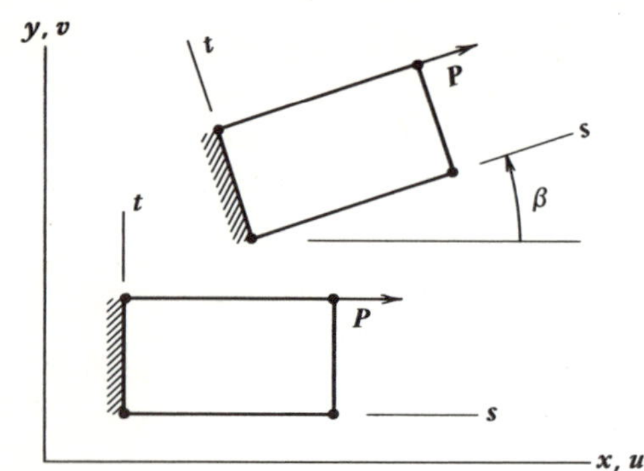

Figure 4.9.1. A one-element structure under load P, shown in two different orientations in global coordinates xy.

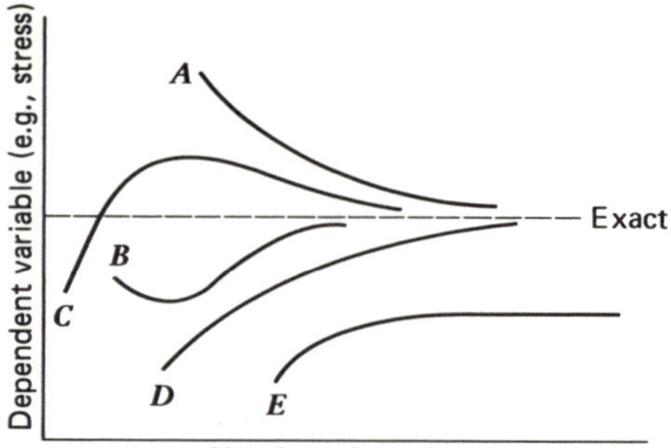

Figure 4.9.2. Types of convergence. A, monotonic from above. B and C, not monotonic. D, monotonic from below. E, monotonic convergence to wrong answer.

Before going on we must describe *completeness*—in the polynomial sense, not in the Rayleigh-Ritz sense of Section 3.6. Let n be the degree of a polynomial and let $A = n + 1$, $B = (n + 2)/2$, and $C = (n + 3)/3$. Up to quadratic terms, coefficients of a complete polynomial in 1, 2, and 3 dimensions are:

	constant ($n = 0$)	linear ($n = 1$)	quadratic ($n = 2$)
$1-D$ (A terms)	1	x	x^2
$2-D$ ($A*B$ terms)	1	x, y	x^2, xy, y^2
$3-D$ ($A*B*C$ terms)	1	x, y, z	$z^2, y^2, z^2, xy, yz, zx$

(4.9.1)

The constant terms provide rigid-body motion capability. A complete linear polynomial includes also the capability for constant strain states. (Analogous expressions in cylindrical coordinates are given in Problem 8.11.) An element based on a complete polynomial of degree n is invariant. Also, the displacement along *any* direction s (Fig. 4.9.1) is also a complete polynomial of degree n.

But most elements are based on *incomplete* polynomials. The displacement field

$$u = a_1 + a_2 x + a_3 y + a_4 xy \tag{4.9.2}$$

is an incomplete quadratic because the x^2 and y^2 terms are absent. This field is the basis for the element of Section 4.6. If formulated in global coordinates xy, element sides must be x parallel or y parallel but, subject to this restriction, the element is invariant. That is, it does not matter whether edge 1-2 is x parallel or y parallel; answers are the same. This happens because the terms retained are symmetric with respect to x and y. No preference is given to either direction. As another example, consider a lateral-displacement polynomial for a thin-plate element.

$$w = a_1 + a_2x + a_3y + a_4x^2 + a_5xy + a_6y^2 + a_7x^3 + a_8x^2y + a_9y^3 \quad (4.9.3)$$

This cubic is incomplete because the xy^2 term is missing. It is not balanced because the x^2y term is retained (see also Section 9.2).

It is worth noting how Requirements 2 and 3 are met by Eqs. 4.9.2 and 4.9.3. In Eq. 4.9.2, a_1 is associated with rigid-body motion and a_2 and a_3 with constant strain. In Eq. 4.9.3, a_1, a_2, and a_3 are associated with rigid-body motion, a_4 and a_6 with constant curvatures $w_{,xx}$ and $w_{,yy}$ and a_5 with constant twist $w_{,xy}$.

Figure 4.9.2 illustrates different types of convergence. Curve *A* might represent a hybrid element (Section 16.1). Curves *B* and *C* are often displayed by incompatible elements. Curve *D* is seen for displacement-field elements that are compatible in any size mesh. Curve *E* might arise if the a_5xy term is omitted from Eq. 4.9.3. Curve *E* also arises from the element of Problem 4.26. It is possible to state conditions under which the convergence of potential energy will be monotonic [4.7]. When convergence is monotonic, results from two different meshes can be extrapolated to a much better answer (see Eq. 15.4.5).

We now state the convergence requirements in broader terms, applicable to problems of continua in general. If a differential equation has derivatives through order $2m$, its corresponding functional has derivatives through order m. Then, in the notation of Section 4.2:

A. The assumed field must be C_m continuous and C_m complete within elements.

B. As the mesh is refined, C_{m-1} continuity between elements must be approached.

Requirement A contains Requirements 1, 2, and 3, and Requirement B contains Requirement 4.

Consider how Requirements A and B apply in structural mechanics. For the problem of Fig. 3.7.2, $2m = 2$. So continuity and constant values of $\epsilon_x = u_{,x}$ are required within elements and continuity of u itself is required between elements. Both requirements are met by Eq. 3.7.2. For thin-plate bending, $2m = 4$. So continuity and constant values of all three second derivatives are required within elements, and continuity of both first derivatives is required between elements. Both requirements are met by Eq. 4.9.3. These requirements are *capabilities* that must be present and do not prevent the addition of still higher-order terms.

Further remarks about convergence appear in Section 15.4.

4.10 STRESS COMPUTATION

Stresses $\{\sigma\}$ in an element can be calculated when its nodal d.o.f. $\{\mathbf{d}\}$ are known. From Eq. 1.6.1, with $\{\epsilon\} = [\mathbf{B}]\{\mathbf{d}\}$,

$$\{\sigma\} = [\mathbf{E}]\left([\mathbf{B}]\{\mathbf{d}\} - \{\epsilon_0\}\right) + \{\sigma_0\} \quad (4.10.1)$$

In general, [**B**] is a function of the coordinates. It must be evaluated at the point where we want the stresses. For computer programming, note that [**E**]([**B**] {**d**}) takes fewer operations than ([**E**] [**B**]){**d**}.

Often we want stresses at a node. They can be found by evaluating [**B**] at the node. But it may be more accurate to find stresses at interior points first, then extrapolate to a node, especially if the element shape is irregular. Stresses at midsides are usually more accurate than stresses at corners. Section 5.11 discusses error-minimal points.

What is the effect of a surface traction or a body force? Consider Fig. 4.10.1. If our model is one bar element, its two nodes must have zero displacement because they are at the fixed supports. Therefore Eq. 4.10.1 predicts zero axial stress all along the length. This incorrect result suggests that we should add to Eq. 4.10.1 the stresses produced by distributed forces when {**d**} = 0. Such an addition is easy to formulate for a framework element but not for a finite element of a continuum. Fortunately, as a mesh is refined, exact answers are approached even when the "extra" stresses are ignored (Figs. 4.10.1*b*, 4.10.1*c*).

Elements that share a node can be expected to predict different stresses at the node. Usually, the *average* stress at a node is more to be trusted than any one of the contributing stresses. The probable error of the average is also of interest (but the statistical sample is small). Too great a probable error suggests a need for reanalysis with a finer mesh.

Various ad hoc procedures may improve accuracy. In a row of simple elements such as constant strain triangles, element stresses may be alternately too high and too low. This requires averaging, with perhaps a weighting scheme that gives more credit to larger elements or to those whose centroids are closer to the node. Averaging between dissimilar materials or across sudden thickness changes must be avoided.

Another ad hoc scheme is to fit a polynomial field to several d.o.f. in the region of interest. Differentiation yields strains: stresses follow from strains in the usual way. Effectively, Eq. 4.10.1 is applied to a region of the structure instead of to a single element. (Take care with a least-squares fit—the algebraic equations tend to be ill conditioned.) Related schemes are available [4.8, 4.9]. Stresses can be calculated indirectly by means of influence surfaces (Problem 16.22).

Stresses are usually greatest at a boundary. We now show another way to calculate them there. Consider the plane problem of Fig. 4.10.2. We have the three equations

$$\{\sigma_s\ \sigma_t\ \tau_{st}\} = [\mathbf{E}]\,(\{\epsilon_s\ \epsilon_t\ \gamma_{st}\} - \{\boldsymbol{\epsilon}_0\}) + \{\boldsymbol{\sigma}_0\} \tag{4.10.2}$$

We define $\epsilon_s = (u_3 - u_4)/L$, where L is the distance between nodes 3 and 4. Now σ_s, ϵ_t, and γ_{st} are the only unknowns in Eq. 4.10.2, so we can find σ_s. Thus the state of stress at point A in local coordinates st is determined—often more accurately than by applying Eq. 4.10.1 in global coordinates and then transforming to st coordinates. In the most common special case, the material is isotropic and $p = f = 0$. Then A is a principal element, and the principal stress of greatest magnitude is simply $\sigma_t = E\epsilon_t = E(u_3 - u_4)/L$. Other schemes also exploit known boundary conditions on stress [4.10]. See also the "Babuska paradox," discussed in Section 15.4.

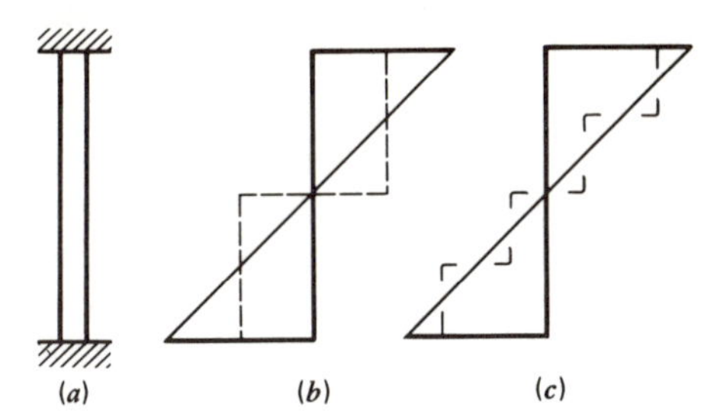

Figure 4.10.1. (*a*) A bar, fixed at upper and lower ends and loaded by its own weight. (*b, c*) Two-and four-element predictions of axial stress (dashed lines) and exact axial stress (solid line).

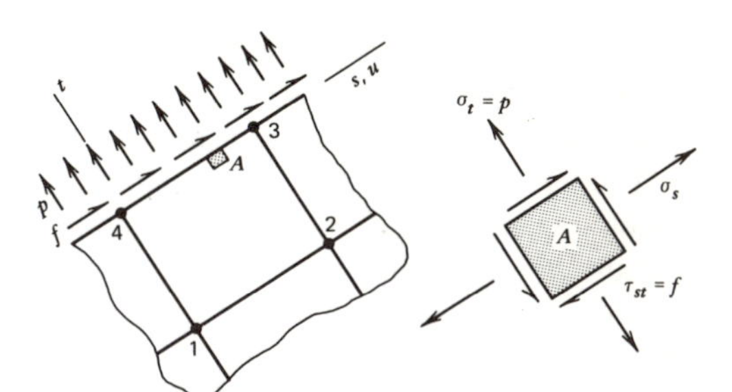

Figure 4.10.2. A typical element on a structure boundary where surface tractions p and f act respectively normal and tangent to the boundary. Stress σ_s is the only unknown on the element at A.

Loubignac [4.11] suggests an iterative scheme that improves both stresses and displacements. We first compute average nodal stresses $\{\bar{\boldsymbol{\sigma}}\}$, perhaps using a weighted average as just suggested. Next let stresses $\{\boldsymbol{\sigma}\}$ in an element be interpolated from the node point average values in the same way that displacements are interpolated from nodal d.o.f. by shape functions $[\mathbf{N}]$. Thus

$$\{\boldsymbol{\sigma}\} = [\mathbf{N}]\{\bar{\boldsymbol{\sigma}}\} \tag{4.10.3}$$

Known stress boundary conditions can be accounted for in Eq. 4.10.3. We recall from Eq. 4.3.5 that the volume integral of $[\mathbf{B}]^T\{\boldsymbol{\sigma}_0\}$ gives element nodal loads caused by initial stresses $\{\boldsymbol{\sigma}_0\}$ and that the sum of these integrals over all *numel* elements is a part of the structure load vector—specifically, the part that balances $\{\boldsymbol{\sigma}_0\}$. If, instead of $\{\boldsymbol{\sigma}_0\}$, we use the *total* stress $\{\boldsymbol{\sigma}\}$, the sum over all m elements should balance the *entire* load vector. So we write

$$[\mathbf{K}]\{\Delta\mathbf{D}\} = \{\mathbf{R}\} - \sum_{n=1}^{\text{numel}} \left(\int_V [\mathbf{B}]^T\{\boldsymbol{\sigma}\}\, dV \right) \tag{4.10.4}$$

where $\{\boldsymbol{\sigma}\}$ comes from Eq. 4.10.3 and $\{\mathbf{R}\}$ is the given load vector (Eq. 4.3.7). If $\{\boldsymbol{\sigma}\}$ is exact, the right side of Eq. 4.10.4 is zero. Otherwise, it is a load unbalance that drives the solution toward a configuration that reduces the unbalance. We compute increments $\{\Delta\mathbf{D}\}$ in the nodal d.o.f. The new configuration is $\{\mathbf{D}\}_{\text{new}} = \{\mathbf{D}\}_{\text{old}} + \{\Delta\mathbf{D}\}$. Stresses $\{\boldsymbol{\sigma}\}$ are computed from $\{\mathbf{D}\}_{\text{new}}$ and Eq. 4.10.3, then Eq. 4.10.4 is reapplied. Note that Eq. 4.10.4 does not require repeated construction and reduction of $[\mathbf{K}]$.

The effectiveness of Loubignac iterations can be illustrated by constructing a beam of constant strain triangles. As seen in Fig. 4.5.3, results can be poor. But with five iterations, the stress field in even the coarse mesh is nearly exact [4.11]. Displacements become more accurate, but are overestimated rather than underestimated. If the iterative scheme is applied to a patch test problem (Fig. 4.8.2), with elements that fail the test, convergence is to exact stresses *and* exact displacements.

Stresses produced by thermal gradients are represented in Eq. 4.10.1 by $\{\boldsymbol{\sigma}_0\}$ (or by $\{\boldsymbol{\epsilon}_0\}$). In many structures, and especially if thermal gradients are sharp, thermally induced strains are largely suppressed by surrounding material or by supports. Thus the contribution to stress in Eq. 4.10.1 comes mostly from $\{\boldsymbol{\sigma}_0\}$ and has little to do with $\{\mathbf{d}\}$. Therefore a coarse mesh is acceptable even if thermal gradients are large, provided that $\{\boldsymbol{\sigma}_0\}$ represents the actual temperature field and not a smoothed field produced by interpolation from node point temperatures [4.12]. This advice runs against the usual suggestion that in thermal stress problems a finite element mesh should be able to display a strain field as complicated as the temperature field.

However, the suggestion is often sound, because expansion can be relatively unrestrained. Consider the element of Eq. 4.6.1. Strain $\epsilon_x = u_{,x}$, for example, depends only on y. But if [N] of Eq. 4.6.2 is used to interpolate temperatures from nodal values, thermal strain $(\epsilon_x)_0$ depends on both x and y. Thus there is a discrepancy between the competence of interpolation fields for mechanical and thermal strains that can lead to errors in nodal stresses. This is especially true in elements with low-order fields. In elements with a linear or a bilinear field, one might do better by fitting the temperature field to the shape function of the strain field or by using a uniform temperature over the entire element.

One might envision ad hoc combinations of the methods of the two foregoing paragraphs, with proportioning according to the relative magnitudes of $\{\boldsymbol{\epsilon}\}$ and $\{\boldsymbol{\epsilon}_0\}$.

Rigid-body motion does not contribute to strains. It can be removed from $\{\mathbf{d}\}$ before applying Eq. 4.10.1. For elements with translational d.o.f., we need only subtract the d.o.f. of (say) node 1 from the d.o.f. of all nodes. If rotational d.o.f. are also present, a $\{\mathbf{d}\}_r$ that expresses rigid-body motion must be constructed and subtracted from $\{\mathbf{d}\}$. The resulting displacement differences can be used to calculate stresses in single-precision arithmethic with confidence that the essential information noted in Section 15.5 has not been lost. *Caution:* In cylindrical coordinates, some strains depend on *total* displacement, not only on a displacement difference.

4.11 CORNER NODES, SIDE NODES, COST, AND DIMENSIONALITY

Elements may have only corner nodes or side nodes as well. Line, plane, and solid elements can be of either type. Here we make a comparison of equation-solving costs under the assumption that cost is proportional to NB^2 (see Section 2.11). Let i be the number of d.o.f. per node.

If k in Fig. 4.11.1*b* is "large" but otherwise unspecified, then the number of equations is $N \approx ik^2$ and the semibandwidth is $B = i(N2 - N1 + 1) \approx ik$. Similarly, if ℓ in Fig. 4.11.1*c* is "large," we find $N \approx 3i\ell^2$ and $B = i(N4 - N3 + 1) \approx 3i\ell$. The latter N and B values reflect the three new nodes brought to the mesh each time a typical eight-node element is added.

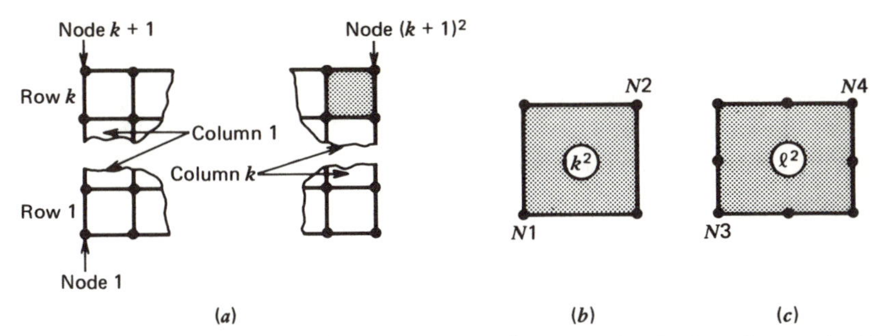

Figure 4.11.1. (*a*) A k-by-k mesh of four-node plane elements. (*b*) The last element in this mesh. (*c*) Similarly, the last element in an ℓ-by-ℓ mesh of eight-node plane elements.

Similar arguments apply in three dimensions. We consider an m-by-m-by-m mesh of 8-node bricks and an n-by-n-by-n mesh of 20-node bricks (Fig. 5.7.1). The first plane of nodes is numbered as in Fig. 4.11.1*a*, the next plane is similarly numbered, and so on. Results of the counting are as follows.

4-node plane	$N \approx ik^2$	$B \approx ik$	$NB^2 \approx i^3k^4$
8-node plane	$N \approx 3i\ell^2$	$B \approx 3i\ell$	$NB^2 \approx 27i^3\ell^4$
8-node solid	$N \approx im^3$	$B \approx im^2$	$NB^2 \approx i^3m^7$
20-node solid	$N \approx 4in^3$	$B \approx 4in^2$	$NB^2 \approx 64i^3n^7$

How are k and ℓ related if meshes of the two plane elements are to have equal cost? We equate the two NB^2 values, assume that i is the same for both element types, and find $k/\ell \approx \sqrt[4]{27} = 2.3$. Thus, for the same cost, we can use a mesh of four-node elements about 2.3 times as fine as a mesh of eight-node elements. In three dimensions, the analogous result is $m/n \approx \sqrt[7]{64} = 1.8$.

These arguments extend to elements with face nodes or with more than one node per edge. The general conclusion is that for efficiency, a typical node should be shared by as many elements as possible.This does not mean that elements with side nodes are to be avoided: they usually perform well and may be the more cost effective when the overall analysis task is considered.

In the preceding NB^2 counts, exponents on the mesh size go from 4 to 7 when we go from two dimensions to three. So if we go from an m-by-m plane mesh to an m-by-m-by-m solid mesh, costs increase by a factor of about m^3. Similarly, we can consider going from one dimension to two. For a beam of m elements with two d.o.f. per node, $NB^2 \approx (2m)\,(4)^2 = 32m$. For an m-by-m mesh of quadrilateral plate bending elements with three d.o.f. per node (Fig. 9.2.1*a*), $NB^2 \approx (3m^2)\,(3m)^2 = 27m^4$. So even in going from one dimension to two, we find again a cost ratio of about m^3. The cost ratio m^3 is large even if m is small, say $m = 10$. This "curse of dimensionality" motivates both the boundary solution process (Section 4.13) and the series treatment (Sections 8.4–8.6, 9.2, 10.10, 10.11, 12.4), in which a series of problems in n dimensions replaces a much more expensive problem in $n + 1$ dimensions.

4.12 FINITE ELEMENTS VERSUS FINITE DIFFERENCES

Both methods discretize a continuum, and both generate simultaneous algebraic equations to be solved for nodal d.o.f. Otherwise, the methods are superficially different. Finite difference molecules overlap one another and sometimes have phantom nodes outside the structure boundary; finite elements are distinct and have no phantom nodes. Finite differences are usually explained as a way to solve diffferential equations: finite elements are explained as a way to minimize a functional.

But a finite difference model *can* be derived from a functional [4.13–4.15]. For example, if Π_p is the functional and $\{\mathbf{D}\}$ are nodal d.o.f., we can write finite difference expressions for the derivatives in Π_p and generate algebraic equations from the stationary condition $\{\partial \Pi_p / \partial \mathbf{D}\} = 0$. The coefficient matrix is symmetric if the finite element method yields a symmetric matrix for the same problem. Also, this *finite difference energy method* is not restricted to a regular node pattern and need not satisfy natural boundary conditions.

Thus the finite difference and finite element methods differ only in the choice of generalized coordinates and in the location of nodes. Indeed, we can say that finite elements are a device for generating finite difference equations. Sometimes the two methods produce identical equations [1.10].

Both methods have about the same accuracy. Computer cost is often less when finite differences are used [4.15]. Inevitably, cost comparisons depend on the type of problem and program organization as well as on the analysis method.

The finite difference energy method is well suited to shells of revolution [4.15]. It is also suited to ''pure'' continua, where there is only one medium, such as a homogeneous solid or a fluid. It is not suited to a structure that must be modeled by a mixture of materials or by different forms, such as a vehicle that combines bar, beam, plate, and shell components. For such a problem the finite element method has no rival.

Finite elements appeal to the structural engineer because they resemble pieces of the actual structure. Physical intuition more than mathematical intuition guides the modeling and treatment of boundary conditions. Improvements in finite element methods have come from both physical and mathematical insight.

4.13 BOUNDARY SOLUTION METHOD

As we go from one dimension to two, or from two to three, the amount of data that must be prepared and processed in a finite element solution increases greatly (Section 4.11). We also have trouble with regions that are infinite in extent. In some problems the *boundary solution method* overcomes these difficulties. Here we use an example problem to outline the method in a physical way and comment on its advantages and limitations.

We begin with a resumé of some theory of elasticity. Stresses in Fig. 4.13.1a are given by the Flamant solution [1.20]. It gives, for example, the stress σ_x at point P as

$$\sigma_x = \frac{F \cos \theta}{4\pi r}[-3 - \nu + 2(1 + \nu)\sin^2 \theta] \tag{4.13.1}$$

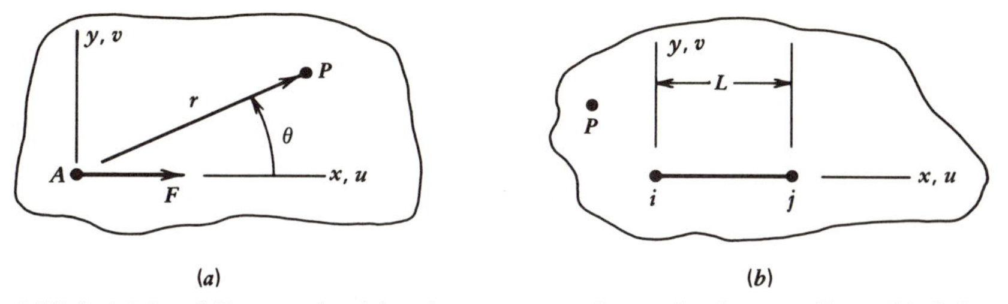

Figure 4.13.1. (*a*) Load F at a point A in a homogeneous isotropic plane medium of infinite extent. (*b*) The infinite plane again, with displacements at i and j used to define a linear variation of displacements u and v over length L.

Displacements u and v at P could also be written in terms of F. In general, the solution can be manipulated to yield stresses or displacements at one point in terms of stresses or displacements at another. Thus, for Fig. 4.13.1*b*, it is possible to write.

$$
\begin{aligned}
\sigma_x &= \lfloor \mathbf{q}_1 \rfloor \{u_i \quad u_j\} + \lfloor \mathbf{q}_2 \rfloor \{v_i \quad v_j\} \\
\sigma_y &= \lfloor \mathbf{q}_3 \rfloor \{u_i \quad u_j\} + \lfloor \mathbf{q}_4 \rfloor \{v_i \quad v_j\} \\
\tau_{xy} &= \lfloor \mathbf{q}_5 \rfloor \{u_i \quad u_j\} + \lfloor \mathbf{q}_6 \rfloor \{v_i \quad v_j\}
\end{aligned}
\tag{4.13.2}
$$

where the q matrices are influence functions that depend on elastic properties and the position of point p relative to line ij.

Now consider Fig. 4.13.2*a*. If the medium was under compressive hydrostatic stress of magnitude p before the tunnel was cut, stresses produced by cutting the tunnel can be found as follows. Approximate displacements of the tunnel wall in piecewise linear fashion and apply Eqs. 4.13.2 with x and y used to represent local directions always normal and tangent to the wall.

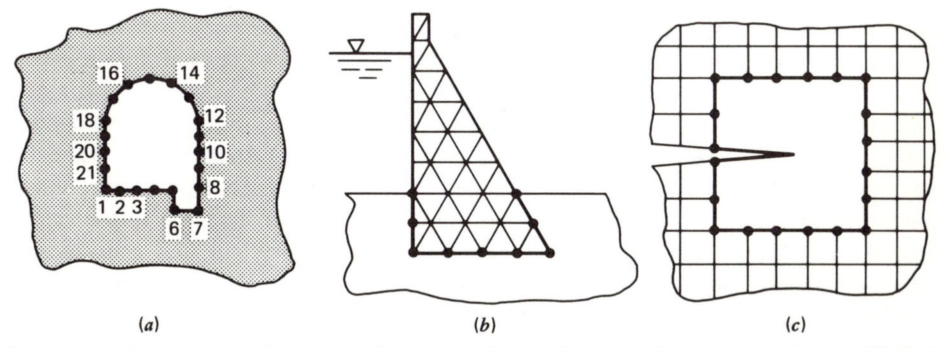

Figure 4.13.2. (*a*) A tunnel in a near-infinite medium. This is a plane strain problem. (*b*) Finite element model of a dam on a boundary solution foundation. (*c*) Boundary solution crack element in a finite element model of the adjacent structure.

$$\begin{aligned} \sigma_x &= p = \lfloor \mathbf{Q}_1 \rfloor \{\mathbf{u}\} + \lfloor \mathbf{Q}_2 \rfloor \{\mathbf{v}\} \\ \tau_{xy} &= 0 = \lfloor \mathbf{Q}_5 \rfloor \{\mathbf{u}\} + \lfloor \mathbf{Q}_6 \rfloor \{\mathbf{v}\} \end{aligned} \tag{4.13.3}$$

where $\{\mathbf{u}\}$ and $\{\mathbf{v}\}$ represent the displacements of all nodes in Fig. 4.13.2*a*, not necessarily in local directions xy. The Q matrices represent the q matrices of Eq. 4.13.2, transformed to local directions and expanded to structure size with summation of terms that belong to the same node. Equations 4.13.3 can be written for each node and solved for $\{\mathbf{u}\}$ and $\{\mathbf{v}\}$. Thus we find nodal d.o.f. that yield normal stress p and zero shear stress all over the tunnel wall. These d.o.f. solve the problem because, when this state of stress is superposed on the original hydrostatic compression, the tunnel wall is completely unloaded. Stress σ_y, and stresses away from the tunnel wall, can now be found from Eqs. 4.13.2 by using the appropriate coordinates in the q matrices and summing contributions from the several linear-displacement segments.

The process just described is associated with the names *boundary integral equation* (BIE) *method* and *boundary element method*. The term "element" is appropriate to Fig. 4.13.2*a* in the sense that an interval between nodes is used as the basis of piecewise interpolation, just as in the finite element method. For a homogeneous *three*-dimensional problem, we might draw four-node or eight-node elements on the surface of the region and interpolate surface displacements over each element from nodal d.o.f. of the element. Unlike a finite element model, there are no nodes *within* the region. We can now see a major advantage of BIE: the number of d.o.f. remains small because an n-dimensional mesh is used for an $(n + 1)$-dimensional problem.

Other advantages of BIE include the following. Accuracy can be high. Bodies with singularities such as cracks or with infinite regions such as in Fig. 4.13.2*a* can be treated if a suitable analytical solution is available. (Finite element methods for these problems are discussed in Sections 7.3, 7.14, and 7.15.) Like the finite element method, BIE is general: it is not limited to stress analysis, regions of special shape, isotropy, or convenient boundary conditions. A boundary solution can be coupled to a finite element solution, as in Figs. 4.13.2*b* and 4.13.2*c*. Such a coupling is attractive in fluid-structure interaction problems, particularly where fluid surrounds a structure and extends to infinity: the structure is modeled by finite elements, the fluid by the boundary method. Similarly, a "messy" structure, not suited to BIE analysis, can be imbedded in a medium for which BIE is appropriate.

Drawbacks of BIE include its greater mathematical complexity and the necessity of having an analytical solution as a starting point. (Analytical solutions are scarce for anisotropy and nonlinearity.) It is best suited to three-dimensional problems: with an increase in the ratio of surface to volume—or, in general, an increase in the ratio of boundary to region—BIE becomes less attractive. "Boundary" in this sense includes internal boundaries, such as holes, joint planes, or where material properties change. A boundary solution would be most inappropriate for an engineering structure built of a combination of bars, plates, and so on.

The boundary solution method is developing rapidly. There has been some application to nonlinear problems. Sample references include Refs. 4.16 and 4.17.

PROBLEMS

4.1 Find $[\mathbf{A}]^{-1}$ in Eq. 4.2.4 and verify Eqs. 4.2.5. *Suggestion*. To find $[\mathbf{A}]^{-1}$, you need not formally invert a matrix. Instead, regard Eq. 4.2.3 as four equations to be solved for the four a_i. The first two are obviously $a_1 = w_1$ and $a_2 = \theta_1$.

4.2 Check the N_i of Eqs. 4.2.5 in the manner suggested below these equations.

4.3 (a) What are the coefficients of u_1, u_5, and u_9 in Eq. 4.2.11, after Eqs. 4.2.10 are substituted, in terms of b, c, x, and y?

(b) Make an isometric sketch, similar to Fig. 5.6.2, that shows u as a displacement normal to the xy plane if $u_1 = 1$ and u_2 through u_8 are zero in Eq. 4.2.11. Make another sketch for $u_5 = 1$ and other $u_i = 0$, and another for $u_9 = 1$ and other $u_i = 0$.

4.4 If element d.o.f. are given infinitesimal displacements $\{\delta \mathbf{d}\}$, strains are changed in the amount $\{\delta \boldsymbol{\epsilon}\} = [\mathbf{B}]\{\delta \mathbf{d}\}$. Applied loads do work, and store strain energy in the amount $\int \{\delta \boldsymbol{\epsilon}\}^T \{\boldsymbol{\sigma}\}\, dV$. Use this virtual work argument to rederive Eqs. 4.3.4, 4.3.5, and 4.3.8.

4.5 Check Eq. 4.3.9 by writing out Eqs. 4.3.4 and 4.3.9 for a hypothetical case where $[\mathbf{E}]$ is 2 by 2 and $[\mathbf{B}]$ is 2 by 3.

4.6 Derive forces P_1 in Fig. 4.3.1*b* from the q_1 distribution in Fig. 4.3.1*a*. Use Eq. 4.3.13 as a guide.

4.7 (a) Verify Eqs. 4.3.15.

(b) Verify the second of Eqs. 4.3.18 by a virtual work argument (see Problem 4.4).

4.8 (a) Verify that Eqs. 3.4.16, 4.3.4, and 4.4.4 yield $[\mathbf{k}]$ of Eq. 4.4.5.

(b) Show that $[\mathbf{k}]\{\mathbf{d}\} = 0$ if $\{\mathbf{d}\}$ represents a small, counterclockwise, rigid-body rotation of the beam about its left end.

4.9 Use elementary mechanics of materials to verify the remark made below Eq. 4.4.5 about Figs. 4.2.1*b* and 4.2.1*c*.

4.10 (a) Verify that Eq. 4.4.6 leads to Eq. 4.4.7.

(b) Verify Eq. 4.4.7 by using elementary beam theory to find fixed-end forces and moments.

(c) It may surprise you that Eq. 4.4.7 agrees with beam theory because the assumed cubic shape is not correct for these loadings. How may the correct results be explained? *Suggestion*. Consider virtual work done by applied loads as they act through displacements produced by a virtual change in nodal d.o.f. Finally, use this virtual work argument to compute the left force $qL/2$ in Fig. 4.4.2*a*.

4.11 If we model the truss of Fig. 2.7.1 by plane frame elements (Fig. 4.4.3) and suppress θ d.o.f. at all seven nodes, do we generate a correct model of the truss? Explain.

4.12 Let the bar of Fig. 4.4.1*a* have mass density ρ and be rotating at constant angular velocity ω about a point b units to the right of node 2. Compute the resulting $\{\mathbf{r}\}$ and check that $\{\mathbf{r}\}$ looks reasonable for two special cases: (*a*) $b >> L$, and (*b*) b

$= -L/2$ (rotation about the midpoint). Is the assumed displacement field of Eq. 4.4.1 exact for this problem? Is it acceptable? Explain.

4.13 Compute {**r**} if the bar of Fig. 4.4.1*a* is heated uniformly an amount T but has a cross-sectional area that varies linearly from A_1 at $x = 0$ to A_2 at $x = L$. Compare this {**r**} with the exact values. Why is there a discrepancy?

4.14 According to Eq. 4.3.5, what is {**r**} if a uniform bar element is heated so that its temperature varies linearly from T_1 at $x = 0$ to T_2 at $x = L$? Is this {**r**} exact?

4.15 According to Eq. 4.3.5, what is {**r**} if the temperature of a uniform beam element varies linearly from $-T_0$ at the lower surface to $+T_0$ at the top surface? Is this {**r**} exact?

4.16 Compute {**r**} for a distributed lateral load on a beam element if the load varies linearly from $q = q_1$ at $x = 0$ to $q = q_2$ at $x = L$.

4.17 Consider a beam element that has the positive directions of nodal d.o.f. and generalized force shown in the sketch. How does this sign convention change the stiffness matrix of Eq. 4.4.5? What is bad about this sign convention?

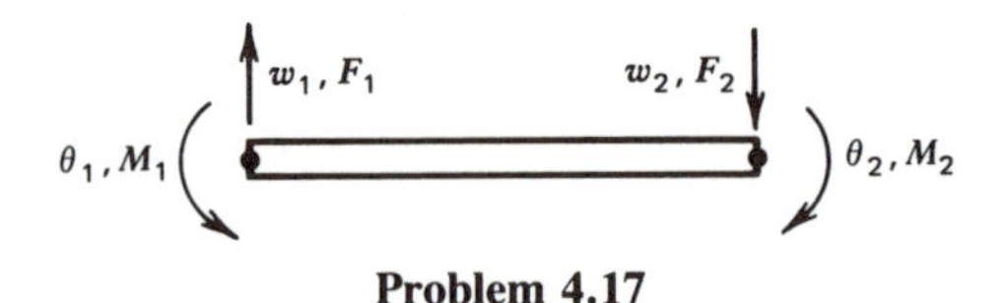

Problem 4.17

4.18 (a) Given Eq. 4.5.4, verify that Eq. 4.5.3 is the inverse of [**A**] in Eq. 4.5.2.
(b) Evaluate [**B**] in Eq. 4.5.8. In other words, verify Eq. 7.9.16.

4.19 (a) Let nodes 1, 2, and 3 of a constant strain triangle have the respective nodal coordinates (0,0), (b,0) and (0,c). Write matrix [**B**] in terms of b and c. Show that [**B**] {**d**} yields the expected strains when nodal displacements {**d**} are assigned values consistent with the displacement field of Eqs. 4.5.1 (that is, show that $\epsilon_x = a_2$, $\epsilon_y = a_6$, and $\gamma_{xy} = a_3 + a_5$ are produced).
(b) What values of the a's describe a counterclockwise rigid-body rotation of 0.0001 radian about node 1?
(c) Find [**k**] in terms of b, c, E, and t if thickness t is constant and $\nu = 0$.
(d) Support nodes 1 and 2 and apply an upward force P to node 3. Use [**k**] from part (c) to compute v_3. Compare v_3 with the stretch of a prismatic bar under tensile force P.

4.20 Outline the formulation of a constant strain tetrahedron in the same way that the formulation of a constant strain triangle is outlined in Section 4.5. You need not invert the 4-by-4 matrix [**A**] (row 4 corresponds to the fourth node and column 4 lists nodal z coordinates). The determinant of this [**A**] is six times the element volume. Nodes 1, 2, and 3 should be numbered counterclockwise when viewed from node 4.

4.21 Verify the N_i in Fig. 4.6.1*b* by use of Lagrange's interpolation formula.

4.22 Element j has stiffness matrix $[\mathbf{k}]$ and nodal displacement vector $\{\mathbf{d}\} = \{u_1 \ v_1 \ u_2 \ v_2 \ . \ . \ . \ v_4\}$. It is to be attached to nodes 19, 20, 30, and 31 of the structure.

(a) If the structure matrix $[\mathbf{K}]$ is to be stored in full (not banded) format and all d.o.f. are retained in $\{\mathbf{D}\}$, what structural subscripts are associated with the respective columns of $[\mathbf{k}]$?

(b) If element j is of the type described in Section 4.6, what is its numerical contribution to the single coefficient in $[\mathbf{K}]$ at the intersection of: (1) row 59 and column 61, (2) row 48 and column 39, (3) row 59 and column 59, and (4) row 37 and column 37?

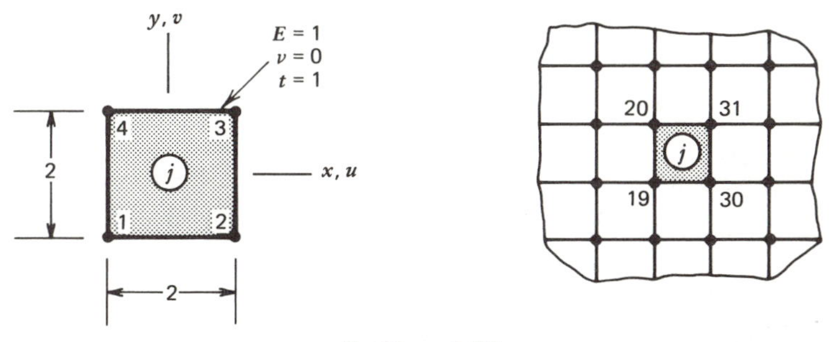

Problem 4.22

4.23 Consider an isotropic material and show that Eqs. 4.6.1 do not satisfy the differential equations of equilibrium unless $a_4 = a_8 = 0$.

4.24 Sketch an assembly of hexahedral solid elements of arbitrary shape, suitable for use as a patch test mesh.

4.25 Divide the lower left element of Fig. 4.8.2 into four elements by connecting mid-points of opposite sides. Subdivide the new elements again and again in the same way. Show that this kind of mesh refinement makes all elements approach parallelograms in shape.

4.26 The bar shown is uniform and has nodes 1 and 2. Let the assumed displacement field be $u = \lfloor 1 \ x^2 \rfloor \{a_1 \ a_2\}$. Find the resulting strain-displacement matrix $\lfloor \mathbf{B} \rfloor$ and stiffness matrix $[\mathbf{k}]$. What defects do you see in u, $\lfloor \mathbf{B} \rfloor$, and $[\mathbf{k}]$?

4.27 Repeat Problem 4.26 using the sketch shown and the assumed displacement field $u = \lfloor x \ x^2 \rfloor \{a_1 \ a_2\}$.

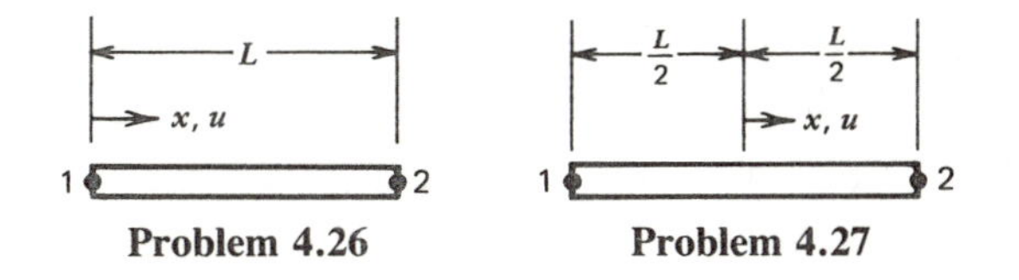

Problem 4.26 **Problem 4.27**

4.28 Consider the displacement field $u = a_1 + a_2x + a_3y$, $v = a_4 + a_5x + a_6y$. Let nodal d.o.f. $\{\mathbf{d}\}$ be consistent with this field of constant strain.
(a) Show that Eq. 4.6.2 yields the given u and v fields.
(b) Show that Eq. 4.6.3 yields the expected strains $\epsilon_x = a_2$, $\epsilon_y = a_6$, and $\gamma_{xy} = a_3 + a_5$.

4.29 A simply supported beam carries a distributed load $q = q(x)$. Nodal moments associated with q are omitted from the load vector $\{\mathbf{r}\}$, but nodal forces associated with q are retained. Will exact results be approached as the beam is divided into more and more elements? Explain why or why not.

4.30 Let the displacement field of Problem 11.8 be used with the beam element of Fig. 4.4.1*b*. Can this field represent the rigid-body motion $v = a + bx$? Does it represent the correct constant curvature $w_{,xx}$ when $w_1 = w_2 = 0$ and $\theta_1 = -\theta_2$? Formulate $[\mathbf{k}]$ using the integral in Eq. 4.4.5 and $w_{,xx} = \lfloor \mathbf{B} \rfloor \{\mathbf{d}\}$. Comment on the shortcomings of $[\mathbf{k}]$ and the assumed w field. (A similar constant-curvature element is used in Problem 9.23.)

4.31 It is proposed that a beam element be based on a cubic polynomial for lateral displacement w but that the d.o.f. are to be only lateral displacements w_i. The nodes, $i = 1, 2, 3, 4$, are to be at either end and at the third points. Why is this element unacceptable? Which criteria of Sections 4.8 and 4.9 does it violate?

4.32 If each coefficient xy in Eq. 4.6.1 is replaced by $x^2 + y^2$, the element becomes incompatible. Why? *Suggestion*. Two adjacent elements have two corner nodes in common. Along the shared boundary, d.o.f. of the corner nodes must produce the same edge displacement in each element if elements are to be compatible. But how many d.o.f. are needed to define a quadratic? And what does this imply?

4.33 (a) Add the cubic terms to Eq. 4.9.1.
(b) In three dimensions, which of the cubic terms would you retain in an incomplete cubic if the polynomial is "balanced" and if the cubic part contains: (1) one term, (2) three terms, (3) four terms, (4) six terms, (5) seven terms, or (6) nine terms?

4.34 Let $\beta = 45°$ in Fig. 4.9.1. Then $x = (s - t)/\sqrt{2}$ and $y = (s + t)/\sqrt{2}$.
(a) Express Eq. 4.9.2 in terms of s and t. Is the element still compatible?
(b) Add $a_5x^2 + a_6y^2$ to Eq. 4.9.2 to make it a complete quadratic field, then express it in terms of s and t. Is it like the field of part (a) or is it still a complete quadratic?

4.35 Let the bar of Fig. 4.10.1 have weight density γ and be divided into m elements, each of length b. Show that as b shrinks and m grows, element stress associated with the weight of the element itself vanishes in comparison with the stress associated with element nodal displacements.

4.36 Consider these one-element beams under uniform load: (*a*) cantilever, right end free, (*b*) cantilever, right end simply supported, and (*c*) simply supported at both ends. Use the nodal load vector from Fig. 4.4.2*a*. Compute nodal d.o.f. and, hence, nodal moments from $M = EIw_{,xx}$. Compare these results with exact answers. (This problem gives some indication of the error in stress calculation introduced by neglecting loads within an element, discussed in connection with Fig. 4.10.1*a*.)

4.37 Imagine that a node is shared by n plane elements, each of area A_i, $i = 1, 2, \ldots, n$. A stress σ_i is computed in each.
(a) Write a formula for average stress at the node, with each σ_i weighted by the area of its element.
(b) Let the elements be triangles, with centroidal coordinates x_i and y_i, and let the node have coordinates x_A and y_A. Write a formula for average stress at the node, with each σ_i weighted by the distance of its centroid from the node.
(c) Combine parts (a) and (b) into a formula that uses both weighting methods.

4.38 Imagine that the tip load in Fig. 15.9.1 is a moment M_0 instead of a force P. How do you expect σ_x to vary with x on the line $y = c$? Show how by a sketch.

4.39 If a temperature field is linear in x and y and the material is homogeneous and isotropic, an unrestrained element should be free of stress. Does this happen with the element of Fig. 4.6.1? In general, how must element temperature and displacement fields be related if $\{\boldsymbol{\sigma}\}$ in Eq. 4.10.1 is to vanish? Or, how does the nature of a given temperature field influence the choice of element used in the analysis?

4.40 Consider the following cases of thermal loading. All nodes of the respective elements are fixed against motion, and T_0 is a constant. In each case, do you think it appropriate to compute stresses as suggested in the paragraph below Eq. 4.10.4? Explain.
(a) The truss element of Fig. 4.4.1 *a*, with $\sigma_0 = -E\alpha T = -E\alpha(x^4/L^4)T_0$.
(b) The rectangular element of Fig. 4.6.1, with $b = 4c$ and $T = T_0 y^2/c^2$.

4.41 A uniform bar extends from $x = 0$ to $x = L$. It is heated so that $T = cx/L$, where c is a constant. Model the bar by a single truss element, fixed only at node 1 (where $x = 0$).
(a) Solve for u_2 and for σ_x.
(b) Replace $T = cx/L$ by an appropriate constant temperature T_0. Again, solve for u_2 and for σ_x.
(c) Why do the results of parts (a) and (b) differ? Is either correct?
(d) Imagine that the element models a bar that is fully restrained along its entire length ($u = 0$ for all x). Is $T = cx/L$ or $T = T_0$ the most appropriate field?
(e) Suggest a way to combine parts (a) and (b) for good stress calculation in a bar that is neither completely free nor completely restrained.

4.42 For a mesh of each of the following hypothetical elements, approximate N, B, and NB^2, as was done to find the operation counts tabulated in Section 4.11.
(a) Plane rectangle, one node per corner and one node in the center of the rectangle (5 nodes in all).
(b) As in part (a), but add a node to the center of each edge (9 nodes in all).
(c) Solid parallelepiped, one node per corner plus one at the center of each face (14 nodes in all).

4.43 Consider the enumerated programming projects listed in the preface of this book. What rigid-body motions and constant strain states are possible in each case?

4.44 (a) Show that the stiffness of a bar element is inversely proportional to its length if the cross section remains constant.
(b) Show that the stiffness of a plane element is independent of size if the shape and thickness are not changed.
(c) Show that the stiffness of a solid element is directly proportional to a linear dimension if its shape is not changed.

4.45 Consider the trapezoidal element shown. Define u as in Eq. 4.6.1. Then follow the procedure of Eqs. 4.5.2 to 4.5.6 to exchange the four a_i for the four nodal u_i. Finally, express u on the right edge as a function of y and show that this is an incompatible element. (The *isoparametric formulation,* Chapter 5, permits us to generate nonrectangular elements and elements with curved sides yet maintain interelement compatibity.)

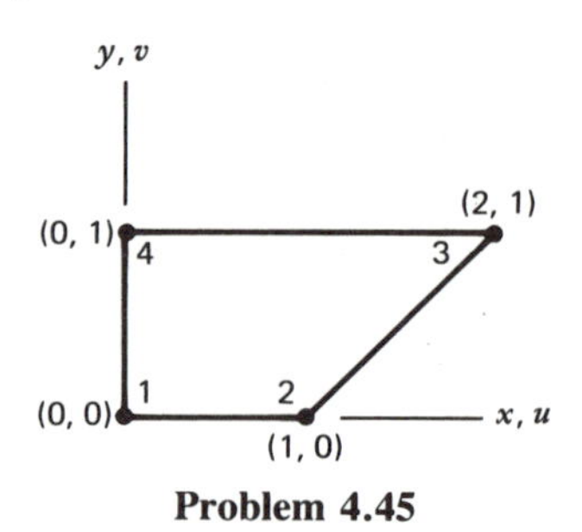

Problem 4.45

CHAPTER 5

THE ISOPARAMETRIC FORMULATION

5.1 ISOPARAMETRIC ELEMENTS

Isoparametric elements were publicly introduced in 1966 [5.1]. They make it possible to have nonrectangular quadrilateral elements. Their most appararent features are sides that may be curved and their special coordinate system ($\xi\eta\zeta$ in Fig. 5.1.1). Isoparametric elements are useful in modeling structures with curved edges and in grading a mesh from coarse to fine. Isoparametric elements are versatile: they have proved effective in two- and three-dimensional elasticity, shell analysis, and nonstructural applications. The formulation procedure looks strange at first, but it is not complicated. And once mastered for one type of element, it is easily applied to most other isoparametric elements. Indeed, a single "shape function subroutine" could generate the **[B]** matrix for a quadrilateral with four to eight nodes, the choice being made by user-supplied control data. The same routine can supply basic matrices needed in nonstructural applications.

Element nodes define two things:

1. Nodal d.o.f. $\{\mathbf{d}\}$ dictate displacements $\{u\ v\ w\}$ of a point in the element. Symbolically, $\{u\ v\ w\} = [\mathbf{N}]\{\mathbf{d}\}$.
2. Nodal coordinates $\{\mathbf{c}\}$ define global coordinates $\{x\ y\ z\}$ of a point in the element. Symbolically, $\{x\ y\ z\} = [\tilde{\mathbf{N}}]\ \{\mathbf{c}\}$.

Matrices $[\mathbf{N}]$ and $[\tilde{\mathbf{N}}]$ are functions of ξ, η, and ζ. *An element is isoparametric if the node sets of Items 1 and 2 are identical and if* $[\mathbf{N}]$ *and* $[\tilde{\mathbf{N}}]$ *are identical.*

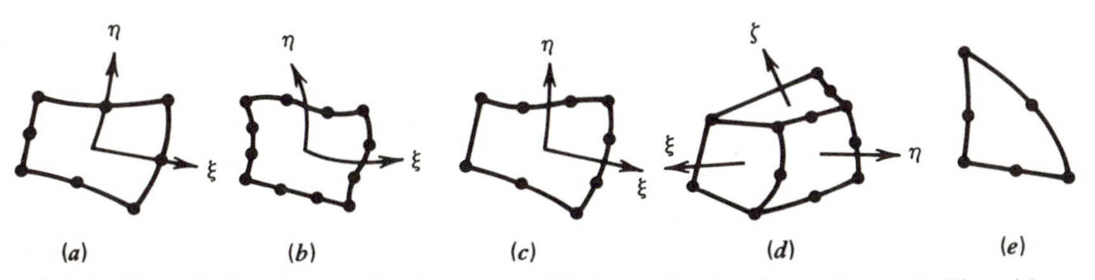

Figure 5.1.1. Sample isoparametric elements. (*a*) A quadratic plane element. The sides can be straight lines or quadratic curves. (*b*) A cubic element whose sides can be first-, second-, or third-degree curves. (*c*) A "degraded" cubic element. The left and lower sides can mate with linear and quadratic elements. (*d*) A solid element with linear and quadratic edges. (*e*) A triangular quadratic element.

5.2 EXAMPLES IN ONE DIMENSION

Our examples are straight bars (Fig. 5.2.1). Coodinate ξ is a *natural* or *intrinsic* coordinate: ends are defined by $\xi = \pm 1$, regardless of the length of the bar; also, ξ is attached to the bar and remains an axial coordinate, regardless of how the bar is oriented in global coordinates xy. For convenience, not necessity, ξ and x are colinear in the present examples.

Shape functions for axial displacement can be found by the usual process (Eqs. 3.7.2-3.7.6). In Fig. 5.2.1 a, let

$$u = \lfloor 1 \quad \xi \rfloor \begin{Bmatrix} a_1 \\ a_2 \end{Bmatrix}, \qquad \text{then} \qquad \begin{Bmatrix} u_1 \\ u_2 \end{Bmatrix} = \begin{bmatrix} 1 & -1 \\ 1 & 1 \end{bmatrix} \begin{Bmatrix} a_1 \\ a_2 \end{Bmatrix} \tag{5.2.1}$$

$$u = \lfloor \mathbf{N} \rfloor \begin{Bmatrix} u_1 \\ u_2 \end{Bmatrix}, \qquad \text{where} \qquad \lfloor \mathbf{N} \rfloor = \left\lfloor \frac{1-\xi}{2} \quad \frac{1+\xi}{2} \right\rfloor \tag{5.2.2}$$

Shape functions $\lfloor \mathbf{N} \rfloor$ are the product of $\lfloor 1 \quad \xi \rfloor$ and the inverse of the square matrix in Eq. 5.2.1. Equation 5.2.2 gives the displacement of a point on the element in terms of nodal d.o.f. when the coordinate ξ of the point is inserted in $\lfloor \mathbf{N} \rfloor$. Similarly, the interpolation

$$x = \lfloor \mathbf{N} \rfloor \{x_1 \quad x_2\} \tag{5.2.3}$$

gives the x coordinate of a point on the element in terms of nodal coordinates x_i when the coordinate ξ of the point is inserted in $\lfloor \mathbf{N} \rfloor$. Since u and x are defined by the same nodes and the same $\lfloor \mathbf{N} \rfloor$, the element is isoparametric.

Next, we generate $[\mathbf{k}]$ for the bar of Fig. 5.2.1a to illustrate the manipulations that isoparametric elements demand. We need a $[\mathbf{B}]$ matrix, so we write $\epsilon_x = du/dx$. But $\lfloor \mathbf{N} \rfloor$ contains ξ instead of x, so the chain rule is invoked.

$$\epsilon_x = \frac{du}{dx} = \frac{d}{dx} \lfloor \mathbf{N} \rfloor \begin{Bmatrix} u_1 \\ u_2 \end{Bmatrix}, \qquad \text{where} \qquad \frac{d}{dx} = \frac{d\xi}{dx} \frac{d}{d\xi} \tag{5.2.4}$$

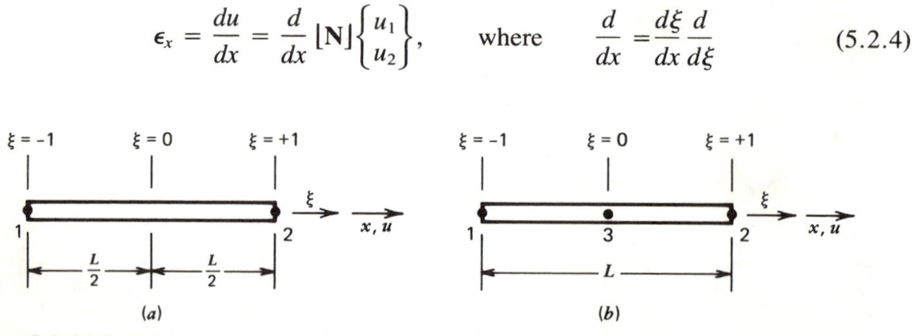

Figure 5.2.1. Bar elements in natural coordinates. (a) Standard two-node bar. (b) Bar with an internal node, node 3.

From Eq. 5.2.3, since $x_2 - x_1 = L$,

$$\frac{dx}{d\xi} = \left\lfloor -\frac{1}{2} \quad \frac{1}{2} \right\rfloor \begin{Bmatrix} x_1 \\ x_2 \end{Bmatrix} = \frac{L}{2}, \qquad \text{so} \qquad \frac{d\xi}{dx} = \frac{2}{L} \tag{5.2.5}$$

In calculus, the scale factor between two coordinate systems is called the "Jacobian." We denote it by J. Here

$$dx = J\, d\xi, \qquad \text{where} \qquad J = dx/d\xi = L/2 \tag{5.2.6}$$

The strain-displacement matrix is

$$\lfloor \mathbf{B} \rfloor = \frac{d}{dx} \lfloor \mathbf{N} \rfloor = \frac{1}{J} \frac{d}{d\xi} \lfloor \mathbf{N} \rfloor = \left\lfloor -\frac{1}{L} \quad \frac{1}{L} \right\rfloor \tag{5.2.7}$$

The stiffness matrix is

$$[\mathbf{k}] = \int_0^L \lfloor \mathbf{B} \rfloor^T AE \lfloor \mathbf{B} \rfloor \, dx = \int_{-1}^1 \lfloor \mathbf{B} \rfloor^T AE \lfloor \mathbf{B} \rfloor J \, d\xi \tag{5.2.8}$$

This $[\mathbf{k}]$ agrees with that of Eq. 3.7.10.

For the element of Fig. 5.2.1b, we start with the assumed field $u = \lfloor 1 \;\; \xi \;\; \xi^2 \rfloor \{u_1 \; u_2 \; u_3\}$. Instead of Eqs. 5.2.2 and 5.2.3, we find

$$u = \lfloor \mathbf{N} \rfloor \{u_1 \; u_2 \; u_3\} \qquad \text{and} \qquad x = \lfloor \mathbf{N} \rfloor \{x_1 \; x_2 \; x_3\} \tag{5.2.9}$$

where

$$\lfloor \mathbf{N} \rfloor = \left\lfloor \frac{-\xi + \xi^2}{2} \quad \frac{\xi + \xi^2}{2} \quad 1 - \xi^2 \right\rfloor$$

The Jacobian, and the strain-displacement matrix for axial strain, are

$$J = \left\lfloor \frac{-1 + 2\xi}{2} \quad \frac{1 + 2\xi}{2} \quad -2\xi \right\rfloor \{x_1 \;\; x_2 \;\; x_3\}$$
$$\lfloor \mathbf{B} \rfloor = \frac{1}{J} \left\lfloor \frac{-1 + 2\xi}{2} \quad \frac{1 + 2\xi}{2} \quad -2\xi \right\rfloor \tag{5.2.10}$$

Now J is a function of ξ—the scale factor J changes as we move along the bar. Only if x_3 represents the midpoint does J reduce to $L/2$ as in Eq. 5.2.6. Matrix $\lfloor \mathbf{B} \rfloor$ is a function of ξ, with polynomials in ξ in its denominator because of the $1/J$ factor. Accordingly,

Eq. 5.2.8 cannot usually be integrated in closed form to yield algebraic expressions that can be programmed. Instead, the computer program must generate **[k]** by numerical integration (Section 5.4).

Additional axial elements are described in Ref. 5.2.

5.3 THE PLANE LINEAR ISOPARAMETRIC ELEMENT

First, we remark again that different isoparametric elements are nevertheless similar. The formulation of this section extends directly to more complicated elements: the essential changes are the addition of nodes and the use of different shape functions.

The subject element resembles that of Fig. 4.6.1, except that it can be an arbitrary quadrilateral. Axes ξ and η in Fig. 5.3.1*c* pass through the midpoints of opposite sides. Axes ξ and η need not be orthogonal, and neither one must be parallel to x- or y-axes. The orientation of $\xi\eta$ coordinates is determined by the node numbers we assign. That is, with the N_i defined as follows, node 1 is at $\xi = \eta = -1$, node 2 is at $\xi = -\eta = 1$, and so on.

Definitions of global coordinates and displacements are

$$\begin{Bmatrix} x \\ y \end{Bmatrix} = [\mathbf{N}]\{\mathbf{c}\} \quad \text{and} \quad \begin{Bmatrix} u \\ v \end{Bmatrix} = [\mathbf{N}]\{\mathbf{d}\} \tag{5.3.1}$$

where

$$\{\mathbf{c}\} = \{x_1 \ y_1 \ x_2 \ y_2 \ x_3 \ y_3 \ x_4 \ y_4\}$$

$$\{\mathbf{d}\} = \{u_1 \ v_1 \ u_2 \ v_2 \ u_3 \ v_3 \ u_4 \ v_4\}$$

$$[\mathbf{N}] = \begin{bmatrix} N_1 & 0 & N_2 & 0 & N_3 & 0 & N_4 & 0 \\ 0 & N_1 & 0 & N_2 & 0 & N_3 & 0 & N_4 \end{bmatrix}$$

Sometimes it is convenient to write Eqs. 5.3.1 in the form

$$\begin{aligned} x &= \Sigma N_i x_i \qquad & u &= \Sigma N_i u_i \\ y &= \Sigma N_i y_i \qquad & v &= \Sigma N_i v_i \end{aligned} \tag{5.3.2}$$

The individual shape functions are

$$\begin{aligned} N_1 &= \tfrac{1}{4}(1 - \xi)(1 - \eta) \qquad & N_2 &= \tfrac{1}{4}(1 + \xi)(1 - \eta) \\ N_3 &= \tfrac{1}{4}(1 + \xi)(1 + \eta) \qquad & N_4 &= \tfrac{1}{4}(1 - \xi)(1 + \eta) \end{aligned} \tag{5.3.3}$$

These N_i are similar to those in Fig. 4.6.1*b*. They can be found by the usual procedure, most recently applied in Eqs. 5.2.1 and 5.2.2 (Problem 5.2).

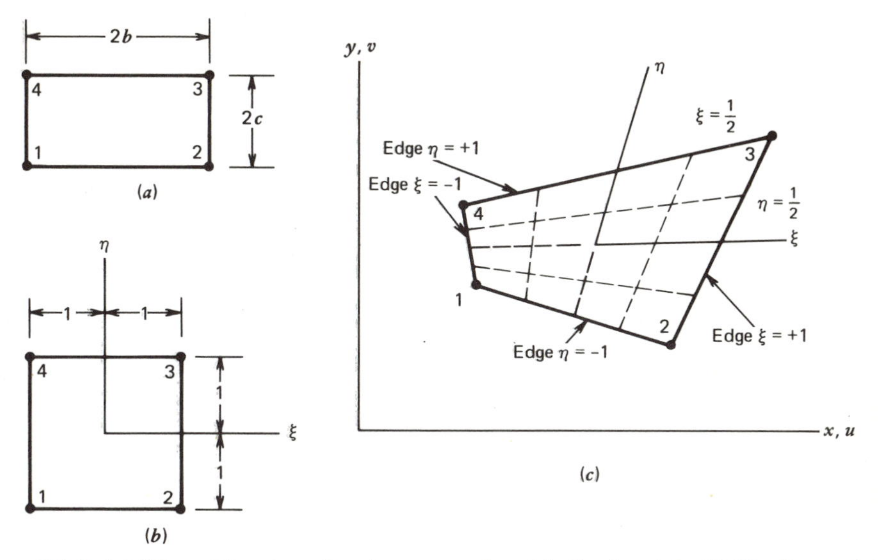

Figure 5.3.1. (*a*) "Parent" rectangular element, analyzed in Section 4.6. (*b*) The linear element in $\xi\eta$ space. (*c*) Mapping of the element into xy space, where nodal coordinates define its shape.

As usual, u and v are x-parallel and y-parallel displacements: in general they are *not* ξ parallel and η parallel. A special case is a rectangle with sides parallel to global coordinates xy (Fig. 4.6.1). Then ξ and η become dimensionless centroidal coordinates, where

$$x = x_c + b\xi \qquad \text{and} \qquad y = y_c + c\eta \tag{5.3.4}$$

and x_c and y_c are global coordinates of the element centroid. It may be helpful to check that subsequent expressions reduce to this special case.

The following development parallels that of Section 5.2. To find [**k**] from Eq. 4.3.4, we need [**B**]. We cannot express [**B**] in terms of x and y, so we write it in terms of ξ and η. This requires that the following coordinate transformation of derivatives be invoked. Let ϕ be some function of x and y. Then the chain rule yields

$$\begin{aligned} \frac{\partial\phi}{\partial\xi} &= \frac{\partial\phi}{\partial x}\frac{\partial x}{\partial\xi} + \frac{\partial\phi}{\partial y}\frac{\partial y}{\partial\xi} \\ \frac{\partial\phi}{\partial\eta} &= \frac{\partial\phi}{\partial x}\frac{\partial x}{\partial\eta} + \frac{\partial\phi}{\partial y}\frac{\partial y}{\partial\eta} \end{aligned} \qquad \text{or} \qquad \begin{Bmatrix} \phi_{,\xi} \\ \phi_{,\eta} \end{Bmatrix} = [\mathbf{J}] \begin{Bmatrix} \phi_{,x} \\ \phi_{,y} \end{Bmatrix} \tag{5.3.5}$$

where [**J**] is the Jacobian matrix.

$$[\mathbf{J}] = \begin{bmatrix} x,_\xi & y,_\xi \\ x,_\eta & y,_\eta \end{bmatrix} = \begin{bmatrix} J_{11} & J_{12} \\ J_{21} & J_{22} \end{bmatrix} \tag{5.3.6}$$

The inverse relations, from Eq. 5.3.5, are

$$\begin{Bmatrix} \phi,_x \\ \phi,_y \end{Bmatrix} = [\boldsymbol{\Gamma}] \begin{Bmatrix} \phi,_\xi \\ \phi,_\eta \end{Bmatrix}, \qquad \text{where} \qquad [\boldsymbol{\Gamma}] = [\mathbf{J}]^{-1} \tag{5.3.7}$$

Equations 5.3.5, 5.3.6, and 5.3.7 are *general*. For the plane elements of this chapter, ϕ is either u or v. Numerical values of coefficients in [**J**] depend on the size, shape, and orientation of the element (dictated by the node coordinates and numbers we assign). For the four-node element at hand, from Eqs. 5.3.2,

$$J_{11} = x,_\xi = N_{1,\xi}x_1 + N_{2,\xi}x_2 + N_{3,\xi}x_3 + N_{4,\xi}x_4 \tag{5.3.8}$$

There are similar expressions for J_{12}, J_{21}, and J_{22}, where

$$N_{1,\xi} = -\frac{1-\eta}{4}, \qquad N_{2,\xi} = \frac{1-\eta}{4}, \text{ etc.} \tag{5.3.9}$$

If $x = \xi$ and $y = \eta$, then $[\mathbf{J}] = [\boldsymbol{\Gamma}] = [\mathbf{I}]$.

Now we are prepared to find [**B**]. We write

$$\{\boldsymbol{\epsilon}\} = \begin{Bmatrix} \epsilon_x \\ \epsilon_y \\ \gamma_{xy} \end{Bmatrix} = \begin{bmatrix} 1 & 0 & 0 & 0 \\ 0 & 0 & 0 & 1 \\ 0 & 1 & 1 & 0 \end{bmatrix} \begin{Bmatrix} u,_x \\ u,_y \\ v,_x \\ v,_y \end{Bmatrix} \tag{5.3.10}$$

$$\begin{Bmatrix} u,_x \\ u,_y \\ v,_x \\ v,_y \end{Bmatrix} = \begin{bmatrix} \Gamma_{11} & \Gamma_{12} & 0 & 0 \\ \Gamma_{21} & \Gamma_{22} & 0 & 0 \\ 0 & 0 & \Gamma_{11} & \Gamma_{12} \\ 0 & 0 & \Gamma_{21} & \Gamma_{22} \end{bmatrix} \begin{Bmatrix} u,_\xi \\ u,_\eta \\ v,_\xi \\ v,_\eta \end{Bmatrix} \tag{5.3.11}$$

$$\begin{Bmatrix} u,_\xi \\ u,_\eta \\ v,_\xi \\ v,_\eta \end{Bmatrix} = \begin{bmatrix} N_{1,\xi} & 0 & N_{2,\xi} & 0 & \text{etc.} \\ N_{1,\eta} & 0 & N_{2,\eta} & 0 & \text{etc.} \\ 0 & N_{1,\xi} & 0 & N_{2,\xi} & \text{etc.} \\ 0 & N_{1,\eta} & 0 & N_{2,\eta} & \text{etc.} \end{bmatrix} \underset{8\times 1}{\{\mathbf{d}\}} \tag{5.3.12}$$

Matrix [**B**] is the product of the three successive rectangular matrices in Eqs. 5.3.10 through 5.3.12. The stiffness matrix is

$$[\mathbf{k}] = \iint [\mathbf{B}]^T [\mathbf{E}][\mathbf{B}]t\, dx\, dy = \int_{-1}^{1}\int_{-1}^{1} [\mathbf{B}]^T[\mathbf{E}][\mathbf{B}]tJ\, d\xi\, d\eta \qquad (5.3.13)$$

where J is the determinant of the Jacobian matrix and t is the element thickness. In two dimensions,

$$J = \det[\mathbf{J}] = J_{11}J_{22} - J_{21}J_{12} \qquad (5.3.14)$$

Jacobian J is a function of position within the element and is the multiplier that yields area $dx\, dy$ from $d\xi\, d\eta$. A typical coefficient in $[\mathbf{B}]$ depends on nodal coordinates and has ξ and η polynomials in both numerator and denominator. Therefore the integration in Eq. 5.3.13 must be done numerically (Section 5.4). Contributions to the element load vector (Eq. 4.3.5) include

$$\int_{-1}^{1}\int_{-1}^{1} ([\mathbf{B}]^T [\mathbf{E}]\{\boldsymbol{\epsilon}_0\} - [\mathbf{B}]^T \{\boldsymbol{\sigma}_0\} + [\mathbf{N}]^T \{\mathbf{F}\})\, t\, d\xi\, d\eta \qquad (5.3.15)$$

This integral must also be evaluated numerically. Surface traction effects are considered in Section 5.8.

5.4 SUMMARY OF GAUSS QUADRATURE

A definite integral can be evaluated numerically by any of several methods. Here we describe only the Gauss method [5.3, 5.4], since it has proved most useful in finite element work.

To approximate the integral

$$I = \int_{-1}^{1} \phi\, d\xi \qquad \text{where} \qquad \phi = \phi(\xi) \qquad (5.4.1)$$

we can sample (evaluate) ϕ at the midpoint of the interval and multiply by the length of the interval, as shown in Fig. 5.4.1*a*. Thus we find $I \approx 2\phi_1$. This result is exact if the function ϕ happens to be a straight line of any slope.

Generalization of Eq. 5.4.1 leads to the formula

$$I = \int_{-1}^{1} \phi\, d\xi \approx W_1\phi_1 + W_2\phi_2 + \cdots + W_n\phi_n \qquad (5.4.2)$$

Thus, to approximate I, we evaluate $\phi = \phi(\xi)$ at each of several locations ξ_i, multiply the resulting ϕ_i by an appropriate weight W_i, and add. Gauss's method locates the sampling points so that for a given number of them, greatest accuracy is achieved.

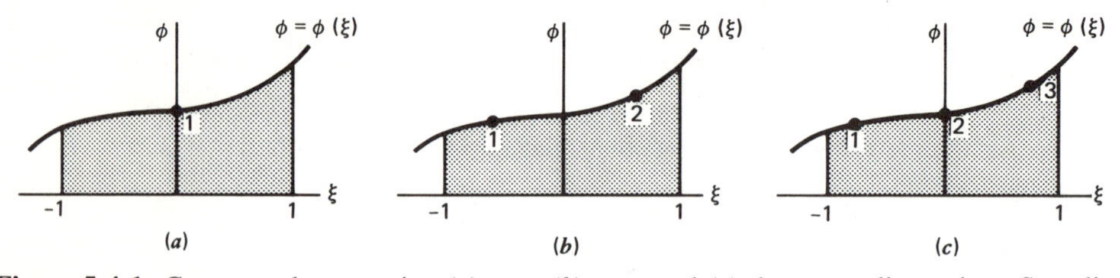

Figure 5.4.1. Gauss quadrature using (*a*) one, (*b*) two, and (*c*) three sampling points. Sampling points are also called Gauss points.

Sampling points are located symmetrically with respect to the center of the interval. Symmetrically paired points have the same weight W_i. Table 5.4.1 gives data for Gauss rules of order $n = 1$ through $n = 6$. As an example application, consider $n = 2$ and Eq. 5.4.2. We find

$$I \approx (1.)(\phi \text{ at } \xi = -0.57735\cdots) + (1.)(\phi \text{ at } \xi = +0.57735\cdots) \tag{5.4.3}$$

In computer work, numerical data for the ξ_i and W_i should be written with as many digits as the machine allows.

Figure 5.4.2*a* illustrates the behavior of Gauss quadrature. The exact area under $\phi = a + c\xi^2$ for $-1 < \xi < 1$ is $I = 2a + 2c/3$. It is clear that ordinates $\phi_1 = \phi_2$ can be chosen such that

$$I_{\text{exact}} = 2a + 2c/3 = \phi_1 + \phi_2 = 2\phi_1 \tag{5.4.4}$$

The $\phi = b\xi$ and $\phi \neq d\xi^3$ curves in Fig. 5.4.2*a* each integrate to zero, since each includes as much positive area on one side as negative area on the other. So whether $\phi = a +$

TABLE 5.4.1. SAMPLING POINTS AND WEIGHTS FOR GAUSS QUADRATURE

Order n	Location ξ_i	Weight W_i
1	0.	2.
2	±0.57735 02691 89626	1.00000 00000 00000
3	±0.77459 66692 41483	0.55555 55555 55556
	0.00000 00000 00000	0.88888 88888 88889
4	±0.86113 63115 94053	0.34785 48451 37454
	±0.33998 10435 84856	0.65214 51548 62546
5	±0.90617 98459 38664	0.23692 68850 56189
	±0.53846 93101 05683	0.47862 86704 99366
	0.00000 00000 00000	0.56888 88888 88889
6	±0.93246 95142 03152	0.17132 44923 79179
	±0.66120 93864 66265	0.36076 15730 48139
	±0.23861 91860 83197	0.46791 39345 72691

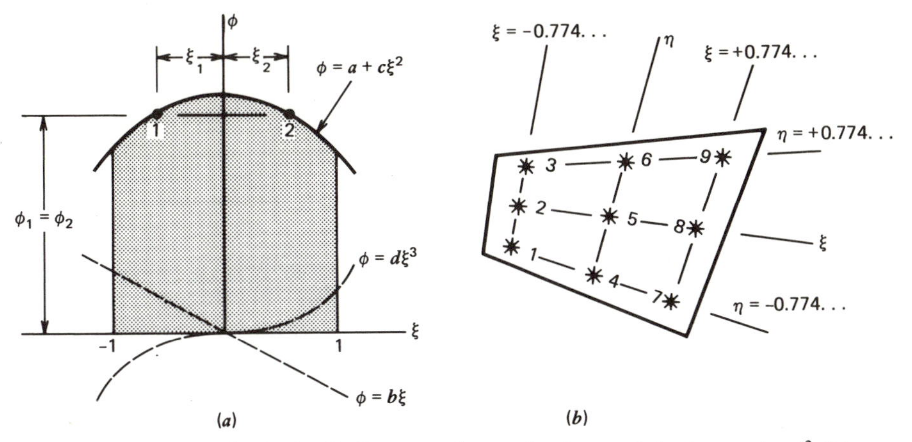

Figure 5.4.2. (*a*) Two-point (order 2) Gauss quadrature for the parabola $\phi = a + c\xi^2$. Linear and cubic curves are shown by dashed lines. Arbitrary constants are denoted by *a*, *b*, *c*, and *d*. (*b*) Nine-point (order 3) Gauss rule applied to a plane region.

$c\xi^2$ or $\phi = a + b\xi + c\xi^2 + d\xi^3$, two Gauss points yield the exact area under the curve. In general, *a polynominal of degree* 2n $-$ *1 is integrated exactly by* n-*point Gauss quadrature*.

In *two* dimensions we find the quadrature formula for $\phi = \phi(\xi, \eta)$ by integrating with respect to ξ and then with respect to η.

$$\begin{aligned} I = \int_{-1}^{1}\int_{-1}^{1} \phi(\xi,\eta)\, d\xi\, d\eta &\approx \int_{-1}^{1}\left[\sum_i W_i\phi(\xi_i,\eta)\right] d\eta \\ &\approx \sum_j W_j\left[\sum_i W_i\phi(\xi_i,\eta_j)\right] = \sum_i\sum_j W_iW_j\phi(\xi_i,\eta_j) \end{aligned} \tag{5.4.5}$$

For example, consider a 3-by-3 (nine-point) rule (Fig. 5.4.2*b*). Let $A = 0.555\ldots$ and $B = 0.888\ldots$. Then

$$\begin{aligned} I \approx A^2\phi_1 + AB\phi_2 + A^2\phi_3 + BA\phi_4 + B^2\phi_5 \\ + BA\phi_6 + A^2\phi_7 + AB\phi_8 + A^2\phi_9 \end{aligned} \tag{5.4.6}$$

where ϕ_i is the numerical value of the function at the *i*th Gauss point.

In *three* dimensions we have

$$I = \int_{-1}^{1}\int_{-1}^{1}\int_{-1}^{1} \phi(\xi,\eta,\zeta)\, d\xi\, d\eta\, d\zeta \approx \sum_i\sum_j\sum_k W_iW_jW_k\phi(\xi_i,\eta_j,\zeta_k) \tag{5.4.7}$$

We need not use the same number of Gauss points in each direction, but this is most common.

In Eq. 5.3.13 *each coefficient* in the integrand matrix $[\mathbf{B}]^T [\mathbf{E}][\mathbf{B}]tJ$ must be integrated as $\phi(\xi, \eta)$ is integrated in Eq. 5.4.5. Then I represents one coefficient k_{ij} in the element stiffness matrix.

5.5 COMPUTER SUBROUTINES FOR THE LINEAR ELEMENT

The stiffness matrix and load vector are generated numerically by a computer program. Subroutines listed here are not unique: other coding may be more compact, more general, or more efficient [5.5], but the subroutines show exactly what must be done. A thorough understanding of these subroutines makes it much easier to understand other isoparametric elements.

The notation, assumptions, and procedures are as follows. Shape functions and their derivatives can be written in the form

$$\begin{aligned} N_i &= (1 + \xi\xi_i)(1 + \eta\eta_i)/4 \\ N_{i,\xi} &= \xi_i(1 + \eta\eta_i)/4 \\ N_{i,\eta} &= \eta_i(1 + \xi\xi_i)/4 \end{aligned} \tag{5.5.1}$$

where i is the number of the shape function, and

$$\begin{aligned} \xi_i &= -1., 1., 1., -1. \quad \text{for} \quad i = 1, 2, 3, 4 \\ \eta_i &= -1., -1., 1., 1. \quad \text{for} \quad i = 1, 2, 3, 4 \end{aligned} \tag{5.5.2}$$

In Fig. 5.5.1, ξ_i and η_i are placed in arrays XII and ETI by DATA and EQUIVALENCE statements. N_i, $N_{i,\xi}$, and $N_{i,\eta}$ are computed and stored in arrays EN, NXI, and NET. Coordinates ξ and η in Eqs. 5.5.1 are called PXI and PET in Fig. 5.5.1 and are transmitted as formal parameters. PXI and PET are Gauss point coordinates if SHAPE is called by QUAD4, but other coordinates could be prescribed by a subsequent calling routine (as, for example, when applying Eq. 4.10.1).

Through statement 40, subroutine SHAPE follows exactly the development in Section 5.3. In the DO 50 loop, initial strains (ϵ_{x0} = EXI, ϵ_{y0} = EYI, γ_{xy0} = ESI) and element thickness (t = THC) are prescribed at the four nodes. Values of these quantities at coordinates PXI and PET are found by interpolation (write $t = \Sigma N_i t_i$, analogous to one of Eqs. 5.3.2). Initial stresses $\{\boldsymbol{\sigma}_0\}$ can be added as an exercise (Problem 5.11).

The foregoing calculations in SHAPE must be carried out at every Gauss point used by QUAD4 (Fig. 5.5.2).

QUAD4 requires as input data the global nodal coordinates XL and YL, nodal thicknesses and initial strains, the material property matrix [E] (presumed full, as for a general material, and constant over the element), body forces F_x = BODYFX and F_y = BODYFY, and the quadrature order NGAUSS. While perhaps less obvious than subroutine SHAPE, subroutine QUAD4 is a straightforward application of Gauss quadrature to the second of Eqs. 5.3.13 and to Eq. 5.3.15. DV represents the product $W_i W_j tJ$, which is a common multiplier of each coefficient to be integrated—that is, of each coefficient in $[\mathbf{B}]^T[\mathbf{E}][\mathbf{B}]$, in $[\mathbf{B}]^T[\mathbf{E}]\{\epsilon_0\}$, and in $[\mathbf{N}]^T\{\mathbf{F}\}$.

Cost is reduced by using quadrature to generate only the upper triangle of $[\mathbf{k}]$, leaving the lower triangle to be completed by symmetry as the last step. Coding of the DO 20 loop is an attempt to exploit the sparsity of $[\mathbf{B}]$. A similar efficiency in the DO 50 loop could be more easily coded if the arrangement of d.o.f. were $\{\mathbf{d}\} = \{u_1\ u_2\ u_3\ u_4\ v_1\ v_2\ v_3\ v_4\}$. Additional economies have been proposed [2.31, 4.2, 5.1, 5.6].

```
      SUBROUTINE SHAPE (PXI,PET,XL,YL,EXI,EYI,ESI,THC,THK,DETJAC)
      IMPLICIT DOUBLE PRECISION (A-H,O-Z)
      DOUBLE PRECISION NXI(4),NET(4),JAC
      REAL XII(4),ETI(5)
      DIMENSION DUP(1),XL(4),YL(4),EXI(4),EYI(4),ESI(4),THC(4)
      COMMON /Q4/ EN(4),JAC(2,2),B(3,9),E(3,3),SE(8,8),RE(8)
      EQUIVALENCE (XII(1),ETI(2)),(DUP(1),JAC(1))
      DATA ETI /-1.,-1.,+1.,+1.,-1./
C --- FIND SHAPE FUNCTIONS (EN) AND THEIR DERIVATIVES (NXI, NET).
      DO 10 L=1,4
      DUM1 = (1. + XII(L)*PXI)/4.
      DUM2 = (1. + ETI(L)*PET)/4.
      EN(L) = 4.*DUM1*DUM2
      NXI(L) = XII(L)*DUM2
   10 NET(L) = ETI(L)*DUM1
C --- CLEAR ARRAYS JAC AND B (WORDS IN COMMON ARE STORED IN SEQUENCE).
      DO 20 L=1,31
   20 DUP(L) = 0.
C --- FIND JACOBIAN JAC AND ITS DETERMINANT. REPLACE JAC BY ITS INVERSE.
      DO 30 L=1,4
      JAC(1,1) = JAC(1,1) + NXI(L)*XL(L)
      JAC(1,2) = JAC(1,2) + NXI(L)*YL(L)
      JAC(2,1) = JAC(2,1) + NET(L)*XL(L)
   30 JAC(2,2) = JAC(2,2) + NET(L)*YL(L)
      DETJAC = JAC(1,1)*JAC(2,2) - JAC(2,1)*JAC(1,2)
      DUM1   = JAC(1,1)/DETJAC
      JAC(1,1) =  JAC(2,2)/DETJAC
      JAC(1,2) = -JAC(1,2)/DETJAC
      JAC(2,1) = -JAC(2,1)/DETJAC
      JAC(2,2) =  DUM1
C --- FORM STRAIN-DISPLACEMENT MATRIX B (ZERO ENTRIES ARE ALREADY SET).
      DO 40 J=1,4
      L = 2*J
      K = L-1
      B(1,K) = JAC(1,1)*NXI(J) + JAC(1,2)*NET(J)
      B(2,L) = JAC(2,1)*NXI(J) + JAC(2,2)*NET(J)
      B(3,K) = B(2,L)
   40 B(3,L) = B(1,K)
C --- INTERPOLATE INITIAL STRAINS FROM CORNER VALUES, AND STORE AS
C --- COL. 9 OF B (ALREADY CLEARED TO ZERO). SIMILARLY, GET THICKNESS.
      THK = 0.
      DO 50 L=1,4
      B(1,9) = B(1,9) + EN(L)*EXI(L)
      B(2,9) = B(2,9) + EN(L)*EYI(L)
      B(3,9) = B(3,9) + EN(L)*ESI(L)
   50 THK    = THK    + EN(L)*THC(L)
      RETURN
      END
```

Figure 5.5.1. Fortran subroutine SHAPE. For the element of Section 5.3, it calculates the shape functions and therir derivatives, the Jacobian matrix, its inverse and determinant, matrix $[\mathbf{B}]$, and initial strains and element thickness, all at the point whose coordinates are PXI and PET.

```
      SUBROUTINE QUAD4 (NGAUSS)
      IMPLICIT DOUBLE PRECISION (A-H,O-Z)
C  THE USER MUST INSERT COMMON AND DIMENSION STATEMENTS HERE. ALSO, MA-
C  TRIX E MUST BE GIVEN, AND DATA AT ELEMENT NODES SUPPLIED AS FOLLOWS.
C  XL,YL         = CARTESIAN COORDINATES.        (USED IN SUBROUTINE SHAPE)
C  EXI,EYI,ESI = INITIAL STRAINS (X,Y,SHEAR). (USED IN SUBROUTINE SHAPE)
C  THC           = THICKNESSES IN Z DIRECTION.  (USED IN SUBROUTINE SHAPE)
C  USE DIMENSION AND DATA STMNTS TO SET UP 3 BY 3 ARRAYS PLACE AND WGT.
C            ( 0.  -.57735---   -.77459--- )          ( 2. 1. .555--- )
C  (PLACE) = ( 0.  +.57735---   0.         ), (WGT) = ( 0. 1. .888--- )
C            ( 0.  0.           +.77459--- )          ( 0. 0. .555--- )
C --- CLEAR ELEMENT LOAD VECTOR AND UPPER TRIANGLE OF EL. STIFFNESS MAT.
      DO 10 K=1,8
      RE(K) = 0.
      DO 10 L=K,8
   10 SE(K,L) = 0.
C --- START GAUSS QUADRATURE LOOP. USE NGAUSS BY NGAUSS RULE.
      DO 90 NA = 1,NGAUSS
      PXI = PLACE(NA,NGAUSS)
         DO 80 NB = 1,NGAUSS
         PET = PLACE(NB,NGAUSS)
         CALL SHAPE (PXI,PET,XL,YL,EXI,EYI,ESI,THC,THK,DETJAC)
         DV = WGT(NA,NGAUSS)*WGT(NB,NGAUSS)*THK*DETJAC
C --------- STORE B-TRANSPOSE TIMES E IN 8 BY 3 WORK-SPACE ARRAY BTE.
            DO 30 J=1,4
            L = 2*J
            K = L-1
C ------------ DO ONLY MULTIPLICATIONS THAT GIVE A NONZERO PRODUCT.
               DO 20 N=1,3
               BTE(K,N) = B(1,K)*E(1,N) + B(3,K)*E(3,N)
   20          BTE(L,N) = B(2,L)*E(2,N) + B(3,L)*E(3,N)
C --------- ADD CONTRIBUTION OF BODY FORCES TO ELEMENT NODAL LOAD ARRAY.
            RE(K) = RE(K) + EN(J)*BODYFX*DV
   30       RE(L) = RE(L) + EN(J)*BODYFY*DV
C --------- LOOP ON ROWS OF ARRAYS SE AND RE.
            DO 70 NROW=1,8
C ------------ ADD CONTRIBUTION OF INITIAL STRAINS TO LOAD ARRAY RE.
               DO 40 J=1,3
   40          RE(NROW) = RE(NROW) + BTE(NROW,J)*B(J,9)*DV
C ------------ LOOP TO ADD CONTRIBUTION TO ELEMENT STIFFNESS ARRAY SE.
               DO 60 NCOL=NROW,8
               DUM = 0.
C --------------- LOOP FOR PRODUCT (B)T*(E)*(B). ZEROS IN B NOT SKIPPED.
                  DO 50 J=1,3
   50             DUM = DUM + BTE(NROW,J)*B(J,NCOL)
   60          SE(NROW,NCOL) = SE(NROW,NCOL) + DUM*DV
   70       CONTINUE
   80    CONTINUE
   90 CONTINUE
C --- FILL IN LOWER TRIANGLE OF ELEMENT STIFFNESS MATRIX BY SYMMETRY.
      DO 100 K=1,7
      DO 100 L=K,8
  100 SE(L,K) = SE(K,L)
      RETURN
      END
```

Figure 5.5.2. Fortran subroutine QUAD4. It generates [**k**] and {**r**} for the element of Section 5.3 by Gauss quadrature. We store [**k**] in array SE and {**r**} in array RE. (Indentations to display the loop hierarchy comprise the "prettyprinted" format.)

5.6 ADDITIONAL PLANE ELEMENTS

Side nodes can be added to the linear element to yield higher-order members of the isoparametric family. In Fig. 5.6.1*a* the shape of an edge and the displacement of an edge can vary as ξ^2 or η^2. Similarly, in Fig. 5.6.1*b*, the edge shapes and displacements can vary as ξ^3 or η^3. These are *capabilities*, not requirements: an edge can be straight if so desired, an edge of the cubic element can have a linear variation of displacement, and so on. Equations 5.3.2 describe the quadratic and cubic elements if we use the appropriate N_i and sum over 8 and 12 terms, respectively.

Interpolation polynomials for shape and displacement in the quadratic element are based on the incomplete cubic polynomial [5.7].

$$a_1 + a_2\xi + a_3\eta + a_4\xi^2 + a_5\xi\eta + a_6\eta^2 + a_7\xi^2\eta + a_8\xi\eta^2 \tag{5.6.1}$$

Instead of generating a matrix $[\mathbf{A}]^{-1}$ to replace the a_i by nodal quantities in the manner of Eqs. 4.5.2 to 4.5.6, we can follow a more direct and physically appealing method [5.8]. We examine various modes and combine them appropriately (Fig. 5.6.2). In Fig. 5.6.2*a*, the disaplacement varies quadratically with ξ and linearly with η. Also, u is zero at all nodes except node 5, where it is unity. Therefore the shape function N_5 is $N_5 = u_{(a)}$. Similarly, $N_8 = u_{(b)}$. In Fig. 5.6.2*c*, u varies linearly with ξ and η and is $\frac{1}{2}$ at nodes 5 and 8. This "excess" displacement at nodes 5 and 8 can be removed from $u_{(c)}$ by subtracting $u_{(a)}/2$ and $u_{(b)}/2$ from $u_{(c)}$. Thus $N_1 = u_{(d)}$, where $u_{(d)}$ is constructed so that $u_{(d)} = 1$ at $\xi = \eta = -1$ and $u_{(d)} = 0$ at other nodes, as is required of an interpolation scheme. This example shows that shape functions can be generated by intuition, trial, and familiarity with simpler elements.

The complete set of shape functions can be written as shown in Table 5.6.1. This format permits us to "degrade" element edges, so that linear and quadratic elements can be joined (see also Section 7.12). It is easy to change Figs. 5.5.1 and 5.5.2 to accommodate an element with four to eight nodes. Input flags could be used to signal which if any side nodes are to be retained. The changes consist largely of redefining shape functions,

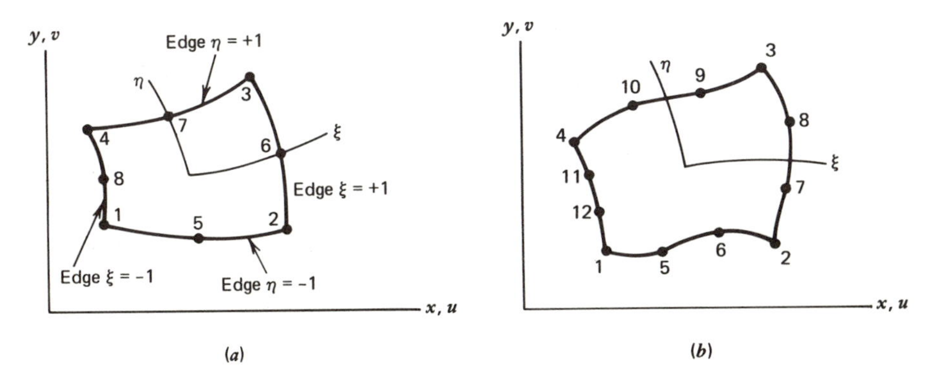

Figure 5.6.1. Plane isoparametric elements. (*a*) Quadratic. (*b*) Cubic.

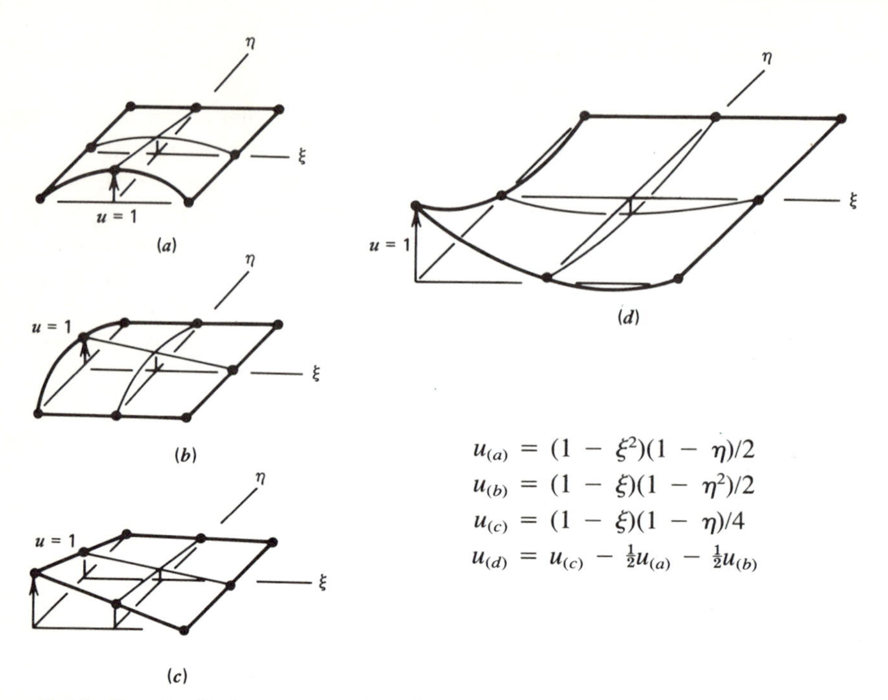

Figure 5.6.2. Sample displacement modes of the quadratic element. For ease of representation only, displacement u is shown normal to the plane of the element.

redimensioning arrays, and increasing the limits on DO loops. For example, Eqs. 5.3.4 through 5.3.7, 5.3.10, 5.3.11, and 5.3.13 through 5.3.15 are unchanged, while all other equations in Section 5.3 include more terms and use other shape functions.

As shown by numerical examples (Table 5.12.1), the quadratic element works far better than the linear element—so much so that an entire book is based on it [2.31].

Shape functions of the cubic element (Fig. 5.6.1*b*) are based on the incomplete quartic polynomial [5.7]

$$\text{(Terms of Eq. 5.6.1)} + a_9\xi^3 + a_{10}\eta^3 + a_{11}\xi^3\eta + a_{12}\xi\eta^3 \tag{5.6.2}$$

For the corner nodes, $i = 1, 2, 3, 4$,

$$N_i = \tfrac{1}{32}(1 + \xi\xi_i)(1 + \eta\eta_i)(-10 + 9\xi^2 + 9\eta^2) \tag{5.6.3}$$

where ξ_i and η_i are given by Eq. 5.5.2. For nodes on sides $\xi = \pm 1$, for which $i = 7, 8, 11, 12$,

$$N_i = \tfrac{9}{32}(1 + \xi\xi_i)(1 + 9\eta\eta_i)(1 - \eta^2) \tag{5.6.4}$$

TABLE 5.6.1. Shape Functions for the Plane Quadratic Isoparametric Element, Written in a Form that Permits a Variable Number of Nodes

	4 Linear Edges	3, 2, 1, or 0 Linear Edges, Others Quadratic			
N_i	Include Nodes 1 to 4	Add Node 5	Add Node 6	Add Node 7	Add Node 8
N_1	$(1-\xi)(1-\eta)/4$	$-N_5/2$			$-N_8/2$
N_2	$(1+\xi)(1-\eta)/4$	$-N_5/2$	$-N_6/2$		
N_3	$(1+\xi)(1+\eta)/4$		$-N_6/2$	$-N_7/2$	
N_4	$(1-\xi)(1+\eta)/4$			$-N_7/2$	$-N_8/2$
N_5		$(1-\xi^2)(1-\eta)/2$			
N_6			$(1+\xi)(1-\eta^2)/2$		
N_7				$(1-\xi^2)(1+\eta)/2$	
N_8					$(1-\xi)(1-\eta^2)/2$

where $\xi_i = \pm 1$ and $\eta_i = \pm\frac{1}{3}$. For nodes on sides $\eta = \pm 1$, for which $i = 5, 6, 9, 10$,

$$N_i = \tfrac{9}{32}(1 + 9\xi\xi_i)(1 + \eta\eta_i)(1 - \xi^2) \tag{5.6.5}$$

where $\eta_i = \pm 1$ and $\xi_i = \pm\frac{1}{3}$.

5.7 SOLID ISOPARAMETRIC ELEMENTS

The isoparametric formulation extends directly from two dimensions to three. We add coordinate ζ and define faces of the element by $\xi, \eta, \zeta = \pm 1$. The formulation of Section 5.3 is adjusted as follows. We add $z = \Sigma N_i z_i$ and $w = \Sigma N_i w_i$ to Eqs. 5.3.2, where w is the z-direction displacement and i ranges over the number of nodes in the element. Derivatives $\phi,_\zeta$ and $\phi,_z$ are added to Eqs. 5.3.5 so that the Jacobian matrix becomes

$$[\mathbf{J}] = \begin{bmatrix} x,_\xi & y,_\xi & z,_\xi \\ x,_\eta & y,_\eta & z,_\eta \\ x,_\zeta & y,_\zeta & z,_\zeta \end{bmatrix} = \begin{bmatrix} J_{11} & J_{12} & J_{13} \\ J_{21} & J_{22} & J_{23} \\ J_{31} & J_{32} & J_{33} \end{bmatrix} \tag{5.7.1}$$

Again, $[\boldsymbol{\Gamma}] = [\mathbf{J}]^{-1}$. Equation 5.3.11 is expanded to include displacement w and derivatives of u, v, and w with respect to z and ζ. Equation 5.3.12 is expanded to yield the 9-by-1 vector $\{u,_\xi \; u,_\eta \; u,_\zeta \; . \; . \; . \; w,_\zeta\}$. In place of Eq. 5.3.10 we have

$$\{\epsilon_x \; \epsilon_y \; \epsilon_z \; \gamma_{xy} \; \gamma_{yz} \; \gamma_{zx}\} = [\mathbf{H}] \, \{u,_x \; u,_y \; u,_z \; \cdots \; w,_z\} \tag{5.7.2.}$$

where

$$\underset{6\times 9}{[\mathbf{H}]} = \begin{bmatrix} 1 & 0 & 0 & 0 & 0 & 0 & 0 & 0 & 0 \\ 0 & 0 & 0 & 0 & 1 & 0 & 0 & 0 & 0 \\ 0 & 0 & 0 & 0 & 0 & 0 & 0 & 0 & 1 \\ 0 & 1 & 0 & 1 & 0 & 0 & 0 & 0 & 0 \\ 0 & 0 & 0 & 0 & 0 & 1 & 0 & 1 & 0 \\ 0 & 0 & 1 & 0 & 0 & 0 & 1 & 0 & 0 \end{bmatrix}$$

Finally, the stiffness matrix is

$$[\mathbf{k}] = \int_{-1}^{1}\int_{-1}^{1}\int_{-1}^{1} [\mathbf{B}]^T [\mathbf{E}][\mathbf{B}]J \, d\xi \, d\eta \, d\zeta \tag{5.7.3}$$

where the Jacobian determinant $J = \det[\mathbf{J}]$ expresses the ratio of the volume element $dx\, dy\, dz$ to $d\xi\, d\eta\, d\zeta$. In general, J is a function of ξ, η, and ζ, as is $[\mathbf{B}]$.

For the linear solid (Fig. 5.7.1*a*) we have simply

$$N_i = \tfrac{1}{8}(1 + \xi\xi_i)(1 + \eta\eta_i)(1 + \zeta\zeta_i) \tag{5.7.4}$$

where $\xi_i, \eta_i, \zeta_i = \pm 1$ and $i = 1, 2, \ldots, 8$, in direct analogy to Eqs. 5.5.1 and 5.5.2.

For the quadratic solid (Fig. 5.7.1*b*), we can allow from 8 to 20 nodes, in analogy to Table 5.6.1. Again, let $\xi_i, \eta_i, \zeta_i = \pm 1$. For nodes on $\xi = 0$, $A = 17, 18, 19, 20$,

$$N_A = (1 - \xi^2)(1 + \eta\eta_i)(1 + \zeta\zeta_i)/4 \tag{5.7.5}$$

For nodes on $\eta = 0$, $B = 10, 12, 14, 16$,

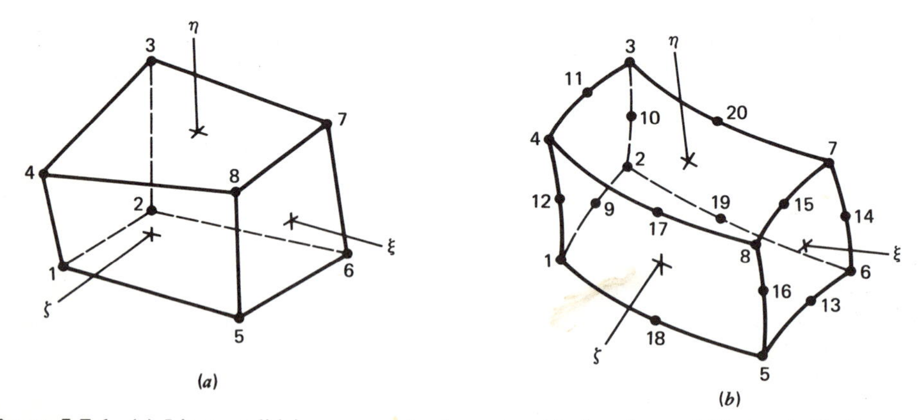

Figure 5.7.1. (*a*) Linear solid isoparametric element, with 8 nodes and 24 d.o.f. (*b*) Quadratic solid isoparametric element, with 20 nodes and 60 d.o.f.

$$N_B = (1 + \xi\xi_i)(1 - \eta^2)(1 + \zeta\zeta_i)/4 \tag{5.7.6}$$

For nodes on $\zeta = 0$, $C = 9, 11, 13, 15$,

$$N_C = (1 + \xi\xi_i)(1 + \eta\eta_i)(1 - \zeta^2)/4 \tag{5.7.7}$$

For the corner nodes, $i = 1, 2, \ldots, 8$,

$$N_i = \frac{(1 + \xi\xi_i)(1 + \eta\eta_i)(1 + \zeta\zeta_i)}{8} - \frac{N_A + N_B + N_C}{2} \tag{5.7.8}$$

where $(N_A + N_B + N_C)/2$ refers to only the three nodes adjacent to node i. For example, if $i = 8$, then $A = 17$, $B = 16$, and $C = 15$. The full quadratic element includes all 20 nodes and shape functions. Nodes and shape functions are omitted for each linear edge. For example, if edge 4-17-8 is to remain straight, we set $N_{17} = 0$ in Eqs. 5.7.5 and 5.7.8.

Figures 1.1.4 and 5.7.2 show example applications of solid elements.

5.8 NODAL LOADS FROM SURFACE TRACTION AND BODY FORCE

Consistent nodal loads are given by Eqs. 4.3.5 and 5.3.15. The effects of initial strains $\{\epsilon_0\}$ and body forces $\{\mathbf{F}\}$ are automatically included in the subroutines of Figs. 5.5.1 and 5.5.2, regardless of element shape. The effects of surface tractions $\{\boldsymbol{\Phi}\}$ must be calculated separately. In this section we treat $\{\boldsymbol{\Phi}\}$ and give numerical examples of loads $\{\mathbf{r}\}$ that come from $\{\mathbf{F}\}$ and $\{\boldsymbol{\Phi}\}$.

Consider the plane *linear* element (Fig. 5.3.1) with a uniform body force F_y in the $+y$ direction. Thus $\{\mathbf{F}\} = \{0 \ \ F_y\}$. If the element is *rectangular,* the volume integral of $[\mathbf{N}]^T\{\mathbf{F}\}$ yields $+y$ direction loads at each node, each equal to one-quarter the total force on the element. This is true regardless of the orientation of the element with respect to the y-axis. Similarly, a rectangular eight-node *solid* element has nodal loads each equal to one-eighth the total force on the element.

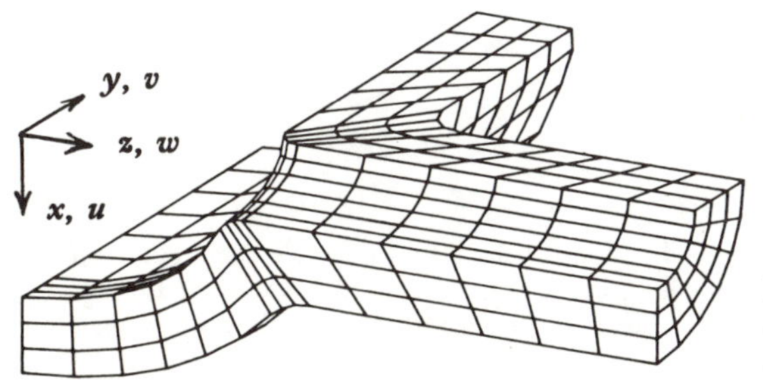

Figure 5.7.2. Solid elements, linear or quadratic, used to model one octant of a cylinder-to-cylinder intersection.

Next consider a plane *quadratic* element. Again, let $\{\mathbf{F}\} = \{0 \ \ F_y\}$, where F_y is constant. Let the element be rectangular and have its side nodes at midsides. Now we are surprised: the forces are unequal, and corner forces are directed *opposite* to the body force (Fig. 5.8.1*a*). Nevertheless, $4P - 4Q = 1$, as required.

If a mesh is fine, we need not use consistent nodal loads. Thus, for example, we could use equal load fractions $P = Q = \frac{1}{8}$ in Fig. 5.8.1*a*, with confidence in convergence is toward exact answers as the mesh is refined. But for greater accuracy in a coarse mesh the consistent formulation is preferred.

Tractions on a *linear* edge are considered in Section 4.3. Consider now a uniform traction σ_y on a *quadratic* edge of overall length L and with a midside node (Fig. 5.8.1*b*). The last integral of Eq. 4.3.5 and the shape functions of Table 5.6.1 yield

$$\underset{3\times 1}{\{\mathbf{r}\}} = \sigma_y \int_{-1}^{1} \{\bar{N}_4 \ \bar{N}_7 \ \bar{N}_3\} \frac{L}{2} t \, d\xi = \sigma_y t L \left\{ \frac{1}{6} \ \ \frac{2}{3} \ \ \frac{1}{6} \right\} \tag{5.8.1}$$

where t is element thickness and overbars indicate that these N_i are evaluated at $\eta = 1$. Fractions of the total force $\sigma_y tL$ are shown in Fig. 5.8.1*b*. Similarly, we can examine one surface, say $\zeta = 1$, on a solid element and derive the load fractions in Fig.5.8.1*c*.

Consider now a curved edge (Fig. 5.8.2*a*). In place of Eq. 5.8.1 we have

$$\underset{3\times 1}{\{\mathbf{r}\}} = \int_{\text{Edge 4-7-3}} \begin{bmatrix} \bar{N}_4 & 0 & \bar{N}_7 & 0 & \bar{N}_3 & 0 \\ 0 & \bar{N}_4 & 0 & \bar{N}_7 & 0 & \bar{N}_3 \end{bmatrix}^T \begin{Bmatrix} \Phi_x \\ \Phi_y \end{Bmatrix} t \, dS \tag{5.8.2}$$

where dS is an increment of edge length and the x- and y-parallel traction components are, from Fig. 1.3.2,

$$\begin{Bmatrix} \Phi_x \\ \Phi_y \end{Bmatrix} dS = \begin{Bmatrix} \tau \, dS \cos\beta - \sigma \, dS \sin\beta \\ \sigma \, dS \cos\beta + \tau \, dS \sin\beta \end{Bmatrix} = \begin{Bmatrix} \tau \, dx - \sigma \, dy \\ \sigma \, dx + \tau \, dy \end{Bmatrix} \tag{5.8.3}$$

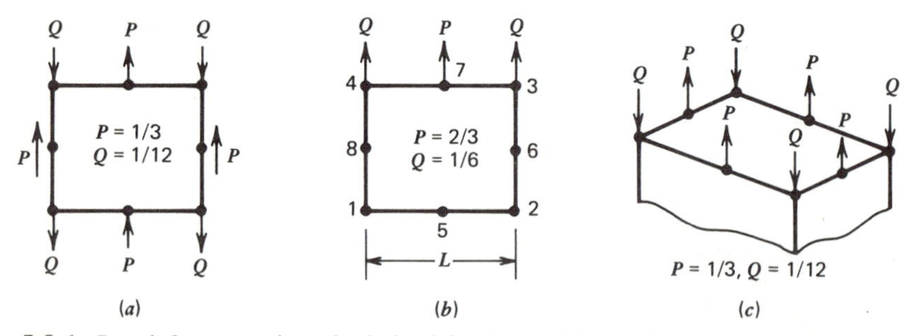

Figure 5.8.1. P and Q are consistently derived fractions of the total force on a quadratic element. In each case the force is uniformly distributed and directed upward, and side nodes are at midsides. (*a*) Body force on a rectangular plane element. (*b*) Traction on the top edge of a plane element. (*c*) Traction on the rectangular top surface of a solid element.

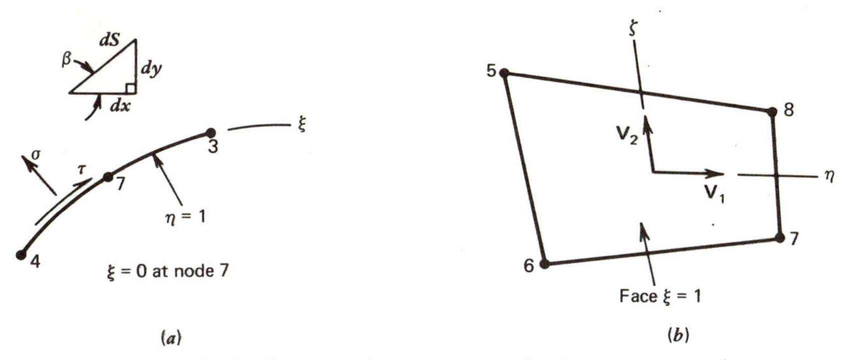

Figure 5.8.2. (*a*) Prescribed edge-normal traction σ and edge-tangent traction τ on a curved quadratic edge. (*b*) The $\xi = 1$ face of the linear solid element of Fig. 5.7.1*a*.

and, from Eq. 5.3.6,

$$dx = x,_\xi \, d\xi = J_{11} \, d\xi, \qquad dy = y,_\xi \, d\xi = J_{12} \, d\xi \tag{5.8.4}$$

So the x and y components of nodal load are

$$\begin{aligned} (r_i)_x &= \int_{-1}^{1} \bar{N}_i(\tau J_{11} - \sigma J_{12}) t \, d\xi \\ (r_i)_y &= \int_{-1}^{1} \bar{N}_i(\sigma J_{11} + \tau J_{12}) t \, d\xi \end{aligned} \tag{5.8.5}$$

where i is 4, 7, or 3, and $\bar{N}_i$, t, and the J_{ij} are evaluated on $\eta = 1$.

Finally, consider the three-dimensional problem (Fig. 5.8.2*b*). To illustrate the method, we consider only a prescribed traction σ normal to the $\xi = 1$ face (which may be warped). This face is defined by nodes 5 through 8. The main task is to evaluate the $\{\Phi\} \, dS$ term in Eq. 4.3.5, where dS is an element of area on the face. For this evaluation we need direction cosines of a normal to the face. So we first form the face-tangent vectors $\mathbf{V}_1$ and $\mathbf{V}_2$.

$$\begin{aligned} \mathbf{V}_1 &= (J_{21}\mathbf{i} + J_{22}\mathbf{j} + J_{23}\mathbf{k}) \, d\eta \\ \mathbf{V}_2 &= (J_{31}\mathbf{i} + J_{32}\mathbf{j} + J_{33}\mathbf{k}) \, d\zeta \end{aligned} \tag{5.8.6}$$

where $\mathbf{i}$, $\mathbf{j}$, and $\mathbf{k}$ are the usual unit vectors in the global x, y, and z directions and

$$J_{21} = x,_\eta, \qquad J_{22} = y,_\eta, \; \cdots, \; J_{33} = z,_\zeta \tag{5.8.7}$$

as in Eq. 5.7.1. A unit vector normal to the face contains the direction cosines ℓ, m, and n.

$$\ell\mathbf{i} + m\mathbf{j} + n\mathbf{k} = \frac{\mathbf{V}_1 \times \mathbf{V}_2}{|\mathbf{V}_1 \times \mathbf{V}_2|} = \frac{\mathbf{V}_1 \times \mathbf{V}_2}{dS} \tag{5.8.8}$$

Thus $\ell = (J_{22}J_{33} - J_{23}J_{32})\, d\eta\, d\zeta/dS$, and so on. Now $\{\Phi\}\, dS = \{\ell\ m\ n\}\, \sigma\, dS$, so the x, y, and z components of load at node i are

$$\begin{Bmatrix} r_{xi} \\ r_{yi} \\ r_{zi} \end{Bmatrix} = \int_{-1}^{1}\int_{-1}^{1} \bar{N}_i \sigma \begin{Bmatrix} J_{22}J_{33} - J_{23}J_{32} \\ J_{23}J_{31} - J_{21}J_{33} \\ J_{21}J_{32} - J_{22}J_{31} \end{Bmatrix} d\eta\, d\zeta \tag{5.8.9}$$

where $\bar{N}_i$ and all the J_{ij} are evaluated at $\xi = 1$, and i ranges over all nodes on the face $\xi = 1$ ($i = 5, 6, 7, 8$ for the linear face in Fig. 5.8.2*b*).

5.9 THE VALIDITY OF ISOPARAMETRIC ELEMENTS

Isoparametric elements are based on continuous fields. They are invariant because they use the intrinsic coordinates $\xi\eta\zeta$. It remains to show that they meet other requirements of Sections 4.8 and 4.9: compatibility, rigid-body modes, and constant strain states.

To demonstrate compatibility, we note that along a common edge, such as *ABC* in Fig. 5.9.1*a*, adjacent elements display the same edge-tangent coordinate: $\eta_1 = \eta_2$. Also, along the common edge the N_i of the two elements are identical functions of η and have as coefficients the same nodal d.o.f. The same argument shows that shared nodes define the same interelement boundary curve, regardless of which element we examine, so elements also match *before* they are strained.

Isoparametric elements discussed in this chapter have no derivative d.o.f. and so have only C_0 compatibility (Section 4.2). They are suited to problems for which $m = 1$ in Requirement A of Section 4.9. In stress analysis terms, Requirement A will be satisfied for $m = 1$ if the element can display rigid-body modes and constant strain states. We now show that Requirement A is indeed satisfied.

Consider the displacement fields

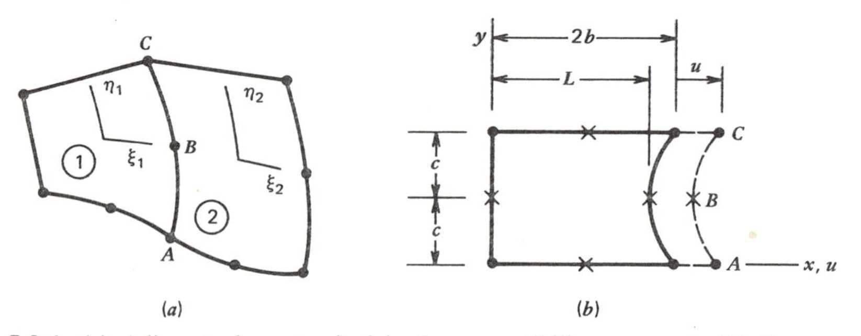

Figure 5.9.1. (*a*) Adjacent elements cited in the compatibility argument. (*b*) Superparametric element, with displacement *u* prescribed at nodes *A* and *C*.

$$u = a_1 + a_2x + a_3y + a_4z$$
$$v = a_5 + a_6x + a_7y + a_8z \tag{5.9.1}$$
$$w = a_9 + a_{10}x + a_{11}y + a_{12}z$$

By suitable choice of the constants a_i, we can produce any required state of constant strain or rigid-body translation or rotation. For example, if all a_i are zero except for $a_9 \neq 0$ and $a_3 = -a_6$, we have the rigid-body motion of z-direction translation and rotation about the z-axis. Therefore, because u, v, and w fields all use the same shape functions N_i, it is sufficient to show that the first of Eqs. 5.9.1 prevails within the element when nodal d.o.f. are assigned values that agree with the field [5.7]. Accordingly, we prescribe the displacement of node i as

$$u_i = a_1 + a_2x_i + a_3y_i + a_4z_i \tag{5.9.2}$$

where i ranges over all nodes of the element. By definition, $u = \Sigma N_iu_i$, so

$$u = a_1\Sigma N_i + a_2\Sigma N_ix_i + a_3\Sigma N_iy_i + a_4\Sigma N_iz_i \tag{5.9.3}$$

But, also by definition,

$$x = \Sigma N_ix_i, \qquad y = \Sigma N_iy_i, \qquad z = \Sigma N_iz_i \tag{5.9.4}$$

Therefore, if $\Sigma N_i = 1$, Eq. 5.9.3 reduces to Eq. 5.9.1, and we have demonstrated constant strain capability.

The relation $\Sigma N_i = 1$ is clearly true at nodes; $N_i = 1$ and $N_j = 0$ for $j \neq i$ is a condition satisfied by the interpolation scheme. If the element is valid, it is also true that $\Sigma N_i = 1$ for arbitrary values of ξ, η, and ζ in the range from -1 to $+1$. Accordingly, a valid shape function subroutine should yield

$$\Sigma N_i = 1, \qquad \Sigma N_{i,\xi} = \Sigma N_{i,\eta} = \Sigma N_{i,\zeta} = 0 \tag{5.9.5}$$

when assigned the ξ, η, and ζ coordinates of any point within the element. These are useful tests for errors in derivation and programming. However, Eqs. 5.9.5 do not apply to nonisoparametric elements, where rotational d.o.f. may be present (see, for example, Eqs. 4.2.5).

The foregoing arguments indicate that an isoparametric element is valid in the patch test sense, but they say nothing about accuracy in a coarse mesh, convergence rate, or how the performance of an element may decline as its shape is made less rectangular.

Isoparametric concepts can be generalized [5.9]. Consider the displacement and shape expansions

$$\{u \ v \ w\} = \Sigma N_i \{u_i \ v_i \ w_i\}$$
$$\{x \ y \ z\} = \Sigma \tilde{N}_j \{x_j \ y_j \ z_j\} \tag{5.9.6}$$

For an isoparametric element, $i = j$ and $N_i \equiv \tilde{N}_j$. For a *subparametric* element, $j < i$ and the N's and $\tilde{N}$'s differ. For a *superparametric* element, $i < j$ and the N's and $\tilde{N}$'s differ.

The plane quadratic element of Fig. 5.6.1 is subparametric if its sides are straight. Then its shape is completely defined by the corner nodes, but edges can still displace quadratically.

The foregoing proof remains valid for subparametric elements if side nodes are at *midsides*. Then in Eq. 5.9.3 we have, for example, $x_5 = (x_1 + x_2)/2$, and with Table 5.6.1 we find that $\Sigma\, N_i x_i$ reduces to $\Sigma\, \tilde{N}_j x_j$, where i and j run from 1 to 8 and 1 to 4, respectively.

Superparametric elements are usually not valid. Consider the element of Fig. 5.9.1*b*, for which $i = 1, 2, 3, 4$ (corner nodes only) and $j = 1, 2, \ldots, 8$ (corner and side nodes). Since edge displacements contain no quadratic terms, if $u_A = u_C = u$, then $u_B = u$ also. Therefore $\epsilon_x = u/2b$ along $y = 0$ and along $y = 2c$, and $\epsilon_x = u/L$ along $y = c$. Unless $L = 2b$ the element does not display constant ϵ_x and is therefore not valid.

If certain restrictions are imposed, valid superparametric elements for plates and shells can be generated (Chapters 9, 10).

5.10 APPROPRIATE ORDER OF QUADRATURE

The best order of quadrature is usually decided after numerical testing, since the profits and pitfalls of a particular rule are hard to forsee. Often it is best to use as low an order as possible without precipitating a numerical disaster. Here we examine the lower limits placed on quadrature rules and suggest rules appropriate to various elements. Theoretical arguments about the interaction of quadrature order and convergence rate are left to other references [4.5, 5.10].

A low-order rule is desirable for two reasons. First, fewer points result in lower computation cost. The expense of numerical integration is proportional to the number of integration points times the square of the number of element d.o.f. Second, a low-order rule tends to soften an element, thus countering the overly stiff behavior associated with an assumed displacement field. Softening comes about when some high-order terms that would otherwise contribute to strain energy happen to vanish at the Gauss points of a low-order rule. In other words, with fewer sampling points, some of the more complicated deformation modes offer less resistance. (However, it is *possible* that certain nodes will offer more resistance.)

There is a lower limit on the number of sampling points because the quadrature rule must be competent enough to integrate the element volume exactly. The argument begins with the observation that as a mesh is refined and a constant strain condition comes to prevail in each element, the strain energy expression assumes the form

$$\iint CJt\, d\xi\, d\eta \qquad \text{or} \qquad \iiint CJ\, d\xi\, d\eta\, d\zeta \tag{5.10.1}$$

in plane and solid elements, respectively, where $C = \{\boldsymbol{\epsilon}\}^T[\mathbf{E}]\{\boldsymbol{\epsilon}\}/2 =$ constant. Each integral is C times the element volume. Accordingly, if the requirements of Sections 4.8 and 4.9 are met, the strain energy of a structure is correctly assessed if the volume of the structure is correctly assessed.

Thus, for elements of *finite* size, we choose a rule that exactly integrates Jt (plane case) or J (solid case). For a plane linear element of constant thickness, Jt is linear in ξ and in η, so only one Gauss point is needed. For a plane quadratic element of constant thickness, Jt contains ξ^3 and η^3, so a four-point rule is needed. The linear solid requires a 2-by-2-by-2 rule (eight points).

However, with indefinitely repeated subdivision, elements in a mesh become straight-sided parallelograms and element thicknesses become constant. Thus J and t cease to be functions of the coordinates and, *in the limit,* a single Gauss point yields element volume. But we cannot accept a one-point rule because it produces zero-energy deformation modes.

A *zero-energy deformation mode* (or *kinematic mode* or *mechanism*) arises when a pattern of nodal d.o.f. produces a strain field that is zero at all quadrature points. Consider, for example, Fig. 5.10.1*a*. Bending about either axis produces no strain at $\xi = \eta = 0$. The elements resist only constant strain states and have no resistance at all to loads that tend to produce "hourglassing." Zero-energy modes can exist in combination with one another and in combination with acceptable modes. Elements need not be square: nodal d.o.f. in elements of arbitrary shape can have values that produce zero strain at the sampling points. If a structure has one or more zero-energy modes, its stiffness matrix $[\mathbf{K}]$ is singular, and a solution for static d.o.f. $\{\mathbf{D}\}$ is not possible.

To avoid a singular $[\mathbf{K}]$ for a mesh of plane linear elements, we can use a 2-by-2 rule in at least one element. Or, if the physical situation permits, we can provide external restraint, say by fixing both d.o.f. at nodes 1, 2, and 3 in Fig. 5.10.1*a*.

In Fig. 5.10.1*b*, the zero-energy deformation mode is

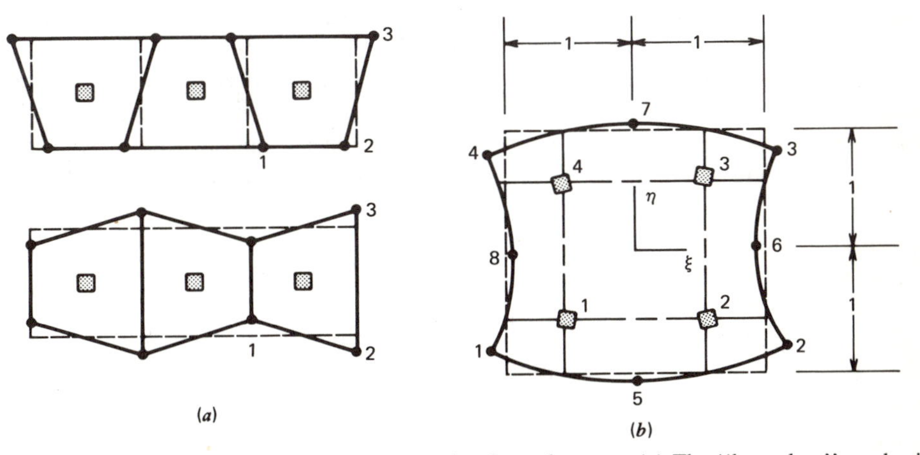

Figure 5.10.1. Zero-energy deformation modes in plane elements. (*a*) The "hourglass" modes in a linear element integrated by a one-point rule. (*b*) A quadratic element integrated by a four-point rule. Differential elements at the Gauss points rotate but do not strain.

$$
\begin{aligned}
-u_1 = u_2 = u_3 = -u_4 = v_1 = v_2 = -v_3 = -v_4 = 2C \\
-v_5 = -u_6 = v_7 = u_8 = C \\
u_5 = v_6 = u_7 = v_8 = 0
\end{aligned}
\tag{5.10.2}
$$

where C is an arbitrary constant, either positive or negative. Fortunately, if adjacent elements are connected, they cannot both have this mode. So we can use a 2-by-2 rule with confidence that $[\mathbf{K}]$ for a structure of two elements or more will not be singular. (However, a stack of quadratic *solid* elements can deform into a prismatic shape whose cross section is as shown by Fig. 5.10.1*b*.)

An element has zero-energy deformation modes if its order N less its rank R exceeds the number of rigid-body motions β; that is, if $N - R > \beta$. It is tempting to say that $R = nr$, where n = the number of sampling points and r = rank of $[\mathbf{E}]$ (the rank and order of $[\mathbf{E}]$ are the same unless special adjustments are made, as in Eq. 10.6.1). But $R < nr$ if a 1-by-5 rule is applied to the quadratic plane element. And hourglass modes are possible in the linear *solid* element if integrated by a six-point rule that has a sampling point at the middle of each face [5.11]. The safest test for R is the eigenvalue test (Section 15.2).

However, a structure can be without zero-energy deformation modes, yet be too soft in resisting them. For example, consider the sketch for Problem 5.32 and imagine that the elements are those of Fig. 4.2.2*b*, integrated by a 2-by-2 rule. Also, let all five nodes along BC be fully fixed to suppress a zero-energy deformation mode of the structure (Problem 7.20). For the bending load shown, results will be good. But if, instead, there is a horizontal load at point A, the restraint of the left end will not be felt enough near the right end. A near-mechanism develops that masks the desired results near the right end [5.24]. This difficulty does not arise if a 3-by-3 rule is used.

What rule *should* we use? For plane elements, either linear or quadratic, the 2-by-2 rule is usually best, except for a markedly nonrectangular or elongated quadratic element, where a 3-by-3 rule may give better accuracy [2.1]. A 2-by-2-by-2 rule seems best for the *linear* solid element but, if applied to the quadratic solid element, it admits zero-energy deformation modes and can misrepresent element volume. But we need not use a 3-by-3-by-3 rule: there is a 14-point rule that is cheaper and also more accurate for extremely thin curved elements [5.11–5.13]. The 14-point rule has the form

$$
\begin{aligned}
&\int_{-1}^{1}\int_{-1}^{1}\int_{-1}^{1} f(\xi, \eta, \zeta)\, d\xi\, d\eta\, d\zeta \\
&\qquad = B_6[f(-b,0,0) + f(b,0,0) + f(0,-b,0) + \ldots \text{(6 terms)}] \\
&\qquad\quad + C_8[f(-c,-c,-c) + f(c,-c,-c) + \cdots \text{(8 terms)}]
\end{aligned}
\tag{5.10.3}
$$

where

$$
\begin{aligned}
B_6 &= 0.88642\ 65927\ 97784, & b &= 0.79582\ 24257\ 54222 \\
C_8 &= 0.33518\ 00554\ 01662, & c &= 0.75878\ 69106\ 39328
\end{aligned}
$$

As expected, $6B_6 + 8C_8 = 8$ (why?). A 13-point rule is also available, but it is unpalatable because the sampling points are not symmetrically located.

5.11 REMARKS ON STRESS COMPUTATION

Stresses follow from Eq. 4.10.1, repeated here as Eq. 5.11.1.

$$\{\boldsymbol{\sigma}\} = [\mathbf{E}]\big([\mathbf{B}]\{\mathbf{d}\} - \{\boldsymbol{\epsilon}_0\}\big) + \{\boldsymbol{\sigma}_0\} \tag{5.11.1}$$

where, in isoparametric elements, $[\mathbf{B}]$ is a function of the natural coordinates, but $\{\boldsymbol{\sigma}\}$ contains stresses in the global Cartesian system. Where should stresses be calculated? Often they are least accurate at element corners, more accurate at midsides (or midface in solid elements), and most accurate at certain interior points. These interior points can be used to define a stress field that can be extrapolated to yield stresses at element boundaries. Here we describe the calculations. Theoretical arguments appear in Refs. 5.14 to 5.17.

Points where stresses are most accurate can be located as follows [5.17]. Impose on the element a complete polynomial displacement field $\{\mathbf{f}'\}$ whose order is one higher than the highest *complete* polynomial in the actual element field $\{\mathbf{f}\}$ (for example, impose a complete cubic field on the quadratic element). Let $\{\mathbf{d}'\}$ be nodal d.o.f. associated with $\{\mathbf{f}'\}$. Now seek locations in the element where stresses calculated from $\{\mathbf{f}'\}$ are identical to stresses calculated from $\{\mathbf{f}\}$ when nodal d.o.f. $\{\mathbf{d}'\}$ are imposed. These locations are found to be Gauss point locations. At these points the stresses may be almost as accurate as nodal displacements $\{\mathbf{d}\}$. (Effectively, optimal stress points define a least-squares fit to a field of higher degree. It is reasonable that the stress field is of lower degree than the displacement field because stresses are proportional to displacement derivatives.)

Table 5.11.1 lists optimal points. Figure 5.11.1*a* shows an example of their benefit [5.14]. We see that stresses away from the optimal points define a parabola that is grossly in error. But optimal points define a straight line that is essentially exact and is a least-squares fit to the parabola.

Table 5.11.1 should be interpreted to mean that stresses are most accurate at the *same* Gauss points used to generate $[\mathbf{k}]$.

To extrapolate stresses from Gauss points, we do as follows. Consider first a one-dimensional situation, the upper part of Fig. 5.11.1*a*. In the span $-P < \xi < P$ we interpolate linearly between the known stresses σ_1 and σ_2 at stations 1 and 2.

$$\sigma = \left\lfloor \frac{1-s}{2} \quad \frac{1+s}{2} \right\rfloor \{\sigma_1 \ \sigma_2\} \tag{5.11.2}$$

where s is the natural coordinate $s = \xi/P$, so that $s = -1$ at $\xi = -P$ and $s = +1$ at $\xi = +P$. Again, $P = 0.57735\ldots$. To extrapolate to points A and B, at $\xi = \pm 1$ we set $s = \pm 1/P$. Thus, with $\sigma = \sigma_A$ and then $\sigma = \sigma_B$ in Eq. 5.11.2,

TABLE 5.11.1. LOCATION OF OPTIMAL POINTS FOR STRESS CALCULATION. $P = 0.57735\ 02691\ 89626$

Element	Locations: Gauss Rule and/or Coordinates
Beam (Fig. 4.4.1*b*)	2 point ($\pm PL/2$ from center)
Linear plane (Fig. 5.3.1*c*)	1 point ($\xi = \eta = 0$)
Quadratic Lagrange (Fig. 4.2.2*b*)	2×2 ($\xi = \pm P$, $\eta = \pm P$)
Quadratic plane (Fig. 5.6.1*a*)	2×2 ($\xi = \pm P$, $\eta = \pm P$)
Cubic plane (Fig. 5.6.1*b*)	3×3 (9 points; see Table 5.4.1)
Linear solid (Fig. 5.7.1*a*)	1 point ($\xi = \eta = \zeta = 0$)
Quadratic solid (Fig. 5.7.1*b*)	$2 \times 2 \times 2$ ($\xi = \pm P$, $\eta = \pm P$, $\zeta = \pm P$)

$$\begin{Bmatrix} \sigma_A \\ \sigma_B \end{Bmatrix} = \frac{1}{2} \begin{bmatrix} 1 + \dfrac{1}{P} & 1 - \dfrac{1}{P} \\ 1 - \dfrac{1}{P} & 1 + \dfrac{1}{P} \end{bmatrix} \begin{Bmatrix} \sigma_1 \\ \sigma_2 \end{Bmatrix} \tag{5.11.3}$$

Stresses σ_A, σ_B, σ_1, and σ_2 each represent one stress component (σ_x, σ_y, or τ_{xy}).

Extrapolation in Fig. 5.11.1*b* is based on four Gauss point values. That the element has eight nodes does not matter. We use the linear shape functions of Eq. 5.3.3 and insert ξ and η values of the location desired, be it inside or outside the quadrilateral defined by the Gauss points. For example, at point A, $\xi = \eta = -1/P$. So, from Eqs. 5.3.3,

$$\sigma_A = \tfrac{1}{4} \lfloor a^2\ ab\ b^2\ ab \rfloor \{\sigma_1\ \sigma_2\ \sigma_3\ \sigma_4\} \tag{5.11.4}$$

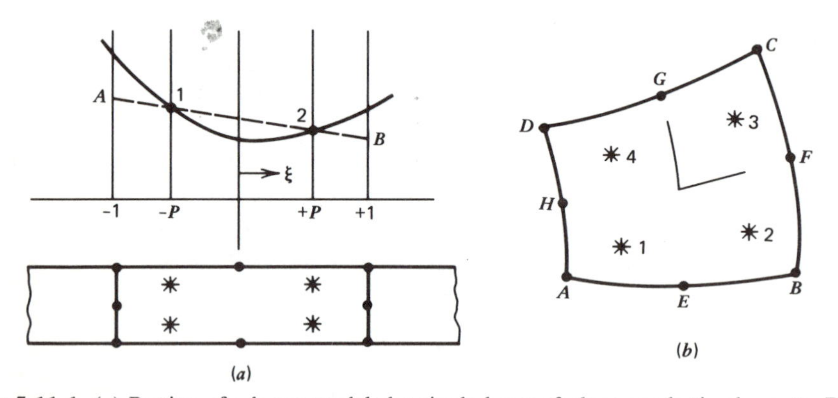

Figure 5.11.1. (*a*) Portion of a beam modeled a single layer of plane quadratic elements. Dashed line: correct transverse shear stress under uniformly distributed load. Solid line: computed shear stress. Optimal (Gauss) points are * symbols, at $\xi = \pm P = \pm\ 0.57735\ldots$. (*b*) Identification of Gauss points and nodes for extrapolation of stresses.

where $a = 1 + 1/P$ and $b = 1 - 1/P$. At point E, $\xi = 0$ and $\eta = -1/P$, so

$$\sigma_E = \tfrac{1}{4} \lfloor a\ a\ b\ b \rfloor \{\sigma_1\ \sigma_2\ \sigma_3\ \sigma_4\} \tag{5.11.5}$$

Note that the Gauss points in Fig. 5.11.1*b* are numbered in the same order as are the nodes in Section 5.3.

The solid element is similarly treated. Extrapolation from eight Gauss points to eight corners yields

$$\{\boldsymbol{\sigma}_{A-H}\} = [\mathbf{Q}]\,\{\boldsymbol{\sigma}_{1-8}\} \tag{5.11.6}$$

where $[\mathbf{Q}]$ contains only four different coefficients (Problem 5.37).

In elements with incompatible modes (Section 7.7), extrapolated stresses at the corners may be inaccurate. Midside or midface stresses are preferred, either extrapolated to these places or calculated there directly.

5.12 CONCLUDING REMARKS. EXAMPLES. ERRORS

Merits of numerically integrated isoparametric elements are now apparent. Element edges can be curved and can have side nodes if necessary. Triangular forms are available (Sections 7.9, 7.10). Thicknesses can vary. So can material properties: like thicknesses, material properties at Gauss points can be interpolated from nodal values. The different element varieties are formulated and programmed in the same way, which makes them easier to learn and remember.

Elements can also be formulated from a displacement field that is a least-squares fit to the nodal d.o.f. This method sometimes yields the same $[\mathbf{k}]$ as given by a reduced order of quadrature. Some triangular elements benefit more from this least-squares method than from a reduced order of quadrature [5.10, 5.14].

Numerical examples of element behavior appear in Tables 5.12.1 and 5.12.2 [5.18–5.20, 5.22]. Stresses in Fig. 5.12.1*d* are as accurate as the displacements. Stresses in Figs. 5.12.1*e* to 5.12.1*g* are not available in the references cited.

TABLE 5.12.1. Tip deflection and horizontal stress in the linear-element cantilever beams of Figs. 5.12.1*a* to 5.12.1*c*. Results are expressed as the ratio of computed result to exact result. Integration is by four-point Gauss quadrature, and stresses are calculated directly at B and C.

Case	Vertical Deflection at A	σ_x at B	σ_x at C
Fig. 5.12.1*a*	0.091	0.048	0.096
Fig. 5.12.1*b*	0.682	0.727	0.727
Fig. 5.12.1*c*	0.494	0.604	0.301

TABLE 5.12.2. TIP DEFLECTION IN THE QUADRATIC-ELEMENT BEAMS OF FIGS. 5.12.1*d* TO 5.12.1*g*. RESULTS ARE EXPRESSED AS THE RATIO OF COMPUTED RESULT TO EXACT RESULT. QUADRILATERAL ELEMENTS ARE INTEGRATED BY 2×2 AND 3×3 GAUSS QUADRATURE. TRIANGULAR ELEMENTS ARE INTEGRATED EXACTLY.

(*d*), 2×2	(*d*), 3×3	(*e*), 2×2	(*e*), 3×3	(*f*)	(*g*)
0.968	0.930	0.366	0.161	0.796	0.801

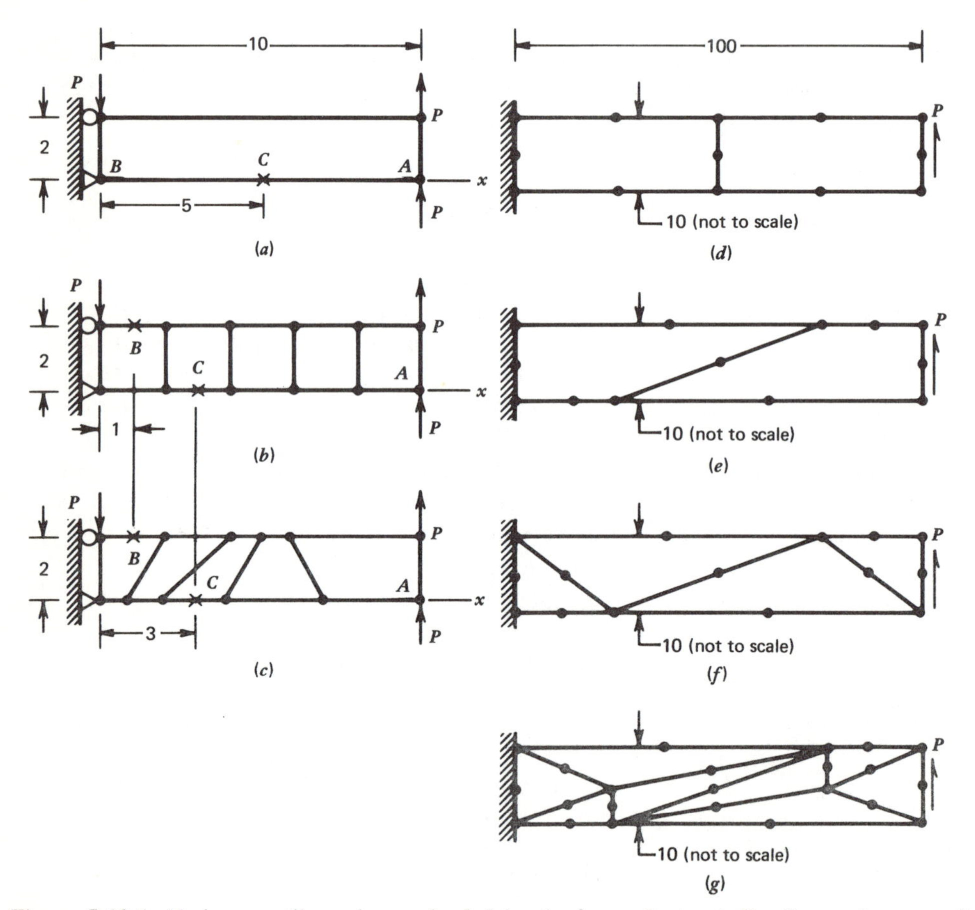

Figure 5.12.1. Various cantilever beams loaded by tip forces P. (*a–c*) The linear elements of Section 5.3, where (*a*) is a one-element case and (*b–c*) are five-element cases. (*d–e*) The quadratic elements of Fig. 5.6.1*a*. (*f–g*) The quadratic elements of Fig. 7.10.1*a*.

Tables 5.12.1 and 5.12.2 show that an elongated linear element is particularly poor (for reasons explained in Section 7.6) but that the quadratic element behaves well even when elongated. More integration points stiffen the elements, as expected. Quadrilateral elements, either linear or quadratic, tend to stiffen and loose accuracy if their form becomes a general quadrilateral rather than a rectangle.

Moving the side nodes of a quadratic or a cubic element, either toward a corner or so that the edge becomes curved, is particularly detrimental (see Ref. 9.9 and the bibliography of Ref. 5.19). As an element departs from a rectangle with uniformly spaced nodes, it tends to lose the ability to represent quadratic and higher polynomials in *xyz* coordinates. That the displacement field may be a complete quadratic or cubic polynomial in $\xi\eta\zeta$ coordinates does not guarantee completeness in *xyz* coordinates. The argument of Section 5.9 shows only that a complete *linear* polynomial in *xyz* will remain. Accordingly, accuracy is improved if elements *within* a mesh are regular, even if some edges are curved to fit the structure boundary.

Clearly, the triangular elements of Section 7.10 can be more accurate than quadrilaterals, and they may be no more expensive to use if element sides are straight.

If adapted to the problem at hand, element distortion can improve answers. But we do not know how to do it when we lay out the mesh. The safest rule is to avoid unnecessary irregularity: keep interior corner angles θ near 90°, and always $0 < \theta < 180°$. Figure 5.12.2 shows some elements likely to encourage poor answers in the immediate neighborhood of the overdistortion. In Fig. 5.12.2*a* the element is made to fold over on itself. In Fig. 5.12.2*c* some of the element is made to fall outside the intended rectangle.

Tests for coding errors are mentioned below Eq. 5.9.5. Another test is to check derivatives by entering the shape function routine at two closely spaced points, for example, at ξ_1 and ξ_2 on the ξ-axis, and see if $N_{,\xi} = (N_2 - N_1)/(\xi_2 - \xi_1)$, where N is any of the shape functions. Other error tests include the patch test and those noted in Section 14.5.

A test for excessive shape distortion is to see if the coordinates of a side node are "close" to the average of the coordinates of the two adjacent corners. The Jacobian J can also be checked. It is constant if the element is a parallelepiped with evenly spaced side nodes, but it varies if other distortions exist. It is positive if *xyz* and $\xi\eta\zeta$ are both right-handed systems, as in Fig. 5.2.1. It becomes zero or negative in the neighborhood of an excessive distortion. We can compute a distortion parameter DP as

Figure 5.12.2. Badly shaped quadratic elements.

$$DP = 4\left(\frac{\min J}{\text{area}}\right) \qquad \text{or} \qquad DP = 8\left(\frac{\min J}{\text{volume}}\right) \tag{5.12.1}$$

for plane and solid elements, respectively. For zero distortion from a parallelogram or parallelepiped shape, $DP = 1.0$. If DP is 0.2 or less at any sampling point, distortion is considered excessive [5.21].

Yet $DP = 1$ for some distortions, such as elongated rectangles or parallelepipeds, and for hourglass or barrel shapes. Elongated plane quadratic elements, if point loaded and integrated with a 2-by-2 rule, can display an oscillating displacement pattern due to activation of a low-energy deformation mode. A 3-by-3 rule is then preferred [5.23].

PROBLEMS

5.1 (a) Start with $u = \lfloor 1 \;\; \xi \;\; \xi^2 \rfloor \{a_1 \; a_2 \; a_3\}$ and derive $\lfloor \mathbf{N} \rfloor$ of Eq. 5.2.9.

(b) The sketch shows how three shape functions vary over the range $-1 < \xi < 1$. Can you combine them to get $\lfloor \mathbf{N} \rfloor$ in Eq. 5.2.9 directly, without working through matrix $[\mathbf{A}]$?

(c) Show that if x_3 is at the middle of a bar of length L, then $J = L/2$ in Eq. 5.2.10. Let node 1 have the arbitrary value x_1.

(d) Find $[\mathbf{k}]$ if $x_1 = 0$, $x_2 = L$, and $x_3 = L/2$ in Fig. 5.2.1*b*. Let A and E be constant.

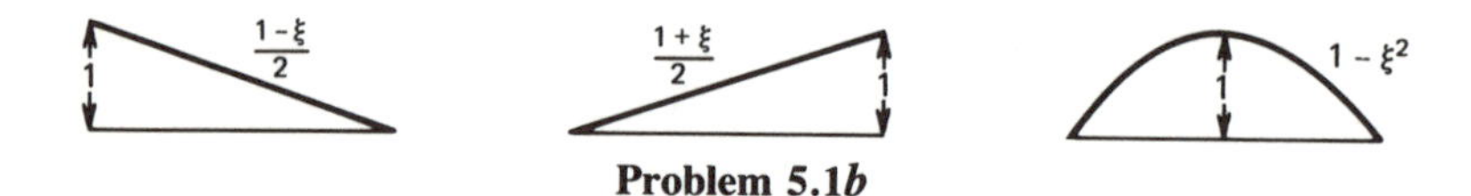

Problem 5.1*b*

5.2 With reference to Fig. 5.3.1*c*, let $x = \lfloor 1 \;\; \xi \;\; \eta \;\; \xi\eta \rfloor \{a_1 \; a_2 \; a_3 \; a_4\}$.

(a) Hence, write $[\mathbf{A}]$ in the relation $\{x_1 \; x_2 \; x_3 \; x_4\} = [\mathbf{A}] \{a_1 \; a_2 \; a_3 \; a_4\}$.

(b) By inspection of Eqs. 5.3.3, write $[\mathbf{A}]^{-1}$ in the relation $x = \lfloor 1 \;\; \xi \;\; \eta \;\; \xi\eta \rfloor [\mathbf{A}]^{-1} \{x_1 \; x_2 \; x_3 \; x_4\}$.

(c) Check your answers by seeing if $[\mathbf{A}] [\mathbf{A}]^{-1} = [\mathbf{I}]$.

5.3 Consider a square linear element, located and oriented such that $\xi = y/2$ and $\eta = -x/2$. Show that the Jacobian matrix provides the proper coordinate transformation. That is, are dx, dy, $d\xi$, and $d\eta$ properly related?

5.4 Use one-, two-, and three-point Gauss quadrature to integrate each of the following functions. Compare these answers with the exact answers.

(a) $\phi = \cos \pi x/2$, for $-1 < x < 1$.

(b) $\phi = x^2 + x^3$, for $-1 < x < 1$.

(c) $\phi = 1/x$, for $1 < x < 7$.

5.5 Write an expression for I, analogous to Eq. 5.4.6, for (*a*) a 2-by-3 quadrature rule, and (*b*) a 3-by-4 quadrature rule.

5.6 For any of the quadrature rules in Table 5.4.1, weights W_i sum to 2 in one dimension, weight products W_iW_j sum to 4 in two dimensions, and weight products $W_iW_jW_k$ sum to 8 in three dimensions. Why? Check this behavior in Eq. 5.4.6.

5.7 With reference to Figs. 5.5.1 and 5.5.2:
(a) Add appropriate COMMON, DIMENSION, and DATA statements to subroutine QUAD4.
(b) QUAD4 would be more efficient if DV were removed from statement 60 and placed elsewhere. Where? And why?
(c) Let $\{\mathbf{d}\} = \{u_1\ u_2\ u_3\ u_4\ v_1\ v_2\ v_3\ v_4.\}$ Revise the subroutines to accommodate this new ordering. Exploit matrix sparsity where possible.

5.8 Use Figs. 5.5.1 and 5.5.2 as a guide and write subroutines that use Eqs. 5.2.10 to generate $[\mathbf{k}]$ for the bar of Fig. 5.2.1*b* by two-point Gauss quadrature. Let area A be linearly interpolated from known values at nodes 1 and 2. Node 3 is not necessarily at the midpoint.

5.9 Use Figs. 5.5.1 and 5.5.2 as a guide and write subroutines that generate $[\mathbf{k}]$ of Eq. 4.4.5 by two-point Gauss quadrature. But allow for the possibility that I varies linearly along the beam element, with values I_1 and I_2 known at nodes 1 and 2.

5.10 Since $[\mathbf{E}]$ is positive definite, it can be decomposed to yield $[\mathbf{E}] = [\mathbf{U}]^T[\mathbf{U}]$ (Eq. 2.12.1). Then

$$[\mathbf{B}]^T[\mathbf{E}][\mathbf{B}] = ([\mathbf{U}][\mathbf{B}])^T([\mathbf{U}][\mathbf{B}])$$

and

$$[\mathbf{B}]^T[\mathbf{E}]\{\boldsymbol{\epsilon}_0\} = ([\mathbf{U}][\mathbf{B}])^T([\mathbf{U}]\{\boldsymbol{\epsilon}_0\})$$

We can exploit the sparsity of $[\mathbf{U}]$ in forming the product $[\mathbf{U}][\mathbf{B}]$ and overwrite $[\mathbf{B}]$ with this product [5.6]. Revise Fig. 5.5.2 to use this formulation method. For simplicity, discard the body forces.

5.11 Revise Figs. 5.5.1 and 5.5.2 to accommodate:
(a) The plane quadratic element of Fig. 5.6.1*a*.
(b) A plane element with four to eight nodes.
(c) An initial stress vector $\{\boldsymbol{\sigma}_0\}$.
(d) Body forces that are functions of the coordinates.
(e) The linear solid element of Fig. 5.7.1*a*.

5.12 (a) Sketch a four-node linear element for which J is a function of ξ but not of η.
(b) Sketch an eight-node *rectangular* quadratic element for which J is a function of η but not of ξ.
(c) Repeat part (b) but make the element nonrectangular. Place side nodes halfway along the sides.

(d) Let *AABB* in Fig. 6.12.1 represent a single plane quadratic element with midside nodes and one 180° angle. If $\xi = \eta = -1$ at node 1 on edge *AA*, sketch the lines $\xi = -0.5$, 0.0, and 0.5 and the lines $\eta = -0.5$, 0.0, and 0.5.
(e) Repeat part (d) but let $\xi = \eta = -1$ at node 9 on edge *AA*.
(f) Repeat part (d) but let $\xi = \eta = -1$ at node 9 on edge *BB*.

5.13 Consider plane *rectangular* elements with evenly spaced side nodes.
(a) The quadratic type (Fig. 5.6.1*a*).
(b) The cubic type (Fig. 5.6.1*b*).
What powers of x and y appear in the polynomials for the strains ϵ_x, ϵ_y, and γ_{xy}? If used to model a cantilever beam by one layer of elements, for what types of tip and distributed loading can this model yield exact results (according to mechanics of materials beam theory)?

5.14 Consider an element whose shape functions are given by Table 5.6.1. Why can we not generate two straight edges by omitting node 1, as shown in the sketch?

5.15 A plane element having 16 external d.o.f. and 2 internal d.o.f. can be formed by combining eight constant strain triangles, as shown. What advantages or disadvantages might this element have in comparison with the plane quadratic isoparametric element?

5.16 What is bad about joining linear and quadratic elements as shown?

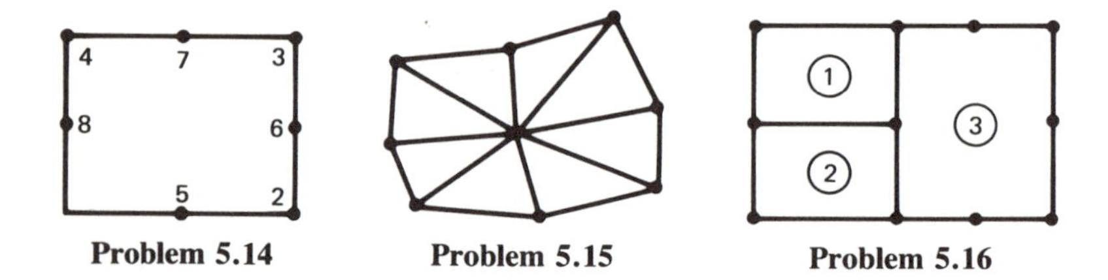

Problem 5.14 **Problem 5.15** **Problem 5.16**

5.17 Verify that nodal loads due to body force on the plane linear element are as stated in the second paragraph of Section 5.8.

5.18 Verify the nodal forces shown in Fig. 5.8.1*c*.

5.19 Let a traction σ_y on the top edge of the element in Fig. 5.8.1*b* vary linearly from σ_A at node 4 to σ_B at node 3. Find the consistent nodal forces $\{\mathbf{r}\}$.

5.20 Show that appropriate specialization reduces Eq. 5.8.5 to Eq. 5.8.1.

5.21 Consider a traction σ that acts normal to the $\zeta = -1$ surface of a linear solid element. Derive equations that correspond to Eqs. 5.8.6 to 5.8.9.

5.22 Consider two adjacent plane elements, as in Fig. 5.9.1*a*. Show from $\{\mathbf{f}\} = [\mathbf{N}]\{\mathbf{d}\}$ that the elements have identical displacements along the common edge. Consider (*a*) linear elements, (*b*) quadratic elements, and (*c*) cubic elements.

5.23 Consider the shape functions of Eqs. 5.2.9.
   (a) Show that at node i, $N_i = 1$ and $N_j = 0$ where $i \neq j$.
   (b) Show that $\Sigma N_i = 1$.
   (c) Show that $\Sigma N_{i,\xi} = \Sigma N_{i,\eta} = 0$.

5.24 Repeat Problem 5.23 for the N_i of Eqs. 5.3.3.

5.25 Repeat Problem 5.23 for the N_i of Table 5.6.1.

5.26 Repeat Problem 5.23 for the N_i of Eq. 5.7.4.

5.27 Let u_1, u_2, and u_3 be prescribed in Fig. 5.2.1*b*. Imagine that strain energy in the element is to be calculated and that AE is constant. (*a*) How many Gauss points are needed if $x_3 = L/2$? (*b*) How many if $x_3 \neq L/2$?

5.28 Verify that a 2-by-2-by-2 Gauss rule is needed to get the exact volume of a linear solid element.

5.29 If element thickness t can vary and is computed by shape-function interpolation from nodal values t_i, what quadrature rule is needed to get the exact volume of (*a*) the plane linear element, and (*b*) the plane quadratic element?

5.30 Let the following elements be rectangular, with side nodes evenly spaced and thicknesses constant. What quadrature rule is needed to get the exact *stiffness matrix*—that is, to integrate each k_{ij} exactly?
   (a) Plane linear element.
   (b) Plane quadratic element.
   (c) Solid linear element.
   (d) Solid quadratic element.

5.31 By substitution into $\{\epsilon\} = [\mathbf{B}]\{\mathbf{d}\}$, verify that the deformation modes of Fig. 5.10.1*a* produce zero strain at the Gauss point. Let elements be rectangular.

5.32 The cantilever beam is built of plane linear elements, each integrated by one-point quadrature.
   (a) Sketch the possible zero-energy deformation modes of the structure.
   (b) If you ask a computer program to solve for static displacements, what do you think will happen?
   (c) What support conditions at the left end will suppress zero-energy deformation modes? Will these supports always prevent trouble?
   (d) Is a zero-energy deformation mode possible if $[\mathbf{k}]$ of the standard beam element (Eq. 4.4.4) is integrated by one-point Gauss quadrature?

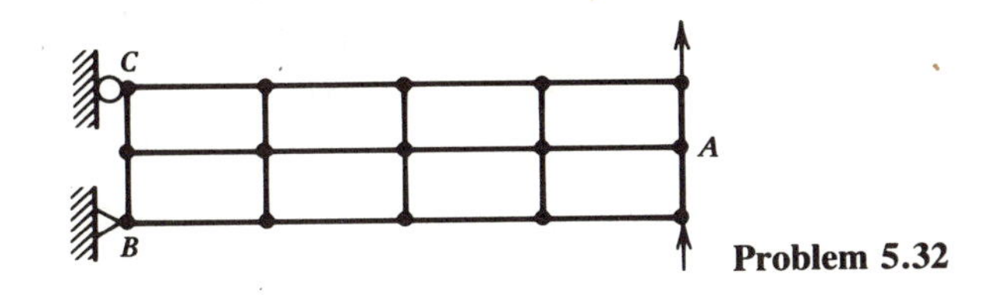

Problem 5.32

5.33 Imagine that we use a 4-by-4 Gauss rule for the cantilever beam of Fig. 5.12.1*d*. In comparison with the result in Table 5.12.2 for a 3-by-3 rule, what result do you expect, and why?

5.34 Let $u = (-1 + 3y^2)x$ and $v = (1 - 3x^2)y$ in a rectangular quadratic plane element, two units along a side and with midside nodes. The element center is at $x = y = 0$. Show that this field is a zero-energy deformation mode under a 2-by-2 Gauss rule. Sketch the deformed element.

5.35 Section 5.10 mentions a six-point quadrature rule for solid elements. Sketch the zero-energy deformation modes that this rule permits in a linear element. Can a *mesh* of solid elements suffer from this mode?

5.36 The rank of **[k]** for a plane quadratic element can be less than 13 if integrated by a 1-by-*m* rule, where *m* is *any* positive integer. Show this by sketching an appropriate deformation mode.

5.37 As in Eq. 5.11.6, stresses $\{\boldsymbol{\sigma}_N\}$ at nodes can be related to stresses $\{\boldsymbol{\sigma}_P\}$ at Gauss points by a matrix **[Q]**. Write **[Q]** for the following cases. Give numerical values of the Q_{ij}.

(a) Equation 5.11.6 as stated, for the linear solid [5.16].

(b) $\{\boldsymbol{\sigma}_N\}$ represents eight nodes and $\{\boldsymbol{\sigma}_P\}$ represents four Gauss points in the plane quadratic element.

5.38 In a plane linear element, stresses can be calculated directly at the nodes or extrapolated to the nodes from values at the four Gauss points. These methods give identical results if the element is a parallelogram. They differ if the element is of general shape. Why?

5.39 We can account for thickness variation in a plane element when **[k]** is generated. Should any adjustment for thickness variation be made during stress computation? Consider axially loaded tapered elements, both linear and quadratic, in answering the question.

5.40 Consider a quadratic bar element (Fig. 5.2.1*b*). How far can node 3 be shifted toward either end before J vanishes at either end?

5.41 Show that $DP = 1$ in Eq. 5.12.1 if:

(a) A plane element is a parallelogram.

(b) A solid element is a parallelepiped.

5.42 In addition to the formulations in Sections 4.5 and 7.9 to 7.12, triangular elements can be formulated as follows [7.50]. We adopt natural coordinates s and t such that a linear element is bounded by $0 \leq s \leq 1$ and $0 \leq t \leq 1$. Shape functions for a linear element are $N_1 = 1-s-t$, $N_2 = s$, and $N_3 = t$. For a quadratic element with midside nodes they are

$$N_1 = 1 - 3s - 3t + 2s^2 + 4st + 2t^2 \qquad N_4 = 4(s - s^2 - st)$$
$$N_2 = -s + 2s^2 \qquad N_5 = 4st$$
$$N_3 = -t + 2t^2 \qquad N_6 = 4(t - t^2 - st)$$

(a) Verify that these elements satisfy Eqs. 5.9.5.
(b) For the special case s parallel to x and t parallel to y, with node 2 at $x = b$ and node 3 at $y = c$, show that the linear element gives the same u and v as Eq. 4.5.7.
(c) Let $x = \Sigma N_i x_i$ and $y = \Sigma N_i y_i$, and let sides be straight. Derive the N_i of the linear element from the N_i of the quadratic element.
(d) What functions of t do the N_i become along edge 2-5-3 of the quadratic element? Do these N_i equal zero or one at the nodes, as expected?
(e) Outline the formation of [**k**] for the quadratic element, in the fashion of Section 5.3.
(f) What changes must be made in Figs. 5.5.1 and 5.5.2 to make them applicable to the quadratic element? Assume that an appropriate quadrature rule can be invoked.

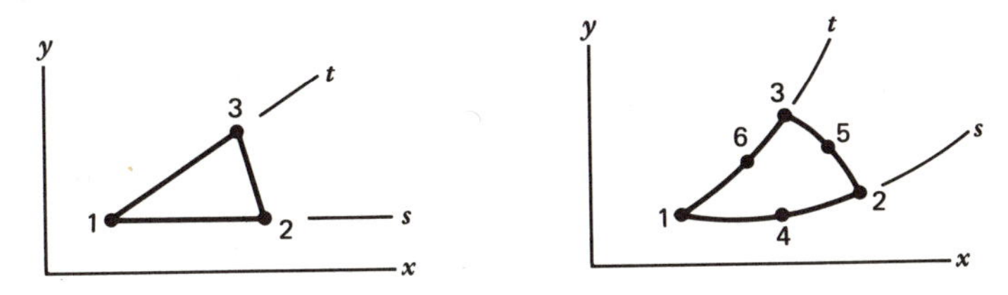

Problem 5.42

CHAPTER 6

COORDINATE TRANSFORMATION

6.1 INTRODUCTION

Common uses of coordinate transformation include the calculation of the elastic property matrix **[E]** and stiffness matrix **[k]** in one coordinate system from corresponding matrices **[E′]** and **[k′]** in another. Other uses include condensation techniques in structural dynamics and imposition of constraints. The reader can study sections of this chapter as the need arises.

The form $[\mathbf{Q}] = [\mathbf{T}]^T [\mathbf{Q}'] [\mathbf{T}]$ appears repeatedly. Here $[\mathbf{Q}']$ is the matrix to be transformed and $[\mathbf{T}]$ is the transformation matrix. The transformed matrix $[\mathbf{Q}]$ is symmetric if $[\mathbf{Q}']$ is symmetric. Matrix $[\mathbf{T}]$ may be rectangular or square. If square it may not be orthogonal. The specific form of $[\mathbf{T}]$ depends on the problem at hand.

One often has the option of taking $[\mathbf{Q}']$ as either an element matrix or a structure matrix. Computer programming is usually easiest when transformations are done *before* elements are assembled, even though we must then transform several small matrices instead of one large one.

Formal matrix multiplication to produce $[\mathbf{T}]^T [\mathbf{Q}'] [\mathbf{T}]$ is usually wasteful, since $[\mathbf{T}]$ is often sparse. Sparsity should be exploited, or terms in the product should be hand-calculated and then coded.

Caution. Transformations modify stiffness matrices. Errors and inconsistencies in stiffness matrices can seriously degrade accuracy (see Chapter 15). It matters little if errors in $[\mathbf{T}]$ produce only a slightly different geometry than intended. But damage is done if errors in $[\mathbf{T}]$ act to falsify equilibrium equations. A safe rule is that transformation matrices and constraint equations must be stated and manipulated with as much precision as is granted to stiffness coefficients K_{ij}.

6.2 TRANSFORMATION OF STRESS, STRAIN, AND MATERIAL PROPERTIES

Transformation of stresses $\{\boldsymbol{\sigma}\}$ and strains $\{\boldsymbol{\epsilon}\}$ in two dimensions leads to the familiar Mohr's circle calcuiations. In this section we consider the problem in three dimensions. We also consider the transformation of material properties $[\mathbf{E}]$. Transformations related to plate bending appear in Section 9.1.

Strain transformations are essentially transformations of displacement derivatives. For example, from Fig. 6.2.1, displacement u' in direction x' can be written in terms of displacements in the unprimed directions.

$$u' = \ell_1 u + m_1 v + n_1 w \tag{6.2.1}$$

By chain rule differentiation,

$$u'_{,x'} = u'_{,x} x_{,x'} + u'_{,y} y_{,x'} + u'_{,z} z_{,x'} \tag{6.2.2}$$

where

$$\begin{aligned} u'_{,x} &= \ell_1 u_{,x} + m_1 v_{,x} + n_1 w_{,x}, \quad \text{etc.}, \\ x_{,x'} &= \ell_1 \qquad y_{,x'} = m_1 \qquad z_{,x'} = n_1 \end{aligned} \tag{6.2.3}$$

By this process we find

$$\{u'_{,x'} \ u'_{,y'} \ u'_{,z'} \ \cdots \ w'_{,z'}\} = [\mathbf{T}]\{u_{,x} \ u_{,y} \ u_{,z} \ \cdots \ w_{,z}\} \tag{6.2.4}$$

where

$$\underset{9\times 9}{[\mathbf{T}]} = \begin{bmatrix} \ell_1 \mathbf{T}_c & m_1 \mathbf{T}_c & n_1 \mathbf{T}_c \\ \ell_2 \mathbf{T}_c & m_2 \mathbf{T}_c & n_2 \mathbf{T}_c \\ \ell_3 \mathbf{T}_c & m_3 \mathbf{T}_c & n_3 \mathbf{T}_c \end{bmatrix}, \qquad \underset{3\times 3}{[\mathbf{T}_c]} = \begin{bmatrix} \ell_1 & m_1 & n_1 \\ \ell_2 & m_2 & n_2 \\ \ell_3 & m_3 & n_3 \end{bmatrix} \tag{6.2.5}$$

Matrix $[\mathbf{T}]$ is orthogonal; that is, $[\mathbf{T}]^{-1} = [\mathbf{T}]^T$. Therefore the inverse relation of Eq. 6.2.4 involves simply $[\mathbf{T}]^T$.

From Eq. 6.2.4 and the strain-displacement relations (Eq. 1.4.4) we write the relation between strains in $x'y'z'$ coordinates and the *same* strains but expressed in xyz coordinates.

$$\{\epsilon'\} = [\mathbf{T}_\epsilon]\{\epsilon\}, \qquad \{\epsilon\} = [\mathbf{T}_\epsilon]^{-1}\{\epsilon'\} \tag{6.2.6}$$

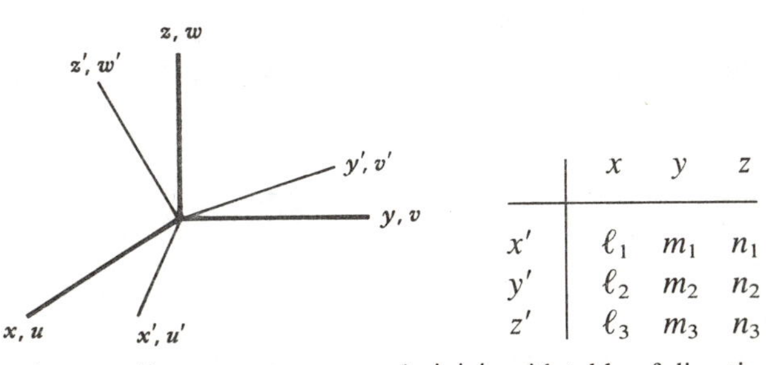

	x	y	z
x'	ℓ_1	m_1	n_1
y'	ℓ_2	m_2	n_2
z'	ℓ_3	m_3	n_3

Figure 6.2.1. Cartesian coordinate systems xyz and $x'y'z'$, with table of direction cosines between axes.

where coefficients in $\{\boldsymbol{\epsilon}\}$ and $\{\boldsymbol{\epsilon}'\}$ are ordered as they are in Eq. 1.4.4, and $[\mathbf{T}_\epsilon]$ is the 6-by-6 matrix

$$\begin{bmatrix} \ell_1^2 & m_1^2 & n_1^2 & \ell_1 m_1 & m_1 n_1 & n_1 \ell_1 \\ \ell_2^2 & m_2^2 & n_2^2 & \ell_2 m_2 & m_2 n_2 & n_2 \ell_2 \\ \ell_3^2 & m_3^2 & n_3^2 & \ell_3 m_3 & m_3 n_3 & n_3 \ell_3 \\ 2\ell_1\ell_2 & 2m_1m_2 & 2n_1n_2 & (\ell_1 m_2 + \ell_2 m_1) & (m_1 n_2 + m_2 n_1) & (n_1 \ell_2 + n_2 \ell_1) \\ 2\ell_2\ell_3 & 2m_2m_3 & 2n_2n_3 & (\ell_2 m_3 + \ell_3 m_2) & (m_2 n_3 + m_3 n_2) & (n_2 \ell_3 + n_3 \ell_2) \\ 2\ell_3\ell_1 & 2m_3m_1 & 2n_3n_1 & (\ell_3 m_1 + \ell_1 m_3) & (m_3 n_1 + m_1 n_3) & (n_3 \ell_1 + n_1 \ell_3) \end{bmatrix} \tag{6.2.7}$$

The engineering definition of shear strain is used, for example, $\gamma_{xy} = u_{,y} + v_{,x}$.

By partitioning $[\mathbf{T}_\epsilon]$ into 3-by-3 submatrices, we define a new matrix $[\mathbf{T}_\sigma]$ from it as follows.

$$\text{Let} \quad \underset{6\times 6}{[\mathbf{T}_\epsilon]} = \begin{bmatrix} \mathbf{T}_Q & \mathbf{T}_R \\ \mathbf{T}_S & \mathbf{T}_T \end{bmatrix}, \qquad \text{define} \quad \underset{6\times 6}{[\mathbf{T}_\sigma]} = \begin{bmatrix} \mathbf{T}_Q & 2\mathbf{T}_R \\ \frac{1}{2}\mathbf{T}_S & \mathbf{T}_T \end{bmatrix} \tag{6.2.8}$$

Thus factors of 2 in $[\mathbf{T}_\epsilon]$ are moved symmetrically about the diagonal to form $[\mathbf{T}_\sigma]$. Useful relations are

$$[\mathbf{T}_\epsilon]^{-1} = [\mathbf{T}_\sigma]^T, \qquad [\mathbf{T}_\sigma]^{-1} = [\mathbf{T}_\epsilon]^T \tag{6.2.9}$$

The *stress* transformation is derived by considering the equilibrium of a differential element. As shown in various elasticity texts, with coefficients in $\{\boldsymbol{\sigma}\}$ and $\{\boldsymbol{\sigma}'\}$ ordered as in Eq. 1.5.2,

$$\{\boldsymbol{\sigma}'\} = [\mathbf{T}_\sigma]\{\boldsymbol{\sigma}\}, \qquad \{\boldsymbol{\sigma}\} = [\mathbf{T}_\sigma]^{-1}\{\boldsymbol{\sigma}'\} \tag{6.2.10}$$

Initial stresses $\{\boldsymbol{\sigma}_0\}$ and initial strains $\{\boldsymbol{\epsilon}_0\}$ transform as do the stresses and strains in Eqs. 6.2.6 and 6.2.10.

A stress-strain relation can be written in either coordinate system, as $\{\boldsymbol{\sigma}\} = [\mathbf{E}]\{\boldsymbol{\epsilon}\}$ or as $\{\boldsymbol{\sigma}'\} = [\mathbf{E}']\{\boldsymbol{\epsilon}'\}$. Suppose that $[\mathbf{E}']$ is known and $[\mathbf{E}]$ is desired. To derive the transformation, we argue that strain energy density U_0 must be independent of the coordinate system in which it is computed. Thus, if $\{\delta\boldsymbol{\epsilon}\}$ and $\{\delta\boldsymbol{\epsilon}'\}$ both describe the strains produced by a given virtual displacement,

$$\delta U_0 = \{\delta\boldsymbol{\epsilon}\}^T\{\boldsymbol{\sigma}\} = \{\delta\boldsymbol{\epsilon}'\}^T\{\boldsymbol{\sigma}'\} \tag{6.2.11}$$

Substitution of Eq. 6.2.6 and $\{\boldsymbol{\sigma}'\} = [\mathbf{E}']\{\boldsymbol{\epsilon}'\}$ into Eq. 6.2.11 yields

$$\begin{aligned} \{\delta\boldsymbol{\epsilon}\}^T\{\boldsymbol{\sigma}\} &= \{\delta\boldsymbol{\epsilon}\}^T[\mathbf{T}_\epsilon]^T[\mathbf{E}']\{\boldsymbol{\epsilon}'\} \\ \{\delta\boldsymbol{\epsilon}\}^T(\{\boldsymbol{\sigma}\} - [\mathbf{T}_\epsilon]^T[\mathbf{E}'][\mathbf{T}_\epsilon]\{\boldsymbol{\epsilon}\}) &= 0 \end{aligned} \tag{6.2.12}$$

The latter equation must hold for *any* $\{\delta\boldsymbol{\epsilon}\}$. So the coefficient of $\{\delta\boldsymbol{\epsilon}\}^T$ must vanish. The coefficient of $\{\boldsymbol{\epsilon}\}$ is identified as the required matrix $[\mathbf{E}]$.

$$[\mathbf{E}] = [\mathbf{T}_\epsilon]^T\,[\mathbf{E}']\,[\mathbf{T}_\epsilon] \tag{6.2.13}$$

In two-dimensional problems we are concerned with $\{\boldsymbol{\epsilon}\} = \{\epsilon_x\ \epsilon_y\ \gamma_{xy}\}$ and with $\{\boldsymbol{\sigma}\} = \{\sigma_x\ \sigma_y\ \tau_{xy}\}$. Accordingly, discarding rows and columns 3, 5, and 6 from Eqs. 6.2.8 and using Fig. 6.2.2, we have

$$[\mathbf{T}_\epsilon] = \begin{bmatrix} c^2 & s^2 & cs \\ s^2 & c^2 & -cs \\ -2cs & 2cs & c^2 - s^2 \end{bmatrix}, \qquad [\mathbf{T}_\sigma] = \begin{bmatrix} c^2 & s^2 & 2cs \\ s^2 & c^2 & -2cs \\ -cs & cs & c^2 - s^2 \end{bmatrix} \tag{6.2.14}$$

where $c = \cos\beta$, and $s = \sin\beta$.

6.3 STIFFNESS TRANSFORMATION IN TWO DIMENSIONS

A plane frame element, defined in Section 4.4, is shown in Fig. 6.3.1. Imagine that its 6-by-6 stiffness matrix $[\mathbf{k}']$ is known in local coordinates $x'y'$. The stiffness matrix $[\mathbf{k}]$ in global coordinates xy is desired. (Coordinate transformation is a good way to derive $[\mathbf{k}]$ for an arbitrarily oriented frame element.)

Nodal rotation vectors maintain their directions, so $\theta_i' = \theta_i$, where $i = 1, 2$. By looking at vector components of u_i and w_i, we find that translational d.o.f. in the two systems are related by the equations

$$u_i' = u_i \cos\beta + w_i \sin\beta, \qquad w_i' = -u_i \sin\beta + w_i \cos\beta \tag{6.3.1}$$

where $i = 1, 2$. Primed and unprimed nodal forces are related in the same way as primed and unprimed nodal d.o.f. Let nodal d.o.f. $\{\mathbf{d}\}$ and nodal forces $\{\mathbf{r}\}$ be arranged in the order $\{\mathbf{d}\} = \{u_1\ w_1\ \theta_1\ u_2\ w_2\ \theta_2\}$ and $\{\mathbf{r}\} = \{p_1\ q_1\ M_1\ p_2\ q_2\ M_2\}$. Primed quantities are similarly arranged. Then, from Eqs. 6.3.1,

$$\{\mathbf{d}'\} = [\mathbf{T}]\{\mathbf{d}\}, \qquad \{\mathbf{r}'\} = [\mathbf{T}]\{\mathbf{r}\} \tag{6.3.2}$$

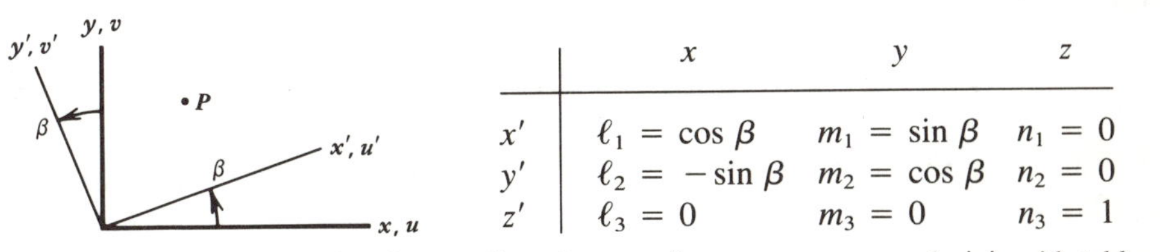

	x	y	z
x'	$\ell_1 = \cos\beta$	$m_1 = \sin\beta$	$n_1 = 0$
y'	$\ell_2 = -\sin\beta$	$m_2 = \cos\beta$	$n_2 = 0$
z'	$\ell_3 = 0$	$m_3 = 0$	$n_3 = 1$

Figure 6.2.2. The two-dimensional case: Cartesian coordinate systems xy and $x'y'$, with table of direction cosines between axes.

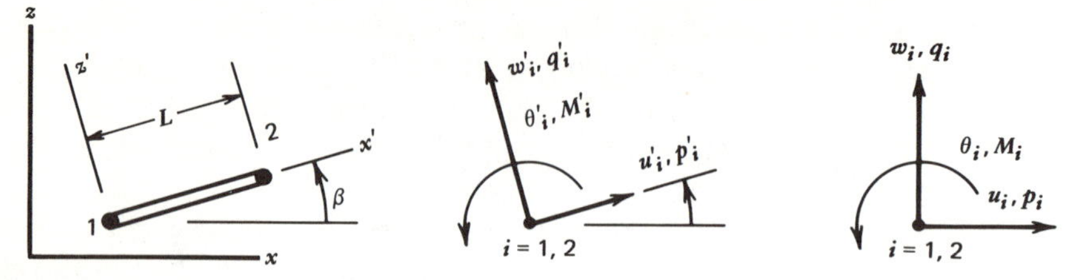

Figure 6.3.1. A plane frame element. Nodal d.o.f. and forces are shown in local and global coordinate systems.

where

$$\underset{6\times 6}{[\mathbf{T}]} = \begin{bmatrix} \boldsymbol{\Lambda} & \mathbf{O} \\ \mathbf{O} & \boldsymbol{\Lambda} \end{bmatrix}, \qquad \underset{3\times 3}{[\boldsymbol{\Lambda}]} = \begin{bmatrix} \cos\beta & \sin\beta & 0 \\ -\sin\beta & \cos\beta & 0 \\ 0 & 0 & 1 \end{bmatrix} \tag{6.3.3}$$

Matrix $[\mathbf{T}]$ is orthogonal; that is, $[\mathbf{T}]^{-1} = [\mathbf{T}]^T$. Therefore

$$\{\mathbf{d}\} = [\mathbf{T}]^T\{\mathbf{d}'\}, \qquad \{\mathbf{r}\} = [\mathbf{T}]^T\{\mathbf{r}'\} \tag{6.3.4}$$

However, the orthogonality of $[\mathbf{T}]$ need not be invoked to obtain the latter equation. Instead, we can argue that since $\{\mathbf{r}\}$ and $\{\mathbf{r}'\}$ describe the same set of forces, the work they do during a virtual displacement must be independent of the coordinate system in which it is computed. Let $\{\delta\mathbf{d}\}$ and $\{\delta\mathbf{d}'\}$ be two ways to describe the same virtual displacement. Virtual work is

$$\{\delta\mathbf{d}\}^T\{\mathbf{r}\} = \{\delta\mathbf{d}'\}^T\{\mathbf{r}'\} = \{\delta\mathbf{d}\}^T[\mathbf{T}]^T\{\mathbf{r}'\}$$

so

$$\{\delta\mathbf{d}\}^T(\{\mathbf{r}\} - [\mathbf{T}]^T\{\mathbf{r}'\}) = 0, \qquad \{\mathbf{r}\} = [\mathbf{T}]^T\{\mathbf{r}'\} \tag{6.3.5}$$

The latter result follows because the equation before it must be true for *any* $\{\delta\mathbf{d}\}$. This argument is useful because it does not demand that $[\mathbf{T}]$ be either orthogonal or square. Transformation matrices exist that have neither property.

To transform the stiffness matrix, we start with $[\mathbf{k}]\{\mathbf{d}\} = \{\mathbf{r}\}$ and substitute from the preceding equations.

$$[\mathbf{k}]\{\mathbf{d}\} = \{\mathbf{r}\} = [\mathbf{T}]^T\{\mathbf{r}'\} = [\mathbf{T}]^T[\mathbf{k}']\{\mathbf{d}'\} = [\mathbf{T}]^T[\mathbf{k}'][\mathbf{T}]\{\mathbf{d}\} \tag{6.3.6}$$

Since this relation is presumed valid for *any* $\{\mathbf{d}\}$, we conclude that the required stiffness transformation is

$$[\mathbf{k}] = [\mathbf{T}]^T[\mathbf{k}'][\mathbf{T}] \tag{6.3.7}$$

Stiffness matrix transformation always has this form, regardless of element type.

Consider next an element that has derivative d.o.f., for example, a plane triangle with $\{u\ v\ u_{,x}\ u_{,y}\ v_{,x}\ v_{,y}\}$ as d.o.f. at each node. The u and v d.o.f. transform as in Eq. 6.3.1. The derivative d.o.f. transform as in Eq. 6.2.4. Thus the [**T**] matrix would contain coefficients seen in Eq. 6.3.3 and coefficients seen in Eq. 6.2.5.

The transformation equation for mass matrices has the same form as Eq. 6.3.7.

$$[\mathbf{m}] = [\mathbf{T}]^T [\mathbf{m}'] [\mathbf{T}] \tag{6.3.8}$$

Mass matrices are discussed in Chapter 11.

Under rotation about the z-axis, coordinates transform by the same rule as displacements in Eq. 6.3.2. That is, any point P in Fig. 6.2.2 can be located by its primed or its unprimed coordinates. The two are related by $\{x_P'\ y_P'\ z_P'\} = [\Lambda]\ \{x_P\ y_P\ z_P\}$, where $[\Lambda]$ is given by Eq. 6.3.3. If rotation about *any* axis is permitted, $[\Lambda]$ is replaced by $[\mathbf{T}_c]$ of Eq. 6.2.5.

6.4 TRANSFORMATION TO TREAT A SKEW SUPPORT

Imagine that structure equations for the plane frame of Fig. 6.4.1 are based on u, w, and θ d.o.f at each node, as usual. The skew support demands that $w_i = -u_i \tan \beta$ at node i. This awkward constraint could be imposed by the methods of Sections 6.8 to 6.10. A better way is to invoke a coordinate transformation that replaces u_i and v_i by U_i and W_i, then set $W_i = 0$ by a standard method (see Section 2.9). The transformation problem is this: given [**K**′] and {**R**′} in the original system [**K**′]{**D**′} = {**R**′}, what are [**K**] and {**R**} in the transformed system [**K**]{**D**} = {**R**}? The transformed equations are to be solved for {**D**}.

Original and transformed displacement vectors are, respectively,

$$\{\mathbf{D}'\} = \{\ldots\ u_i\ w_i\ \theta_i\ \ldots\}, \qquad \{\mathbf{D}\} = \{\ldots\ U_i\ W_i\ \theta_i\ \ldots\} \tag{6.4.1}$$

D.o.f. not at node i are not transformed.

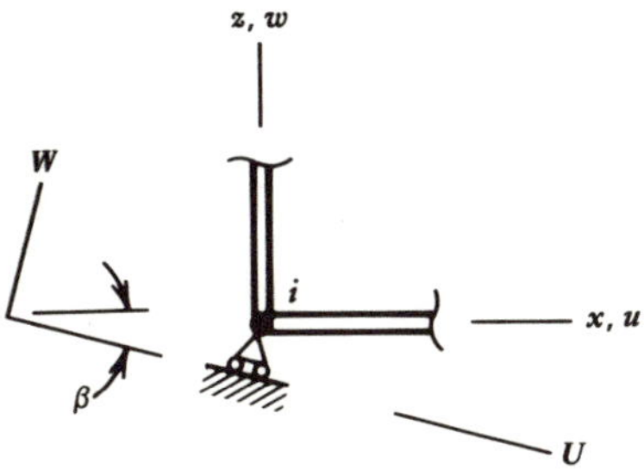

Figure 6.4.1. Node i of a plane frame. The support condition allows displacement U but prohibits displacement W.

Nodal d.o.f. and nodal force vectors here are written on the structure level, but otherwise they are the same as the corresponding vectors in Section 6.3. Accordingly, the transformation equations have the same form as Eqs. 6.3.2, 6.3.4, and 6.3.7.

$$\{\mathbf{D}'\} = [\mathbf{T}]\{\mathbf{D}\}, \qquad \{\mathbf{R}\} = [\mathbf{T}]^T\{\mathbf{R}'\}, \qquad [\mathbf{K}] = [\mathbf{T}]^T[\mathbf{K}'][\mathbf{T}] \tag{6.4.2}$$

where $[\mathbf{T}]$ is a unit matrix except for the appearance of $[\Lambda]$ from Eq. 6.3.3 in the position that corresponds to node i.

$$[\mathbf{T}] = \begin{bmatrix} \mathbf{I} & \mathbf{O} & \mathbf{O} \\ \mathbf{O} & \Lambda & \mathbf{O} \\ \mathbf{O} & \mathbf{O} & \mathbf{I} \end{bmatrix} \tag{6.4.3}$$

Transformed arrays $[\mathbf{K}]$ and $\{\mathbf{R}\}$ can themselves be transformed if there is another skew support. Conceivably, all nodes of the frame could be skew and all translational d.o.f. in $\{\mathbf{D}\}$ could have different directions. If n successive transformations are used, original d.o.f. $\{\mathbf{D}'\}$ are related to final d.o.f. $\{\mathbf{D}\}$ by the equation

$$\{\mathbf{D}'\} = [\mathbf{T}_1][\mathbf{T}_2] \ldots [\mathbf{T}_n]\{\mathbf{D}\} \tag{6.4.4}$$

One might need $\{\mathbf{D}'\}$ to do stress calculation and therefore invoke Eq. 6.4.4.

In the preceding explanation, transformation is done at the structure level. This approach requires that we construct and use $[\mathbf{T}]$ in a manner consistent with whatever compact storage format has been adopted for the structure stiffness matrix. It also requires that W_i be present in $\{\mathbf{D}\}$. If, instead, the separate element matrices are transformed *before* assembly, the scheme of Fig. 2.9.2 can be used to exclude W_i from $\{\mathbf{D}\}$. The required transformation matrix for a plane frame element is

$$\underset{6\times 6}{[\mathbf{T}]} = \begin{bmatrix} \Lambda & \mathbf{O} \\ \mathbf{O} & \mathbf{I} \end{bmatrix} \qquad \text{or} \qquad \underset{6\times 6}{[\mathbf{T}]} = \begin{bmatrix} \mathbf{I} & \mathbf{O} \\ \mathbf{O} & \Lambda \end{bmatrix} \tag{6.4.5}$$

depending on whether node 1 or node 2 of the element coincides with node i of the frame. This transformation must be applied to every element that frames into node i.

6.5 TRANSFORMATION IN SPACE. NONRECTANGULAR [T] MATRICES

In local coordinates $x'y'z'$, stiffness equations of the truss element in Fig. 6.5.1 are $[\mathbf{k}']\{\mathbf{d}'\} = \{\mathbf{r}'\}$, where

$$[\mathbf{k}'] = \frac{AE}{L}\begin{bmatrix} 1 & -1 \\ -1 & 1 \end{bmatrix}, \qquad \{\mathbf{d}'\} = \begin{Bmatrix} u_1' \\ u_2' \end{Bmatrix}, \qquad \{\mathbf{r}'\} = \begin{Bmatrix} p_1' \\ p_2' \end{Bmatrix} \tag{6.5.1}$$

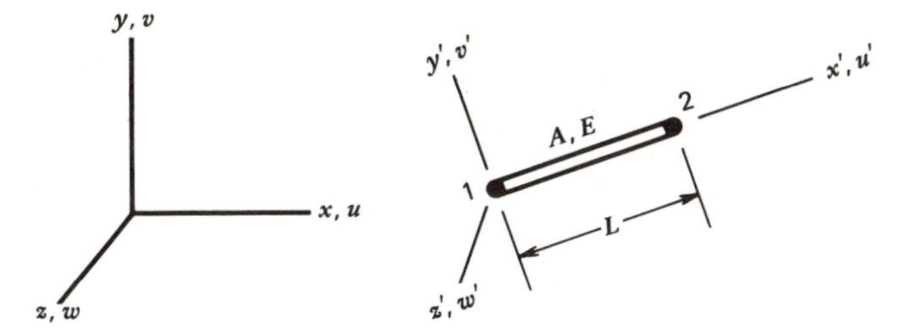

Figure 6.5.1. Truss element on the x'-axis of local coordinates $x'y'z'$.

The **[k]** and **{r}** matrices in global coordinates *xyz* are desired.

A straightforward but inefficient approach begins with expansion of $[\mathbf{k}']$ to 6 by 6, and $\{\mathbf{r}'\}$ to 6 by 1 by adding zeros. Thus

$$\{\mathbf{d}'\} = \{u_1' \; v_1' \; w_1' \; u_2' \; v_2' \; w_2'\}$$

$$\{\mathbf{r}'\} = \{p_1' \; 0 \; 0 \; p_2' \; 0 \; 0\} \tag{6.5.2}$$

$$k_{ij} = 0 \qquad \text{except} \qquad k_{11} = k_{44} = -k_{14} = -k_{41} = \frac{AE}{L}$$

Matrices in global coordinates *xyz* are, as usual, $\{\mathbf{r}\} = [\mathbf{T}]^T\{\mathbf{r}'\}$ and $[\mathbf{k}] = [\mathbf{T}]^T[\mathbf{k}'][\mathbf{T}]$, where

$$\underset{6\times 6}{[\mathbf{T}]} = \begin{bmatrix} \mathbf{\Lambda} & \mathbf{O} \\ \mathbf{O} & \mathbf{\Lambda} \end{bmatrix}, \qquad [\mathbf{\Lambda}] = \begin{bmatrix} \ell_1 & m_1 & n_1 \\ \ell_2 & m_2 & n_2 \\ \ell_3 & m_3 & n_3 \end{bmatrix} \tag{6.5.3}$$

We see that $[\mathbf{\Lambda}]$ is simply the table of direction cosines in Fig. 6.2.1.

Upon carrying out the transformations, we find that direction cosines with subscripts 2 and 3 do not contribute to **{r}** or to **[k]**. Figure 6.5.1 shows that this is physically reasonable, since the orientation of axis x' is all that matters. Accordingly, the transformation can be written as follows. Let

$$\underset{2\times 6}{[\mathbf{T}]} = \begin{bmatrix} \ell_1 & m_1 & n_1 & 0 & 0 & 0 \\ 0 & 0 & 0 & \ell_1 & m_1 & n_1 \end{bmatrix} \tag{6.5.4}$$

where ℓ_1, m_1, and n_1 can be calculated from *xyz* coordinates of nodes 1 and 2. Then

$$\underset{2\times 1}{\{\mathbf{d}'\}} = [\mathbf{T}]\underset{6\times 1}{\{\mathbf{d}\}}, \qquad \underset{6\times 1}{\{\mathbf{r}\}} = [\mathbf{T}]^T\underset{2\times 1}{\{\mathbf{r}'\}}, \qquad \underset{6\times 6}{[\mathbf{k}]} = [\mathbf{T}]^T\underset{2\times 2}{[\mathbf{k}']}[\mathbf{T}] \tag{6.5.5}$$

where the primed matrices are defined by Eq. 6.5.1. Clearly, Eqs. 6.5.5 are more economical of storage and effort than operations with full-sized matrices. The rank of $[\mathbf{k}']$ and $[\mathbf{k}]$ are the same (rank one, in this case).

If the element in Fig. 6.5.1 were a space beam, y' and z' would probably be principal axes of the cross section. Also, there would be six d.o.f. per node, $[\mathbf{k}']$ would be 12 by 12, and $[\mathbf{T}]$ must also be 12 by 12. Rotational d.o.f., being small, can be treated as vectors, so they transform in the same way as translational d.o.f.

6.6 JOINING DISSIMILAR ELEMENTS TO ONE ANOTHER

Often it is necessary to join elements that are not of the same type. An instance is the reinforcement of a plate by an edge beam. No transformation is needed if the elements have nodes at the same locations and if the $\{\mathbf{d}\}$ of one element is contained in the $\{\mathbf{d}\}$ of the other. (*Eccentric* stiffeners are discussed in Section 6.7.) For example, let a beam element span edge 2–3 of the plate element in Fig. 9.2.1*a*. Neutral axes of plate and beam coincide. Thus, at node 3, the beam d.o.f. are w_3 and $w_{,y3}$. Let the elements be part of a general-purpose program that allows up to six d.o.f. per node, in the order $\{u_i \; v_i \; w_i \; w_{,xi} \; w_{,yi} \; \theta_z\}$. Turning now to the ID array produced by Fig. 2.9.2, we locate the column that represents node 3 (this is column 3 if 3 is the *structure* node number). In this column, equation numbers for the plate node are in rows 3, 4, and 5, and equation numbers for the beam node are in rows 3 and 5. These equation numbers are placed in the KK array of the version of Fig. 2.9.3 appropriate to plate and beam elements, respectively. The d.o.f. u_i, v_i, and θ_z are suppressed.

An element match more aptly termed "dissimilar" is depicted in Fig. 6.6.1*a*. Node numbers are arbitrary. One end (node 5) of a plane frame element is to be attached at an arbitrary location along an edge of a four-node quadrilateral element. The quadrilateral has linear edges, no side nodes, and no rotational d.o.f.; the plane frame element has the usual three d.o.f. per node. Let $\{\mathbf{r}'\}$ and $[\mathbf{k}']$ refer to the frame element. The problem is to replace the usual frame d.o.f. by d.o.f. of the quadrilateral and hence convert $\{\mathbf{r}'\}$ to $\{\mathbf{r}\}$ and $[\mathbf{k}']$ to $[\mathbf{k}]$.

A suitable transformation states that translational motion of node 5 is linearly interpolated from translational d.o.f. at nodes 2 and 3, and rotation of node 5 is the same as rotation of edge 2–3. Thus

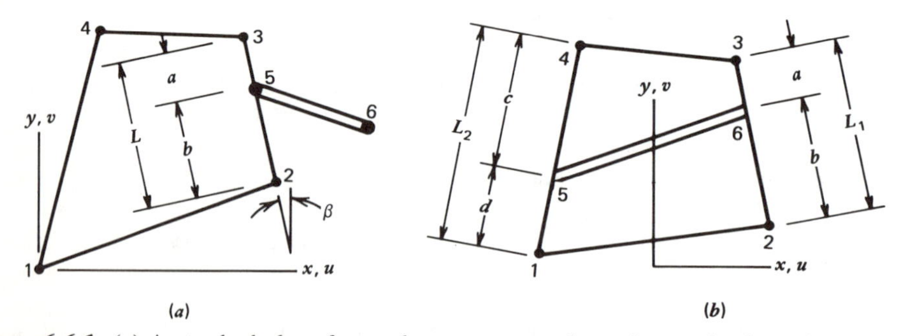

Figure 6.6.1. (*a*) A standard plane frame element connected to a four-node plane element. (*b*) A two-force member connected to a four-node plane element.

$$\{u_5 \; v_5 \; \theta_5\} = [\Lambda] \{u_2 \; v_2 \; u_3 \; v_3\}$$

where (6.6.1)

$$[\Lambda] = \frac{1}{L}\begin{bmatrix} a & 0 & b & 0 \\ 0 & a & 0 & b \\ \cos\beta & \sin\beta & -\cos\beta & -\sin\beta \end{bmatrix}$$

In other symbols, $\theta_5 = (n_2 - n_3)/L$, where n_2 and n_3 are nodal displacements normal to edge 2–3. The required transformation matrix is

$$\underset{6\times 7}{[\mathbf{T}]} = \begin{bmatrix} \mathbf{\Lambda} & \mathbf{O} \\ \mathbf{O} & \mathbf{I} \end{bmatrix} \tag{6.6.2}$$

where $[\mathbf{I}]$ is a 3-by-3 unit matrix. The transformed load vector and stiffness matrix of the frame element are, by the same arguments that lead to Eqs. 6.3.5 and 6.3.6,

$$\underset{7\times 1}{\{\mathbf{r}\}} = [\mathbf{T}]^T \underset{6\times 1}{\{\mathbf{r}'\}}, \qquad \underset{7\times 7}{[\mathbf{k}]} = [\mathbf{T}]^T \underset{6\times 6}{[\mathbf{k}']} [\mathbf{T}] \tag{6.6.3}$$

where $[\mathbf{k}]$ operates on the vector $\{\mathbf{d}\} = \{u_2 \; v_2 \; u_3 \; v_3 \; u_6 \; v_6 \; \theta_6\}$. Matrices of the two elements can be assembled in the usual way, whereupon node 5 and its d.o.f. disappear.

As a second example, consider the problem of Fig. 6.6.1*b*. A two-force member, perhaps a reinforcing bar in concrete, is connected to arbitrary points on edges of a linear element. Matrices $\{\mathbf{r}'\}$ and $[\mathbf{k}']$ of the bar are associated with d.o.f. $\{u_5 \; v_5 \; u_6 \; v_6\}$ but, through transformation, are to be associated instead with d.o.f. u_i and v_i of the four corner nodes. Thus

$$\underset{4\times 1}{\{\mathbf{r}'\}} \text{ becomes } \underset{8\times 1}{\{\mathbf{r}\}}, \qquad \underset{4\times 4}{[\mathbf{k}']} \text{ becomes } \underset{8\times 8}{[\mathbf{k}]} \tag{6.6.4}$$

The $\{\mathbf{r}\}$ and $[\mathbf{k}]$ arrays can then be directly added to the corresponding arrays of either the quadrilateral or the finite element structure. Clearly, the required $[\mathbf{T}]$ matrix involves coefficients like those in the first two rows of $[\Lambda]$ in Eq. 6.6.1.

Some structures can be modeled as one or more axisymmetric parts joined to one or more nonaxisymmetric parts. A cylinder-to-cylinder intersection is such a structure. A full three-dimensional analysis is expensive. A cheaper way is to analyze the axisymmetric part under forces applied by the nonaxisymmetric part [6.1]. This method requires a force estimate and ignores the changes in force that accompany deformations of the interface. A third approach is to write constraint equations that couple d.o.f. along an interface shared by axisymmetric and nonaxisymmetric parts [6.2]. By writing the constraint relations in the homogeneous form $(\ldots) = 0$, they can be imposed by the Lagrange multiplier method (Section 6.9).

6.7 RIGID LINKS. RIGID ELEMENTS. EXTRA HINGES

Consider a plate stiffened by an edge beam. Usually the neutral surfaces of plate and beam are not coincident: the stiffener is on one side of the plate. Therefore the procedure described at the outset of Section 6.6 cannot be applied directly. A standard preliminary treatment is to connect adjacent plate and beam nodes by a rigid link, so that d.o.f. of the beam are replaced by d.o.f. of the plate. The usual assembly is then possible. The necessary transformation is now described.

The plane frame element in Fig. 6.7.1*a* has the usual six d.o.f.—three at node i and three at node j. With reference to these d.o.f., element load and stiffness matrices are $\{\mathbf{r}'\}$ and $[\mathbf{k}']$. Similar d.o.f. are used at nodes 1 and 2 of the rigid links $i1$ and $j2$. The "master" d.o.f. at node 1 and "slave" d.o.f. at node i have the relation

$$\begin{Bmatrix} u_i \\ w_i \\ \theta_i \end{Bmatrix} = [\Lambda_1] \begin{Bmatrix} u_1 \\ w_1 \\ \theta_1 \end{Bmatrix}, \qquad \text{where} \qquad [\Lambda_1] = \begin{bmatrix} 1 & 0 & b_1 \\ 0 & 1 & -a_1 \\ 0 & 0 & 1 \end{bmatrix} \tag{6.7.1}$$

where $a_1 = x_1 - x_i$ and $b_1 = y_1 - y_i$.

A similar expression is written for link $j2$ by replacing subscripts i and 1 by j and 2. In Eq. 6.7.1, $b_1\theta_1$ and $-a_1\theta_1$ are the x and y displacement components of the (small) displacement $\theta_1 L_{i1}$, where $L_{i1} = (a_1^2 + b_1^2)^{1/2}$. The transformed arrays $\{\mathbf{r}\}$ and $[\mathbf{k}]$, associated with d.o.f. at nodes 1 and 2, are

$$\begin{aligned} \{\mathbf{r}\} &= [\mathbf{T}]^T \{\mathbf{r}'\} \\ [\mathbf{k}] &= [\mathbf{T}]^T [\mathbf{k}'][\mathbf{T}] \end{aligned} \qquad \text{where} \qquad \underset{6\times 6}{[\mathbf{T}]} = \begin{bmatrix} \Lambda_1 & \mathbf{O} \\ \mathbf{O} & \Lambda_2 \end{bmatrix} \tag{6.7.2}$$

If the frame element reinforces a plate, we would probably use $\beta = a_1 = a_2 = 0$ in Fig. 6.7.1*a*.

The foregoing transformation introduces the following error. For simplicity consider $\beta = a_1 = a_2 = 0$ in Fig. 6.7.1*a*. Now u is linear in x, but w is cubic and therefore θ

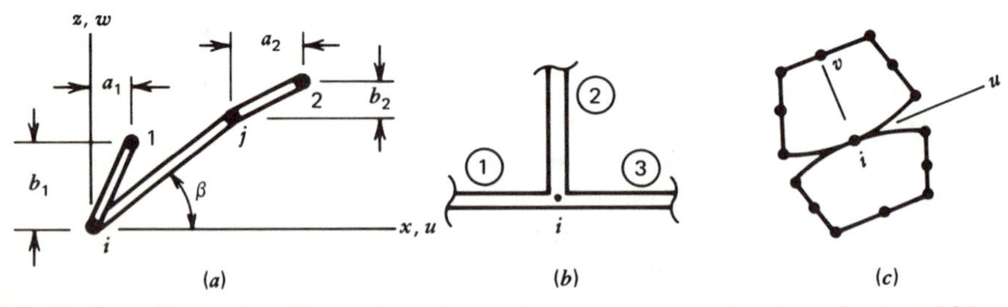

Figure 6.7.1. (*a*) Nodes i and j of a plane frame element are made slave to nodes 1 and 2 by rigid links $i1$ and $j2$. (*b*) Typical node i of a plane frame. (*c*) Sliding contact at node i of two elements in a larger plane mesh (not shown).

is quadratic. The transformation makes u depend on θ and so introduces an unwanted quadratic field into u. It is found that a beam-stiffened plate is overly flexible in a coarse mesh. Mesh refinement helps: error is quartered if the number of elements is doubled. Error is *eliminated* if a nodeless d.o.f. is added [6.3].

Another difficulty of rigid links is that the transformation converts a diagonal mass matrix $[\mathbf{m}']$ to a full matrix $[\mathbf{m}]$. Ad hoc adjustments of $[\mathbf{m}]$ can make it diagonal.

From rigid links it is a short step to rigid *elements*. Suppose, for example, that the triangle of Fig. 4.5.1*a* is infinitely stiff. Then three d.o.f. completely define its motion and, by kinematics, we write

$$\{u_1\ v_1\ u_2\ v_2\ u_3\ v_3\} = \left[\frac{\mathbf{I}}{\mathbf{T}}\right] \{u_1\ v_1\ u_2\} \tag{6.7.3}$$

Unit matrix $[\mathbf{I}]$ is 3 by 3 and expresses the identity $\{u_1\ v_1\ u_2\} \equiv \{u_1\ v_1\ u_2\}$. Matrix $[\mathbf{T}]$ is 3 by 3 and is a function of the nodal coordinates. The choice of $\{u_1\ v_1\ u_2\}$ on the right is not unique, and it is even unacceptable if $y_1 = y_2$.

The transformation of Eq. 6.7.3 differs from the transformations of Eq. 6.7.2 and Section 6.6 in that we are not preparing to add an elastic member to the structure. Instead, we are using geometry alone to impose a relation among existing d.o.f. of the structure. Equation 6.7.3 is a *constraint condition*, and it is in the form of Eq. 6.8.4 (see also Problem 6.24). A constraint *decreases* the number of d.o.f.

A hinge *increases* the number of d.o.f. In Fig. 6.7.1*b*, θ_i would ordinarily denote the rotation of all three bars at node i. But if (say) bar 1 is to be hinged at i, we need two independent rotational d.o.f.: θ_{1i} for bar 1 at i and θ_i for the remaining two bars. We can avoid carrying θ_{1i} into $\{\mathbf{D}\}$ by the simple expedient of condensing θ_{1i} from element 1 before assembly using the procedure of Eq. 7.2.3. Thus θ_{1i} "floats," and no moment is transferred between the bar and the joint at node i. The condensed $[\mathbf{k}]$ remains 6 by 6, but its θ_{1i} row and column are filled with zeros so that no rotational stiffness is contributed to the structure by element 1. (This treatment is called a *release* of θ_{1i}.) If n bars terminate at a plane frame node, the treatment can be applied to $n - 1$ of the bars without affecting the number of d.o.f. in $\{\mathbf{D}\}$. If all frame nodes are so treated, the structure becomes a pin-jointed truss.

The sliding contact in Fig. 6.7.1*c* can also be introduced by condensation. Before assembly, d.o.f. u_i in one element is condensed and the corresponding row and column of $[\mathbf{k}]$ are filled with zeros. The v_i d.o.f. are not treated. Thus u_i is released and the elements can slide, but they neither separate nor interpenetrate at node i.

6.8 CONSTRAINTS AND TRANSFORMATION EQUATIONS

Thus far we have regarded transformations as ways to change directions and introduce new element d.o.f. In this section we regard transformation as a way to impose relations among d.o.f. already present in the structure.

Examples in Sections 6.6 and 6.7 can be viewed as constraint problems. Additional examples that involve constraints are shown in Fig. 6.8.1. In Figure 6.8.1*a* nodes B and C are separate but are to have the same coordinates and the same displacement, $v_B = v_D$. Nodes A and B in Figure 6.8.1*b* are to be pin-connected by forcing them to have the same translational d.o.f. In a plane frame (Figure 6.8.1*c*) members usually have much less bending stiffness than axial stiffness. Therefore it is reasonable to say that nodes A and B have the same horizontal displacement. (This is a case of a stiff member supported by members that are much more flexible. Numerical error is likely unless equality of d.o.f. is explicitly imposed. See Section 15.5.)

For each equation of constraint, one d.o.f. can be eliminated. Accordingly, formal constraint transformations reduce the number of structural equations. However, considerable matrix manipulation is needed. The Lagrange multiplier method of treating constraints (Section 6.9) *adds* to the number of equations but requires less manipulation.

The theory of transformation equations is as follows. Equations that couple d.o.f. in $\{\mathbf{D}\}$ are

$$[\mathbf{C}]\,\{\mathbf{D}\} = \{\mathbf{Q}\} \tag{6.8.1}$$

where $[\mathbf{C}]$ is a matrix of constants and $\{\mathbf{Q}\}$ is a vector of constants. There are more d.o.f. than constraint equations, so $[\mathbf{C}]$ has more columns than rows. We consider the common case $\{\mathbf{Q}\} = 0$. Let the equations be partitioned so that

$$[\mathbf{C}_e\ \mathbf{C}_r]\,\{\mathbf{D}_e\ \mathbf{D}_r\} = 0 \tag{6.8.2}$$

where $\{\mathbf{D}_e\}$ are $\{\mathbf{D}_r\}$ are, respectively, d.o.f. to be eliminated and d.o.f. to be retained. Matrix $[\mathbf{C}_r]$ is mostly null if d.o.f. $\{\mathbf{D}_e\}$ are coupled to only a few of the d.o.f. $\{\mathbf{D}_r\}$. Since there are as many constraint equations as d.o.f. $\{\mathbf{D}_e\}$, matrix $[\mathbf{C}_e]$ is square. Next we solve for $\{\mathbf{D}_e\}$.

$$\{\mathbf{D}_e\} = [\mathbf{C}_{er}]\,\{\mathbf{D}_r\}, \qquad \text{where} \qquad [\mathbf{C}_{er}] = -\,[\mathbf{C}_e]^{-1}\,[\mathbf{C}_r] \tag{6.8.3}$$

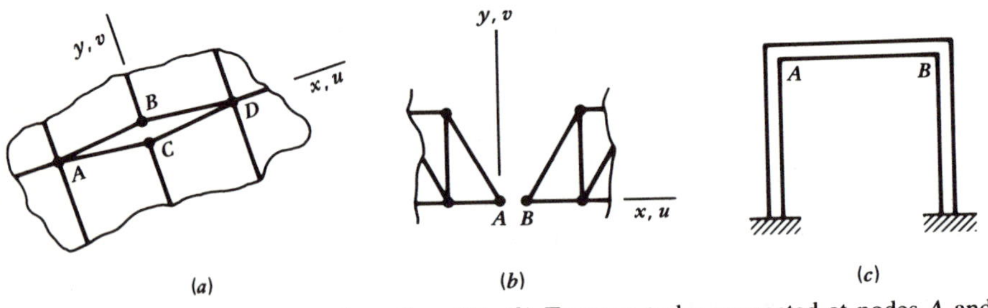

Figure 6.8.1. (*a*) Sliding contact along line *AD*. (*b*) Trussses to be connected at nodes *A* and *B*. (*c*) Plane frame.

We now write as one expression the relation $\{\mathbf{D}_r\} = \{\mathbf{D}_r\}$ and Eq. 6.8.3.

$$\begin{Bmatrix} \mathbf{D}_e \\ \mathbf{D}_r \end{Bmatrix} = [\mathbf{T}]\{\mathbf{D}_r\}, \qquad \text{where} \qquad [\mathbf{T}] = \begin{bmatrix} \mathbf{C}_{er} \\ \mathbf{I} \end{bmatrix} \tag{6.8.4}$$

With the transformation matrix $[\mathbf{T}]$ now defined, the familiar operations $\{\mathbf{R}\} = [\mathbf{T}]^T\{\mathbf{R}'\}$ and $[\mathbf{K}] = [\mathbf{T}]^T[\mathbf{K}'][\mathbf{T}]$ can be carried out. After solution of the structural equations has yielded $\{\mathbf{D}_r\}$, $\{\mathbf{D}_e\}$ can be recovered by use of Eq. 6.8.3.

The choice of $[\mathbf{C}_e]$ is not unique. One method of forming $[\mathbf{C}_e]$ is to choose the first e linearly independent columns of $[\mathbf{C}]$. Regardless of the method, finding that $[\mathbf{C}_e]$ is nonsingular checks that the constraint equations are linearly independent. Too many constraints will "lock" the system and require that $\{\mathbf{D}\} = 0$.

Example

Consider the one-dimensional truss of Fig. 6.8.2. Let each of the three members have the same stiffness, $AE/L = k$. The structure equations are

$$\begin{bmatrix} k & -k & 0 \\ -k & 2k & -k \\ 0 & -k & 2k \end{bmatrix} \begin{Bmatrix} u_1 \\ u_2 \\ u_3 \end{Bmatrix} = \begin{Bmatrix} P \\ P \\ P \end{Bmatrix} \tag{6.8.5}$$

Suppose that the constraint $u_1 = u_2$ is to be imposed. Eqs. 6.8.2 and 6.8.3 become

$$[1 \quad -1 \quad 0]\,\{u_1 \; u_2 \; u_3\} = 0, \qquad \{u_1\} = [1 \quad 0]\{u_2 \; u_3\} \tag{6.8.6}$$

where u_3 is multiplied by zero because it does not enter the constraint condition $u_1 = u_2$. Equation 6.8.4 and the reduced system are

$$\begin{Bmatrix} u_1 \\ u_2 \\ u_3 \end{Bmatrix} = \begin{bmatrix} 1 & 0 \\ 1 & 0 \\ 0 & 1 \end{bmatrix} \begin{Bmatrix} u_2 \\ u_3 \end{Bmatrix}, \qquad \begin{bmatrix} k & -k \\ -k & 2k \end{bmatrix} \begin{Bmatrix} u_2 \\ u_3 \end{Bmatrix} = \begin{Bmatrix} 2P \\ P \end{Bmatrix} \tag{6.8.7}$$

Solution yields $u_2 = 5P/k$ and $u_3 = 3P/k$, as expected.

In Section 6.12 we note that two nodes can be forced to have the same d.o.f. by giving them the same node number. Then external loads $\{\mathbf{P}\}$ on d.o.f. $\{\mathbf{D}_e\}$ must be transferred to d.o.f. $\{\mathbf{D}_r\}$ if they are not to be lost. When applied to Fig. 6.8.2, this method adds all four coefficients in $[\mathbf{k}]$ of the left element, for a sum of zero at node 2, and immediately yields the second of Eqs. 6.8.7.

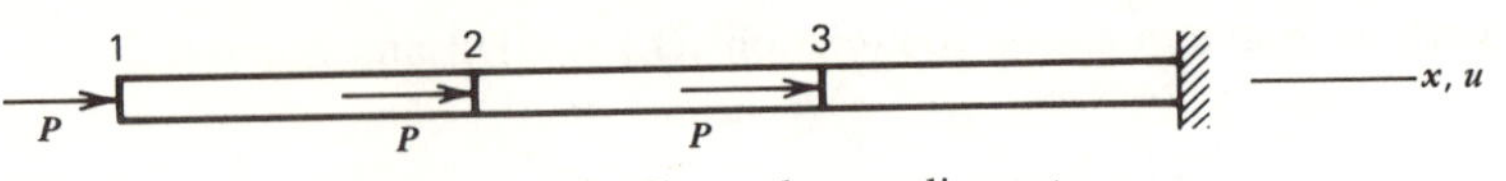

Figure 6.8.2. Three-element linear truss.

The formal operations of constraint equations can be restructed as follows [6.24]. Retain $\{\mathbf{Q}\}$ in Eq. 6.8.2 and write it as

$$-[\mathbf{H}]\{\mathbf{D}_r\} + \{\mathbf{D}_e\} = \{\bar{\mathbf{Q}}\}, \tag{6.8.8}$$

$$\text{where} \qquad [\mathbf{H}] = -[\mathbf{C}_e]^{-1}[\mathbf{C}_r] \qquad \text{and} \qquad \{\bar{\mathbf{Q}}\} = [\mathbf{C}_e]^{-1}\{\mathbf{Q}\}$$

Prior to constraint, the (symmetric) structural equations are

$$\begin{bmatrix} \mathbf{K}_{rr} & \mathbf{K}_{re} \\ \mathbf{K}_{er} & \mathbf{K}_{ee} \end{bmatrix} \begin{Bmatrix} \mathbf{D}_r \\ \mathbf{D}_e \end{Bmatrix} = \begin{Bmatrix} \mathbf{R}_r \\ \mathbf{R}_e \end{Bmatrix} \qquad \begin{matrix} (6.8.9) \\ (6.8.10) \end{matrix}$$

We substitute Eq. 6.8.8 into Eq. 6.8.9 and replace Eq. 6.8.10 by Eq. 6.8.8. Thus

$$\begin{bmatrix} (\mathbf{K}_{rr} + \mathbf{K}_{re}\mathbf{H}) & \mathbf{O} \\ -\mathbf{H} & \mathbf{I} \end{bmatrix} \begin{Bmatrix} \mathbf{D}_r \\ \mathbf{D}_e \end{Bmatrix} = \begin{Bmatrix} \mathbf{R}_r - \mathbf{K}_{re}\bar{\mathbf{Q}} \\ \bar{\mathbf{Q}} \end{Bmatrix} \qquad \begin{matrix} (6.8.11) \\ (6.8.12) \end{matrix}$$

To restore symmetry, we premultiply Eq. 6.8.10 by $[\mathbf{H}]^T$ and add it to Eq. 6.8.11, and we premultiply Eq. 6.8.12 by $-[\mathbf{K}_{ee}]$. Thus

$$\begin{bmatrix} (\mathbf{K}_{rr} + \mathbf{K}_{re}\mathbf{H} + \mathbf{H}^T\mathbf{K}_{er}) & \mathbf{H}^T\mathbf{K}_{ee} \\ \mathbf{K}_{ee}\mathbf{H} & -\mathbf{K}_{ee} \end{bmatrix} \begin{Bmatrix} \mathbf{D}_r \\ \mathbf{D}_e \end{Bmatrix} = \begin{Bmatrix} \mathbf{H}^T\mathbf{R}_e + \mathbf{R}_r - \mathbf{K}_{re}\bar{\mathbf{Q}} \\ -\mathbf{K}_{ee}\bar{\mathbf{Q}} \end{Bmatrix} \tag{6.8.13}$$

In this form the number of equations is preserved, and all d.o.f. in $\{\mathbf{D}\} = \{\mathbf{D}_r\ \mathbf{D}_e\}$ are produced by a standard equation solver (however, the coefficient matrix is not generally positive definite, so Fig. 2.11.2 will encounter negative pivots).

The reordering, partitioning, and matrix multiplications in Eq. 6.8.13 can be avoided by applying the constraints of Eq. 6.8.8 serially instead of all at once. The necessary algebra is given in Ref. 6.24.

6.9 CONSTRAINTS AND LAGRANGE MULTIPLIERS

Lagrange's method of undetermined multipliers is used to find the maximum or minimum of a function whose variables are not independent but have some prescribed relation. In our context the function is total potential energy Π_p and the variables are those in $\{\mathbf{D}\}$. System unknowns become $\{\mathbf{D}\}$ *and* the Lagrange multipliers.

The theory is easy to apply. We write the constraint equation (Eq. 6.8.1) as the homogeneous equation $[\mathbf{C}]\{\mathbf{D}\} - \{\mathbf{Q}\} = 0$ and multiply its left side by a row vector $\{\boldsymbol{\lambda}\}^T$ that contains as many Lagrange multipliers as there are constraint equations. Next we add the result to the total potential expression, Eq. 4.3.6.

$$\Pi_p = \tfrac{1}{2}\{\mathbf{D}\}^T[\mathbf{K}]\{\mathbf{D}\} - \{\mathbf{D}\}^T\{\mathbf{R}\} + \{\boldsymbol{\lambda}\}^T\big([\mathbf{C}]\{\mathbf{D}\} - \{\mathbf{Q}\}\big) \tag{6.9.1}$$

The expression in parentheses is zero, so we have added nothing to Π_p. Next we make Π_p stationary by writing the equations $\{\partial\Pi_p/\partial\mathbf{D}\} = \{\partial\Pi_p/\partial\boldsymbol{\lambda}\} = 0$, following differentiation rules in Section 3.3. The result is

$$\begin{bmatrix} \mathbf{K} & \mathbf{C}^T \\ \mathbf{C} & \mathbf{O} \end{bmatrix} \begin{Bmatrix} \mathbf{D} \\ \boldsymbol{\lambda} \end{Bmatrix} = \begin{Bmatrix} \mathbf{R} \\ \mathbf{Q} \end{Bmatrix} \tag{6.9.2}$$

The second of Eqs. 6.9.2 is Eq. 6.8.1, the equation of constraint. Equations 6.9.2 are solved for both $\{\mathbf{D}\}$ and $\{\boldsymbol{\lambda}\}$. The λ_i may be interpreted as forces of constraint (see the following example).

Strict partitioning — $\{\mathbf{D}\}$ followed by $\{\boldsymbol{\lambda}\}$ in Eq. 6.9.2 — increases bandwidth to the maximum. If the D_i and λ_i are interlaced, bandwidth can be much less, although not so small as when the λ_i are absent. However, in a Gauss elimination solution with pivoting on the diagonal, a zero pivot appears if a constraint equation is processed before any of the d.o.f. to which it is coupled. Otherwise, the null submatrix fills in and the solution proceeds normally if the stiffness matrix $[\mathbf{K}]$ is by itself positive definite.

The Lagrange multiplier method is more attractive than the transformation method of Section 6.8 if there are few constraint equations that couple many d.o.f. However, Lagrange multipliers are active at the structure level, while transformation equations can be applied at either the structure level or element by element. The latter has the appeal of disposing of constraints at an early stage, when the matrices are small and more manageable.

Another use of Lagrange multipliers is to enforce interelement compatibility. Consider, for example, two plane elements A and B that are adjacent but unconnected and that share a boundary of length L. Let u's and v's represent displacements along the common boundary. If we add the expressions

$$\lambda_1 \int_0^L (u_A - u_B)\, ds \qquad \text{and} \qquad \lambda_2 \int_0^L (v_A - v_B)\, ds \tag{6.9.3}$$

to Π_p, we enforce the compatibility conditions $u_A = u_B$ and $v_A = v_B$ in an average sense. System d.o.f. include the separate d.o.f. (or generalized coordinates) of elements A and B plus the Lagrange multipliers λ_1 and λ_2 that couple the d.o.f. If applied to all elements and edges, this procedure replaces element assembly [3.6, 6.4, 6.5, 7.62].

Example

We again solve the problem of Fig. 6.8.2. The constraint equation is Eq. 6.8.6. Equation 6.9.2 becomes

$$\begin{bmatrix} k & -k & 0 & 1 \\ -k & 2k & -k & -1 \\ 0 & -k & 2k & 0 \\ 1 & -1 & 0 & 0 \end{bmatrix} \begin{Bmatrix} u_1 \\ u_2 \\ u_3 \\ \lambda \end{Bmatrix} = \begin{Bmatrix} P \\ P \\ P \\ 0 \end{Bmatrix} \tag{6.9.4}$$

The solution is the same as that found in Section 6.8.

$$\{u_1 \; u_2 \; u_3 \; \lambda\} = \{5P/k \;\; 5P/k \;\; 3P/k \;\; P\} \tag{6.9.5}$$

Two comments are in order. First, the arrangement of unknowns is poor in that the bandwidth in Eq. 6.9.4 is made as large as possible. Second, $\lambda = P$ can be regarded as the force of constraint applied to node 2 through the now rigid link 1–2.

6.10 CONSTRAINTS AND PENALTY FUNCTIONS

If the constraint equation, Eq. 6.8.1, is written in the form

$$\{\mathbf{t}\} = [\mathbf{C}]\{\mathbf{D}\} - \{\mathbf{Q}\} \tag{6.10.1}$$

then $\{\mathbf{t}\} = 0$ implies satisfaction of the constraint. The usual potential Π_p of the system can be augmented by a *penalty function* $\{\mathbf{t}\}^T\lceil\boldsymbol{\alpha}\rfloor\{\mathbf{t}\}/2$, where $\lceil\boldsymbol{\alpha}\rfloor$ is a diagonal matrix of "penalty numbers" α_i. Thus

$$\Pi_p = \tfrac{1}{2}\{\mathbf{D}\}^T[\mathbf{K}]\{\mathbf{D}\} - \{\mathbf{D}\}^T\{\mathbf{R}\} + \tfrac{1}{2}\{\mathbf{t}\}^T\lceil\boldsymbol{\alpha}\rfloor\{\mathbf{t}\} \tag{6.10.2}$$

If $\{\mathbf{t}\} = 0$ the constraints are satisfied and we have added nothing to Π_p. If $\{\mathbf{t}\} \neq 0$ the penalty of constraint violation becomes more prominent as $\lceil\boldsymbol{\alpha}\rfloor$ increases. Next we substitute Eq. 6.10.1 into Eq. 6.10.2 and write the stationary condition $\{\partial\Pi_p/\partial\mathbf{D}\} = 0$; we find

$$[\mathbf{K} + \mathbf{C}^T\boldsymbol{\alpha}\mathbf{C}]\{\mathbf{D}\} = \{\mathbf{R}\} + [\mathbf{C}]^T\lceil\boldsymbol{\alpha}\rfloor\{\mathbf{Q}\} \tag{6.10.3}$$

If $\lceil\boldsymbol{\alpha}\rfloor = 0$, the constraints are ignored. As $\lceil\boldsymbol{\alpha}\rfloor$ grows, $\{\mathbf{D}\}$ changes in such a way that the constraint equations are more nearly satisfied.

However, if $\lceil\boldsymbol{\alpha}\rfloor$ is too large, accuracy declines. The reason can be seen through an example that also conveys a physical meaning of the penalty function. Consider again the problem of Fig. 6.8.2. We want u_1 and u_2 to be equal, so Eq. 6.10.1 becomes

$$\{\mathbf{t}\} = \lfloor 1 \quad -1 \quad 0 \rfloor \{u_1 \quad u_2 \quad u_3\} \tag{6.10.4}$$

The $[\mathbf{C}]^T \lceil \boldsymbol{\alpha} \rfloor [\mathbf{C}]$ term in Eq. 6.10.3 is found to be essentially the stiffness of a truss element that connects nodes 1 and 2. This is the physical interpretation of the constraint. Section 15.5 shows that numerical error is likely if a very stiff element lies entirely within a structure. In the present context, a very stiff element appears when $\lceil \boldsymbol{\alpha} \rfloor$ is large.

As suggested in Section 15.5, numerical trouble is reduced if violation-of-constraint d.o.f. are used. In the foregoing example, either u_1 or u_2 could be replaced by u_1', where $u_1' = u_1 - u_2$. A penalty function then imposes the constraint $u_1' = 0$.

Miscellaneous applications of penalty functions are many. An early one is the use of a very stiff member to support a structure (Fig. 2.9.5). Another is the suppression of transverse shear deformation in a plate or beam by means of high shear stiffness (Section 9.5). A third is that of making a material nearly incompressible by letting Poisson's ratio approach 0.5 (Section 16.6). In the latter two applications, $[\mathbf{C}]^T \lceil \boldsymbol{\alpha} \rfloor [\mathbf{C}]$ is effectively assembled element by element and involves all d.o.f. in $\{\mathbf{D}\}$. Other penalty functions are noted below Eq. 9.6.4 and in Eq. 10.2.3.

If $[\mathbf{C}]$ has many more columns than rows, $[\mathbf{C}]^T \lceil \boldsymbol{\alpha} \rfloor [\mathbf{C}]$ is a singular matrix. In the shear stiffness and incompressible material problems noted in the preceding paragraph, the effective $[\mathbf{C}]$ has as many d.o.f. as $\{\mathbf{D}\}$, and $[\mathbf{C}]^T \lceil \boldsymbol{\alpha} \rfloor [\mathbf{C}]$ may not be singular. But we *want* it to be singular, as the following argument shows.

For simplicity let all α_i in $\lceil \boldsymbol{\alpha} \rfloor$ be the same number, say α. Also, let $\{\mathbf{Q}\} = 0$ in Eq. 6.10.3. Then, as α becomes large, Eq. 6.10.3 becomes

$$[\mathbf{C}]^T [\mathbf{C}]\{\mathbf{D}\} = \frac{1}{\alpha}\{\mathbf{R}\} \tag{6.10.5}$$

Equation 6.10.5 shows that if $[\mathbf{C}]^T [\mathbf{C}]$ is nonsingular, as α grows the mesh "locks," giving $\{\mathbf{D}\} = 0$. Only if $[\mathbf{C}]^T[\mathbf{C}]$ is singular can $\{\mathbf{D}\}$ be nonzero. Then the number of independent nonzero $\{\mathbf{D}\}$'s that solve Eq. 6.10.5 is equal to the difference between the order of $[\mathbf{C}]^T [\mathbf{C}]$ and its rank. The import of this is discussed in connection with Eq. 9.5.5.

In comparison with Lagrange multipliers, penalty functions have the advantages of introducing no new variables and requiring no care to order unknowns for successful equation solving. Their implementation can be as easy as assigning a high modulus to an element already in the program. Penalty functions have the disadvantage that the penalty numbers must be chosen in an allowable range: big enough to be effective but not so big as to provoke numerical problems [5.10, 6.6].

6.11 MAKING USE OF STRUCTURAL SYMMETRY

Figure 6.11.1*a* represents a homogeneous, isotropic, uniformly loaded, thin square plate, with all edges simply supported. The *x*-axis, the *y*-axis, and the two diagonals are all axes of symmetry. So we need not analyze the plate as a whole. Analysis of a typical

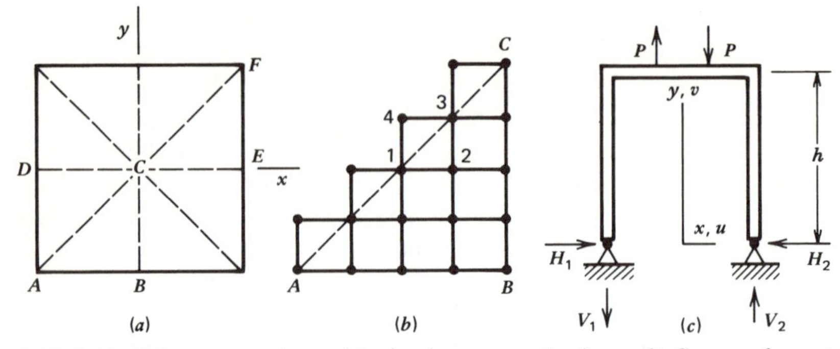

Figure 6.11.1 (*a*) Thin, square plate with simply supported edges. (*b*) Square elements in one octant. (*c*) Symmetric frame with antisymmetric loads.

octant *ABC* tells us all there is to know. If the material were orthotropic, with principal axes x and y, we would have to analyze only a typical quadrant *ABCD*. In either case the benefit is a reduction of effort.

How can symmetry be recognized? To be symmetric, a structure must have symmetry of shape, elastic properties, and support conditions. Symmetry can be classed as reflective or rotational. A symmetric structure is one for which one or more reflections and/or rotations brings the structure to a configuration indistinguishable from the original configuration with respect to shape, elastic properties, and support conditions. In the plate of Fig. 6.11.1*a*, each dashed line is an axis of reflective symmetry. A z-axis through point C is an axis of rotational symmetry, since successive 90° rotations bring the structure into coincidence with itself. Other examples of rotational symmetry include solids of revolution (Chapter 8) and cyclic symmetry (Section 6.12). Before a large job is run, a model with few d.o.f. can be studied to check that anticipated symmetries indeed exist, and perhaps discover unanticipated symmmetries.

A symmetric structure can carry symmetric or antisymmetric loads. Antisymmetry of loads exists if a single reflection of the structure, *with* its loads, followed by reversal of all loads, results in self-coincidence.

Symmetric loads, when acting on symmetric structures, produce symmmetric effects [6.7]. To exploit this rule, we analyze a part of the structure with appropriate boundary conditions. Boundary conditions for symmetric loading are:

S1. Zero translation perpendicular to a plane of symmetry.

S2. Hence all rotation vectors are perpendicular to a plane of symmetry.

Boundary conditions for antisymmetric loading are:

A1. Zero translation in a plane of structural symmetry.

A2. Hence all rotation vectors lie in a plane of structural symmetry.

Consider, for example, a uniform load on the plate of Fig. 6.11.1*a*. Let nodal d.o.f. be lateral displacement w and rotations $w_{,x}$ and $w_{,y}$ about $-y$ and $+x$ axes, respectively.

We can analyze one quadrant *ABCD* by setting $w_{,y} = 0$ along *DC* and $w_{,x} = 0$ along *BC*. We can analyze one octant *ABC* by setting $w_{,x} = 0$ along *BC* and rotations about *AC* to zero along the diagonal *AC*. The latter d.o.f. are not present in the original formulation and must be established by coordinate transformation.

Analysis of an octant *ABC* is easy with triangular elements because the mesh can fit in the octant exactly. But square elements can also be used. Consider the typical symmetrically placed points 2 and 4 in Fig. 6.11.1*b*. Symmetry about *AC* is achieved by imposing the constraints

$$w_2 = w_4, \qquad w_{,x4} = w_{,y2}, \qquad w_{,y4} = w_{,x2} \tag{6.11.1}$$

Further arguments are given by Glockner [6.7] and Noor [6.8]. Glockner also considers antisymmetric loads. If the frame of Fig. 6.11.1*c* is reflected or rotated about the *y*-axis, the structure is brought into self-coincidence, but the loads are not. But a load reversal would achieve coincidence and would also reverse the displacements and the support reactions. From this Glockner concludes that $V_1 = V_2$ and $H_1 = H_2 = 0$. In a displacement analysis, we could analyze half the frame with $v = 0$ as the displacement boundary condition at the point $x = 0$, $y = h$.

A *skew*-symmetric problem requires a rotation or more than one reflection to achieve self-coincidence. The plate problem of Fig. 6.11.1*a* would be called skew-symmetric if (say) octants *ACD* and *FCE* carried equal uniform loads and other octants were unloaded. Then the lateral displacement w is $w(r) = w(-r)$, where r is a radial coordinate measured from the center *C*. We would have $w(r) = -w(-r)$ in the skew-*anti*symmetric case, where (say) the load on *ACD* acts up and the load on *FCE* acts down. Boundary conditions for skew problems are given in Ref. 6.9.

Occasionally it is expedient to express a load as the sum of symmetric and antisymmetric parts. Then a half-size structure under two load conditions replaces the complete structure under a single load case. Analysis results from the two load cases must be superposed.

Caution is needed if deflections become large, since initial symmetries may then disappear.

6.12 CYCLIC SYMMETRY

In some structures, such as impellers in centrifugal pumps, we can recognize a repetition of geometry and loading even though no axis of reflective symmetry exists. This circumstance is called *cyclic symmetry*. Figure 6.12.1*a* depicts a plane structure with cyclic symmetry. Each triangle is identified as a "repeating substructure." Stresses and deformations also display cyclic symmetry. Therefore all that need be analyzed is one repeating substructure, such as *AABB* in Fig. 6.12.1. The necessary equations are developed as follows. They represent a transformation, not an approximation.

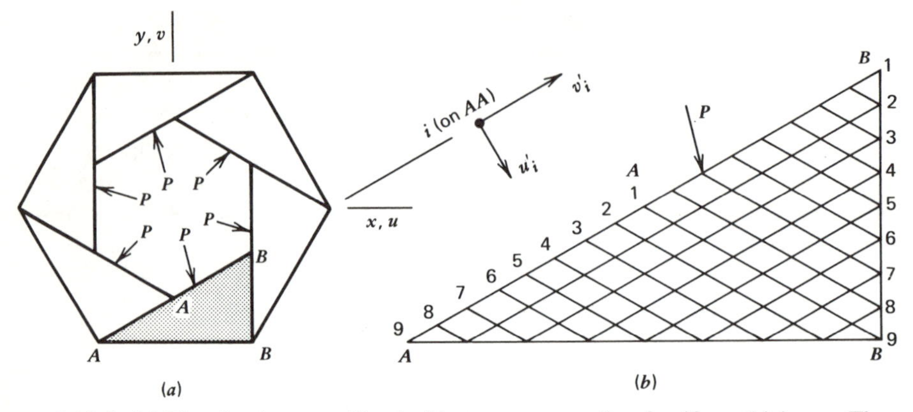

Figure 6.12.1. (*a*) The six plates are identical but not necessarily of uniform thickness. They are joined to form a structure with cyclic symmetry. (*b*) Typical repeating substructure, showing node numbers for automatic production of Eq. 6.12.4.

The stiffness equations $[\mathbf{K}]\{\mathbf{D}\} = \{\mathbf{R}\}$ for a typical substructure (Fig. 6.12.1*b*) are

$$\begin{bmatrix} \mathbf{K}_{II} & \mathbf{K}_{IA} & \mathbf{K}_{IB} \\ \mathbf{K}_{IA}^T & \mathbf{K}_{AA} & \mathbf{K}_{AB} \\ \mathbf{K}_{IB}^T & \mathbf{K}_{AB}^T & \mathbf{K}_{BB} \end{bmatrix} \begin{Bmatrix} \mathbf{D}_I \\ \mathbf{D}_A \\ \mathbf{D}_B \end{Bmatrix} = \begin{Bmatrix} \mathbf{R}_I \\ \mathbf{R}_A \\ \mathbf{O} \end{Bmatrix} + \begin{Bmatrix} \mathbf{O} \\ \mathbf{F}_A \\ \mathbf{F}_B \end{Bmatrix} \tag{6.12.1}$$

where $\{\mathbf{D}_A\}$ and $\{\mathbf{D}_B\}$ represent d.o.f. on edges AA and BB, and $\{\mathbf{D}_I\}$ represents all other d.o.f. of the substructure. Forces $\{\mathbf{F}_A\}$ and $\{\mathbf{F}_B\}$ are applied along AA and BB by neighboring substructures. Forces $\{\mathbf{R}_I\}$ and $\{\mathbf{R}_A\}$ represent applied loads (concentrated, distributed, thermal, and so on). Loads $\{\mathbf{R}_B\}$ are absent because, as subsequent discussion shows, external loads on substructure boundaries must be placed in either $\{\mathbf{R}_A\}$ or $\{\mathbf{R}_B\}$: if mistakenly placed in both, the substructure receives twice the load intended.

All substructures are identical. Therefore (see the following caution)

$$\{\mathbf{D}_B\} = \{\mathbf{D}_A\} \qquad \text{and} \qquad \{\mathbf{F}_B\} = -\{\mathbf{F}_A\} \tag{6.12.2}$$

Using techniques described in Section 6.8, we write

$$\{\mathbf{D}_I\ \mathbf{D}_A\ \mathbf{D}_B\} = [\mathbf{T}]\{\mathbf{D}_I\ \mathbf{D}_A\} \tag{6.12.3}$$

to express the first of Eqs. 6.12.2. Then we apply the constraint transformation and find

$$\begin{bmatrix} \mathbf{K}_{II} & \mathbf{K}_{IA} + \mathbf{K}_{IB} \\ \mathbf{K}_{IA}^T + \mathbf{K}_{IB}^T & \mathbf{K}_{AA} + \mathbf{K}_{AB} + \mathbf{K}_{AB}^T + \mathbf{K}_{BB} \end{bmatrix} \begin{Bmatrix} \mathbf{D}_I \\ \mathbf{D}_A \end{Bmatrix} = \begin{Bmatrix} \mathbf{R}_I \\ \mathbf{R}_A \end{Bmatrix} \tag{6.12.4}$$

The transformation incidentally produces the sum $\{\mathbf{F}_A\} + \{\mathbf{F}_B\}$, which vanishes according to Eq. 6.12.2.

Equation 6.12.4 can be produced *automatically* by the element assembly process, thus avoiding the constraint transformation. The trick, shown in Fig. 6.12.1*b*, is to assign the same node number to cyclically symmetric nodes along *AA* and *BB*. This numbering is only for element assembly: actual node point coordinates must be used in the generation of element matrices. (See the example problem in Section 6.8.)

Caution. Node patterns must match on boundaries shared by substructures. In Fig. 6.12.1 the number and spacing of nodes along *AA* must be the same as along *BB*. Furthermore, $\{\mathbf{D}_A\}$ and $\{\mathbf{D}_B\}$ must match. For example, let u_i and v_i be d.o.f. of the ith node along *BB* in Fig. 6.12.1*b*. Then corresponding d.o.f. normal and tangential to edge *AA* are needed at the ith node along *AA*. These are the u_i' and v_i' d.o.f. in the figure. Accordingly, if elements are generated with the usual u and v d.o.f., a transformation must be applied to elements along *AA*. The transformation must be done before the renumbering and assembly described in the preceding paragraph.

Configurations with a z-axis of rotational symmetry, such as Fig. 6.12.1, are a special case of cyclic symmetry called *sectorial symmetry* or *rotational periodicity*. In the xy plane, rotation about the z-axis is the only rigid-body motion possible for the substructure of Eq. 6.12.4. In three dimensions, translation along the z-axis is also possible. There must be enough supports to prevent these motions.

Reference 6.11 discusses the dynamics of rotationally periodic structures.

6.13 SUBSTRUCTURING

The division of a single structure into component substructures is called *substructuring* [6.12]. The procedure has other names in other contexts: *diakoptics* or *tearing* when used by electrical engineers, and *blocking* or *dissection* when used by numerical analysts. Figure 6.13.1 shows a structural application. The division of a substructure into further substructures can be repeated indefinitely.

The computational procedure is now outlined. Later we remark on why substructuring is done. A typical substructure is built of finite elements and has arbitrary loads on any or all nodes, as usual. The first analysis step determines nodal loads and stiffnesses seen by nodes on boundaries where the substructure will join with others. At this stage, the structure becomes a "superelement," all of whose nodes are on interface boundaries (such as boundary *AA* in Fig. 6.13.1*a*). Other substructures are similarly treated. In the second analysis step, substructures are assembled to yield $[\mathbf{K}]$. At this stage, all structure d.o.f. $\{\mathbf{D}\}$ lie on the interfaces. Internal d.o.f. of substructures do not appear. The third analysis step solves $[\mathbf{K}]\{\mathbf{D}\} = \{\mathbf{R}\}$ for $\{\mathbf{D}\}$. The fourth analysis step extracts from $\{\mathbf{D}\}$ the interface d.o.f. $\{\mathbf{d}\}$ of the substructure at hand and analyzes this substructure under the combination of prescribed boundary d.o.f. $\{\mathbf{d}\}$ and known loads on the nonboundary d.o.f. Hence stresses in individual elements are found in the usual way.

The first and fourth analysis steps amount to condensation and recovery of internal d.o.f. (Sections 7.2–7.4) It does not matter whether the substructure is a single element or many.

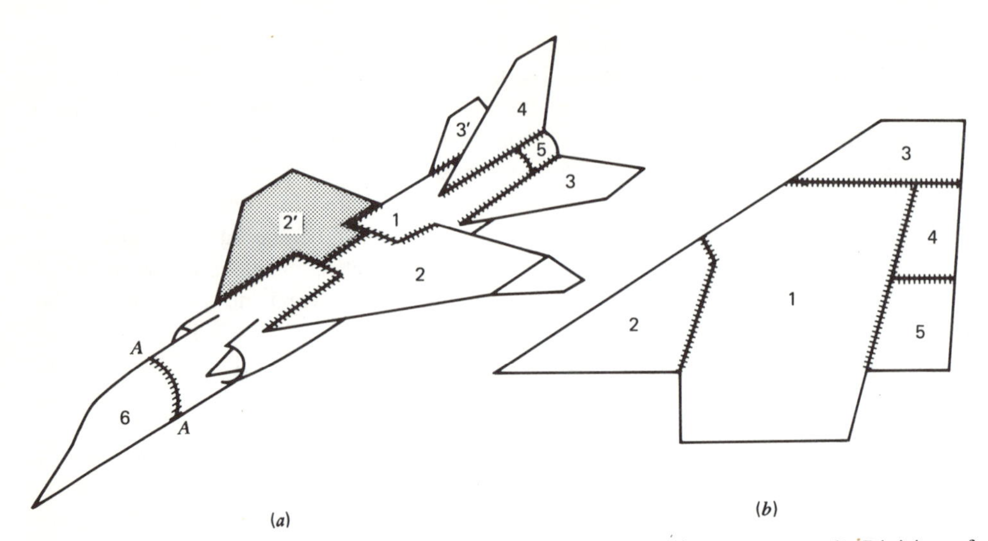

Figure 6.13.1. (*a*) An aircraft divided into substructures 1, 2, 2′, and so on. (*b*) Division of substructure 2′ into substructures.

In the "reduced" substructure method, only a fraction of the interface d.o.f. are used to connect substructures together. The rest are constrained to follow along in such a way that the interface remains a smooth curve (see boundary *AB* in Fig. 7.13.4).

No two substructures need be alike. But there is special advantage if a structure can be idealized as a repetition of a single substructure, as in Fig. 6.13.2*a*. When the first analysis step has produced the substructure [**k**] with respect to d.o.f. on *AA* and *BB*, this [**k**] need only be replicated with different node numbers to form the structure matrix [**K**] = Σ[**k**]. Indeed, the stiffness matrix of a substructure quadrant (Fig. 6.13.2*b*) can be transformed to the three other orientations, then the four matrices added to yield [**k**] of substructure *AABB*. Figure 6.12.1*a* can be similarly treated if the loads lack cyclic symmetry.

Boundaries between substructures should be easy to generate and, if possible, should be far from regions that may be refined or redesigned [6.13]. Boundary nodes on mating substructures must agree in topology and in orientation of their d.o.f. A user-oriented program should permit node numbers in one substructure to be independent of node numbers in others. In dynamic analysis, it may be desirable to retain some internal d.o.f. of substructures and carry them into the structure equations.

Advantages of substructuring include the following. Mass storage files may be inadequate for analysis of a nonsubstructured system with a huge number of d.o.f. Even if such an analysis is possible, a long residency time makes it more likely that the computer will crash before the job is finished. Substructuring replaces the long job by shorter jobs. A repetition of structural form can be exploited, as noted already. Separate substructures may be treatable by special programs of high quality (such as one for plates, another for shells, and so on). There is the managerial advantage of breaking a large problem into

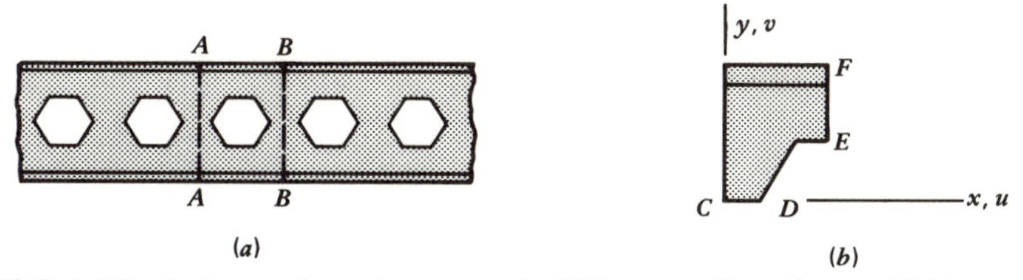

Figure 6.13.2. (*a*) Typical repeating substructure *AABB* in a castellated beam. (*b*) Upper left quarter of the typical substructure *AABB*.

smaller and more tractable parts. The different substructures can be simultaneously studied by different design groups. The work of one group can be almost independent of others if interaction between substructures is small. Design changes or analysis nonlinearities, if confined to a single substructure, leave matrices of other substructures unchanged. Thus computer cost is greatly reduced. The results of substructure analyses can be checked separately and revised if necessary before substructures are combined to form the complete structure.

Disadvantages of substructuring include the following. As compared with the computational cost of a *single-pass, single-load* analysis, the computational cost of substructuring is greater if there are no repeating substructures. (In practice, additional load cases and design changes are common, so a single analysis is rare.) Substructuring requires more computer runs. This is a disadvantage if turnaround is slow. The coding is more complicated, and more effort is needed to achieve efficiency in data structure and file handling. Bookkeeping and overhead expense increase, and the program becomes less portable. Indeed, successful implementation of substructuring is primarily a programming problem [6.12–6.16].

Substructuring can be used in a dynamic context. Then it is called *modal synthesis* or *component mode synthesis*. The procedure involves finding mode shapes and frequencies of the substructures, by experiment or by analysis, and assembling this information to find mode shapes and frequencies of the whole structure. Modal synthesis shares the advantages and disadvantages of static substructuring. See Ref. 2.1 and 6.17 to 6.22; Ref. 6.22 includes a literature survey. Application to transient heat conduction appears in Ref. 6.23.

PROBLEMS

6.1 Write matrix $[\mathbf{T}_\epsilon]$ of Section 6.2 for the layered orthotropic material shown in Fig. 8.2.1.

6.2 The material of the element in the sketch consists of very stiff fibers in a very weak matrix. Some fibers lie in direction β_1. Others lie in direction β_2. What are $[\mathbf{E}']$ matrices for the β_1 and β_2 layers of fibers? Does this suggest a way to get $[\mathbf{k}]$ for the complete element?

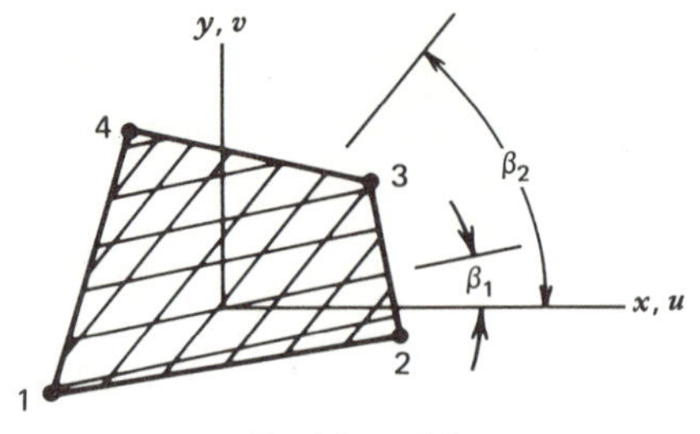

Problem 6.2

6.3 (a) Derive row 5 of **[T]** in Eq. 6.2.5. Use the procedure of Eqs. 6.2.1 to 6.2.3.
(b) Similarly, derive row 9 of **[T]** in Eq. 6.2.5.
(c) Show that **[T]** of Eq. 6.2.5 is orthogonal.
(d) Derive Eq. 6.2.7 by using Eqs. 6.2.5 and 1.4.4.
(e) Show that $[\mathbf{T}_\epsilon]$ of Eq. 6.2.7 is *not* orthogonal. *Suggestion*. Write $[\mathbf{T}_\epsilon] = [\mathbf{Q}]\,[\mathbf{T}]$, where **[Q]** contains 1's and 0's, instead of tackling $[\mathbf{T}_\epsilon]$ directly.
(f) Use Eqs. 6.2.14 to show that $[\mathbf{T}_\epsilon]^T\,[\mathbf{T}_\sigma] = [\mathbf{I}]$.
(g) Use Eqs. 6.2.6, 6.2.10, and 6.2.13 to obtain Eq. 6.2.9.
(h) Use the expressions $\{\boldsymbol{\sigma}'\} = [\mathbf{E}']\,\{\boldsymbol{\epsilon}'\}$, $\{\boldsymbol{\sigma}\} = [\mathbf{T}_\epsilon]^T\{\boldsymbol{\sigma}'\}$, and $\{\boldsymbol{\epsilon}'\} = [\mathbf{T}_\epsilon]\,\{\boldsymbol{\epsilon}\}$ to obtain Eq. 6.2.13.

6.4 Let $[\mathbf{E}']$ be 3 by 3, as for plane stress. Verify that Eq. 6.2.13 yields $[\mathbf{E}] = [\mathbf{E}']$ if the material is isotropic.

6.5 Imagine that we want to calculate the stress σ_n normal to edge 2-3 of a four-node plane quadrilateral as follows. First, set up local *xy*-axes such that the *x*-axis passes through diagonally opposite nodes 1 and 3. Then calculate stresses $\{\boldsymbol{\sigma}\} = [\mathbf{E}]\,[\mathbf{B}]\,\{\mathbf{d}\}$, with **[B]** referred to the local axes. Finally, extract σ_n from $\{\boldsymbol{\sigma}\}$. Explain in detail the necessary transformations. *Note*. This is an exercise, not a recommended procedure.

6.6 (a) Verify the **[k]** found in Problem 2.7 by coordinate transformation of Eq. 2.4.3.
(b) Obtain the same result by coordinate transformation of **[k]** in Eq. 2.4.5.
(c) For an arbitrarily oriented plane frame element (Fig. 6.3.1), find the stiffness matrix **[k]** produced by Eq. 6.3.7, where $[\mathbf{k}']$ operates on local (primed) d.o.f. Write the result compactly, so that it could be effectively coded.

6.7 (a) Derive Eq. 6.3.7 by arguing that strain energy must be the same regardless of the coordinate system in which a virtual displacement is expressed.
(b) Similarly, use a kinetic energy argument to derive Eq. 6.3.8.

6.8 Verify Eq. 6.4.4.

6.9 Consider the plane analogue of the problem in Fig. 6.5.1. Now only a rotation of the element in the *xy* plane is required. Write equations analogous to Eqs. 6.5.4 and 6.5.5. Thus convert Eq. 2.4.5 to Eq. 2.4.3.

6.10 If *A*, *E*, and nodal coordinates are given in Fig. 6.5.1, how many Fortran statements are needed to generate **[k]** in global coordinates? (Write these statements.)

6.11 The size, shape, and orientation in space of a plane triangular element are defined by the known global coordinates of its three corner nodes.

(a) Let node 3 be at the origin of local coordinates $x'y'z'$. Also, let nodes 3 and 1 define the x'-axis and let the plane of the element define the $x'y'$ plane. Describe how to compute the direction cosines from the given information.

(b) If the element is a constant strain triangle, write the matrix [**T**] that will produce the 9-by-9 global matrix [**k**] from the 6-by-6 local matrix [**k**′].

6.12 The 6-by-6 stiffness matrix [**k**′] of the constant strain triangle 1-2-3 in the sketch operates on d.o.f. $\{u_1\ w_1\ u_2\ w_2\ u_3\ w_3\}$. The element is to be rotated about the x-axis to position ABC, where its 9-by-9 stiffness matrix [**k**] is to operate on d.o.f. $\{u_A\ v_A\ w_A\ u_B\ v_B\ w_B\ u_C\ v_C\ w_C\}$. Write the appropriate transformation matrix.

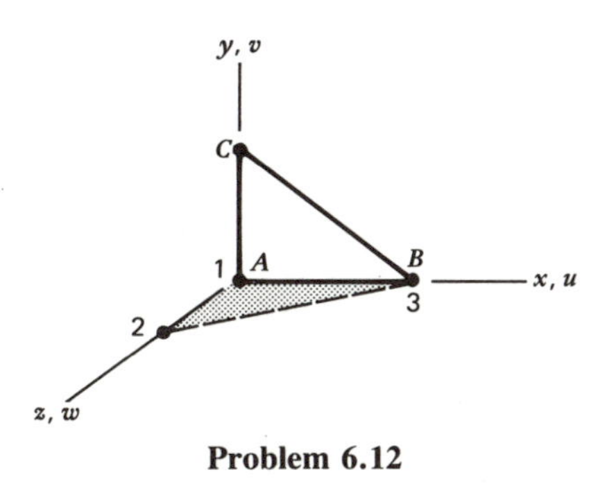

Problem 6.12

6.13 Let node 5 of the frame element in Fig. 6.6.1a lie at an arbitrary point *within* the four-node element rather than at an edge. Suggest a procedure to find a transformation matrix that yields the three frame d.o.f. from the eight d.o.f. of the plane element.

6.14 (a) Write the transformation matrix for the problem described in connection with Eq. 6.6.4. What is [**k**′] for this problem?

(b) Imagine the analogous problem in space: a layer of parallel bars is to be modeled as a quadrilateral whose four nodes are to be attached to edges of an eight-node solid. How would you formulate [**k**] for the quadrilateral? On what would its [**E**] depend?

6.15 Plane element 1 in the sketch has the usual two d.o.f. per node (Fig. 4.6.1). Plane *frame* element 2 has the usual three d.o.f. per node (u_i, w_i, θ_i). Consider the element stiffness matrices [$\mathbf{k}_1$] and [$\mathbf{k}_2$].

(a) Write a transformation matrix [$\mathbf{T}_1$] that could be used to convert [$\mathbf{k}_1$] so that it operates on the d.o.f. of element 2.

(b) Write a transformation matrix [$\mathbf{T}_2$] that could be used to convert [$\mathbf{k}_2$] so that it operates on the d.o.f. of element 1.

(c) Should [$\mathbf{T}_1$] [$\mathbf{T}_2$] and [$\mathbf{T}_2$] [$\mathbf{T}_1$] be unit matrices? Find an argument that says so.

(d) Evaluate the products [$\mathbf{T}_1$] [$\mathbf{T}_2$] and [$\mathbf{T}_2$] [$\mathbf{T}_1$]. How can the results be explained?

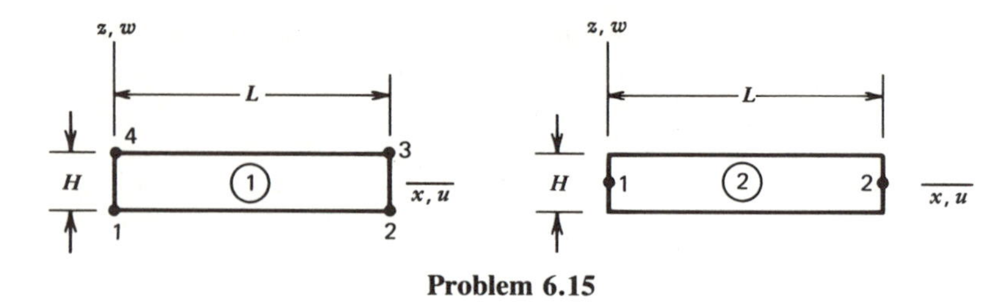

Problem 6.15

6.16 Imagine that Fig. 6.7.1*a* represents a frame element in space. Let the element and the rigid links have arbitrary orientations. Then $\{\mathbf{d}'\} = \{u_i \; v_i \; w_i \; \theta_{xi} \; \theta_{yi} \; \theta_{zi}\}$, where the θ d.o.f. vectors point in positive global coordinate directions. Write the appropriate form of $[\Lambda_1]$ (see Eq. 6.7.1).

6.17 In Section 6.7 we note that rigid links create a nondiagonal mass matrix $[\mathbf{m}]$. Suggest at least one way to make $[\mathbf{m}]$ diagonal.

6.18 (a) Write the coefficients in $[\mathbf{T}]$ of Eq. 6.7.3.
(b) In part (a), why is the condition $y_1 = y_2$ unacceptable? What can be done if $y_1 \approx y_2$?

6.19 Consider the discussion associated with Fig. 6.7.1*b*. What happens if θ_i is condensed in *all* elements that terminate at node i? What if the bars remain rigidly joined, but the joint itself is connected through a hinge to a rigid base?

6.20 In the 10 d.o.f. element of Fig. 7.2.1*a*, imagine that d.o.f. u_5 and v_5 are to be replaced by d.o.f. u_{5r} and v_{5r}, which represent the motion of node 5 *relative* to the average motions $\Sigma\, u_i/4$ and $\Sigma\, v_i/4$, where $i = 1, 2, 3, 4$. Write the arrays in the transformation $\{\mathbf{d}'\} = [\mathbf{T}]\{\mathbf{d}\}$, where $[\mathbf{T}]$ is 10 by 10.

6.21 Consider the eight-node plane element of Fig. 5.6.1*a*. Let sides be straight and side nodes be at midsides. Imagine that each d.o.f. of each side node is to be converted to motion *relative* to the average motion dictated by the two adjacent corner nodes. (See the caution in Section 7.12.)
(a) Write a transformation matrix that would convert the "usual" stiffness matrix so that it operates on the new nodal displacement vector.
(b) In a mesh of such elements, how would you enforce linearity of displacement along all sides?

6.22 Consider the 60 d.o.f. solid element of Fig. 5.7.1*b*. Imagine that we want to obtain from it a 48 d.o.f. element, in which nodes 9, 11, 13, and 15 do not appear because displacements along edges that join the $\zeta = 1$ and $\zeta = -1$ surfaces are constrained to be linear. Let the nodes to be eliminated be at midedge. By means of a transformation, express the 16 shape functions $\bar{N}_i$ of the 48 d.o.f. element in terms of the 20 shape functions N_i of the 60 d.o.f. element.

6.23 Assume that $\{\mathbf{Q}\} \neq 0$ in Eq. 6.8.1. Start with equations $[\mathbf{K}']\{\mathbf{D}_e \; \mathbf{D}_r\} = \{\mathbf{R}_e' \; \mathbf{R}_r'\}$, in which $[\mathbf{K}']$ is partitioned into four parts that correspond to partitioning in the vectors. Find expressions for $[\mathbf{K}]$ and $\{\mathbf{R}_r\}$ in the transformed and condensed equations $[\mathbf{K}]\{\mathbf{D}_r\} = \{\mathbf{R}_r\}$. Express answers in terms of submatrices in the partitioned equation.

6.24 A constraint condition is written in the form

$$\{\mathbf{D}_r\ \mathbf{D}_e\} = \begin{bmatrix} \mathbf{I} \\ \mathbf{T} \end{bmatrix} \{\mathbf{D}_r\}$$

in Eq. 6.7.3. Devise a formula that converts this constraint condition to the form $[\mathbf{C}_e\ \mathbf{C}_r]\{\mathbf{D}_e\ \mathbf{D}_r\} = 0$, as in Eq. 6.8.2.

6.25 Write Eq. 6.7.3 in the form of Eq. 6.8.1 (see also Problem 6.24). How many equations of constraint are there? How many would there be if the element were a rigid eight d.o.f. quadrilateral?

6.26 (a) Consider the eight-node plane element of Fig. 5.6.1*a*. Let sides be straight and side nodes be at midsides. Write a constraint equation in the form of Eq. 6.8.1 that constrains u and v along each edge to be linear in the edge-parallel coordinate.

(b) Let nodes 1 to 8 of this element be the first eight nodes in a mesh of n nodes, where $n >> 8$. Imagine that the constraint of part (a) is to be applied to only this one element in the mesh. What change would you make in the element constraint of part (a) to make the constraint expression applicable on the structure level?

6.27 (a) Write the specific form of Eqs. 6.8.1 appropriate to Fig. 6.8.1*b*.

(b) For the same problem, write [**T**] in the relation $\{u_A\ v_A\ u_B\ v_B\} = [\mathbf{T}]\{u_A\ v_A\ u_r\ v_r\}$, where $u_r = u_B - u_A$ and $v_r = v_B - v_A$. Outline the complete procedure of generating the transformed equations, how u_r and v_r are set to zero, and possible advantages and disadvantages of this method of enforcing the constraints $u_A = u_B$ and $v_A = v_B$.

6.28 Three nodes lie on an x-axis at coordinates x_1 to x_3. Write a relation that constrains their x-direction displacements to be directly proportional to x.

6.29 Element 1 is to be connected to elements 2 and 3 at the two nodes indicated. Node 3 is to have no d.o.f. in the structure vector {**D**}. All three elements have linear edges. How would you treat the {**r**′} and [**k**′] matrices of elements 2 and 3?

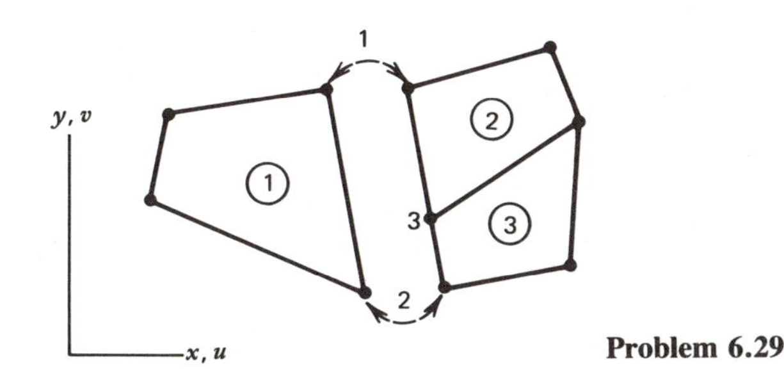

Problem 6.29

6.30 The sketch shows two plane truss members. With $i = 1, 2$, element force and displacement vectors are $\{\mathbf{r}_i\}$ and $\{\mathbf{d}_i\}$, each 4 by 1. Element stiffness matrices are $[\mathbf{k}_i]$, each 4 by 4. A two-element, six d.o.f. structure is produced by joining nodes B and C. (This problem illustrates the use of transformation matrices to assemble elements. The procedure was popular in early descriptions of matrix structural analysis.)
  (a) Write $\{\mathbf{d}_1\}$ and $\{\mathbf{d}_2\}$ consecutively in an 8-by-1 vector $\{\mathbf{d}_1 \ \mathbf{d}_2\}$. Then write $[\mathbf{T}]$ in the relation $\{\mathbf{d}_1 \ \mathbf{d}_2\} = [\mathbf{T}]\{\mathbf{D}\}$, where $\{\mathbf{D}\} = \{u_1 \ v_1 \ u_2 \ v_2 \ u_3 \ v_3\}$.
  (b) Similarly, write the 8-by-1 vector of nodal force $\{\mathbf{r}_1 \ \mathbf{r}_2\}$ and relate it to the 6-by-1 vector of structure forces $\{\mathbf{R}\}$.
  (c) Write the unassembled stiffness matrix $[\mathbf{k}_1 \ \mathbf{k}_2]$, which is 8 by 8 and has two 4-by-4 null matrices off-diagonal. Find the relation between $[\mathbf{k}_1 \ \mathbf{k}_2]$ and the 6-by-6 structure matrix $[\mathbf{K}]$.
  (d) Carry out the transformations. Verify the correctness of the resulting $\{\mathbf{R}\}$ and $[\mathbf{K}]$.

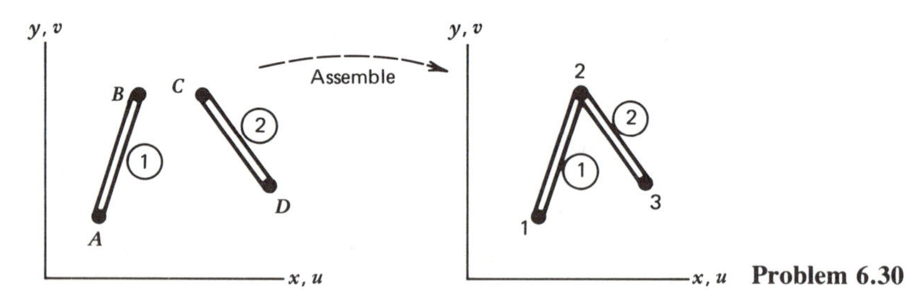

Problem 6.30

6.31 (a) Solve the example problem of Section 6.8 by the "same node number" trick suggested below Eq. 6.8.7.
  (b) If Fig. 6.8.2 were to represent three *beam* elements under *lateral* load, what sort of constraint would be enforced by the "same node number" trick applied to nodes 1 and 2? Would $[\mathbf{k}]$ of the left element contribute to $[\mathbf{K}]$ of the structure? If so, what is it contribution?

6.32 (a) Derive Eq. 6.8.13 from Eqs. 6.8.8 through 6.8.10.
  (b) Solve the example problem of Section 6.8 by the method of Eq. 6.8.13.

6.33 An m-by-n rectangular mesh is built from mn plane stress rectangular elements like the one in Fig. 4.6.1. Imagine that no volume change is permitted in any element; that is, constraints $\int(\epsilon_x + \epsilon_y)\,dx\,dy = 0$ are imposed. What constraint equations would you write? By what method would you impose constraints? Would you work element by element or at the structure level?

6.34 (a) Derive Eq. 6.10.3 from Eqs. 6.10.1 and 6.10.2.
  (b) Revise the argument associated with Eq. 6.10.5 without making the simplifying assumptions $[\boldsymbol{\alpha}] = [\mathbf{I}]\alpha$ and $\{\mathbf{Q}\} = 0$.

6.35 A spring of unit stiffness lies along the x-axis. Its right end carries a force $P = 3$ in the $+x$ direction. Its left end is to be displaced two units toward the right. Impose the displacement condition $u_1 = 2$ and solve for u_2 by (*a*) Lagrange multipliers, and (*b*) a penalty function. In part (b), tabulate u_1 and u_2 versus α for $\alpha = 1, 4,$ 10, and 100.

6.36 (a) Solve Problem 6.35 by using transformation equations (see Problem 6.23).
(b) Solve Problem 6.35 by using Eq. 6.8.13.

6.37 Imagine that a prescribed axial strain $\bar{\epsilon}$ in bar 2-4 of Fig. 2.7.1 is imposed by adding a bar of high stiffness AE/L between nodes 2 and 4 and loading it by opposing axial forces $F = EA\bar{\epsilon}$. In the symbolism of Section 6.10, what are $[\mathbf{C}]$, $[\boldsymbol{\alpha}]$, and $\{\mathbf{Q}\}$?

6.38 The pin-jointed plane hexagon is loaded by equal forces P, each radial from center O. All bars are uniform and identical. Consider bar BC.
(a) If $\{\mathbf{d}'\} = \{u_B\ v_B\ u_C\ v_C\}$, what $[\mathbf{T}]$ relates $\{\mathbf{d}'\}$ to four d.o.f. $\{\mathbf{d}\}$ in the r and θ directions at B and C?
(b) What boundary conditions and constraints can be imposed on $\{\mathbf{d}\}$ so that bar BC and a single d.o.f. in $\{\mathbf{d}\}$ are representative of the entire hexagon? Write the transformation matrix.
(c) Carry out the operations. Find the radial displacement of a typical node. Check your result by elementary theory.

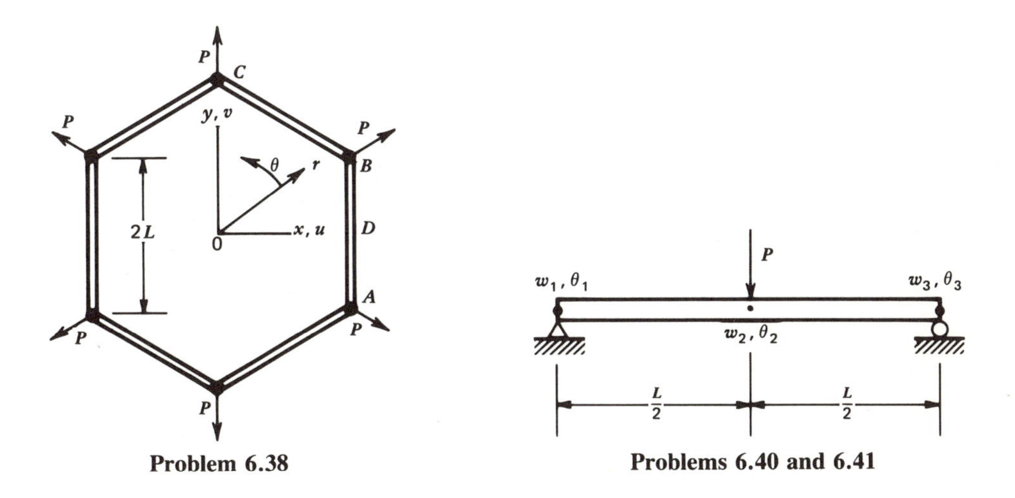

Problem 6.38 **Problems 6.40 and 6.41**

6.39 If the pipe intersection in Fig. 5.7.2 is symmetric with respect to each of the three coordinate planes, what displacement boundary conditions should be imposed on the mesh shown?

6.40 The uniform, simply supported beam is modeled by two beam elements, each of length $L/2$. Load P acts at midspan.
(a) Write the 6-by-6 structure stiffness matrix and the 6-by-1 load vector.
(b) Reduce the order of the matrices by imposing the essential boundary conditions.
(c) Impose symmetry about the midpoint. (After this step, there are only two equations left.)
(d) Solve for the midspan displacement w_2.

6.41 Let P in Problem 6.40 be replaced by a counterclockwise moment M_0 at midspan. Repeat Problem 6.40, but impose antisymmetry in part (c) and solve for θ_2 in part (d).

6.42 (a) Make sketches and add detail and complete the symmetry argument outlined in Section 6.11 with reference to Fig. 6.11.1*c*.

(b) A skew-symmetric thin-plate problem is mentioned at the end of Section 6.11. For this problem, how do slopes, moments, and shears at $+r$ relate to those at $-r$? Repeat for the skew-antisymmetric case.

(c) Imagine that nodes 2 and 4 in Fig. 6.11.1*b* must have equal rotations about direction BD and equal but opposite rotations about direction AC. Derive the slope relations (Eq. 6.11.1).

6.43 A simply supported beam is modeled by two elements of equal length. A uniform lateral load acts along the left half. Describe how the problem can be analyzed as the sum of symmetric and anti-symmetric states. Does this approach have any advantage over a single analysis?

6.44 Sketch an example structure that has cyclic symmetry but lacks rotational symmetry.

6.45 In Eq. 6.12.1, let m be the number of d.o.f. in $\{\mathbf{D}_I\}$ and let n be the number of d.o.f. in $\{\mathbf{D}_A\ \mathbf{D}_B\}$. Write $[\mathbf{T}]$ of Eq. 6.12.3 in terms of unit matrices. In terms of m and n, how many rows and columns are in its submatrices and in the submatrices of Eq. 6.12.1? Show that the constraint transformation yields Eq. 6.12.4 from Eq. 6.12.1.

6.46 Consider the structure used in Problem 6.38. Add a node at D so as to model link AB by two bars of equal length and stiffness AE/L. Solve for the radial displacement at B by taking advantage of cyclic symmetry. Give nodes A and B the same number when assembling the two bars so as to obtain Eq. 6.12.4 directly, with a 4-by-4 stiffness matrix. Restrict d.o.f. at node D to impose boundary conditions.

6.47 Let d.o.f. in Fig. 6.13.2*b* be u_i and v_i at all nodes. If $[\mathbf{k}]$ is known for this portion, what constraint or relation must its d.o.f. satisfy if $[\mathbf{T}]^T\ [\mathbf{k}]\ [\mathbf{T}]$ is to yield $[\mathbf{k}]$ for one of the other three parts? (There are three answers, one for each part.) No specific number or location of nodes need be assumed, and $[\mathbf{T}]$ need not be written out for any of the three parts.

CHAPTER 7

TOPICS IN ELEMENT FORMULATION AND USE

7.1 THE ELASTIC KERNEL AND THE GENERAL ELEMENT

Some nodal d.o.f. in the element vector $\{\mathbf{d}\}$ can be used to define rigid-body motion of the element. Let these d.o.f. be called $\{\mathbf{d}_R\}$. The choice of d.o.f. to be called $\{\mathbf{d}_R\}$ is not unique. Let the remaining d.o.f. be $\{\mathbf{d}_E\}$, so that $\{\mathbf{d}\} = \{\mathbf{d}_R \; \mathbf{d}_E\}$. D.o.f. $\{\mathbf{d}_E\}$ define straining modes. A stiffness matrix $[\mathbf{k}_{EE}]$ that operates on only $\{\mathbf{d}_E\}$ can be transformed to one that operates on $\{\mathbf{d}\}$ and includes rigid-body motion capability. In this section we formulate the procedure and comment on why it is useful.

First, the element stiffness equation $[\mathbf{k}]\{\mathbf{d}\} = \{\bar{\mathbf{r}}\}$ is partitioned.

$$\begin{bmatrix} \mathbf{k}_{RR} & \mathbf{k}_{RE} \\ \mathbf{k}_{RE}^T & \mathbf{k}_{EE} \end{bmatrix} \begin{Bmatrix} \mathbf{d}_R \\ \mathbf{d}_E \end{Bmatrix} = \begin{Bmatrix} \bar{\mathbf{r}}_R \\ \bar{\mathbf{r}}_E \end{Bmatrix} \tag{7.1.1}$$

The lower partition is solved for $\{\mathbf{d}_E\}$. Also, this expression for $\{\mathbf{d}_E\}$ is substituted into the upper partition. Thus

$$\begin{bmatrix} (\mathbf{k}_{RR} - \mathbf{k}_{RE}\mathbf{k}_{EE}^{-1}\mathbf{k}_{RE}^T) & \mathbf{k}_{RE}\mathbf{k}_{EE}^{-1} \\ -\mathbf{k}_{EE}^{-1}\mathbf{k}_{RE}^T & \mathbf{k}_{EE}^{-1} \end{bmatrix} \begin{Bmatrix} \mathbf{d}_R \\ \bar{\mathbf{r}}_E \end{Bmatrix} = \begin{Bmatrix} \bar{\mathbf{r}}_R \\ \mathbf{d}_E \end{Bmatrix} \tag{7.1.2}$$

Matrix $[\mathbf{k}_{EE}]$ is symmetric and is invertible because rigid-body motion is prevented. If there is no elastic distortion, then $\{\bar{\mathbf{r}}_E\} = 0$ and, therefore,

$$-[\mathbf{k}_{EE}]^{-1}[\mathbf{k}_{RE}]^T\{\mathbf{d}_R\} = \{\mathbf{d}_E\} \tag{7.1.3}$$

With $\{\bar{\mathbf{r}}_E\} = 0$, only rigid-body motion is possible. Therefore nodal d.o.f. are related strictly by kinematics, expressed by a matrix $[\mathbf{T}]$ of element dimensions.

$$\{\mathbf{d}_E\} = [\mathbf{T}]\{\mathbf{d}_R\} \tag{7.1.4}$$

Usually $[\mathbf{T}]$ has more rows than columns. Because Eqs. 7.1.3 and 7.1.4 must be true for *any* $\{\mathbf{d}_R\}$, we conclude that

$$[\mathbf{k}_{RE}]^T = -[\mathbf{k}_{EE}][\mathbf{T}] \tag{7.1.5}$$

Next imagine that $\{\bar{\mathbf{r}}_E\} \neq 0$. Because $\{\mathbf{d}_R\}$ contains only enough d.o.f. to prevent rigid-body motion, $\{\bar{\mathbf{r}}_R\}$ can be computed from $\{\bar{\mathbf{r}}_E\}$ entirely by equations of statics, independently of $\{\mathbf{d}_R\}$. Therefore the upper partition of Eq. 7.1.2 and Eq. 7.1.5 yields

$$[\mathbf{k}_{RR}] = [\mathbf{k}_{RE}][\mathbf{k}_{EE}]^{-1}[\mathbf{k}_{RE}]^T = [\mathbf{T}]^T[\mathbf{k}_{EE}][\mathbf{T}] \tag{7.1.6}$$

The stiffness matrix that operates on *all* nodal d.o.f, $\{\mathbf{d}\} = \{\mathbf{d}_R \ \mathbf{d}_E\}$, is therefore

$$[\mathbf{k}] = \begin{bmatrix} \mathbf{T}^T\mathbf{k}_{EE}\mathbf{T} & -\mathbf{T}^T\mathbf{k}_{EE} \\ -\mathbf{k}_{EE}\mathbf{T} & \mathbf{k}_{EE} \end{bmatrix} \tag{7.1.7}$$

Matrix $[\mathbf{k}_{EE}]$ is called the *elastic kernel*.

The principal merit of Eq. 7.1.7 is that it permits the user to employ an element of his or her own divising. This is the *general element*. Its properties can be found by theory or by experiment. It is introduced into a computer program by prescribing as input data the coefficients in $[\mathbf{k}_{EE}]$. Alternatively, coefficients in the flexibility matrix $[\mathbf{k}_{EE}]^{-1}$ can be prescribed, and $[\mathbf{k}_{EE}]$ can be found by inversion. Equation 7.1.7 then produces a $[\mathbf{k}]$ that *requires no force to produce rigid-body motion*. If the user were required to prescribe the *entire* $[\mathbf{k}]$, the individual k_{ij} must be of full computer-word accuracy to avoid the possibility of introducing serious errors (Section 15.5). By use of Eq. 7.1.7, slight errors in the $(k_{EE})_{ij}$ produce only slight defects in elastic response; they do not cause rigid-body motion to be misrepresented.

Example

Consider the standard beam element of Fig. 4.4.1*b*. Let $\{\mathbf{d}_R\} = \{w_1 \ \theta_1\}$. Then $\{\mathbf{d}_E\} = \{w_2 \ \theta_2\}$. From Eq. 4.4.4, the curvature

$$w_{,xx} = \left\lfloor \frac{6}{L^2} - \frac{12x}{L^3} \quad -\frac{2}{L} + \frac{6x}{L^2} \right\rfloor \{\mathbf{d}_E\} \tag{7.1.8}$$

is used to construct $[\mathbf{k}_{EE}]$, which is found to be the lower right 2-by-2 submatrix in Eq. 4.4.5. Equation 7.1.4 becomes

$$\begin{Bmatrix} w_2 \\ \theta_2 \end{Bmatrix} = [\mathbf{T}]\begin{Bmatrix} w_1 \\ \theta_1 \end{Bmatrix}, \quad \text{where} \quad [\mathbf{T}] = \begin{bmatrix} 1 & L \\ 0 & 1 \end{bmatrix} \tag{7.1.9}$$

Equation 7.1.7 then yields $[\mathbf{k}]$ of Eq. 4.4.5.

a-Basis Formulation. Now we can see that matrix $[\mathbf{k}_a]$ in Eq. 4.3.19 is an elastic kernel. Again, using the beam element as an example, we find from Eqs. 4.2.1 that

$$w_{,xx} = \lfloor \mathbf{B}_a \rfloor \{a_3 \; a_4\}, \qquad \text{where} \qquad \lfloor \mathbf{B}_a \rfloor = \lfloor 2 \quad 6x \rfloor \tag{7.1.10}$$

Coefficients a_1 and a_2 describe only rigid-body motion and need not appear. The stiffness matrix that operates on $\{a_3 \; a_4\}$ is

$$[\mathbf{k}_a] = EI \begin{bmatrix} 4L & 6L^2 \\ 6L^2 & 12L^3 \end{bmatrix} \tag{7.1.11}$$

If a_1 and a_2 were retained to make $[\mathbf{k}_a]$ a 4-by-4 matrix, additional zeros would appear in $\lfloor \mathbf{B}_a \rfloor$ and $[\mathbf{k}_a]$.

To convert $[\mathbf{k}_a]$ of Eq. 7.1.11 to $[\mathbf{k}_{EE}]$, we write $w = a_3x^2 + a_4x^3$ for a cantilever fixed at the left end and evaluate w_2 and θ_2. Thus

$$\begin{Bmatrix} w_2 \\ \theta_2 \end{Bmatrix} = [\mathbf{A}] \begin{Bmatrix} a_3 \\ a_4 \end{Bmatrix}, \qquad \text{where} \qquad [\mathbf{A}] = \begin{bmatrix} L^2 & L^3 \\ 2L & 3L^2 \end{bmatrix} \tag{7.1.12}$$

Hence $[\mathbf{k}_{EE}] = [\mathbf{A}]^{-T}[\mathbf{k}_a][\mathbf{A}]^{-1}$ is found to be identically the $[\mathbf{k}_{EE}]$ that results from Eq. 7.1.8. Then Eqs. 7.1.7 and 7.1.9 yield $[\mathbf{k}]$.

7.2 NODES WITHIN ELEMENTS. CONDENSATION

In this section and in Section 7.3 we consider the purpose and theory of internal d.o.f. Computational algorithms are discussed in Section 7.4.

Assemblages of elements are shown in Fig. 7.2.1. Each can be regarded as a substructure, or simply as an element with subelements and internal nodes. These examples have

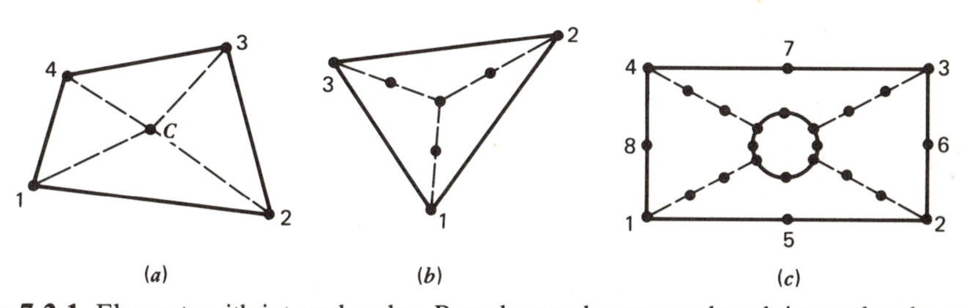

Figure 7.2.1. Elements with internal nodes. Boundary nodes are numbered; internal nodes are not. (*a*) A plane quadrilateral built from four constant strain triangles. (*b*) A plate bending element that has three subtriangles. (*c*) A plane element with a hole, modeled by four degenerate cubic elements.

2, 12, and 32 internal d.o.f., respectively. In each case it is expected that the user of a computer program will provide data about only the external (numbered) nodes. The program will locate the internal nodes, generate and combine matrices of the subelements, and produce a stiffness matrix and load vector associated with d.o.f. of only the external nodes. The motivation is to improve accuracy by using more d.o.f., without burdening the user with the task of defining subelements.

The element of Fig. 7.2.1*a* was suggested long ago [1.7]. It is much better than a two-triangle quadrilateral but no better than the linear isoparametric element of Section 5.3. (Similarly, in three dimensions, combinations of tetrahedra are no better than a linear isoparametric solid [7.1]). The element in Fig. 7.2.1*b* is described in [9.3]. Elements such as that of Fig. 7.2.1*c* are described in Section 7.13.

Let d.o.f. $\{\mathbf{d}\}$ of an assemblage be partitioned so that $\{\mathbf{d}\} = \{\mathbf{d}_r\ \mathbf{d}_e\}$, where $\{\mathbf{d}_r\}$ are boundary d.o.f. to be retained and $\{\mathbf{d}_e\}$ are internal d.o.f. to be eliminated. Element stiffness equations are then

$$\begin{bmatrix} \mathbf{k}_{rr} & \mathbf{k}_{re} \\ \mathbf{k}_{er} & \mathbf{k}_{ee} \end{bmatrix} \begin{Bmatrix} \mathbf{d}_r \\ \mathbf{d}_e \end{Bmatrix} = \begin{Bmatrix} \mathbf{r}_r \\ \mathbf{r}_e \end{Bmatrix} \tag{7.2.1}$$

We regard Eq. 7.2.1 as a "fragment" of the structural equations, so the right side represents loads applied *to* nodes *by* elements. The lower partition of Eq. 7.2.1 is solved for $\{\mathbf{d}_e\}$.

$$\{\mathbf{d}_e\} = -[\mathbf{k}_{ee}]^{-1}([\mathbf{k}_{er}]\{\mathbf{d}_r\} - \{\mathbf{r}_e\}) \tag{7.2.2}$$

This expression for $\{\mathbf{d}_e\}$ is substituted into the upper partition of Eq. 7.2.1. Thus

$$\begin{aligned} [\mathbf{k}]\{\mathbf{d}_r\} &= \{\mathbf{r}\}, \qquad \text{in which} \\ [\mathbf{k}] &= [\mathbf{k}_{rr}] - [\mathbf{k}_{re}][\mathbf{k}_{ee}]^{-1}[\mathbf{k}_{er}] \\ \{\mathbf{r}\} &= \{\mathbf{r}_r\} - [\mathbf{k}_{re}][\mathbf{k}_{ee}]^{-1}\{\mathbf{r}_e\} \end{aligned} \tag{7.2.3}$$

The process symbolized in Eq. 7.2.3 is called *condensation* or *static condensation*. It is a manipulation, not an approximation.

The condensed element is now treated like any other: $[\mathbf{k}]$ and $\{\mathbf{r}\}$ in Eq. 7.2.3 are assembled into the structure in the usual way. When $\{\mathbf{d}_r\}$ is known because structural equations have been solved for $\{\mathbf{D}\}$, Eq. 7.2.2 yields $\{\mathbf{d}_e\}$. This process is called *recovery*. The $\{\mathbf{d}_e\}$ may be needed for stress calculation. Equation 7.2.2 serves as a constraint relation and, indeed, can be cast in the form of Eq. 6.8.1.

It is clear, especially if $\{\mathbf{d}_e\}$ contains a single d.o.f., that Eqs. 7.2.3 represent the Gauss elimination method of solving equations. But in Eqs. 7.2.3 we stop when only the internal d.o.f. $\{\mathbf{d}_e\}$ have been eliminated. Elimination of the remaining d.o.f. $\{\mathbf{d}_r\}$ is done on the structural level, after all elements have been assembled. Thus condensation is simply the first step in solving the structural equations $[\mathbf{K}]\{\mathbf{D}\} = \{\mathbf{R}\}$. The same solution vector $\{\mathbf{D}\}$ would result if *all* d.o.f. were carried into $\{\mathbf{D}\}$. Then there would be more structural equations. However, the effort required to solve them would be no greater if the d.o.f. in $\{\mathbf{D}\}$ were ordered so that all $\{\mathbf{d}_e\}$ were processed first—but this is accomplished when the $\{\mathbf{d}_e\}$ are condensed before assembly. Computer programming for condensation and recovery is simpler than programming that would be needed if the $\{\mathbf{d}_e\}$ were carried into $\{\mathbf{d}\}$.

Equation 7.2.2 shows that $\{\mathbf{d}_e\}$ depends on internal loads $\{\mathbf{r}_e\}$, stiffness coefficients k_{ij}, and element boundary d.o.f. $\{\mathbf{d}_r\}$. In other words, condensed d.o.f. "ride along" in the same way as d.o.f. processed by Gauss elimination, as described in Section 2.11. Accordingly, if d.o.f. of a node on the element boundary were condensed and the affected rows and columns of $[\mathbf{k}]$ filled with zeros before assembly, the boundary node would be detached from the structure. Its d.o.f. would be dictated by element coefficients in Eq. 7.2.2.

7.3 NODELESS D.O.F. GLOBAL-LOCAL FORMULATION

Conventionally, stiffness matrix $[\mathbf{k}]$ operates on nodal d.o.f. $\{\mathbf{d}\}$. Section 4.3 also describes a stiffness matrix $[\mathbf{k}_a]$ that operates on generalized coordinates $\{\mathbf{a}\}$. The two concepts are not mutually exclusive: an element can have d.o.f. $\{\mathbf{d}\}$ and $\{\mathbf{a}\}$ at the same time. However, in this context $\{\mathbf{a}\}$ represents *additional* element d.o.f. and is not another way of representing $\{\mathbf{d}\}$. These a_i are called *internal* or *nodeless* d.o.f.

For example, consider a four-node plane quadrilateral with the displacement field

$$u = \sum_1^4 N_i u_i + N_5 a_1, \qquad v = \sum_1^4 N_i v_i + N_5 a_2 \tag{7.3.1}$$

where N_1 to N_4 are given by Eqs. 5.3.3 and N_5 is the "bubble function"

$$N_5 = (1 - \xi^2)(1 - \eta^2) \tag{7.3.2}$$

The fifth mode has no effect on displacements along element edges. The d.o.f. a_1 and a_2 are internal to the element and are not associated with a node. They can be regarded as u and v displacements at $\xi = \eta = 0$ *relative* to the displacements at $\xi = \eta = 0$ produced by the corner d.o.f. u_i and v_i.

Although $\{\mathbf{a}\}$ is now present, arguments associated with Eqs. 4.3.1 through 4.3.5 require no change. But now displacement and strain fields are augmented.

$$\{\mathbf{f}\} = [\mathbf{N} \;\; \mathbf{N}_a]\{\mathbf{d} \;\; \mathbf{a}\}, \qquad \{\boldsymbol{\epsilon}\} = [\mathbf{B} \;\; \mathbf{B}_a]\{\mathbf{d} \;\; \mathbf{a}\} \tag{7.3.3}$$

We adopt the notation

$$[\mathbf{k}_{rr}] = \int_V [\mathbf{B}]^T [\mathbf{E}] [\mathbf{B}] \, dV$$

$$[\mathbf{k}_{er}]^T = [\mathbf{k}_{re}] = \int_V [\mathbf{B}]^T [\mathbf{E}] [\mathbf{B}_a] \, dV \tag{7.3.4}$$

$$[\mathbf{k}_{ee}] = \int_V [\mathbf{B}_a]^T [\mathbf{E}] [\mathbf{B}_a] \, dV$$

Matrix $[\mathbf{k}_{rr}]$ is the usual stiffness matrix that would appear if $\{\mathbf{a}\}$ were absent. Nodal loads $\{\mathbf{r}_r\}$ are as stated by Eq. 4.3.5. Loads $\{\mathbf{r}_e\}$ are those associated with the use of $[\mathbf{N}_a]$ and $[\mathbf{B}_a]$ instead of $[\mathbf{N}]$ and $[\mathbf{B}]$ in Eq. 4.3.5. The element stiffness equation is

$$[\mathbf{k}']\{\mathbf{d} \;\; \mathbf{a}\} = \{\mathbf{r}_r \;\; \mathbf{r}_e\} \tag{7.3.5}$$

where $[\mathbf{k}']$ is given by assembling Eqs. 7.3.4 in the format of Eq. 7.2.1.

The d.o.f. $\{\mathbf{a}\}$ are now treated like d.o.f. $\{\mathbf{d}_e\}$ in Section 7.2. Condensed matrices are found by applying Eqs. 7.2.3. The a_i are recovered later for stress calculation. Indeed, d.o.f. $\{\mathbf{d}_e\}$ and $\{\mathbf{a}\}$ serve the same purpose: to improve accuracy by adding more d.o.f. Specifically, we may want to add modes that make a polynomial complete, since the convergence rate with mesh refinement depends on the degree of the highest complete polynomial in the element displacement field (Section 15.4).

Computer programming to accommodate nodeless d.o.f. $\{\mathbf{a}\}$ is not difficult. For example, to use Eqs. 7.3.1, we must introduce N_5, $N_{5,\xi}$, and $N_{5,\eta}$. Arrays that contain $[\mathbf{B}]$ and the uncondensed $[\mathbf{k}]$ must be enlarged, and condensation and recovery algorithms must be added. Jacobian calculations for an isoparametric element are unchanged, because d.o.f. $\{\mathbf{a}\}$ are used only for displacements; they are not used to modify element geometry. Accordingly, the element can be classed as subparametric.

The four-node linear element is only slightly improved by adding the N_5 mode in Eq. 7.3.1. Yet it is instructive to consider nodal loads in this element under body-force loading (Fig. 7.3.1). Load $\{\mathbf{r}_a\}$ is not associated with a specific point but, for convenience, we place it at the center. Because corner loads in Fig. 7.3.1 *a* already sum to 4.0, the internal load seems incorrect. Also, rows and columns of the uncondensed stiffness matrix do not sum to zero, as would seem required for force-free, rigid-body motion. These seeming defects do not appear in the condensed matrix and load vector. The confusion arises because a_1 and a_2 are *relative* displacements. Conversion to *absolute* internal d.o.f. removes the seeming defects (Problem 7.16).

Nodeless d.o.f. can provoke zero-energy deformation modes, particularly if the a_i are used in profusion. If zero-energy modes are avoided, increasing the number of a_i without limit would not produce convergence to exact answers unless the ability of element *boundaries* to model arbitrary displacements were simultaneously increased. Simultaneous

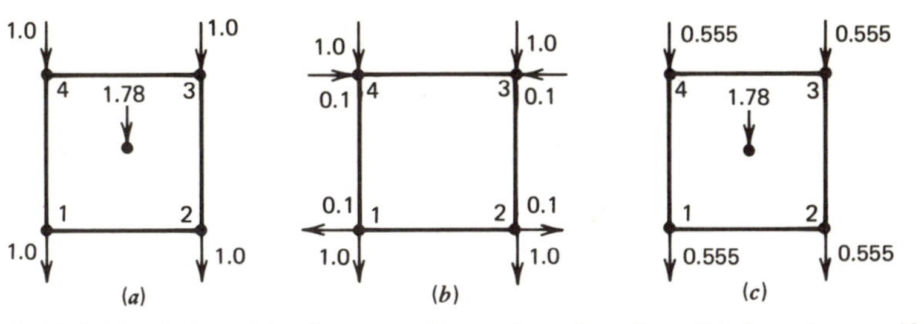

Figure 7.3.1. Nodal loads in a 2-by-2 square element based on Eqs. 7.3.1, under a uniform body force that totals 4.0 units. Poisson's ratio = 0.3. (*a*) Before condensation. (*b*) After condensation. (*c*) Loads if the internal d.o.f. represents absolute instead of relative motion.

improvement is possible: one can generate a convergent sequence of solutions by improving elements of a given mesh instead of by using more and more elements of a given type [7.2].

Some elements are not improved if nodeless variables are added. The constant strain triangle, for example, has too few d.o.f. to interpret a strain field properly. Consider Fig. 7.3.2*a*. The triangle cannot decide whether it should represent constant strain $\epsilon_x = 2\bar{u}/L$ (which it can do exactly) or bending strain $\epsilon_x = 2y\bar{u}/HL$ (for which its stiffness coefficients are too large). Another argument is that adding d.o.f. a_i enables an element to satisfy the differential equations of equilibrium more exactly. Since the triangle *already* satisfies equilibrium equations exactly, adding a_i can make it no better. Neither argument applies to quadrilateral elements, which *are* improved by adding nodeless variables.

A *Lagrange element* is introduced in Fig. 4.2.2 b and Eqs. 4.2.9 and is applied in a plate bending context in Tables 9.5.1 and 9.6.1. This element can be produced by adding the bubble function (Eq. 7.3.2) to the eight N_i of the quadratic isoparametric element of Section 5.6. It does not matter whether a_1 and a_2 are "relative" or "absolute" d.o.f. (Problem 7.19). Either way, the element displays the same behavior and has an 18-by-18 uncondensed stiffness matrix. It requires 3-by-3 Gauss quadrature to avoid a zero-energy deformation mode (Problem 7.20).

The benefit of internal d.o.f. must be weighed against their computational expense. Cost can be high in nonlinear problems if internal freedoms are condensed and recovered in each cycle of an iterative solution.

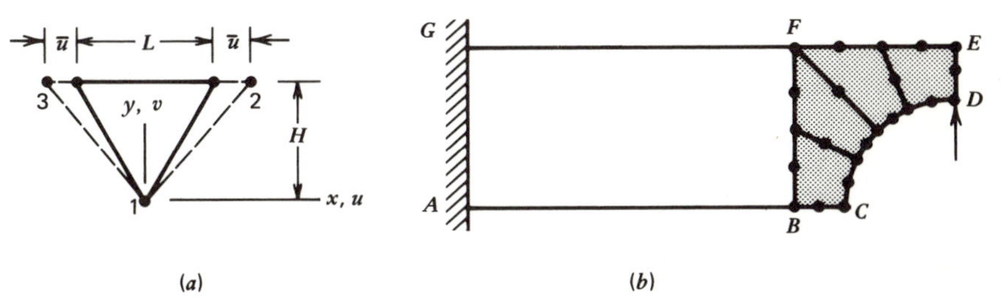

Figure 7.3.2. (*a*) A constant strain triangle, with deformed configuration shown by dashed lines. (*b*) The global-local approach. Ritz functions span region *ABCDEFG* and finite elements are added in region *BCDEF*.

Until now our discussion has tacitly assumed that the structure is entirely divided into elements and that each nodeless variable is added to a single element. The *global-local* procedure is the reverse process: displacements of the entire structure are modeled by the a_i, and finite elements are added, usually only in regions of particular interest [7.3]. In Fig. 7.3.2 *b* elements happen to be restricted to the right end. More could be added to cover the entire structure.

Briefly, the global-local method treats the entire structure by the classical Rayleigh-Ritz method but adds finite elements in regions of particular interest. The Ritz functions need not satisfy as many boundary conditions as would be required if they were used alone because the finite elements "patch things up" in the regions of interest.

Stiffness equations of the global-local approach have the form of Eqs. 7.2.1, but on the *structure* level. Thus $[\mathbf{K}_{rr}]$ would be the stiffness matrix of the finite element portion by itself, $[\mathbf{K}_{ee}]$ would be the stiffness matrix of a classical Rayleigh-Ritz analysis, and the off-diagonal matrices couple the two together. Therefore the structure equations have a larger bandwidth than that of a purely finite element structure.

The global-local approach is useful when a large portion of the structure is regular enough for a classical Rayleigh-Ritz treatment. Then, for a given level of accuracy, it requires fewer d.o.f. than a strictly finite element analysis. The global-local method has concepts in common with boundary, infinite, and singularity elements (Sections 4.13, 7.14, 7.15).

7.4 CONDENSATION AND RECOVERY ALGORITHMS

Our first algorithm is a straightforward coding of Gauss elimination to produce the condensed arrays $[\mathbf{k}]$ and $\{\mathbf{r}\}$ of Eqs. 7.2.3 from the uncondensed arrays of Eq. 7.2.1. Let there be NSIZE d.o.f. in the uncondensed system, and let it be required to condense the last NUM of these d.o.f. In Fig. 7.4.1 condensation is done in place, so that after completion of the algorithm, $[\mathbf{k}]$ resides in the upper left NSIZE – NUM rows and columns of array SE, and $\{\mathbf{r}\}$ resides in the upper NSIZE – NUM rows of array RE.

After structural equations have been solved, d.o.f. $\{\mathbf{d}_r\}$ reside in the structure solution vector $\{\mathbf{D}\}$. It remains to recover $\{\mathbf{d}_e\}$ in Eq. 7.2.2. This can be done by back-substitution, thus completing the Gauss elimination solution that was begun by condensation. Figure 7.4.2 gives the Fortran coding. In Fig. 7.4.2, arrays SM and RM contain the last NUM rows of arrays SE and RE exactly as they stand after completion of the routine in Fig. 7.4.1. The first NSIZE – NUM rows of array DE contain the known d.o.f. $\{\mathbf{d}_r\}$. Internal d.o.f. $\{\mathbf{d}_e\}$ are computed and stored in the last NUM rows of array DE.

In the foregoing algorithm, mass storage is probably used to store the last NUM rows of arrays SE and RE from Fig. 7.4.1. Node point coordinates, and $\{\boldsymbol{\epsilon}_0\}$ and $\{\boldsymbol{\sigma}_0\}$, are also stored. After $\{\mathbf{D}\}$ is known, these data are recalled, and Fig. 7.4.2 used to compute $\{\mathbf{d}_e\}$. At each point where stresses are needed, $[\mathbf{B}]$ is reconstructed from the node point coordinates. Finally, strains are $\{\boldsymbol{\epsilon}\} = [\mathbf{B}]\,\{\mathbf{d}_r\ \mathbf{d}_e\} - \{\boldsymbol{\epsilon}_0\}$ and stresses are $\{\boldsymbol{\sigma}\} = [\mathbf{E}]\,\{\boldsymbol{\epsilon}\} + \{\boldsymbol{\sigma}_0\}$.

```
C --- DO CONDENSATION OPERATIONS ON THE LOWER TRIANGLE OF SE.
      DO 30 K=1,NUM
      LL = NSIZE - K
      KK = LL + 1
      DO 20 L=1,LL
      IF (SE(KK,L) .EQ. 0.)  GO TO 20
      DUM = SE(KK,L)/SE(KK,KK)
      DO 10 M=1,L
   10 SE(L,M) = SE(L,M)  - SE(KK,M)*DUM
      RE(L)   = RE(L)    - RE(KK)   *DUM
   20 CONTINUE
   30 CONTINUE
C --- FILL IN THE UPPER TRIANGLE OF SE BY SYMMETRY.
      DO 40 K=1,LL
      DO 40 L=1,K
   40 SE(L,K) = SE(K,L)
```

Figure 7.4.1. Fortran condensation algorithm that accepts a *symmetric* matrix, stored in full instead of banded format, does Gauss elimination from the bottom with diagonal pivots and yields [**k**] and {**r**} of Eqs. 7.2.3.

In an alternative algorithm [7.5], explicit recovery of $\{\mathbf{d}_e\}$ is avoided. At each point where stresses are needed, we define

$$[\mathbf{H}] = [\mathbf{E}]\,[\mathbf{B}] \qquad \text{and} \qquad \{\tau_0\} = [\mathbf{E}]\,\{\epsilon_0\} \tag{7.4.1}$$

Hence, after partitioning [**H**], stresses are

$$\{\sigma\} = [\mathbf{H}_r \;\; \mathbf{H}_e]\,\{\mathbf{d}_r \;\; \mathbf{d}_e\} - \{\tau_0\} + \{\sigma_0\} \tag{7.4.2}$$

Now substitute $\{\mathbf{d}_e\}$ from Eq. 7.2.2.

$$\begin{aligned} \{\sigma\} &= [\mathbf{H}^*]\,\{\mathbf{d}_r\} + \{\rho\}, \qquad \text{where} \\ [\mathbf{H}^*] &= [\mathbf{H}_r] - [\mathbf{H}_e]\,[\mathbf{k}_{ee}]^{-1}[\mathbf{k}_{er}] \\ \{\rho\} &= [\mathbf{H}_e]\,[\mathbf{k}_{ee}]^{-1}\,\{\mathbf{r}_e\} - \{\tau_0\} + \{\sigma_0\} \end{aligned} \tag{7.4.3}$$

Let there be m stresses in $\{\sigma\}$ and n stress points in the element. Then the n matrices [**H**] can be stacked one atop the other in an array with NS $= mn$ rows. To implement the algorithm of Eqs. 7.4.3, we combine it with condensation as follows [7.5]. To compute [**H***], insert just before statement 20 in Fig. 7.4.1 the Fortran statements

```
      DO 60 J=1,NUM
      JJ = NSIZE - NUM + J
      DUM = 0.
      K = JJ - 1
      DO 50 L=1,K
   50 DUM = DUM + SM(J,L)*DE(L)
   60 DE(JJ) = (RM(J) - DUM)/SM(J,JJ)
```

Figure 7.4.2. Recovery of eliminated d.o.f. $\{\mathbf{d}_e\}$ when $\{\mathbf{d}_r\}$ and $\{\mathbf{r}_e\}$ are known.

```
      DO 15 M = 1, NS
   15 H(M,L) = H(M,L) - H(M,KK)*DUM
```

To compute $\{\boldsymbol{\rho}\}$, insert just after statement 20 in Fig. 7.4.1

```
      DUM = RE(KK)/SE(KK,KK)
      DO 25 M = 1, NS
   25 RHO(M) = RHO(M) + H(M,KK)*DUM
```

where it is presumed that array RHO contains $\{\boldsymbol{\sigma}_0\} - \{\boldsymbol{\tau}_0\}$ before the algorithm is begun. After condensation, arrays H and RHO are stored. When $\{\mathbf{d}_r\}$ is known, arrays H and RHO are recalled and used to compute stresses $\{\boldsymbol{\sigma}\}$ according to Eq. 7.4.3.

The most significant differences between the two methods are as follows. The first method explicitly recovers $\{\mathbf{d}_e\}$ and reconstructs a (somewhat sparse) matrix $[\mathbf{B}]$ at each stress point. The second method generates, condenses, and stores an array $[\mathbf{H}^*]$ that is smaller than $[\mathbf{B}]$ but not sparse. Both methods yield the same stresses. The relative cost of the two methods depends on billing charges for computing and for secondary storage, the size of $\{\mathbf{d}_r\}$ in relation to $\{\mathbf{d}_e\}$, the number of load conditions, the number of stress points, and other less important factors. The method of Figs. 7.4.1 and 7.4.2 tends to be cheaper if $\{\mathbf{d}_e\}$ is small in relation to $\{\mathbf{d}_r\}$, if the number of load cases is small, or if the number of stress points is large. Further comparison appears in Ref. 7.4.

References 2.20, 2.21, and 6.16 consider condensation in the specific context of substructuring.

7.5 HIGHER DERIVATIVES AS NODAL D.O.F.

Here we define a "higher derivative" as one that is not needed to define interelement compatibility. Thus, in the stretching of a bar or in a plane stress, *all* derivatives of u and v would be considered "higher." In the bending of thin plates, higher derivatives are second and greater derivatives of lateral displacement. When used as nodal d.o.f., higher derivatives are also called "extra" or "excessive." In common usage, the term "higher-order element" includes elements with extra d.o.f. as well as elements with both corner and side nodes and only the essential d.o.f. at each node.

Higher-order elements have advantages. They are based on fields with many generalized coordinates, so they provide good accuracy in a coarse mesh. The element d.o.f. can be confined largely to element corners if higher derivatives are used. Thus the bandwidth of $[\mathbf{K}]$ is not seriously increased. By using higher derivatives as nodal d.o.f., the strains (or curvatures) needed for stress and moment calculation appear in $\{\mathbf{D}\}$. So, being primary unknowns, strains may be computed more accurately than those found from $[\mathbf{B}]\{\mathbf{d}\}$ when $\{\mathbf{d}\}$ contains only essential d.o.f. Plane elements can accommodate a nodal moment in $\{\mathbf{r}\}$ if $\{\mathbf{d}\}$ includes the average rotation $(v_{,x} - u_{,y})/2$ as a nodal d.o.f.

Higher derivatives as nodal d.o.f. also have disadvantages. Input data that describes element shape can be cumbersome. By enforcing a higher order of interelement continuity than is necessary, they stiffen the structure, which somewhat negates their benefit. At an elastic-plastic boundary, or where there is an abrupt change in stiffness or material properties, continuity of higher d.o.f. must *not* be enforced. For example, if two beam elements of different stiffness are joined, they have the same moment but different curvature at the node they share. A maneuver appropriate to such a circumstance is to release the curvature d.o.f. in one of the elements before assembly (Section 6.7). But, by doing so, we reduce the benefit of these d.o.f. where it is most needed—near a high stress gradient.

An alternative maneuver is illustrated by the following example. Imagine that elements in Fig. 3.7.1 have u and $u_{,x}$ as d.o.f. at each node. For convenience, we denote nodal d.o.f. of elements 1 and 2 by the symbols

$$\{\mathbf{d}\}_1 = \{u_1 \ \epsilon_1 \ u_2 \ \epsilon_2\}, \qquad \{\mathbf{d}\}_2 = \{u_2 \ \epsilon_2 \ u_3 \ \epsilon_3\} \tag{7.5.1}$$

Let elements 1 and 2 have different stiffness, so that $(AE)_1/(AE)_2 = g$. We can use ϵ_1 as a global d.o.f. in $\{\mathbf{D}\}$, avoid condensation of ϵ_2, and yet incorporate the relation $\epsilon_2 = g\epsilon_1$. We do so by transforming the matrices $[\mathbf{k}]_2 \{\mathbf{d}\}_2 = \{\mathbf{r}\}_2$ of element 2 as follows (before assembly). The original equations are

$$[\mathbf{k}]_2 \{u_2 \ \epsilon_2 \ u_3 \ \epsilon_3\} = \{r_2 \ \ \sigma_2 \ \ r_3 \ \ \sigma_3\} \tag{7.5.2}$$

By transformation, these equations become

$$[\mathbf{k}_g]_2 \{u_2 \ \epsilon_1 \ u_3 \ \epsilon_3\} = \{r_2 \ \ g\sigma_2 \ \ r_3 \ \ \sigma_3\} \tag{7.5.3}$$

where the latter matrices follow from Eqs. 6.3.4 and 6.3.7, with $[\mathbf{T}] = [1 \ g \ 1 \ 1]$. After solving equations, the nodal strain $g\epsilon_1$ is used for stress calculation in element 2.

Release of higher derivative d.o.f. is again required and the benefit of these d.o.f. is again reduced if elements with derivative d.o.f. must be used in combination with elements that have only essential nodal d.o.f. Indeed, most computer programs allow up to six d.o.f. per node (three translations and three rotations) and so may be unable to accommodate a higher-order element without basic changes.

The physical meaning of higher-derivative d.o.f. and their associated nodal loads is often obscure. Boundary conditions become awkward. For example, if a plane element includes derivatives $u_{,x}$, $u_{,y}$, $v_{,x}$, and $v_{,y}$ as nodal d.o.f., a stress-free boundary dictates a constraint relation among these d.o.f. but does not dictate the numerical value of any of them.

In summary, higher-derivative d.o.f. make the finite element method awkward in application to problems for which it is most powerful—to structures built of different element types and involving thickness changes, stiffeners, and parts that join with sharp angles instead of curves.

7.6 PARASITIC SHEAR IN LINEAR ELEMENTS

Linear elements, such as those of Figs. 5.3.1 and 5.7.1a, are attractive because they are simple and have only corner nodes. But in bending they are too stiff. Consider Fig. 7.6.1. The element is rectangular, so $\xi = 2x/L$ and $\eta = 2y/H$. If displacements $\bar{u}$ are imposed, the element *must* respond as in Fig. 7.6.1*b* because its sides must remain straight. Its deformation field is

$$u = \bar{u}\xi\eta, \qquad v = 0 \tag{7.6.1}$$

The *correct* shape under pure bending (Fig. 7.6.1*c*), is [1.20]

$$u = \bar{u}\xi\eta, \qquad v = \frac{L\bar{u}}{2H}(1 - \xi^2) + \nu\frac{H\bar{u}}{2L}(1 - \eta^2) \tag{7.6.2}$$

Equations 7.6.2 yield the correct value of shear strain ($\gamma_{xy} = 0$), but Eqs. 7.6.1 do not. So, to impose displacements $\bar{u}$ on the element, we must apply a moment big enough to overcome shear resistance as well as bending resistance. Thus the element is too stiff in a bending mode. We call this effect *parasitic shear*. Its influence is disastrous if L/H is large.

In the general plane problem, *two* bending modes are possible: the one shown in Fig. 7.6.1 and one where the *vertical* sides should become curved ($v = \bar{v}\xi\eta$).

Many schemes that overcome parasitic shear have been proposed. We outline some of them in the following paragraphs. Our goal is more to provide insight into element behavior than to advocate a particular scheme. The topic is continued in Section 7.7. The concepts apply to solid as well as plane elements.

An improvement scheme may destroy invariance. It can be restored by using local axes and coordinate transformation, as explained below Item 5 in Section 4.9.

Consider the element of Eqs. 7.3.1 and 7.3.2 and the bending mode of Fig. 7.6.1. If a_2 in Eq. 7.3.1 is large enough, γ_{xy} is zero at the four Gauss points. However, the element

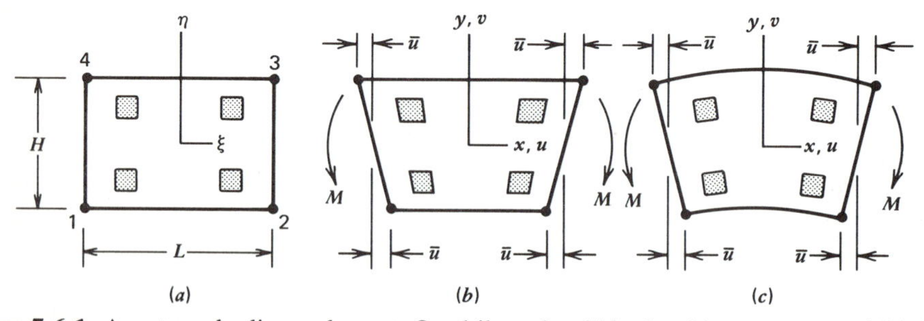

Figure 7.6.1. A rectangular linear element. Quadrilaterals within the element represent initial and deformed shapes at Gauss points of a 2-by-2 quadrature rule. (*a*) Undeformed shape. (*b*) Prescribed d.o.f. $\bar{u}$ deform the element in a bending mode. (*c*) The correct shape of a beam segment in bending.

is then too stiff because of a parasitic *normal* strain $\epsilon_y = (1 - \xi^2)(-4\eta/H)a_2$. To avoid it, we decouple the bubble function from ϵ_x and ϵ_y and allow it to affect only γ_{xy}. Thus, in the last two columns of the strain-displacement matrix $\mathsf{B} = [\mathbf{B}\ \mathbf{B}_a]$ that operates on $\{u_1\ v_1\ u_2\ .\ .\ .\ v_4\ a_1\ a_2\}$, only B(3,9) and B(3,10) are nonzero [7.6].

Imagine that the *a*-basis matrix $[\mathbf{k}_a]$ is known, by direct use of Eqs. 4.3.19 and 4.6.1 if the element is rectangular, otherwise by coordinate transformation [7.7]. The mode in Fig. 7.6.1*b* is $u = a_4xy$, and the other (not shown) is $v = a_8xy$. To soften these overly stiff bending modes, we can reduce the magnitudes of the associated coefficients in $[\mathbf{k}_a]$. Accordingly, $(k_a)_{4,4}$, $(k_a)_{4,8}$, $(k_a)_{8,4}$, and $(k_a)_{8,8}$ are multiplied by reduction factors before $[\mathbf{k}_a]$ is transformed to $[\mathbf{k}]$. But too much reduction can make the structure unstable [7.7].

A similar softening can be applied to the element of Eqs. 7.3.1. If diagonal coefficients of $[\mathbf{k}]$ associated with a_1 and a_2 are multiplied by $(1 + e)$, where $e < 1$, we have added negative springs that "boost" these d.o.f. instead of restraining them. Thus the element becomes more flexible. The four-triangle element of Fig. 7.2.1*a* can be treated similarly if d.o.f. u_5 and v_5 are first converted to relative motions a_1 and a_2 [7.8, 16.15] (see also Problem 7.16).

The bending modes $u = \bar{u}\xi\eta$ and $v = \bar{v}\xi\eta$ both display the correct strain $\gamma_{xy} = 0$ at $\xi = \eta = 0$ in Fig. 7.6.1. Accordingly, when generating $[\mathbf{k}]$ numerically, we can base the γ_{xy} row of $[\mathbf{B}]$ on conditions at $\xi = \eta = 0$ regardless of the actual coordinates of the sampling point [7.9]. "Ligitimate" shear strain, $\gamma_{xy} \neq 0$ at $\xi = \eta = 0$, is still correctly computed. This technique can be applied to the basic 8 d.o.f. element and to the 10 d.o.f. element of Eqs. 7.3.1, but both elements fail the patch test unless they are rectangular. An alternative definition of average γ_{xy} can be written in terms of nodal d.o.f. and nodal coordinates (Problem 7.28). This element passes the patch test.

Yet another method makes u and v both depend on all eight d.o.f., so $[\mathbf{N}]$ becomes a full matrix [7.6]. This allows the quadratic v in Fig. 7.6.1*c* to be activated by d.o.f. $\bar{u}$.

Hybrid elements, (see Section 16.1) easily avoid parasitic shear by adopting a constant τ_{xy} in their assumed stress field.

An eight-node solid element of quality comparable to the QM6 solid element of Section 7.7 can be formed by added physically motivated higher modes to a constant strain element. Space does not permit its description here [7.10, 7.11].

7.7 THE QM6 ELEMENTS. INCOMPATIBLE ELEMENTS

In Eqs. 7.6.1 and 7.6.2 we saw that in the bending mode $u = \bar{u}\xi\eta$, the plane element errs by omitting the v modes $(1 - \xi^2)$ and $(1 - \eta^2)$. In the bending mode $v = \bar{v}\xi\eta$, the element errs by omitting the u modes $(1 - \xi^2)$ and $(1 - \eta^2)$. The linear *solid* element has similar defects. Linear elements, whether retangular or not, can be improved by adding these modes as internal freedoms. We write [7.12]

Eight-node solid element →
Four-node plane element →

$$
\begin{aligned}
u &= N_i u_i + (1 - \xi^2)a_1 + (1 - \eta^2)a_2 \quad + (1 - \zeta^2)a_7 \\
v &= N_i v_i + (1 - \xi^2)a_3 + (1 - \eta^2)a_4 \quad + (1 - \zeta^2)a_8 \\
w &= N_i w_i + (1 - \xi^2)a_5 + (1 - \eta^2)a_6 \quad + (1 - \zeta^2)a_9
\end{aligned}
\tag{7.7.1}
$$

For the plane element, $i = 1,2,3,4$ and the N_i are given by Eq. 5.3.3. For the solid element, $i = 1,2, \ldots, 8$ and the N_i are given by Table 5.6.1. These elements are usually called QM6 elements. If integrated as described next, they are invariant and without zero-energy deformation modes. If rectangular, they model pure bending exactly, regardless of aspect ratio. In contrast, the element of Fig. 7.6.1*b* is about 10 times too stiff in bending if $L/H = 10$.

The QM6 elements are *incompatible*. For example, as suggested by Fig. 7.7.1, the mode $u = (1 - \eta^2)a_2$ might be activated in one element but not in its neighbors to the left and right, thus producing a gap on one side and an overlap on the other. But incompatible elements are still valid if incompatibilities disappear and a constant strain state is approached as the mesh is refined. Accordingly, we must ask if incompatible elements can pass a patch test.

Straightforward coding of Eqs. 7.7.1 (as outlined in Section 7.3) yields elements that pass the patch test only if they are parallelograms of constant thickness, or parallelepipeds in the solid case. We next consider a modified numerical integration scheme that removes this shape restriction [7.13].

Let forces be applied to element nodes but not to internal d.o.f., as in a patch test. Then the element equations are

$$
\begin{bmatrix} \mathbf{k}_{rr} & \mathbf{k}_{re} \\ \mathbf{k}_{er} & \mathbf{k}_{ee} \end{bmatrix} \begin{Bmatrix} \mathbf{d} \\ \mathbf{a} \end{Bmatrix} = \begin{Bmatrix} \mathbf{r} \\ \mathbf{O} \end{Bmatrix} \tag{7.7.2}
$$

where the submatrices are defined by Eq. 7.3.4. We also require that $\{\mathbf{a}\} = 0$, so that incompatibilities vanish, as they must if a patch test is to be passed. Let $\{\mathbf{d}\} = \{\mathbf{d}_c\}$,

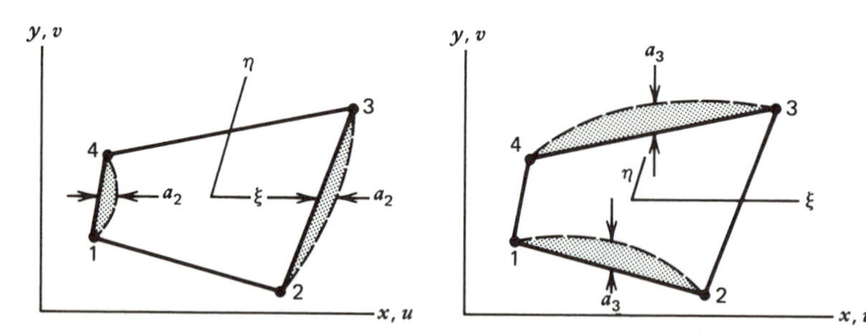

Figure 7.7.1. Dashed lines show edge displacements associated with the incompatible modes $u = (1 - \eta^2)a_2$ and $v = (1 - \xi^2)a_3$ in a plane QM6 element.

where $\{\mathbf{d}_c\}$ represents nodal d.o.f. consistent with rigid-body motion or a constant strain state. Thus, from the lower partition, $[\mathbf{k}_{er}]\{\mathbf{d}_c\} = 0$. By substitution from Eq. 7.3.4.,

$$\int_V [\mathbf{B}_a]^T[\mathbf{E}][\mathbf{B}]\, dV \{\mathbf{d}_c\} = \int_V [\mathbf{B}_a]^T\, dV \{\boldsymbol{\sigma}_c\} = 0 \qquad (7.7.3)$$

where $\{\boldsymbol{\sigma}_c\}$ is the constant stress state $\{\boldsymbol{\sigma}_c\} = [\mathbf{E}]\{\boldsymbol{\epsilon}_c\}$. We know that $\{\boldsymbol{\epsilon}_c\} = [\mathbf{B}]\{\mathbf{d}_c\}$ is a constant strain state because of arguments in Section 5.9. So, if an incompatible element is to pass a patch test, we must have, for plane and solid elements, respectively,

$$\int_{-1}^{1}\int_{-1}^{1} [\mathbf{B}_a]^T tJ\, d\xi\, d\eta = 0, \qquad \int_{-1}^{1}\int_{-1}^{1}\int_{-1}^{1} [\mathbf{B}_a]^T J\, d\xi\, d\eta\, d\zeta = 0 \qquad (7.7.4)$$

where t is element thickness and J is the Jacobian determinant. For QM6 elements, $[\mathbf{B}_a]$ contains first powers of ξ and η (and ζ for solids). Accordingly, Eqs. 7.7.4 are automatically satisfied if t and J are constant, as for parallelograms of constant thickness and parallelepipeds. For elements of arbitrary shape, we can satisfy Eqs. 7.7.4 by *forcing* the integrand to be linear in the isoparametric coordinates. We do this by evaluating $[\mathbf{B}_a]$ and J from a Jacobian matrix $[\mathbf{J}]$ constructed at $\xi = \eta = \zeta = 0$, regardless of the actual coordinates of the integration point. The values of J and $[\mathbf{B}]$ used in the first of Eqs. 7.3.4. are not modified, so this equation still defines $[\mathbf{k}]$ of the eight-node linear element of Section 5.3. This integration scheme allows the element to pass the patch test, but it scarcely improves accuracy in a coarse mesh.

Figure 5.5.1 can be modified to produce the plane QM6 element as follows. Enlarge the EN, NXI, and NET arrays by two words each to accommodate the $(1 - \xi^2)$ and $(1 - \eta^2)$ modes. Compute $[\mathbf{J}_c]$ and J_c, the Jacobian matrix and its determinant at $\xi = \eta = 0$. Form the first eight columns of array B as usual. Let the last four columns of array B contain $[\mathbf{B}_a]$; form them using coefficients in $[\mathbf{J}_c]^{-1} J_c/J$, where J is DETJAC of Fig. 5.5.1. Also multiply the $[\mathbf{B}_a]$ terms by t_c/t, where t_c is the thickness at $\xi = \eta = 0$ (for a solid of revolution, multiply by r_c/r, where r_c is the radius to $\xi = \eta = 0$). Then the only changes needed in Fig. 5.5.2 are to enlarge arrays, increase limits on DO loops, and condense the internal freedoms.

Figure 7.7.2 gives numerical evidence of the quality of the QM6 element. It has near-quadratic accuracy, but only if it is a parallelogram. In plane problems, which are comparatively cheap, the eight-node quadratic element may be preferred. Solid QM6 elements may be preferred over 20-node bricks because of the expense of three-dimensional analysis. But if a solid is thin enough that one layer of elements seems acceptable, the element of Fig. 10.4.2*b* should be considered, with the integration rule of Eq. 5.10.3.

If a QM6 element is homogeneous, isotropic, and rectangular, analytical integration and an explicit $[\mathbf{k}]$ are possible (Fig. 7.7.3).

How are incompatible modes used in stress calculation? Imagine that the element in Fig. 7.6.1*a* is a QM6 element, supported so that $u_1 = u_4 = v_4 = 0$ on its left edge and

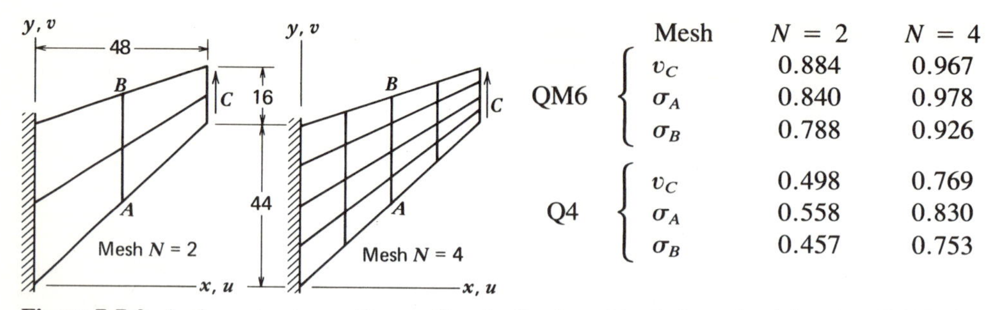

	Mesh	$N = 2$	$N = 4$
QM6	v_C	0.884	0.967
	σ_A	0.840	0.978
	σ_B	0.788	0.926
Q4	v_C	0.498	0.769
	σ_A	0.558	0.830
	σ_B	0.457	0.753

Figure 7.7.2. A plane structure with a uniformly distributed load along the right edge [7.13]. Q4 = element of Section 5.3. v_C = deflection at C, σ_A = maximum stress at A, and σ_B = minimum stress at B, all reported as the ratio of computed value to that given by an $N = 16$ mesh of hybrid elements.

$$
\underset{8\times 8}{[\mathbf{k}]} = \frac{Qt}{12(1 - m^2)} \begin{bmatrix} A_1 & C_1 & A_2 & -C_2 & A_4 & -C_1 & A_3 & C_2 \\ & B_1 & C_2 & B_3 & -C_1 & B_4 & -C_2 & B_2 \\ & & A_1 & -C_1 & A_3 & -C_2 & A_4 & C_1 \\ & & & B_1 & C_2 & B_2 & C_1 & B_4 \\ & & & & A_1 & C_1 & A_2 & -C_2 \\ & \text{Symmetric} & & & & B_1 & C_2 & B_3 \\ & & & & & & A_1 & -C_1 \\ & & & & & & & B_1 \end{bmatrix}
$$

In plane stress: $Q = E, \qquad m = \nu$

In plane strain: $Q = \dfrac{E}{1 - \nu^2}, \qquad m = \dfrac{\nu}{1 - \nu}$

$$
\begin{aligned}
A_1 &= (4 - m^2)c/b + 1.5\,(1 - m)b/c \\
A_2 &= -(4 - m^2)c/b + 1.5\,(1 - m)b/c \\
A_3 &= (2 + m^2)c/b - 1.5\,(1 - m)b/c \\
A_4 &= -(2 + m^2)c/b - 1.5\,(1 - m)b/c \\
C_1 &= 1.5\,(1 + m) \\
C_2 &= 1.5\,(1 - 3m)
\end{aligned}
$$

$B_1 - B_4$ are obtained from $A_1 - A_4$ by interchanging b and c

Figure 7.7.3. The condensed stiffness matrix of a plane, isotropic, rectangular QM6 element of constant thickness t [16.63, with sign errors corrected]. Element dimensions and orientation are shown by Fig. 4.6.1a, and $\{\mathbf{d}\} = \{u_1\ v_1\ u_2\ v_2\ u_3\ v_3\ u_4\ v_4\}$. This example indicates that $[\mathbf{k}]$ of a plane element depends on its aspect ratio c/b but not on its size.

loaded by a uniform traction Φ_x along its right edge. According to Eq. 4.3.5, Φ_x produces loads associated with u_2, u_3, and a_2 of Eqs. 7.7.1. As shown by Fig. 7.3.1, condensation redistributes the load associated with a_2. But unless recovery of internal d.o.f. yields $a_2 = 0$, the element will bulge to the right, as in Fig. 7.7.1. It will not display the expected constant strain state. This is an instance of a general empirical rule: *do not load incompatible modes during recovery of internal d.o.f. and stress calculation.*

A similar problem arises if a QM6 element is used in its solid-of-revolution form (see Section 8.2).

7.8 WARPED MEMBRANE ELEMENTS

On occasion we want to use a four-node quadrilateral to represent a slightly curved surface. The skin of an airplane can be modeled this way. But now the four nodes of an element are in general not coplanar. If not adjusted for warping, the stiffness matrix of a plane element behaves strangely: different meshes can give wildly different answers [7.14]. The corrective adjustments expand the 8-by-8 stiffness matrix to 12 by 12 to allow for three displacement d.o.f. per node; they are summarized as follows [15.27].

A warped quadrilateral 1234 is shown in Fig. 7.8.1*a*. In making corrective adjustments, the first step is to set up local axes *xyz* so that the four nodes are alternately *H* units above and *H* units below the *xy* plane. The projection of the element on the *xy* plane is *ABCD*. In the usual way, we generate a stiffness matrix $[\mathbf{k}']$ for element *ABCD* that operates on the eight d.o.f. $\{u_A \ v_A \ u_B \ . \ . \ . \ v_D\}$. Next we need a transformation matrix $[\mathbf{T}]$ to expand $[\mathbf{k}']$ to a 12-by-12 matrix that operates on d.o.f. $\{u_1 \ v_1 \ w_1 \ u_2 \ . \ . \ . \ w_4\}$.

To find $[\mathbf{T}]$, we consider a force transformation that satisfies the three out-of-plane equilibrium equations and is insensitive to a permutation of node numbers. At each

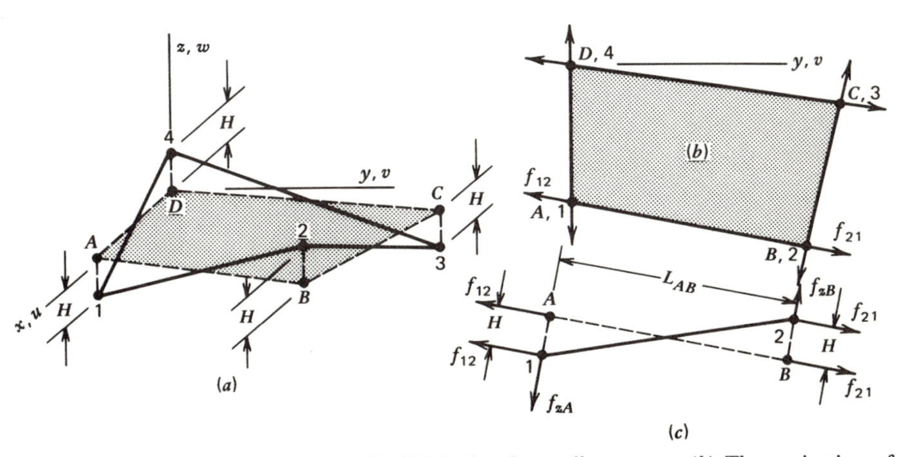

Figure 7.8.1. (*a*) A warped quadrilateral 1234 in local coordinates *xyz*. (*b*) The projection of the element onto the *xy* plane. (*c*) A view parallel to the *xy* plane and normal to edge *AB*.

lettered node, x-parallel and y-parallel nodal forces are resolved into their edge-parallel components, such as f_{12} and f_{21} in Fig. 7.8.1b, parallel to edge AB. The z-direction "kickoff" forces are found by requiring moment equilibrium to prevail when f_{12} and f_{21} are transferred from lettered to numbered nodes. From Fig. 7.8.1c,

$$f_{zA} = f_{zB} = \frac{H}{L_{AB}}(f_{12} + f_{21}) \tag{7.8.1}$$

Summing results from all four sides,

$$f_{z1} = f_{zD} - f_{zA}, \qquad f_{z2} = f_{zA} - f_{zB}, \qquad f_{z3} = f_{zB} - f_{zC}, \qquad f_{z4} = f_{zC} - f_{zD} \tag{7.8.2}$$

The force transformation $\{\mathbf{f}\} = [\mathbf{T}]^T\{\mathbf{f}'\}$ is

$$\{f_{x1}\ f_{y1}\ f_{z1}\ f_{x2} \cdots f_{z4}\} = [\mathbf{T}]^T \{f_{xA}\ f_{yA}\ f_{xB} \cdots f_{yD}\} \tag{7.8.3}$$

Rows 3, 6, 9, and 12 of $[\mathbf{T}]^T$ express the transformation summarized by Eq. 7.8.2. Coefficients in these rows [15.27] are functions of H and local xy coordinates of the lettered nodes. They are derived by tedious algebra and careful attention to signs. The remaining rows of $[\mathbf{T}]^T$ each contain a single 1 and seven 0's and express the conditions $f_{x1} = f_{xA}$, $f_{y1} = f_{yA}$, and so on. The siffness matrix of the warped element is $[\mathbf{k}] = [\mathbf{T}]^T[\mathbf{k}'][\mathbf{T}]$ (as in Eq. 6.3.7, for example).

The stiffness matrix $[\mathbf{k}]$ has order 12 but rank 5 because of six rigid-body modes and unrestrained out-of-plane warping. Warping would still be unrestrained if the quadrilateral were replaced by two constant strain triangles (whose stiffness matrices need not be adjusted as described in this section). To restrain warping, nodes of the membrane must be attached to other structures, such as to nodes of the spar model in an airplane wing. Warping of the membrane element is also restrained if the quadrilateral is replaced by four overlapping triangles (123, 234, 341, 412) because the four triangles form a tetrahedral box that cannot warp. The two- and four-triangle schemes avoid the erratic behavior displayed by a warped but unadjusted quadrilateral, but they force us to use the rather crude constant strain triangle.

7.9 TRIANGULAR ELEMENTS AND AREA COORDINATES

In Fig. 7.9.1a, dashed lines are drawn between an arbitrary point P and the vertices of a triangle. Thus we divide the triangle into the three subareas A_1, A_2, and A_3. Let A be the total area of the triangle, $A = A_1 + A_2 + A_3$. Then, by definition,

$$\alpha = \frac{A_1}{A}, \qquad \beta = \frac{A_2}{A}, \qquad \gamma = \frac{A_3}{A} \tag{7.9.1}$$

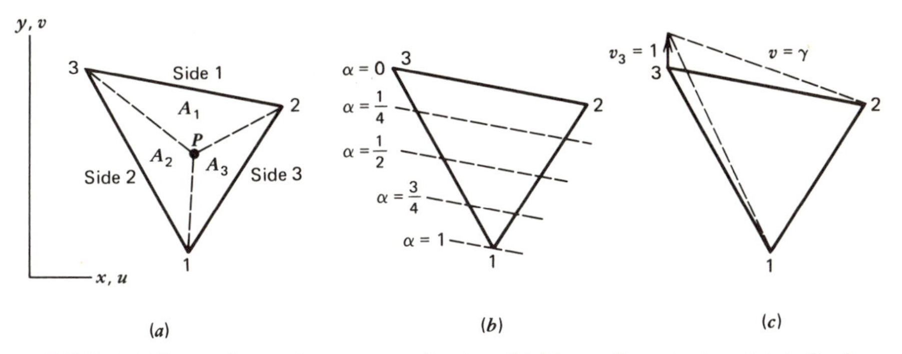

Figure 7.9.1. (*a*) Nomenclature for area coordinates. (*b*) Lines of constant α. (*c*) A displacement mode of the constant strain triangle is simple to state in area coordinates.

Consequently,

$$\alpha + \beta + \gamma = 1 \tag{7.9.2}$$

The dimensionless quantities α, β, and γ are called *area coordinates*. The location of P is defined by any two of them. Area coordinates could also be defined as ratios of length; for example, $\alpha = s/h$, where s is the distance from side 1 to point P and h is the distance from side 1 to vertex 1. The latter interpretation is suggested by Fig. 7.9.1*b*. The centroid of a straight-sided triangle is at $\alpha = \beta = \gamma = \frac{1}{3}$.

Other names for area coordinates are *areal, triangular,* and *trilinear* coordinates. To mathematicians they are instances of *simplex* and *barycentric* coordinates. They are not new [7.15, 7.16] but seem to have been independently devised when finite elements suggested a need for them.

Why should area coordinates be used instead of Cartesian coordinates? Area coordinates are natural for a triangle, just as $\xi\eta$ coordinates are natural for isoparametric quadrilaterals in Chapter 5. Indeed, when the shape and displacement of a triangular element are defined by the same area coordinate shape functions, the triangular element *is, in fact, isoparametric*. It can have side nodes and curved sides, and the argument in Section 5.9 establishes the validity of such an element. Like other isoparametrics, triangular elements formulated in area coordinates are invariant. Section 9.2 notes difficulties of plate-element formulation that were overcome by changing from Cartesian to area coordinates.

The relation between the Cartesian and area coordinates of a point is simply stated but not obvious. For straight-sided triangles, it is

$$\begin{Bmatrix} 1 \\ x \\ y \end{Bmatrix} = \begin{bmatrix} 1 & 1 & 1 \\ x_1 & x_2 & x_3 \\ y_1 & y_2 & y_3 \end{bmatrix} \begin{Bmatrix} \alpha \\ \beta \\ \gamma \end{Bmatrix} \tag{7.9.3}$$

Of these three equations, the first restates Eq. 7.9.2, and the last two are analogous to Eqs. 5.3.2, where α, β, and γ play the role of shape functions N_i. Indeed, area coordinates *are* the shape functions of the constant strain triangle (Fig. 7.9.1*c*):

$$u = \alpha u_1 + \beta u_2 + \gamma u_3, \qquad v = \alpha v_1 + \beta v_2 + \gamma v_3 \tag{7.9.4}$$

The inverse coordinate transformation is

$$\{\alpha\ \beta\ \gamma\} = [\mathbf{A}]^{-T}\{1\ x\ y\} \tag{7.9.5}$$

Matrix $[\mathbf{A}]^{-1}$ is stated in Eq. 4.5.3 (see also Problem 7.35).

Fields expressed in terms of area coordinates must be differentiated with respect to Cartesian coordinates during element formulation. Consider a field $\phi = \phi(\alpha, \beta, \gamma)$. By the chain rule,

$$\phi_{,x} = \phi_{,\alpha}\alpha_{,x} + \phi_{,\beta}\beta_{,x} + \phi_{,\gamma}\gamma_{,x} \tag{7.9.6}$$

and similarly for $\phi_{,y}$. For straight-sided triangles, from Eq. 7.9.5,

$$\alpha_{,x} = \frac{y_2 - y_3}{2A} \qquad \beta_{,x} = \frac{y_3 - y_1}{2A} \qquad \gamma_{,x} = \frac{y_1 - y_2}{2A} \tag{7.9.7}$$

Integrals of polynomials in area coordinates over a triangle of area A are given by a simple formula that involves factorials of integer exponents k, ℓ, and m [7.17]. It is

$$\int_{\text{area}} \alpha^k \beta^\ell \gamma^m \, dA = 2A \frac{k!\ \ell!\ m!}{(2 + k + \ell + m)!} \tag{7.9.8}$$

(Remember that $0! = 1$.) Or, by Eq. 7.9.2, we can eliminate one coordinate from a polynomial in α, β, and γ. If (say) γ is eliminated, we simply omit γ and m from Eq. 7.9.8.

Along a side of length L, say side 3 in Fig. 7.9.1,

$$\int_{\text{length}} \alpha^k \beta^\ell \, dL = L \frac{k!\ \ell!}{(1 + k + \ell)!} \tag{7.9.9}$$

If area coordinates appear in the denominator of a term that must be integrated, as happens with solids of revolution and with elements whose sides are curved, the terms to be integrated do not have the simple form $\alpha^k \beta^\ell \gamma^m$. Then Eqs. 7.9.8 must be replaced by numerical integration (Table 7.10.1).

Equations 7.9.8 and 7.9.9 are the two-space instance of formulas for a *simplex*—a figure in n-dimensional space that has $n + 1$ vertices and is bounded by $n + 1$ surfaces of dimensionality $n - 1$ [7.18]. Accordingly, we could invoke analogous *volume* coordinates to formulate tetrahedron elements for solids [7.1, 7.19].

There are various approaches to generating shape functions. One is to form products of Lagrange formulas, as in Eq. 4.2.12 [7.19]. This method yields elements with internal d.o.f. for cubic and higher expansions, as in Fig. 7.9.2*b*.

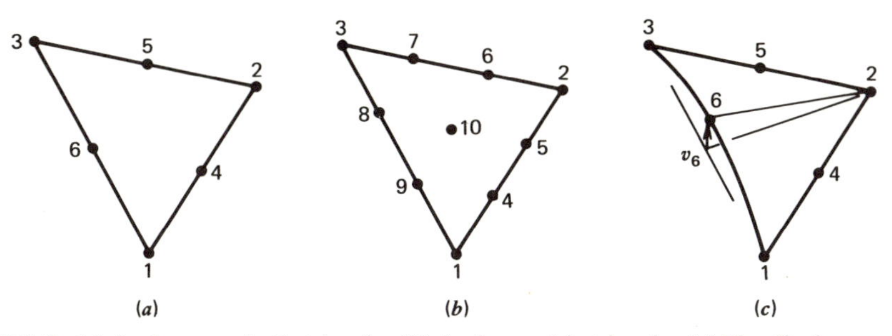

Figure 7.9.2. (*a*) A plane quadratic triangle. (*b*) A plane cubic triangle. (*c*) The displacement mode $v = 4\alpha\gamma v_6$ in a quadratic triangle.

A second approach is to start with a field expressed in terms of generalized coordinates a_i as, for example, in Eq. 4.5.1, and then convert from a-basis to d-basis. Thus, for the quadratic and cubic elements, respectively, in Fig. 7.9.2, we start with

$$\phi = \alpha^2 a_1 + \beta^2 a_2 + \gamma^2 a_3 + \alpha\beta a_4 + \beta\gamma a_5 + \gamma\alpha a_6 \tag{7.9.10}$$

$$\phi = \alpha^3 a_1 + \beta^3 a_2 + \gamma^3 a_3 + \alpha^2\beta a_4 + \alpha\beta^2 a_5 + \beta^2\gamma a_6 + \beta\gamma^2 a_7 + \gamma^2\alpha a_8 + \gamma\alpha^2 a_9 + \alpha\beta\gamma a_{10} \tag{7.9.11}$$

Equation 7.9.5 shows that the coefficient of each a_i is a polynomial in x and y of degree 2 and 3 for the quadratic and cubic elements, respectively. Indeed, the expansion

$$\phi = \sum_{i=1}^{n} \alpha^p \beta^q \gamma^r a_i \tag{7.9.12}$$

where p, q, and r range over the n possible combinations for which $(p + q + r) = m$, yields a complete polynomial of degree m [7.19]. As noted in Eq. 4.9.1, $n = (m + 1)(m + 2)/2$. Mode $\alpha\beta\gamma$ in Eq. 7.9.11 is a bubble function like Eq. 7.3.2. Without it the ϕ expansion in Cartesian coordinates xy would be an incomplete cubic, so its inclusion is well worthwhile.

A third approach to generating shape functions is to combine modes in a visual way, as in Fig. 5.6.2. For example, in Fig. 7.9.1*c*, $v = 0.5$ at $\alpha = \gamma = 0.5$. To achieve $v = 0$ at all nodes but node 3, we can subtract half the shape function $4\alpha\gamma$ in Fig. 7.9.2*c* and half the corresponding mode $4\beta\gamma$ for node 5. Thus, using Eq. 7.9.2 to eliminate α and β,

$$N_3 = \gamma - 2\alpha\gamma - 2\beta\gamma = \gamma(2\gamma - 1) \tag{7.9.13}$$

This mode is zero at $\gamma = 0$ and at $\gamma = 0.5$, but unity at $\gamma = 1$, as it should be. The displacement field of the quadratic triangle is therefore

$$u = \sum N_i u_i \qquad v = \sum N_i v_i \tag{7.9.14}$$

where

$$\begin{array}{lll} N_1 = \alpha(2\alpha - 1) & N_2 = \beta(2\beta - 1) & N_3 = \gamma(2\gamma - 1) \\ N_4 = 4\alpha\beta & N_5 = 4\beta\gamma & N_6 = 4\gamma\alpha \end{array} \tag{7.9.15}$$

Shape functions for other arrangements of nodal d.o.f. and for cubic elements appear, for example, in Ref. 11.13.

Example

Consider the constant strain triangle, first discussed in Section 4.5. In area coordinates, the shape functions in Eq. 4.5.7 are simply $N_1 = \alpha$, $N_2 = \beta$, and $N_3 = \gamma$. The strain-displacement relation $\{\boldsymbol{\epsilon}\} = [\mathbf{B}]\{\mathbf{d}\}$, from Eqs. 7.9.6 and 7.9.7, is

$$\begin{Bmatrix} \epsilon_x \\ \epsilon_y \\ \gamma_{xy} \end{Bmatrix} = \frac{1}{2A} \begin{bmatrix} y_{23} & 0 & y_{31} & 0 & y_{12} & 0 \\ 0 & x_{32} & 0 & x_{13} & 0 & x_{21} \\ x_{32} & y_{23} & x_{13} & y_{31} & x_{21} & y_{12} \end{bmatrix} \begin{Bmatrix} u_1 \\ v_1 \\ \vdots \\ v_3 \end{Bmatrix} \tag{7.9.16}$$

where $y_{23} = y_2 - y_3$, $x_{32} = x_3 - x_2$, and so on. Substitution of this $[\mathbf{B}]$ matrix into Eq. 4.5.8 yields the element stiffness matrix.

7.10 QUADRATIC TRIANGLES. NUMERICAL INTEGRATION

First, we formulate the stiffness matrix of a straight-sided element with midside nodes (Fig. 7.10.1*a*). Later we remove these restrictions and consider the element of Fig. 7.10.1*b*.

Shape functions for the triangle of Fig. 7.10.1*a* are given by Eq. 7.9.15. Straightforward use of Eqs. 7.9.6 and 7.9.7, where first $\phi = u$ and then $\phi = v$, yields the strain-displacement relation $\{\boldsymbol{\epsilon}\} = [\mathbf{B}]\{\mathbf{d}\}$.

$$\begin{Bmatrix} \epsilon_x \\ \epsilon_y \\ \gamma_{xy} \end{Bmatrix} = \begin{bmatrix} \mathbf{B}_x & \mathbf{O} \\ \mathbf{O} & \mathbf{B}_y \\ \mathbf{B}_y & \mathbf{B}_x \end{bmatrix} \underset{12 \times 1}{\{u_1 \; u_2 \; \cdots \; u_6 \; v_1 \; v_2 \; \cdots \; v_6\}} \tag{7.10.1}$$

where, for example, with $y_{23} = y_2 - y_3$ and so on,

$$\underset{1\times 6}{[\mathbf{B}_x]} = \frac{1}{2A}\Big\lfloor (4\alpha - 1)y_{23} \quad (4\beta - 1)y_{31} \quad (4\gamma - 1)y_{12} \quad 4\alpha y_{31} + 4\beta y_{23} \quad 4\beta y_{12} + 4\gamma y_{31} \quad 4\gamma y_{23} + 4\alpha y_{12} \Big\rfloor \tag{7.10.2}$$

Because the displacement field of this element is a complete quadratic polynomial, the strain field is a complete linear polynomial. This circumstance permits the following trick that simplifies the formulation. Strains $\{\boldsymbol{\epsilon}\}$ at any point can be linearly interpolated from strains $\{\boldsymbol{\epsilon}_c\}$ at the corners. Thus

$$\{\boldsymbol{\epsilon}\} = [\mathbf{Q}]\{\boldsymbol{\epsilon}_c\} \tag{7.10.3}$$

where

$$\{\boldsymbol{\epsilon}_c\} = \{\epsilon_{x1}\ \epsilon_{x2}\ \epsilon_{x3}\ \epsilon_{y1}\ \epsilon_{y2}\ \epsilon_{y3}\ \gamma_{xy1}\ \gamma_{xy2}\ \gamma_{xy3}\} \tag{7.10.4}$$

and

$$[\mathbf{Q}] = \begin{bmatrix} \alpha & \beta & \gamma & 0 & 0 & 0 & 0 & 0 & 0 \\ 0 & 0 & 0 & \alpha & \beta & \gamma & 0 & 0 & 0 \\ 0 & 0 & 0 & 0 & 0 & 0 & \alpha & \beta & \gamma \end{bmatrix} \tag{7.10.5}$$

Corner strains $\{\boldsymbol{\epsilon}_c\}$ are written in terms of nodal d.o.f. $\{\mathbf{d}\}$ by evaluating Eq. 7.10.1 at the corners. We abbreviate this relation as

$$\{\boldsymbol{\epsilon}_c\} = [\mathbf{H}]\{\mathbf{d}\} \tag{7.10.6}$$

where $[\mathbf{H}]$ is a 9-by-12 matrix of element dimensions. Therefore $[\mathbf{B}] = [\mathbf{Q}][\mathbf{H}]$, and the element stiffness matrix is

$$[\mathbf{k}] = [\mathbf{H}]^T \int_A [\mathbf{Q}]^T[\mathbf{E}][\mathbf{Q}]t\, dA[\mathbf{H}] \tag{7.10.7}$$

The expression to be integrated is an elastic kernel, as described in Section 7.1. If material properties $[\mathbf{E}]$ and element thickness t are constant, the integral is easy to evaluate by Eq. 7.9.8, and a purely algebraic expression for $[\mathbf{k}]$ is produced (Fig. 7.10.2). Numerical examples in Section 5.12 show how well this element behaves.

If the element has curved sides (Fig. 7.10.1*b*), or if side nodes are not at midsides, the N_i of Eqs. 7.9.14 and 7.9.15 still apply. If we also define element shape in the same way as displacements, that is,

$$\{x\ \ u\} = \sum N_i\{x_i\ \ u_i\}, \qquad \{y\ \ v\} = \sum N_i\{y_i\ \ v_i\} \tag{7.10.8}$$

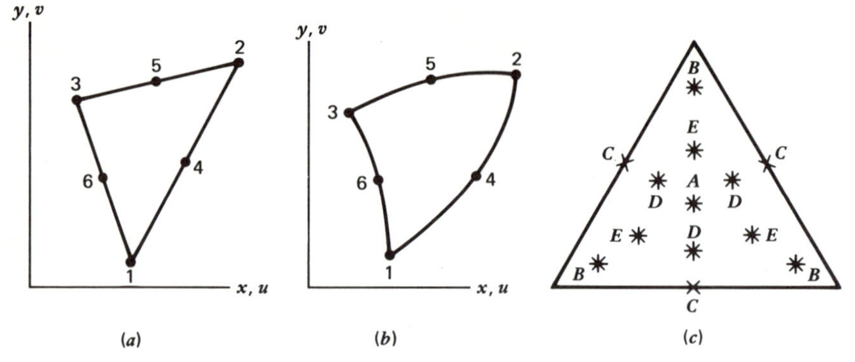

Figure 7.10.1. (*a*) Quadratic triangle with straight sides and midside nodes. (*b*) General quadratic triangle. (*c*) Sampling points used in quadrature formulas of Table 7.10.1.

$$[\mathbf{k}] = \frac{t}{48A}[\mathbf{H}_1]^T[\mathbf{Q}_1][\mathbf{H}_1], \quad \text{where}$$

$$\underset{9\times 9}{[\mathbf{Q}_1]} = \begin{bmatrix} 2E_{11} & E_{11} & E_{11} & 2E_{12} & E_{12} & E_{12} & 2E_{13} & E_{13} & E_{13} \\ & 2E_{11} & E_{11} & E_{12} & 2E_{12} & E_{12} & E_{13} & 2E_{13} & E_{13} \\ & & 2E_{11} & E_{12} & E_{12} & 2E_{12} & E_{13} & E_{13} & 2E_{13} \\ & & & 2E_{22} & E_{22} & E_{22} & 2E_{23} & E_{23} & E_{23} \\ & & & & 2E_{22} & E_{22} & E_{23} & 2E_{23} & E_{23} \\ & & & & & 2E_{22} & E_{23} & E_{23} & 2E_{23} \\ \text{Symmetric} & & & & & & 2E_{33} & E_{33} & E_{33} \\ & & & & & & & 2E_{33} & E_{33} \\ & & & & & & & & 2E_{33} \end{bmatrix}$$

$$\underset{9\times 12}{[\mathbf{H}_1]} = \begin{bmatrix} 3b_1 & -b_2 & -b_3 & 4b_2 & 0 & 4b_3 & 0 & 0 & 0 & 0 & 0 & 0 \\ -b_1 & 3b_2 & -b_3 & 4b_1 & 4b_3 & 0 & 0 & 0 & 0 & 0 & 0 & 0 \\ -b_1 & -b_2 & 3b_3 & 0 & 4b_2 & 4b_1 & 0 & 0 & 0 & 0 & 0 & 0 \\ 0 & 0 & 0 & 0 & 0 & 0 & 3a_1 & -a_2 & -a_3 & 4a_2 & 0 & 4a_3 \\ 0 & 0 & 0 & 0 & 0 & 0 & -a_1 & 3a_2 & -a_3 & 4a_1 & 4a_3 & 0 \\ 0 & 0 & 0 & 0 & 0 & 0 & -a_1 & -a_2 & 3a_3 & 0 & 4a_2 & 4a_1 \\ 3a_1 & -a_2 & -a_3 & 4a_2 & 0 & 4a_3 & 3b_1 & -b_2 & -b_3 & 4b_2 & 0 & 4b_3 \\ -a_1 & 3a_2 & -a_3 & 4a_1 & 4a_3 & 0 & -b_1 & 3b_2 & -b_3 & 4b_1 & 4b_3 & 0 \\ -a_1 & -a_2 & 3a_3 & 0 & 4a_2 & 4a_1 & -b_1 & -b_2 & 3b_3 & 0 & 4b_2 & 4b_1 \end{bmatrix}$$

and

$$a_1 = x_3 - x_2 \qquad a_2 = x_1 - x_3 \qquad a_3 = x_2 - x_1$$
$$b_1 = y_2 - y_3 \qquad b_2 = y_3 - y_1 \qquad b_3 = y_1 - y_2$$

Figure 7.10.2. The stiffness matrix of the linear strain triangle (LST) of Fig. 7.10.1*a* expressed in terms of elastic constants E_{ij} and corner node coordinates [11.13]. Sides are straight, side nodes are at midsides, thickness t is constant, area A is given by Eq. 4.5.4, and nodal d.o.f. have the order $\{\mathbf{d}\} = \{u_1\ u_2\ u_3\ u_4\ u_5\ u_6\ v_1\ v_2\ v_3\ v_4\ v_5\ v_6\}$.

then we have again written Eqs. 5.3.2, and the triangular element is *isoparametric*. With curved sides, Eqs. 7.9.7 no longer apply, and the procedure of Section 5.3 must be used as follows.

Because $\alpha + \beta + \gamma = 1$, we can eliminate (say) γ from the shape functions and write them in terms of only α and β. Now equations 5.3.5 through 5.3.7 apply to the triangle if we replace ξ by α and η by β. The equation analogous to Eq. 5.3.8 is

$$J_{11} = x,_{\alpha} = N_{1,\alpha}x_1 + N_{2,\alpha}x_2 + \cdots + N_{6,\alpha}x_6 \tag{7.10.9}$$

Equations 5.3.10 through 5.3.12 require only that α and β replace ξ and η to be applicable to the triangle. Hence [**B**] is again the product of three rectangular matrices. The element stiffness matrix is

$$[\mathbf{k}] = \iint_A [\mathbf{B}]^T [\mathbf{E}][\mathbf{B}]tJ \, d\alpha \, d\beta \tag{7.10.10}$$

The integrand is a function of α and β and must be evaluated numerically.

The procedure outlined here can be applied to higher-order curved triangles and to prismatic wedges that involve both area coordinates and the coordinates of Chapter 5 (Fig. 7.15.2*c*, Problem 7.40). Again, it is clear that the manipulations of isoparametric elements apply to a broad class of element types.

Gauss quadrature formulas for triangular regions have a form analogous to Eq. 5.4.2.

$$\iint_A f(\alpha, \beta, \gamma) \, d\alpha \, d\beta = \sum_{i=1}^{n} W_i f(\alpha_i, \beta_i, \gamma_i) \tag{7.10.11}$$

where n is the number of sampling points. Different choices of weights and sampling points are sometimes possible for a given n [5.4, 7.20–7.23]. Table 7.10.1 lists some quadrature formulas. For points B through E in Fig. 7.10.1*c*, there are three points that have the same weight W_i. Area coordinates are given in Table 7.10.1 for only one of the points B through E. Coordinates of the other two are obtained by cyclic permutation. The center point A has multiplicity one. Points *ns* (not shown) have multiplicity 6; that is, all six possible permutations of the coordinates in row *ns* of Table 7.10.1 are used. Thus the integration formulas are symmetric in the area coordinates.

The "degree of precision" in Table 7.10.1 is the highest-degree polynomial that the formula integrates exactly. The seven-point formula is used in Reference 8.2, where it is coded in Cartesian instead of area coordinates.

If $f(\alpha, \beta, \gamma)$ is unity, Eq. 7.10.11 becomes

$$\int_A d\alpha \, d\beta = \sum_{i=1}^{n} W_i = 1 \tag{7.10.12}$$

TABLE 7.10.1. Gauss quadrature formulas for integration over a triangle. Approximate locations of entries in the "points" column are shown in Fig. 7.10.1c. ns = points of multiplicity 6, not shown in Fig. 7.10.1c

Points	Multiplicity	Area Coordinates α_i, β_i, γ_i			Weights W_i
		1-point formula	degree of precision 1		
A	1	0.33333 33333 33333	0.33333 33333 33333	0.33333 33333 33333	1.00000 00000 00000
		3-point formula	degree of precision 2		
B	3	0.66666 66666 66667	0.16666 66666 66667	0.16666 66666 66667	0.33333 33333 33333
		3-point formula	degree of precision 2		
C	3	0.50000 00000 00000	0.50000 00000 00000	0.00000 00000 00000	0.33333 33333 33333
		4-point formula	degree of precision 3		
A	1	0.33333 33333 33333	0.33333 33333 33333	0.33333 33333 33333	−0.56250 00000 00000
B	3	0.60000 00000 00000	0.20000 00000 00000	0.20000 00000 00000	0.52083 33333 33333
		6-point formula	degree of precision 4		
B	3	0.81684 75729 80459	0.09157 62135 09771	0.09157 62135 09771	0.10995 17436 55322
D	3	0.10810 30181 68070	0.44594 84909 15965	0.44594 84909 15965	0.22338 15896 78011
		7-point formula	degree of precision 5		
A	1	0.33333 33333 33333	0.33333 33333 33333	0.33333 33333 33333	0.22500 00000 00000
B	3	0.79742 69853 53087	0.10128 65073 23456	0.10128 65073 23456	0.12593 91805 44827
D	3	0.47014 20641 05115	0.47014 20641 05115	0.05971 58717 89770	0.13239 41527 88506
		12-point formula	degree of precision 6		
B	3	0.87382 19710 16996	0.06308 90144 91502	0.06308 90144 91502	0.05084 49063 70207
E	3	0.50142 65096 58179	0.24928 67451 70910	0.24928 67451 70911	0.11678 62757 26379
ns	6	0.63650 24991 21399	0.31035 24510 33785	0.05314 50498 44816	0.08285 10756 18374
		13-point formula	degree of precision 7		
A	1	0.33333 33333 33333	0.33333 33333 43333	0.33333 33333 33333	−0.14957 00444 67670
E	3	0.47930 80678 41923	0.26034 59660 79038	0.26034 59660 79038	0.17561 52574 33204
B	3	0.86973 97941 95568	0.06513 01029 02216	0.06513 01029 02216	0.05334 72356 08839
ns	6	0.63844 41885 69809	0.31286 54960 04875	0.04869 03154 25316	0.07711 37608 90257

For the element of Fig. 7.10.1*a*, the Jacobian J is constant so, by Eq. 7.10.12, the element area is

$$A = \iint_A J \, d\alpha \, d\beta = J \sum_{i=1}^{n} W_i = J \tag{7.10.13}$$

7.11 OTHER OPTIONS IN ELEMENT FORMULATION

Exact integration of polynomials in x and y over rectangles and *straight-sided* triangles is straightforward but tedious. Fortunately, the results are given by simple formulas. By using the formulas with a polynomial field in Cartesian coordinates, we can easily generate the a-basis stiffness matrix of Eq. 4.3.19. Each term that must be integrated has the form

$$I = Q \int_A x^m y^n \, dA \tag{7.11.1}$$

where m and n are integers and Q contains elastic constants, element thickness, and numbers produced by differentiation of the displacement field. For elements in Figs. 7.11.1*a* and 7.11.1*b* [15.5, 7.24, 7.25],

$$\text{Triangle:} \quad I = Qc^{n+1} \left[a^{m+1} - (-b)^{m+1} \right] \frac{m! \, n!}{(m + n + 2)!}$$

$$\text{Rectangle:} \quad I = Q \frac{a^{m+1} \, b^{n+1}}{(m + 1)(n + 1)} \tag{7.11.2}$$

Elements formulated in local xy coordinates can be transformed to global coordinates by Eq. 6.3.7.

A computer program based on this approach need only be supplied with element dimensions, elastic properties, and the degree of polynomial to be used. It then can

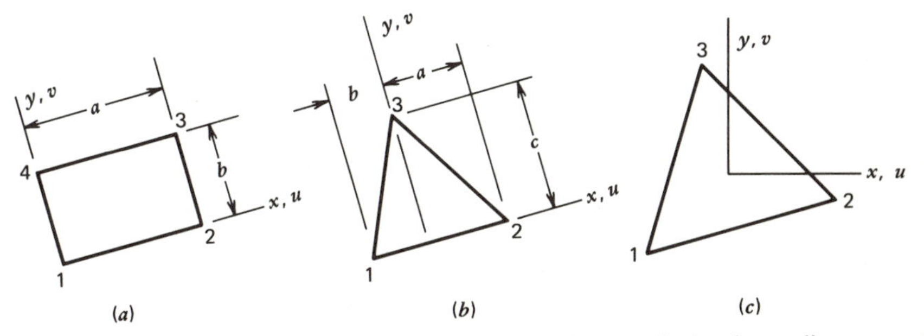

Figure 7.11.1. Rectangular and straight-sided triangular elements in local coordinates xy. (*a, b*) Coordinates for Eqs. 7.11.2. (*c*) Centroidal coordinates xy for Eq. 7.11.3.

generate $[\mathbf{k}_a]$. For example, Ref. 7.25 contains two Fortran subroutines of about 65 statements each, capable of generating mass, initial stress, and $[\mathbf{k}_a]$ matrices for eight plane elements and nine plate bending elements. The transformation from a-basis to d-basis (Eq. 4.3.18), must be coded by the user for each element.

Integration over the triangle can be done in other coordinates. In centroidal coordinates of arbitrary orientation, $x_1 + x_2 + x_3 = y_1 + y_2 + y_3 = 0$ in Fig. 7.11.1*c*. Then the integral of Eq. 7.11.1 is

$$I = QAc_{m+n}(x_1{}^m y_1{}^n + x_2{}^m y_2{}^n + x_3{}^m y_3{}^n) \tag{7.11.3}$$

where A is the triangle area and $c_0 = \frac{1}{3}$, $c_1 = 0$, $c_2 = \frac{1}{12}$, $c_3 = c_4 = \frac{1}{30}$, and $c_5 = \frac{2}{105}$ [15.5]. Integrals for $m + n = 6$ are more complicated, but they are tabulated [7.26]. Still other local coordinates are possible [7.27, 7.28].

For a straight-sided triangle, instead of working in the a-basis, we can work in the d-basis by starting with a displacement field in terms of area coordinates and nodal d.o.f., then transforming the field to Cartesian coordinates by Eq. 7.9.5. In this way we achieve invariance if the polynomial is symmetric in area coordinates. The algebra of this process—and of some others—can be vast. Can these symbolic manipulations be automated?

Indeed they can. Available programs can, among other things, do the basic operations on polynomials: addition, multiplication, differentiation, and integration. Input data required by such a program include the geometry of the element, the type and placement of d.o.f., and the terms in the displacement field polynomial. Output includes *algebraic* expressions for coefficients in the element stiffness matrix. These expressions can be cast in the format of Fortran statements that can be incorporated in another program that will evaluate the coefficients numerically [7.28–7.30].

The obvious advantage of a symbol-processing program is that the developer of an element can avoid a great deal of pencil-and-paper algebra. Also, the numerical program may become more efficient, because the final algebraic expressions for stiffness coefficients are often simple and because the customary numerical-integration formulation spends time calculating zeros and other quantities already known. The numerical approach remains preferable when stiffness coefficients are not simple polynomials, as with isoparametric elements.

7.12 DEGRADING OF ELEMENTS

The four-to-eight node plane element of Table 5.6.1 can have one to four linear edges. With one linear edge it makes a useful transition element between quadratic and linear elements. In this form it might be called a "once-degraded" quadratic element. Similar remarks apply to the quadratic brick of Section 5.7.

When an element is to be degraded, it is convenient to use *departures from linearity* as side d.o.f. For example, v_5' in Fig. 7.12.1*b* is such a d.o.f. The total y-direction displacement of node 5 is

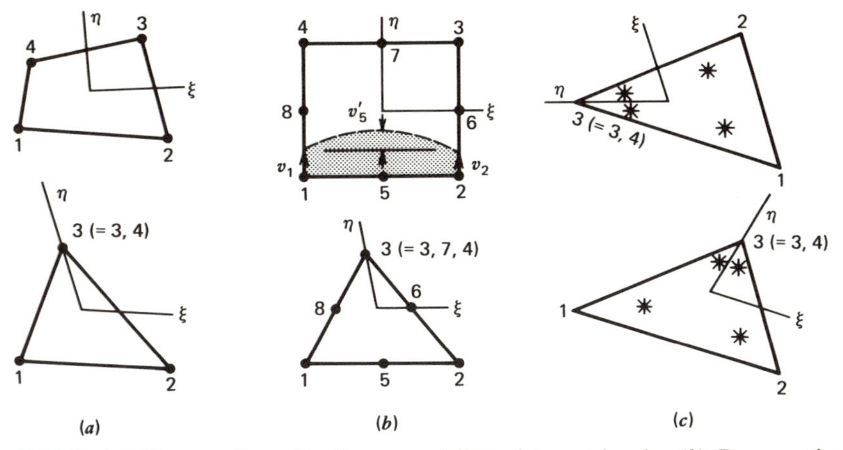

Figure 7.12.1. (*a*) Degeneration of a linear quadrilateral to a triangle. (*b*) Degeneration of a quadratic quadrilateral to a triangle. (*c*) Gauss point locations of a 2-by-2 rule, produced by different node numberings.

$$v_5 = \frac{v_1 + v_2}{2} + v_5' \tag{7.12.1}$$

Thus v_5' is the displacement *relative* to a midside displacement linearly interpolated between nodes 1 and 2. Substitution of equations such as Eq. 7.12.1 into the equation $v = \Sigma N_i v_i$, with the eight N_i of Table 5.6.1, shows that in terms of departures from linearity, the element basis is

$$\begin{Bmatrix} x \\ y \\ u \\ v \end{Bmatrix} = \sum_4^1 N_i \begin{Bmatrix} x_i \\ y_i \\ u_i \\ v_i \end{Bmatrix} + \sum_5^8 N_i \begin{Bmatrix} x_i' \\ y_i' \\ u_i' \\ v_i' \end{Bmatrix} \tag{7.12.2}$$

where N_1 to N_4 are given by Eqs. 5.3.3 and N_5 to N_8 remain as stated by Table 5.6.1. Linearity of an edge is enforced simply by setting the appropriate x_i' and y_i' to zero and omitting the corresponding u_i' and v_i' from the vector of element d.o.f. $\{\mathbf{d}\}$. (Another advantage of this format is that if an edge is quadratic but straight, so that u_i' and v_i' are retained but $x_i' = y_i' = 0$, the user need not supply x_i' and y_i' as input data.)

Caution. In elements such as this, side node *coordinates* must also be expressed as departures from linearity if [**N**] and [**J**] are to be the same for shape as for displacement. Also, a "departure" edge of one element cannot be attached to an "ordinary" edge of another element.

Quadrilaterals can be degraded to triangles. In Fig. 7.12.1*a*, nodes 3 and 4 coalesce to yield a three-node triangle. The shape functions of the triangle are

$$N_1 = \frac{(1 - \xi)(1 - \eta)}{4}, \qquad N_2 = \frac{(1 + \xi)(1 - \eta)}{4}, \qquad N_3 = \frac{1 + \eta}{2} \tag{7.12.3}$$

where N_3 is the sum of N_3 and N_4 in Eqs. 5.3.3. This element is a constant strain triangle. Similarly, an eight-node solid element yields a constant strain tetrahedron by coalescing its corners [7.31].

Unfortunately, *quadratic* elements do not neatly degenerate to triangles and tetrahedra. Consider a rectangular element (Fig. 7.12.1*b*). If degenerated to a straight-sided triangle with midside nodes, its shape (not its displacement field) is

$$x = N_1x_1 + N_2x_2 + N_3x_3 = \alpha x_1 + \beta x_2 + \gamma x_3 \tag{7.12.4}$$

and similarly for y, where N_1 to N_3 are given by Eq. 7.12.3 and α, β, and γ are the area coordinates discussed in Section 7.9. Therefore, from Eqs. 7.12.3 and 7.12.4,

$$\alpha = \frac{(1 - \xi)(1 - \eta)}{4}, \qquad \beta = \frac{(1 + \xi)(1 - \eta)}{4}, \qquad \gamma = \frac{1 + \eta}{2} \tag{7.12.5}$$

Turning now to the displacement field, Eqs. 7.12.2 through 7.12.5 yield

$$N_5 = \frac{(1 - \xi^2)(1 - \eta)}{2} = \frac{4\alpha\beta}{1 - \gamma} = \frac{4\alpha\beta}{(1 - \eta)/2} \tag{7.12.6}$$

We see that N_5 becomes infinite at node 3. (In special circumstances, such a singularity is desirable; see Section 7.15.) The desired shape function, according to Eq. 7.9.15, is $N_5 = 4\alpha\beta$, which can be obtained by multiplying Eq. 7.12.6 by $(1 - \eta)/2$. Generalizing, it is recommended that when a side node represents departure from linearity, the side node shape function N_5 to N_8 in Eq. 7.12.2 be multiplied by $(1 - \eta)/2$, $(1 + \xi)/2$, $(1 + \eta)/2$, or $(1 - \xi)/2$, respectively, if the side opposite is to be degraded to a point. Similar but more complicated considerations apply to quadratic solids [7.31, 7.32].

In the preceding descriptions, nodes are made to coalesce by giving them the same coordinates and modified shape functions. If input data to a program give nodes the same coordinates but the original shape functions are not changed, nodes are superposed but can move independently. Then, if $[\mathbf{k}]$ is integrated exactly, stiffness coefficients between the superposed nodes become infinite, and this is disastrous (Section 15.5). If the element is numerically integrated but none of the Gauss points coincide with the superposed nodes, performance declines near the superposed nodes but the element does not fail. An attempt to calculate strains directly at superposed nodes will fail because of division by a zero Jacobian determinant.

Figure 7.12.1*c* shows that in $\xi\eta$ coordinates, different node numberings of a given figure can lead to a different distribution of Gauss points. In this sense the element looses invariance. Area coordinates and the quadrature formulas of Table 7.10.1 yield Gauss point locations that are insensitive to node point numbering.

7.13 MODELING, MESH LAYOUT, AND GRADING

An inexperienced user finds it hard to decide how to lay out a mesh. What element types should be used? How many elements? How are meshes graded from coarse to fine? What are the limits of element aspect ratio and shape distortion?

Modeling is an art based on the ability to visualize physical interactions. Little is published; practitioners learn by doing and by talking with others. Many aspects of finite element behavior bear on modeling and mesh-layout decisions. (Sections 4.8–4.11, 5.12, 6.12, 6.13, 7.12, 9.2, 10.2, 10.4, 12.5, 14.3–14.5, 15.3–15.5, and Appendix B contain pertinent observations.) This section includes additional remarks, suggestions, and references.

It is clear that the *arrangement* of a given number of elements influences results, most strongly if the mesh is coarse. For example, in Fig. 7.13.1 elements with dashed lines are utterly inactive because the boundary conditions fully restrain them. The plate is modeled by only the solid-outline elements. The first mesh is most accurate. The second is too stiff. The third and fourth do not predict a symmetric response with respect to symmetry axes of the actual structure [7.33].

Figure 7.13.2 shows some mesh-grading schemes. Triangular elements such as 1, 2, and 3 can be replaced by a single four-node quadrilateral if edge *ABC* is constrained to remain straight. Problem 6.29 considers this constraint problem. Other approaches to the ''side node transition'' problem appear in Refs. 7.34, 7.35, and 7.62.

Meshes in Figs. 7.13.3*a* and 7.13.3*b* also involve the side node transition problem. All elements of these meshes have good aspect ratios. Internal d.o.f. of the hole elements are to be condensed before the patch with eight external nodes is assembled into the structure. Alternative versions of hole elements appear in Figs. 7.2.1*c* and 7.13.3*c*. Both might be based on assumed displacement fields. But hybrid elements seem preferable. They incorporate an assumed stress field that can represent the sinusoidal variations of stress that theory demands [7.36].

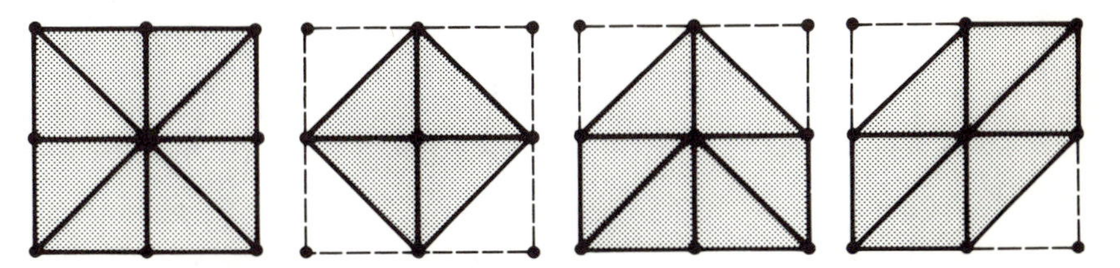

Figure 7.13.1. Arrangements of eight triangular elements to model a uniformly loaded rectangular plate with clamped edges.

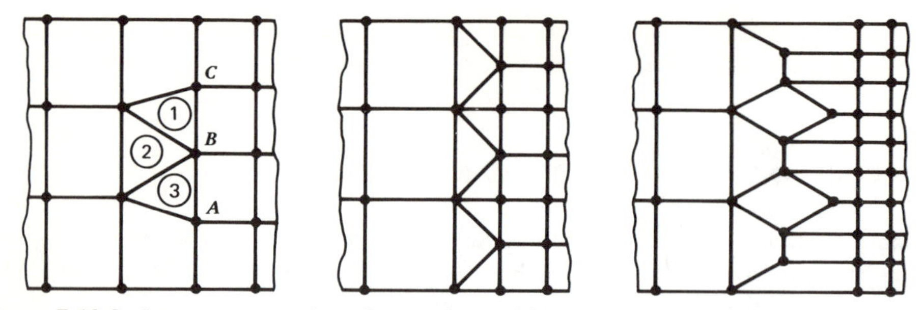

Figure 7.13.2. Some coarse-mesh to fine-mesh transition schemes that use three- and four-node plane elements.

A "staged" analysis can be used to study areas of particular interest. In Fig. 7.13.4, analysis begins with the coarse mesh of Fig. 7.13.4*b*. Nodal displacements along *AB* and *CD* are computed. Then the region of interest *only* is modeled with a finer mesh (Fig. 7.13.4*c*). Boundaries *AB* and *CD* of the finer mesh are forced to have the displacements computed in the coarse-mesh analysis. Thus neither the coarse-mesh analysis nor the fine-mesh analysis involves a large number of d.o.f. However, it is questionable whether this approach is better than using a fine-mesh substructure to represent region *ABCD* (Section 6.13).

Some fine-mesh nodes along *AB* and *CD* in Fig. 7.13.4 do not appear in the coarser mesh. D.o.f. of these nodes must be computed by interpolation from the known d.o.f. [7.37].

A source of error in the staged analysis lies in the finer mesh being more flexible than the coarser mesh (if elements are based on assumed displacement fields). Therefore displacements computed in Fig. 7.13.4*b* are underestimated, so stresses in Fig. 7.13.4*c* are underestimated. An ad hoc way to improve results is as follows. For coarse and fine meshes, respectively, compute the boundary forces $\{\mathbf{R}\}_c$ and $\{\mathbf{R}\}_f$ associated with the prescribed displacements along *AB* and *CD*. Add absolute magnitudes in these vectors

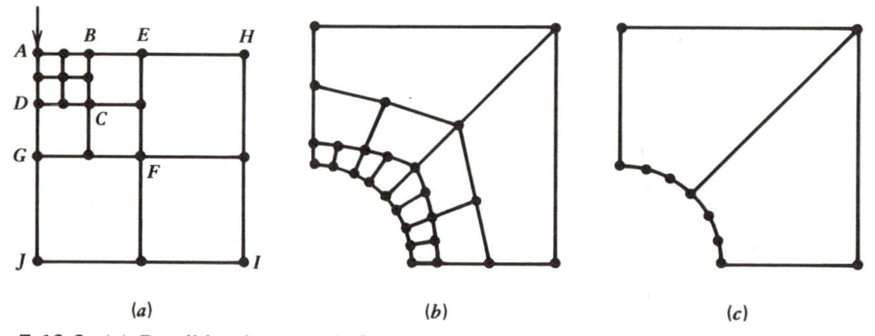

Figure 7.13.3. (*a*) Possible plane mesh for analysis of high stress gradients near a point load at *A*. (*b*) One-quarter of a "hole element"—an eight-node patch that models a hole in a plate. (*c*) Alternative arrangement for a hole element.

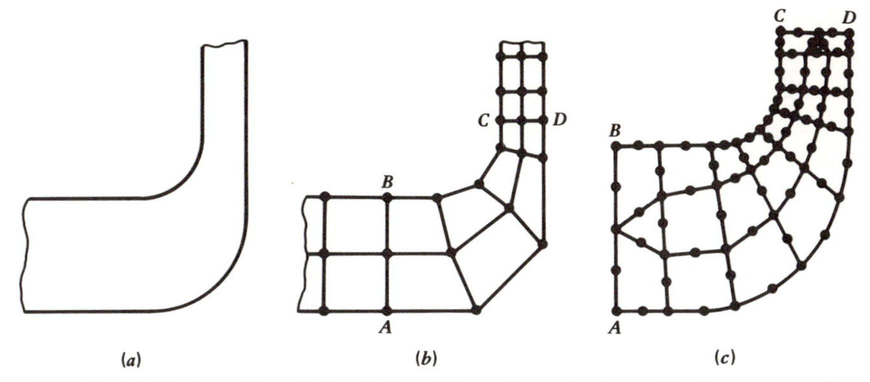

Figure 7.13.4. (*a*) Region of particular interest in a plane structure. (*b*) A coarse-mesh model. (*c*) An enlarged view of a finer mesh in the region where stresses and stress gradients are high.

(or use some other norm) and call the respective sums F_c and F_f. Finally, stresses computed from the fine mesh are multiplied by the ratio F_c/F_f. This ratio exceeds unity because the coarse mesh is stiffer. This device is reminiscent of Loubignac's stress-improvement scheme (Eq. 4.10.4).

The preceding staged analysis is an instance of the general dictum *start simply*. The first analysis step might be to model the structure by a few beam elements to see the overall response and find out where high stresses are likely. Next, a coarse mesh of simple elements might be used and, finally, a finer mesh of more competent elements. Often a mesh fine enough for accurate stress analysis is automatically produced by good modeling of the structure geometry. Many simple elements may model the geometry better than a few higher-order elements. (However, a curved boundary should not be represented by straight line segments; see Section 15.4.)

Various effects can reduce the stiffness of structures and elements: joints (riveted, bolted, glued, and so on), holes, notches, corrugations, swages, and curvature of membranes [7.38]. These effects should be modeled, either explicitly or by an empirical reduction of element stiffness. Unfortunately, the needed empirical data are often lacking.

Advice that supplements remarks found elsewhere in this book includes the following [7.38, 7.39].

1. Use the smallest number and the simplest types of elements compatible with stress gradients and structure geometry.
2. Include all real structure in the model. Analysis may show that "nonload-carrying" members are, in fact, overstressed.
3. When possible, align element boundaries with structure members and with principal loading trajectories.
4. Element aspect ratios should not exceed roughly 7 for good displacement results and roughly 3 for good stress results.

5. Mesh grading should be done in such a way that abrupt changes in element size are minimized.
6. If a direct-integration transient analysis is contemplated, a mass less than roughly one-thousandth of the average mass should be eliminated by combining it with other masses.
7. The mesh layout (or the time step in a dynamic analysis) is probably adequate if alterations do little to change the results.
8. Only if computed displacements are deemed agreeable should computed stresses be taken seriously. However, a mesh that gives good displacements (or vibration frequencies) may be too coarse to yield accurate stresses (or vibration modes).

These items are suggestions more than rules, and for each one there may be an important exception. The items should be tempered by broad knowledge and experience.

7.14 ELEMENTS THAT ARE SPECIALLY SHAPED OR INFINITE

Until now we have emphasized elements that are not restricted to any particular shape. But less geometric generality sometimes permits the use of particularly good displacement fields. Also, the elements can model the geometry of certain structural shapes exactly. Examples include those of Fig. 7.14.1*a*: annular sector elements for plane problems and plate bending, parallelogram elements for skew plates, and elements with 90° corner angles for circular cylindrical shells [7.40–7.44, 10.11, 10.19].

In some problems the region of interest is contained in a medium that extends to infinity. Examples include a load on a semiinfinite plate and wave action on an offshore

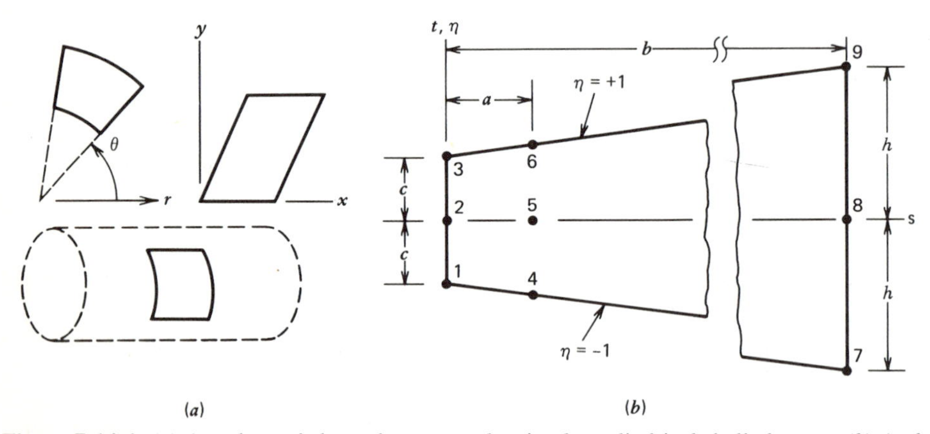

Figure 7.14.1. (*a*) Annular and skew elements and a circular cylindrical shell element. (*b*) A plane quadratic "infinite" element in local coordinates *st*. Typically b is set at about 15 times a. The element is "semiparametric" in that η is a natural coordinate but s is Cartesian.

structure: in the respective problems, stresses and the velocity potential vanish at infinity. It is awkward to approximate an infinite region by a large mesh of standard elements. Also, in the fluid problem, we do not want the outer boundary to reflect a wave.

Infinite elements model the effect of a distant region on the domain of interest. They do not pretend to describe accurately what happens in the distant region itself. An "infinite" element is actually finite in size but incorporates exponential decay in its shape functions.

The infinite element of Fig. 7.14.1*b* is a Lagrange element, similar to the element of Fig. 4.2.2*b*. Let a function ϕ be interpolated from nodal values ϕ_i, where $i = 1, 2, \ldots, 9$. Interpolation in the η direction is done in standard isoparametric fashion. For example, along $s = 0$, we have $\phi = L_1\phi_1 + L_2\phi_2 + L_3\phi_3$, where

$$L_1 = \frac{\eta^2 - \eta}{2}, \qquad L_2 = 1 - \eta^2, \qquad L_3 = \frac{\eta^2 + \eta}{2} \tag{7.14.1}$$

In the perpendicular direction s (for example, along $\eta = -1$), we have $\phi = M_1\phi_1 + M_2\phi_4 + M_3\phi_7$, where

$$M_1 = \frac{a - s}{a}\exp\left(-\frac{s}{L}\right), \qquad M_2 = \frac{s}{a}\exp\left(\frac{a - s}{L}\right), \qquad M_3 = 1 - M_1 - M_2 \tag{7.14.2}$$

and L is an arbitrary length, perhaps in the range $a < L < 4a$, that controls the severity of decay. All M_i satisfy the "zero-one" rule for shape functions; for example,

$$M_1 = 1 \quad \text{at} \quad s = 0, \qquad M_1 = 0 \quad \text{at} \quad s = a \quad \text{and at} \quad s = b \tag{7.14.3}$$

provided that the exponential function is nearly zero at $s = b$. In practice, M_3 is not needed because d.o.f. at "infinity" ($s = b$) are set to zero. Thus, in effect, the element has only six nodes. Accordingly, for the two-dimensional region, $\phi = N_1\phi_1 + N_2\phi_2 + \cdots + N_6\phi_6$, where

$$N_1 = L_1M_1, \qquad N_2 = L_2M_1, \quad \cdots, \qquad N_6 = L_3M_2 \tag{7.14.4}$$

If ϕ represents first u and then v, we can generate the stiffness matrix of a plane element. Here we need not restrict ϕ to mean displacement: we speak only of an element matrix, a typical coefficient of which is

$$\int_A f(s, \eta)\, dA = \int_0^b \int_{-1}^1 f(s, \eta) J \, d\eta \, ds \tag{7.14.5}$$

The Jacobian determinant J is easy to calculate because of the simple element geometry.

$$t = \left(c + \frac{h - c}{b} s\right)\eta, \qquad \text{hence} \qquad J = \frac{\partial t}{\partial \eta} = c + \frac{h - c}{b} s \tag{7.14.6}$$

Derivatives in the integrand are $\phi,_t = \phi,_\eta / J$ and simply $\phi,_s$. The standard Gauss quadrature of Table 5.4.1 is used in the η direction. A special six-point formula is recommended for the s direction [7.45].

After formulation, matrices of the infinite element can be transformed from local st coordinates to global xy coordinates by the standard operations (Eqs. 6.3.4, 6.3.7).

The element of Fig. 7.14.1*b* is simple but more restricted in shape and in shape function than necessary. Nevertheless, it is useful because infinite elements bound a region of standard elements that can be shaped as needed to make the transition to these infinite elements. Further elaboration and greater generality are found in Refs. 7.45 and 7.46.

7.15 FRACTURE MECHANICS. SINGULARITY ELEMENTS

Figure 7.15.1 illustrates the three deformation modes that can exist near a crack. Theory provides expressions for linearly elastic stresses $\{\boldsymbol{\sigma}\}$ and displacements $\{\mathbf{f}\}$ in the immediate vicinity of a crack tip. These expressions are independent of *overall* geometry and have the form

$$\{\boldsymbol{\sigma}\} = \frac{1}{\sqrt{r}}\left(K_{\mathrm{I}}\{\boldsymbol{\sigma}_{\mathrm{I}}\} + K_{\mathrm{II}}\{\boldsymbol{\sigma}_{\mathrm{II}}\} + K_{\mathrm{III}}\{\boldsymbol{\sigma}_{\mathrm{III}}\}\right) \tag{7.15.1}$$

$$\{\mathbf{f}\} = \sqrt{r}\left(K_{\mathrm{I}}\{\mathbf{f}_{\mathrm{I}}\} + K_{\mathrm{II}}\{\mathbf{f}_{\mathrm{II}}\} + K_{\mathrm{III}}\{\mathbf{f}_{\mathrm{III}}\}\right) \tag{7.15.2}$$

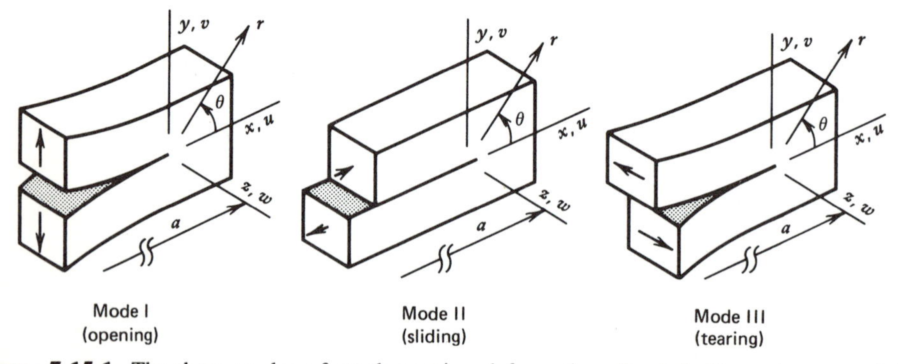

Figure 7.15.1. The three modes of crack-opening deformation. By definition, stresses and displacements are independent of z. Also, $r = \sqrt{x^2 + y^2}$ and $\theta = \arctan(y/x)$.

where r is the distance from the crack tip (Fig. 7.15.1). Vectors $\{\boldsymbol{\sigma}_i\}$ are functions of θ only, and vectors $\{\mathbf{f}_i\}$ are functions of θ and material properties. Factors K_{I}, K_{II}, and K_{III} are *stress intensity factors* for the respective modes in Fig. 7.15.1. Their magnitude depends on the geometry of the specimen and crack and on the distribution and magnitude of loading. Stress intensity factors are tabulated in handbooks for several geometries and loadings. In cases that are not tabulated the factors can be found by the finite element methods described in this section.

A crack propagates when the stress intensity factor reaches a critical value. The critical value is a material property (if the specimen is thick enough to avoid a size effect). Accordingly, by comparing the computed intensity factor with its critical value, we can decide whether or not the applied loading will make the crack propagate.

We summarize two ways that a finite element analysis can be used to compute a stress intensity factor. The first is an energy-release analysis. The second involves comparing finite element results with Eq. 7.15.1 or 7.15.2. In either method, calculations are simplified or made more accurate by elements that incorporate the $\sqrt{1/r}$ singularity for stress. Isoparametric elements can easily be made to display this singularity, as we will see. For brevity, we will consider only Mode I. It is the most common and the most studied.

Let a crack propagate an incremental distance da in a plane structure of unit thickness. The energy release rate is $d\Pi_p/da$. It can be shown that for plane stress and plane strain, respectively,

$$\frac{d\Pi_p}{da} = \frac{1}{E} K_{\mathrm{I}}^2 \qquad \text{or} \qquad \frac{d\Pi_p}{da} = \frac{1 - \nu^2}{E} K_{\mathrm{I}}^2 \tag{7.15.3}$$

In a finite element context we model the structure with its crack and write (Eq. 3.3.6)

$$\Pi_p = \tfrac{1}{2}\{\mathbf{D}\}^T[\mathbf{K}]\{\mathbf{D}\} - \{\mathbf{D}\}^T\{\mathbf{R}\} \tag{7.15.4}$$

Next we consider the same structure and loads but with the crack extended an amount Δa. Then all quantities in Eq. 7.15.4 but $\{\mathbf{R}\}$ are incremented. Thus we have an expression for $\Pi_p + \Delta\Pi_p$. We take the difference between the two potential expressions, use the equation $[\mathbf{K}]\{\mathbf{D}\} = \{\mathbf{R}\}$, ignore second order quantities, and find

$$\Delta\Pi_p = (\Pi_p + \Delta\Pi_p) - \Pi_p = \tfrac{1}{2}\{\mathbf{D}\}^T [\Delta\mathbf{K}]\{\mathbf{D}\} \tag{7.15.5}$$

To compute $\Delta\Pi_p$, we solve $[\mathbf{K}]\{\mathbf{D}\} = \{\mathbf{R}\}$ for $\{\mathbf{D}\}$, then use Eq. 7.15.5. Matrix $[\Delta\mathbf{K}]$ is very sparse because extending the crack an amount Δa changes only the mesh immediately adjacent to the crack tip. Accordingly, $\Delta\Pi_p$ can be found by adding contributions from only the elements that surround the crack tip. Finally, we approximate $d\Pi_p/da$ as $\Delta\Pi_p/t\ \Delta\mathrm{a}$, where t is the thickness of the finite element model, and use Eq. 7.15.3 to compute K_{I} [7.47].

To estimate a stress intensity factor by comparing finite element computations with Eq. 7.15.1 or 7.15.2, we can simply insert a computed stress or displacement into the left side, the r and θ coordinates of the stress point into the right side, and solve for K_{I}, K_{II}, or K_{III}. For example, consider a crack along the x-axis of a plate loaded by forces along the y-axis. We might use finite element results for either σ_y on $\theta = 0$ or v on $\theta = \pi$ to solve for K_{I}. By doing this at several values of r, we can extrapolate a plot of K_{I} versus r to $r = 0$. Thus we find a best value of K_{I}. Because displacements are usually more accurate than stresses, it is preferable to use Eq. 7.15.2 and the crack-opening displacement, v on $\theta = \pi$ (see Eq. 17.15.6).

Singularity or *crack-tip* elements incorporate the theoretical solution in the element itself. Hybrid elements employ an assumed stress field within the element and therefore use Eq. 7.15.1 [7.48]. Displacement-based singularity elements add Eq. 7.15.2 to the assumed displacement field [7.49]. Both element types can have a stress intensity factor as one of the primary unknowns, so that the factor becomes part of the solution vector $\{\mathbf{D}\}$. Other singularity elements augment their fields without explicitly including a stress intensity factor [7.50]. The elements are then better able to approximate Eqs. 7.15.1 and 7.15.2, but they yield a stress intensity factor only by additional calculations (such as the energy-release or equation-matching methods outlined previously). Singularity elements can be attached to standard elements that model the remainder of the structure away from the crack.

In whatever form, singularity elements can promote accuracy and efficiency. It is better to surround the crack tip by a few good singularity elements than by an extravagantly refined mesh of standard elements.

Some singularity elements fail to be compatible with their neighbors or lack rigid-body motion and constant strain capabilities. They are therefore suspect. Isoparametric elements do not have these failings. Also, they can be made to display the $\sqrt{r}$ displacement field and $1/\sqrt{r}$ stress field of Eqs. 7.15.1 and 7.15.2 simply by locating their side nodes in the proper places [7.51, 7.52]. Thus a standard computer program will contain singularity elements if the right input data are presented to it.

A singularity element is produced simply by moving the side nodes of a quadratic element toward the crack tip until they are at the quarter points (Fig. 7.15.2). The patch of elements that surrounds the crack tip is embedded in a mesh of standard elements that models the rest of the structure. For Mode I conditions, if $v_A = 0$ and displacements are symmetric about the x-axis, then

$$K_{\text{I}} = \frac{2G\sqrt{2\pi}}{\kappa + 1}\left(\frac{4v_B - v_c}{\sqrt{L}}\right) \tag{7.15.6}$$

where G = shear modulus and $\kappa = 3 - 4\nu$ for plane strain and $\kappa = (3 - \nu)/(1 + \nu)$ for plane stress [7.53].

The quadrilateral element of Fig. 7.15.2*a* displays the $1/\sqrt{r}$ stress field only along its edges and therefore should probably not be used. A triangular element (Fig. 7.15.2*b*) displays the $1/\sqrt{r}$ stress field throughout [7.54]. A triangular element can be formed

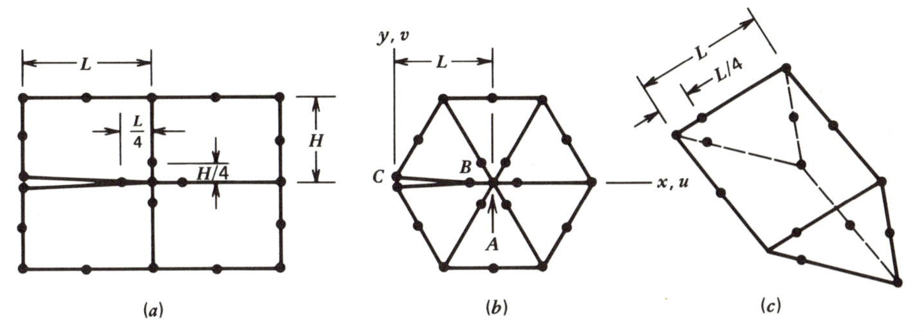

Figure 7.15.2. (*a*, *b*) A crack tip surrounded by plane quadratic isoparametric elements. The elements are singularity elements because side nodes adjacent to the crack tip are at quarter points of their respective sides. (*c*) A quadratic singularity element for solids.

either directly, as in Section 7.10, or by coalescing the three nodes on one side of a quadrilateral. Both triangles are excellent elements but, in the latter guise, element sides must be straight to avoid significant errors [7.55]. Making sides straight and placing side nodes exactly at quarter points requires care in data preparation and mesh generation. If a triangle or a triangular face is formed by collapsing one side of a quadrilateral, the stress field varies as $1/\sqrt{r}$ if the three nodes are constrained to have the same displacements. It varies as $1/r$ if the three nodes are left independent. The former is suited to linearly elastic conditions, the latter to perfectly plastic conditions near the crack tip [7.54].

Cubic elements can also be used. If side nodes are placed at $\frac{1}{9}$ and $\frac{4}{9}$ of the length of the side from the crack tip, the $1/\sqrt{r}$ stress field is displayed [7.58]. Indeed, an *n*th-order element can also display a stress field of the form $r^{-1/n}$, depending on where the side nodes are placed [7.59].

Quarter-point elements are not limited to plane problems. They also find application in general solids, solids of revolution, and plates and shells [7.54, 7.56, 7.57]. If a solid element models a *curved* crack front, the element overlaps itself a bit, and accuracy declines somewhat.

Accuracy is enhanced if transition elements *adjacent* to crack-tip elements (elements 1 to 4 in Fig. 7.15.3) have their side nodes in positions that imply singularity at the crack-tip location. Thus transition elements indicate a singularity *outside* their boundaries. For a singularity at a distance of qh units from the transition element (Fig. 7.15.3*b*), the side node location must be [7.60]

$$\frac{s}{h} = \frac{1 - q + \sqrt{q^2 + 2q}}{2} \tag{7.15.7}$$

For $q = 0$, this reduces to the expected quarter-point value $s/h = 0.5$. For the mesh in Fig. 7.15.3 and with $L/a = 0.1$, K_I is in error by only about 2%. Error increases if L/a is made larger or smaller than approximately 0.1.

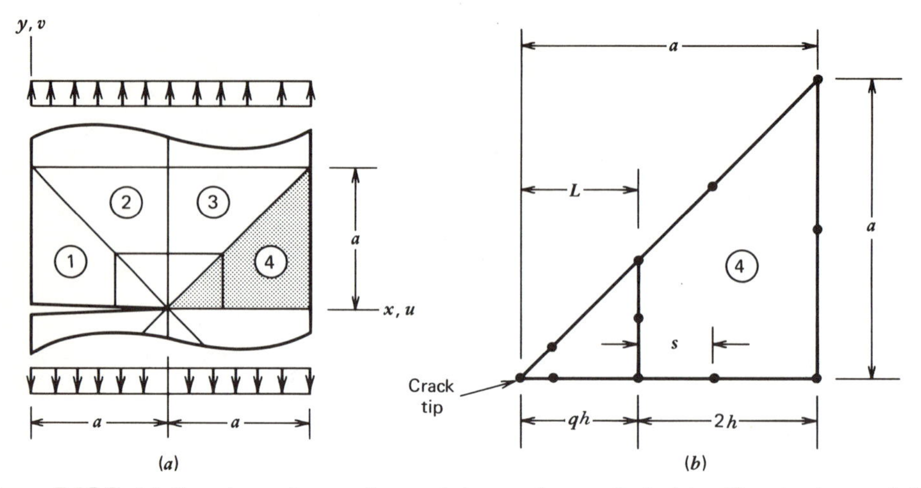

Figure 7.15.3. (*a*) Sample mesh near the crack in an edge-cracked plate. The structure and the mesh are symmetric about $x = 2a$ and about $y = 0$. (*b*) Quarter-point crack-tip element and transition element number 4.

PROBLEMS

7.1 Verify Eqs. 7.1.2, 7.1.5, and 7.1.6.

7.2 (a) Fill in details of the example associated with Eqs. 7.1.8 and 7.1.9 and verify that [**k**] is indeed as stated by Eq. 4.4.5.

(b) Repeat the derivation of [**k**] for the beam element, but this time use $\{\mathbf{d}_R\} = \{w_1 \; w_2\}$.

(c) Imagine that the leading diagonal coefficient of $[\mathbf{k}_{EE}]$ in part (a) is in error by an amount e. Show that [**k**] still represents rigid-body motion correctly.

7.3 (a) Verify that Eq. 7.1.11 follows from Eqs. 4.3.19 and 7.1.10.

(b) Show that $[\mathbf{k}_a]$ yields $[\mathbf{k}_{EE}]$ (last paragraph of Section 7.1).

(c) Show that $[\mathbf{k}_a]$ can be transformed directly to [**k**] by using [**T**] from the relation $\{a_3 \; a_4\} = [\mathbf{T}] \{w_1 \; \theta_1 \; w_2 \; \theta_2\}$.

7.4 Find the 2-by-2 stiffness matrix of a truss element that lies on the x-axis and has d.o.f. u_1 and u_2. Use the a-basis method described at the end of Section 7.1.

(a) Use a 1-by-2 matrix [**T**] to get [**k**] from $[\mathbf{k}_a]$.

(b) Convert $[\mathbf{k}_a]$ to $[\mathbf{k}_{EE}]$, then find [**k**] from Eq. 7.1.7.

7.5 (a) A constant strain triangle is supported against rigid-body motion as shown in the sketch. Write the u and v displacement fields in terms of three generalized coordinates a_i. Is there a nodal arrangement for which the expressions are invalid?

(b) For the specific element geometry of Problem 4.19, generate the 3-by-3 $[\mathbf{k}_{EE}]$ and, finally, the 6-by-6 [**k**] from Eq. 7.1.7. Let $\nu = 0$.

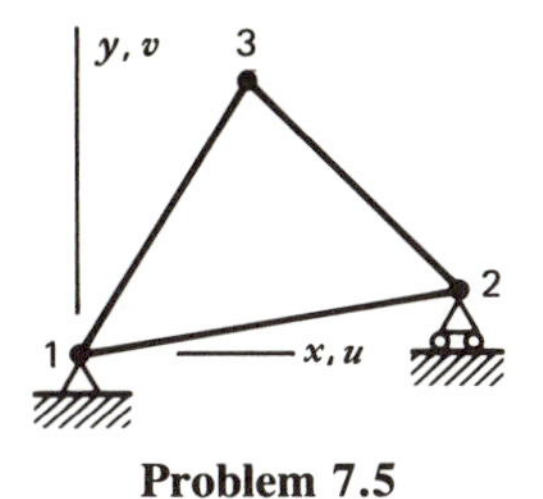

Problem 7.5

7.6 A flat disc has inside and outside radii r_1 and r_2. Nodal d.o.f. are v_1 and v_2 in the circumferential direction θ. When the inside edge is fixed, the ratio of the torque T_2 on the outside edge to its corresponding angle of twist θ_2 is a number C. Find the 2-by-2 element stiffness matrix $[\mathbf{k}]$ that operates on $\{v_1\ v_2\}$. Verify that $[\mathbf{k}]\{\mathbf{d}\} = 0$ if $\{\mathbf{d}\}$ represents rigid-body motion.

7.7 (a) Express Eq. 7.2.2 in the form of Eq. 6.8.1.
(b) Apply Eq. 7.2.3 to the problem of Fig. 2.11.1. Condense D_1 and D_2 and show that the equation for D_3 in Fig. 2.11.1*d* is obtained.

7.8 Write computer program statements that will place the internal node in Fig. 7.2.1*a* at the average of the corner node coordinates and assemble matrices of subtriangles to yield the 10-by-10 $[\mathbf{k}]$ and 10-by-1 $\{\mathbf{r}\}$ of the quadrilateral. Given information consists of $[\mathbf{E}]$ matrices and corner node coordinates. Assume that matrices for any triangle are generated merely by calling a suitable subroutine. *Suggestion.* Let the internal node be associated with d.o.f. 9 and 10 and set up an indexing array that relates the 6 d.o.f. of each triangle to the 10 d.o.f. of the quadrilateral.

7.9 Instead of computing stresses in each subtriangle of Fig. 7.2.1*a*, it is much faster and at least as accurate to find average stresses for the quadrilateral by adapting linear fields $u = a_1 + a_2x + a_3y$ and $v = a_4 + a_5x + a_6y$ to d.o.f. of the four corners by a least-squares fit [7.61]. We write

$$\begin{Bmatrix} u_1 \\ u_2 \\ u_3 \\ u_4 \end{Bmatrix} - \begin{bmatrix} 1 & x_1 & y_1 \\ 1 & x_2 & y_2 \\ 1 & x_3 & y_3 \\ 1 & x_4 & y_4 \end{bmatrix} \begin{Bmatrix} a_1 \\ a_2 \\ a_3 \end{Bmatrix} = \begin{Bmatrix} e_1 \\ e_2 \\ e_3 \\ e_4 \end{Bmatrix}$$

or $\{\mathbf{u}\} - [\mathbf{Q}]\{\mathbf{a}\} = \{\mathbf{e}\}$. There is a similar expression for v. Residuals $\{\mathbf{e}\}$ are present because three a_i cannot exactly satisfy four equations. Place local axes so that $\Sigma x_i = \Sigma y_i = 0$. Show that minimization of $\{\mathbf{e}\}^T\{\mathbf{e}\}$ with respect to the a_i yields $\{\mathbf{a}\} = [\mathbf{Q}^T\mathbf{Q}]^{-1}\ [\mathbf{Q}]^T\{\mathbf{u}\}$. Find expressions for average stains ϵ_x, ϵ_y, and γ_{xy} in terms of the x_i, y_i, u_i, and v_i ($i = 1, 2, 3, 4$).

7.10 Imagine that θ_1 and θ_2 of the beam element of Fig. 4.4.1*b* are to be condensed. What sort of stiffness matrix do you think will result? Verify your prediction.

7.11 Let a plane frame element have a rotational spring at each end, with respective coefficients k_1 and k_2 (moment per radian). Let β_1 and β_2 represent *structure* node rotations. Elements are joined to structure nodes through the springs, so that in Fig. 4.4.1*b* $\theta_1 \neq \beta_1$ and $\theta_2 \neq \beta_2$ unless k_1 and k_2 approach infinity.

(a) Describe a method that involves condensation and yields **[k]** of a frame element, ready for assembly. Thus **[k]** is to operate on $\{u_1\ w_1\ \beta_1\ u_2\ w_2\ \beta_2\}$.

(b) How could the desired 6-by-6 **[k]** be generated without condensation?

7.12 Element 2 of the propped cantilever beam carries a concentrated load P at $x = 3L/2$. Imagine that the θ d.o.f. at node A of element 2 is condensed; then the θ_A row and column of $[\mathbf{k}]_2$ are filled with zeros. Next, elements 1 and 2 are assembled. Use elementary beam theory to compute the expected displacement at A in terms of P, L, E, and I.

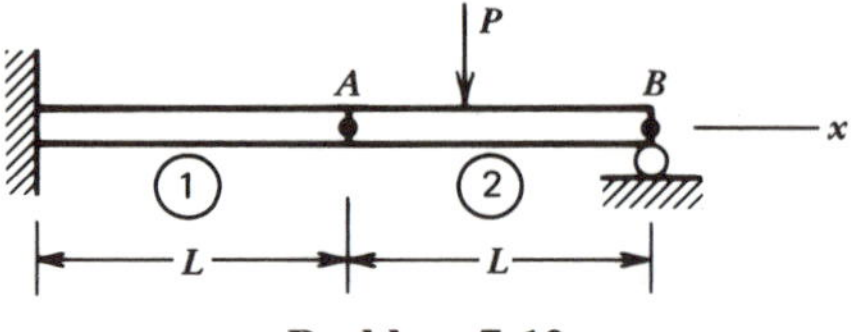

Problem 7.12

7.13 Consider the beam of Problem 7.12, but omit the condensation of θ_A. How would you find the force P needed to produce a prescribed θ_B?

7.14 Consider the uniform, two-force member of Fig. 4.4.1*a*, but let its displacement field be

$$u = \lfloor (1 - x/L) \qquad x/L \qquad x(x - L) \rfloor \{u_1\ u_2\ a\}$$

where a is a nodeless d.o.f. Let the bar carry a uniform body force that totals F units in the $+x$ direction.

(a) Find the uncondensed stiffness matrix and nodal load vector.

(b) Transform the results of part (a) so that d.o.f. a is replaced by the actual displacement at $x = L/2$.

(c) Similarly, transform the given displacement field u.

(d) Condense the internal d.o.f. in the matrices of parts (a) and (b). Do the results agree?

7.15 (a) Verify the "center" force of 1.78 units in Fig. 7.3.1*a*.

(b) How do the results in Fig. 5.8.1*b* change if the bubble function of Eq. 7.3.2 is added to the element displacement field?

7.16 Consider the element of Eqs. 7.3.1. Conversion from "relative" d.o.f. a_1 to "absolute" d.o.f. u_5 can be written

$$\{u_1\ u_2\ u_3\ u_4\ a_1\} = [\mathbf{T}]\{u_1\ u_2\ u_3\ u_4\ u_5\}$$

and similarly for the v's, a_2, and v_5. Write the matrix **[T]**. Verify that it produces Fig. 7.3.1*c* from Fig. 7.3.1*a*. Can you write $[\mathbf{T}]^{-1}$ by inspection?

7.17 Could N_5 in Eq. 7.3.2 be replaced by $N_5 = \cos(\pi\xi/2)\cos(\pi\eta/2)$? Would this be advantageous? What "center" load replaces 1.78 in Fig. 7.3.1*a*? Why are trigonometric functions not used for N_1 to N_4?

7.18 Under what circumstances will the addition of nodeless d.o.f. be of benefit to a standard, uniform beam element? And to a tapered beam element? Suggest an appropriate polynomial bubble function.

7.19 Start with Eq. 4.2.12 and write the nine shape functions N_i of a quadratic plane Lagrange element of arbitrary shape. Use Eq. 7.3.2 for node 9. Verify that the N_i assume values of zero and one where they should. *Suggestion.* Set $x/b = \xi$ and $y/c = \eta$; then set $b = c = 1$ to get an isoparametric-type formulation.

7.20 Consider the Lagrange element in Fig. 4.2.2*b*. Under 2-by-2 Gauss quadrature, zero-energy deformation modes can appear that make the *structure* stiffness matrix singular. Let $b = c = 1$ in Fig. 4.2.2*b* and consider these values of nodal d.o.f. on the element boundary: $u_1 = u_2 = u_3 = u_4 = -0.4$, $u_5 = u_6 = u_7 = u_8 = 0.2$, and all $v_i = 0$. Sketch this mode. Also, find the d.o.f. at node 9, in both relative and absolute forms, that make the mode zero energy.

7.21 Modify Figs. 7.4.1 and 7.4.2 to allow for NL load cases rather than only one. Also modify the suplementary statements below Eq. 7.4.3 to allow for NL load cases.

7.22 Verify that Eq. 7.4.3 follows from Eqs. 7.2.2 and 7.4.2.

7.23 Imagine that element rigid-body motion is removed, as suggested in the last paragraph of Section 4.10. Next, internal d.o.f. are recovered. Are the computed internal d.o.f. changed by the rigid-body motion treatment?

7.24 A two-node frame element lies on a local x'-axis and has nodal d.o.f. u', v', $u'_{,x'}$ and $v'_{,x'}$. Its local stiffness matrix $[\mathbf{k}']$ is known.
(a) What transformation matrix $[\mathbf{T}]$ must be used to find $[\mathbf{k}]$ for an element of arbitrary orientation in the xy plane?
(b) Two elements join at node i, as shown. What d.o.f. from node i should appear in the structure vector $\{\mathbf{D}\}$? What condensation or constraints are necessary?
(c) Repeat part (b), but consider node j.

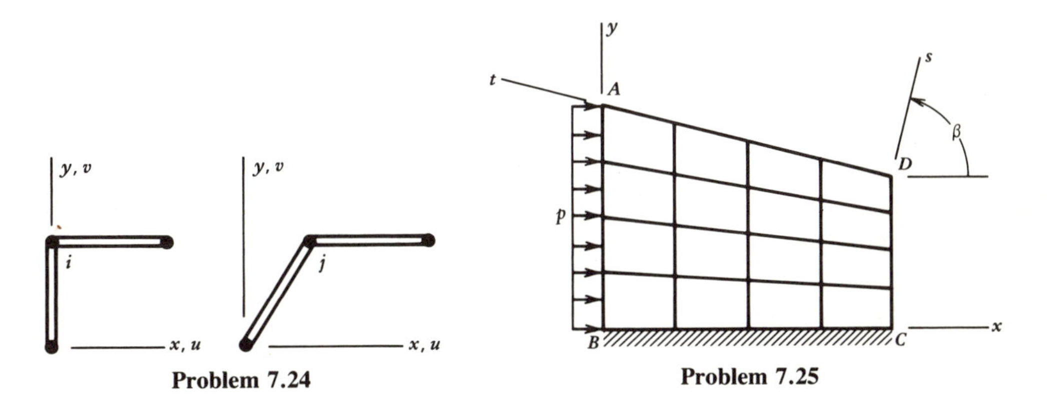

Problem 7.24 **Problem 7.25**

7.25 The structure is composed of plane elements. Nodal d.o.f. are u, v, $u_{,x}$, $v_{,x}$, $u_{,y}$, and $v_{,y}$. Pressure p acts along AB. Edge BC is fixed. What boundary conditions should be imposed on the structural equations $[\mathbf{K}]\{\mathbf{D}\} = \{\mathbf{R}\}$ along edges AB, BC, CD, and DA? Assume that the material is isotropic (what happens if it is not?).

7.26 (a) Show that Eqs. 7.6.2 yield $\gamma_{xy} = 0$ for all x and y and that $\epsilon_x = -\epsilon_y/\nu = \eta\epsilon_0$, where ϵ_0 is the strain at $\eta = 1$.
(b) Show that Eqs. 7.6.2 predict that points at the middepth move above the neutral axis when the element is bent. Is this prediction physically correct?
(c) Let displacements $\bar{u}$ be imposed, as in Fig. 7.6.1, and let t be element thickness. Find an expression for R, the ratio of strain energy in the linear element (Fig. 7.6.1*b*) to strain energy in a beam segment of like dimensions.
(d) Let $\nu = 0$ and compute R for $L/H = 0.25$, 0.50, 1.00, 2.00, and 10.0.

7.27 Consider 2-by-2 Gauss quadrature, as in Fig. 7.6.1*a*, but let γ_{xy} be evaluated at $\xi = \eta = 0$, as suggested in Section 7.6. Show that strain energy in the bending mode falls (by one-half if $\nu = 0$) when the rectangular element is rotated so that $\xi\eta$ axes are 45° to the xy-axes. That is, demonstrate lack of invariance. *Suggestion.* Consider Mohr's circle for strain.

7.28 By drawing diagonals in a four-node plane element, we can identify four triangles (123, 234, 341, and 412 in Fig. 5.3.1*c*). To eliminate parasitic shear, a weighted average strain can be defined as $\bar{\gamma}_{xy} = (\Sigma A_i\gamma_i)/(\Sigma A_i)$ [7.8]. Here A_i is the triangle area and γ_i it its γ_{xy} according to formulas for the constrant strain triangle.
(a) Show that

$$\bar{\gamma}_{xy} = \frac{1}{x_{31}y_{42} + x_{24}y_{31}} \lfloor x_{42} \; y_{24} \; x_{13} \; y_{31} \; x_{24} \; y_{42} \; x_{31} \; y_{13} \rfloor \{\mathbf{d}\}$$

where $x_{31} = x_3 - x_1$, and so on, and $\{\mathbf{d}\} = \{u_1 \; v_1 \; \cdots \; v_4\}$.
(b) Show that $\bar{\gamma}_{xy} = 0$ for the mode of Eq. 7.6.1 in a rectangular element.

7.29 A cantilever beam is to be modeled by one layer of the elements defined by Table 5.6.1. If there are to be only two side nodes per element, where should the side nodes be located?

7.30 In the sketch, the respective four-element asemblages contain constant strain triangles in part (a) and QM6 quadrilaterals in part (b). Corner loads P are consistent with pure bending.
(a) Imagine that d.o.f. at node 5 are condensed *before* elements are assembled (thus a total of eight d.o.f. are condensed). Sketch the deformed shape of the assemblage, especially near node 5.
(b) Will exact stresses be computed along AB? Will the elements even be compatible?

7.31 (a) Modify Figs. 5.5.1 and 5.5.2 so that they will generate matrices for the QM6 plane element.

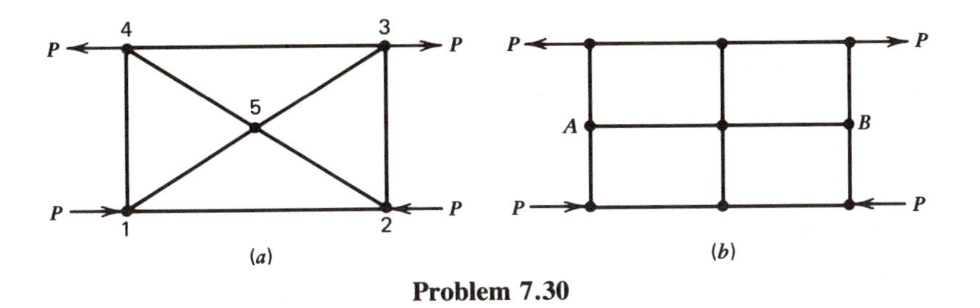

Problem 7.30

(b) Equation 7.7.3 demands only that $[\mathbf{k}_{re}]$ be modified. No such demand is placed on $[\mathbf{k}_{ee}]$. Perform numerical tests to decide if $[\mathbf{k}_{ee}]$ should be modified in the manner described below Eq. 7.7.4.

(c) Similar programming projects can be devised for the elements of Eqs. 7.3.1 and Section 7.6.

7.32 (a) Are the a_i of Eq. 7.7.1 relative or absolute motions?

(b) If, after computation of nodal d.o.f. in a mesh of QM6 elements, the nodeless d.o.f. a_i are omitted from stress computation, what consequences do you expect? Consider the test case of Fig. 5.12.1*b*.

7.33 (a) Describe how local coordinates xyz in Fig. 7.8.1 can be established from the global coordinates of nodes 1 to 4. The x-axis need not coincide with DA.

(b) If $\{\mathbf{f}\}$ in Eq. 7.8.3 satisfies equilibrium of z-direction force and equilibrium of moments about x- and y- axes, we can write $\lfloor \mathbf{Q}_i \rfloor [\mathbf{T}]^T = 0$, where $i = 1, 2, 3$. (This provides a check on the derivation of $[\mathbf{T}]$.) What is $\lfloor \mathbf{Q}_i \rfloor$ for the three equilibrium cases in terms of dimensions in Fig. 7.8.1?

7.34 Review the validity proof for isoparametric elements in Section 5.9. If shape functions are expressed in area coordinates instead of $\xi\eta\zeta$ coordinates, what changes are needed?

7.35 If x_1 and y_1 are replaced by x and y, Eq. 4.5.4 yields A_1 of Fig. 7.9.1. Hence $\alpha = A_1/A$, and β and γ are similarly expressed. Derive Eq. 7.9.5 by this procedure.

7.36 (a) Consider Eq. 7.9.8 and its simplified form without γ (see below Eq. 7.9.8). Verify that both yield the same result if $k = \ell = m = 1$.

(b) By inspection, write equations analogous to Eqs. 7.9.8 and 7.9.9 for tetrahedral coordinates α, β, γ, and δ. *Suggestion.* Consider the special case where all exponents are zero.

7.37 (a) Sketch the displacement mode of Eq. 7.9.13.

(b) What N_i replace those in Eq. 7.9.15 if d.o.f. of side nodes are departures from linearity? (See Eq. 7.12.1.)

7.38 (a) Write $\lfloor \mathbf{B}_y \rfloor$, analogous in form to $\lfloor \mathbf{B}_x \rfloor$ in Eq. 7.10.2.

(b) Show that Eq. 7.10.1 yields $\{\boldsymbol{\epsilon}\} = 0$ if all nodal u_i are equal and all nodal v_i are zero.

(c) Write the upper left 3-by-6 submatrix in $[\mathbf{H}]$ of Eq. 7.10.6.

(d) To generate matrices for the element of Fig. 7.10.1*b*, what specific changes are needed in Figs. 5.5.1 and 5.5.2?
(e) Sketch the differential area $d\alpha\, d\beta$ used in Eq. 7.10.11.
(f) By Eq. 7.10.12, $\Sigma W_i = 1$ over the n sampling points. What similar check can be made on other entires in Table 7.10.1?

7.39 Consider a quadratic rectangle, with four corner nodes and four midside nodes. Can the corner strain device of Eqs. 7.10.3 through 7.10.7 be applied to this rectangle?

7.40 Using area coordinates and an isoparametric coordinate ζ that has values $\zeta = +1$ and $\zeta = -1$ on the top and bottom triangular faces of a solid wedge element (see sketch), write the shape functions for (*a*) the linear element, and (*b*) the quadratic element, in terms of α, β, γ, and ζ.

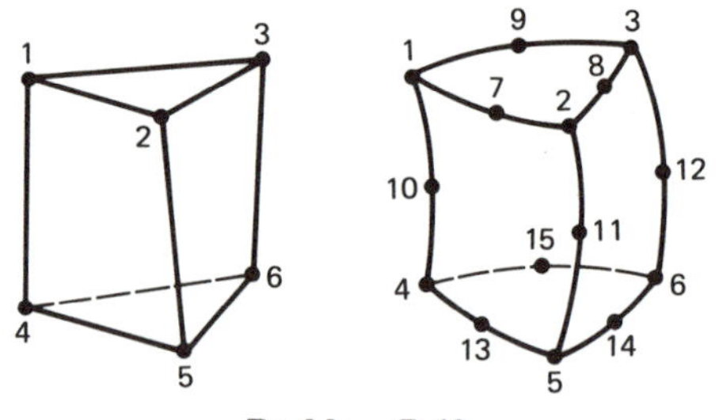

Problem 7.40

7.41 Consider replacing x and y in Fig. 7.11.1*b* by nonorthogonal coordinates s and t [7.27]. The triangle sides are straight. The origin is at corner 1; s runs along side 1–2 and t runs along side 1–3. Also, $s = 1$ at corner 2 and $t = 1$ at corner 3. Then $x = (1 - s - t)(-b) + sa$ and $y = tc$.
(a) How are s and t related to area coordinates?
(b) If $\phi = \phi\,(s, t)$, what are $\phi_{,x}$ and $\phi_{,y}$ in terms of a, b, c, $\phi_{,s}$ and $\phi_{,t}$?

7.42 Let $a = b$ in Fig. 7.11.1*b* and let $Q = 1$ in Eq. 7.11.1. For both the rectangle and the triangle, check that Eqs. 7.11.2 yield the expected I for (*a*) $m = n = 0$, (*b*) $m = 0$, $n = 1$, (*c*) $m = n = 1$, (*d*) $m = 0$, $n = 2$, and (*e*) $m = 2$, $n = 0$.

7.43 Let the triangle of Fig. 7.11.1*c* be isosceles, with y the axis of symmetry. Check that Eq. 7.11.3 yields the expected I for (*a*) $m = n = 0$, (*b*) $m = 0$, $n = 1$, (*c*) $m = n = 1$, and (*d*) $m = 2$, $n = 0$.

7.44 Let the xy corner coordinates in the triangle of Fig. 7.12.1*a* be $(-1, -1)$, $(1, -1)$, and $(0, 1)$, respectively. Use Eqs. 7.12.3 and:
(a) Write the Jacobian matrix $[\mathbf{J}]$.
(b) Write the strain-displacement matrix $[\mathbf{B}]$.
(c) If nodal d.o.f. $u_i = a_1 + a_2x_i + a_3y_i$ and $v_i = a_4 + a_5x_i + a_6y_i$ are imposed, show that $\{\boldsymbol{\epsilon}\} = [\mathbf{B}]\{\mathbf{d}\}$ yields the correct strains.

7.45 Coalesce corners 1–2–4, 3–7, and 5–6 of the linear brick in Fig. 5.7.1. Sketch the resulting element and write its shape functions in terms of ξ, η, and ζ. What are the $\xi\eta\zeta$ coordinates of the four corners?

7.46 (a) Show that by combining shape functions from Table 5.6.1 in the manner stated below Eq. 7.12.1, N_1 to N_4 become as stated by Eq. 5.3.3.

(b) Let a uniform traction σ_y act on edge $\eta = -1$ of the quadrilateral in Fig. 7.12.1*b*. Use v_5' rather than v_5 as a d.o.f. at node 5. Also, let $x_1 = -L$ and $x_2 = L$. What $\{\mathbf{r}\}$ is produced?

(c) Transform the $\{\mathbf{r}\}$ of part (b) so that v_5 replaces v_5' in $\{\mathbf{d}\}$. Check your result against Fig. 5.8.1*b*.

7.47 Verify Eq. 7.12.6.

7.48 Imagine that a report about a plane stress analysis states that the mesh contains 302 nodes and 192 triangular elements but does not describe the element type. Are they constant strain or linear strain triangles?

7.49 Let the element of fig. 7.14.1*b* have corner nodes only. Write the apropriate froms of Eqs. 7.14.1, 7.14.2, and 7.14.6 for such an element.

7.50 (a) Let $a = 1$ and $b = 15$ in Fig. 7.14.1*b*. Sketch M_1, M_2, and M_3 versus s for $1 < s < 4$. Let $L = 1$ in Eqs. 7.14.2.

(b) Repeat part (*a*), but let $L = 3$.

(c) Plot $\partial M_3/\partial s$ versus s for the conditions of part (a).

7.51 (a) Verify Eq. 7.15.5.

(b) What are the dimensions (units) of a stress intensity factor?

(c) Show that Eq. 7.15.7 yields $s = h$ as q becomes infinite.

CHAPTER 8

SOLIDS OF REVOLUTION

8.1 INTRODUCTION

A solid of revolution is axially symmetric if its geometry and material properties are independent of the circumferential coordinate θ. The problem is physically three dimensional but mathematically two dimensional: subject to restrictions noted above Eq. 8.2.1, material points have only u (radial) and w (axial) displacement components. The analysis procedure is essentially that of plane stress, so a program based on (say) the plane quadratic element in Chapter 5 can be used almost as is. Essential changes consist of adding more terms to the [**B**] and [**E**] matrices.

If the solid is axially symmetric but the loading is not, a Fourier series method can be used. The given loading is expressed as the sum of several component loadings, and an analysis is done for each component. According to the principle of superposition, the original problem is solved by adding the solutions of the component problems. Each analysis remains two dimensional. Thus the original three-dimensional problem is exchanged for a series of two-dimensional problems. The exchange is worthwhile because three-dimensional problems are so expensive to set up and run (see Section 4.11).

Bodies of revolution, whether solids or shells, have nodal circles, not nodal points. Accordingly, it is difficult to attach bodies of revolution to finite elements that have only nodal points (Section 6.6). So programs for solids or shells of revolution tend to be "stand-alone" programs.

Finite elements for axially symmetric solids were published in 1965 [8.1]. Programs are readily available; see, for example, Ref. 8.2.

8.2 FORMULATION FOR AXIALLY SYMMETRIC LOADING

The finite element is a ring of constant cross section (Fig. 8.2.1). Centers of all nodal circles lie on the z-axis, which is the axis of revolution. The body and each of its elements are solids of revolution about the z-axis. Figure 1.1.3 depicts a solid of revolution modeled by triangular finite elements.

Let θ be a principal material direction. Thus, in this section, we exclude anistropic materials with layering that spirals upward or inward (see Section 8.6). Under axially symmetric loading, no point has θ direction displacement. The displacement field {**f**} has only r and z components.

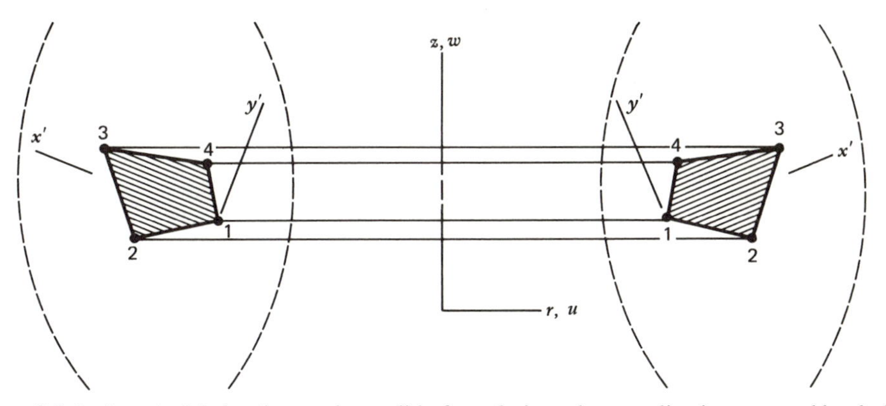

Figure 8.2.1. A typical finite element in a solid of revolution whose outline is suggested by dashed lines. Cross-hatching suggests a layered or orthotropic material whose principal directions are x', y', and the circumferential direction θ. (This is called *polar orthotropy*.)

$$\{\mathbf{f}\} = \{u \; w\} = [\mathbf{N}]\{\mathbf{d}\} \tag{8.2.1}$$

where $[\mathbf{N}]$ is a function of the coordinates and $\{\mathbf{d}\}$ contains the u's and w's of the nodal circles. For example, $[\mathbf{N}]$ and $\{\mathbf{d}\}$ could be defined by Eqs. 5.3.1 and 5.3.3. However, hoop strain ϵ_θ is neither zero nor uniquely determined by the other strains. So ϵ_θ must appear in the stress-strain relation.

$$\{\sigma_r \; \sigma_\theta \; \sigma_z \; \tau_{zr}\} = [\mathbf{E}]\left(\{\epsilon_r \; \epsilon_\theta \; \epsilon_z \; \gamma_{zr}\} - \{\epsilon_0\}\right) + \{\boldsymbol{\sigma}_0\} \tag{8.2.2}$$

where, if the material is isotropic,

$$[\mathbf{E}] = \frac{E}{(1+\nu)(1-2\nu)}\begin{bmatrix} 1-\nu & \nu & \nu & 0 \\ \nu & 1-\nu & \nu & 0 \\ \nu & \nu & 1-\nu & 0 \\ 0 & 0 & 0 & \dfrac{1-2\nu}{2} \end{bmatrix} \tag{8.2.3}$$

Except for hoop strain $\epsilon_\theta = u/r$, the strain-displacement relation has the form of Eq. 5.3.10. Thus

$$\begin{Bmatrix} \epsilon_r \\ \epsilon_\theta \\ \epsilon_z \\ \gamma_{zr} \end{Bmatrix} = \begin{bmatrix} \partial/\partial r & 0 \\ 1/r & 0 \\ 0 & \partial/\partial z \\ \partial/\partial z & \partial/\partial r \end{bmatrix}\begin{Bmatrix} u \\ w \end{Bmatrix} = \begin{bmatrix} 1 & 0 & 0 & 0 & 0 \\ 0 & 0 & 0 & 0 & 1/r \\ 0 & 0 & 0 & 1 & 0 \\ 0 & 1 & 1 & 0 & 0 \end{bmatrix}\begin{Bmatrix} u_{,r} \\ u_{,z} \\ w_{,r} \\ w_{,z} \\ u \end{Bmatrix} \tag{8.2.4}$$

If $[\mathbf{N}]$ in Eq. 8.2.1 is a function of r and z, the strain-displacement matrix $[\mathbf{B}]$ is produced by applying the 4-by-2 operator in Eq. 8.2.4 to $[\mathbf{N}]$. If $[\mathbf{N}]$ is a function of ξ and η, as it is for isoparametric elements, we follow the pattern of Eqs. 5.3.11 and 5.3.12 and write

$$\begin{Bmatrix} u,_r \\ u,_z \\ w,_r \\ w,_z \\ u \end{Bmatrix} = \begin{bmatrix} \Gamma_{11} & \Gamma_{12} & 0 & 0 & 0 \\ \Gamma_{21} & \Gamma_{22} & 0 & 0 & 0 \\ 0 & 0 & \Gamma_{11} & \Gamma_{12} & 0 \\ 0 & 0 & \Gamma_{21} & \Gamma_{22} & 0 \\ 0 & 0 & 0 & 0 & 1 \end{bmatrix} \begin{Bmatrix} u,_\xi \\ u,_\eta \\ w,_\xi \\ w,_\eta \\ u \end{Bmatrix} \tag{8.2.5}$$

$$\underset{5\times 1}{\{u,_\xi \, u,_\eta \cdots u\}} = \begin{bmatrix} \text{(First 4 rows in Eq. 5.3.12)} \\ N_1 \quad 0 \quad N_2 \quad 0 \quad \cdots \quad \text{etc.} \end{bmatrix} \underset{n\times 1}{\{\mathbf{d}\}} \tag{8.2.6}$$

where the last row in Eq. 8.2.6 expresses the relation $u = N_1u_1 + N_2u_2 + \cdots + N_nu_n$. Matrix $\{\mathbf{B}\}$ is then the product of successive rectangular matrices in Eqs. 8.2.4, 8.2.5, and 8.2.6.

All ingredients are now at hand, and element matrices follow from the usual formulas (Eqs. 4.3.4 and 4.3.5). The differential volume element is $dV = r\,dr\,d\theta\,dz$, and the differential surface element is $dS = r\,d\theta\,d\ell$, where $d\ell$ is an increment of meridional length. The stiffness matrix of an isoparametric element becomes

$$\underset{n\times n}{[\mathbf{k}]} = \int_{-1}^{1}\int_{-1}^{1}\int_{-\pi}^{\pi} \underset{n\times 4}{[\mathbf{B}]^{\mathrm{T}}}\,\underset{4\times 4}{[\mathbf{E}]}\,\underset{4\times n}{[\mathbf{B}]}\, r\,d\theta\,J\,d\xi\,d\eta \tag{8.2.7}$$

The radius at a Gauss point is $r = [N_1\,N_2 \cdots N_n]\,\{r_1\,r_2 \cdots r_n\}$, where the r_i are nodal radii and the N_i are evaluated at the ξ and η coordinates of the Gauss point. To avoid the superfluous multiplier 2π on all k_{ij} and r_i, circumferential integration can span zero to one radian.

Some coefficients in the integrand of Eq. 8.2.7 have $1/r$ as a multiplier. With Gauss quadrature these terms remain finite because there are no Gauss points $r = 0$. If $[\mathbf{k}]$ is formed explicitly, we can produce "core" elements by using a displacement field that sets nodal u's to zero for nodes on the z-axis and evaluating indeterminate forms 0/0 by the l'Hospital rule.

During stress computation, the indeterminate form $\epsilon_\theta = u/r = 0/0$ arises for points on the z-axis. We can avoid this trouble by finding ϵ_θ slightly away from the axis or by extrapolating Gauss point strains to the axis. Another option is to exploit the theoretical requirement that $\epsilon_r = \epsilon_\theta$ at $r = 0$. Thus, for stress computation at $r = 0$, we merely replace the ϵ_θ row of $[\mathbf{B}]$ by the ϵ_r row.

Because of axial symmetry, z-direction translation is the only possible rigid-body motion. It can be restrained by prescribing w on a single nodal circle. The radial displacement $u = 0$ should be prescribed at all nodes that lie on the z-axis.

As shown, elements devised for plane problems require few additions to serve for solids of revolution. But on occasion there are unexpected troubles, as with element QM6 of Section 7.7: hoop strain ϵ_θ activates the internal d.o.f. and creates a spurious radial bulge that in turn creates spurious shear strain γ_{zr} except at $\xi = \eta = 0$. The effect is more prominent for elements nearer the z-axis. One remedy, which slightly degrades element performance, is to decouple ϵ_θ from the incompatible modes before generating [**k**]. This is done by placing zeros in the ϵ_θ row of $[\mathbf{B}_a]$. Specifically, in Eq. 8.2.7 we set B_{2j} to zero, where $j = 9, 10, 11, 12$. Another remedy is to form [**k**] with all [**B**] coefficients included but, during stress computation, let γ_{zr} at $\xi = \eta = 0$ apply all over the element [8.3].

The patch test in axially symmetric solids deserves comment. If numerically integrated, [**k**] will not be exact because of the $1/r$ term in [**B**]. Hence, for example, a uniform radial traction on the outer surface of a solid cylinder canot be expected to yield the *correct* constant strain state. Yet exact results are approached as a mesh is refined.

8.3 REMARKS ON FOURIER SERIES

Fourier series represent functions that are periodic. A Fourier series for a dependent variable $\phi = \phi(\theta)$ can be written

$$\phi = \sum_{n=0}^{\infty} p_n \cos n\theta + \sum_{n=1}^{\infty} q_n \sin n\theta \tag{8.3.1}$$

where n is an integer.

Sine terms are called *odd* or *antisymmetric*, as $\phi(\theta) = -\phi(-\theta)$. Cosine terms are called *even* or *symmetric*, as $\phi(\theta) = \phi(-\theta)$. Coefficients p_n and q_n are functions of n but not of θ. The following integrals, where m and n are integers, are useful.

$$\int_{-\pi}^{\pi} \sin m\theta \sin n\theta \, d\theta = \begin{cases} \pi & \text{for} \quad m = n \neq 0 \\ 0 & \text{for} \quad m \neq n \text{ and for } m = n = 0 \end{cases}$$

$$\int_{-\pi}^{\pi} \cos m\theta \cos n\theta \, d\theta = \begin{cases} 2\pi & \text{if} \quad m = n = 0 \\ \pi & \text{if} \quad m = n \neq 0 \\ 0 & \text{if} \quad m \neq n \end{cases} \tag{8.3.2}$$

$$\int_{-\pi}^{\pi} \sin m\theta \cos n\theta \, d\theta = 0 \qquad \text{for} \quad \text{all } m \text{ and } n$$

How do we find p_n and q_n to represent a known function $\phi = \phi(x)$ as a Fourier series? We integrate the series, then multiply it by the single term $\cos n\theta$ and integrate, then multiply it by the single term $\sin n\theta$ and integrate. Thus, from Eqs. 8.3.1 and 8.3.2,

$$\int_{-\pi}^{\pi} Q \, \phi \, d\theta = \begin{cases} 2\pi p_0 & \text{if} \quad Q = 1 \\ \pi p_n & \text{if} \quad Q = \cos n\theta \\ \pi q_n & \text{if} \quad Q = \sin n\theta \end{cases} \tag{8.3.3}$$

Integrals of ϕ, $\phi \cos n\theta$, and $\phi \sin n\theta$, where ϕ is the known function, can *also* be found by graphical or analytical integration over the range $-\pi < \theta < \pi$. Thus we obtain three equations that yield p_0, p_n, and q_n.

Example

Consider Fig. 8.3.1. Here $\theta = \pi x/L$, and ϕ_0 can be regarded as a uniform load on a span of length L. Integration of ϕ, $\phi \cos n\theta$, and $\phi \sin n\theta$ over $-\pi < \theta < \pi$ yields the respective results 0, 0, and 0 (n even) or $4\phi_0/n$ (n odd). Hence, from Eqs. 8.3.3, $p_0 = 0$, $p_n = 0$, and $q_n = 0$ (n even). For n odd,

$$q_n = \frac{4\phi_0}{n\pi}, \qquad \text{so} \qquad \phi = \sum_{n=1,3,\cdots} \frac{4\phi_0}{n\pi} \sin \frac{n\pi x}{L} \tag{8.3.4}$$

As suggested by Figs. 8.3.1*b* to 8.3.1*d*, the square wave is matched more and more closely as more terms are used.

Concentrated loads yield an interesting series. In the coordinates of Fig. 8.3.1*a*, consider P units downward at $x = -L/2$ and P units upward at $x = +L/2$. We find $p_0 = p_n = 0$ and

$$q_n = \frac{2P}{L} \sin \frac{n\pi}{2}, \qquad \phi = \sum_{n=1,2,3,\cdots} \left(\frac{2P}{L} \sin \frac{n\pi}{2}\right) \sin \frac{n\pi x}{L} \tag{8.3.5}$$

The series for ϕ does not converge but, when used in stress analysis, it yields convergent results for displacement and stress (see Problems 8.7 and 8.8).

To show the application of Fourier series to stress analysis, consider the beam of Fig. 8.3.2*a*. It displays the essential concepts that apply to bodies of revolution but does not

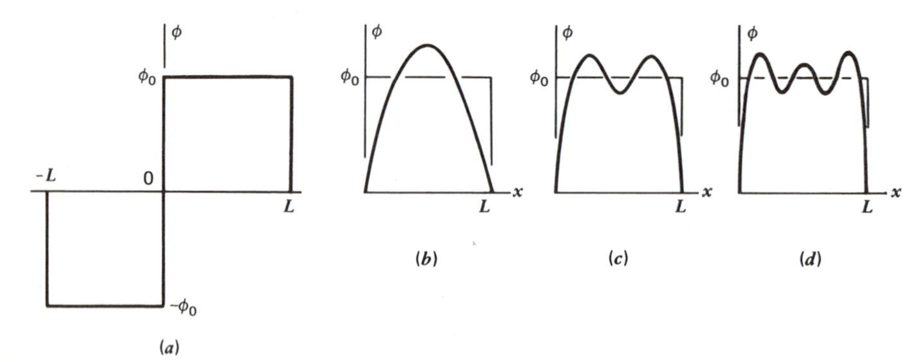

Figure 8.3.1. (*a*) Square wave and its Fourier series representation using (*b*) one, (*c*) two, and (*d*) three Fourier harmonics.

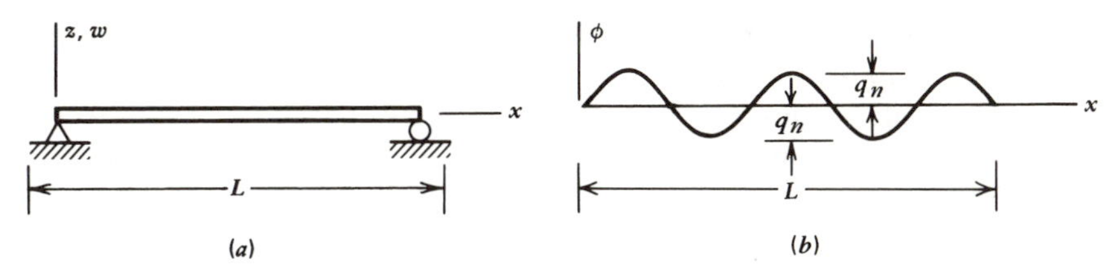

Figure 8.3.2. (*a*) Simply supported beam. (*b*) The sine-wave loading $\phi = q_n \sin n\pi x/L$. The case $n = 5$ is pictured.

require us to write so many equations. The equilibrium and moment-curvature relations of a beam are

$$V_{,x} = \phi \qquad M_{,x} = V \qquad EIw_{,xx} = M \tag{8.3.6}$$

where V is transverse shear, M is bending moment, and ϕ is distributed load. When combined, Eqs. 8.3.6 yield, for constant bending stiffness EI,

$$EIw_{,xxxx} = \phi \tag{8.3.7}$$

Consider the loading of Fig. 8.3.2*b*.

$$\phi = q_n \sin \frac{n\pi x}{L} \tag{8.3.8}$$

where q_n does not depend on x. Assume that the displacement is the admissible function

$$w = w_n \sin \frac{n\pi x}{L} \tag{8.3.9}$$

where w_n does not depend on x. We substitute w and ϕ into Eq. 8.3.7 and find

$$\left[EI\left(\frac{n\pi}{L}\right)^4 w_n - q_n \right] \sin \frac{n\pi x}{L} = 0 \tag{8.3.10}$$

This can be true for all x only if the bracketed expression vanishes. Hence

$$w_n = \frac{q_n}{EI}\left(\frac{L}{n\pi}\right)^4 \tag{8.3.11}$$

Equations 8.3.9 and 8.3.11 define the correct and unique solution, since they satisfy all requirements of equilibrium, compatibility, and boundary conditions (Section 1.3). Note that a sine wave of loading produces a corresponding sine wave of deflection and bending moment, regardless of n. That is, the different harmonics are uncoupled—the ith wave does not interact with the jth wave.

Now that w_n is known, Eq. 8.3.9 defines w for any x and for any n. If several sine-wave loads act simultaneously, each associated with a different n, the total deflection w is found by superposition.

$$w = \sum \frac{q_n}{EI}\left(\frac{L}{n\pi}\right)^4 \sin\frac{n\pi x}{L} \tag{8.3.12}$$

Moments are found by substitution of Eq. 8.3.12 into the expression $M = EIw_{,xx}$.

What we have done is find the response of the beam to a single load component from an equation (Eq. 8.3.11) that does not contain the independent variable x. This reduction is dimensionality is obtained cheaply. The cost is merely the summation required by Eq. 8.3.12. The loading is arbitrary: if uniform, q_n is given by Eq. 8.3.4; if concentrated at midspan, q_n is given by Eq. 8.3.5.

These concepts apply to bodies of revolution in the manner described next.

8.4 GENERAL LOADING: INTRODUCTION

Except for Section 8.6, our discussion is restricted to materials described by the following stress-strain law $\{\boldsymbol{\sigma}\} = [\mathbf{E}]\{\boldsymbol{\epsilon}\}$.

$$\begin{Bmatrix} \sigma_r \\ \sigma_\theta \\ \sigma_z \\ \tau_{zr} \\ \tau_{r\theta} \\ \tau_{\theta z} \end{Bmatrix} = \begin{bmatrix} E_{11} & E_{12} & E_{13} & E_{14} & 0 & 0 \\ & E_{22} & E_{23} & E_{24} & 0 & 0 \\ & & E_{33} & E_{34} & 0 & 0 \\ & & & E_{44} & 0 & 0 \\ & \text{Symmetric} & & & E_{55} & E_{56} \\ & & & & & E_{66} \end{bmatrix} \begin{Bmatrix} \epsilon_r \\ \epsilon_\theta \\ \epsilon_z \\ \gamma_{zr} \\ \gamma_{r\theta} \\ \gamma_{\theta z} \end{Bmatrix} \tag{8.4.1}$$

The layered material of Fig. 8.2.1 is described by such a relation. So is an isotropic material, but then $E_{14} = E_{24} = E_{34} = E_{56} = 0$.

In this section we introduce the analysis procedure and attempt to make it plausible. Let the loading be expressed as Fourier series

$$\begin{aligned} \{F_r\, F_z\, \Phi_r\, \Phi_z\, T\} &= \sum \{\bar{F}_{rn}\, \bar{F}_{zn}\, \bar{\Phi}_{rn}\, \bar{\Phi}_{zn}\, \bar{T}_n\} \cos n\theta \\ \{F_\theta\, \Phi_\theta\} &= \sum \{\bar{F}_{\theta n}\, \bar{\Phi}_{\theta n}\} \sin n\theta \end{aligned} \tag{8.4.2}$$

where T is temperature, and the F's and Φ's are, respectively, body forces per unit volume and surface tractions in the r, θ, and z directions. Similarly, let

$$\begin{aligned} \text{Radial displacement} &= u = \sum \bar{u}_n \cos n\theta \\ \text{Circumferential displacement} &= v = \sum \bar{v}_n \sin n\theta \\ \text{Axial displacement} &= w = \sum \bar{w}_n \cos n\theta \end{aligned} \tag{8.4.3}$$

All three displacements are needed because the physical problem is three dimensional. In Eqs. 8.4.2 and 8.4.3, n is an integer, and all barred quantities are functions of r, z, and n but not of θ. Thus the barred terms are amplitudes. Equations 8.4.2 and 8.4.3 represent a state of symmetry with respect to θ about the plane $\theta = 0$ (antisymmetric Fourier terms are considered subsequently).

The strain-displacement relations in cylindrical coordinates are [1.20]

$$\begin{Bmatrix} \epsilon_r \\ \epsilon_\theta \\ \epsilon_z \\ \gamma_{zr} \\ \gamma_{r\theta} \\ \gamma_{\theta z} \end{Bmatrix} = \begin{bmatrix} \partial/\partial r & 0 & 0 \\ 1/r & \partial/r\partial\theta & 0 \\ 0 & 0 & \partial/\partial z \\ \partial/\partial z & 0 & \partial/\partial r \\ \partial/r\partial\theta & (\partial/\partial r - 1/r) & 0 \\ 0 & \partial/\partial z & \partial/r\partial\theta \end{bmatrix} \begin{Bmatrix} u \\ v \\ w \end{Bmatrix} \tag{8.4.4}$$

These relations are independent of material properties and of whether or not u, v, and w are described by series.

Consider now a typical single harmonic of displacement, say the nth. If we substitute Eqs. 8.4.3 into Eq. 8.4.4 and the resulting strains into Eq. 8.4.1, we find that stresses are

$$\begin{aligned} \{\sigma_{rn}\ \sigma_{\theta n}\ \sigma_{zn}\ \tau_{zrn}\} &= \{\bar{\sigma}_{rn}\ \bar{\sigma}_{\theta n}\ \bar{\sigma}_{zn}\ \bar{\tau}_{zrn}\} \cos n\theta \\ \{\tau_{r\theta n}\ \tau_{\theta zn}\} &= \{\bar{\tau}_{r\theta n}\ \bar{\tau}_{\theta zn}\} \sin n\theta \end{aligned} \tag{8.4.5}$$

where the barred quantities are functions of r, z, and n but not of θ. If Eqs. 8.4.2 and 8.4.5 are substituted into the three differential equations of equilibrium [1.20], we find that they have the forms

$$Q_1 \cos n\theta = 0, \qquad Q_2 \cos n\theta = 0, \qquad Q_3 \sin n\theta = 0 \tag{8.4.6}$$

where the Q_i are functions of r, z, and n but not of θ. Equations 8.4.6 are analogous to Eq. 8.3.10. Equations 8.4.6 prevail for all θ, so $Q_1 = Q_2 = Q_3 = 0$. In a finite element context the equations $Q_1 = Q_2 = Q_3 = 0$ are the equilibrium equations

$$[\mathbf{K}_n]\{\mathbf{D}_n\} - \{\mathbf{R}_n\} = 0 \tag{8.4.7}$$

Equations 8.4.7 are analogous to Eq. 8.3.11. Their solution yields the nodal d.o.f. $\{\mathbf{D}_n\} = \{\bar{u}_{1n}\ \bar{v}_{1n}\ \bar{w}_{1n}\ \bar{u}_{2n}\ \ldots\}$, which are displacement amplitudes of the nodal circles. Matrix $[\mathbf{K}_n]$ depends on n. The load coefficients in $\{\mathbf{R}_n\}$ correspond to the q_n of Eq. 8.3.1.

We see that n circumferential waves of loading are associated with n circumferential waves of stress and of displacement. *The Fourier harmonics are not coupled*. Different numerical values of n present different problems that do not interact. Thus the need for a division into finite elements in the circumferential direction is replaced by the need to superpose separate solutions for a structure divided into finite elements in only its cross section. A single mesh is used for all the separate solutions. In most practical problems only a few load harmonics need be analyzed. A computer program can automatically cycle through a user-specified number of harmonics and superpose the separate solutions.

The preceding discussion invoked only loads and displacements that have $\theta = 0$ as a plane of symmetry. In general, antisymmetric terms are also present. Thus Eqs. 8.4.2 are augmented to read

$$\begin{aligned} \{F_r\ F_z\ \Phi_r\ \Phi_z\ T\} &= \sum \{\bar{\mathbf{Q}}_{An}\} \cos n\theta + \sum \{\bar{\bar{\mathbf{Q}}}_{An}\} \sin n\theta \\ \{F_\theta\ \Phi_\theta\} &= \sum \{\bar{\mathbf{Q}}_{Bn}\} \sin n\theta - \sum \{\bar{\bar{\mathbf{Q}}}_{Bn}\} \cos n\theta \end{aligned} \tag{8.4.8}$$

where $\{\bar{\mathbf{Q}}_{An}\}$ and $\{\bar{\mathbf{Q}}_{Bn}\}$ represent the amplitudes already present in Eq. 8.4.2, and $\{\bar{\bar{\mathbf{Q}}}_{An}\}$ and $\{\bar{\bar{\mathbf{Q}}}_{Bn}\}$ represent the additional amplitudes. Similarly, the displacement field (Eqs. 8.4.3) is augmented and becomes

$$\begin{aligned} u &= \sum \bar{u}_n \cos n\theta + \sum \bar{\bar{u}}_n \sin n\theta \\ v &= \sum \bar{v}_n \sin n\theta - \sum \bar{\bar{v}}_n \cos n\theta \\ w &= \sum \bar{w}_n \cos n\theta + \sum \bar{\bar{w}}_n \sin n\theta \end{aligned} \tag{8.4.9}$$

The motivation for the arbitrarily chosen negative signs in Eqs. 8.4.8 and 8.4.9 is explained below Eq. 8.5.9. Arguments that precede Eq. 8.4.8 can be repeated with "sin" and "cos" interchanged to justify use of the double-barred terms.

Axially symmetric problems are represented by the $n = 0$ terms of the single-barred series. Use of only the even terms $n = 0, 2, 4, \cdots$ of the single-barred series corresponds to loading and deformation that have both $\theta = 0$ and $\theta = 90°$ as planes of symmetry with respect to θ.

Antisymmetric problems (Fig. 8.4.1) are represented by the double-barred series. Pure torque is represented by the $n = 0$ terms of the double-barred series. Thus we can study the twist of shafts of variable diameter. In the torsion problem u and w are everywhere zero, so a finite element solution based on a stress function is also possible [8.4].

When $n = 0$, for any node i, nodal d.o.f. $\bar{v}_{ni}$, $\bar{\bar{u}}_{ni}$, and $\bar{\bar{w}}_{ni}$ have no stiffness associated with them, so these d.o.f. must be suppressed to avoid a singular stiffness matrix.

The simplest displacement boundary condition is zero displacement on a nodal circle. This requires that displacement amplitudes on the circle be zero in every harmonic. If

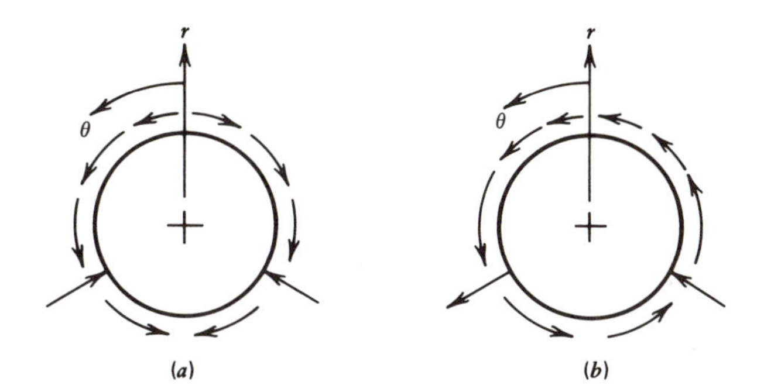

Figure 8.4.1. Examples of (*a*) symmetric, and (*b*) antisymmetric loading.

loads are symmetric about both the $\theta = 0$ and $\theta = 90°$ planes, then u and v are zero at $r = 0$ in all harmonics. Nonzero and asymmetric displacement conditions can be represented as Fourier series and the separate amplitude coefficients used as prescribed displacements in the separate analyses.

Additional constraints on the Fourier displacement amplitudes are deduced from the condition that strains remain finite at $r = 0$ [8.5]. If these constraints are not imposed, numerical integration makes some stiffness coefficients significantly larger than others. This circumstance is not likely to be troublesome in static analysis, but it may require a prohibitively small time step if explicit integration is applied to transient problems.

8.5 GENERAL LOADING: ELEMENT MATRICES

The starting point for stiffness matrix derivation is an element displacement field. Equation 8.4.9 can be written in the form

$$u = \sum u_n, \qquad \text{where} \qquad u_n = \bar{u}_n \cos n\theta + \bar{\bar{u}}_n \sin n\theta \tag{8.5.1}$$

and similarly for the v's and w's. Taking, for example, the single-barred series, we can write the displacement field for harmonic n as

$$\{u_n \; v_n \; w_n\} = \lceil \cos n\theta \;\; \sin n\theta \;\; \cos n\theta \rfloor \{\bar{u}_n \; \bar{v}_n \; \bar{w}_n\} \tag{8.5.2}$$

Amplitudes $\bar{u}_n$, $\bar{v}_n$, and $\bar{w}_n$ are independent of θ. they are interpolated from nodal amplitudes:

$$\{\bar{u}_n \; \bar{v}_n \; \bar{w}_n\} = [\mathbf{N}] \{\bar{u}_{1n} \; \bar{v}_{1n} \; \bar{w}_{1n} \; \bar{u}_{2n} \; \text{etc.}\} \tag{8.5.3}$$

where

$$[\mathbf{N}] = \begin{bmatrix} N_1 & 0 & 0 & N_2 & 0 & 0 & \text{etc.} \\ 0 & N_1 & 0 & 0 & N_2 & 0 & \text{etc.} \\ 0 & 0 & N_1 & 0 & 0 & N_2 & \text{etc.} \end{bmatrix} \tag{8.5.4}$$

If we select, for example, the four-node element of Fig. 5.3.1, there are 12 nodal amplitudes and the N_i are defined by Eq. 5.3.3.

If we apply Eq. 8.4.4 to Eqs. 8.4.9, we find that strains in harmonic n can be written in the form

$$\begin{aligned} \{\epsilon_{rn}\ \epsilon_{\theta n}\ \epsilon_{zn}\ \gamma_{zrn}\} &= [\mathbf{H}_A]\left(\begin{Bmatrix} \bar{u}_n \\ \bar{v}_n \\ \bar{w}_n \end{Bmatrix} \cos n\theta + \begin{Bmatrix} \bar{\bar{u}}_n \\ \bar{\bar{v}}_n \\ \bar{\bar{w}}_n \end{Bmatrix} \sin n\theta\right) \\ \{\gamma_{r\theta n}\ \gamma_{\theta zn}\} &= [\mathbf{H}_B]\left(\begin{Bmatrix} \bar{u}_n \\ \bar{v}_n \\ \bar{w}_n \end{Bmatrix} \sin n\theta - \begin{Bmatrix} \bar{\bar{u}}_n \\ \bar{\bar{v}}_n \\ \bar{\bar{w}}_n \end{Bmatrix} \cos n\theta\right) \end{aligned} \tag{8.5.5}$$

where

$$[\mathbf{H}_A] = \begin{bmatrix} \partial/\partial r & 0 & 0 \\ 1/r & n/r & 0 \\ 0 & 0 & \partial/\partial z \\ \partial/\partial z & 0 & \partial/\partial r \end{bmatrix} \tag{8.5.6}$$

$$[\mathbf{H}_B] = \begin{bmatrix} -n/r & \partial/\partial r - 1/r & 0 \\ 0 & \partial/\partial z & -n/r \end{bmatrix} \tag{8.5.7}$$

By applying operators $[\mathbf{H}_A]$ and $[\mathbf{H}_B]$ to matrix $[\mathbf{N}]$ of Eq. 8.5.3, then including the $\cos n\theta$ and $\sin n\theta$ coefficients, we produce a strain-displacement matrix $[\mathbf{B}_n]$ for harmonic n. For example, the first column of $[\mathbf{B}_n]$ is, from the single-barred series,

$$\left\{\frac{\partial N_1}{\partial r}\cos n\theta \quad \frac{N_1}{r}\cos n\theta \quad 0 \quad \frac{\partial N_1}{\partial z}\cos n\theta \quad -\frac{nN_1}{r}\sin n\theta \quad 0\right\} \tag{8.5.8}$$

We see that $[\mathbf{B}_n]$ is a function of r, z, n, and θ. If the element is isoparametric, the usual transformation must be included:

$$N_{i,r} = \Gamma_{11}N_{i,\xi} + \Gamma_{12}N_{i,\eta} \qquad N_{i,z} = \Gamma_{21}N_{i,\xi} + \Gamma_{22}N_{i,\eta} \tag{8.5.9}$$

Continuing the development, we find that for harmonic n, integrands of the element stiffness matrix and load vector (Eqs. 4.3.4 and 4.3.5) contain $\sin^2 n\theta$ and/or $\cos^2 n\theta$ in every term. Integration with respect to θ is done explicitly by means of Eqs. 8.3.2. The resulting factor of π (or 2π for $n = 0$) can be canceled from every term. Integration with respect to r and z (or ξ and η) is done as if the problem were axially symmetric. The resulting stiffness matrix $[\mathbf{k}_n]$ is the same whether we consider the single-barred or the double-barred series. (This is the motivation for the negative sign in Eq. 8.4.9.) Coefficients in $[\mathbf{k}_n]$ have the forms $A + Bn^2$ and Cn, where A, B, and C are independent of n. Accordingly, A, B, and C need be generated only once, regardless of the number of harmonics used. Assembly of elements yields the structural equations $[\mathbf{K}]\{\mathbf{D}\} = \{\mathbf{R}\}$:

$$\lceil \mathbf{K}_0 \;\; \mathbf{K}_1 \;\; \mathbf{K}_2 \;\; \cdots \rfloor \{\mathbf{D}_0 \;\; \mathbf{D}_1 \;\; \mathbf{D}_2 \;\; \cdots\} = \{\mathbf{R}_0 \;\; \mathbf{R}_1 \;\; \mathbf{R}_2 \;\; \cdots\} \tag{8.5.10}$$

where the diagonal nature of $[\mathbf{K}]$ reflects the decoupling into separate problems for $n = 0$, $n = 1$, $n = 2$, and so on. In computation, matrices are not built for all harmonics at once. The harmonics are treated in sequence, as suggested by Eq. 8.4.7.

Loads are treated by the usual formulas. Consider, for example, a surface traction in harmonic n.

$$\{\mathbf{r}_n\} = \int_{-\pi}^{\pi} [\mathbf{N}_n]^T \{\boldsymbol{\Phi}_n\}\, r\, d\theta\, d\ell \tag{8.5.11}$$

where $d\ell$ is an increment of meridional length. For the single-barred series, from Eqs. 8.5.2 and 8.5.3, $[\mathbf{N}_n] = [\cos n\theta \quad \sin n\theta \quad \cos n\theta]\,[\mathbf{N}]$, where $[\mathbf{N}]$ is evaluated on the loaded surface. From Eq. 8.4.8, for the single-barred series, $\{\Phi_n\} = \{\bar{\Phi}_{rn}\cos n\theta \quad \bar{\Phi}_{\theta n}\sin n\theta \quad \bar{\Phi}_{zn}\cos n\theta\}$, where the barred Φ_n's are amplitudes of the nth Fourier harmonic. When $n = 0$, $\bar{\Phi}_{rn}$ yields a uniform radial "force" to be treated like any other component of $\{\mathbf{R}_n\}$. Being axially symmetric, this force has no radial resultant from a physical point of view.

8.6 GENERAL LOADING AND GENERAL PROPERTIES

Consider that the solid is axially symmetric in its geometry and elastic properties but that $[\mathbf{E}]$ of Eq. 8.4.1 is a full matrix. We now find that the stresses on the left side of Eq. 8.4.5 each contain both $\cos n\theta$ and $\sin n\theta$. Consequently, the simple decoupling of harmonics expressed by Eqs. 8.4.6 and 8.4.7 does not happen. A decoupling occurs, but it is more complicated, as explained next.

With all terms in Eqs. 8.4.9 retained, the strain-displacement relation $\{\epsilon\} = [\mathbf{B}]\{\mathbf{d}\}$ can be written

$$\{\epsilon\} = [\bar{\mathbf{B}}_0\; \bar{\bar{\mathbf{B}}}_0\; \bar{\mathbf{B}}_1\; \bar{\bar{\mathbf{B}}}_1\; \bar{\mathbf{B}}_2 \cdots]\, \{\bar{\mathbf{d}}_0\; \bar{\bar{\mathbf{d}}}_0\; \bar{\mathbf{d}}_1\; \bar{\bar{\mathbf{d}}}_1\; \bar{\mathbf{d}}_2 \cdots\} \tag{8.6.1}$$

If the element has M nodes and three d.o.f. per node, each $[\mathbf{B}_n]$ is 6 by $3M$ and each $\{\mathbf{d}_n\}$ is $3M$ by 1. Tilde overbars indicate the source of terms from the single- or double-barred series. Except for $[\tilde{\mathbf{B}}_0]$ and $[\tilde{\tilde{\mathbf{B}}}_0]$, the $[\tilde{\mathbf{B}}_n]$'s and $[\tilde{\tilde{\mathbf{B}}}_n]$'s contain cos $n\theta$ *and* sin $n\theta$ coefficients. Subscripts indicate the harmonic number.

For the sake of explanation, imagine that $[\mathbf{E}]$ is decomposed into its Choleski factor $[\mathbf{U}]$ (Eq. 2.12.1). Then, in the integrand of the element stiffness matrix,

$$[\mathbf{B}]^T\,[\mathbf{E}][\mathbf{B}] = [\mathbf{b}]^T[\mathbf{b}], \qquad \text{where} \qquad [\mathbf{b}] = [\mathbf{U}]\,[\mathbf{B}] \tag{8.6.2}$$

and

$$[\mathbf{b}]^T[\mathbf{b}] = \begin{bmatrix} \tilde{\mathbf{b}}_0{}^T\tilde{\mathbf{b}}_0 & \tilde{\mathbf{b}}_0{}^T\tilde{\tilde{\mathbf{b}}}_0 & \tilde{\mathbf{b}}_0{}^T\tilde{\mathbf{b}}_1 & \tilde{\mathbf{b}}_0{}^T\tilde{\tilde{\mathbf{b}}}_1 & \tilde{\mathbf{b}}_0{}^T\tilde{\mathbf{b}}_2 & \cdots \\ & \tilde{\tilde{\mathbf{b}}}_0{}^T\tilde{\tilde{\mathbf{b}}}_0 & \tilde{\tilde{\mathbf{b}}}_0{}^T\tilde{\mathbf{b}}_1 & \tilde{\tilde{\mathbf{b}}}_0{}^T\tilde{\tilde{\mathbf{b}}}_1 & \tilde{\tilde{\mathbf{b}}}_0{}^T\tilde{\mathbf{b}}_2 & \cdots \\ & & \tilde{\mathbf{b}}_1{}^T\tilde{\mathbf{b}}_1 & \tilde{\mathbf{b}}_1{}^T\tilde{\tilde{\mathbf{b}}}_1 & \tilde{\mathbf{b}}_1{}^T\tilde{\mathbf{b}}_2 & \cdots \\ \text{Symmetric} & & & \tilde{\tilde{\mathbf{b}}}_1{}^T\tilde{\tilde{\mathbf{b}}}_1 & \tilde{\tilde{\mathbf{b}}}_1{}^T\tilde{\mathbf{b}}_2 & \cdots \end{bmatrix} \tag{8.6.3}$$

Each matrix product in Eq. 8.6.3, such as $[\tilde{\mathbf{b}}_0]^T[\tilde{\mathbf{b}}_0]$, is $3M$ by $3M$. As can be seen by expanding these products, those that have *unlike* subscripts vanish after integration with respect to θ because of Eqs. 8.3.2. If $[\mathbf{E}]$ were as given by Eq. 8.4.1, each *off-diagonal* matrix product with *like* subscripts would contain only sin $n\theta$ times cos $n\theta$ and would also vanish after θ integration. But now these terms contain $\sin^2 n\theta$ and $\cos^2 n\theta$ and remain, as do the on-diagonal matrix products.

So again the problem can be symbolized by Eq. 8.5.10. But now each $[\mathbf{K}_n]$ is $6M$ by $6M$, not $3M$ by $3M$ as when Eq. 8.4.1 prevails. The harmonics are again uncoupled, but there is coupling between the single- and double-barred series of Eq. 8.4.9.

Matrix $[\mathbf{K}_0]$ couples the amplitudes $\bar{u}_0$, $\bar{w}_0$, and $\bar{\bar{v}}_0$ but is null (and therefore singular) in coefficients associated with $\bar{v}_0$, $\bar{\bar{u}}_0$, and $\bar{\bar{w}}_0$. Thus, if $[\mathbf{E}]$ is full, as for a layered material where θ is not a principal material direction, loading that does not vary with θ ($n = 0$) requires only $\bar{u}_0$, $\bar{w}_0$, and $\bar{\bar{v}}_0$ as d.o.f. at each node.

Another approach to the problem of solids with a general $[\mathbf{E}]$ starts by writing the displacements as

$$\{u \quad v \quad w\} = \sum \{u_n \quad v_n \quad w_n\}\, e^{jn\theta} \tag{8.6.4}$$

where $j = \sqrt{-1}$. Each $[\mathbf{K}_n]$ remains $3M$ by $3M$ but is Hermitian, not symmetric. The force and displacement vectors are complex variables [8.7, 8.8].

The problem is yet more complicated when elastic properties depend on θ. This happens when temperature varies circumferentially and $[\mathbf{E}]$ is temperature dependent, or when the material has plane layers that are not perpendicular to the axis of revolution. By our definition, such a solid is not axially symmetric. But the Fourier series attack can still be used. However, *all* harmonics are then coupled, and the order and bandwidth of $[\mathbf{K}]$ are proportional to the number of harmonics used [8.9–8.11].

8.7 CONCLUDING REMARKS

The Fourier series treatment described in Sections 8.4 through 8.6 is also known as the *semianalytical method* and the *separation of variables method*. When used for plates, it is called the *finite strip method* (Section 9.2).

Besides its application to plates and to solids and shells of revolution, the Fourier series method can be applied to prismatic solids [8.12, 8.13]. Then the name *finite prism method* may be used [9.12]. The solid, and its elements, are prismatic (Fig. 8.7.1). The displacement field is again Eq. 8.4.9, except that $\pi y/L$ replaces θ. If only the single-barred series are used, deformation and loading are symmetric about the xz plane, with $v = 0$ at $y = 0$ and at $y = \pm L$. As usual, arbitrary loads and displacements are treated by finding their Fourier coefficients and making a separate analysis for each, then superposing results. Problems such as that of Fig. 8.7.1 may require 9 to 19 Fourier coefficients.

In a physical sense, what has been done in Fig. 8.7.1 is to take a toroidal solid that extends from $-\pi$ to π and straighten it out to form a prismatic solid that extends from $-L$ to L. The straightening can be faked by moving the z-axis far to the left and making it an axis of revolution. Then the almost-prismatic solid can be analyzed by a program for solids of revolution. However, elements of large radius and small cross section invite numerical trouble because of inaccuracies in numerical integration [8.15]. Elements with complicated displacement fields are more susceptible than simple elements.

Some geometries are ''almost'' axially symmetric: a large portion is axisymmetric, or a certain geometry repeats itself around the circumference. Some aspects of these problems are noted in Sections 6.6, 6.12, and 6.13.

The problem of Fig. 8.7.2 is axially symmetric in geometry and material properties. But it is not obvious that an axisymmetric analysis can deal with a moment loading. Reference 8.14 describes how. The trick is to use a thermal load to simulate the strains produced by M_0. Note that if the beam is a thin-walled pipe elbow its cross section may become oval in response to M_0. Pipe elbow elements that include this effect have been developed [8.16].

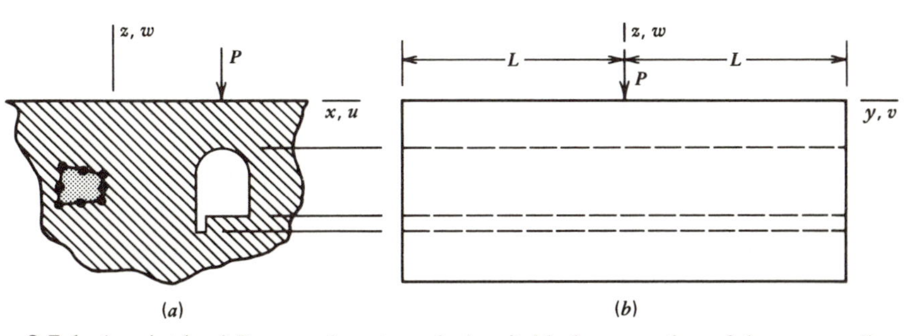

Figure 8.7.1. A point load P over a long tunnel. A suitably large portion of the surrounding earth on rock is modeled by finite elements, one of which is shown and shaded. (*a*) Front view. (*b*) Right side view.

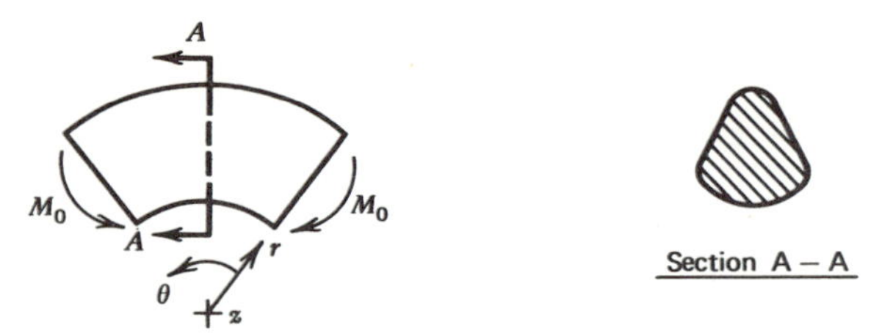

Figure 8.7.2. A curved beam of arbitrary cross section, loaded by a bending moment M_0.

PROBLEMS

8.1 Dashed lines represent displacement modes of a four-node element whose shape functions are given by Eq. 5.3.3. The original cross section (solid lines) is square, and the axis of revolution is to the left of each. Loads and displacements are axially symmetric.
   (a) If **[k]** is generated using one-point Gauss quadrature, which of the eight modes are rigid-body modes? Which are zero-energy deformation modes?
   (b) Let the Gauss quadrature be 2 by 2 and repeat part (a).
   (c, d) Let the element be one for *plane stress*, not axial symmetry, and repeat parts (a) and (b).

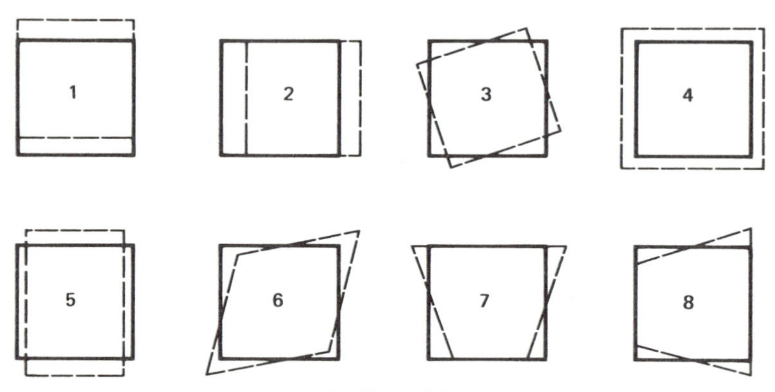

Problem 8.1

8.2 Section 5.10 presents and discusses the number of Gauss points needed for correct volume calculation and for correct convergence. Reference is to plane elements. How should these arguments be amended if reference is to axially symmetric elements?

8.3 Justify the theoretical requirement that $\epsilon_r = \epsilon_\theta$ at $r = 0$ in an axially symmetric problem.

8.4 Nodes 4–7–3 represent a z = constant face of an axially symmetric quadratic isoparametric element (Fig. 5.6.1*a*). Find the 3-by-1 vector of consistent nodal loads {r} produced by the z-direction surface traction $\Phi_z = p$ (a uniform "tensile pressure" p).

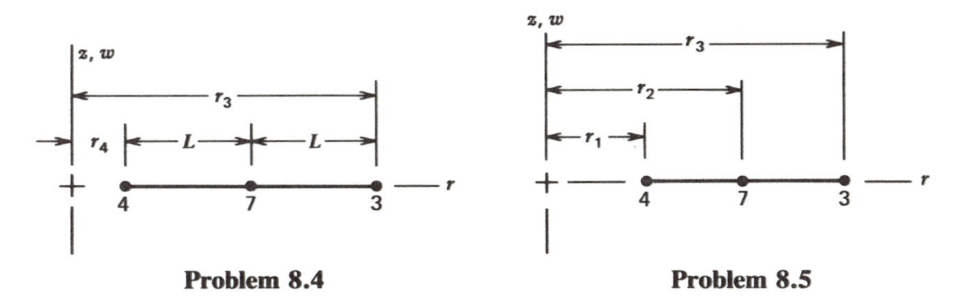

Problem 8.4 **Problem 8.5**

8.5 The sketch represents equally spaced nodes of an axially symmetric quadratic element. Consider a *parabolic* traction; that is, $\Phi_z = (\xi^2 - \xi)\, p_4/2 + (1 - \xi^2)\, p_7 + (\xi^2 + \xi) p_3/2$. Find the consistent nodal loads {r}.

8.6 In Problem 8.4, nodal load associated with node 4 is zero if $r_4 = 0$. This is not physically reasonable. Explain why the polynomial for w leads to this result. *Suggestion*. Consider $w_4 > 0$, $w_3 = w_7 = 0$.

8.7 (a) Use Eqs. 8.3.2 to verify the Q's in 8.3.3.
(b) Verify Eq. 8.3.4.
(c) Verify Eq. 8.3.5. *Suggestion*. Imagine that a concentrated force is produced by large stress on a small area.
(d) Find a series that represents concentrated radial forces P on a disc, one outward at $\theta = -\pi/2$ and another outward at $\theta = \pi/2$.

8.8. Use one, then two, terms of Eq. 8.3.12 to evaluate the deflection and bending moment at midspan in Fig. 8.3.2*a*. Compare exact and approximate answers.
(a) Consider a uniform load.
(b) Consider a concentrated load at midspan.

8.9 Consider a view down the z-axis onto a disc. Sketch the distribution (the variation with θ) of the following tractions on the outer edge of the disc.
(a) $\Phi_r = \bar{\Phi}_r \cos n\theta$ with $n = 0$.
(b) $\Phi_r = \bar{\Phi}_r \cos n\theta$ with $n = 2$.
(c) $\Phi_\theta = \bar{\Phi}_\theta \sin n\theta$ with $n = 1$.

8.10 Assume that a certain cylindrical shell can be adequately modeled by 20 four-node elements arranged in a 20-by-1 mesh so that one element spans the thickness. Also assume that the loading is described by Eq. 8.4.2, with $n = 1, 2, 3, 4, 5,$ and 6. Alternatively, one could contemplate a fully three-dimensional analysis with a 20-by-1-by-m mesh, where m is the number of elements around the circumference.
(a) Estimate m so that the three-dimensional model would be adequate in accuracy.
(b) Estimate the cost ratio of the two alternatives. Base your estimate on the expense of a banded equation solver (Fig. 2.11.2).

8.11 If $\{\epsilon\} = 0$, Eqs 8.4.4 have the solution

$$u = (a_1 + a_2 z) \sin \theta + (b_1 + b_2 z) \cos \theta$$
$$v = (a_1 + a_2 z) \cos \theta - (b_1 + b_2 z) \sin \theta + c_2 r$$
$$w = -a_2 r \sin \theta - b_2 r \cos \theta + c_1$$

where the a's, b's, and c's are constants. An assumed field must contain these terms if the element is to exhibit rigid-body motion without strain. Show that this field does, in fact, yield $\{\epsilon\} = 0$.

8.12 Specialize Eqs. 8.4.9 to represent the following rigid-body motions [8.6].
(a) Axial translation.
(b) Translation perpendicular to the z-axis in the $\theta = 0$ plane.
(c) Translation perpendicular to the z-axis in the $\theta = 90°$ plane.
(d) Rotation about the z-axis.
(e) Rotation about the line $\theta = z = 0$.
(f) Rotation about the line $\theta = 90°$, $z = 0$.

8.13 Show that element matrix integrands contain $\sin^2 n\theta$ or $\cos^2 n\theta$ in every term, as noted in the paragraph that contains Eq. 8.5.10.

8.14 Show that coefficients in $[\mathbf{k}_n]$ have the forms $A + Bn^2$ and Cn, as noted above Eq. 8.5.10.

8.15 (a) Revise Figs. 5.5.1 and 5.5.2 so that they apply to a linear isoparametric element for a solid of revolution with axially symmetric loads.
(b) Similarly, revise these subroutines so that they apply to loads without axial asymmetry. Follow the development of Section 8.5. Consider only single-barred terms and assume that $[\mathbf{E}]$ is given by Eq. 8.4.1.

8.16 Consider an isotropic, plane stress element of constant thickness t that has only radial displacement u (see sketch). Element nodal d.o.f. are u_1 and u_2.
(a) Formulate matrices $[\mathbf{N}]$ and $[\mathbf{B}]$.
(b) Incorporate the element in a computer program. For simplicity, one-point Gauss quadrature can be used to generate $[\mathbf{k}]$. As test cases, consider radial tension on the outer edge of a disc divided into several of these elements, with and without a central hole.

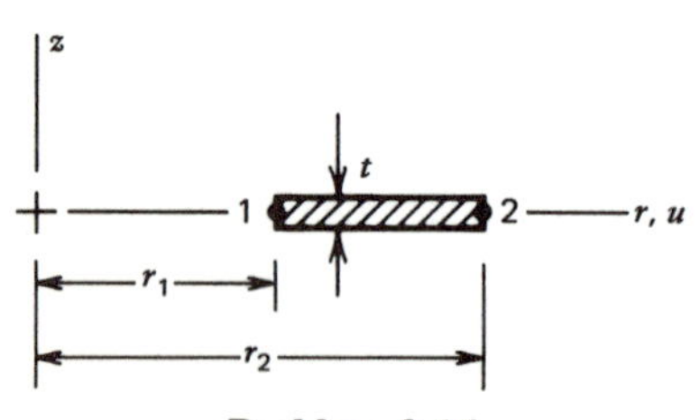

Problem 8.16

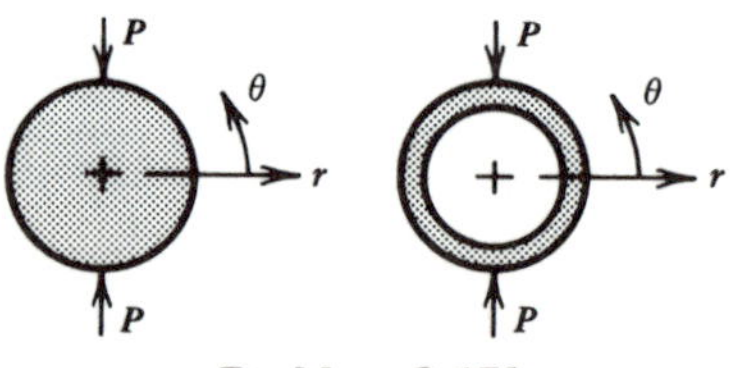

Problem 8.17b

8.17 (This problem is suitable as a term project.) Repeat Problem 8.16 but also allow the circumferential displacement v so that loads without axial symmetry can be treated. Element nodal d.o.f. are u_1, v_1, u_2, and v_2. Let $\theta = 0$ be a plane of symmetry. Additional test cases include (*a*) a hole in a wide plate under uniaxial tension, and (*b*) pinching loads P on a disc or ring, as shown. The program should treat a user-specified number of harmonics and superpose results for stress and displacement.

8.18 What Fourier harmonics of surface tractions $\{\boldsymbol{\Phi}\}$, and what nodal loads in $\{\mathbf{R}\}$, are associated with Problem 8.17*a*?

8.19 (a) Let Poisson's ratio $\nu = 0$ in Problem 8.16. Generate the coefficients in $[\mathbf{k}]$ by explicit integration.

(b) Specialize this $[\mathbf{k}]$ to represent a core element, that is, an element for which $u_1 = r_1 = 0$.

CHAPTER 9

BENDING OF FLAT PLATES

9.1 BEHAVIOR OF PLATES AND SHELLS

A flat plate, like a straight beam, supports transverse loads by bending action. Figure 9.1.1*a* shows stresses that act on vertical cross sections. Stresses σ_x and σ_y vary linearly with z, and τ_{zx} and τ_{yz} vary quadratically with z. Shear stress τ_{xy} varies linearly with z and is associated with twisting (Fig. 9.1.3*c*). Transverse load q includes both surface traction and body force. Unless stated otherwise, "plate bending" means that there are no loads in the plane of the plate.

Stresses in Fig. 9.1.1*a* produce moments and shear forces.

$$M_x = \int_{-t/2}^{t/2} \sigma_x z \, dz, \qquad M_y = \int_{-t/2}^{t/2} \sigma_y z \, dz, \qquad M_{xy} = \int_{-t/2}^{t/2} \tau_{xy} z \, dz$$
$$Q_x = \int_{-t/2}^{t/2} \tau_{zx} \, dz, \qquad Q_y = \int_{-t/2}^{t/2} \tau_{yz} \, dz \tag{9.1.1}$$

The M's are moments *per unit length* and the Q's are shear forces *per unit length*. Differential *total* moments and forces are $M_x\,dy$, $Q_x\,dy$, and so on. Moments M_x and M_y resist the curvatures $w_{,xx}$ and $w_{,yy}$. Twisting moment M_{xy} resists the twist $w_{,xy}$.

In a small lateral displacement w of a homogeneous, linearly elastic plate, points on the midsurface $z = 0$ have only z-direction displacements. Points *not* on $z = 0$ also have x and y motions. In a *thin* plate—one with negligible transverse shear deformation—a line normal to the midsurface of the undeformed plate is assumed to remain normal to the deformed midsurface. This is the Kirchhoff approximation [9.1]. So we have, from Eqs. 1.4.4 and Fig. 9.1.2,

$$\begin{aligned} u &= -zw_{,x} & & & \epsilon_x &= -zw_{,xx} \\ v &= -zw_{,y} & & \text{hence} & \epsilon_y &= -zw_{,yy} \\ & & & & \gamma_{xy} &= -2zw_{,xy} \end{aligned} \tag{9.1.2}$$

These are the strain-displacement relations for a thin plate.

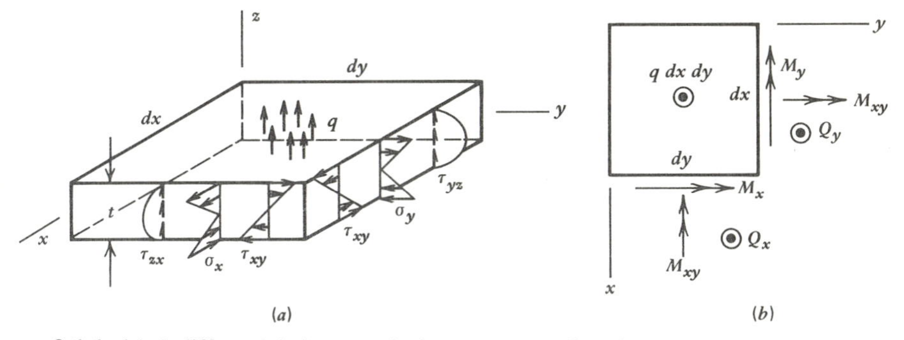

Figure 9.1.1. (*a*) A differential element of a homogeneous, linearly elastic plate. The lateral load is q (force per unit area). (*b*) The same element, seen from above.

Moments and curvatures are related as follows. Let x and y be principal directions of an orthotropic material. Stress σ_z is considered negligible in comparison with σ_x and σ_y. Then, in the notation of Ref. 9.1, Eqs. 1.6.1 state that

$$\begin{Bmatrix} \sigma_x \\ \sigma_y \\ \tau_{xy} \end{Bmatrix} = \begin{bmatrix} E'_x & E'' & 0 \\ E'' & E'_y & 0 \\ 0 & 0 & G \end{bmatrix} \left(\begin{Bmatrix} \epsilon_x \\ \epsilon_y \\ \gamma_{xy} \end{Bmatrix} - \begin{Bmatrix} \alpha_x T \\ \alpha_y T \\ 0 \end{Bmatrix} \right) \tag{9.1.3}$$

where $\{\boldsymbol{\sigma}_0\} = 0$, and initial strains $\{\boldsymbol{\epsilon}_0\}$ are presumed caused by thermal expansion with principal coefficients α_x and α_y. For an isotropic material,

$$E'_x = E'_y = \frac{E''}{\nu} = \frac{E}{1 - \nu^2} \qquad G = \frac{E}{2(1 + \nu)} \tag{9.1.4}$$

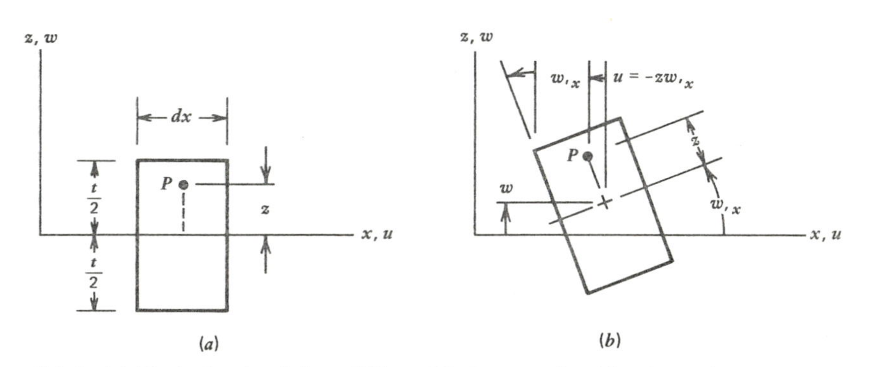

Figure 9.1.2. (*a*) Typical point P in a differential element of a thin plate. (*b*) Point P displaces w units up and $zw,_x$ units to the left because of midsurface motions w and $w,_x$.

Let $T = 2zT_0/t$, which is a linear temperature variation from $-T_0$ at $z = -t/2$ to T_0 at $z = t/2$. Then, if we substitute Eqs. 9.1.2 into Eq. 9.1.3 and substitute the result into Eqs. 9.1.1, we find

$$\{M_x \ M_y \ M_{xy}\} = -\underset{3\times 3}{[\mathfrak{D}]}\left(\{w,_{xx} \ w,_{yy} \ 2w,_{xy}\} + \left\{\frac{2\alpha_x T_0}{t} \quad \frac{2\alpha_y T_0}{t} \quad 0\right\}\right) \tag{9.1.5}$$

where $[\mathfrak{D}]$ is symmetric, $\mathfrak{D}_{13} = \mathfrak{D}_{23} = 0$, and

$$\mathfrak{D}_{11} = \frac{E'_x t^3}{12}, \qquad \mathfrak{D}_{12} = \frac{E'' t^3}{12}, \qquad \mathfrak{D}_{22} = \frac{E'_y t^3}{12}, \qquad \mathfrak{D}_{33} = \frac{G t^3}{12} \tag{9.1.6}$$

For an isotropic material we have

$$\begin{aligned} \mathfrak{D}_{11} &= \mathfrak{D}_{22} = \frac{\mathfrak{D}_{12}}{\nu} = \mathfrak{D}, \qquad \text{where } \mathfrak{D} = \frac{Et^3}{12(1 - \nu^2)} \\ \mathfrak{D}_{33} &= \frac{1}{2(1 + \nu)}\frac{Et^3}{12} = \frac{1 - \nu}{2}\mathfrak{D} \end{aligned} \tag{9.1.7}$$

The "flexural rigidity" $\mathfrak{D}$ is analogous to the bending stiffness EI of a beam. Indeed, if the plate has unit width and $\nu = 0$, then $\mathfrak{D} = EI$.

We are still talking about *thin* plates. Additional terms that account for transverse shear effects are introduced when needed (in Eqs. 9.1.9–9.1.11 and in Section 9.3).

Equation 9.1.5 shows that actions in the x and y directions are coupled, even for an isotropic plate. In Fig. 9.1.3*a*, $w,_{xx} = w,_{xy} = 0$ in the center portion, but M_x still exists because of the Poisson effect. But $M_x = 0$ at the free edges $x = \pm a$, so the edges curl a bit (Fig. 9.1.3*b*). Only if $a \approx t$, so that $M_x \approx 0$ throughout, does the plate act like a beam, displaying the familiar anticlastic surface. The pure twist of Fig. 9.1.3*c* is associated with moments $-M_{xy}$ alone ($M_x = M_y = 0$) if the plate is isotropic.

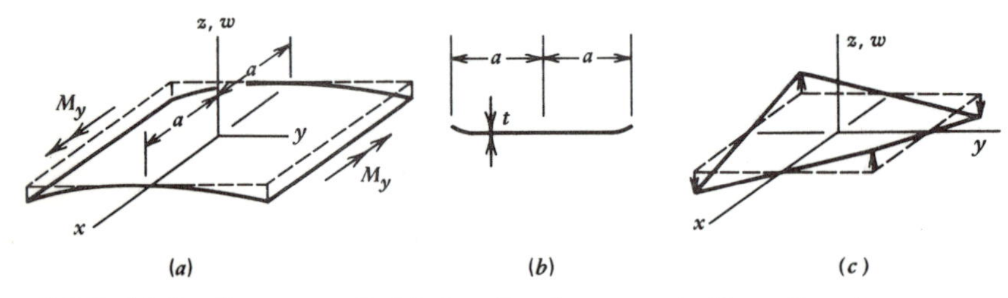

Figure 9.1.3. (*a*) Bending to a cylindrical surface by moments M_y on the edges y = constant. (*b*) Cross section cut by the xz plane. (*c*) The $w = xy$ state of pure twist: $w,_{xx} = w,_{yy} = 0$, $w,_{xy} > 0$.

Suppose that $x'y'$ are principal axes of an orthotropic material (Fig. 9.1.4*a*). Then Eqs. 9.1.6 pertain to the primed axes, and $[\mathfrak{D}]$ in Eq. 9.1.5 becomes a full matrix. It is obtained from $[\mathfrak{D}']$ by coordinate transformation as in Section 6.2. The transformations are

$$
\begin{aligned}
\{M_{x'} \;\; M_{y'} \;\; M_{x'y'}\} &= [\mathbf{T}_\sigma]\,\{M_x \;\; M_y \;\; M_{xy}\} \\
\{w,_{x'x'} \;\; w,_{y'y'} \;\; 2w,_{x'y'}\} &= [\mathbf{T}_\epsilon]\,\{w,_{xx} \;\; w,_{yy} \;\; 2w,_{xy}\} \\
[\mathfrak{D}] &= [\mathbf{T}_\epsilon]^T[\mathfrak{D}'][\mathbf{T}_\epsilon]
\end{aligned}
\tag{9.1.8}
$$

where $[\mathbf{T}_\epsilon]$ and $[\mathbf{T}_\sigma]$ are given by Eqs. 6.2.14.

Thick plates and sandwich plates must account for transverse shear deformation. In principal coordinates $x'y'$ the shear force versus shear strain relation is written

$$
\{Q_{y'} \; Q_{x'}\} = \lceil \mathbf{S}' \rfloor \{\gamma_{y'z} \; \gamma_{zx'}\}
\tag{9.1.9}
$$

where $\lceil \mathbf{S}' \rfloor$ is a diagonal matrix that depends on shear moduli and plate thickness (see Eq. 9.4.5). Coordinate transformations are

$$
\begin{aligned}
\{Q_{y'} \;\; Q_{x'}\} &= [\mathbf{T}]\{Q_y \;\; Q_x\} \\
\{\gamma_{y'z} \;\; \gamma_{zx'}\} &= [\mathbf{T}]\{\gamma_{yz} \;\; \gamma_{zx}\}
\end{aligned}
\tag{9.1.10}
$$

where $T_{11} = T_{22} = \cos\beta$ and $T_{21} = -T_{12} = \sin\beta$. In global coordinates, the analogue of Eq. 9.1.9 is

$$
\{Q_y \;\; Q_x\} = ([\mathbf{T}]^T\lceil \mathbf{S}' \rfloor [\mathbf{T}])\,\{\gamma_{yz} \; \gamma_{zx}\}
\tag{9.1.11}
$$

The total potential expression for a thin plate is found by specializing Eq. 3.4.9. We ignore ϵ_z, γ_{yz}, γ_{zy} and all $x-$ and $y-$ direction loads, substitute Eqs. 9.1.2, and integrate from $z = -t/2$ to $z = t/2$. The result is analogous to Eq. 3.4.16:

$$
\begin{aligned}
\Pi_p = &\iint \{\boldsymbol{\kappa}\}^T \left[[\mathfrak{D}] \left(\frac{1}{2}\{\boldsymbol{\kappa}\} - \{\boldsymbol{\kappa}_0\} \right) + \{\mathbf{m}_0\} \right] dx\,dy \\
&- \iint qw\,dx\,dy - \{\mathbf{D}\}^T\{\mathbf{P} + \mathbf{M}\}
\end{aligned}
\tag{9.1.12}
$$

where $\{\boldsymbol{\kappa}\} = \{w,_{xx} \; w,_{yy} \; 2w,_{xy}\}$, and $[\mathfrak{D}]$ is the same matrix as seen in Eq. 9.1.5. Loads $\{\mathbf{P}\}$ and $\{\mathbf{M}\}$ represent concentrated forces and moments. These include discrete nodal loads and contributions from prescribed shears and moments on edges; for example

$$
\int \lfloor \mathbf{N} \rfloor^T \bar{Q}_x\,dy, \qquad \int \lfloor \mathbf{N},_x \rfloor^T \bar{M}_x\,dy, \qquad \int \lfloor \mathbf{N},_x \rfloor^T \bar{M}_{xy}\,dx
\tag{9.1.13}
$$

where barred quantities are prescribed, and $[\mathbf{N}]$ is the shape function matrix in $w = \lfloor \mathbf{N} \rfloor \{\mathbf{d}\}$. Nodal d.o.f. $\{\mathbf{d}\}$ and $\{\mathbf{D}\}$ include nodal displacements and nodal rotations. The stiffness matrix of a thin-plate element is

$$[\mathbf{k}] = \iint [\mathbf{B}]^T [\mathfrak{D}] [\mathbf{B}] \, dx \, dy,$$

where (9.1.14)

$$[\mathbf{B}] = \left\{ \frac{\partial^2}{\partial x^2} \quad \frac{\partial^2}{\partial y^2} \quad 2\frac{\partial^2}{\partial x \partial y} \right\} \lfloor \mathbf{N} \rfloor$$

Usually, as a plate is loaded, it wants to assume a nondevelopable shape (a *developable* surface is made by folding or bending a flat sheet without straining its midsurface). Therefore a loaded plate tends to develop in-plane strains and forces. These are *membrane forces*. They carry part of the load. As load increases, so do the membrane forces, and the plate seems to become stiffer (Fig. 9.1.4*b*). The problem is nonlinear. To solve it, higher-order terms must be added to Eq. 9.1.2. Linear theory may err by 50% if deflection w equals the thickness t.

Shells always develop membrane forces because they are initially curved. Let z be a midsurface-normal coordinate. If the material is linearly elastic, σ_x, σ_y, and τ_{xy} of Fig. 9.1.1*a* still vary linearly with z, but they do not vanish at the midsurface. Therefore they generate the membrane forces

$$N_x = \int_{-t/2}^{t/2} \sigma_x \, dz, \qquad N_y = \int_{-t/2}^{t/2} \sigma_y \, dz, \qquad N_{xy} = \int_{-t/2}^{t/2} \tau_{xy} \, dz \qquad (9.1.15)$$

These forces are shown in Fig. 12.2.2*b*. If σ_x, σ_y, and τ_{xy} are independent of z, the moments of Eqs. 9.1.1 vanish, and we have a membrane shell, which represents a very efficient use of material. But most shells carry *some* of their load by bending.

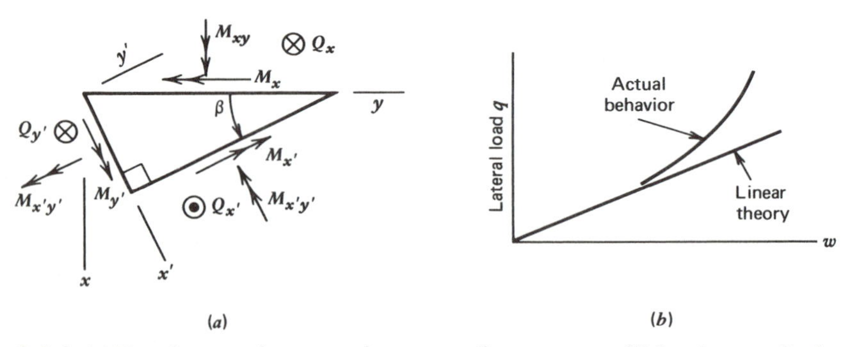

Figure 9.1.4. (*a*) Plate forces and moments in two coordinate systems. (*b*) Load versus displacement behavior of a typical plate.

Plates can also develop membrane forces if they are nonhomogeneous (see Eqs. 9.4.14–9.4.16).

9.2 FINITE ELEMENTS AND FINITE STRIPS

So many plate elements have been proposed that we do not even attempt to catalog them all. Here we discuss some difficulties of plate elements and some options in element formulation. Unless stated otherwise, the elements are thin (without transverse shear deformation) and are based on assumed displacement fields.

The deformation of a thin-plate element is completely described by its lateral displacement, $w = w(x, y) = \lfloor \mathbf{N} \rfloor \{\mathbf{d}\}$. An acceptable element must display C_1 continuity in the limit of mesh refinement. This is Requirement 4 of Section 4.9. Here it means that w and its first derivatives must be continuous across interelement boundaries. Furthermore, we must satisfy Requirement 2 of Section 4.8. According to Eqs. 9.1.2, it now means that the element must be able to display constant curvatures $w_{,xx}$ and $w_{,yy}$ and constant twist $w_{,xy}$. The element should also be invariant. It is hard to satisfy all these requirements at once.

In what follows we name the elements by symbols in parentheses so that they can be identified in Fig. 9.2.3.

An early rectangular element (element ACM) is based on a 12-term polynomial [4.1, 9.2]. It is $w = \lfloor \mathbf{X} \rfloor \{\mathbf{a}\}$, where

$$\lfloor \mathbf{X} \rfloor = \lfloor 1, x, y, x^2, xy, y^2, x^3, x^2y, xy^2, y^3, x^3y, xy^3 \rfloor \tag{9.2.1}$$

which is an incomplete quartic because the terms x^4, x^2y^2, and y^4 are absent. The element has three d.o.f. at each corner i: displacement w_i and slopes $w_{,xi}$ and $w_{,yi}$ (Fig. 9.2.1*a*). Expressed in terms of nodal d.o.f., the displacement field is [4.1]

$$\begin{aligned}
w &= \lfloor \mathbf{N} \rfloor \{w_1 \;\; w_{,x1} \;\; w_{,y1} \;\; w_2 \cdots w_{,y4}\}, \qquad \text{where} \\
\lfloor \mathbf{N} \rfloor &= \lfloor \mathbf{N}_1 \;\; \mathbf{N}_2 \;\; \mathbf{N}_3 \;\; \mathbf{N}_4 \rfloor, \qquad \text{and} \\
\lfloor \mathbf{N}_1 \rfloor &= \frac{TV}{32} \lfloor 2TV - 2SU - 4ST - 4UV, \;\; -4aST, \;\; -4bUV \rfloor \\
\lfloor \mathbf{N}_2 \rfloor &= \frac{SV}{32} \lfloor 2SV - 2TU + 4ST + 4UV, \;\; -4aST, \;\; 4bUV \rfloor \\
\lfloor \mathbf{N}_3 \rfloor &= \frac{SU}{32} \lfloor 2SU - 2TV - 4ST - 4UV, \;\; 4aST, \;\; 4bUV \rfloor \\
\lfloor \mathbf{N}_4 \rfloor &= \frac{TU}{32} \lfloor 2TU - 2SV + 4ST + 4UV, \;\; 4aST, \;\; -4bUV \rfloor \\
&\quad S = (x + a)/a \qquad T = (x - a)/a \\
&\quad U = (y + b)/b \qquad V = (y - b)/b
\end{aligned} \tag{9.2.2}$$

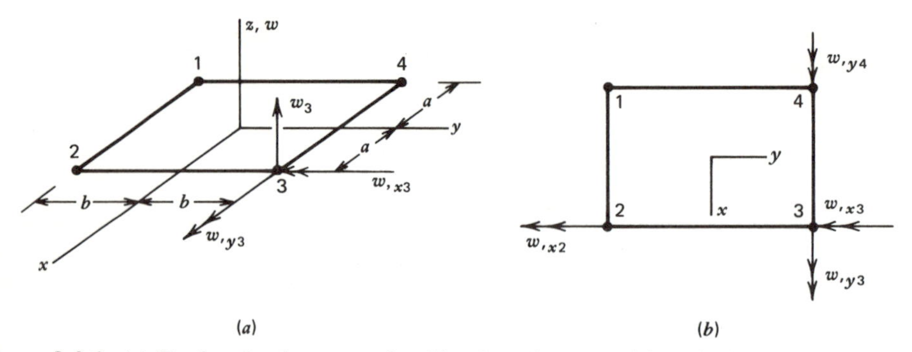

Figure 9.2.1. (*a*) Twelve d.o.f. rectangular thin-plate element, with typical nodal d.o.f. shown at node 3. (*b*) D.o.f. used to explain the uniqueness-compatibility conflict.

The stiffness matrix follows from Eq. 9.1.14 and is explicitly stated in Refs. 9.2, 9.3, and others. This element satisfies all convergence requirements and performs well, but it is incompatible (Problem 9.4).

The "uniqueness-compatibility" conflict can be explained with reference to the foregoing element and Fig. 9.2.1*b*. *If adjacent elements are to be compatible,* $w,_x$ along edge 2–3 must depend on only $w,_{x2}$ and $w,_{x3}$, while $w,_y$ along edge 3–4 must depend on only $w,_{y3}$ and $w,_{y4}$. Therefore $w,_{xy}$ along 2–3 is influenced by $w,_{x2}$, and $w,_{xy}$ along 3–4 is influenced by $w,_{y4}$. Since $w,_{x2}$ and $w,_{y4}$ are independent, the two values of $w,_{xy}$ at node 3 usually differ. *In general,* if nodal d.o.f. in a thin-plate element consist of only w and its first derivatives, then uniquely defined twist at the corners and interelement compatibility are mutually exclusive properties [9.4]. This annoyance is avoided by adding $w,_{xy}$ to the list of nodal d.o.f. But elements with twist and curvature as nodal d.o.f. have the drawbacks noted in Section 7.5.

Next, consider triangular elements, with corner nodes only and three d.o.f. at each. A complete cubic contains 10 terms.

$$w = \lfloor 1, x, y, x^2, xy, y^2, x^3, x^2y, xy^2, y^3 \rfloor \{\mathbf{a}\} \tag{9.2.3}$$

We need only 9 terms. Terms through y^2 must be kept if the element is to have rigid-body motion and constant-curvature capabilities. If we discard the xy term, we obtain an element that converges to the wrong answer as the mesh is refined. If we discard any one of the last four terms, we favor x or y, and so lose invariance. Equating coefficients to yield the cubic terms

$$a_7x^3 + a_8(x^2y + xy^2) + a_9y^3 \tag{9.2.4}$$

is specious: the polynomial is rendered incomplete, so the displacement along an edge is a function of its orientation. Indeed, $[\mathbf{k}]$ is then singular for some element shapes (element T). If all 10 terms are kept and one d.o.f. is condensed after $[\mathbf{k}]$ is formed, we obtain an invariant element—but one that converges to wrong answers [9.3], for reasons yet to be made clear (element T10). A satisfactory nine d.o.f. plate element (element BCIZ) was at last generated with the aid of area coordinates [9.5]. It is incompatible but works much better than a nine d.o.f. compatible element. (Reference 9.21 explains why some of these triangular elements have behaved badly, and suggests a simple improvement.)

When we speak of *thick* plates and *sandwich* plates instead of thin plates, it is understood that transverse shear deformations are to be included. Let θ_y and θ_x be the rotation components, about $+x$ and $-y$ directions, respectively, of a straight line that was normal to the midsurface of the undeformed plate. Shear strains in the plate are

$$\gamma_{yz} = w_{,y} - \theta_y \qquad \gamma_{zx} = w_{,x} - \theta_x \tag{9.2.5}$$

(This is called "Mindlin plate theory.") In general, $w_{,y} \neq \theta_y$ and $w_{,x} \neq \theta_x$: equality prevails only when $\gamma_{yz} = \gamma_{zx} = 0$, that is, when the plate is thin. When γ_{zx} and γ_{yz} do not vanish, it is θ_x and θ_y that must match between elements and must be used as nodal d.o.f., not $w_{,x}$ and $w_{,y}$. Surface slopes $w_{,x}$ and $w_{,y}$ may change discontinuously if adjacent elements have different shear moduli (Fig. 9.2.2*a*).

One way to construct a thick-plate element is to assume independent fields for w, θ_x, and θ_y. From Fig. 9.2.2*a* we see that the displacement components of a point P, which is z units above the midsurface, are

$$w, \qquad u = -z\theta_x \qquad \text{and} \qquad v = -z\theta_y \tag{9.2.6}$$

It is assumed that w is independent of z. From Eq. 1.4.4 we can find [**B**] and hence [**k**]. Subsequent sections explain the procedure in more detail. A merit of this approach is that assumed fields need be only C_0 continuous, not C_1 continuous as for thin plates. (An element of this type is LR in Fig. 9.2.3 and Table 9.5.1 [9.10]. It uses four Gauss points for in-plane actions and one Gauss point for transverse shear actions.)

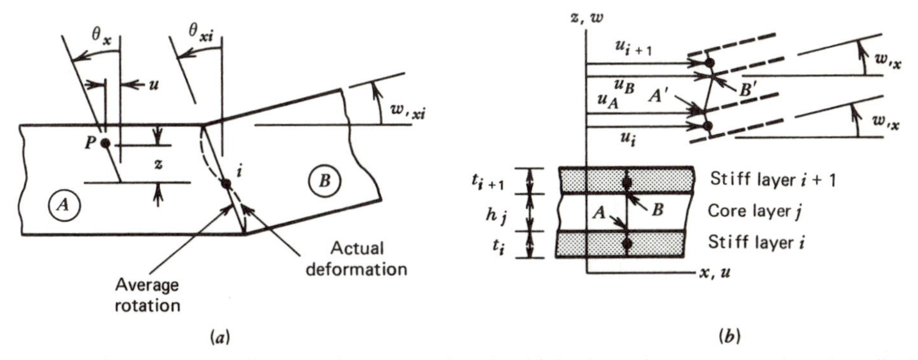

Figure 9.2.2. (*a*) Compatible rotations at node i in thick-plate elements A and B that display transverse shear deformation. (*b*) Deformations in three typical layers of a layered plate.

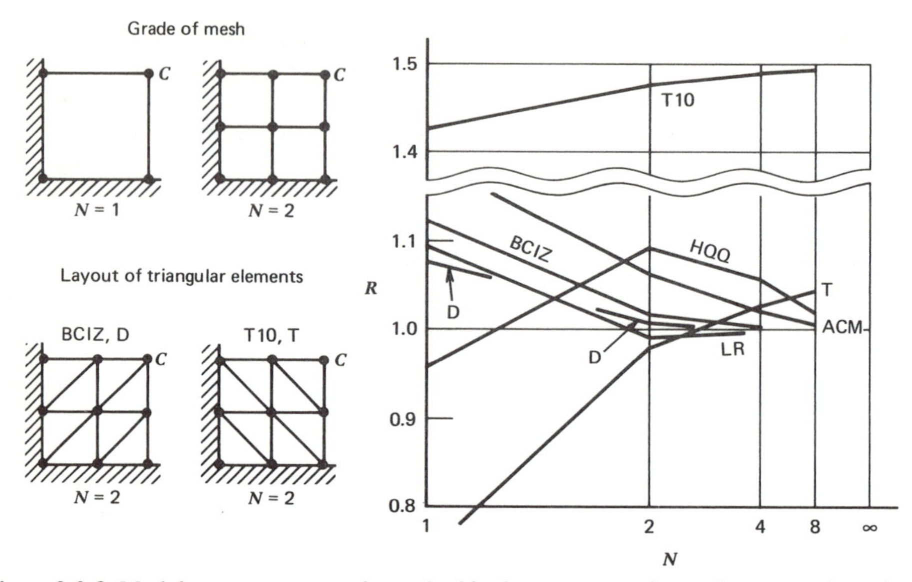

Figure 9.2.3. Mesh layouts on one quadrant of a thin, homogeneous, isotropic, square plate, simply supported and with a concentrated load at C. There are N^2 quadrilaterals or $2N^2$ triangles per quadrant. In the plot of R versus N, R is the ratio of computed deflection at C to the exact deflection at C.

Discrete Kirchhoff elements (element D) start with three independent fields, as in the foregoing paragraph. Then constraints $\gamma_{yz} = \gamma_{zx} = 0$ are imposed at a finite number of points to produce a thin-plate element [9.6]. Each constraint removes one d.o.f., so the initial displacement fields must contain more d.o.f. than are associated with the final $[\mathbf{k}]$.

In a layered plate, soft layers alternate with stiff layers. A special case, the sandwich plate, has one soft layer (the core) between two stiff layers (the facings). Each stiff layer resists both membrane and bending strains and so requires d.o.f. u_i, v_i, w_i, $w_{,xi}$, and $w_{,yi}$. In each layer, membrane action is treated as a plane stress problem in the xy plane, and bending action is accounted for by the thin-plate equation (Eq. 9.1.14). Core layers are assumed to resist only the transverse shear strains γ_{yz} and γ_{zx} where, for example, from Fig. 9.2.2*b*,

$$\gamma_{zxj} = w_{,x} + \frac{u_B - u_A}{h_j} = w_{,x} + \frac{1}{h_j}\left[\left(u_{i+1} + \frac{t_{i+1}}{2}w_{,x}\right) - \left(u_i - \frac{t_i}{2}w_{,x}\right)\right] \tag{9.2.7}$$

The u's, v's, and w's are interpolated from nodal d.o.f., and element matrices are derived in the usual way [9.11].

In the foregoing description there would be $5n$ d.o.f. per node for a plate with n stiff layers. But all layers are assumed to have the same $w = w(x, y)$. Thus there are $2n + 3$ d.o.f. per node: $\{w_1 \; w_{,x1} \; w_{,y1} \; u_1 \; v_1 \; u_2 \; v_2 \cdots u_n \; v_n\}$. If all t_i are small in comparison with the h_j, Eq. 9.2.7 simplifies. Also, if the t_i are small in comparison with the total thickness, the bending stiffness of the individual stiff layers becomes negligible, and $w,_{x1}$ and $w,_{y1}$ are not needed as nodal d.o.f.

It is possible to analyze a layered plate as if it were a thick plate with modified stiffness coefficients. We need only three d.o.f. per node (w, θ_{xi}, and θ_{yi}). But we find only an average shear stress in each soft layer if there is more than one.

Elements need not be based on assumed displacement fields. Mixed and hybrid variational principles are available (Section 16.1). They produce good elements for plates with and without transverse shear deformation (element HQQ of Ref. 16.16).

Which element is best? Attempts to answer [9.7, 9.8] lead us into a morass. Shall we judge stresses or deflections? Since performance is problem dependent, what will the test cases be? Should every element work for thin plates *and* for thick plates? Can we say that triangles, quadrilaterals, and elements with side node or higher-order d.o.f. are equally desirable? Are mixed and hybrid elements less worthy because their formulation is unfamiliar? Do some elements have shape restrictions, mesh layout restrictions, or other quirks? Shall we compare accuracy versus number d.o.f., versus number of elements, versus formulation time, versus NB^2 for [**K**], or versus something else? For what it may be worth, the behavior of some elements is displayed in Fig. 9.2.3. We see at least that elements may converge from above, from below, or not at all.

The *finite strip method* is the analysis procedure of Sections 8.4 through 8.6, applied to plates. If edges $y = 0$ and $y = L$ in Fig. 9.2.4*a* are simply supported, a typical strip has displacement

$$w = \sum \lfloor \mathbf{N} \rfloor \{\mathbf{d}\} \sin \frac{n\pi y}{L} \tag{9.2.8}$$

which is a trimmed-down form of Eqs. 8.5.1 through 8.5.4. Beam deflection functions serve in the x direction, so that the N_i are given by Eqs. 4.2.5, but with span a instead

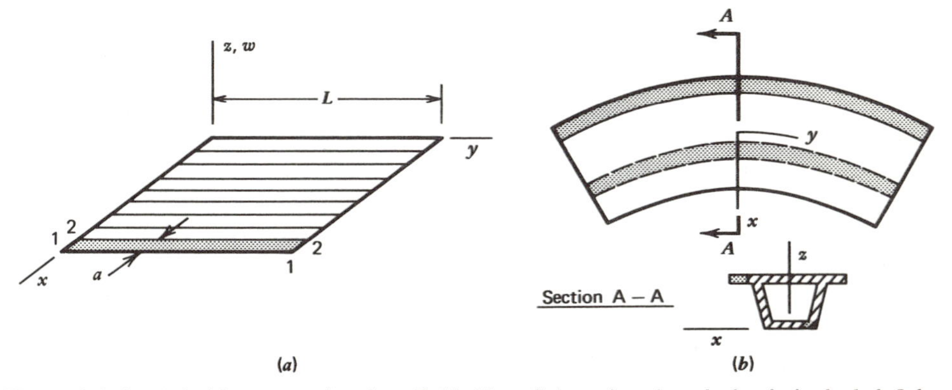

Figure 9.2.4. (*a*) A thin rectangular plate divided into finite strips. A typical strip is shaded. It has nodal lines 1–1 and 2–2. (*b*) A curved box-girder bridge. Typical finite strip elements are shaded.

of L and with $\{\mathbf{d}\} = \{w_1 \ w_{,x1} \ w_2 \ w_{,x2}\}$. The load is expressed as Fourier series. An analysis is made for each load component, and the results are superposed. In Fig. 9.2.4*b*, both bending and stretching are present, so $\{\mathbf{d}\}$ represents eight d.o.f.: $\{u_1 \ v_1 \ w_1 \ \theta_{x1} \ u_2 \ v_2 \ w_2 \ \theta_{x2}\}$.

In more general terms, the name finite strip is used for prismlike structures, whether they are plates, straight or curved folded plates (Fig. 9.2.4*b*), or solids (Fig. 8.7.1). Geometry and material properties are not functions of y. A deflection component, say w, is assumed to have the form

$$w = \sum X_n(x)\, Y_n(y) \tag{9.2.9}$$

where $X_n(x)$ is a polynomial in x, and $Y_n(y)$ is a function that satisfies boundary conditions. In Eq. 9.2.8, $Y_n(y) = \sin n\pi y/L$. Thus, analyses for separate load harmonics uncouple, as in Eq. 8.5.10. Less fortunate choices for $Y_n(y)$ lead to the coupling of all harmonics.

Advantages of the finite strip method include its modest requirements for computer resources and input data. Disadvantages, relative to finite elements, include in inability to cope with arbitrary geometry, boundary conditions, and material variation. Finite strips are discussed in several papers and in a book [9.12].

9.3 ISOPARAMETRIC PLATE ELEMENTS

We discuss elements that can be used for thin, thick, or sandwich plates [9.13, 9.14]. Transverse shear deformation is automatically accounted for, even if its influence is negligible. This section treats the displacement field and the $[\mathbf{B}]$ matrix. Subsequent sections explain the remainder of the element formulation.

Three-dimensional elements such as those of Section 5.7 can be made thin in one dimension to simulate a plate (Fig. 9.3.1*a*). But, as the element becomes thin, stiffness coefficients associated with thickness-direction strain ϵ_z become much larger than other coefficients. The discrepancy produces ill-conditioned equations (Section 15.5). Fortunately, d.o.f. that define strains ϵ_z are "excess," since ϵ_z is ignored in plate theory.

Solid elements are satisfactorily specialized to model plate bending by saying that (1) in-plane displacements u and v vary linearly with distance from the midsurface, and (2) on any thickness-direction line, all points have the same lateral displacement w. Thus, lines originally normal to the midsurface are assumed to remain straight and inextensible after deformation (but not normal to the midsurface unless transverse shear deformation is negligible). The deformation state is defined by the lateral displacement of the lines and their two rotation components. On the midsurface $z = 0$, membrane displacements u and v are zero.

Accordingly, after specialization, the elements appear as in Figs. 9.3.1*b* and 9.3.1*c*. Each node has three d.o.f.—one displacement and two rotations, as shown in Fig. 9.3.2*a*. (This direction of θ_{xi} is adopted to simplify notation and make explanations easier to

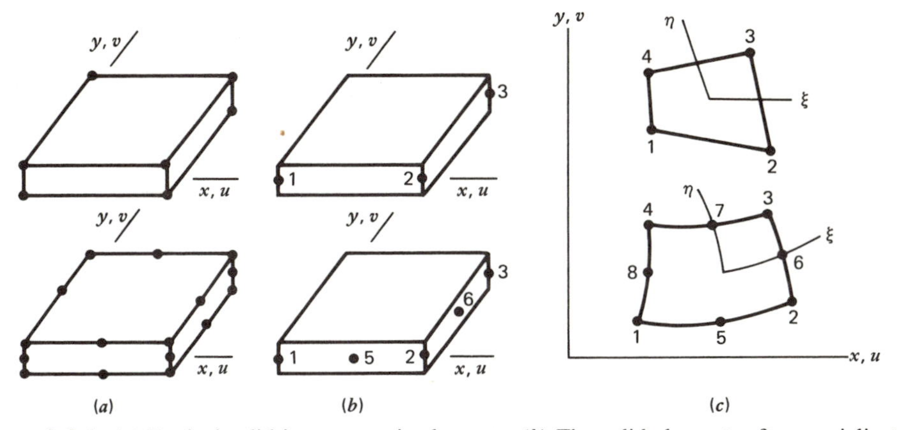

Figure 9.3.1. (*a*) Typical solid isoparametric elements. (*b*) The solid elements after specialization to plate elements. (*c*) Plan view, showing isoparametric coordinates $\xi\eta$ and elements of general shape.

follow. A general-purpose computer program will probably define positive nodal rotation vectors as pointing in the positive coordinate directions. This would necessitate changing signs of the θ_x columns of [**N**] and [**B**] and of the corresponding rows and columns of [**k**].)

Displacement w and rotations θ_x and θ_y within the element are interpolated from nodal d.o.f., as usual.

$$w = \sum N_i w_i \qquad \theta_x = \sum N_i \theta_{xi} \qquad \theta_y = \sum N_i \theta_{yi} \tag{9.3.1}$$

As in Fig. 9.2.2*a*, θ_x and θ_y are average rotation components of a line that was originally normal to the midsurface. The N_i are functions of ξ and η and are given by Eq. 5.3.3 for the four-node element and by Table 5.6.1 for the eight-node element. Thickness t can

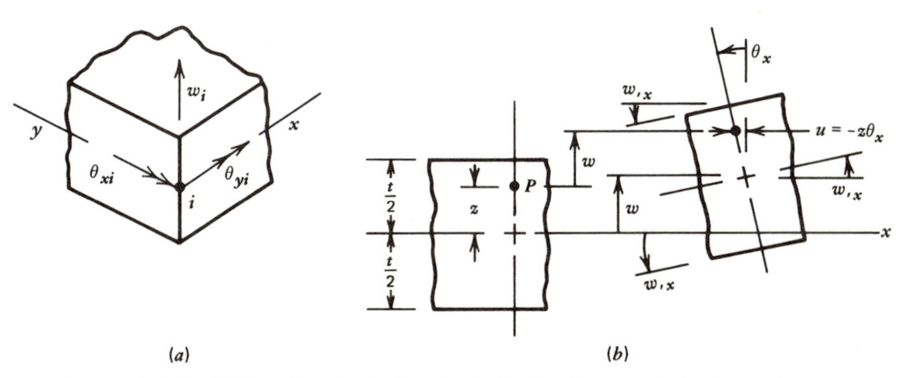

Figure 9.3.2. (*a*) D.o.f. at typical node *i*. (*b*) Displacements in the *xz* plane.

also be interpolated from nodal t_i: $t = \Sigma N_i t_i$. From Fig. 9.3.2*b*, displacements of any point P in the plate are

$$w, \qquad u = -z\theta_x \qquad \text{and} \qquad v = -z\theta_y \tag{9.3.2}$$

These displacements suffice to define all strains but ϵ_z, which is not needed.

The strain-displacement matrix **[B]** is derived by the same steps as seen in Eqs. 5.3.10 through 5.3.12. The strains are

$$\{\epsilon_x \; \epsilon_y \; \gamma_{xy} \; \gamma_{yz} \; \gamma_{zx}\} = [\mathbf{H}] \{u_{,x} \; u_{,y} \; u_{,z} \; v_{,x} \; v_{,y} \; v_{,z} \; w_{,x} \; w_{,y} \; w_{,z}\} \tag{9.3.3}$$

where **[H]** is 5 by 9 and is given by Eq. 5.7.2, but with row three deleted (because $\epsilon_z = 0$) and column 9 null (because $w_{,z} = 0$). The analogue of Eq. 5.3.11 is

$$\underset{9\times 1}{\{u_{,x} \; u_{,y} \cdots w_{,z}\}} = \underset{9\times 9}{[\Gamma \;\; \Gamma \;\; \Gamma]} \{u_{,\xi} \; u_{,\eta} \; u_{,z} \; v_{,\xi} \cdots w_{,\eta} \; w_{,z}\} \tag{9.3.4}$$

where

$$[\Gamma] = [\mathbf{J}]^{-1} \qquad \text{and} \qquad [\mathbf{J}] = \begin{bmatrix} x_{,\xi} & y_{,\xi} & 0 \\ x_{,\eta} & y_{,\eta} & 0 \\ 0 & 0 & 1 \end{bmatrix}$$

Finally, the analogue of Eq. 5.3.12 is

$$\begin{Bmatrix} u_{,\xi} \\ u_{,\eta} \\ u_{,z} \\ v_{,\xi} \\ v_{,\eta} \\ v_{,z} \\ w_{,\xi} \\ w_{,\eta} \\ w_{,z} \end{Bmatrix} = \sum_{i=1}^{N} \begin{bmatrix} 0 & -zN_{i,\xi} & 0 \\ 0 & -zN_{i,\eta} & 0 \\ 0 & -N_i & 0 \\ 0 & 0 & -zN_{i,\xi} \\ 0 & 0 & -zN_{i,\eta} \\ 0 & 0 & -N_i \\ N_{i,\xi} & 0 & 0 \\ N_{i,\eta} & 0 & 0 \\ 0 & 0 & 0 \end{bmatrix} \begin{Bmatrix} w_i \\ \theta_{xi} \\ \theta_{yi} \end{Bmatrix} \tag{9.3.5}$$

where N is the number of nodes in the element. The product of the rectangular matrices in Eqs. 9.3.3 through 9.3.5 yields the strain-displacement relation $\{\epsilon\} = [\mathbf{B}] \{\mathbf{d}\}$.

$$\begin{Bmatrix} \epsilon_x \\ \epsilon_y \\ \gamma_{xy} \\ \gamma_{yz} \\ \gamma_{zx} \end{Bmatrix} = \sum_{i=1}^{N} \left(\begin{bmatrix} 0 & 0 & 0 \\ 0 & 0 & 0 \\ 0 & 0 & 0 \\ b_i & 0 & -N_i \\ a_i & -N_i & 0 \end{bmatrix} + \begin{bmatrix} 0 & -za_i & 0 \\ 0 & 0 & -zb_i \\ 0 & -zb_i & -za_i \\ 0 & 0 & 0 \\ 0 & 0 & 0 \end{bmatrix} \right) \begin{Bmatrix} w_i \\ \theta_{xi} \\ \theta_{yi} \end{Bmatrix} \tag{9.3.6}$$

where

$$a_i = \Gamma_{11} N_{i,\xi} + \Gamma_{12} N_{i,\eta}$$
$$b_i = \Gamma_{21} N_{i,\xi} + \Gamma_{22} N_{i,\eta} \tag{9.3.7}$$

The complete matrix [**B**] is 5 by 12 for a four-node element, 5 by 24 for an eight-node element, and so on. The reason for arbitrarily splitting [**B**] so as to segregate terms linear in z is seen in what follows.

9.4 ISOPARAMETRIC PLATE ELEMENTS (CONTINUED)

We continue the development by integrating through the thickness. Integration in the xy plane is considered in Section 9.5.

In the stress-strain relation $\{\sigma\} = [\mathbf{E}]\{\epsilon\}$, we have

$$\{\sigma\} = \{\sigma_x \ \sigma_y \ \tau_{xy} \ \tau_{yz} \ \tau_{zx}\}$$

$$\{\epsilon\} = \{\epsilon_x \ \epsilon_y \ \gamma_{xy} \ \gamma_{yx} \ \gamma_{zx}\}$$

$$\underset{5\times 5}{[\mathbf{E}]} = \begin{bmatrix} E'_x & E'' & 0 & 0 & 0 \\ E'' & E'_y & 0 & 0 & 0 \\ 0 & 0 & G & 0 & 0 \\ 0 & 0 & 0 & G_{yz} & 0 \\ 0 & 0 & 0 & 0 & G_{zx} \end{bmatrix} \tag{9.4.1}$$

This is basically Eq. 9.1.3, augmented by shear moduli G_{yz} and G_{zx} to account for transverse shear deformation. If z is a principal material axis but x and y are not, then the only zeros in [**E**] are $E_{ij} = E_{ji} = 0$ for $i = 1, 2, 3$ and $j = 4, 5$. We presume that z is always a principal material axis. Then [**E**] is always composed of a 3-by-3 "in-plane" matrix and a 2-by-2 "transverse shear" matrix. We ignore ϵ_z and presume that $\sigma_z = 0$, as in a state of plane stress.

As suggested by Eq. 9.3.6, [**B**] can be split into parts $[\mathbf{B}_0]$ and $z[\mathbf{B}_1]$ that are respectively independent of z and linear in z. The stiffness matrix is given by Eq. 4.3.4, as usual.

$$[\mathbf{k}] = \int_V [\mathbf{B}]^T [\mathbf{E}][\mathbf{B}]\, dV = \iiint [\mathbf{B}_0 + z\mathbf{B}_1]^T [\mathbf{E}]\, [\mathbf{B}_0 + z\mathbf{B}_1]\, dx\, dy\, dz \tag{9.4.2}$$

Because of the distribution of zeros in $[\mathbf{B}_0]$, $[\mathbf{B}_1]$, and [**E**], the products $z[\mathbf{B}_0]^T [\mathbf{E}]\, [\mathbf{B}_1]$ and $z[\mathbf{B}_1]^T [\mathbf{E}]\, [\mathbf{B}_0]$ both vanish, $[\mathbf{B}_0]^T [\mathbf{E}]\, [\mathbf{B}_0]$ involves only the transverse shear portion of [**E**], and $z^2[\mathbf{B}_1]^T [\mathbf{E}][\mathbf{B}_1]$ involves only the in-plane portion of [**E**]. This is not a fortuitious circumstance: it is the natural result of our assumptions about the displacement field and the material properties. After integration from $z = -t/2$ to $z = t/2$,

$$[\mathbf{k}] = \iint [\mathcal{B}]^T [\mathcal{D}] [\mathcal{B}] \, dx \, dy = \iint [\mathcal{B}]^T [\mathcal{D}][\mathcal{B}] J \, d\xi \, d\eta \tag{9.4.3}$$

where J is the determinant of the Jacobian matrix, and $[\mathcal{B}]$ is

$$\underset{5 \times N}{[\mathcal{B}]} = \begin{bmatrix} 0 & -a_1 & 0 & 0 & -a_2 & 0 & \cdots \\ 0 & 0 & -b_1 & 0 & 0 & -b_2 & \cdots \\ 0 & -b_1 & -a_1 & 0 & -b_2 & -a_2 & \cdots \\ b_1 & 0 & -N_1 & b_2 & 0 & -N_2 & \cdots \\ a_1 & -N_1 & 0 & a_2 & -N_2 & 0 & \cdots \end{bmatrix} \tag{9.4.4}$$

where the a's and b's are defined by Eq. 9.3.7. If x and y are principal material axes, the 5-by-5 matrix $[\mathcal{D}]$ of a homogeneous plate is

$$\begin{aligned} &\mathcal{D}_{ij} \quad \text{are given by Eq. 9.1.6 or 9.1.7 for} \quad i, j = 1, 2, 3 \\ &\mathcal{D}_{ij} = \mathcal{D}_{ji} = 0 \quad \text{for } i = 1, 2, 3 \quad \text{and} \quad j = 4, 5 \\ &\mathcal{D}_{44} = G_{yz}t \qquad \mathcal{D}_{55} = G_{zx}t \end{aligned} \tag{9.4.5}$$

In the notation of Eq. 9.1.9, $\lceil \mathbf{S}' \rfloor = \lceil \mathcal{D}_{44} \; \mathcal{D}_{55} \rfloor$. If thickness t varies, $[\mathcal{D}]$ is a function of x and y.

Mechanics of materials theory shows that a shear force, say Q_x, produces the strain energy $1.2Q_x^2 \, dx \, dy/2G_{zx}t$ in a differential element $t \, dx \, dy$ of a homogeneous plate. Our plate formulation yields the strain energy $\gamma_{zx}^2 \mathcal{D}_{55} \, dx \, dy/2 = Q_x^2 \, dx \, dy/2G_{zx}t$. Accordingly, to get the same strain energy and to analyze thick plates more accurately, we should use $\mathcal{D}_{44} = G_{yz}t/1.2$ and $\mathcal{D}_{55} = G_{zx}t/1.2$ instead of the values given in Eq. 9.4.5.

Nonhomogeneous plates, either sandwich or layered, can be analyzed by adopting the appropriate $[\mathcal{D}]$. For example, consider an isotropic sandwich plate where h is the thickness of each facing, c is the thickness of the core, G is the shear modulus of the core, and E and ν are the elastic modulus and Poisson ratio of each facing [9.15]. Then

$$\begin{aligned} &\mathcal{D}_{11} = \mathcal{D}_{22} = \frac{\mathcal{D}_{12}}{\nu} = \frac{\mathcal{D}_{21}}{\nu} = \frac{Eh(c + h)^2}{2(1 - \nu^2)} \\ &\mathcal{D}_{33} = \frac{Eh(c + h)^2}{4(1 + \nu)}, \qquad \mathcal{D}_{44} = \mathcal{D}_{55} = \frac{G(c + h)^2}{c} \end{aligned} \tag{9.4.6}$$

and the remaining $\mathcal{D}_{ij}$ are zero [16.12]. This formulation presumes that a facing by itself has only membrane stiffness.

Let xyz be principal material axes, and let

$$\{\epsilon_0\} = \frac{2zT_0}{t}\{\alpha_x \;\; \alpha_y \;\; 0 \;\; 0 \;\; 0\} \tag{9.4.7}$$

where T_0 is the temperature at the surface $z = t/2$, and α_x and α_y are principal coefficients of thermal expansion. Nodal loads are given by Eq. 4.3.5 as

$$\{\mathbf{r}\} = \int_V [\mathbf{B}_0 + z\mathbf{B}_1]^T [\mathbf{E}] \{\epsilon_0\}\, dV = \iint [\mathcal{B}]^T [\mathcal{D}]\{\bar{\epsilon}\}\, dx\, dy \tag{9.4.8}$$

where $[\mathcal{D}]$ and $[\mathcal{B}]$ are given by Eqs. 9.4.4 and 9.4.5, and

$$\{\bar{\epsilon}\} = \frac{2T_0}{t}\{\alpha_x \;\; \alpha_y \;\; 0 \;\; 0 \;\; 0\} \tag{9.4.9}$$

If x and y are not principal material axes, the third terms in $\{\epsilon_0\}$ and $\{\bar{\epsilon}\}$ are also nonzero. If the plate is a sandwich plate, $[\mathcal{D}]$ is given by Eq. 9.4.6, and t in Eq. 9.4.9 is regarded as the effective thickness $c + h$.

When nodal d.o.f. have been calculated, strains can be found from Eq. 9.3.6, and stresses follow from Eqs. 1.6.1 and 9.4.1. Transverse shear stresses calculated in this way are average values. In a homogeneous (nonlayered) plate, their maximum values, at $z = 0$, are

$$\tau_{yz} = \frac{3}{2}(\tau_{yz})_{\text{ave}} = \frac{3}{2}\frac{Q_y}{t}, \qquad \tau_{zx} = \frac{3}{2}(\tau_{zx})_{\text{ave}} = \frac{3}{2}\frac{Q_x}{t} \tag{9.4.10}$$

Bending and twisting moments can be calculated as follows. Equations 9.1.2 and 9.3.2 indicate the following correspondence.

$$w_{,xx} = \theta_{x,x} \qquad w_{,yy} = \theta_{y,y} \qquad 2w_{,xy} = \theta_{x,y} + \theta_{y,x} \tag{9.4.11}$$

These are equations only in the sense that the derivatives of w and of θ_x and θ_y have the same effect on ϵ_x, ϵ_y, and γ_{xy} (and hence the same effect on M_x, M_y, and M_{xy}). So we can find moments in an isoparametric plate element by substituting "Eqs." 9.4.11 into Eqs. 9.1.5 where, from Eqs. 9.3.1 and 9.3.7,

$$\begin{aligned} \theta_{x,x} &= \sum_{i=1}^{N} a_i\theta_{xi} \qquad \theta_{y,y} = \sum_{i=1}^{N} b_i\theta_{yi} \\ \theta_{x,y} + \theta_{y,x} &= \sum_{i=1}^{N} (b_i\theta_{xi} + a_i\theta_{yi}) \end{aligned} \tag{9.4.12}$$

If the plate is a sandwich plate, t in Eq. 9.1.5 represents the effective thickness $c + h$.

Transverse shear forces (for example, Q_x) follow from Eqs. 9.1.1, 9.3.6, 9.4.1, and 9.4.5.

$$Q_x = \int_{-t/2}^{t/2} \tau_{zx}\, dz = tG_{zx}\gamma_{zx} = \mathcal{D}_{55} \sum_{i=1}^{N} (a_i w_i - N_i \theta_{xi}) \tag{9.4.13}$$

Coupling of Bending and Stretching. To make a point, we temporarily depart from strict plate bending and the isoparametric element. Imagine that stresses σ_x, σ_y, and τ_{xy} may have nonzero resultants in the xy plane. Forces and moments are then, from Eqs. 9.1.1 and 9.1.15,

$$\underset{6\times 1}{\{N_x\ N_y\ N_{xy}\ M_x\ M_y\ M_{xy}\}} = \int_{-t/2}^{t/2} \begin{Bmatrix} \boldsymbol{\sigma} \\ z\boldsymbol{\sigma} \end{Bmatrix} dz \tag{9.4.14}$$

where $\{\boldsymbol{\sigma}\} = \{\sigma_x\ \sigma_y\ \tau_{xy}\}$. Next introduce

$$\{\boldsymbol{\sigma}\} = [\mathbf{E}]\{\boldsymbol{\epsilon}\} \qquad \text{and} \qquad \{\boldsymbol{\epsilon}\} = \{\boldsymbol{\epsilon}_m\} - z\{\boldsymbol{\kappa}\} \tag{9.4.15}$$

where curvatures $\{\boldsymbol{\kappa}\}$ are those used in Eq. 9.1.12 and $\{\boldsymbol{\epsilon}_m\}$ are strains of the surface $z = 0$. Thus Eq. 9.4.14 can be written

$$\underset{6\times 1}{\begin{Bmatrix} \mathbf{N} \\ \mathbf{M} \end{Bmatrix}} = \int_{-t/2}^{t/2} \begin{bmatrix} \mathbf{E}(\boldsymbol{\epsilon}_m - z\boldsymbol{\kappa}) \\ z\mathbf{E}(\boldsymbol{\epsilon}_m - z\boldsymbol{\kappa}) \end{bmatrix} dz = \int_{-t/2}^{t/2} \begin{bmatrix} \mathbf{E} & -z\mathbf{E} \\ z\mathbf{E} & -z^2\mathbf{E} \end{bmatrix} dz \begin{Bmatrix} \boldsymbol{\epsilon}_m \\ \boldsymbol{\kappa} \end{Bmatrix} \tag{9.4.16}$$

Equation 9.4.16 shows that forces $\{\mathbf{N}\}$ are produced by $\{\boldsymbol{\epsilon}_m\}$ *and* by $\{\boldsymbol{\kappa}\}$ unless the integral of $z[\mathbf{E}]\ dz$ is zero. The same is true of moments $\{\mathbf{M}\}$; indeed, $\{\mathbf{M}\}$ can depend on $\{\boldsymbol{\epsilon}_m\}$ even if $\{\mathbf{N}\} = 0$. Membrane and bending actions decouple if $z[\mathbf{E}]\ dz$ integrates to zero, which happens only if material properties are either independent of z or symmetric with respect to the reference surface $z = 0$ (as is tacitly assumed in Eq. 9.4.2). Otherwise, $\{\mathbf{N}\}$ and $\{\mathbf{M}\}$ interact, and in-plane displacement u and v must be added to the list of nodal d.o.f. Then $\{\mathbf{d}\}$ contains all the d.o.f. needed for shell analysis. The coupling effect is pronounced in two-layer laminated plates [9.16].

If all E_{ij} in $[\mathbf{E}]$ vary with z in the same way, a reference surface $\mathfrak{z} = 0$ can be located, a distance b *away* from the midsurface, so that $z = b + \mathfrak{z}$. Hence

$$\int \mathfrak{z} E_{ij}\, d\mathfrak{z} = 0 \qquad \text{yields} \qquad b = \int zE_{ij}\, dz \Big/ \int E_{ij}\, dz \tag{9.4.17}$$

If $\mathfrak{z}$ replaces z in preceding equations, then the coupling terms again disappear. In elastic-plastic problems the $\mathfrak{z} = 0$ surface moves as yielding progresses. However, it is an ad hoc simplification to say that the elastic-plastic E_{ij} all vary with z in the same way.

9.5 ISOPARAMETRIC ELEMENTS FOR THICK OR THIN PLATES

The formulation in Sections 9.3 and 9.4 applies to thin plates as well as to thick plates but only if the sampling points for integration in the *xy* plane are properly chosen. To locate them, we begin with an argument that applies to the *quadratic* element and has physical appeal. Subsequent arguments are more general.

Consider a beam loaded by a linearly varying moment (Fig. 9.5.1*a*). Because of bending moment $M = M_0x/a$, end cross sections rotate an amount θ_B. Because of the constant transverse shear $Q_x = dM/dx = M_0/a$, *all* cross sections rotate an additional amount θ_S. Usually $t >> a$, so that $\theta_B >> \theta_S$. Now impose the preceding rotations as d.o.f. at nodes *A*, *M*, and *C* of an isoparametric element (Fig. 9.5.1*b*). Lateral displacement *w* vanishes for all *x*, because it can be only quadratic in *x* and must vanish at *A*, *M*, and *C*. Rotation θ is also quadratic in *x*:

$$\theta = -\left(\frac{\theta_B}{2} - \theta_S\right) + \frac{3\theta_B}{2}\left(\frac{x}{a}\right)^2 \tag{9.5.1}$$

In the isoparametric element, $\gamma_{zx} = w_{,x} - \theta$ and, since $w = 0$, $\gamma_{zx} = -\theta$ for all *x*. But in the actual beam, θ is much larger in magnitude than γ_{zx}. Therefore, under the prescribed d.o.f., the element stores an excess of strain energy in shear. If $a >> t$, the element is much too stiff and gives poor answers. But the element works well, even when thin, if quadrature points are located at places where θ happens to match the correct shear strain. So we set $\theta = \theta_S$ in Eq. 9.5.1 and find

$$x/a = \pm\sqrt{3}/3 = \pm 0.57735\cdots \tag{9.5.2}$$

which are precisely the Gauss point locations of an order 2 rule.

We introduce more general arguments by looking first at the beam analogue of a linear isoparametric plate element (Fig. 9.5.2). The following discussion complements the "parasitic shear" discussion in Section 7.6. Strains in the element of Fig. 9.5.2 are

$$\epsilon_x = u_{,x} = -z\theta_{,x} \qquad \text{and} \qquad \gamma_{zx} = w_{,x} - \theta \tag{9.5.3}$$

Strain energy in a beam of width *b* is

$$U = \frac{1}{2}\left[\frac{Ebt^3}{12}\int_0^L \theta_{,x}^2\,dx + \frac{Gbt}{1.2}\int_0^L (w_{,x} - \theta)^2\,dx\right] \tag{9.5.4}$$

In the usual way, we obtain from *U* a stiffness matrix $[\mathbf{k}] = [\mathbf{k}_b] + [\mathbf{k}_s]$, where $[\mathbf{k}_b]$ resists bending strain and $[\mathbf{k}_s]$ resists transverse shear strain. With this notation, for a structure,

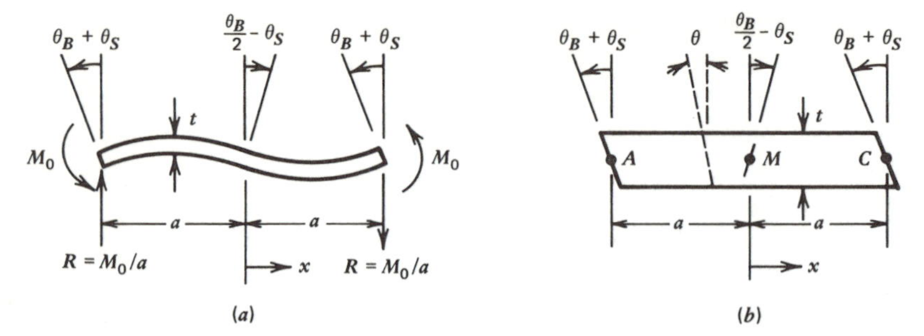

Figure 9.5.1. (*a*) A beam, simply supported at $x = -a$ and at $x = a$, shown deflected by end moments M_0. (*b*) A quadratic isoparametric beam element, shown deflected by the nodal d.o.f. of part (a). This element can also be regarded as the quadratic plate element of Fig. 9.3.1, seen in edge view.

$$([\mathbf{K}_b] + [\mathbf{K}_s]) \{\mathbf{D}\} = \{\mathbf{R}\} \tag{9.5.5}$$

Deflections $\{\mathbf{D}\}$ of a *thin* beam should be governed by only $[\mathbf{K}_b]$ because transverse shear deformation is negligible. We cannot simply discard $[\mathbf{K}_s]$ because it is all that couples the w and θ d.o.f. But, as a beam becomes slender, $[\mathbf{K}_s]$ grows in relation to $[\mathbf{K}_b]$. So $[\mathbf{K}_s]$ acts as a penalty function that yields $\{\mathbf{D}\} = 0$ unless $[\mathbf{K}_s]$ is singular, as described in Section 6.10.

In other words, when $Gbt/1.2$ becomes much larger than $Ebt^3/12$ in Eq. 9.5.4, the second integral enforces the constraint $\gamma_{zx} = 0$. Now, every additional element brings two nodal d.o.f. with it. If integrated by a two-point rule, it also brings two $\gamma_{zx} = 0$ constraints. So the mesh locks because the addition of nodes and elements does not increase the number of "free" d.o.f. But if elements are integrated by a one-point rule, each element adds only one constraint: $[\mathbf{K}_s]$ is singular and the element works well. (Since $\theta_{,x}$ is constant, one-point quadrature is exact for $[\mathbf{K}_b]$ of the beam element.) These arguments [9.17, 11.53] are illustrated by Problems 9.22 to 9.24.

The preceding arguments, made with reference to beams, also apply to isotropic plates. Reduced integration for transverse shear terms makes $[\mathbf{K}_s]$ singular (*general* material

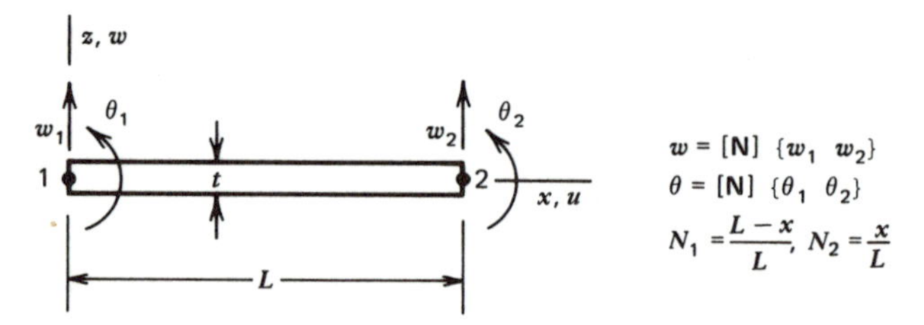

Figure 9.5.2. Beam analogue of a linear isoparametric plate element. This beam can model constant moment but not linearly varying moment.

properties may couple transverse shear with other modes, as in Eq. 9.4.16). Of the many possible elements [9.10, 9.17, 9.18], three are cited in Table 9.5.1. Despite the mixture of quadrature rules, element 3 is invariant [9.19].

As usual, low-order quadrature rules may trouble us with zero-energy deformation modes. In Table 9.5.1, their presence is indicated if (order) - (rank) $>$ 3, since three rigid-body modes are possible. Element 3 has none. Element 2 has one, of the type described by Eq. 5.10.2, so it is of no consequence. Element 1 has two. We see from Fig. 9.5.3 that the first is of no consequence because, as in Eq. 5.10.2, two elements that share an edge cannot have this mode at the same time, so the *structure* $[\mathbf{K}]$ has no such zero-energy mode. The hourglass mode can cause trouble. For example, if a rectangular plate has only corner supports, its $[\mathbf{K}]$ is singular.

Despite reduced integration, very thin elements can still fail by becoming too stiff or too flexible because of numerical difficulties. (If the plate is thin, we implicitly generate a discrete Kirchhoff element by imposing the constraints $\gamma_{yz} = \gamma_{zx} = 0$ at Gauss points by penalty functions, with their attendant difficulties.) A remedy is to arbitrarily reduce the transverse shear stiffness if elements are too thin [9.10, 9.17]. Thus, computed transverse shear deformation is bigger than strict theory requires, yet negligibly small. Equation 9.5.4 shows that the ratio of $[\mathbf{k}_s]$ to $[\mathbf{k}_b]$ is of order (L^2/t^2), where t is thickness and L is element length or width, whichever is larger. If, for example, results deteriorate when element dimensions exceed $L/t = 1000$, the shear moduli G_{yz} and G_{zx} of all elements for which $L/t > 1000$ can be reduced by multiplying these moduli by $(1000t/L)^2$. Whether or not $L/t = 1000$ is "too thin" depends on the test case and the number of digits carried.

TABLE 9.5.1. EFFECTS OF REDUCED INTEGRATION. ELEMENTS 1 AND 2 APPEAR IN FIG. 9.3.1. ELEMENT 1 IS THE SAME AS ELEMENT LR OF FIG. 9.2.3. ELEMENT 3 [9.18] USES NODAL D.O.F. w_i, θ_{xi}, and θ_{yi} AT THE EIGHT BOUNDARY NODES BUT ONLY θ_{x9} AND θ_{y9} AT THE CENTER. SHAPE FUNCTIONS FOR θ_x AND θ_y FOLLOW FROM EQS. 4.2.12 (SEE PROBLEM 7.19).

Element	Gauss Quadrature	d.o.f.[a]	Constraints[b]	$[\mathbf{k}_b] + [\mathbf{k}_s]$
1.	2×2 for $[\mathbf{k}_b]$ 1×1 for $[\mathbf{k}_s]$	3	2×1 $= 2$	Order 12 Rank 7
2.	2×2 for both $[\mathbf{k}_b]$ and $[\mathbf{k}_s]$	9	2×4 $= 8$	Order 24 Rank 20
3.	3×3 for $[\mathbf{k}_b]$ 2×2 for $[\mathbf{k}_s]$	11	2×4 $= 8$	Order 24[c] Rank 21[c]

[a]The number of nodal d.o.f. each element brings with it when added to a large mesh.
[b]The number of shear constraints $\gamma_{yz} = \gamma_{zx} = 0$ for a thin element.
[c]After condensation of internal d.o.f. θ_{x9} and θ_{y9}.

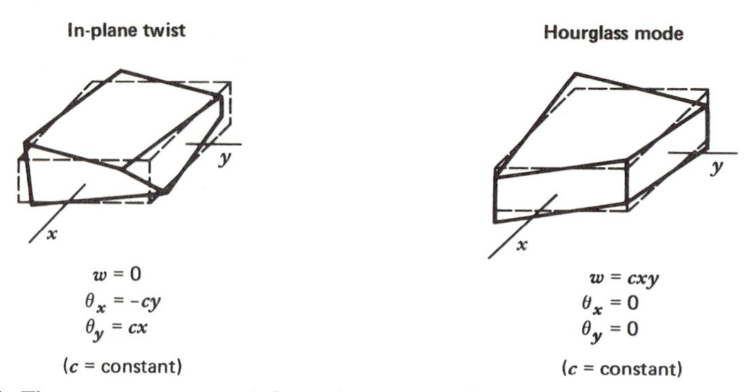

Figure 9.5.3. The two zero-energy deformation modes of element 1 in Table 9.5.1. Dashed lines show the undeformed configuration.

9.6 TEST CASES FOR PLATE BENDING

The behavior of a new or unfamiliar element should be studied by the patch test. Elements for *thin* plates must be able to assume constant states of $w_{,xx}$, $w_{,yy}$, and $w_{,xy}$. These states produce constant strains ϵ_x, ϵ_y, and γ_{xy} in any z = constant surface. Analogous constant states in plates that account for transverse shear deformation are constant values of [9.19]

$$\begin{aligned} &\theta_{x,x} \quad \theta_{y,y} \quad \text{and} \quad \theta_{x,y} + \theta_{y,x} \\ &\qquad \text{(yield constant values of } \epsilon_x, \epsilon_y, \text{ and } \gamma_{xy} \text{ on } z = \text{constant)} \\ &w_{,y} - \theta_y \quad \text{and} \quad w_{,x} - \theta_x \\ &\qquad \text{(yield constant values of } \gamma_{yz} \text{ and } \gamma_{zx}\text{)} \end{aligned} \tag{9.6.1}$$

Elements that satisfy these conditions, but not the patch test for constant $w_{,xx}$, $w_{,yy}$, and $w_{,xy}$, still converge properly when used for thin plates.

Boundary conditions for plates are as follows (see Fig. 9.6.1*a*). If the plate is thin, θ_n and θ_s are replaced by $w_{,n}$ and $w_{,s}$, respectively.

$$\begin{array}{cccc} \textit{clamped} & \textit{free} & \textit{simply supported (1)} & \textit{simply supported (2)} \\ w = 0 & Q = 0 & w = 0 & w = 0 \\ \theta_s = 0 & M_{ns} = 0 & M_{ns} = 0 & \theta_s = 0 \\ \theta_n = 0 & M_n = 0 & M_n = 0 & M_n = 0 \end{array} \tag{9.6.2}$$

Quantities not prescribed are unknowns to be computed. At nodes on a clamped edge we prescribe $w = \theta_s = \theta_n = 0$ (or its equivalent, $w = \theta_x = \theta_y = 0$). After solving for unknown d.o.f., we can compute shear force Q and moments M_s and M_n. On a free edge we prescribe *none* of the d.o.f. w, θ_s, and θ_n. We hope to compute $Q = M_{ns} = M_n = 0$ but actually will only approach this condition with mesh refinement.

Simply supported conditions 2 in Eqs. 9.6.2 are demanded by classical thin-plate theory. But in finite element work, if we use thin-plate elements (that demand C_1 continuity) and a curved edge is approximated by straight line segments, the condition

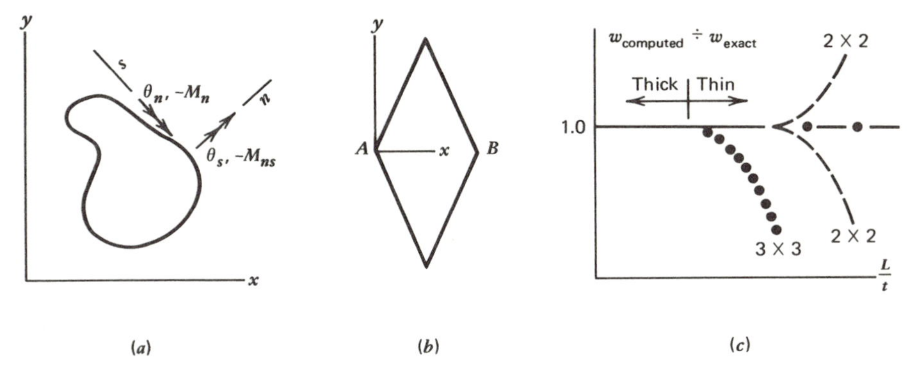

Figure 9.6.1. (*a*) Rotations θ_n and θ_s and moments M_n and M_{ns}, where n and s are directions normal and tangent to the plate boundary. (*b*) Simply supported rhombic plate. (*c*) Behavior as the span-to-thickness ratio L/t is increased, while the number of quadratic isoparametric plate elements is held constant.

$w_{,s} = 0$ implies $w_{,n} = 0$ as well. Thus mesh refinement converts a simple support to a clamped edge! This pitfall is avoided by using instead elements that demand only C_0 continuity and simply supported conditions 1 in Eqs. 9.6.2, where we prescribe only $w_i = 0$ for nodes i on a simple support [9.19]. (Other options are to include a singularity element or higher-derivative nodal d.o.f.) A case in point is a rhombic plate (Fig. 9.6.1*b*). At obtuse corners A and B, theory predicts the M_x approaches negative infinity while M_y approaches positive infinity. Conditions 1 yield this behavior but conditions 2 often do not [9.19]. If corner angles are 90°, either set of simply supported conditions in acceptable. See also the Babuska paradox (Section 15.4).

Figure 9.6.1c shows the effect of making isoparametric plate elements thinner. Consider, for example, element 2 of Table 9.5.1. If integrated with a 3-by-3 Gauss rule, the element fails as soon as transverse shear deformation in the plate becomes negligible. If integrated with a 2-by-2 Gauss rule, the element works into the thin-plate regime but eventually diverges up or down. As shown by the dot-and-dash line, correct thin-plate behavior can be maintained by using 2-by-2 integration and the shear stiffness modification suggested in the last paragraph of Section 9.5.

Table 9.6.1 gives numerical examples [9.10, 9.17, 9.18]. Elements 1, 2, and 3 are those of Table 9.5.1. Element 4 is element 2 augmented by a single bubble function internal d.o.f., so its displacement field is [9.20]

$$\theta_x = \sum_1^8 N_i \theta_{xi} \qquad \theta_y = \sum_1^8 N_i \theta_{yi} \qquad w = \sum_1^8 N_i w_i + (1 - \xi^2)(1 - \eta^2) w_9 \tag{9.6.3}$$

where the N_i are the usual quadratic shape functions. Condensation removes w_9 before elements are assembled. This element produces a more realistic $\{\mathbf{r}\}$ caused by lateral pressure than an element that lacks w_9 (see Problem 9.30).

The element of Eqs. 9.6.3 admits a zero-energy deformation mode under 2-by-2 Gauss quadrature. Consider, for example, a rectangular element, with

TABLE 9.6.1. RATIO OF COMPUTED CENTER DEFLECTION TO EXACT CENTER DEFLECTION. THE PLATE IS SQUARE, HOMOGENEOUS, ISOTROPIC, WITH A SPAN-TO-THICKNESS RATIO OF 100, AND HAS CLAMPED EDGES AND A CONCENTRATED CENTER LOAD. N = NUMBER OF (SQUARE) ELEMENTS ALONG ONE SIDE OF A QUADRANT, AS IN FIG. 9.2.3. SEE TEXT FOR ELEMENT NOTATION.

N	element 1	element 2	element 3	element 4
1	≈ 0	1.10	1.10	1.12
2	0.87	0.66	0.97	1.01
4	0.97	0.98	1.00	1.00
8	0.99	—	—	—

$$\theta_x = \theta_y = 0, \qquad w = c(1 + \xi^2 + \eta^2 - 3\xi^2\eta^2) \tag{9.6.4}$$

where c is an arbitrary constant. Like the $w = cxy$ mode of Fig. 9.5.3, this mode can be troublesome in a corner-supported plate. The mode can be resisted by multiplying the twenty-fifth diagonal coefficient of $[\mathbf{k}]$—the one associated with w_9—by $(1 + e)$ before condensation, where e is a small number [9.20]. This is a penalty function technique that physically amounts to adding a soft spring, of stiffness $ek_{25,25}$, to resist w_9. Note that w_9 is the displacement at $\xi = \eta = 0$ *relative* to the value of w at $\xi = \eta = 0$ dictated by the eight-term summation in Eq. 9.6.3. This "e-device" does not inhibit the element's response to the original 24 d.o.f. But it inhibits w_9 enough to restrict the amplitude of the unwanted mode in the rare practical cases where there are so few supports that it is possible. Limited tests indicate that bending moments are accurately computed when $e = 0.004$.

As noted in Section 5.11, it is good practice to compute stresses at quadrature points, then extrapolate or interpolate to other points. Transverse shear strains γ_{yz} and γ_{zx} are apt to be wildly in error, except at $\xi = \eta = 0$ in element 1 of Table 9.6.1, and at the Gauss points of a 2-by-2 rule in elements 2, 3, and 4.

Which of these elements is best? The "e-modified" form of element 4 seems quite good. So does element 3—it is more expensive to generate but more foolproof to use. See also Problem 9.30.

PROBLEMS

9.1 (a) Verify Eqs. 9.1.5.
(b) Verify Eqs. 9.1.8.
(c) Verify Eqs. 9.1.10.
(d) Verify Eqs. 9.1.11.

9.2 (a) In Fig. 9.1.1*b* presume that the M's and Q's are functions of x and y, so that (for example) M_x acts along the edge $y = 0$ and $M_x + M_{x,x}\,dx$ acts along the parallel edge. Show that the equilibrium equations are

$$Q_{x,x} + Q_{y,y} + q = 0, \qquad M_{x,x} + M_{xy,y} = Q_x, \qquad M_{xy,x} + M_{y,y} = Q_y$$

(b) Hence, show that $M_{x,xx} + 2M_{xy,xy} + M_{y,yy} + q = 0$.

(c) Invoke Eqs. 9.1.5, with $T_0 = 0$ and isotropy assumed. Show that $\nabla^4 w = q/\mathfrak{D}$.

9.3 In polar coordinates the curvatures and twist of a thin plate are $\{\boldsymbol{\kappa}\} = \{\kappa_r \; \kappa_\theta \; 2\kappa_{r\theta}\}$, where

$$\kappa_r = w_{,rr} \qquad \kappa_\theta = \frac{w_{,\theta\theta}}{r^2} + \frac{w_{,r}}{r} \qquad \kappa_{r\theta} = \frac{w_{,r\theta}}{r} - \frac{w_{,\theta}}{r^2}$$

(a) Write an expression for $w = w(r, \theta)$ that contains three independent constants and expresses rigid-body motion. Show that $\{\boldsymbol{\kappa}\} = 0$ for this w.

(b) Write an expression for [**B**], analogous to that in Eq. 9.1.14.

9.4 Consider the 12 d.o.f. rectangular element of Eq. 9.2.2.

(a) Show that $w = w_i$ when $\lfloor \mathbf{N} \rfloor$ is evaluated at node i, where $i = 1, 2, 3, 4$.

(b) From Eq. 9.2.2, write the expression for $w_{,x}$ along $x = a$. Show that it yields $w_{,x} = w_{,x3}$ at $y = b$ and $w_{,x} = w_{,x2}$ at $y = -b$.

(c) Let there be an element in front of the element in Fig. 9.2.1*a*, so that the two elements share edge 2–3. By activating d.o.f. at nodes 1 and 4 only, show that the normal slope $w_{,x}$ is not compatible between elements.

(d) Equation 9.2.1 also implies that normal slopes $w_{,x}$ are not compatible between the two elements in part (c). Explain how.

(e) Now consider the *tangential* slope $w_{,y}$ between the two elements in part (c). What nodal d.o.f. should affect it? What degree polynomial is it? What does Eq. 9.2.1 imply about compatibility of $w_{,y}$?

9.5 A certain thin-plate element has w_i, $w_{,xi}$, $w_{,yi}$, and $w_{,xyi}$ as nodal d.o.f. at corner nodes i. It is to be used to model a plate with a step change in thickness. What treatment do you suggest to avoid enforcing too much continuity?

9.6 Imagine that the lateral displacement w of a triangular thin-plate element is taken as a complete quintic (21 terms). For each element shown, and without calculation, allocate d.o.f. to the nodes in a way that seems acceptable. Consider higher-order d.o.f. as needed. Is interelement compatibility achieved? Consider compatibility conditions on the edge $x = 0$.

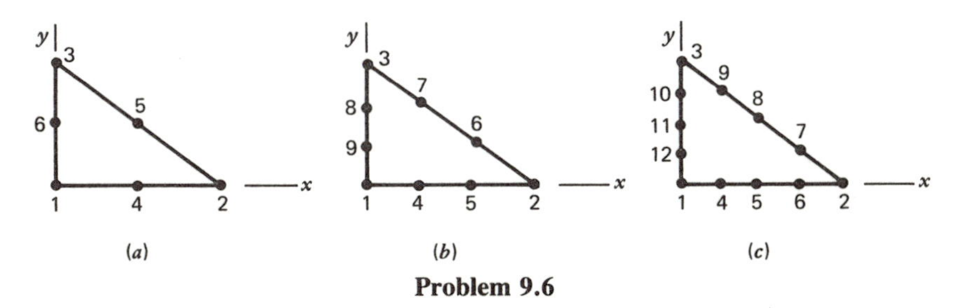

Problem 9.6

9.7 Imagine that the quadratic element (element 2 of Table 9.5.1) is to be given the "discrete Kirchhoff" treatment by explicitly enforcing zero shear strain at the Gauss points. How many d.o.f. do these constraints eliminate? What d.o.f. do you think appropriate to retain in $\{\mathbf{d}\}$, and why?

9.8 As an exercise, imagine that we need a stiffness matrix that represents only the resistance to γ_{zxj} of Eq. 9.2.7. If

$$u_i = \lfloor \mathbf{N}_u \rfloor \{\mathbf{u}_i\}, \qquad u_{i+1} = \lfloor \mathbf{N}_u \rfloor \{\mathbf{u}_{i+1}\}, \qquad w = \lfloor \mathbf{N}_w \rfloor \{\mathbf{w}\}$$

write $[\mathbf{N}]$ in the expression $\{u_i \ u_{i+1} \ w\} = [\mathbf{N}]\{\mathbf{d}\}$, in terms of $\lfloor \mathbf{N}_u \rfloor$ and $\lfloor \mathbf{N}_w \rfloor$. Then write $\lfloor \mathbf{Q} \rfloor$ in the expression $\gamma_{zxj} = \lfloor \mathbf{Q} \rfloor \{u_i \ u_{i+1} \ w\}$. What are the symbolic expressions for $\lfloor \mathbf{B} \rfloor$ and $[\mathbf{k}]$?

9.9 In Fig. 9.2.3, which subdivision of a square into triangles do you think is better? Why?

9.10 The elements of Fig. 9.3.1*b* are "superparametric." Why?

9.11 Show that if z is parallel to ζ and normal to the $\xi\eta$ plane, and line 7-8 in Fig. 5.7.1*a* is constrained to remain straight and inextensible, with $u = 0$ at $z = 0$, then the N_i of Section 5.7 yield $u = \ldots -zN_3\theta_{x3} + \ldots$ and $w = \ldots + N_3w_3 + \ldots$, where N_3 is given by Table 5.6.1. (This exercise uses N_3 as an example that illustrates the conversion of a solid element to a plate element.)

9.12 Verify that Eqs. 9.3.3 through 9.3.5 yield Eq. 9.3.6.

9.13 By inspection (of Fig. 9.3.2*b*, for example), write expressions for γ_{yz} and γ_{zx} in terms of $w_{,x}$, $w_{,y}$, θ_x, and θ_y. Then show that Eqs. 9.3.2 and 9.3.3 yield the same expressions for γ_{yz} and γ_{zx}.

9.14 Equations 9.3.2 suggest that surface-parallel strains ϵ_x, ϵ_y, and γ_{xy} depend only on θ_x and θ_y. Can it be true that lateral displacement w has no effect on these strains? Explain.

9.15 (a) In a sandwich plate, the average transverse shear strain γ is related to the core shear strain γ_c by $(c + h)\gamma = c\gamma_c$. Derive this expression.
(b) Derive the expression for $\mathcal{D}_{44}$ in Eq. 9.4.6. *Suggestion*. Consider strain energy.
(c) Work from the bending stiffness EI of a beam and derive the expression for $\mathcal{D}_{11}$ in Eq. 9.4.6, except for the $(1 - \nu^2)$ factor.
(d) Use Eq. 9.1.7 to get $\mathcal{D}_{33}$ in Eq. 9.4.6 from $\mathcal{D}_{11}$.

9.16 (a) Verify that Eq. 9.4.3 follows from Eq. 9.4.2.
(b) Verify that the second integral in Eq. 9.4.8 follows from the first.
(c) Why can you say, by inspection, that $[\mathbf{B}_0]$ has no effect on $\{\mathbf{r}\}$ when $\{\boldsymbol{\epsilon}_0\}$ is given by Eq. 9.4.7?
(d) What are the dimensions (units) of the coefficients in $\{\mathbf{r}\}$, as determined from the second integral in Eq. 9.4.8?
(e) Does Eq. 9.4.7 cause nodal forces to appear in $\{\mathbf{r}\}$? Explain.

9.17 (a) Verify that Eqs. 9.4.12 follow from Eqs. 9.3.1 and 9.3.7.
(b) Express Q_y in a form that corresponds to Q_x of Eq. 9.4.13.

9.18 Consider a beam that extends along the y-axis and is of width c in the x direction. Let $E = a + bz$ in Eq. 9.4.16, and evaluate N_y and M_y in terms of ϵ_y, $\theta_{y,y}$, a, b, c, and thickness t.

9.19 Let the six nodal d.o.f. $\{\mathbf{d}\} = \{w_A \ \theta_A \ w_M \ \theta_M \ w_C \ \theta_C\}$ in Fig. 9.5.1*b* be arbitrary, not prescribed. Of what degree in x are the polynomials for bending moment M and transverse shear Q_x? How *should* the polynomial degrees be related, according to elementary beam theory? What degree polynomial for Q_x is implied by two-point quadrature?

9.20 (a) Write the displacement fields $w = \lfloor \mathbf{N} \rfloor \{\mathbf{d}\}$ and $\theta = \lfloor \mathbf{N} \rfloor \{\mathbf{d}\}$ for a quadratic beam element (Fig. 9.5.1*b*). Let $\{\mathbf{d}\} = \{w_A \ \theta_A \ w_M \ \theta_M \ w_C \ \theta_C\}$, and let $\xi = x/a$.
(b) Impose $w_A = w_C = 0$ and $\theta_A = \theta_C = \theta_B + \theta_S$. Form expressions for ϵ_x and γ_{zx}, then use explicit integration to find Π_p in terms of E, G, θ_B, θ_S, w_M, θ_M, t, a, and b, where b is the width of the beam.
(c) Show that the equations $\partial \Pi_p / \partial w_M = \partial \Pi_p / \partial \theta_M = 0$ yield $w_M = 0$ and $\theta_M = \theta_S$ if $E >> G$, and $\theta_M = -\theta_B/4$ if $G >> E$.
(d) Finally, compute Π_p by a two-point Gauss rule and show that the stationary condition of Π_p yields $w_M = 0$ and $\theta_M = \theta_S - \theta_B/2$.

9.21 Consider the element of Fig. 9.5.2, integrated by one-point quadrature. What plane stress element is likely to have similar accuracy if a single layer of plane elements models a beam?

9.22 Use Eq. 9.5.4 to evaluate stiffness matrices for the beam element of Fig. 9.5.2 in the following ways.
(a) Evaluate $[\mathbf{k}_b]$ by one-point Gauss quadrature.
(b) Evaluate $[\mathbf{k}_s]$ by one-point Gauss quadrature. Call the result $[\mathbf{k}_s]_1$.
(c) Evaluate $[\mathbf{k}_s]$ by two-point Gauss quadrature. Call the result $[\mathbf{k}_s]_2$.

9.23 Use $[\mathbf{k}_b]$ and $[\mathbf{k}_s]_1$ from Problem 9.22 to model a one-element cantilever beam fixed at its left end. Load the right end by force P and moment. M. Solve for w_2 and θ_2. Investigate what happens as L becomes much larger than t.

9.24 Repeat Problem 9.23 using $[\mathbf{k}_s]_2$ instead of $[\mathbf{k}_s]_1$.

9.25 Suppose that the condition $\gamma_{zx} = 0$ is to be *explicitly* imposed at Gauss points of the beam elements in Figs. 9.5.1 and 9.5.2.
(a) What two d.o.f. can be removed from $\{\mathbf{d}\}$ for the element of Fig. 9.5.1? Write an appropriate 6-by-4 constraint matrix $[\mathbf{C}]$.
(b) If one Gauss point is used in Fig. 9.5.2, what constraint relation can be written? Can any d.o.f. be eliminated?
(c) Does the constraint relation of part (b) agree with the results of Problem 9.23?

9.26 Consider element 2 of Table 9.5.1. Under what circumstances or deformation states will both transverse shear strains be correctly evaluated (*a*) along $\xi = 0$, and (*b*) at $\xi = \eta = 0$? For simplicity, assume that $\xi = x$ and $\eta = y$.

9.27 Model a rectangular plate by six identical square elements, arranged in a 2-by-3 mesh. Sketch the mesh deformed into the hourglass mode of Fig. 9.5.3.

9.28 If element 1 of Table 9.5.1 uses one-point quadrature for *all* terms, two more zero-energy deformation modes appear. Sketch them. (Mathematical arguments are not needed.)

9.29 Let a constant z-direction traction p act on the lateral surface of a rectangular element. Compute the nodal loads $\{\mathbf{r}\}$ that result.
(a) Consider the four-node element of Fig. 9.3.1.
(b) Consider the eight-node element of Fig. 9.3.1. Let side nodes be evenly spaced.
(c) Consider the element of Eq. 9.6.3. Let side nodes be evenly spaced.

9.30 (a) In what way do the $\{\mathbf{r}\}$'s of Problem 9.29 seem unpromising? Consider, for example, using a single element to model a simply supported plate.
(b) How does the element of Eq. 9.6.3 alleviate the problem?
(c) Does element 3 of Table 9.5.1 also alleviate the problem?

9.31 (a) Show that Eq. 9.6.4 yields zero strains at the Gauss points of a 2-by-2 quadrature rule.
(b) What value of w_9 in Eq. 9.6.3 does this mode demand?
(c) Sketch an element in this mode and show by a sketch that a rectangular mesh with only corner supports is susceptible to it.

9.32 (a) Why should you be able to predict that actual calculation will yield the ≈ 0 entry in Table 9.6.1?
(b) For each entry in Table 9.6.1, how many d.o.f. are there in $\{\mathbf{D}\}$ after the d.o.f. that are fixed have been struck out? Include internal d.o.f. of elements in your count.
(c) Plot the ratios in Table 9.6.1 versus the number of d.o.f. counted in part (b).

9.33 How wide can a beam be before it should be regarded as a plate? Suggest at least one criterion. How might finite elements be used to answer the question?

CHAPTER 10

SHELLS

10.1 INTRODUCTION

Shells are curved, so it is reasonable that shell elements be curved. It makes as much sense, or as little, to model a shell by flat elements as to model a flat plate by curved elements. A curved shell element must model shell geometry and must combine membrane and bending actions in its displacement behavior. The element should contain rigid-body motion and constant strain capability, be mixable with other types of elements, and be accurate in a coarse mesh. It should also be without strains and curvatures as nodal d.o.f. and be easy to describe (say by only nodal coordinates and thicknesses). An element with so many merits is a chimera.

There are three options in shell element formulation.

1. A flat element, made by combining a membrane element with a plate bending element.
2. A curved element, formulated by shell theory.
3. A solid or degenerate solid element, analogous to the plate elements of Section 9.3.

Option 1 is easiest for both formulation and use, but accuracy is often mediocre and sometimes unacceptable. Option 2 is difficult, and the element may have many d.o.f., some of them higher-displacement derivatives (Section 7.5). Option 3 occupies the middle ground. It is discussed in Sections 10.5 and 10.6. Another product of Option 3 is the SemiLoof element for thin shells [10.1, 10.2]. Its 32 d.o.f. are u, v, and w at corner and midside nodes, and rotations (such as θ_n in Fig. 9.6.1) at two Gauss points on each edge. Its performance to date is good, but the unusual allocation of d.o.f. requires that edge beams, if present, be specially formulated. Option 2 has two drawbacks: shell theory is complicated, and theoreticians have proposed many different strain-displacement relations for the same geometry of shell because there is room for argument about which of many terms can be discarded as negligible. Shallow shell theory uses simple relations but, again, different theories are available, and the resulting elements may or may not work for deep shells. Elements for plane arches are worth studying because they display in a simpler context some of the difficulties associated with general shells [10.3, 10.4].

After choosing a shell theory, analysis options available for plates are again available for shells: discrete Kirchhoff, hybrid and mixed variational principles, thick or thin, and so on. (A thin shell, like a thin plate, is one for which transverse shear deformation effects are negligible.)

In testing and using shell elements, there are again special problems. Because thin shells are much stiffer in stretching than in bending, the stiffness matrix **[K]** has a high condition number. Numerical troubles are possible (Section 15.5). For example [10.2], if the diameter-to-thickness ratio of a shell is 1000, the condition number may exceed 10^{15}.

In this chapter we look at some options and problems associated with shell elements. We emphasize isoparametric shell elements that resemble the plate elements in Section 9.3. Some analysis details for shells of revolution are also presented.

10.2 FLAT ELEMENTS

Curved shell geometry can be approximated by a surface of triangular facets. It is easy to generate **[k]** for a flat element. Let the 18 nodal d.o.f. of the element in Fig. 10.2.1 be

$$\{\mathbf{d}\} = \{u_1\ u_2\ u_3\ v_1\ v_2\ v_3\ w_1\ w_2\ w_3\ \theta_{x1}\ \theta_{x2}\ \theta_{x3}\ \theta_{y1}\ \theta_{y2}\ \theta_{y3}\ \theta_{n1}\ \theta_{n2}\ \theta_{n3}\} \qquad (10.2.1)$$

We take the 6-by-6 membrane stiffness matrix $[\mathbf{k}_m]$ of a constant strain triangle, which operates on u and v d.o.f., and combine it with the 9-by-9 bending stiffness matrix $[\mathbf{k}_b]$ of a flat plate element, which operates on w, θ_x, and θ_y d.o.f. The result is

$$\underset{18\times 18}{[\mathbf{k}]} = \begin{bmatrix} \mathbf{k}_m & \mathbf{O} & \mathbf{O} \\ \mathbf{O} & \mathbf{k}_b & \mathbf{O} \\ \mathbf{O} & \mathbf{O} & \mathbf{O} \end{bmatrix} \qquad (10.2.2)$$

The three rows and three columns of zeros correspond to θ_n d.o.f., which are included to facilitate the next step: coordinate transformation of **[k]** from local coordinates to global. The transformed **[k]** is associated with nodal d.o.f. in global directions (Fig. 10.2.1*c*).

The flat element is simple to formulate, easy to describe by input data, easy to mix with other element types, and capable of rigid-body motion without strain. Thus it seems preferable to a doubly curved element for a general shell, whose geometric description alone may require 54 data items [10.2].

But a flat element has limited accuracy and several drawbacks. Membrane and bending actions are uncoupled within a single element, simply because it is flat. The necessary coupling for the entire shell comes about only because a membrane force N in one element exerts an element-normal force component on its neighbor (Fig. 10.2.1*c*). Thus, apart from requiring many elements to obtain accuracy, flat elements may display bending moments where there should be none, for example, in a cylinder under pressure (this

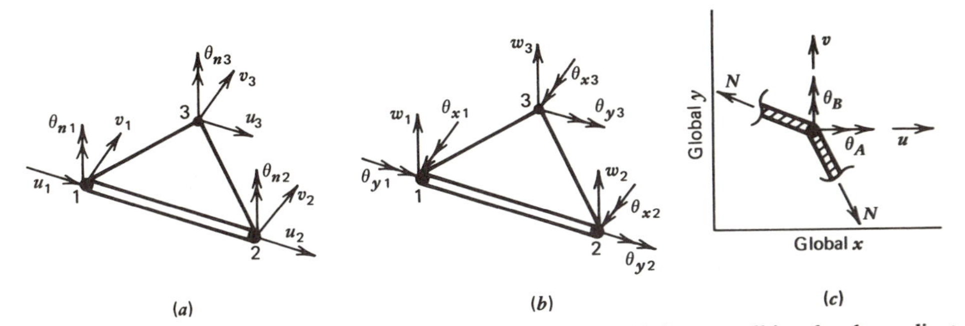

Figure 10.2.1 (*a*) Membrane and (*b*) bending d.o.f. of a flat shell element, all in a local coordinate system *xyz*. The element lies in the local *xy* plane. θ_n is a rotation about a normal to the element. (*c*)A typical structure node, showing global d.o.f. that lie in the global *xy* plane. *N* represents a membrane force.

discrepancy is greatly reduced if nodal moments are omitted from $\{\mathbf{r}\}$). Indeed, as the diameter-to-thickness ratio increases, the cylinder displays correct membrane forces but bizarre displacements! And the element utterly fails to model a slit cylinder under torsion (see Ref. 10.2 and Section 10.7).

By modifying the membrane response of the element just described, we obtain another 18 d.o.f. flat triangular element [10.34]. It cannot represent states of constant membrane strain and so fails a patch test. But it seems to provide results of engineering accuracy at reasonable cost.

The four-node quadrilateral element of Ref. 10.35 is the shell analogue of the four-node plate element in Fig. 9.3.1 and Table 9.5.1. It can be warped yet, like triangular elements, it provides a poor fit to shell geometry and has fold lines between elements. Nevertheless, it is simple and accurate (see Section 10.7).

What if all elements happen to be coplanar at a node? Then Eq. 10.2.2 indicates that the structure stiffness matrix will be singular: none of the elements that share the node resist θ_n, so $[\mathbf{K}]$ will have a null row and a null column. An obvious treatment is to change the diagonal coefficient from zero to some number α. Computed results will then show $\theta_n = 0$. Other d.o.f. will be independent of α, since θ_n is unloaded and not connected to any other d.o.f.

If elements at a node are *almost* coplanar, θ_n is not quite normal to any of the elements it shares. Accordingly, θ_n is weakly coupled to other d.o.f., and the near-singularity of $[\mathbf{K}]$ may lead to inaccuracy. So it is advisable to add a small number α to the appropriate diagonal coefficient in $[\mathbf{K}]$. Physically, this amounts to adding a soft torsional spring that helps resist θ_n. Numerical experiment aids in defining "almost coplanar" and "a small number α."

Another way to avoid singularity when elements are coplanar is to modify element $[\mathbf{k}]$ matrices before assembly. We replace the on-diagonal null matrix in Eq. 10.2.2 by the 3-by-3 matrix in the following equation [10.5].

$$\begin{Bmatrix} M_{n1} \\ M_{n2} \\ M_{n3} \end{Bmatrix} = \alpha EV \begin{bmatrix} 1.0 & -0.5 & -0.5 \\ -0.5 & 1.0 & -0.5 \\ -0.5 & -0.5 & 1.0 \end{bmatrix} \begin{Bmatrix} \theta_{n1} \\ \theta_{n2} \\ \theta_{n3} \end{Bmatrix} \tag{10.2.3}$$

where E is elastic modulus, V is element volume, and α is again a "small number." Equation 10.2.3 is a penalty function that provides each θ_n with a fictitious stiffness but offers no resistance to the mode $\theta_{n1} = \theta_{n2} = \theta_{n3}$ or to any other rigid-body motion. The lower limit of α is determined by the length of a computer word. Typically, $10^{-6} < \alpha < 10^{-2}$. Values as large as $\alpha = 1.0$ do not change results very much [10.5]. Reference 10.35 describes another penalty function method, not restricted to flat elements (that should be used with *four* Gauss points).

The possibility of a "θ_n singularity" can be avoided by not using θ_n as a d.o.f. Such a formulation needs only five d.o.f. per node: three translations and two rotations. The two rotation vectors are perpendicular to a shell-normal direction that must be defined at each node. The absence of θ_n does not imply a restraint of rotation about normals to the shell any more than the absence of rotational d.o.f. θ_z in the plane truss analysis of Chapter 2 implies restraint of rigid-body rotation in the xy plane.

10.3 RIGID-BODY MOTION

Displacement fields that are easy to write may not allow strainfree motion if an element is curved. For the elements of Fig. 10.3.1, let the displacements of a point P on the midsurface be

$$\begin{aligned} w &= a_1 + a_2 s + a_3 s^2 + a_4 s^3 \\ u &= a_5 + a_6 s \end{aligned} \tag{10.3.1}$$

where the a_i are generalized coordinates. Now consider a rigid-body translation, H units downward. For the element that spans 90°, we have

$$\begin{aligned} &\text{Actual displacement } u\text{:} \quad u = H \sin \frac{\pi s}{2L} \\ &\text{Predicted by Eq. 10.3.1:} \quad u = Hs/L \end{aligned} \tag{10.3.2}$$

The two expressions for u cannot be the same, so this arch element cannot model rigid-body motion. Shell elements present similar difficulties.

However, as angle θ in Fig. 10.3.1b decreases to zero, the element becomes straight, and Eqs. 10.3.1 yield the correct values: $u = H$ and $w = 0$. An element for thin shells of revolution is also based on Eqs. 10.3.1. As the angle θ subtended by each shell element decreases, there is convergence toward correct results. In other words, rigid-body motion capability is recovered by mesh refinement [10.6].

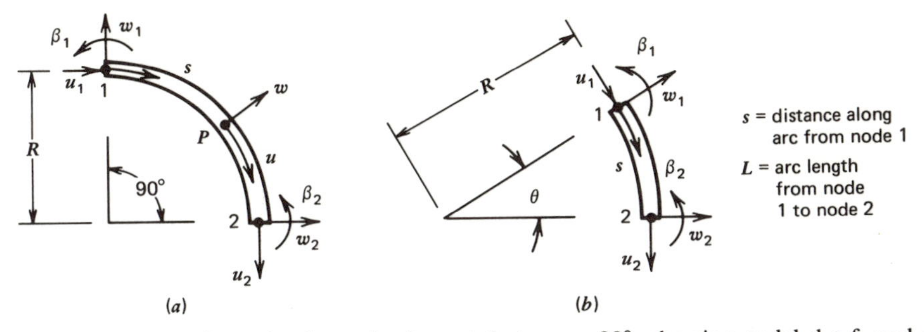

Figure 10.3.1 (*a*) A plane circular arch element that spans 90°, showing nodal d.o.f. and displacements u and w, tangent and normal to the element. (*b*) An arch element that spans $\theta°$ from the equator, showing nodal d.o.f. u_i, w_i, and β_i.

Additional modes make the arch element (and the axisymmetric shell element) more nearly able to represent rigid-body motion. Let the second of Eqs. 10.3.1 be replaced by

$$u = a_5 + a_6 s + a_7 s(L - s)$$

or by (10.3.3)

$$u = a_5 + a_6 s + a_7 s(L - s) + a_8 s^2(L - s)$$

where a_7 and a_8 are nodeless d.o.f. to be condensed before assembly. Thus the finite element solution approximates the sine curve of Eq. 10.3.2 by a parabola or by a cubic in s instead of by a straight line. Performance improves, as shown by Table 10.3.1 [10.7]. The eigenvalue quoted, which should be zero, is proportional to the strain energy in the element when its d.o.f. are assigned values consistent with rigid-body motion. The lowest eigenvalue associated with elastic distortion is on the order of 10^6.

An element that lacks rigid-body motion capability can be persuaded to behave properly by modifying its stiffness matrix [10.8]. However, the treatment is numerically sensitive and may hurt rather than help [10.9]. Fortunately, exact inclusion of rigid-body modes by trigonometric functions is not essential because approximate polynomial treatments, such as Eqs. 10.3.3 and the proper coupling of normal and tangential displacements [10.37], are so helpful.

TABLE 10.3.1. EIGENVALUES OF **[k]** ASSOCIATED WITH RIGID-BODY MOTION OF AN ARCH ELEMENT.

θ in Fig. 10.3.1*b*	$u = f(a_5, a_6)$	$u = f(a_5, a_6, a_7)$	$u = f(a_5, a_6, a_7, a_8)$
2°	2013.	0.05688	0.00005
6°	12177.	0.02387	0.00019
12°	26794.	0.21930	0.00618
20°	44388.	2.46182	0.08052

10.4 CHOICE OF SHELL THEORY AND DISPLACEMENT FIELD

In this section we discuss some of the theories on which a finite element formulation can be based.

An efficient shell carries load mainly by membrane action. Membrane theory ignores bending action. The displacement field need define only the membrane strains ϵ_r, ϵ_s, and γ_{rs}, where r and s coordinates lie in the midsurface. Rotations need not appear as nodal d.o.f. in a membrane shell (or in a portion of a shell where membrane action dominates). But a membrane is sensitive to its geometric description: if it contains (or approaches) a flat spot or a straight line, it will probably display a mechanism or a near-mechanism.

Shallow shell theory is simpler than general shell theory because calculations are done in a reference plane onto which the shell is projected. But shell curvature is not discarded, so membrane and bending actions interact, as they should.

Some concepts of shallow shell theory are easier to explain by way of a shallow *arch* element (Fig. 10.4.1*a*). Arch geometry is defined by $z = f(x)$. The arch is shallow if $z_{,x}^2 << 1$, where $z_{,x} = dz/dx$. Then the radius of curvature is $R = -1/z_{,xx}$. In Fig. 10.4.1*a*, $z_{,xx} < 0$ because the arch is concave down. For a thin arch ($t << \ell$), the midsurface strain ϵ_m and midsurface-normal rotation β are [10.4]

$$\epsilon_m = u_{,x} + \frac{w}{R} \qquad \beta = w_{,x} - \frac{u}{R} \tag{10.4.1}$$

Strain ϵ at a point $\mathfrak{z}$ units above the midsurface is

$$\epsilon = \epsilon_m - \mathfrak{z}\beta_{,x} = u_{,x} - wz_{,xx} - \mathfrak{z}(w_{,xx} + u_{,x}z_{,xx} + uz_{,xxx}) \tag{10.4.2}$$

where $\beta_{,x}$ is change in curvature produced by u and w. The strain equation simplifies if the arch is parabolic, since then $z_{,xxx} = 0$. Assumed fields for u and w may be taken as polynomials in x.

Equations 10.4.1 yield correct results for deep arches if each element is shallow. If the term u/R is discarded from the expression for β, the *entire* arch must be shallow. If

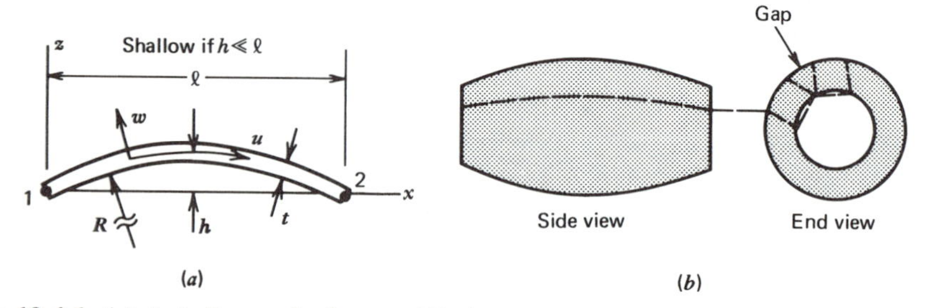

Figure 10.4.1. (*a*) A shallow arch element. Displacements u and w are, respectively, tangent and normal to the arch. (*b*) Gap created by projecting shell elements onto different reference planes.

we use simply $\beta = w_{,x}$ for a deep arch, mesh refinement produces convergence to wrong answers [10.2]. These remarks apply also to shallow shells, where there is another coordinate, another displacement component, another principal radius of curvature, and additional strains and curvature changes.

Figure 10.4.1*b* shows a minor problem in the shallow shell theory of doubly curved elements. Interelement gaps appear, but they shrink as the mesh is refined [10.10].

Cylindrical shells are closely related to arches. Elements for cylindrical shells are often restricted to circular cylinders, right angles at corner nodes, and two opposite edges parallel to the axis of the cylinder. A very effective element of this type starts with a strain field instead of a displacement field (see Ref. 10.11 and Table 10.7.1). This approach is not simple, but it permits exact representation of rigid-body and constant strain modes.

Let u and v be membrane displacements (here abbreviated uv), and let w be the surface-normal displacement of a general shell. In a thin-shell element, with a polynomial rather than a trigonometric displacement field, what should be the relative degrees of the polynominals for uv and for w? Different arguments yield different conclusions.

1. Convergence requirements A and B of Section 4.9 suggest that uv should be one degree less than w because strain energy is proportional to first derivatives of uv and to second derivatives of w.
2. If interelement compatibility is to prevail, all polynomials should be of the same degree. (Imagine a 90° fold line of a folded plate. Displacement w of one element edge is displacement u or v of the neighboring element edge.)
3. Hoop strain ϵ_θ in a cylindrical shell of radius r is $\epsilon_\theta = (v_{,\theta} + w)/r$. In order to assure the possibility that ϵ_θ can be constant, uv polynomials should be one degree higher than the w polynomial.

Limited numerical evidence suggests that doubly curved shells are modeled better by flat elements based on fields of equal degree than by doubly curved elements in which uv is of appreciably lower degree than w [10.12]. Also, an efficient shell carries most of the load by membrane action, which again suggests that uv fields be at least as competent as the w field.

Novozhilov, Love, Reissner, Vlasov, Sanders, Donnell, and others have proposed shell theories. The theories use different strain-displacement relations and therefore yield different finite elements. Some theories do not yield zero strain when a rigid-body motion is prescribed [10.13]. We elect a simple theory if we do not know which is best.

Figure 10.4.2 depicts the middle course between flat elements and elements derived by shell theory. A particular brick element is shown, but the concepts apply also to triangular elements and to elements with a different allocation of nodes. In Fig. 10.4.2*b*, surplus d.o.f. on thickness-direction edges have been eliminated (see Eqs. 5.7.5 through 5.7.8). The element is nevertheless able to model thickness-direction strains ϵ_ζ, which can be important in a very thick shell. Its stiffness matrix is formulated as if it were a solid element.

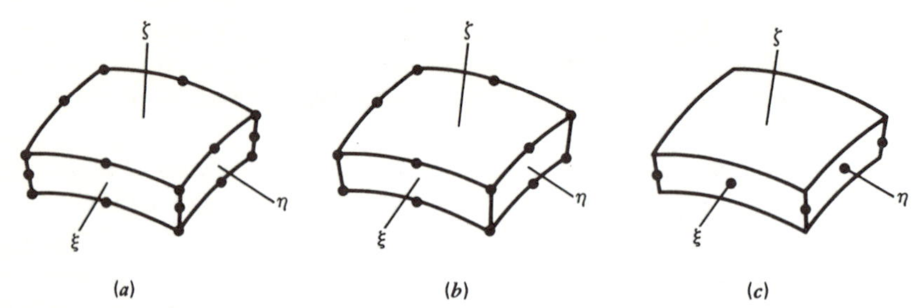

Figure 10.4.2. (*a*) The 60 d.o.f. solid element of Section 5.7. (*b*) Specialization accomplished by eliminating four side nodes to yield a 48 d.o.f. element. (*c*) Further specialization to yield a 40 d.o.f. element: each node has three translational and two rotational d.o.f.

The element of Fig. 10.4.2*b* presents two difficulties. First, making the element thin in the ζ direction invites numerical troubles, as noted in Section 9.3. Numerical trouble is much less likely, but still possible, if we define d.o.f. on (say) the outer surface as nodal displacements *relative* to nodal displacements on the inner surface (see Section 15.5). The second difficulty is that when an element is bent, its normal strain ϵ_ζ should vary with ζ, but the element yields $\epsilon_\zeta = 0$. Therefore the stress-strain relation yields too much bending stress for a given bending strain, and the element is too stiff [10.5]. This difficulty can be overcome by using modified elastic constants when the bending contribution to the element stiffness matrix is computed (see Eqs. 10.6.1 and 10.6.2).

The troubles of the 48 d.o.f. element are avoided, but information about ϵ_ζ is lost, by adopting the 40 d.o.f. element of Fig. 10.4.2*c*. This element retains an ability to account for transverse shear deformation, yet it can be used for thin shells. It strongly resembles the plate elements of Sections 9.3 through 9.5. Some details of its formulation are given in the following.

10.5 ISOPARAMETRIC SHELL ELEMENTS

The following development resembles that of Section 9.3 [9.13, 9.14, 10.5, 10.35]. One result, the eight-node element of Fig. 10.4.2*c*, is widely used. But our arguments are not restricted to four edges or to eight nodes. Strictly, the elements are "superparametric" according to definitions in Section 5.9.

Coordinates ξ and η lie in the midsurface, which is therefore the $\zeta = 0$ surface. Coordinate ζ and element edges that span the thickness are normal to the midsurface. Let subscripts p and q be appended to *nodal* quantities to indicate nodes on the $\zeta = -1$ and $\zeta = +1$ surfaces, respectively, of the element in Fig. 10.4.2*b*. Element geometry is

$$\begin{Bmatrix} x \\ y \\ z \end{Bmatrix} = \sum N_i \frac{1-\zeta}{2} \begin{Bmatrix} x_{ip} \\ y_{ip} \\ z_{ip} \end{Bmatrix} + \sum N_i \frac{1+\zeta}{2} \begin{Bmatrix} x_{iq} \\ y_{iq} \\ z_{iq} \end{Bmatrix} \tag{10.5.1}$$

Shape functions N_i contain ξ and η but not ζ (see, for example, Table 5.6.1). In Fig. 10.4.2*b* index i ranges from 1 to 8. If there are nodes on the midsurface, as in Fig. 10.4.2*c*, their coordinates are

$$\{x_i\ y_i\ z_i\} = (\{x_{ip}\ y_{ip}\ z_{ip}\} + \{x_{iq}\ y_{iq}\ z_{iq}\})/2 \qquad (10.5.2)$$

At each node i we also define a vector $\mathbf{C}_i$ that spans the thickness t_i and is normal to the midsurface (Fig. 10.5.1*a*).

$$\mathbf{C}_i = \{x_{iq}\ y_{iq}\ z_{iq}\} - \{x_{ip}\ y_{ip}\ z_{ip}\} \qquad (10.5.3)$$

Equation 10.5.1 can now be written

$$\{x\ y\ z\} = \sum N_i \{x_i\ y_i\ z_i\} + \sum N_i \frac{\zeta}{2} \mathbf{C}_i \qquad (10.5.4)$$

The elements of Figs. 10.4.2*b* and 10.4.2*c* have the same geometry. Therefore the Jacobian coefficients needed to formulate either element can be obtained from Eq. 10.5.4.

The displacement field of an element with only midsurface nodes is

$$\begin{Bmatrix} u \\ v \\ w \end{Bmatrix} = \sum N_i \begin{Bmatrix} u_i \\ v_i \\ w_i \end{Bmatrix} + \sum N_i \zeta \frac{t_i}{2} [\boldsymbol{\mu}_i] \begin{Bmatrix} \alpha_i \\ \beta_i \end{Bmatrix} \qquad (10.5.5)$$

(The positive sense of β_i is as shown in Fig. 10.5.1*c* for the same reason that θ_{xi} is similarly directed in Fig. 9.3.2*a*.) The first summation is the contribution of the d.o.f. in Fig. 10.5.1*b* to the displacement of any point in the element. The second summation is the contribution of the rotation of a midsurface-normal line and requires the following explanation.

An arbitrary point P on one of the vectors $\mathbf{C}_i$ is $\zeta t_i/2$ units distant from node i. *Small* rotations α_i and β_i produce displacements u_i' and v_i' of point P.

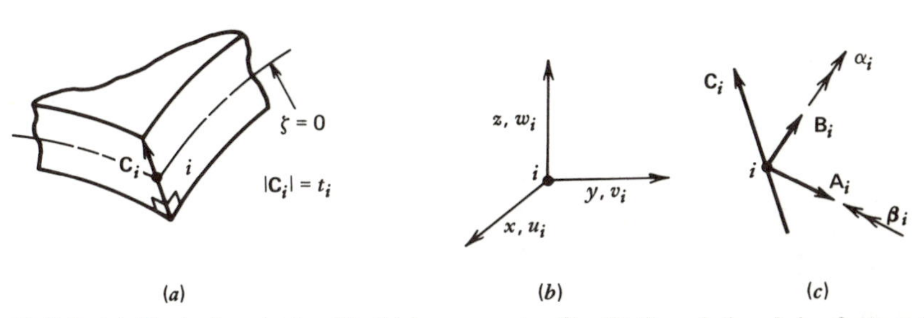

Figure 10.5.1. (*a*) Typical node i, with thickness vector $\mathbf{C}_i$. (*b*) Translational d.o.f. at node i, in global directions x, y, *and* z. (*c*) Rotational d.o.f. α_i and β_i at node i, with arrow directions defined by the right-hand rule. Vectors $\mathbf{A}_i$ and $\mathbf{B}_i$ are normal to each other and to $\mathbf{C}_i$ but are otherwise arbitrarily directed.

$$u_i' = (\zeta t_i/2)\,\alpha_i \qquad\qquad v_i' = (\zeta t_i/2)\,\beta_i \tag{10.5.6}$$

where u_i' and v_i' are, respectively, parallel to $\mathbf{A}_i$ and $\mathbf{B}_i$ in Fig. 10.5.1*c*. Vectors $\mathbf{A}_i$ and $\mathbf{B}_i$ have direction cosines

$$\{\ell_{Ai}\; m_{Ai}\; n_{Ai}\} = \frac{\mathbf{A}_i}{A_i} \qquad\qquad \{\ell_{Bi}\; m_{Bi}\; n_{Bi}\} = \frac{\mathbf{B}_i}{B_i} \tag{10.5.7}$$

where A_i and B_i are the magnitudes of $\mathbf{A}_i$ and $\mathbf{B}_i$. By resolving u_i' and v_i' into *xyz* components, we find that the globally directed displacement components caused by α_i and β_i are

$$\begin{Bmatrix} u_i \\ v_i \\ w_i \end{Bmatrix}_{\alpha\beta} = [\boldsymbol{\mu}_i] \begin{Bmatrix} u_i' \\ v_i' \end{Bmatrix}, \qquad \text{where} \qquad [\boldsymbol{\mu}_i] = \begin{bmatrix} \ell_{Ai} & \ell_{Bi} \\ m_{Ai} & m_{Bi} \\ n_{Ai} & n_{Bi} \end{bmatrix} \tag{10.5.8}$$

The total displacement of an arbitrary point in the element is

$$\{u\; v\; w\} = \sum N_i \left(\{u_i\; v_i\; w_i\} + \{u_i\; v_i\; w_i\}_{\alpha\beta}\right) \tag{10.5.9}$$

where, in the case of the quadratic element of Fig. 10.4.2*c*, index *i* ranges from 1 to 8. Substitution of Eqs. 10.5.6 into Eqs. 10.5.8, and that result into Eq. 10.5.9, yields Eq. 10.5.5.

Equation 10.5.5 describes displacements in global directions, not displacements tangent and normal to the shell. So far, the element is a three-dimensional solid, specialized only by the assumptions that thickness-direction strains need not be computed and that thickness-direction edges are straight and normal to the midsurface. These assumptions permit the use of only midsurface nodes, with five d.o.f. at each. (*Deformed* edges need not remain normal, so transverse shear effects are not excluded.)

At any structure node *i*, vectors $\mathbf{A}_i$, $\mathbf{B}_i$, and $\mathbf{C}_i$ are shared by all elements that share node *i*. Input data define any two of $\mathbf{A}_i$, $\mathbf{B}_i$, and $\mathbf{C}_i$. The third is computed as a cross product. (Alternatively, the $\mathbf{C}_i$ direction can be defined by $\mathbf{v}_{1i} \times \mathbf{v}_{2i}$, the latter two vectors being surface-tangents computed as in Eq. 5.8.6.) If the shell is orthotropic, it is convenient to let $\mathbf{A}_i$ and $\mathbf{B}_i$ coincide with principal material directions. Since $\mathbf{A}_i$ and $\mathbf{B}_i$ need be only midsurface-tangent and normal to $\mathbf{C}_i$, it is possible that no two of the α_i and β_i vectors in a structure will be parallel.

Accuracy of the finite element solution declines if *none* of the vectors $\mathbf{C}_i$ in an element are midsurfce normals (Fig. 10.5.2). Similarly, the element of Fig. 10.4.2*c* has trouble at a fold line in a shell (Problems 10.12 and 10.16). If, as in Ref. 10.35, we were to use three rotational d.o.f. per node instead of two, the rotational d.o.f. need have no particular orientation with respect to the shell. Then a fold line would present no difficulty. Also, input data to a program could omit $\mathbf{C}_i$; nodal coordinates and thicknesses would suffice (Problem 10.15).

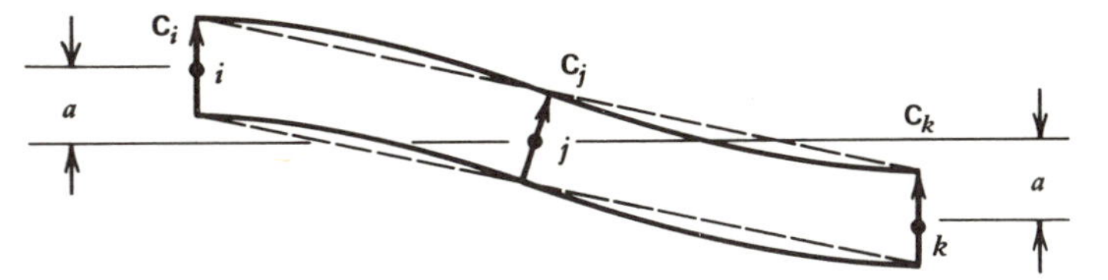

Figure 10.5.2. Intended cubic shape (solid lines) and actual linear shape (dashed lines) of a quadratic element. The **C** vectors are normal to the solid line but not to the dashed line because a quadratic curve fitted to nodal ordinates (a, 0, $-a$) is a straight line.

Finally, we write the strain-displacement relation. Equations 5.7.1 and 5.7.2 for solid elements apply without any change to the present shell elements.

$$\begin{aligned} \{\epsilon_x\ \epsilon_y\ \epsilon_z\ \gamma_{xy}\ \gamma_{yz}\ \gamma_{zx}\} &= [\mathbf{H}]\ \{u_{,x}\ u_{,y}\ \cdots\ w_{,z}\} \\ \{u_{,x}\ u_{,y}\ \cdots\ w_{,z}\} &= [\boldsymbol{\Gamma}\ \boldsymbol{\Gamma}\ \boldsymbol{\Gamma}]\ \{u_{,\xi}\ u_{,\eta}\ \cdots\ w_{,\zeta}\} \end{aligned} \tag{10.5.10}$$

where again [**H**] is 6 by 9, and each Jacobian inverse [$\boldsymbol{\Gamma}$] is 3 by 3. For now, all six strains are included because the shell surface has no particular orientation with respect to global axes xyz. From Eq. 10.5.5 we write

$$\begin{Bmatrix} u_{,\xi} \\ u_{,\eta} \\ u_{,\zeta} \end{Bmatrix} = \sum \begin{bmatrix} N_{i,\xi} & 0 & 0 \\ N_{i,\eta} & 0 & 0 \\ 0 & 0 & 0 \end{bmatrix} \begin{Bmatrix} u_i \\ v_i \\ w_i \end{Bmatrix} + \sum \frac{t_i}{2} \begin{bmatrix} \zeta N_{i,\xi} & 0 & 0 \\ \zeta N_{i,\eta} & 0 & 0 \\ N_i & 0 & 0 \end{bmatrix} [\boldsymbol{\mu}_i] \begin{Bmatrix} \alpha_i \\ \beta_i \end{Bmatrix} \tag{10.5.11}$$

Similar expressions are written for the derivatives of v and w. Then, from Eqs. 10.5.10 and 10.5.11, we find

$$\{\epsilon_x\ \epsilon_y\ \epsilon_z\ \gamma_{xy}\ \gamma_{yz}\ \gamma_{zx}\} = \sum\ [\mathbf{B}_i]\ \{u_i\ v_i\ w_i\ \alpha_i\ \beta_i\} \tag{10.5.12}$$

The complete strain-displacement matrix [**B**] is built of as many 6-by-5 blocks [$\mathbf{B}_i$] as there are nodes in the element.

Isoparametric beam elements are available [10.32, 10.36.]. They can be used as stand-alone elements or as stiffeners for plate and shell elements. Their formulation procedure is similar to the preceding one.

10.6 ISOPARAMETRIC SHELL ELEMENTS (CONTINUED)

We continue the development of elements with five d.o.f. per node by integrating through the thickness. The element stiffness matrix formula is Eq. 5.7.3, except that the 6-by-6 matrix [**E**] must pertain to a shell, not to a solid. That is, [**E**] must reflect the condition of zero normal stress in the thickness direction. We define directions A and B tangent to the midsurface and direction C normal to the midsurface. In these directions

$$\{\sigma_A \ \sigma_B \ \sigma_C \ \tau_{AB} \ \tau_{BC} \ \tau_{CA}\} = [\mathbf{E}'] \{\epsilon_A \ \epsilon_B \ \epsilon_C \ \gamma_{AB} \ \gamma_{BC} \ \gamma_{CA}\} \tag{10.6.1}$$

In $[\mathbf{E}']$ row three is null, so that $\sigma_C = 0$, and column three is null to decouple all stresses from ϵ_C. In the case of isotropy and homogeneity, nonzero terms in the upper triangle of $[\mathbf{E}']$ are

$$E'_{11} = E'_{22} = \frac{E'_{12}}{\nu} = \frac{E}{1 - \nu^2} \qquad E'_{44} = 1.2E'_{55} = 1.2E'_{66} = G \tag{10.6.2}$$

where the shear modulus is $G = E/2(1 + \nu)$ and factors 1.2 account for transverse shear effects as explained in Section 9.4.Transformation of $[\mathbf{E}']$ to $[\mathbf{E}]$ is given by Eq. 6.2.13. This is required because $[\mathbf{E}]$ in Eq. 5.7.3 refers to *global* directions, not to the *local* directions of Eq. 10.6.1. The null row and null column in $[\mathbf{E}']$ permit us to discard the third row of $[\mathbf{T}_\epsilon]$ to yield a 5-by-6 matrix $[\mathbf{T}_\epsilon^*]$ and to compact the 6-by-6 matrix $[\mathbf{E}']$ to yield a 5-by-5 matrix $[\mathbf{E}^*]$. Thus the transformation can be written

$$\underset{6\times6}{[\mathbf{E}]} = \underset{6\times5}{[\mathbf{T}_\epsilon^*]^T} \ \underset{5\times5}{[\mathbf{E}^*]} \ \underset{5\times6}{[\mathbf{T}_\epsilon^*]} \tag{10.6.3}$$

The N-by-N stiffness matrix can be written in either of two forms.

$$[\mathbf{k}] = \int_V \underset{N\times6}{[\mathbf{B}]^T} \ \underset{6\times6}{[\mathbf{E}]} \ \underset{6\times N}{[\mathbf{B}]} \, dV = \int_V \underset{N\times5}{[\mathbf{B}^*]^T} \ \underset{5\times5}{[\mathbf{E}^*]} \ \underset{5\times N}{[\mathbf{B}^*]} \, dV \tag{10.6.4}$$

where $dV = J \, d\xi \, d\eta \, d\zeta$ and $[\mathbf{B}^*] = [\mathbf{T}_\epsilon^*] [\mathbf{B}]$. The second form is obtained by associating $[\mathbf{T}_\epsilon^*]$ with $[\mathbf{B}]$ rather than with $[\mathbf{E}^*]$. Programming skill determines which option is more efficient.

Either form of Eq. 10.6.4 requires that direction cosines of the local directions in Eq. 10.6.1 be known at each numerical quadratic point. Vectors **A**, **B**, and **C** at any point can be found by shape-function interpolation from nodal values. For example,

$$A_x = \sum N_i A_{xi} \qquad A_y = \sum N_i A_{yi} \qquad A_z = \sum N_i A_{zi} \tag{10.6.5}$$

where $\mathbf{A} = \{A_x \ A_y \ A_z\}$. Hence the nine direction cosines are given by $\mathbf{A}/|\mathbf{A}|$, $\mathbf{B}/|\mathbf{B}|$, and $\mathbf{C}/|\mathbf{C}|$.

Integration of $[\mathbf{k}]$ can be done by an n-by-n-by-2 rule, or by an n-by-n-by-1 rule (all points in the $\zeta = 0$ surface). The n-by-n-by-1 option requires explicit integration in the ζ direction, as described shortly. There is evidence that as element geometry becomes more sharply curved, explicit thickness integration leads to appreciable error, particularly in nodal forces and moments computed by the operation $[\mathbf{k}]\{\mathbf{d}\}$ from known d.o.f. $\{\mathbf{d}\}$. Related numerical evidence appears in Table 10.7.1. The n-by-n-by-2 rule is twice as costly but less risky.

To do explicit thickness integration, we first examine $[\mathbf{J}]$, calculated from Eqs. 5.7.1 and 10.5.4. Certain terms in $[\mathbf{J}]$ have ζ as a coefficient. Reference 9.14 suggests discarding these terms if the ratio t/R is small, where t is thickness and R is the least radius of curvature of the shell element. Thus $[\mathbf{J}]$ and its determinant become functions of only ξ and η. Next, as in Section 9.4, we split $[\mathbf{B}]$ into a "membrane part" $[\mathbf{B}_0]$ and a "bending part" $\zeta[\mathbf{B}_1]$. Thus, $[\mathbf{B}] = [\mathbf{B}_0] + \zeta[\mathbf{B}_1]$, where $[\mathbf{B}_0]$ and $[\mathbf{B}_1]$ are independent of ζ (the same split is possible if we use the second form in Eq. 10.6.4). Integration of $\zeta\, d\zeta$ yields zero. Integration of $d\zeta$ and $\zeta^2\, d\zeta$ yields 2 and $\frac{2}{3}$, respectively. Accordingly,

$$[\mathbf{k}] = \int_{-1}^{1}\int_{-1}^{1} \left(2[\mathbf{B}_0]^T[\mathbf{E}][\mathbf{B}_0] + \tfrac{2}{3}[\mathbf{B}_1]^T[\mathbf{E}][\mathbf{B}_1]\right) J\, d\xi\, d\eta \tag{10.6.6}$$

As in Section 9.4, we have again assumed that $[\mathbf{E}]$ is either independent of ζ or symmetric with respect to the surface $\zeta = 0$. Note also that although ζ is absent from Eq. 10.6.6, J is still calculated from a 3-by-3 Jacobian matrix.

Section 9.5 presents arguments about what quadrature rules to use in the $\zeta = 0$ surface and what to do if the element is extremely thin. The same arguments apply qualitatively to shell elements, but detailed studies and numerical tests have not been done. For the eight-node element (Fig. 10.4.2*c*), an additional argument is available: unless the element is flat, two Gauss points must be used in the ξ direction and two in the η direction if membrane strains are to be correctly computed (as zero) when a field of pure bending prevails [10.15]. Other quadrature rules overestimate membrane stiffness. Fortunately, an order two rule is also indicated by Eq. 9.5.2.

The implication for stress calculation is that for accuracy, transverse shear strains and membrane strains tangent to curved sections must be evaluated at Gauss points of a 2-by-2-by-2 rule or a 2-by-2-by-1 rule, then extrapolated elsewhere if so desired.

A shell of sandwich construction can be treated as follows [10.14]. Let $[\mathbf{E}']$ of Eq. 10.6.1 describe properties of the facings, except for E'_{55} and E'_{66}, which describe transverse shear stiffnesses of the core. Formulate $[\mathbf{k}]$ as if the shell has thickness $t = 2h$, where h = thickness of each facing. Membrane stiffness is therefore properly accounted for. However, bending stiffness is underestimated. It must be corrected by multiplying the contribution of $[\mathbf{B}_1]^T[\mathbf{E}][\mathbf{B}_1]$ in Eq. 10.6.6 by a flexural rigidity ratio. For isotropic materials, the ratio is

$$\frac{\mathcal{D}_{\text{sandwich}}}{\mathcal{D}_{\text{homogeneous}}} = \frac{Eh(c + h)^2/2(1 - \nu^2)}{Et^3/12(1 - \nu^2)} = \frac{3(c + h)^2}{4h^2} \tag{10.6.7}$$

where the two $\mathcal{D}$'s are stated in Eqs. 9.1.7 and 9.4.6 and core thickness c must be much greater than h. Also, transverse shear stiffnesses E'_{55} and E'_{66} must be multiplied by $c/2h$

to compensate for the reduction in thickness of the shear-resisting layer from c to $2h$. Nodal moments caused by a temperature gradient also require a correction factor (Problem 10.23).

In computing element nodal loads, column vectors in Eq. 4.3.5 must be referred to *global* directions because [**N**] and [**B**] refer to displacements and strains in global directions. Grouping terms in $\{\boldsymbol{\sigma}_0\}$ and $\{\boldsymbol{\epsilon}_0\}$ according to the exponent on ζ can be helpful, as it is in Eq. 10.6.6.

Strains $\{\boldsymbol{\epsilon}\} = [\mathbf{B}]\{\mathbf{d}\}$ are referred to global directions. Strains $\{\boldsymbol{\epsilon}'\}$, referred to directions A, B, and C of Eq. 10.6.1, are $\{\boldsymbol{\epsilon}'\} = [\mathbf{T}_\epsilon]\{\boldsymbol{\epsilon}\}$. The corresponding stresses are $\{\boldsymbol{\sigma}'\} = [\mathbf{E}']\{\boldsymbol{\epsilon}'\}$. If a sandwich shell is analyzed by an equivalent section of thickness $2h$, bending strains must be increased by the factor $(c + 2h)/2h$ to yield correct strains in the facings.

10.7 TEST CASES FOR SHELL ELEMENTS

It is hard to find a shell element that does well in a variety of test cases. Indeed, the available test cases may be few. Analytical solutions are not plentiful. Constant strain states, desired for patch tests, cannot be stated for arbitrarily curved shells. Computed results, especially buckling loads, may depend strongly on geometry [10.16]. Thus the description of element shape can be as important as the displacement field.

In shells analyzed by elements based on displacement fields, a "sensitive" problem is one that has significant bending stress but negligible membrane stress [10.17]. A slit cylinder under torsional loading is a sensitive problem. (It is also a patch test that should yield constant twisting moment.) Mesh refinement may not improve results given by an element that performs badly in a sensitive problem [10.10].

A dilemma in developing a shell element is that to model rigid-body motion exactly, the displacement field must contain trigonometric terms, but to model a sensitive problem exactly, the field must contain polynomials. An element can model one behavior or the other, but not both [10.10].

A cylinder under internal pressure is not a sensitive problem, but it is a problem that flat elements may fail, as noted in Section 10.2.

Theory indicates that there is a bending moment concentration where three fold lines meet, as at a corner of a box [10.18]. This is a severe test case, since it requires either many small elements near the corner or larger elements that somehow avoid displaying too large a moment well away from the corner.

Figure 10.7.1 depicts a popular test case. Table 10.7.1 reports some of the published results. Except for the element of Fig. 10.4.2c, all elements in the table are restricted to circular cylindrical shells. We see that one of the elements works for the thicker shell but fails for the thinner shell.

Another popular test case is depicted in Fig. 10.7.2. Computed results are in Table 10.7.2 [5.21]. The 16-node elements use modified elastic constants, as suggested near

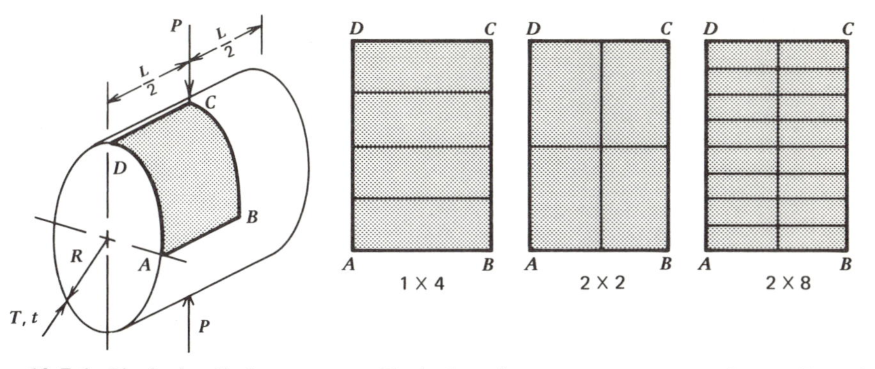

Figure 10.7.1. Pinched cylinder test case. Typical meshes on one octant are shown. $R = 4.953$ in., $T = 0.094$ in. (thicker) or $t = 0.01548$ in. (thinner), $L = 10.35$ in., $E = 10.5(10)^6$ psi, $\nu = 0.3125$, $P = 100$ lb (thicker) or $P = 0.10$ lb (thinner).

the end of Section 10.4. A 2-by-2-by-2 Gauss quadrature rule in used for all elements. Element nodal loads $\{\mathbf{r}\}$ from self-weight are formulated consistently (Eq. 4.3.5). Stresses are extrapolated from Gauss points to nodes. Analytical solutions of the shell roof problem are based on approximations, so no "exact" value is quoted in Table 10.7.2. A 4-node isoparametric shell element [10.35] gives the following values for w_B: -4.518 (2×2), -3.472 (4×4), and -3.546 (8×8).

The quadratic element is somewhat improved if three nodeless d.o.f. are added to permit u, v, and w to have the bubble function mode $(1 - \xi^2)(1 - \eta^2)$ [10.14].

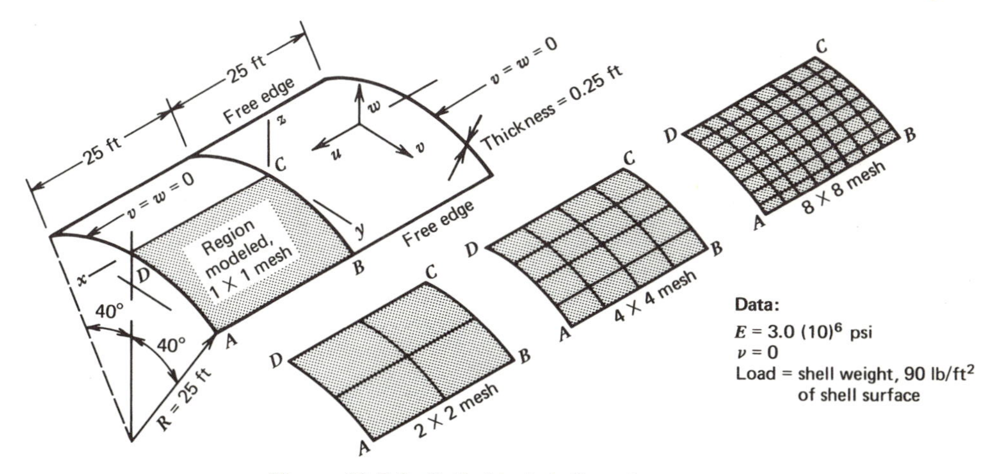

Figure 10.7.2. Cylindrical shell roof test case.

TABLE 10.7.1. Deflection (inches) under one load in the pinched cylinder problem of Fig. 10.7.1. The number of d.o.f. before constraints are applied is given in parentheses. The last two columns pertain to the element of Fig. 10.4.2c, with and without explicit thickness integration.

Mesh, and shell thickness (T = thicker, t = thinner)	20 d.o.f. per element [10.11]	24 d.o.f. per element [10.19]	40 d.o.f. element 2-by-2-by-1 rule [10.20]	40 d.o.f. element 2-by-2-by-2 rule [10.20]
$1 \times 1, T$	0.1040 (20)	—	0.0897 (40)	0.0943 (40)
$1 \times 4, T$	0.1106 (50)	0.0769 (72)[a]	—	—
$1 \times 8, T$	—	0.1057 (120)[b]	—	—
$2 \times 2, T$	0.1103 (45)	0.1073 (180)[c]	0.1032 (105)	0.1107 (105)
$4 \times 4, T$	0.1129 (125)	—	0.1051 (325)	0.1133 (325)
$8 \times 8, T$	0.1137 (405)	0.1129 (1200)[d]	0.1055 (1125)	0.1139 (1125)
$1 \times 1, t$	0.02301 (20)	0.00001 (24)	0.02022 (40)	0.02022 (40)
$1 \times 4, t$	0.02403 (50)	0.00074 (60)	—	—
$1 \times 8, t$	0.02406 (90)	0.00700 (108)	0.02286 (105)[e]	0.02453 (105)[e]
$2 \times 8, t$	0.02414 (135)	0.00699 (162)	0.02313 (325)[f]	0.02499 (325)[f]
$8 \times 8, t$	0.02431 (405)	0.00708 (486)	0.02322 (1125)	0.02511 (1125)

[a] 1×5 mesh.
[b] 1×9 mesh.
[c] 2×9 mesh.
[d] 3×49 mesh.
[e] 2×2 mesh.
[f] 4×4 mesh.

TABLE 10.7.2. Computed results for the shell roof of Fig. 10.7.2 [5.21]. Element type (8) is the 8-node element of Fig. 10.4.2*c*. Element type (16) is the 16-node element of Fig. 10.4.2*b*.

Mesh and element type	Displacements (inches)		Stresses, averaged at shared nodes			
			At node B		At node C	
	w_B	v_B	σ_x top	σ_x bottom	σ_y top	σ_y bottom
1 × 1 (8)	−3.672	−2.057	−32	1820	1592	−1408
1 × 1 (16)	−3.794	−2.139	−133	1858	1370	−1438
2 × 2 (8)	−3.523	−1.865	1084	2026	1416	−1619
2 × 2 (16)	−3.510	−1.861	1052	2084	1410	−1586
4 × 4 (8)	−3.623	−1.911	1497	2366	1328	−1519
4 × 4 (16)	−3.627	−1.913	1497	2370	1325	−1517
8 × 8 (8)	−3.619	−1.908	1623	2489	1292	−1482
8 × 8 (16)	−3.617	−1.907	1624	2488	1290	−1430

10.8 THIN SHELLS OF REVOLUTION

A shell of revolution resembles a solid of revolution in that elements are symmetric with respect to an axis and node points are cross sections of nodal circles. Like any other shell, it requires careful attention to geometric description, strain-displacement relations, and assumed displacement field.

Figure 10.8.1 shows the geometry and notation. Displacement v exists because we make no restriction that loads must be axially symmetric. The following geometric relations can be written.

$$R_\theta = \frac{r}{\cos\phi} \qquad \frac{d\phi}{ds} = -\frac{1}{R_s} \qquad \frac{dr}{ds} = \sin\phi \qquad \frac{dz}{ds} = -\cos\phi \tag{10.8.1}$$

where R_θ and R_s are principal radii of curvature. Shell geometry and displacements could be expressed in terms of x, as in the arch element of Fig. 10.4.1*a*. However, published papers use arc length s instead. In what follows we concentrate on the formulations described in Ref. 10.4 and 10.21 to 10.24.

Subsequent expressions require that R_s, R_θ, r, and ϕ be expressed in terms of s. So, to describe the shape of an element meridian, we write

$$\phi = \alpha_0 + \alpha_1 s + \alpha_2 s^2 \tag{10.8.2}$$

which permits elements to have a common tangent where they meet. Coefficients α_0, α_1, and α_2 are evaluated from the conditions $\phi = \phi_1$ at $s = 0$, $\phi = \phi_2$ at $s = L$, and

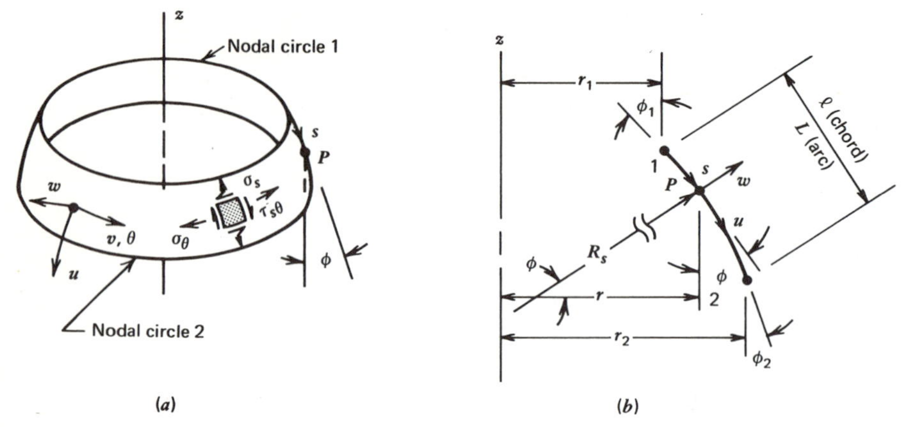

Figure 10.8.1. An axisymmetric shell element, showing geometry and notation. P is an arbitrary point on the midsurface, a distance s along the meridian from node 1. Displacements u, v, and w are mutually orthogonal, with v and w in circumferential and shell-normal directions, respectively. R_s is the radius of curvature of the meridian.

$$0 = \int_0^L \sin(\phi - \phi_c)\, ds \approx \int_0^L (\phi - \phi_c)\, ds \tag{10.8.3}$$

where ϕ_c is the angle of the chord, $\tan \phi_c = (r_2 - r_1)/(z_1 - z_2)$. The results are (Problem 10.26)

$$\alpha_0 = \phi_1 \qquad \alpha_1 = \frac{6\phi_c - 4\phi_1 - 2\phi_2}{L} \qquad \alpha_2 = \frac{3\phi_1 + 3\phi_2 - 6\phi_c}{L^2} \tag{10.8.4}$$

A first approximation for L is simply $L = \ell$, where chord length $\ell = [(r_2 - r_1)^2 + (z_2 - z_1)^2]^{1/2}$. A better approximation for L is found by using Eq. 4.2.4 to fit the meridian by a "deflected beam" of length ℓ that has the nodal d.o.f.

$$w_1 = w_2 = 0 \qquad \theta_1 = \phi_1 - \phi_c \qquad \theta_2 = \phi_2 - \phi_c \tag{10.8.5}$$

Thus $\theta = w_{,x}$ (in radians) is a function of x, where $x = s\ell/L$ is the chord distance that corresponds to s. (In this paragraph θ is *not* the circumferential coordinate of the shell). The arc length of the beam is [13.10]

$$L \approx \ell + \int_0^\ell \frac{w_{,x}^2}{2}\, dx = \ell + \ell\, \frac{2\theta_1^2 - \theta_1\theta_2 + 2\theta_2^2}{30} \tag{10.8.6}$$

Similarly, with $\phi = \phi_c + \theta$, radius r can be written

$$r = r_1 + \int_0^s \sin \phi \, ds \approx r_1 + s \sin \phi_c + \cos \phi_c \int_0^x \theta \, dx \qquad (10.8.7)$$

The first form of Eq. 10.8.7 can be integrated numerically. The second form follows from a trigonometric formula and the approximations $\cos \theta \approx 1$ and $\sin \theta \approx \theta$.

Strain at any point a distance z outward from the midsurface is

$$\{\epsilon_s \ \epsilon_\theta \ \gamma_{s\theta}\} = \{\bar{\epsilon}_s \ \bar{\epsilon}_\theta \ \bar{\gamma}_{s\theta}\} - z \{\kappa_s \ \kappa_\theta \ \kappa_{s\theta}\} \qquad (10.8.8)$$

where the barred quantities are midsurface strains and the κ's are curvatures. In writing Eq. 10.8.8 we imply a thin shell: midsurface-normals remain straight *and normal* during deformation, and plane stress conditions prevail in any z = constant layer. There are many strain-displacement expressions from which to choose. Theoreticians disagree as to what $\kappa_{s\theta}$ should be. Sanders' theory, as stated in Ref. 10.24, is

$$\begin{aligned}
\bar{\epsilon}_s &= u_{,s} + \frac{w}{R_s} + \frac{1}{2}(X^2 + Z^2) \\
\bar{\epsilon}_\theta &= \frac{v_{,\theta}}{r} + \frac{u}{r} r_{,s} + \frac{w}{R_\theta} + \frac{1}{2}(Y^2 + Z^2) \\
\bar{\gamma}_{s\theta} &= \frac{u_{,\theta}}{r} + v_{,s} - \frac{v}{r} r_{,s} + XY \\
\kappa_s &= X_{,s} \qquad \kappa_\theta = \frac{1}{r} Y_{,\theta} + \frac{X}{r} r_{,s} \\
\kappa_{s\theta} &= \frac{1}{r} X_{,\theta} + Y_{,s} - \frac{Y}{r} r_{,s} + \left(\frac{1}{R_\theta} - \frac{1}{R_s}\right) Z
\end{aligned} \qquad (10.8.9)$$

where

$$X = w_{,s} - \frac{u}{R_s} \qquad Y = \frac{w_{,\theta}}{r} - \frac{v}{R_\theta} \qquad Z = \frac{1}{2}\left(\frac{u_{,\theta}}{r} - v_{,s} - \frac{v}{r} r_{,s}\right)$$

The X, Y, and Z terms in the midsurface strains are nonlinear terms. They can be discarded from the midsurface strains unless displacements are large or buckling is involved. If meridians are straight, R_s becomes infinite. If displacements are axially symmetric, $v = 0$ and all derivatives with respect to θ are zero (see Eqs. 10.11.6).

Stresses caused by Eq. 10.8.8 are

$$\{\sigma_s \ \sigma_\theta \ \tau_{s\theta}\} = [\mathbf{E}] \{\epsilon_s \ \epsilon_\theta \ \gamma_{s\theta}\} \qquad (10.8.10)$$

where, in the case of isotropy and homogeneity,

$$E_{11} = E_{22} = \frac{E_{12}}{\nu} = \frac{E_{21}}{\nu} = \frac{E}{1 - \nu^2} \qquad E_{33} = G = \frac{E}{2(1 + \nu)} \tag{10.8.11}$$

and the other E_{ij} are zero.

To analyze a membrane shell, we drop the curvatures from Eq. 10.8.8 and omit nodal rotations from the assumed displacement field.

10.9 THIN SHELLS OF REVOLUTION (CONTINUED)

The displacement field for the nth component of a Fourier series displacement field, symmetric about a $\theta = 0$ plane, can be written

$$\begin{aligned} w &= [a_1 + a_2 s + a_3 s^2 + a_4 s^3] \cos n\theta \\ u &= [a_5 + a_6 s + a_9 s(s - L) + a_{10} s^2(s - L)] \cos n\theta \\ v &= [a_7 + a_8 s + a_{11} s(s - L) + a_{12} s^2(s - L)] \sin n\theta \end{aligned} \tag{10.9.1}$$

where the a_i are generalized coordinates that depend on n. With $n = 0$ we have the axially symmetric case. Clearly, we are beginning a discussion much like Sections 8.3 through 8.6, so here we consider few details.

Freedoms a_9 through a_{12} are nodeless. They are condensed before elements are assembled. As noted with Eqs. 10.3.3, these a_i serve mainly to provide rigid-body motion capability. Rigid-body modes are contained in only the $n = 0$ and $n = 1$ Fourier terms. *For* n > 1, a_9 *through* a_{12} *are omitted from Eqs. 10.9.1* in order to avoid a too-flexible stiffness matrix [10.25].

The element stiffness matrix and element nodal load vector can be written in the a-basis of Eqs. 4.3.19. To find the strain-displacement matrix $[\mathbf{B}_a]_n$ for a typical harmonic n, we substitute Eqs. 10.8.1, 10.8.2, and 10.9.1 into Eqs. 10.8.9 and substitute those results into Eq. 10.8.8. Then

$$[\mathbf{k}_a]_n = \int_{-t/2}^{t/2} \int_0^L \int_{-\pi}^{\pi} [\mathbf{B}_a]_n^T \underset{3\times 3}{[\mathbf{E}]} [\mathbf{B}_a]_n \, r \, d\theta \, ds \, dz \tag{10.9.2}$$

The integrand is a function of n, θ, s, and z. The thickness integration, $-t/2 < z < t/2$, can be done explicitly. If $[\mathbf{E}]$ is independent of z, the results include t^3 terms (bending action) and t terms (membrane action). Integration with respect to θ is done by Eqs. 8.3.2. Integration with respect to s is done by numerical quadrature.

Next, condensation is applied if $n = 0$ or if $n = 1$, to remove a_9 through a_{12}. Thus we obtain an 8-by-8 $[\mathbf{k}_a]$ and an 8-by-1 $\{\mathbf{r}_a\}$ (or a 6-by-6 $[\mathbf{k}_a]$ and a 6-by-1 $\{\mathbf{r}_a\}$ if the problem is axially symmetric, since then $v_1 = v_2 = 0$). Before assembly of elements, two coordinate transformations remain, as described now.

The first transformation exchanges generalized coordinates {**a**} for the nodal d.o.f. {**d**} in Fig. 10.9.1*a*, where

$$\begin{aligned} \{\mathbf{d}\} &= \{w_1\ u_1\ v_1\ \beta_1\ w_2\ u_2\ v_2\ \beta_2\} \\ \beta_i &= w_{i,s} - \frac{u_i}{R_{si}}, \qquad i = 1, 2 \end{aligned} \tag{10.9.3}$$

Like β in Eq. 10.4.1, β_i is the rotation of a cross section. As in Eq. 4.2.3, we write the relation $\{\mathbf{d}\} = [\mathbf{A}]\{\mathbf{a}\}$. But we omit cos $n\theta$ and sin $n\theta$ terms because {**d**} and {**a**} now represent *amplitudes* of harmonic n. Thus [**A**] is produced by evaluating the bracketed expressions in Eqs. 10.9.1 at $s = 0$ and at $s = L$. Coordinates a_9 through a_{12} disappear at $s = 0$ and at $s = L$, so [**A**] is 8 by 8 in the general case or 6 by 6 in the axially symmetric case. Now we can apply Eqs. 4.3.18 to obtain [**k**] and {**r**} from $[\mathbf{k}_a]$ and $\{\mathbf{r}_a\}$.

But we probably do not want to use the nodal d.o.f. {**d**} in Fig. 10.9.1*a*. If there is a cusp at a node, as where a cylinder is joined to a cone, the u_i and w_i d.o.f. in adjacent elements do not have common directions. The directions *are* common if we use the d.o.f. $\{\bar{\mathbf{d}}\}$ in Fig. 10.9.1*b*. The transformation is

$$\begin{aligned} \{\mathbf{d}\} &= \underset{8\times 8}{[\boldsymbol{\Gamma}_1\ \mathbf{I}\ \boldsymbol{\Gamma}_2\ \mathbf{I}]}\ \{\bar{\mathbf{d}}\}, \qquad \text{where} \\ \{\bar{\mathbf{d}}\} &= \{\bar{w}_1\ \bar{u}_1\ v_1\ \beta_1\ \bar{w}_2\ \bar{u}_2\ v_2\ \beta_2\} \\ [\boldsymbol{\Gamma}_i] &= \begin{bmatrix} \sin\phi_i & \cos\phi_i \\ -\cos\phi_i & \sin\phi_i \end{bmatrix}, \qquad i = 1, 2 \end{aligned} \tag{10.9.4}$$

and [**I**] is a 2-by-2 unit matrix. Accordingly, the transformation from a-basis to $\bar{d}$-basis in Fig. 10.9.1*b* is

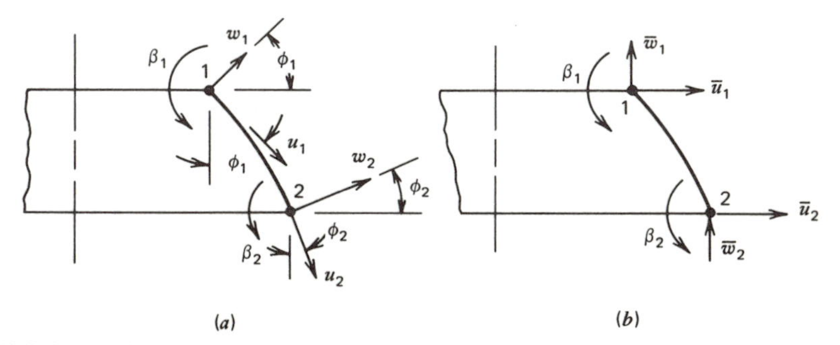

Figure 10.9.1. D.o.f. at nodes, in (*a*) local, and (*b*) global directions. D.o.f. v_1 and v_2, not shown, are perpendicular to the paper.

$$\begin{aligned} [\bar{\mathbf{k}}] &= [\mathbf{T}]^T [\mathbf{k}_a][\mathbf{T}], \qquad \{\bar{\mathbf{r}}\} = [\mathbf{T}]^T [\mathbf{r}_a] \\ \text{where} \qquad [\mathbf{T}] &= [\mathbf{A}]^{-1} [\boldsymbol{\Lambda}_1 \, \mathbf{I} \, \boldsymbol{\Lambda}_2 \, \mathbf{I}] \end{aligned} \tag{10.9.5}$$

10.10 CONCLUDING REMARKS ABOUT THIN SHELLS OF REVOLUTION

At a pole, infinite terms appear in the strain-displacement relations (Eqs. 10.8.9) because radius r is zero. The same problem is noted in Sections 8.2 and 8.4. Special strain-displacement relations can be produced by discarding terms that become infinite. These expressions show that a pole acts as a clamped edge if $n > 1$ in Eqs. 10.9.1 [10.21, 10.23, 10.26].

Special shapes of shells can be identified. If $\phi = c$, a constant, then $c = 0$ is a circular cylinder, $0 < c < 90°$ is a cone, and $c = 90°$ is a circular plate. These configurations can be analyzed by the preceding formulation. Loads need not be axially symmetric.

The Fourier series method can be applied to problems of buckling (Section 12.4). Vibration problems can be treated similarly. In problems of vibration and bifurcation buckling, important deflection modes may be antisymmetric with respect to planes about which the geometry and loading are symmetric [10.27].

Cylindrical shells that are not circular or not axially symmetric or have longitudinal stiffeners can be analyzed in the same way as prismatic solids (Section 8.7). If the shell is slightly bent to form a torus of large radius, a small piece of the torus models an almost straight piece of the original cylinder with simply supported ends [10.28]. The axial coordinate of the cylinder becomes the circumferential coordinate of the torus.

A practical shell may be built of component shells joined to one another by circumferential rings. Midsurfaces of adjacent shells may not intersect at the joint. To assume such an intersection, or to connect the midsurfaces together rigidly, makes the analytical model stiffer than the actual structure [10.27].

A shell of revolution may be unexpectedly sensitive to a change in boundary conditions. When a displacement mode lacks axial symmetry, boundary effects propagate much further along the shell than effects that are axially symmetric [10.27].

Shells of revolution are well suited to analysis by finite differences. A versatile computer program is based on the finite difference energy method [10.29]. This method should be comprehensible to students of finite elements, since the concepts and procedures are much the same.

10.11 ISOPARAMETRIC ELEMENTS FOR SHELLS OF REVOLUTION

A shell of revolution, whether thick, thin, or sandwich, can be analyzed by isoparametric elements. The approach is similar to that in Sections 10.5 and 10.6 [10.30]. Figure 10.11.1 shows some of the possible elements. The linear-quadratic element is analogous

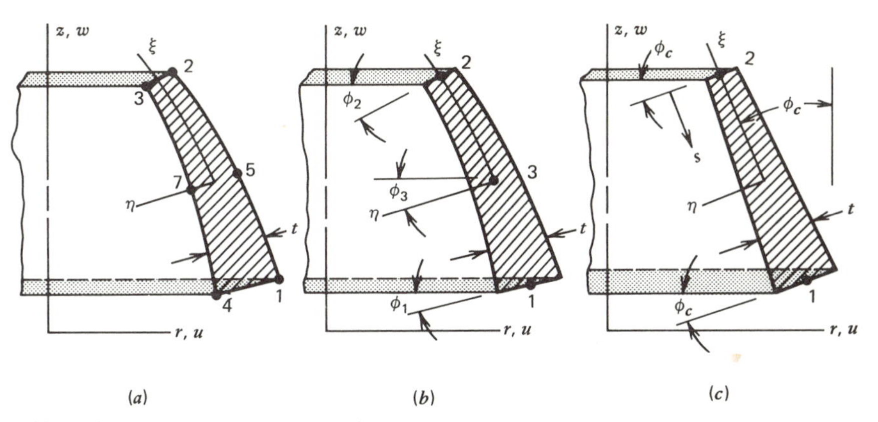

Figure 10.11.1. Isoparametric elements for thick shells of revolution. (*a*) Linear-quadratic. Node numbers correspond to those in Table 5.6.1, which yields the shape functions. (*b*) Midsurface quadratic. (*c*) Midsurface linear.

to the element of Fig. 10.4.2*b*. The midsurface quadratic element is analogous to the element of Fig. 10.4.2*c* and is discussed in more detail in the following, as is the element of Fig. 10.11.1*c*.

Geometry of the midsurface elements is defined by an equation analogous to Eq. 10.5.4.

$$\begin{Bmatrix} r \\ z \end{Bmatrix} = \sum N_i \begin{Bmatrix} r_i \\ z_i \end{Bmatrix} - \sum N_i \eta \frac{t_i}{2} \begin{Bmatrix} \cos \phi_i \\ \sin \phi_i \end{Bmatrix} \tag{10.11.1}$$

The second summation carries a negative sign because η points inward. The N_i for the three-node element are given by Eq. 5.2.9. For the two-node element, the N_i are $N_1 = (1 - \xi)/2$ and $N_2 = (1 + \xi)/2$.

For axially symmetric behavior, the displacement field is

$$\begin{Bmatrix} u \\ w \end{Bmatrix} = \sum N_i \begin{Bmatrix} u_i \\ w_i \end{Bmatrix} + \sum N_i \eta \frac{t_i}{2} \begin{Bmatrix} \sin \phi_i \\ -\cos \phi_i \end{Bmatrix} \beta_i \tag{10.11.2}$$

which corresponds to Eq. 10.5.5. Nodal rotations β_i are directed as in Fig. 10.9.1. However, the u's and w's are now displacements in *global* directions, not in directions tangent and normal to the shell.

If loads are not axially symmetric, we must add the circumferential displacement v, as in Fig. 10.8.1*a*. We must also add α, the rotation of a midsurface-normal line about the ξ-axis, positive when its vector is in the positive ξ direction. Rotation α is needed to provide bending stiffness in the θ direction. For the nth Fourier harmonic of a displacement field symmetric about the $\theta = 0$ plane,

$$\begin{Bmatrix} u \\ v \\ w \end{Bmatrix} = \underset{3\times 3}{\lceil \mathbf{Q} \rfloor} \left(\sum N_i \begin{Bmatrix} \bar{u}_i \\ \bar{v}_i \\ \bar{w}_i \end{Bmatrix}_n + \sum N_i \eta \frac{t_i}{2} \begin{bmatrix} 0 & \sin \phi_i \\ -1 & 0 \\ 0 & -\cos \phi_i \end{bmatrix} \begin{Bmatrix} \bar{\alpha}_i \\ \bar{\beta}_i \end{Bmatrix}_n \right) \tag{10.11.3}$$

$$\text{where} \qquad \lceil \mathbf{Q} \rfloor = \lceil \cos n\theta \;\; \sin n\theta \;\; \cos n\theta \rfloor$$

The barred nodal quantities are amplitudes of harmonic n.

Strain-displacement relations are given by Eqs. 8.2.4 for the axially symmetric case and by Eqs. 8.4.4 for the general case. Hence we find $[\mathbf{B}]_n$. Then, for harmonic n,

$$[\mathbf{k}]_n = \int_{-\pi}^{\pi} \int_{-1}^{1} \int_{-1}^{1} [\mathbf{B}]_n^T [\mathbf{E}] \, [\mathbf{B}]_n \, J \, d\xi \, d\eta \, r \, d\theta \tag{10.11.4}$$

where J is computed from Eq. 10.11.1. Matrix $[\mathbf{E}]$ is 4 by 4 in the axially symmetric case and 6 by 6 in the general case. *However,* $[\mathbf{E}]$ must be obtained as described in connection with Eqs. 10.6.1 through 10.6.3. For example, in the case of isotropy, homogeneity, and axial symmetry, Eq. 8.2.3 is modified to become

$$[\mathbf{E}'] = \frac{E}{1 - \nu^2} \begin{bmatrix} 1 & 0 & \nu & 0 \\ 0 & 0 & 0 & 0 \\ \nu & 0 & 1 & 0 \\ 0 & 0 & 0 & (1 - \nu)/2k \end{bmatrix} \tag{10.11.5}$$

where $k = 1.2$, as explained in Section 9.4. Matrix $[\mathbf{E}']$ operates on $\{\epsilon_\xi \; \epsilon_\eta \; \epsilon_\theta \; \gamma_{\xi\eta}\}$ and reflects the assumption that $\sigma_\eta = 0$ in any surface $\eta =$ constant. Whether axial symmetry prevails or not, a coordinate transformation of $[\mathbf{E}']$ through angle ϕ yields $[\mathbf{E}]$ for Eq. 10.11.4. Again, either of the two forms for $[\mathbf{k}]$ in Eq. 10.6.4 can be used.

In Eq. 10.11.4, we use Eqs. 8.3.2 for integration with respect to θ and follow the discussion in Sections 9.5 and 10.6 for integration in the other two directions.

Elements in Fig. 10.11.1 are analogous to shell elements discussed in Section 10.6. Accordingly, explicit thickness integration is possible but perhaps unwise. In the ξ direction, two points are best for the quadratic element and one point is best for the linear element.

An alternative to the linear element of Fig. 10.11.1*c* has been proposed [10.31]. For axial symmetry, Eqs. 10.8.1 and 10.8.9 yield

$$\begin{aligned} \bar{\epsilon}_s &= \frac{du}{ds} + \frac{w}{R_s} & \qquad \bar{\epsilon}_\theta &= \frac{w \cos \phi + u \sin \phi}{r} \\ \kappa_s &= \frac{d^2 w}{ds^2} - \frac{d(u/R_s)}{ds} & \qquad \kappa_\theta &= \frac{\sin \phi}{r} \left(\frac{dw}{ds} - \frac{u}{R_s} \right) \end{aligned} \tag{10.11.6}$$

The symbols are defined in Figs. 10.8.1 and 10.9.1 (note that w is normal to the shell). By saying that the transverse shear strain γ is $\gamma = dw/ds - \beta$, we can rewrite the curvatures as

$$\kappa_s = \frac{d\beta}{ds} - \frac{d(u/R_s)}{ds}, \qquad \kappa_\theta = \frac{\sin\phi}{r}\left(\beta - \frac{u}{R_s}\right) \tag{10.11.7}$$

where γ is assumed small in comparison with dw/ds and β. Including $\gamma = dw/ds - \beta$, the strain vector of the problem is, with $\mathfrak{z}$ an outward distance normal to the midsurface,

$$\{\epsilon_s \ \ \epsilon_\theta \ \ \gamma\} = \{\bar{\epsilon}_s \ \ \bar{\epsilon}_\theta \ \ \gamma\} - \mathfrak{z}\,\{\kappa_s \ \ \kappa_\theta \ \ 0\} \tag{10.11.8}$$

These strain-displacement relations involve only first derivatives, so the assumed field need be only linear in s (Fig. 10.8.1).

$$\{u\ w\ \beta\} = \sum_{i=1,2} N_i\,\{u_i\ w_i\ \beta_i\} \quad \text{where} \quad N_1 = \frac{L-s}{L} \quad \text{and} \quad N_2 = \frac{s}{L} \tag{10.11.9}$$

Nodal d.o.f. are as shown in Fig. 10.9.1*a*. The element meridian is straight, so $R_s = \infty$, and some terms in Eqs. 10.11.8 disappear. The same is true of the element in Fig. 10.11.1*c*. Thickness integration is done explicitly, and the remaining integration is done by a single Gauss point at the element center, $s = L/2$. Test cases show good performance, but doubts about the quality of straight-sided elements remain (Section 10.2 and Ref. 10.31).

Interelement compatibility prevails when one edge of an axially symmetric quadratic element (Fig. 5.6.1*a*) is attached to the $\eta = 1$ surface of the element of Fig. 10.11.1*a*. The quadratic element is *not* compatible with the shell element of Fig. 10.9.1. These considerations find application in the analysis of rocket motors, where the propellant is bonded to the case.

PROBLEMS

10.1 A doubly curved shell is modeled by the elements of Eq. 10.2.2. Is interelement compatibility present in a coarse mesh? Or in a fine mesh?

10.2 Write an equation, similar to Eq. 10.2.3, appropriate to a flat element that has four nodes. Why is this equation not suited to a *warped* quadrilateral?

10.3 (a) Derive the expression for u in Eqs. 10.3.2 and show the two curves on a plot of u versus s.
(b) What are the corresponding expressions if the angle is not 90° but θ, as in Fig. 10.3.1*b*? Do these expressions agree as θ approaches zero?

10.4 Let a thin, spherical shell be modeled by three elements. The mesh is absurdly coarse: a cylinder of length $2R$, capped by two discs of radius R. Let internal pressure be represented by nodal forces, without nodal moments. Argue why computed results will display bending moments. Would the addition of nodal moment loads or a refinement of the mesh change the qualitative conclusion?

10.5 Modes a_7 and a_8 in Eq. 10.3.3 are beneficial to curved beam and axially symmetric shell elements. Would they also benefit the elements of Problems 4.26 and 4.27? Explain. (You could test your prediction by making a few changes in a standard truss analysis computer program.)

10.6 Consider finding a_7 and a_8 in Eq. 10.3.3 so as to fit the sine curve in Eq. 10.3.2. Describe how this might be done *prior* to formulating element matrices.

10.7 Let nodal d.o.f. for the arch element of Fig. 10.4.1*a* be $\{\mathbf{d}\} = \{u_1 \; w_1 \; w_{,x1} \; u_2 \; w_2 \; w_{,x2}\}$. Let u be linear in x (as for a truss element) and w be cubic in x (as for a beam element). Also let $z = 4hx(\ell - x)/\ell^2$, which describes a parabolic arch with a rise of h units.

(a) Show that rigid-body translations in the x and z directions, respectively, produce zero ϵ_m and zero β if R is infinite.

(b) Write a simple $\{\mathbf{d}\}$ that yields $\epsilon_m = 0$.

10.8 (a) Rotational d.o.f. suggested in Problem 10.7 are $w_{,x1}$ and $w_{,x2}$. Instead, we might use β_1 and β_2 (Eq. 10.4.1). How then would you determine $[\mathbf{N}]$ in the displacement field $\{u \; w\} = [\mathbf{N}] \{u_1 \; w_1 \; \beta_1 \; u_2 \; w_2 \; \beta_2\}$? Explain, but do not carry out all the algebra.

(b) Consider rigid-body rotation about the center of curvature of the arch element in Problem 10.7. Do either the w_{xi} d.o.f. or the β_i d.o.f. allow strain-free motion? (See also Problem 10.34.)

10.9 This problem is a programming project. The strain energy U for the arch element of Fig. 10.4.1*a* and Eqs. 10.4.1 is

$$U = \frac{1}{2}\int EA\,\epsilon_m^{\,2}\,ds + \frac{1}{2}\int EI\,\beta_{,x}^{\,2}\,ds$$

where ds is an increment of arc length (assume that $ds \approx dx$). From U, generate and program $[\mathbf{k}]$. Study some tests cases, such as rigid-body motions, the effect of mesh refinement, and others in References 10.3 and 10.4. You might also study the effect of using arc length L instead of chord length ℓ in the integrations and the effect of adding one or two nodeless d.o.f. to u (Eqs. 10.3.3). The performance of straight and curved elements can be compared. At the cost of reformulation, alternate nodal d.o.f. can be used (Problem 10.8).

10.10 What difficulties of element formulation do you foresee if the flat element of Fig. 10.2.1 is to use displacement fields of the same degree for all three displacement components?

10.11 Consider a folded plate whose component slabs join at right angles. Let it be modeled by finite elements that have three translational and three rotational d.o.f. per node. What d.o.f. can probably be set to zero along the line and why? Should finite elements be used to analyze this structure?

10.12 Consider elements that meet along the fold line of a folded plate. Is interelement compatibility provided by the elements of Eq. 10.2.2? Or by the elements of Fig. 10.4.2*c*?

10.13 (a) Write the displacement field for the element of Fig. 10.4.2*b* in a form analogous to Eq. 10.5.1.

(b) Convert the field of part (a) to a form that uses nodal d.o.f. u_{ip}, v_{ip}, w_{ip} (on the inner surface) and u_{ipq}, v_{ipq}, w_{ipq} (displacements on the outer surface *relative* to u_{ip}, v_{ip}, w_{ip}).

10.14 Why should the shell elements of Section 10.5 be called "superparametric"? Why are they nevertheless valid?

10.15 (a) Is it necessary that rotation vectors α_i and β_i in Eq. 10.5.5 be mutually perpendicular? Explain.

(b) Imagine that instead of α_i and β_i in Fig. 10.5.1 we elect to use α_1, and α_2, and α_3, which are nodal rotations (in the right-hand sense) about global axes x, y, and z. Let z' be a local normal to the shell. Write the new form of Eq. 10.5.5 and define the new matrix $[\boldsymbol{\mu}_i]$.

(c) How could z' be defined? Consider a four-node element in your answer.

(d) Elements 1 and 2 lie in the xz and yz planes, respectively. They join along the global z-axis. Imagine that the nodal rotation α_3 takes place at node i. Show that matrix $[\boldsymbol{\mu}_i]$ of part (b) yields the expected motion at corners A and B.

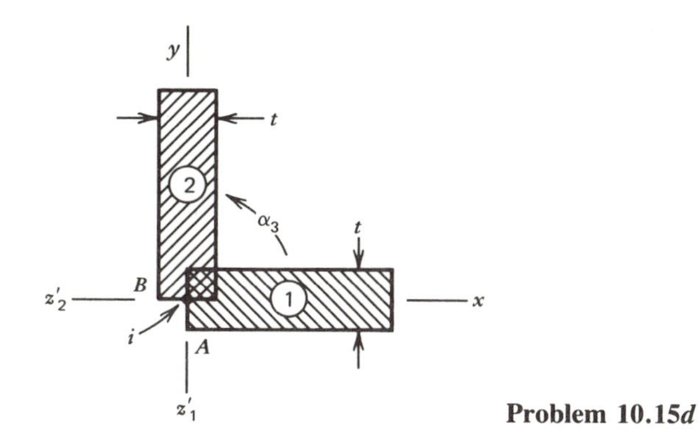

Problem 10.15*d*

10.16 Two elements of the type shown in Fig. 10.4.2*c* are shown in cross section, where they meet along a fold line in a shell. In sketch (a), vector $\mathbf{C}_i$ is not normal to the midsurface of either element. In sketch (b), the elements lack a common surface-parallel tangent. What are the advantages and disadvantages of the two types of junction? Is there a procedure that can make either type acceptable?

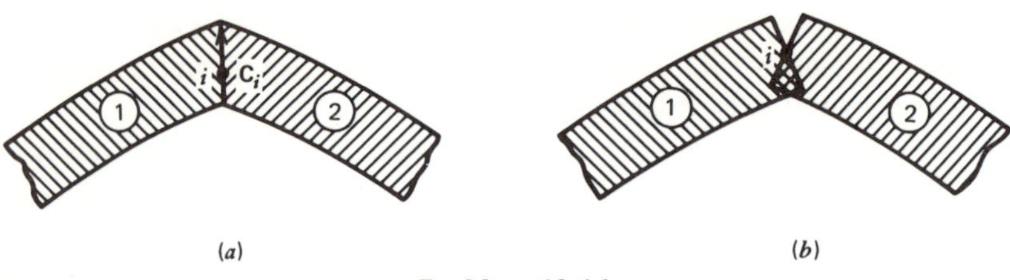

Problem 10.16

10.17 Imagine that $\mathbf{A}_i$ and $\mathbf{C}_i$ of Fig. 10.5.1*c* are provided as input data at all *n* nodes of a structure. Write about 15 Fortran statements that will, for all *n* nodes: check that $\mathbf{A}_i$ and $\mathbf{C}_i$ are nearly orthogonal, compute $\mathbf{B}_i$, recompute $\mathbf{A}_i$ so that the three vectors are exactly orthogonal, and compute the nine direction cosines.

10.18 Let a typical $[\mathbf{B}_i]$ in Eq. 10.5.12 have the form $[\mathbf{B}_i] = [\mathbf{H}]\,[\mathbf{B}_i^*]$, where $[\mathbf{B}_i^*]$ is 9 by 5. Express $[\mathbf{B}_i^*]$ as a function of ζ, t_i, N_i, $N_{i,\xi}$, $N_{i,\eta}$, the Γ_{ij}, and the μ_{ij}.

10.19 Consider the beam element cited at the end of Section 10.5. Let nodal rotation vectors α_i, β_i, and γ_i point in the $+\eta$, $-\xi$, and $-\zeta$ directions, respectively, and let vectors **B** and **C** span the width *b* and height *c* of a beam of rectangular cross section. Coordinate ξ is axial.
   (a) Write the equation that corresponds to Eq. 10.5.4
   (b) Write N_1, N_2, and N_3 (in terms of ξ).
   (c) Write the equation that corresponds to Eq. 10.5.5.
   (d) If torsional stiffness of the beam is neglected and the material is isotropic, what $[\mathbf{E}']$ is appropriate? (See Eq. 10.6.1.)

10.20 Compare the number of multiplications needed to produce the two matrix triple products in the two forms of Eq. 10.6.4. Exploit symmetry but not the possible skipping of products that are zero.

10.21 (a) Write expressions for the nine coefficients in the Jacobian matrix of an isoparametric shell element (see Eqs. 5.7.1 and 10.5.4 and Fig. 10.4.2*c*).
   (b) If the element is flat, of constant thickness, and in the *xy* plane, what becomes of the Jacobian coefficients that contain ζ?

10.22 What Gauss quadrature rule is needed to yield the exact volume of the element in Fig. 10.4.2*c*? Is this the rule we are *obliged* to use?

10.23 Consider a sandwich beam of thickness $c + 2h$ and a homogeneous beam of thickness $2h$. For each, compute the moments, M_h and M_s, caused by the temperature gradient $T = \zeta$ if curvature is prohibited. Also compute the adjustment factor that yields M_s from M_h.

10.24 Consider a membrane shell (a shell that has no bending stiffness). Write the appropriate forms of equations that simplify in Sections 10.5 and 10.6 and briefly justify the changes you make.

10.25 If the shell element of Fig. 10.4.2*c* is very thin, the conditions $\gamma_{BC} = \gamma_{CA} = 0$ can be explicitly imposed at $\xi = \pm\sqrt{3}/3$ and $\eta = \sqrt{3}/3$ (see Eq. 10.6.1 for notation). Simultaneously, two d.o.f. can be removed at each side node—displacement normal to the shell and rotation about a normal to the edge—to yield an element with the 32 d.o.f. $\{\mathbf{d}_r\}$ instead of the original 40 d.o.f. $\{\mathbf{d}\}$. Consider the transformation $\{\gamma\ \mathbf{d}_r\} = [\mathbf{T}]\ \{\mathbf{d}\}$, where $\{\gamma\}$ contains the eight strains in question. How can $[\mathbf{T}]$ be constructed? And how is the condition $\{\gamma\} = 0$ enforced and the 32-by-32 stiffness matrix $[\mathbf{k}_r]$ obtained?

10.26 (a) Let a coordinate x run along the chord in Fig. 10.8.1 from node 1 to node 2. Let y be the elevation of the element meridian above the x-axis. Describe $y = f(x, \ell, \phi_1, \phi_2, \phi_c)$ as a beam deflection (Eqs. 4.2.5). Show that this expression for y contains the constants in Eq. 10.8.4.
(b) Hence, what is the physical meaning of Eq. 10.8.3?

10.27 (a) Verify that the α_i in Eq. 10.8.4 satisfy the conditions imposed on ϕ of Eq. 10.8.2.
(b) Verify Eq. 10.8.6.
(c) Verify Eq. 10.8.7.

10.28 Let $\theta_1 = -\theta_2$ in Eq. 10.8.6. Compute the ratio of approximate L to exact L for (*a*) $\theta_1 = 2°$, (*b*) $\theta_1 = 4°$, (*c*) $\theta_1 = 8°$, and (*d*) $\theta_1 = 30°$.

10.29 Write the form of Eqs. 10.8.9 appropriate to an arch (Fig. 10.4.1*a*).

10.30 Specialize Eqs. 10.8.9 to the case of (*a*) a flat circular plate, and (*b*) a circular cylindrical shell.

10.31 Express Eqs. 10.8.9 in terms of u, v, w, r, ϕ, and derivatives with respect to s and θ. Omit X, Y, and Z from the membrane strains in this exercise.

10.32 Consider the rigid-body motion that leads to Eq. 10.3.2. Use Eq. 10.3.1 and the definition of β_i in Eq. 10.9.3 and find the polynomial for w in terms of H, L, R, and s.

10.33 Let a constant internal pressure p act on a spherical shell element for which $r_1 = r_2$ in Fig. 10.8.1. Write the integral that yields a-basis nodal loads $\{\mathbf{r}_a\}$, with s the only independent variable. Also show, on a sketch like Fig. 10.9.1, how the d-basis loads $\{\mathbf{r}\}$ would be directed.

10.34 A circular cylindrical shell is capped by a segment of a spherical shell. Consider the nodal circle where the two shells meet. Show that an axial rigid-body motion yields $\beta_i = 0$ in Eq. 10.9.3, whether calculated from the cylindrical part or from the spherical part.

10.35 Write the terms in matrix $[\mathbf{A}]$ of Eq. 10.9.5.

10.36 In place of u in Eq. 10.9.1, we could write

$$u = [a_5 + a_6 s + a_9 s^2 + a_{10} s^3] \cos n\theta$$

and similarly for v. How does this choice alter the development outlined in Section 10.9?

10.37 Consider axially symmetric bending of a thin circular cylindrical shell ($v = 0$, axial membrane stress is zero, and all θ derivatives are zero). Let the material be isotropic and the thickness constant. Generate the 4-by-4 element stiffness matrix. (The result is essentially the sum of matrices for beam bending and an elastic foundation.)

10.38 The sketch shows the cross section of an axially symmetric conical shell element loaded by axially symmetric line loads that are tangent to its meridian.

(a) Why is it specious to model the element by applying a coordinate transformation through angle ϕ to the element of Problem 8.16?

(b) Write [**N**] and {**d**} in an assumed displacement field {**f**} = [**N**] {**d**} that will probably be satisfactory for this element. Also write the appropriate form of [**E**] and expressions for $\epsilon_{r'}$ and ϵ_θ in terms of u_1, w_1, u_2, w_2, s, r, L, and ϕ.

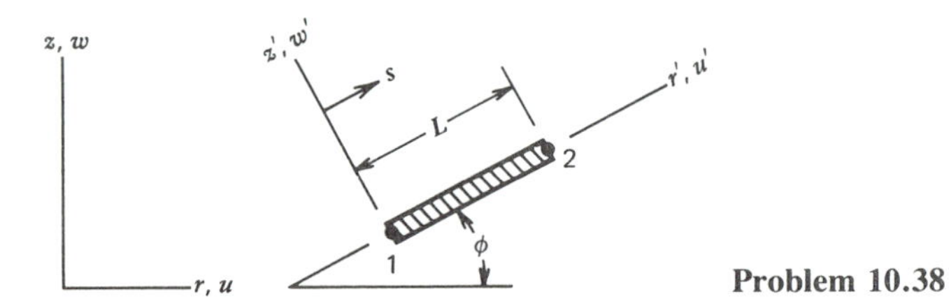

Problem 10.38

10.39 Show that $\bar{\alpha}_i$ is needed in Eq. 10.11.3 if the shell is to have bending stiffness in the circumferential direction. What other purpose does $\bar{\alpha}_i$ serve? Why is $\bar{\alpha}_i$ not needed as a d.o.f. in Eq. 10.9.3?

10.40 Consider numerical integration to find [**k**] for the element of Fig. 10.11.1*c*. If two Gauss points are used, where must they be placed?

10.41 (a) Start with Eq. 10.11.2 and write the component matrices that combine to yield [**B**], as is done, for example, in Eqs. 5.3.10, 5.3.11, and 5.3.12.

(b) Again, for the axially symmetric case, write the transformation matrix needed to convert [**E**′] in Eq. 10.11.5 to [**E**] in Eq. 10.11.4.

10.42 Consider the element described in connection with Eqs. 10.11.6 through 10.11.9. It is tempting to try to improve the element by using R_s of the actual shell instead of setting $R_s = \infty$ because the element meridian is straight. What is bad about this idea?

10.43 Use Eqs. 10.11.8 and 10.11.9 to write expressions for the strains ϵ_s, ϵ_θ, and γ. Express answers in terms of s, r, ϕ_c, L, and the nodal d.o.f. Group terms in the form $\{\epsilon\} = ([\mathbf{B}_0] + z[\mathbf{B}_1])\,\{\mathbf{d}\}$.

10.44 A flat, circular disc (radius r, thickness t) is mounted on a circular shaft (radius ρ). A load parallel to the shaft is uniformly distributed around the outer edge of the disc. What sort of finite element solution do you propose if the disc is thick? And what sort if the disc is thin ($r >> t$)?

10.45 A flange is modeled by four-node solid of revolution elements. The flange is to be attached to a structure of thin-shell elements. At the juncture, nodes A and B are on the flange and node C is on the shell. What transformation matrix [**T**] should be written so that d.o.f. of node C can be replaced by the d.o.f. of nodes A and B?

(a) Let the problem be axially symmetric.

(b) Let the problem be without axial symmetry.

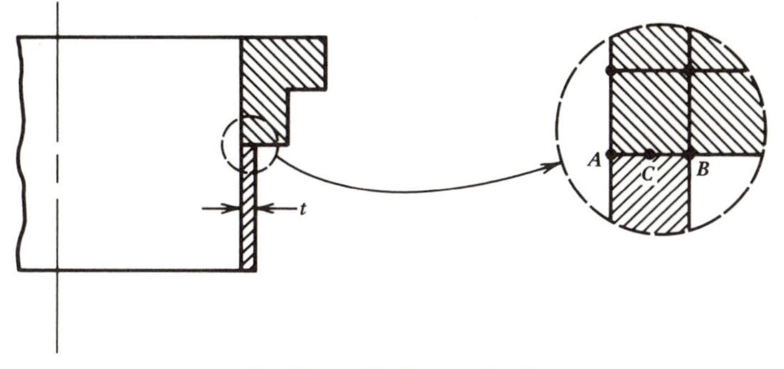

Problems 10.45 and 10.46

10.46 Repeat Problem 10.45 but do not write [**T**]. Instead, generate a constraint matrix [**C**], as in Section 6.8.

CHAPTER 11

FINITE ELEMENTS IN DYNAMICS AND VIBRATIONS

11.1 INTRODUCTION

If forces shake a structure at less than roughly one-third its lowest natural frequency of vibration, we can treat the problem as static. That is, equations $[\mathbf{K}]\{\mathbf{D}\} = \{\mathbf{R}\}$ are accurate enough, even though loads $\{\mathbf{R}\}$ vary with time. More rapid shaking makes the inertia of the structure important, and the problem must be considered dynamic. Inertia is accounted for by the *mass matrix,* written as $[\mathbf{m}]$ for an element and as $[\mathbf{M}]$ for the structure. (If damping is significant, analogous *damping matrices* $[\mathbf{c}]$ and $[\mathbf{C}]$ are introduced.)

Inertia loads produced by gravity and centrifugal force are sometimes called "inertia relief" forces. They do not vary with time, in magnitude or direction. They are accounted for by $\{\mathbf{F}\}$ in Eq. 4.3.5 but could also be computed by an integral over $[\mathbf{m}]\{\ddot{\mathbf{d}}\}$, where $\{\ddot{\mathbf{d}}\} = \partial^2\{\mathbf{d}\}/\partial t^2$.

In contrast to a static problem, $[\mathbf{K}]$ can be singular in a dynamic problem, since (depending on the solution algorithm) mass "holds the structure together."

Problems in structural dynamics fit into two broad classes. In one we ask for the natural frequencies of vibration and the corresponding mode shapes. In the other we ask how the structure moves with time under a prescribed load, impulse, or ground acceleration.

Structural dynamics has an extensive literature and good textbooks [11.1, 11.52]. The methods of structural dynamics are largely independent of finite element analysis because these methods presume the availability of stiffness, mass, and damping matrices but do not demand that the matrices arise from a finite element discretization. Solution algorithms are numerous and proliferating. Accordingly, here we do not attempt to be comprehensive. Techniques we discuss are commonly used with finite elements and the banded equation systems they produce.

11.2 MASS AND DAMPING MATRICES. DYNAMIC EQUATIONS

D'Alembert's principle, while distasteful to some, easily leads to a definition of the mass matrix and conveys its physical meaning. In the usual notation, an assumed element displacement field $\{\mathbf{f}\} = \{u\ v\ w\}$ and its first two time drivatives are

$$\{\mathbf{f}\} = [\mathbf{N}]\{\mathbf{d}\}, \qquad \{\dot{\mathbf{f}}\} = [\mathbf{N}]\{\dot{\mathbf{d}}\}, \qquad \{\ddot{\mathbf{f}}\} = [\mathbf{N}]\{\ddot{\mathbf{d}}\} \tag{11.2.1}$$

where $\{\dot{\mathbf{d}}\}$ and $\{\ddot{\mathbf{d}}\}$ are, respectively, nodal velocities and nodal accelerations. An acceleration field $\{\ddot{\mathbf{f}}\}$ produces d'Alembert body forces $\{\mathbf{F}\}$ in the opposite direction, so $\{\mathbf{F}\} = -\rho\,\{\ddot{\mathbf{f}}\}$, where ρ is mass density. Nodal loads $\{\mathbf{r}\}_\rho$ associated with $\{\mathbf{F}\}$ are given by Eq. 4.3.5. Thus, with Eq. 11.2.1,

$$\{\mathbf{r}\}_\rho = \int_V [\mathbf{N}]^T\{\mathbf{F}\}\,dV = -\int_V \rho[\mathbf{N}]^T[\mathbf{N}]\,dV\,\{\ddot{\mathbf{d}}\} = -[\mathbf{m}]\{\ddot{\mathbf{d}}\} \qquad (11.2.2)$$

where the *element mass matrix* $[\mathbf{m}]$ is defined as

$$[\mathbf{m}] = \int_V \rho[\mathbf{N}]^T[\mathbf{N}]\,dV \qquad (11.2.3)$$

This $[\mathbf{m}]$ is known as the *consistent* mass matrix because it is formulated from the same shape functions $[\mathbf{N}]$ that are used to formulate the stiffness matrix $[\mathbf{k}]$. In general, $[\mathbf{m}]$ is full but symmetric.

The field defined by $[\mathbf{N}]$ is rarely exact in a dynamic problem. For example, a uniform, simply supported beam vibrates as a sine wave, so the cubic beam polynomial does not permit a coarse mesh to yield an exact natural frequency.

The structure mass matrix $[\mathbf{M}]$ is built by the conceptual expansion of element matrices $[\mathbf{m}]$ to "structure size," followed by addition of the overlapping coefficients, exactly as explained for stiffness matrices in Sections 2.5 and 2.6. Similarly, for damping matrices—which we introduce now and comment on later—we symbolize the assembly process as $[\mathbf{C}] = \Sigma\,[\mathbf{c}]$.

Following D'Alembert, we place nodes in equilibrium under all loads applied to them, including the dynamic loads. Dynamic loads include inertia forces $\Sigma\,\{\mathbf{r}\}_\rho$ from Eq. 11.2.2 and damping forces $\Sigma\,[\mathbf{c}]\{\dot{\mathbf{d}}\}$. Thus the governing equation of motion for the discretized system is

$$\begin{aligned} \{\mathbf{R}\} - [\mathbf{K}]\{\mathbf{D}\} - [\mathbf{C}]\{\dot{\mathbf{D}}\} - [\mathbf{M}]\{\ddot{\mathbf{D}}\} &= 0 \\ \text{or} \qquad [\mathbf{K}]\{\mathbf{D}\} + [\mathbf{C}]\{\dot{\mathbf{D}}\} + [\mathbf{M}]\{\ddot{\mathbf{D}}\} &= \{\mathbf{R}\} \end{aligned} \qquad (11.2.4)$$

A column j of $[\mathbf{M}]$ (or of $[\mathbf{m}]$) represents the vector of generalized forces applied to the structure (or to an element) when generalized coordinate j has unit acceleration while all other d.o.f. have zero acceleration. A similar remark applies to $[\mathbf{C}]$ and $[\mathbf{c}]$ if we say "velocity" instead of "acceleration."

Damping matrix $[\mathbf{c}]$ is easily evaluated for a Newtonian fluid. Its terms are given by Rayleigh [11.2]. But in structures we are less interested in viscous damping than in dry friction and hysteresis loss. These energy-loss mechanisms are not well understood, so they are approximated [11.1]. A popular scheme is to combine a fraction α of the stiffness matrix with a fraction β of the mass matrix. Thus

$$[\mathbf{C}] = \alpha[\mathbf{K}] + \beta[\mathbf{M}] \qquad (11.2.5)$$

This is *Rayleigh* or *proportional* damping, and [**C**] is an *orthogonal* damping matrix because it permits modes to decouple (Section 11.7). With proportional damping, eigenvalues are real; with nonproportional damping, they are complex. If $\alpha = 0$, the higher modes are lightly damped; if $\beta = 0$, the higher modes are heavily damped [11.1].

11.3 MASS MATRICES, CONSISTENT AND DIAGONAL

Development of the consistent mass matrix is credited to Archer [11.3]. A simpler and historically earlier formulation is the *lumped* mass matrix. A lumped mass matrix is diagonal. Conceptually, lumping implies that translational motion of a node is shared by the immediately adjacent part of the element, while the remainder of the element is unaffected (Fig. 11.3.1). Or, we can say that the distributed element mass is replaced by particle masses at the nodes. These "lumps" have no rotary inertia.

Examples

Consider the uniform truss element of Fig. 11.3.1. Shape functions to be used in Eq. 11.2.3 are given by Eq. 3.7.6. The mass increment $\rho\, dV$ in Eq. 11.2.3 can be written $(m/L)\, dx$, where m is the total mass of the element. The consistent and lumped mass matrices are, respectively,

$$[\mathbf{m}] = \frac{m}{6}\begin{bmatrix} 2 & 0 & 1 & 0 \\ 0 & 2 & 0 & 1 \\ 1 & 0 & 2 & 0 \\ 0 & 1 & 0 & 2 \end{bmatrix}, \qquad \lceil \mathbf{m} \rfloor = \frac{m}{2}\begin{bmatrix} 1 & 0 & 0 & 0 \\ 0 & 1 & 0 & 0 \\ 0 & 0 & 1 & 0 \\ 0 & 0 & 0 & 1 \end{bmatrix} \tag{11.3.1}$$

For the uniform beam element of Fig. 11.3.1, shape functions are given by Eq. 4.2.5. Equation 11.2.3 yields the consistent mass matrix

$$[\mathbf{m}] = \frac{m}{420}\begin{bmatrix} 156 & 22L & 54 & -13L \\ 22L & 4L^2 & 13L & -3L^2 \\ 54 & 13L & 156 & -22L \\ -13L & -3L^2 & -22L & 4L^2 \end{bmatrix} \tag{11.3.2}$$

where $m = \rho AL$ is the total element mass. The lumped mass matrix of the beam element is

$$\lceil \mathbf{m} \rfloor = \frac{m}{2}\lceil 1 \quad 0 \quad 1 \quad 0 \rfloor \tag{11.3.3}$$

If an element is tapered rather than uniform, its mass and stiffness matrices change, but matrix $[\mathbf{k}_\sigma]$ of Eqs. 12.2.6 and 12.2.13 is unaffected [11.4]. (Further examples appear in Fig. 11.3.2.)

A plane frame has six d.o.f. per node. Its mass matrix is formed by expanding and then combining the truss and beam mass matrices. An element arbitrarily oriented in xy

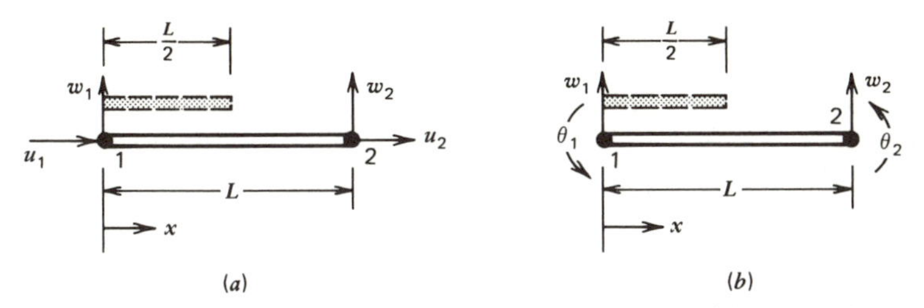

Figure 11.3.1. Uniform truss and beam elements, with the respective $\{\mathbf{d}\}$ vectors $\{u_1\ w_1\ u_2\ w_2\}$ and $\{w_1\ \theta_1\ w_2\ \theta_2\}$. Dashed lines show the (conceptual) deformed shapes used to find the lumped mass coefficients associated with d.o.f. w_1.

Constant strain triangle (Fig. 4.5.1). With a linear displacement field in each direction, and $\{\mathbf{d}\} = \{u_1\ u_2\ u_3\ v_1\ v_2\ v_3\ w_1\ w_2\ w_3\}$,

$$\underset{9\times 9}{[\mathbf{m}]} = \lceil \mathbf{Q}\ \mathbf{Q}\ \mathbf{Q} \rfloor, \qquad \text{where} \qquad [\mathbf{Q}] = \frac{\rho At}{12}\begin{bmatrix} 2 & 1 & 1 \\ 1 & 2 & 1 \\ 1 & 1 & 2 \end{bmatrix}$$

Linear rectangle (Fig. 4.6.1). With the same displacement field in each direction, and $\{\mathbf{d}\} = \{u_1\ u_2\ u_3\ u_4\ v_1\ v_2\ v_3\ v_4\ w_1\ w_2\ w_3\ w_4\}$,

$$\underset{12\times 12}{[\mathbf{m}]} = \lceil \mathbf{Q}\ \mathbf{Q}\ \mathbf{Q} \rfloor, \qquad \text{where} \qquad [\mathbf{Q}] = \frac{\rho At}{36}\begin{bmatrix} 4 & 2 & 1 & 2 \\ 2 & 4 & 2 & 1 \\ 1 & 2 & 4 & 2 \\ 2 & 1 & 2 & 4 \end{bmatrix}$$

Linear strain triangle (Fig. 7.9.2*a*). With straight sides and midside nodes, the same displacement field in each direction, and

$$\{\mathbf{d}\} = \{u_1 \ldots u_6\ v_1 \ldots v_6\ w_1 \ldots w_6\}, \quad \underset{18\times 18}{[\mathbf{m}]} = \lceil \mathbf{Q}\ \mathbf{Q}\ \mathbf{Q} \rfloor,$$

where [11.13]

$$[\mathbf{Q}] = \frac{\rho At}{180}\begin{bmatrix} 6 & -1 & -1 & 0 & -4 & 0 \\ -1 & 6 & -1 & 0 & 0 & -4 \\ -1 & -1 & 6 & -4 & 0 & 0 \\ 0 & 0 & -4 & 32 & 16 & 16 \\ -4 & 0 & 0 & 16 & 32 & 16 \\ 0 & -4 & 0 & 16 & 16 & 32 \end{bmatrix}$$

Figure 11.3.2. Consistent mass matrices for plane elements allowed to move in three dimensions. A = surface area, ρ = constant mass density, and t = constant thickness.

coordinates requires the coordinate transformation of Eq. 6.3.8. A similar transformation applies to damping matrices.

With any mass matrix, the product $[\mathbf{m}]\{\ddot{\mathbf{d}}\}$ or $\lceil \mathbf{m} \rfloor\{\ddot{\mathbf{d}}\}$ must yield the correct total force $\mathbf{F}$ on the element according to Newton's law $\mathbf{F} = m\mathbf{a}$ when $\{\ddot{\mathbf{d}}\}$ represents a rigid-body translational acceleration. The rationale is that for convergence to correct results, this kind of motion must be correctly represented because it is the only motion experienced by an element when a mesh is indefinitely refined.

Consistent mass matrices $[\mathbf{m}]$ and $[\mathbf{M}]$ are positive definite. A lumped mass matrix is positive semidefinite if zeros appear on-diagonal. The zeros may (or may not) make some operations awkward, depending on the algorithm. A lumped mass matrix requires little storage space. It is cheaper to form and manipulate than a consistent mass matrix. If the mesh layout correctly represents the structure volume, elements are compatible and not softened by low-order integration rules, and mass matrices are consistent, then computed natural frequencies are upper bounds to the exact values. If any of these restrictions is violated, a bound cannot be guaranteed [11.5]. The upper bound is computed with an error of order $h^{2(q-1)}$, where h and q are defined in Section 15.4 [11.6]. The upper-bound property is illustrated by the numerical example that closes Section 11.5.

We cannot say that either lumped or consistent mass matrices are best for all problems. Consistent matrices are more accurate for flexural problems, such as beams and shells, but negligibly so if the wavelength of the mode spans more than about four elements [11.7]. Lumped matrices are simpler to form, cheaper to use, and usually yield natural frequencies that are less than the exact values. For a bar element, MacNeal [15.27] finds that $[\mathbf{m}]$ and $\lceil \mathbf{m} \rfloor$ of Eq. 11.3.1 yield natural frequency errors of order h^2 in opposite directions and that an $[\mathbf{m}]$ that is the *average* of the two yields a natural frequency error of only order h^4. In wave propagation problems using elements with linear displacement fields, lumped and consistent mass matrices predict that waves propagate too slow and too fast, respectively. Lumped masses give greater accuracy and fewer spurious oscillations than a consistent mass matrix. So here again an average mass matrix may be best [11.8].

The preceding paragraph suggests that accuracy is only slightly damaged if mass is omitted from element d.o.f. that are to be condensed before assembly. If such mass coefficients are present, mass condensation is required, as described in Section 11.5.

A diagonal mass matrix that is more sophisticated than a lumped mass matrix can be derived from a consistent mass matrix as follows [11.9, 11.10]. The following procedure applies to elements whose translational d.o.f. are mutually parallel, such as beam and plate elements.

1. Compute only the diagonal coefficients of the consistent mass matrix.
2. Compute m, the total mass of the element.
3. Compute a number s by adding the diagonal coefficients m_{ii} associated with translation (but not rotation).
4. Scale the diagonal coefficients m_{ii} by multiplying them by the ratio m/s, thus preserving the translational mass of the element.

Example

For the bar and beam elements of Fig. 11.3.1, the preceding four steps yield, respectively, the diagonal matrices

$$\lfloor \mathbf{m} \rfloor = \frac{m}{2} \lfloor 1 \quad 1 \quad 1 \quad 1 \rfloor \tag{11.3.4}$$

$$\lfloor \mathbf{m} \rfloor = \frac{m}{78} \lfloor 39 \quad L^2 \quad 39 \quad L^2 \rfloor \tag{11.3.5}$$

Test cases to date show that the accuracy of this form of diagonal mass matrix is excellent, often surpassing that of the consistent mass matrix (Table 11.3.1). (In Table 11.3.1, the consistent mass matrix does not guarantee an upper bound because the stiffness matrix is based on reduced quadrature.)

A simpler and more general procedure is to assign particle masses m_i to the nodes, where

$$m_i = \int_V N_i \, \rho \, dV \tag{11.3.6}$$

Shape functions N_i represent the lowest order field that spans the nodes (linear for *both* elements in Fig. 11.3.1). Since $\Sigma N_i = 1$, Σm_i is the total element mass, and no scaling is needed. Mass coefficients associated with rotational d.o.f. are much less important, and if needed can be approximated, even by a crude displacement field. Rotary inertias from thick elements can be approximated as $m_i t_i^2/4$, where t_i is the element thickness at node i [11.10].

TABLE 11.3.1. NATURAL FREQUENCIES OF A SIMPLY SUPPORTED THICK SQUARE PLATE [11.9]. HALF THE PLATE WAS MODELED BY A 4-BY-2 MESH OF THE EIGHT-NODE, 24 d.o.f. ELEMENTS OF FIG. 9.3.1*c*. THE LUMPED MASS MATRIX HAS EQUAL MASS PARTICLES AT EACH NODE. THIS TABLE REPORTS PERCENTAGE ERRORS WITH RESPECT TO THE THEORY OF ELASTICITY SOLUTION.

Mode		Type of mass matrix used		
m	*n*	Consistent	Diagonal [11.9]	Lumped
1	1	− 0.11	+ 0.32	+ 0.32
2	1	− 0.40	+ 0.45	− 0.45
2	2	− 0.35	− 2.75	− 4.12
3	1	+ 5.18	+ 0.05	− 5.75
3	2	+ 4.68	− 2.96	−10.15
3	3	+13.78	− 5.18	−19.42
4	2	+16.88	+ 1.53	+31.70

Nodes of Lagrangian elements coincide with sampling points of the Lobatto quadrature rule. Accordingly, Eq. 11.2.3 and the Lobatto rule produce a lumped mass matrix for a Lagrangian element.

If $[\mathbf{N}]$ in Eq. 11.2.3 is of lower degree than the $[\mathbf{N}]$ used to construct $[\mathbf{k}]$, then $[\mathbf{m}]$ is neither consistent nor lumped. It may be called "distributed." Like a lumped mass matrix, a distributed $[\mathbf{m}]$ does not guarantee a bound on computed frequencies, but it provides correct convergence with mesh refinement. Also, it may be cheaper to form and use than a consistent $[\mathbf{m}]$.

11.4 NATURAL FREQUENCIES. THE EIGENVALUE PROBLEM

We consider natural frequencies of vibration, with damping $[\mathbf{C}]$ and external forces $\{\mathbf{R}\}$ both zero. Each d.o.f. executes harmonic motion in phase with all other d.o.f. Therefore

$$\{\mathbf{D}\} = \{\bar{\mathbf{D}}\} \sin \omega t, \qquad \{\ddot{\mathbf{D}}\} = -\omega^2 \{\bar{\mathbf{D}}\} \sin \omega t \tag{11.4.1}$$

where $\{\bar{\mathbf{D}}\}$ = amplitudes of nodal d.o.f. and ω = *circular frequency* (radians per second). The *cyclic frequency* f (in hertz) is $f = \omega/2\pi$. The *period* is $T = 1/f$ (seconds). In common parlance and in this book, both ω and f are called simply "frequency." Also, ω and f pertain to undamped motion unless stated otherwise.

Equations 11.2.4 and 11.4.1 yield

$$([\mathbf{K}] - \lambda[\mathbf{M}])\{\bar{\mathbf{D}}\} = 0, \qquad \text{where } \lambda = \omega^2 \tag{11.4.2}$$

This is the basic statement of the vibration problem. If there is neither geometric nor material nonlinearity, then neither $[\mathbf{K}]$ nor $[\mathbf{M}]$ is a function of ω, and Eq. 11.4.2 is called a *linear eigenvalue problem.*

Equation 11.4.2 has the trivial solution $\{\bar{\mathbf{D}}\} = 0$. If $\{\bar{\mathbf{D}}\} \neq 0$, only certain values λ_i satisfy Eq. 11.4.2. These λ_i are the *eigenvalues* (or *characteristic numbers* or *latent roots*). To each there corresponds an *eigenvector* $\{\bar{\mathbf{D}}_i\}$, which is called a *natural* (or *normal,* or *characteristic*, or *principal*) mode. The lowest nonzero ω_i is called the *fundamental* vibration frequency. The *eigenproblem* is to extract the solution pairs λ_i and $\{\bar{\mathbf{D}}_i\}$.

If $[\mathbf{K}]$ is real, symmetric, and nonsingular, the number of nonzero independent eigenvalues that satisfy Eq. 11.4.2 is equal to the rank of $[\mathbf{M}]$. Accordingly, if $[\mathbf{M}]$ is a consistent mass matrix there are as many natural frequencies as there are unrestrained nodal d.o.f. If $[\mathbf{M}]$ is a lumped matrix some of its coefficients M_{ii} may be zero. Each zero M_{ii} is associated with an infinite frequency in Eq. 11.4.2. Ways to avoid this difficulty are described in the paragraph that follows Eq. 11.4.8.

A structure that is partly or completely unsupported has a singular stiffness matrix and displays one zero eigenvalue for each possible rigid-body motion. A singular [**K**] prohibits certain operations we may want to use (see following). To avoid a singular [**K**], we can substitute $\lambda = \lambda_0 + c$ into Eq. 11.4.2, where c is a negative number chosen by the analyst. Thus

$$([\mathbf{K} - c\mathbf{M}] - \lambda_0 [\mathbf{M}]) \{\bar{\mathbf{D}}\} = 0 \tag{11.4.3}$$

Now we solve for λ_{0i} and hence for λ_i. (Some eigenvalue routines calculate first the λ_{0i} values that correspond to $\lambda_i = 0$, which amounts to some wasted effort.) Eigenvectors are not changed by the transformation. Matrix $[\mathbf{K} - c\mathbf{M}]$ is positive definite if [**M**] is such that kinetic energy appears for each possible rigid-body motion. This will always happen unless [**M**] is not properly constructed. If $|c|$ is too small, $[\mathbf{K} - c\mathbf{M}]$ is nearly singular; if $|c|$ is too large, convergence is slow. A possible choice is $c = -0.01r$, where r is the ratio of the trace of [**K**] to the trace of [**M**] [11.11, 11.12].

Next, we consider how to set up the eigenproblem (Eq. 11.4.2). A change in the form of these equations, if made, should be suited to the computer algorithm. In turn, the choice of algorithm depends on the nature of the matrices and on how much information is to be computed. Literature in this area is so vast that here we can only present a few basic ideas.

Equation 11.4.2 is satisfied for nonzero $\{\bar{\mathbf{D}}\}$ only if the determinant of $([\mathbf{K}] - \lambda[\mathbf{M}])$ vanishes. Hence we obtain the *characteristic polynomial* in λ whose roots are the eigenvalues λ_i. But this method is useful only for small problems, say of order three or less. For larger problems, even when done by computer, the method is hopelessly lengthy and subject to error.

If we premultiply Eq. 11.4.2 by $\{\bar{\mathbf{D}}\}^T$ and solve for λ, we obtain the *Rayleigh quotient*

$$\lambda = \frac{\{\bar{\mathbf{D}}\}^T [\mathbf{K}] \{\bar{\mathbf{D}}\}}{\{\bar{\mathbf{D}}\}^T [\mathbf{M}] \{\bar{\mathbf{D}}\}} \tag{11.4.4}$$

Let $\{\bar{\mathbf{D}}_i\}$ approximate the ith eigenvector $\{\bar{\mathbf{D}}_i\}$ with a first-order error. Then λ_i approximates the corresponding eigenvalue with a second-order error. Also, the lowest λ_i is overestimated and the highest λ_i is underestimated. The quotient is, in fact, an extreme value when $\{\bar{\mathbf{D}}\}$ varies in the neighborhood of a exact eigenvector [2.1, 11.22]. (Accordingly, the extraction of an eigenvalue can be approached as a minimization problem.)

Eq. 11.4.2 is a *generalized* eigenvalue problem. The *standard* or *special* eigenvalue problem has the form

$$([\mathbf{H}] - \Lambda [\mathbf{I}]) \{\mathbf{Y}\} = 0 \tag{11.4.5}$$

where [**H**] is a *symmetric* matrix, [**I**] is a unit matrix, and {**Y**} is an eigenvector. Eigenvalues Λ that satisfy this equation are called eigenvalues of [**H**]. Some algorithms require the standard form. We can convert from generalized to standard form but must be careful, as noted in the following.

Premultiplication of Eq. 11.4.2 by $[\mathbf{M}]^{-1}$ yields

$$([\mathbf{M}]^{-1}[\mathbf{K}] - \lambda[\mathbf{I}])\{\bar{\mathbf{D}}\} = 0 \tag{11.4.6}$$

This is not the standard form because $[\mathbf{M}]^{-1}[\mathbf{K}]$ is in general not symmetric. But we can define

$$[\mathbf{M}] = [\mathbf{L}][\mathbf{L}]^T \qquad \text{and} \qquad \{\bar{\mathbf{D}}\} = [\mathbf{L}]^{-T}\{\mathbf{Y}\} \tag{11.4.7}$$

where $[\mathbf{L}]$ is a lower triangular matrix, obtained by Choleski decomposition. Substitution of Eq. 11.4.7 into Eq. 11.4.6 and premultiplication by $[\mathbf{L}]^T$ yields the standard form of Eq. 11.4.5, in which

$$[\mathbf{H}] = [\mathbf{L}]^{-1}[\mathbf{K}][\mathbf{L}]^{-T} \qquad \text{and} \qquad \Lambda = \lambda = \omega^2 \tag{11.4.8}$$

The theory of linear algebra shows that the transformations of Eqs. 11.4.6 through 11.4.8 do not change the eigenvalues of the problem [2.1, 15.24]. Natural modes $\{\bar{\mathbf{D}}_i\}$ are recovered from eigenvectors $\{\mathbf{Y}\}$ by Eq. 11.4.7. If $[\mathbf{M}]$ is a band matrix, so is $[\mathbf{L}]$, but $[\mathbf{L}]^{-1}$ is a *full* matrix and, therefore, so is $[\mathbf{H}]$. But if the mass matrix is *diagonal,* so are $[\mathbf{L}]$ and $[\mathbf{L}]^{-1}$, and the ith coefficient in $[\mathbf{L}]^{-1}$ is simply $1/\sqrt{M_{ii}}$. Thus $[\mathbf{H}]$ has the same band structure as $[\mathbf{K}]$.

The foregoing transformation fails if $[\mathbf{M}]$ is singular. This commonly happens if $[\mathbf{M}]$ is a lumped mass matrix, with rotational d.o.f. present in $\{\bar{\mathbf{D}}\}$ but not associated with rotary inertia. In such a case we can take the preliminary step of removing the rotational d.o.f. by condensation. That is, we follow the procedure of Eqs. 7.2.2 and 7.2.3, where $\{\mathbf{d}_e\}$ now represents the rotational d.o.f. of the *structure*. Alternatively, we can use Eq. 11.4.9: eigenvalues Λ that would be infinite in Eq. 11.4.8 become zero in Eq. 11.4.9.

Clough and Penzien [11.1] remark that the substitution $[\mathbf{M}] = [\mathbf{L}][\mathbf{L}]^T$ may produce an eigenproblem that is numerically sensitive and hard to solve accurately. They suggest a more reliable but more expensive transformation.

An alternative transformation involves decomposing $[\mathbf{K}]$ instead of $[\mathbf{M}]$. We premultiply Eq. 11.4.2 by $[\mathbf{K}]^{-1}$, substitute $[\mathbf{K}] = [\mathbf{L}][\mathbf{L}]^T$ and $\{\bar{\mathbf{D}}\} = [\mathbf{L}]^{-T}\{\mathbf{Y}\}$, and then premultiply by $[\mathbf{L}]^T$. As before, we obtain the standard form, but now with

$$[\mathbf{H}] = [\mathbf{L}]^{-1}[\mathbf{M}][\mathbf{L}]^{-T} \qquad \text{and} \qquad \Lambda = 1/\lambda = 1/\omega^2 \tag{11.4.9}$$

The highest eigenvalues now represent the lowest frequencies. This is an advantage if the algorithm extracts eigenvalues in descending order. Again, $[\mathbf{L}]^{-1}$ and therefore $[\mathbf{H}]$ are not sparse, and the transformation fails if $[\mathbf{K}]$ is singular. References 2.1 and 11.13 consider efficient programming techniques.

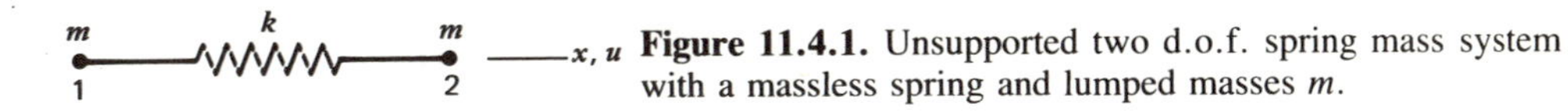

Figure 11.4.1. Unsupported two d.o.f. spring mass system with a massless spring and lumped masses m.

Example

We close this section with an elementary numerical example of eigenvalue calculation. Consider x-direction vibrations in Fig. 11.4.1. Equation 11.4.2 becomes

$$\left(\begin{bmatrix} k & -k \\ -k & k \end{bmatrix} - \omega^2 \begin{bmatrix} m & 0 \\ 0 & m \end{bmatrix}\right)\begin{Bmatrix} \bar{u}_1 \\ \bar{u}_2 \end{Bmatrix} = 0 \tag{11.4.10}$$

For a nontrivial solution to exist, the determinant of the expression in parentheses must vanish. This condition yields

$$\omega^2 (\omega^2 m - 2k) = 0, \qquad \omega_1 = 0, \qquad \omega_2 = \sqrt{2k/m} \tag{11.4.11}$$

An eigenvector can be found by substituting the eigenvalue into Eq. 11.4.10, setting one amplitude to an arbitrary number, and solving for the other amplitudes. If we set $\bar{u}_1 = 1$, we obtain

$$\begin{array}{lll} \text{For } \omega = \omega_1, & \bar{u}_1 = 1, & \bar{u}_2 = 1 \\ \text{For } \omega = \omega_2, & \bar{u}_1 = 1, & \bar{u}_2 = -1 \end{array} \tag{11.4.12}$$

The former mode is rigid-body translation. The latter mode is a motion in which the masses alternately approach and recede from one another. (If node 1 were fixed in Fig. 11.4.1, Eq. 11.4.10 would be replaced by $(k - \omega^2 m)\, \bar{u}_2 = 0$.)

11.5 CONDENSATION TO REDUCE THE NUMBER OF D.O.F.

Calculations associated with dynamics and vibrations are more costly than those of static analysis. A structure may have too many d.o.f. for economical treatment. The number of d.o.f. can be reduced, to reduce the expense of subsequent calculations, by invoking a condensation technique. The technique is not limited to dynamic problems. It is detrimental to accuracy, but only slightly so if properly used. Again, the literature is large, so we omit advanced concepts and longer arguments [2.1, 11.1].

We present a condensation algorithm usually known as *Guyan reduction* [11.14] in the context of a vibration problem. In this context it is also called an "eigenvalue economizer" and "mass condensation." We partition Eq. 11.4.2 and write it in the form

$$\left(\begin{bmatrix} \mathbf{K}_{mm} & \mathbf{K}_{ms} \\ \mathbf{K}_{ms}^T & \mathbf{K}_{ss} \end{bmatrix} - \lambda \begin{bmatrix} \mathbf{M}_{mm} & \mathbf{M}_{ms} \\ \mathbf{M}_{ms}^T & \mathbf{M}_{ss} \end{bmatrix}\right)\begin{Bmatrix} \bar{\mathbf{D}}_m \\ \bar{\mathbf{D}}_s \end{Bmatrix} = 0 \tag{11.5.1}$$

where the m "master" d.o.f. $\{\bar{\mathbf{D}}_m\}$ are to be retained and the s "slave" d.o.f. $\{\bar{\mathbf{D}}_s\}$ are to be removed by condensation. Later we discuss how to choose the masters. Now we make the principal assumption: that *for the lower frequencies*, inertia forces on slave d.o.f. are much less important than elastic forces transmitted by the master d.o.f. That is, matrix [**K**] alone dictates how slaves will follow masters. So we discard all mass but $[\mathbf{M}_{mm}]$ from Eq. 11.5.1 and find, from the lower partition,

$$\underset{s\times 1}{\{\bar{\mathbf{D}}_s\}} = -\underset{s\times s}{[\mathbf{K}_{ss}]^{-1}}\,\underset{s\times m}{[\mathbf{K}_{ms}]^T}\,\underset{m\times 1}{\{\bar{\mathbf{D}}_m\}} \tag{11.5.2}$$

Accordingly, with $n = m + s$ and [**I**] an m-by-m unit matrix,

$$\underset{n\times 1}{\begin{Bmatrix} \bar{\mathbf{D}}_m \\ \bar{\mathbf{D}}_s \end{Bmatrix}} = \underset{n\times m}{[\mathbf{T}]}\,\underset{m\times 1}{\{\bar{\mathbf{D}}_m\}}, \qquad \text{where} \qquad [\mathbf{T}] = \begin{bmatrix} \mathbf{I} \\ -\mathbf{K}_{ss}^{-1}\mathbf{K}_{ms}^T \end{bmatrix} \tag{11.5.3}$$

Substitution of Eq. 11.5.3 into Eq. 11.5.1 and premultiplication by $[\mathbf{T}]^T$ yields the condensed system

$$([\mathbf{K}_r] - \lambda[\mathbf{M}_r])\{\bar{\mathbf{D}}_m\} = 0 \tag{11.5.4}$$

where the condensed matrices are symmetric and are given by

$$\underset{m\times m}{[\mathbf{K}_r]} = [\mathbf{T}]^T\,\underset{n\times n}{[\mathbf{K}]}\,[\mathbf{T}], \qquad \underset{m\times m}{[\mathbf{M}_r]} = [\mathbf{T}]^T\,\underset{n\times n}{[\mathbf{M}]}[\mathbf{T}] \tag{11.5.5}$$

Note that even if [**K**] is banded and [**M**] is diagonal, $[\mathbf{K}_r]$ and $[\mathbf{M}_r]$ are in general full, and $[\mathbf{M}_r]$ is a combination of mass and stiffness coefficients. If a damping matrix [**C**] appears in the full system, the reduced system has a condensed damping matrix $[\mathbf{C}_r] = [\mathbf{T}]^T[\mathbf{C}][\mathbf{T}]$. If there is a right side, as in Eq. 11.2.4, the condensed loads are $[\mathbf{T}]^T\{\mathbf{R}\}$, as in Chapter 6.

In Eq. 11.5.1, [**C**] could also be present, or [**M**] could be replaced by $[\mathbf{K}_\sigma]$ of Chapter 12, and so on. Equation 11.5.3 can be applied to any of these forms if we accept that [**K**] alone defines the master-slave relationship. In general terms, Eq. 11.5.3 is an *equation of constraint*, which we discussed formally in Section 6.8. A similarity with Eqs. 7.2.3 is also apparent. There are computational advantages to generating [**T**] in terms of flexibility instead of stiffness (Problem 11.19 and References 2.1, 11.13, and 11.18).

If [**M**] in Eq. 11.5.1 is lumped, not full, and $[\mathbf{M}_{mm}]$ represents *all* the nonzero masses, then Eq. 11.5.2 follows without benefit of our principal assumption and is, indeed, the "static" condensation of Eqs. 7.2.3. This suggests another approach to condensation: the analyst can lump mass at only the d.o.f. to be retained as masters. This approach requires experience and is less accurate than condensation with a more populated mass matrix.

In going from Eq. 11.5.1 to Eq. 11.5.4, we raise the lower eigenvalues because constraints are imposed [11.15]. This behavior is seen in Fig. 11.5.1. The higher eigenvalues of the original system are absent because several d.o.f. have been discarded. The

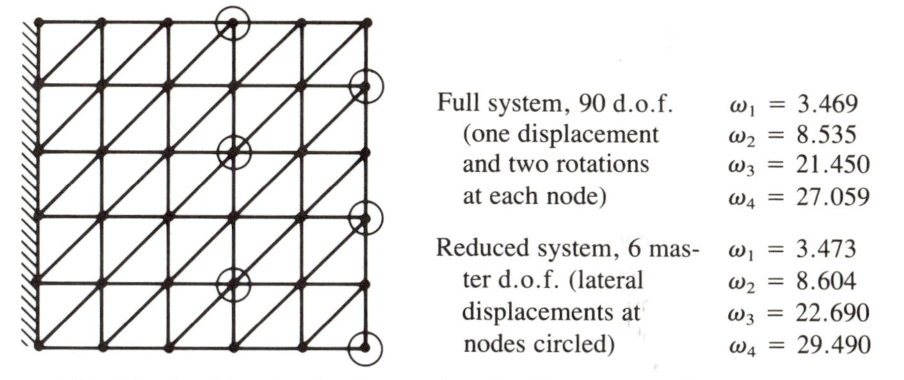

Figure 11.5.1. The first four vibration frequencies of a thin square cantilever plate [11.16]. The analysis uses triangular elements and consistent mass matrices.

higher eigenvalues of the condensed system interlace eigenvalues of the original system [11.16]. Eigenvalues obtained from the reduced system can be improved by substituting the associated vector $\{\bar{\mathbf{D}}_m \ \bar{\mathbf{D}}_s\}$ in the Rayleigh quotient (Eq. 11.4.4).

When eigenvalues λ_i and eigenvectors $\{\bar{\mathbf{D}}_{mi}\}$ of the reduced system are known, slave modes $\{\bar{\mathbf{D}}_{si}\}$ can be recovered by use of Eq. 11.5.2. However, it is more accurate to recover $\{\bar{\mathbf{D}}_{si}\}$ from the lower partition of Eq. 11.5.1. Thus

$$\{\bar{\mathbf{D}}_{si}\} = -[\mathbf{K}_{ss} - \lambda_i \mathbf{M}_{ss}]^{-1} [\mathbf{K}_{ms}^T - \lambda_i \mathbf{M}_{ms}^T] \{\bar{\mathbf{D}}_{mi}\} \tag{11.5.6}$$

Reference 11.17 includes another form of this expression that has better computational efficiency.

The following observations guide us in the choice of master d.o.f. The structure displacement field is dictated by $\{\bar{\mathbf{D}}_m \ \bar{\mathbf{D}}_s\}$ which, in turn, is formed by adding columns of $[\mathbf{T}]$ in proportions prescribed by coefficients in $\{\bar{\mathbf{D}}_m\}$ (see Eq. 11.5.3). Accordingly, columns of $[\mathbf{T}]$ are displacement modes that, when superposed, must be able to represent accurately the important (lower) vibration modes of the structure. Our principal assumption, below Eq. 11.5.1, implies that either terms in $[\mathbf{M}_{mm}]$ are large or terms in $[\mathbf{K}_{mm}]$ are small. Accordingly, master d.o.f. should have a large mass-to-stiffness ratio. Thus rotational d.o.f. rarely appear as masters.

Master d.o.f. should not be clustered in one area. If they are, some vibration modes may be almost linearly dependent. This is not a major caveat but, if ignored, can lead to the disconcerting appearance of negative eigenvalues in the highest modes of the reduced problem [11.20].

The selection of master and slave d.o.f. can be automated as follows [11.19]. Diagonal coefficients of $[\mathbf{K}]$ and $[\mathbf{M}]$ are scanned, and the d.o.f. i for which K_{ii}/M_{ii} is largest is selected as the first slave. In case of a tie, the first d.o.f. encountered is taken as slave.

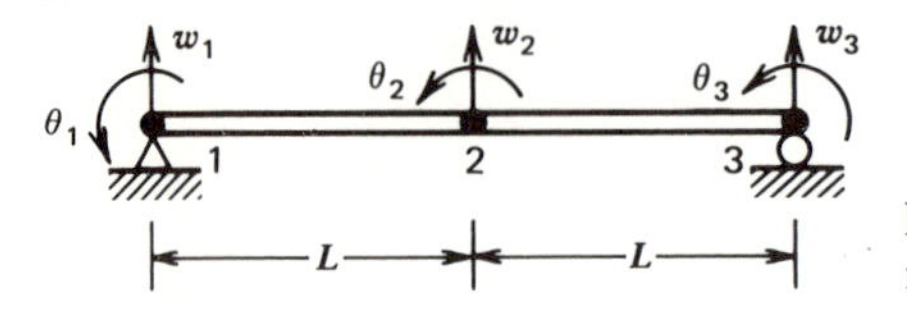

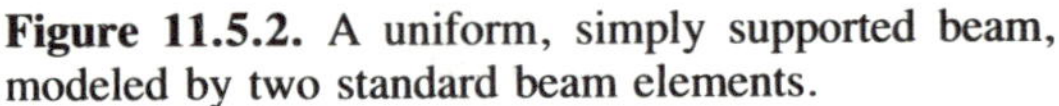

Figure 11.5.2. A uniform, simply supported beam, modeled by two standard beam elements.

Then [**K**] and [**M**] are condensed (by one order). The *condensed* matrices are now scanned, the largest K_{rii}/M_{rii} is selected as the next slave, and another condensation is performed. The process repeats until a user-specified number of d.o.f. remain. These are the masters, chosen in a near-optimal way.

The number of master d.o.f. may be as low as $\frac{1}{10}$ or $\frac{1}{20}$ of the total number of d.o.f. If there are three times as many master d.o.f. as eigenvalues to be computed, the highest computed eigenvalue may err by less than 10%. These estimates are rough and strongly problem dependent [11.16, 11.19].

An objection to mass condensation is that the accuracy obtainable depends on the choice of master d.o.f. The subspace and Lanczos methods of eigenvalue extraction incorporate transformations that may be more accurate than Eq. 11.5.3.

Example

Consider the vibration problem of Fig. 11.5.2. Essential boundary conditions are $w_1 = w_3 = 0$. To keep things simple, we also set $\theta_2 = 0$, which has the effect of eliminating an antisymmetric mode. Thus, from Eqs. 4.4.5, 11.3.2, and 11.4.2,

$$\left(\frac{EI}{L^3}\begin{bmatrix} 4L^2 & -6L & 0 \\ -6L & 24 & 6L \\ 0 & 6L & 4L^2 \end{bmatrix} - \frac{\omega^2 m}{420}\begin{bmatrix} 4L^2 & 13L & 0 \\ 13L & 312 & -13L \\ 0 & -13L & 4L^2 \end{bmatrix}\right)\begin{Bmatrix} \bar{\theta}_1 \\ \bar{w}_2 \\ \bar{\theta}_3 \end{Bmatrix} = 0 \qquad (11.5.7)$$

where $m = \rho AL$ is the mass of half the beam. As a further simplification (one already implied by the first and third of Eqs. 11.5.7), we impose complete symmetry about node 2 by setting $\bar{\theta}_1 = -\bar{\theta}_3$. So we write the constraint equation

$$\begin{Bmatrix} \bar{\theta}_1 \\ \bar{v}_2 \\ \bar{\theta}_3 \end{Bmatrix} = [\mathbf{T}']\begin{Bmatrix} \bar{w}_2 \\ \bar{\theta}_3 \end{Bmatrix}, \qquad \text{where} \qquad [\mathbf{T}'] = \begin{bmatrix} 0 & -1 \\ 1 & 0 \\ 0 & 1 \end{bmatrix} \qquad (11.5.8)$$

Then applying $[\mathbf{T}']^T[\mathbf{K}][\mathbf{T}']$ and $[\mathbf{T}']^T[\mathbf{M}][\mathbf{T}']$ to Eq. 11.5.7, we find

$$\left(\frac{4EI}{L^3}\begin{bmatrix} 6 & 3L \\ 3L & 2L^2 \end{bmatrix} - \frac{\omega^2 m}{210}\begin{bmatrix} 156 & -13L \\ -13L & 4L^2 \end{bmatrix}\right)\begin{Bmatrix} \bar{w}_2 \\ \bar{\theta}_3 \end{Bmatrix} = 0 \qquad (11.5.9)$$

The eigenvalues of this two d.o.f. system are

$$\omega_1^2 = 6.1362 \frac{EI}{mL^3} \qquad \omega_2^2 = 758.17 \frac{EI}{mL^3} \tag{11.5.10}$$

The exact values for these modes are lower, as expected. From beam theory, they are

$$\omega_1^2 = 6.0881 \frac{EI}{mL^3} \qquad \omega_2^2 = 493.13 \frac{EI}{mL^3} \tag{11.5.11}$$

With $\bar{w}_2 = 1$, the amplitude of $\bar{\theta}_3$ in mode 1, from Eq. 11.5.9 and beam theory, respectively, is

$$\bar{\theta}_3 = -1.5704/L \text{ (approximate)}, \qquad \bar{\theta}_3 = -1.5708/L \text{ (exact)} \tag{11.5.12}$$

To condense $\bar{\theta}_3$, we use Eqs. 11.5.3 and 11.5.5 with $K_{ss} = 8EI/L$ and $K_{ms} = 12EI/L^2$. Hence

$$[\mathbf{T}] = \begin{bmatrix} 1 \\ -3/2L \end{bmatrix} \quad \text{and} \quad \left(\frac{6EI}{L^3} - \omega^2 \frac{102m}{105} \right) \bar{w}_2 = 0 \tag{11.5.13}$$

from which $\omega_1^2 = 6.1765EI/mL^3$. This is the only eigenvalue obtainable. As expected, it is higher than ω_1^2 in Eq. 11.5.10. With $\bar{w}_2 = 1$, recovery of $\bar{\theta}_3$ from Eqs. 11.5.2 and 11.5.6, respectively, yields

$$\bar{\theta}_3 = -1.5000/L \quad \text{and} \quad \bar{\theta}_3 = -1.5709/L \tag{11.5.14}$$

An improved ω_1^2 can be found by substituting $\{\bar{\mathbf{D}}\} = \{1 \quad -1.5709/L\}$ into the Rayleigh quotient (Eq. 11.4.4), with $[\mathbf{K}]$ and $[\mathbf{M}]$ taken from Eq. 11.5.9. The resulting ω_1^2 is the same as ω_1^2 in Eq. 11.5.10.

A lumped mass solution is easily found by placing a particle mass m at the center. From beam theory, $w_2 = P_2(2L)^3/48EI$, so $K = P_2/w_2 = 6EI/L^3$. Then $\omega^2 = K/m = 6EI/mL^3$. Note that this value is not an upper bound.

11.6 SOLUTION TECHNIQUES FOR EIGENPROBLEMS

Algorithms that extract eigenvalues and eigenvectors are abundant. Here we introduce a few concepts and cite the literature for details, extensions, and program listings [2.1–2.3, 11.13, 11.21–11.25, 11.55, 15.27].

All algorithms are iterative if there are more than four eigenvalues. This is because in solving Eq. 11.4.2 we are in effect finding roots of the polynomial det$[\mathbf{K} - \lambda\mathbf{M}] = 0$. There exist no closed-form expressions for the λ_i if there are more than four roots. At computer installations, standard library packages offer methods such as those of Jacobi and Householder. These "transformation methods" are best suited to extracting *all* eigenvalues when matrices are small and full.

Other methods, outlined next, are more suited to finite element work, where we usually want the few lowest eigenvalues of a system with sparse matrices and a great many d.o.f. Each method can be used as a stand-alone algorithm. More often they are used in combination.

Expansion of the characteristic polynomial $\det[\mathbf{K} - \lambda\mathbf{M}] = 0$ and extraction of its roots is costly and inaccurate. But we can effectively do the same thing by *determinant tracking* [2.1, 15.27]. Here we estimate a root λ, compute an array $[\mathbf{S}]$ as $[\mathbf{S}] = [\mathbf{K} - \lambda\mathbf{M}]$, and enter Fig. 2.11.2. The determinant of $[\mathbf{S}]$ is given by Eq. 2.12.5. By using additional trial values of λ and interpolation, we can find a λ for which the determinant vanishes. This λ is an eigenvalue. The corresponding eigenvector can be found by the method used in Eq. 11.4.12. Determinant tracking is efficient if $[\mathbf{S}]$ is narrowly banded and if only one eigenvalue is needed, as in a buckling problem. It requires no special form for $[\mathbf{K}]$ or $[\mathbf{M}]$. But the determinant must be scaled to prevent overflow, and it is easy to miss a root.

The *Sturm sequence property* can be used to bracket a root λ as closely as desired, by much the same procedure as used in determinant tracking. We start as in the preceding paragraph: choose λ, evaluate $[\mathbf{S}] = [\mathbf{K} - \lambda\mathbf{M}]$, and complete the loop on statement 790 in Fig. 2.11.2. At this time the number of eigenvalues *exceeded* by the chosen λ is equal to the number of negative diagonal coefficients in $[\mathbf{S}]$; that is, they are equal to the number of negative terms in the first column of band-form array $[\mathbf{S}]$. We can use the Sturm sequence property to find the number of eigenvalues in an arbitrary range, say between λ_a and λ_b. We can also check that another extraction method has not missed a root: if roots purport to be $\lambda_1, \lambda_2, \ldots, \lambda_n$ then, for an accuracy of one part in a thousand, setting $\lambda = 1.001\lambda_n$ in a Sturm sequence calculation should result in exactly n negative coefficients. By itself, the Sturm sequence method is not efficient for eigenvalue extraction, particularly if bandwidths are large [2.1, 2.2, 11.21–11.23, 15.27].

The *inverse iteration* or *inverse power* method permits extraction of the lowest eigenvalue, or of the eigenvalue closest to a given number. We write Eq. 11.4.3 in an iterative form but, in contrast to Eq. 11.4.3, we will prescribe λ_0 and calculate c.

$$[\mathbf{K} - \lambda_0 \mathbf{M}]\{\bar{\mathbf{D}}\}_k = [\mathbf{M}]\{\bar{\mathbf{D}}'\}_{k-1} \tag{11.6.1}$$

$$\{\bar{\mathbf{D}}'\}_k = \{\bar{\mathbf{D}}\}_k / C_k \tag{11.6.2}$$

In Eq. 11.6.2, C_k is the element of $\{\bar{\mathbf{D}}\}_k$ that has the largest magnitude (C_k may be positive or negative). The ratio $1/C_k$ converges to the "shifted eigenvalue" c closest to the "shift point" λ_0, and $\{\bar{\mathbf{D}}\}_k$ converges to the corresponding eigenvector. Hence, $\lambda = \lambda_0 + c$ is the desired root. (To find the *lowest* eigenvalue of Eq. 11.4.2, we set $\lambda_0 = 0$.) To begin the solution, we assume a vector $\{\bar{\mathbf{D}}'\}_0$, solve Eq. 11.6.1 to get $\{\bar{\mathbf{D}}\}_1$, scale with Eq. 11.6.2 to get $\{\bar{\mathbf{D}}'\}_1$, go back to Eq. 11.6.1 with $\{\bar{\mathbf{D}}'\}_1$ on the right side, and so on. The mass matrix $[\mathbf{M}]$ can have zeros on-diagonal. As in the determinant-tracking method, matrix $[\mathbf{K} - \lambda_0\mathbf{M}]$ is symmetric but not necessarily positive definite. Unless the shift point λ_0 is changed to speed convergence, $[\mathbf{K} - \lambda_0 \mathbf{M}]$ need be reduced for equation solving only once. If λ_0 happens to be exactly an eigenvalue, matrix $[\mathbf{K} - \lambda_0\mathbf{M}]$ is singular and reduction fails: precautions are advised [2.1, 2.2, 11.21, 11.22, 15.27].

Inverse iteration, and other algorithms used in dynamics, require matrix multiplication of the form $[\mathbf{E}]\{\mathbf{F}\} = \{\mathbf{R}\}$, where $[\mathbf{E}]$ is symmetric and stored in a compact format. Figure 11.6.1 gives appropriate coding. Modification to accommodate matrices $[\mathbf{F}]$ and $[\mathbf{R}]$ of several columns each is straightforward.

In *subspace iteration* we begin with two steps that are done only once: triangularize $[\mathbf{K}]$ for equation solving, and choose $[\mathbf{R}]_1$. Here $[\mathbf{R}]_1$ is an n-by-m matrix of load patterns, chosen by the analyst so that a linear combination of the resulting deformation patterns will probably represent the important modes. (The "subspace" is the reduced set of m d.o.f. $\{\bar{\mathbf{D}}_m\}$, where $m << n$.) Now solve Eq. 11.6.4, with $k = 0$, to obtain $[\mathbf{T}]_1$. Next use $[\mathbf{T}]_1$ in Eqs. 11.5.5. Thus we obtain an m-by-m eigenvalue problem with full matrices. It is appropriately solved by a transformation method. (The Jacobi method is a good choice because $[\mathbf{K}_r]$ and $[\mathbf{M}_r]$ tend to become diagonal as iteration progresses.) The resulting normalized eigenvectors $[\mathbf{V}_r]$ are entered into Eqs. 11.6.3 and 11.6.4 to compute the following: improved eigenvectors $[\boldsymbol{\phi}]$ of the full system, then a new $[\mathbf{R}]$, and then a new $[\mathbf{T}]$.

$$\underset{n\times m}{[\boldsymbol{\phi}]_k} = \underset{n\times m}{[\mathbf{T}]_k}\,\underset{m\times m}{[\mathbf{V}_r]_k}, \qquad \underset{n\times m}{[\mathbf{R}]_{k+1}} = \underset{n\times n}{[\mathbf{M}]}\underset{n\times m}{[\boldsymbol{\phi}]_k} \tag{11.6.3}$$

```
C --- UPPER TRIANGLE OF E TIMES F, INCLUDING DIAGONAL OF E.
      DO 20 I=1,NEQ
      R(I) = 0.
      L = I - 1
      K = MINO(L+MBAND,NEQ)
      N = 0
      DO 20 J=I,K
      N = N + 1
   20 R(I) = R(I) + E(I,N)*F(J)
C --- NOW PICK UP TERMS BELOW THE DIAGONAL OF E.
      IF (MBAND .EQ. 1)  RETURN
      LL = MBAND - 1
      DO 30 I=2,NEQ
      L = I - 1
      KK = MINO(L,LL)
      JJ = I - KK
      N = JJ - 1
      II = KK + 2
      DO 30 J=JJ,L
      N = N + 1
      II = II - 1
   30 R(I) = R(I) + E(N,II)*F(J)
      RETURN
      END
```

Figure 11.6.1. Fortran coding for the matrix multiplication $[\mathbf{E}]\{\mathbf{F}\} = \{\mathbf{R}\}$. $[\mathbf{E}]$ is an NEQ-by-NEQ symmetric matrix of semibandwidth MBAND, stored in the band format of Fig. 2.7.2*b*. $\{\mathbf{F}\}$ and $\{\mathbf{R}\}$ are NEQ-by-1 column vectors. MINO is a Fortran function that selects the minimum value in the argument list.

$$\underset{n\times n}{[\mathbf{K}]}\ \underset{n\times m}{[\mathbf{T}]_{k+1}} = \underset{n\times m}{[\mathbf{R}]_{k+1}}, \qquad \text{solve for } [\mathbf{T}]_{k+1} \tag{11.6.4}$$

Now $k = 1$. We generate and solve a new reduced problem, and so on, until convergence. A colum R_j of $[\mathbf{R}]$ in Eq. 11.6.3 is seen to be a nodal force vector that results from applying column ϕ_j of $[\boldsymbol{\phi}]$ as a nodal acceleration vector [2.1, 11.22].

Clearly, there is a marked similarity between mass condensation and subspace iteration. Subspace iteration chooses initial master d.o.f. in a different way but, more important, *it iteratively improves them*, so a poor initial choice is not fatal. The size of the reduced problem can be chosen as $m = 2p$ or $m = p + 8$, whichever is less, where p is the number of accurate eigenvalues required [2.1]. If $n = 1$ and $\lambda_0 = 0$ the method becomes inverse iteration with convergence to $\lambda = K_r/M_r$, which is the Rayleigh quotient of Eq. 11.4.4.

The *Lanczos method* shares the accuracy and automatic mass condensation features of subspace iteration, but it seems to be even more efficient. Apparently, it is the best method now available for use with large, sparse, symmetric matrices. A brief explanation is inadequate, so references are cited [11.22, 11.24, 11.25, 15.27; also *NASTRAN* User's Colloquia of recent years].

11.7 DYNAMIC RESPONSE. MODAL METHODS

The linear *dynamic response problem* is that of solving Eq. 11.2.4 to obtain $\{\mathbf{D}\}$, $\{\dot{\mathbf{D}}\}$, and $\{\ddot{\mathbf{D}}\}$ as functions of time, when $[\mathbf{K}]$, $[\mathbf{C}]$, and $[\mathbf{M}]$ are known and time independent, $\{\mathbf{R}\}$ is a known function of time, and initial values of $\{\mathbf{D}\}$ and $\{\dot{\mathbf{D}}\}$ are prescribed. The *modal* or *mode superposition* method transforms Eq. 11.2.4 so that $\{\mathbf{D}\}$ and its time derivatives are replaced by $\{\mathbf{z}\}$ and its time derivatives, where $\{\mathbf{z}\}$ is a vector of modal amplitudes. If damping is orthogonal, the transformed equations are independent. Solutions of these modal equations are superposed to yield the solution of the original problem [2.1, 11.1, 11.11, 11.23]. Other solution methods for dynamic response are considered in Section 11.8.

One usually has much greater confidence in the computational methods than in the correctness of the forcing function $\{\mathbf{R}\}$ or the appropriateness of the structural model.

Some preliminaries are necessary. Eigenvectors of Eq. 11.4.2 are orthogonal with respect to the (symmetric) mass and stiffness matrices. In symbols, this means that

$$\{\bar{\mathbf{D}}_i\}^T[\mathbf{M}]\{\bar{\mathbf{D}}_j\} = \{\bar{\mathbf{D}}_i\}^T[\mathbf{K}]\{\bar{\mathbf{D}}_j\} = 0 \qquad \text{when} \qquad i \neq j \tag{11.7.1}$$

If eigenvectors are normalized with respect to the mass matrix, then the stiffness matrix yields undamped frequencies. That is, by Eq. 11.4.4,

$$\text{If } \{\bar{\mathbf{D}}_i\}^T[\mathbf{M}]\{\bar{\mathbf{D}}_i\} = 1 \qquad \text{then} \qquad \{\bar{\mathbf{D}}_i\}^T[\mathbf{K}]\{\bar{\mathbf{D}}_i\} = \omega_i^2 \tag{11.7.2}$$

If $[\boldsymbol{\phi}]$ is the *modal matrix* (a matrix whose columns are the normalized eigenvectors), then

$$[\boldsymbol{\phi}]^T[\mathbf{M}][\boldsymbol{\phi}] = [\mathbf{I}] \qquad \text{and} \qquad [\boldsymbol{\phi}]^T[\mathbf{K}][\boldsymbol{\phi}] = \lceil \boldsymbol{\omega}^2 \rfloor \tag{11.7.3}$$

where $[\mathbf{I}]$ is a unit matrix and $\lceil \boldsymbol{\omega}^2 \rfloor$ is the *spectral matrix*, a diagonal matrix of the squared natural frequencies.

To derive modal equations, we note that an arbitrary displacement can be expressed as the sum of normal modes in appropriate proportions. Thus

$$\{\mathbf{D}\} = [\boldsymbol{\phi}]\,\{\mathbf{z}\} \tag{11.7.4}$$

where $\{\mathbf{z}\}$ contains weight factors for the superposition of mass-normalized modes in $[\boldsymbol{\phi}]$. In other words, the z_i are modal amplitudes. Like $\{\mathbf{D}\}$, $\{\mathbf{z}\}$ is a function of time. Now substitute Eq. 11.7.4 into Eq. 11.2.4, premultiply by $[\boldsymbol{\phi}]^T$, and take note of Eqs. 11.7.3. If $[\mathbf{C}]$ is given by Eq. 11.2.5, or by other orthogonal forms [2.1, 11.1], then the coefficient matrix of $\{\dot{\mathbf{z}}\}$ becomes diagonal. Then, following custom, diagonal damping coefficients are written as $2\xi_i\omega_i$, where ξ_i is the ratio of damping in mode i to critical damping in mode i. Thus

$$\ddot{z}_i + 2\xi_i\omega_i\dot{z}_i + \omega_i^2 z_i = p_i, \qquad \text{where} \qquad p_i = \lfloor \boldsymbol{\phi}_i \rfloor \{\mathbf{R}\} \tag{11.7.5}$$

and $\lfloor \boldsymbol{\phi}_i \rfloor$ is the ith row of $[\boldsymbol{\phi}]^T$. There are as many of these uncoupled ordinary differential equations as there are d.o.f. In each equation, p_i is a known function of time. Each equation can be solved for z_i as a function of time. Then Eq. 11.7.4 yields $\{\mathbf{D}\}$ as a function of time. Similarly, $\{\dot{\mathbf{D}}\} = [\boldsymbol{\phi}]\{\dot{\mathbf{z}}\}$ and $\{\ddot{\mathbf{D}}\} = [\boldsymbol{\phi}]\,\{\ddot{\mathbf{z}}\}$.

Initial values z_i and $\dot{z}_i$ at $t = 0$, needed in the integration of Eqs. 11.7.5, can be found from known initial nodal quantities $\{\mathbf{D}\}$ and $\{\dot{\mathbf{D}}\}$ as follows. Premultiply both sides of Eq. 11.7.4 by $[\boldsymbol{\phi}]^T[\mathbf{M}]$ and take note of Eqs. 11.7.3. Thus

$$\{\mathbf{z}\} = [\boldsymbol{\phi}]^T[\mathbf{M}]\{\mathbf{D}\} \qquad \text{and} \qquad \{\dot{\mathbf{z}}\} = [\boldsymbol{\phi}]^T[\mathbf{M}]\{\dot{\mathbf{D}}\} \tag{11.7.6}$$

Integration of Eqs. 11.7.5 can be accomplished by various methods [11.23]. One is the direct integration method (Section 11.8), which is very efficient when applied to Eqs. 11.7.5 because the equations are uncoupled. If p_i is piecewise linear in time, exact solutions can be written for z_i and $\dot{z}_i$ in terms of $e^{-\xi\omega t}\sin\omega t$ and $e^{-\xi\omega t}\cos\omega t$. These exact solutions are a good way to solve the *harmonic response problem*, which asks for the steady-state response of a system to loads of known amplitude and frequency.

Finding $[\boldsymbol{\phi}]$ in Eq. 11.7.4 at first glance seems to be a prohibitive computational expense. But, in many problems, *only a few modes are needed*. Thus $[\boldsymbol{\phi}]$ is not n by n but n by m, where $m << n$. Also, there are only m equations in Eqs. 11.7.5. Further comments appear in Section 11.9.

In a *non*linear dynamic response problem, [**K**] and possibly [**C**] are time dependent. Usually [**M**] does not change. The modal method is a superposition technique, so it is often assumed to be inapplicable to nonlinear problems. However, nonlinear problems can be accommodated if all nonlinearities are transferred to the right side of Eq. 11.2.4 and incorporated in the load vector {**R**} (see, for example, Eqs. 13.6.2 and 13.11.9). Then, in Eq. 11.7.5, the prescribed loads p_i are augmented by pseudoforces that are updated at every time step in the integration of Eqs. 11.7.5. Updating requires that modes be superposed to find {**D**} (and {$\dot{\mathbf{D}}$}, if needed), hence the new nonlinear terms in {**R**} and, finally, the revised p_i in Eq. 11.7.5. The computational expense of this approach compares favorably with the expense of direct integration (Section 11.8). Strictly, [$\boldsymbol{\phi}$] does not represent normal modes when nonlinearities are present. Equation 11.7.4 should then be viewed simply as a coordinate transformation [11.26, 11.27].

Elaboration of the modal method includes *modal synthesis* (Section 6.13).

11.8 DYNAMIC RESPONSE. DIRECT INTEGRATION

This section introduces the *direct integration methods* for finding dynamic response. Here the structure equations are attacked directly, without the modal decoupling explained in Section 11.7. Modal and direct methods are compared in Section 11.9. The literature is vast. Our remarks are extracted from [2.1, 11.7, 11.28–11.41].

Direct integration equations are either *explicit* or *implicit*. In *explicit* (or *open*, or *predictor*) equations, neither {**D**} nor {$\dot{\mathbf{D}}$} at time t is a function of {$\ddot{\mathbf{D}}$} at time $t + \Delta t$, where Δt is the time step or time increment. In *implicit* (or *closed*, or *corrector*) equations, there *is* this functional relationship. Stated another way, explicit methods find $\{\mathbf{D}\}_{t+\Delta t}$ by use of $\mathbf{F} = m\mathbf{a}$ equations written at time t, while implicit methods find $\{\mathbf{D}\}_{t+\Delta t}$ from expressions that involve $\mathbf{F} = m\mathbf{a}$ equations written at time $t + \Delta$t. Usually, explicit methods require a small Δt but produce equations that are cheap to solve, while implicit methods allow a large Δt but produce equations that are expensive to solve.

Most explicit methods are *conditionally stable*. If Δt exceeds a certain fraction of the smallest period of the structure, computed displacements and velocities grow without limit. If Δt is too large but damping is appreciable, the process may not "blow up" but may contain errors that are not obvious. *Unconditionally stable* methods are those for which the size of Δt is governed by considerations of accuracy rather than stability. Most unconditionally stable methods are implicit.

Finite element structures produce so-called *stiff equations*. In the present context, stiff equations characterize a structure whose highest natural vibration frequencies are much greater than the lowest ($\omega_{max} >> \omega_{min}$). Especially "stiff" structures therefore include those with a very fine mesh, a thin shell structure (whose membrane stiffness greatly exceeds its bending stiffness), and a structure with near-rigid support members (Fig. 2.9.5). If a conditionally stable algorithm is used for these structures, Δt must be very small, even if the highest modes are of no physical interest. If any $M_{ii} = 0$ in a diagonal mass matrix, a preliminary condensation of the associated d.o.f. is needed in order to avoid infinite frequencies and an allowable Δt of zero.

To construct the *central difference algorithm*, we begin with finite difference expressions in time for nodal velocities and accelerations at time t.

$$\{\dot{\mathbf{D}}\}_t = \frac{1}{2(\Delta t)}\{\mathbf{D}_{t+\Delta t} - \mathbf{D}_{t-\Delta t}\} \tag{11.8.1}$$

$$\{\ddot{\mathbf{D}}\}_t = \frac{1}{(\Delta t)^2}\{\mathbf{D}_{t+\Delta t} - 2\mathbf{D}_t + \mathbf{D}_{t-\Delta t}\} \tag{11.8.2}$$

Equations 11.8.1 and 11.8.2 are substituted into Eq. 11.2.4, written at time t. Therefore the method is explicit. We find

$$\left[\frac{\Delta t}{2}\mathbf{C} + \mathbf{M}\right]\{\mathbf{D}\}_{t+\Delta t} = (\Delta t)^2\{\mathbf{R}\}_t - [(\Delta t)^2\mathbf{K} - 2\mathbf{M}]\{\mathbf{D}\}_t - \left[\mathbf{M} - \frac{\Delta t}{2}\mathbf{C}\right]\{\mathbf{D}\}_{t-\Delta t} \tag{11.8.3}$$

Equation 11.8.2 has an error of order $(\Delta t)^2$, so the error in $\{\ddot{\mathbf{D}}\}$ is quartered when Δt is halved.

The algorithm operates as follows. We start at $t = 0$. Initial conditions prescribe $\{\mathbf{D}\}_0$ and $\{\dot{\mathbf{D}}\}_0$. From these and Eq. 11.2.4 we find $\{\ddot{\mathbf{D}}\}_0$, if $\{\ddot{\mathbf{D}}\}_0$ is not prescribed. Hence Eqs. 11.8.1 and 11.8.2 yield the displacements $\{\mathbf{D}\}_{-\Delta t}$ needed to start the computations;

$$\{\mathbf{D}\}_{-\Delta t} = \{\mathbf{D}\}_0 - (\Delta t)\{\dot{\mathbf{D}}\}_0 + \frac{(\Delta t)^2}{2}\{\ddot{\mathbf{D}}\}_0 \tag{11.8.4}$$

Next, Eqs. 11.8.1, 11.8.2, and 11.8.3 are used repeatedly: Eq. 11.8.3 yields $\{\mathbf{D}\}_{\Delta t}$, then Eqs. 11.8.1 and 11.8.2 yield $\{\dot{\mathbf{D}}\}_{\Delta t}$ and $\{\ddot{\mathbf{D}}\}_{\Delta t}$ (if needed), then Eq. 11.8.3 yields $\{\mathbf{D}\}_{2(\Delta t)}$, and so on. The solution of Eqs. 11.8.3 is trivial if [**M**] is diagonal and [**C**] is either diagonal or zero.

The central difference method is conditionally stable. If [**C**] is null or if [**C**] is diagonal, the limiting time step for stability is $\Delta t_{cr} = 2/\omega_{max}$, whether the problem is linear or not [11.39]. (For reasons that are poorly understood, the practical limit on Δt is roughly 25% less than $2/\omega_{max}$. Nondiagonal damping reduces Δt_{cr}.) In other words, Δt_{cr} is the time required for an acoustic wave to traverse an element, for the element with the least traversal time. Approximate formulas for Δt_{cr} have been offered [11.36]. Alternatively, one can solve an eigenvalue problem to find ω_{max}, or estimate $\omega^2_{max}$from Eq. 15.6.8, where [$\mathbf{K}_s$] in Eq. 15.6.8 is the dame as [**H**] in Eq. 11.4.8.

Newmark's method is based on the assumptions

$$\{\mathbf{D}\}_{t+\Delta t} = \{\mathbf{D}\}_t + (\Delta t)\,\{\dot{\mathbf{D}}\}_t + (\Delta t)^2 \left\{ (\tfrac{1}{2} - \beta)\ddot{\mathbf{D}}_t + \beta\ddot{\mathbf{D}}_{t+\Delta t} \right\} \tag{11.8.5}$$

$$\{\dot{\mathbf{D}}\}_{t+\Delta t} = \{\dot{\mathbf{D}}\}_t + (\Delta t) \left\{ (1 - \gamma)\ddot{\mathbf{D}}_t + \gamma\ddot{\mathbf{D}}_{t+\Delta t} \right\} \tag{11.8.6}$$

where β and γ are numbers that the analyst can choose. Equation 11.8.5 is solved for $\{\ddot{\mathbf{D}}\}_{t+\Delta t}$, then substituted into Eq. 11.8.6. These acceleration and velocity vectors are substituted into Eq. 11.2.4, written at time $t + \Delta$t. Therefore the method is implicit. We find

$$\begin{aligned} \left[\mathbf{K} + \frac{\gamma}{\beta(\Delta t)}\mathbf{C} + \frac{1}{\beta(\Delta t)^2}\mathbf{M} \right] \{\mathbf{D}\}_{t+\Delta t} &= \{\mathbf{R}\}_{t+\Delta t} \\ &+ [\mathbf{C}] \left\{ \frac{\gamma}{\beta(\Delta t)}\mathbf{D}_t + \left(\frac{\gamma}{\beta} - 1\right)\dot{\mathbf{D}}_t + (\Delta t)\left(\frac{\gamma}{2\beta} - 1\right)\ddot{\mathbf{D}}_t \right\} \\ &+ [\mathbf{M}] \left\{ \frac{1}{\beta(\Delta t)^2}\mathbf{D}_t + \frac{1}{\beta(\Delta t)}\dot{\mathbf{D}}_t + \left(\frac{1}{2\beta} - 1\right)\ddot{\mathbf{D}}_t \right\} \end{aligned} \tag{11.8.7}$$

The algorithm operates as follows. We start at $t = 0$. Initial conditions prescribe $\{\mathbf{D}\}_0$ and $\{\dot{\mathbf{D}}\}_0$. From these and Eq. 11.2.4 we find $\{\ddot{\mathbf{D}}\}_0$, if $\{\ddot{\mathbf{D}}\}_0$ is not prescribed. Then Eq. 11.8.7 is solved for $\{\mathbf{D}\}_{\Delta t}$, Eq. 11.8.5 is solved for $\{\ddot{\mathbf{D}}\}_{\Delta t}$, and Eq. 11.8.6 is solved for $\{\dot{\mathbf{D}}\}_{\Delta t}$. Then Eq. 11.8.7 yields $\{\mathbf{D}\}_{2(\Delta t)}$, and so on. The solution of Eq. 11.8.7 for $\{\mathbf{D}\}_{t+\Delta t}$ is not trivial, but the coefficient matrix need be reduced only once if Δt is not changed.

Newmark's method is unconditionally stable if $\gamma \geqq 0.5$ and $\beta \geqq (2\gamma + 1)^2/16$. Artificial positive damping is introduced if $\gamma > 0.5$ and artificial negative damping if $\gamma < 0.5$. If $\gamma = 0.5$ and $\beta = 0$ it can be shown that Newmark's method reduces to the central difference method. A good choice of parameters for an implicit method that is unconditionally stable (in linear problems) is $\gamma = 0.5$ and $\beta = 0.25$. Then the method is also called the *constant average acceleration method,* or the *trapezoidal method,* since Eq. 11.8.6 becomes the so-called trapezoidal formula. With $\gamma = 0.5$ and $\beta = 0.25$ there are no amplitude errors in any sine-wave motion, regardless of its frequency. But periods of sine-wave motion are overestimated. The error increases as $\Delta t/T$ increases, where T is the (correct) period of the motion.

We note that unconditional stability in linear problems is by itself no guarantee that an algorithm will remain unconditionally stable when used in nonlinear problems.

A central difference algorithm that is particularly useful for nonlinear problems can be summarized as follows. Elastic resisting forces provided by the structure are, from equations in Section 4.3,

$$[\mathbf{K}]\,\{\mathbf{D}\} = \sum \int_V [\mathbf{B}]^T\,[\mathbf{E}]\,[\mathbf{B}]\,dV\,\{\mathbf{d}\} = \sum \int_V [\mathbf{B}]^T\,\{\boldsymbol{\sigma}\}\,dV \tag{11.8.8}$$

where $\{\boldsymbol{\sigma}\}$ represents the total stress in an element. The latter integral continues to express the resistance of the structure even when the material is nonlinear, say elastic-plastic. Then, if $[\mathbf{C}] = 0$, Eq. 11.2.4 yields

$$\{\ddot{\mathbf{D}}\}_t = [\mathbf{M}]^{-1}\left(\{\mathbf{R}\}_t - \sum \int_V [\mathbf{B}]^T \{\sigma\}_t \, dV\right) \tag{11.8.9}$$

With $\gamma = 0.5$ and $\beta = 0$ to make them central difference equations, Eqs. 11.8.5 and 11.8.6 yield $\{\mathbf{D}\}_{t+\Delta t}$, Eq. 11.8.9 can now be used to find $\{\ddot{\mathbf{D}}\}_{t+\Delta t}$, and so on.

Many other algorithms are available for dynamic response. The *fourth-order Runge-Kutta method* is often recommended by mathematicians. It has an error of order $(\Delta t)^5$ but requires four equation solutions per time step, and the time step must be small in order to get accurate results from stiff equations. *Houbolt's method* [11.38] damps a higher mode appreciably when Δt exceeds $T/10$ or perhaps even $T/50$, where T is the time period of the mode. But in nonlinear problems it is stable to some three times the Δt_{cr} of Newmark's method with $\alpha = 0.5$ and $\beta = 0.25$, and it is therefore preferred. *Trujillo* [11.40] proposes an explicit algorithm with unconditional stability. *Hilber* et al. [11.37] extend Newmark's method by adding a third parameter that permits greater control over artificial damping. *Park* [11.33] offers a method that seems to be better than Houbolt's method when applied to nonlinear problems. *Hughes* [11.56] considers novel combinations of implicit and explicit methods, including the restriction of each to that part of the structure for which it is best suited. Use of different Δt values in different parts is also possible.

11.9 DYNAMIC RESPONSE. REMARKS ABOUT METHODS

Generally, modal and implicit direct methods are more economical in inertial problems, while explicit direct methods are more economical in shock loading and wave propagation problems. Other considerations that influence the choice of method are much the same as those listed in Section 13.14 [11.34].

In linear problems, modal methods are favored if only a few modes are needed to describe the response. For example, an earthquake may excite only the lowest modes of a building, and a vibrating machine may excite only the nearby frequencies of its support structure. In wave propagation and shock loading problems, many modes are excited and the response may be required for only a short time. Here an explicit direct method would be favored.

The modal matrix is available free, in a sense, because response analysis will probably be preceded by a vibration analysis to see how the current design compares with similar previous designs. If the structure is changed, or more modes are thought necessary, existing eigenvectors can be used as initial vectors in the new eigenproblem to speed its solution. With mode superposition, loading histories after the first are analyzed cheaply.

With direct integration, especially the explicit method, the second and subsequent load histories do not decline in cost so greatly.

With the modal method, the user is burdened with deciding how many columns are needed in the modal matrix, which is perhaps more difficult than assigning the Δt to be used in direct integration. (Ideally, *any* method would demand only that the user specify the accuracy needed, such as "find $\{\mathbf{D}\}$ as a function of time, correct to two digits.") Accelerations computed by a direct method are roughly as accurate as the computed displacements. However, if accelerations must be accurately computed by a modal method and there is no damping, many modes are needed. Shock and blast loading also require many modes.

In the modal method, use a few modes may give good displacements but poor stresses. Displacements, and therefore stresses, can be improved by using in place of Eq. 11.7.4 the equation

$$\{\mathbf{D}\} = [\boldsymbol{\phi}]\{\mathbf{z}\} + \{\mathbf{D}\}_c \tag{11.9.1}$$

where $\{\mathbf{D}\}_c$ is an approximate corrective vector of nodal d.o.f., given by the static equation [11.54]

$$[\mathbf{K}]\{\mathbf{D}\}_c = \{\mathbf{R}\} - [\mathbf{K}][\boldsymbol{\phi}]\{\mathbf{z}\} \tag{11.9.2}$$

The right side of Eq. 11.9.2 is the unbalance between given nodal loads $\{\mathbf{R}\}$ on the structure and nodal loads predicted by the (approximate) modal analysis.

In nonlinear problems, the modal method is again suitable if it would be favored if the nonlinearities were absent, and if nonlinearities are small and confined to a few regions of the structure. Advocates of modal methods have made more glowing claims of efficiency in nonlinear problems than advocates of direct integration.

The form of the matrices influences the choice of algorithm. Damping matrix $[\mathbf{C}]$ must be orthogonal for the modal method of Section 11.7. It must be diagonal or null for an efficient explicit algorithm but need have no special form for an efficient implicit algorithm. If $[\mathbf{M}]$ is not diagonal, the cost of an explicit algorithm increases by a much greater factor than the cost of an implicit algorithm. For greater accuracy, lumped masses should be used with the central difference method because a consistent $[\mathbf{M}]$ overestimates frequencies. However, a consistent $[\mathbf{M}]$ should be used in Newmark's method with $\gamma = 0.5$ and $\beta = 0.25$, because lumped masses underestimate frequencies [11.42].

Equation 11.8.9 is preferred over Eq. 11.8.3 as a format for the central difference method, even in linear problems. Thus $[\mathbf{K}]$ need be neither assembled nor stored. In Eq. 11.8.9 most of the expense lies in computing the integrals, but this is cheaper than using Eq. 11.8.3. In example linear problems, Eq. 11.8.9 requires one-half to one-tenth the number of operations per time step that Eq. 11.8.7 requires [11.43, 11.44].

In nonlinear problems, the modal method may require many modes if plastic action is to be followed properly. If implicit direct integration is used, placing nonlinearities on the right side as pseudoforces is much cheaper than updating $[\mathbf{K}]$ at each time step.

However, the explicit approach is often best for nonlinear problems. Here the stability limit on Δt may be of little concern, since Δt must be small anyway to follow wave propagation or to account for rapidly changing stiffness (especially in gap-closure problems). Residual forces, discussed in Section 13.11, for example, should be included to reduce drift from the correct path [11.43].

An efficient and versatile computer program should be able to change automatically from an explicit method at an early time, where Δt is small to follow transients, to an implicit method at a later time, where a large Δt is sufficiently accurate.

In addition to the methods discussed in Sections 11.7 and 11.8, it is becoming more popular to analyze structures under harmonic loading by the frequency domain approach. Fourier series expansions eliminate the time variable, and the Fast Fourier Transform can yield computational efficiency [2.1, 11.1].

11.10 MISCELLANEOUS DYNAMIC PROBLEMS

Rotation alters the vibration behavior and dynamic response of a structure. Thus, for example, fans and turbomachinery have critical speeds. Finite element analyses are in Refs. 11.45, 11.46, and 11.52.

A structure may be loaded by vehicles or forces that move across it. The analysis of these problems is surveyed in Refs. 11.47 and 11.52. If w represents the vertical deflection of a mass that moves across a beam with velocity $\dot{x}$, it is tempting to say the vertical acceleration of the mass is simply $\ddot{w}$. But the vertical acceleration includes three additional terms and is $w,_{xx}\dot{x}^2 + 2\dot{w},_x\dot{x} + w,_x\ddot{x} + \ddot{w}$.

In large-amplitude vibration, the conventional stiffness matrix $[\mathbf{K}]$ is augmented by a stress stiffness matrix $[\mathbf{K}_\sigma]$ (Chapter 13). Furthermore, $[\mathbf{K}_\sigma]$ is amplitude dependent, so the frequency of nonlinear vibration depends on the amplitude of vibration [11.48, 11.49].

A panel flutters when aerodynamic loading reaches a critical value. Similarly, flutter identifies the critical load in a nonconservative stability problem, such as a column under a tangential tip load. The solution of these problems requires repeated eigenvalue extraction as a loading parameter is varied [11.50, 11.51].

Substructuring of a system with a great many d.o.f. can be done in a dynamic context. The procedures resemble mass condensation (see also the last paragraph in Section 6.13).

PROBLEMS

11.1 A single d.o.f. spring-mass system has natural frequency $\omega_1 = \sqrt{k/m}$. It is excited by a force $P_0 \sin \omega_2 t$. What is the limiting value of the ratio ω_2/ω_1 such that the amplitude of motion differs from the static displacement by less that 10%?

11.2 Show that under constant acceleration $\{\ddot{\mathbf{d}}\}$, nodal loads $[\mathbf{m}]\{\ddot{\mathbf{d}}\}$ given by Eq. 11.2.3 are the same as the body-force nodal loads given by Eq. 4.3.5.

11.3 Under what circumstances might the semibandwidth of [**C**] or [**M**] exceed the semibandwidth of [**K**]?

11.4 (a) Can a diagonal coefficient in a consistent mass matrix ever be negative? Explain.
(b) Imagine that nodes of a vibration mode coincide with nodes of a finite element mesh. Would you prefer a consistent or a lumped mass matrix? Why?
(c) How does your answer to part (b) relate to an element with many d.o.f. but no internal d.o.f.? Give a hypothetical example.

11.5 (a) Derive the consistent mass matrix given by Eq. 11.3.1.
(b) Derive the consistent mass matrix given by Eq. 11.3.2.
(c) Show that Eq. 11.3.2 yields the correct nodal forces and moments under a rigid-body translational acceleration.
(d) How would you change the formulation procedures for [**m**] and for ⌈**m**⌋ if the bars in Fig. 11.3.1 are tapered instead of uniform?
(e) Show that the consistent [**m**] of Eq. 11.3.1 is not changed by the coordinate transformation of Eq. 6.3.8. Would transformation change [**m**] of a plane frame element? Why?
(f) Derive Eq. 6.3.8.

11.6 (a) Find the consistent mass matrix of the constant strain triangle (Fig. 4.5.1). It has constant density and thickness. Arrange d.o.f. in the order $\{\mathbf{d}\} = \{u_1\ u_2\ u_3\ v_1\ v_2\ v_3\}$.
(b) If this form of [**m**] is used with a triangular plate-bending element, with [**m**] reduced to 3 by 3 and $\{\mathbf{d}\} = \{w_1\ w_2\ w_3\}$, what displacement field is implied?

11.7 Replace the zeros in Eq. 11.3.3 by coefficients $m_{22} = m_{44}$ such that the beam has the correct kinetic energy under rigid-body rotation about its center. Then find an example problem for which this ⌈**m**⌋ gives unacceptable results. Can you draw a lesson from this problem?

11.8 For the beam element of Fig. 11.3.1*b*, consider the displacement field

$$w = \left\lfloor \left(1 - \frac{x}{L}\right) \quad \left(\frac{x}{2} - \frac{x^2}{2L}\right) \quad \frac{x}{L} \quad \left(-\frac{x}{2} + \frac{x^2}{2L}\right) \right\rfloor \{w_1\ \theta_1\ w_2\ \theta_2\}$$

(a) Show that w is linear in x if $\theta_1 = \theta_2$. Also show that the field yields the correct displacement and curvature under pure bending.
(b) Find [**m**] using this field and Eq. 11.2.3.

11.9 Consider the procedure described above Eq. 11.3.4 and:
(a) Verify Eqs. 11.3.4 and 11.3.5.
(b) Find ⌈**m**⌋ for the element of Problem 11.6(a). Consider x- and y-direction accelerations separately.
(c) Find ⌈**m**⌋ for a four-node rectangular plane element having eight d.o.f. It has constant density and thickness. Consider x- and y-direction accelerations separately.
(d) Find ⌈**m**⌋ for the element of Problem 11.8.

(e) Propose a rule that predicts when the resulting $\lceil \mathbf{m} \rfloor$ will differ from a simple lumped mass matrix.

11.10 Consider the stiffness and mass matrices

$$[\mathbf{K}] = \begin{bmatrix} 2 & 2 \\ 2 & 5 \end{bmatrix}, \qquad [\mathbf{M}] = \begin{bmatrix} 1 & 0 \\ 0 & 1 \end{bmatrix}$$

Thus, exact eigensolutions of Eq. 11.4.2 are $\lambda_1 = 1$, $\lambda_2 = 6$, $\{\bar{\mathbf{D}}_1\} = \{-2 \quad 1\}$, and $\{\bar{\mathbf{D}}_2\} = \{1 \quad 2\}$. Consider approximate eigenvectors $\{-1.7 \quad 1.0\}$ and $\{1.2 \quad 2.0\}$ and show that Eq. 11.4.4 estimates λ_1 and λ_2 in the predicted way.

11.11 Suppose that prescribed zero d.o.f. are imposed by treating $[\mathbf{K}]$ and $[\mathbf{M}]$ as in Fig. 2.9.6. How does this treatment affect the eigenvalue problem?

11.12 (*a*) Verify Eq. 11.4.8.
(*b*) Verify Eq. 11.4.9.
(*c*) Remove the mass at node 1 in the sketch for Problems 11.24 and 11.25, but retain u_1 and u_2 as d.o.f. Find the natural frequencies and show that Eq. 11.4.9 succeeds while Eq. 11.4.8 fails.

11.13 Solve the example problem that closes Section 11.4 by the following methods, where possible. To make the problem numerical, let $k = 1$ and $m = 2$.
(a) Use Eq. 11.4.3 and a shift of $c = -1$.
(b) Convert to the standard form, using Eq. 11.4.8.
(c) Convert to the standard form, using Eq. 11.4.9.

11.14 Consider axial vibrations of a uniform bar of length L and mass m, free at one end and fixed at the other. Represent the bar first by one element, then by two elements of equal length $L/2$. In each case, compute the lowest natural frequency using:
(a) The consistent mass matrix $[\mathbf{m}]$.
(b) The lumped mass matrix $\lceil \mathbf{m} \rfloor$.
(c) The average mass matrix $([\mathbf{m}] + \lceil \mathbf{m} \rfloor)/2$.
The square of the exact lowest natural frequency is $\pi^2 AE/4Lm$.

11.15 In Problem 11.14, consider the convergence rates as the mesh is refined. Are they in accord with the rates predicted in Section 11.3?

11.16 Model a simply supported uniform beam of length $2L$ by a single element. Find the natural frequencies of vibration, where possible, by using the mass matrices cited here. Base the stiffness matrix on Eq. 4.4.5. The exact fundamental frequency is $\omega_1 = \pi^2 \sqrt{EI/16\rho AL^4}$, where ρ is the mass density and A is the cross-sectional area.
(a) Matrix $[\mathbf{m}]$ is given by the first of Eqs. 11.3.1, with d.o.f. u_1 and u_2 discarded.
(b) Matrix $[\mathbf{m}]$ is given by Eq. 11.3.2.
(c) Matrix $\lceil \mathbf{m} \rfloor$ is given by Eq. 11.3.3.
(d) Matrix $\lceil \mathbf{m} \rfloor$ is given by Eq. 11.3.5.
(e) Matrix $[\mathbf{m}]$ is given by Problem 11.8.

11.17 Model the beam of Problem 11.16 by two elements, each of length L as in Fig. 11.5.2. Impose symmetry about the midpoint, so that θ_1 and w_2 are the only active d.o.f. and only the left element need be considered. Again, find the natural frequencies using the various mass matrices.

11.18 Mass condensation, starting from Eq. 11.5.1, would be more accurate if the assumption $[\mathbf{M}_{ms}] = [\mathbf{M}_{ss}] = 0$ were not made. What is an objection to this approach?

11.19 If $\lambda[\mathbf{M}]\{\bar{\mathbf{D}}\}$ in Eq. 11.4.2 is regarded as a vector of inertia loads $\{\mathbf{R}\}$, and $[\mathbf{K}]$ is inverted to become the flexibility matrix $[\mathbf{F}]$, we can write $[\mathbf{F}]\{\mathbf{R}\} = \{\mathbf{D}\}$.

(a) Partition this equation into m and s d.o.f. as in Eq. 11.5.1 and let $\{\mathbf{R}_s\} = 0$. Derive the transformation

$$\begin{Bmatrix} \bar{\mathbf{D}}_m \\ \bar{\mathbf{D}}_s \end{Bmatrix} = [\mathbf{T}]\{\bar{\mathbf{D}}_m\}, \qquad \text{where} \qquad [\mathbf{T}] = \begin{bmatrix} \mathbf{I} \\ \mathbf{F}_{ms}^T \ \mathbf{F}_{mm}^{-1} \end{bmatrix}$$

(b) Show that this transformation is mathematically the same as that of Eq. 11.5.3.

(c) How can $[\mathbf{F}_{mm}]$ be computed, and what is its physical meaning?

(d) Why is the transformation of part (a) likely to yield greater computational efficiency than the form used in Eq. 11.5.3?

11.20 (a) Show that Eq. 11.5.5 yields $[\mathbf{K}_r] = [\mathbf{K}_{mm}] - [\mathbf{K}_{ms}]\,[\mathbf{K}_{ss}]^{-1}\,[\mathbf{K}_{ms}]^T$. Where has this relation been seen before?

(b) Similarly, work out the expression for $[\mathbf{M}_r]$ in Eq. 11.5.5.

11.21 (a) In the example problem that closes Section 11.5, is the choice of $\bar{w}_2$ as master and $\bar{\theta}_3$ as slave consistent with the rule of largest M_{ii}/K_{ii}?

(b) Make the other choice, $\bar{\theta}_3$ as master and $\bar{w}_3$ as slave, and compute the frequency and mode shape (analogous to Eqs. 11.5.13 and 11.5.14).

11.22 Extend the structure of Problems 11.24 and 11.25 so that five identical massless springs alternate with five identical particle masses. Consider the procedure for automatic selection of master d.o.f. described in Section 11.5. Recall from Section 2.11 that condensation of a node places the two adjacent springs in series. To simplify this problem, but for no sound theoretical reason, assume that condensation of a mass m effectively adds mass $m/2$ to the two adjacent masses, so that $[\mathbf{M}]$ remains diagonal.

(a) Choose two masters by making three left-to-right sweeps.

(b) Choose two masters by making three right-to-left sweeps.

(c) Find the frequencies ω_1 and ω_2 in part (a) and in part (b) and compare results.

11.23 (a) The frequency $\omega_1^2 = 6.1765\, EI/mL^3$ is computed below Eq. 11.5.13. Improve this estimate, if possible, by using $\bar{w}_2 = 1$ and the first $\bar{\theta}_3$ is Eq. 11.5.14 in the Rayleigh quotient (Eq. 11.4.4). Take $[\mathbf{K}]$ and $[\mathbf{M}]$ from Eq. 11.5.9.

(b) Repeat part (a) using the second θ_3 in Eq. 11.5.14.

(c) Improve the ω calculated in Problem 11.21(*b*) by using the mode shape in the Rayleigh quotient.

11.24 Apply mass condensation to the system shown. For simplicity, let $k = 1$ and $m = 2$. Find the fundamental frequency of the reduced system and compare it with the exact answer. Find the corresponding mode, using first Eq. 11.5.3 and then Eq. 11.5.6. Finally, improve the estimate of ω_1 by using the latter mode in the Rayleigh quotient.

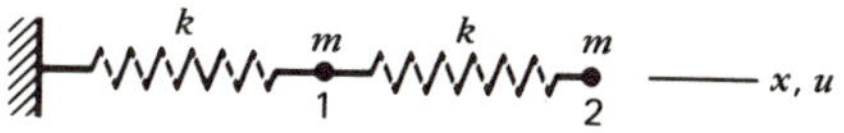

Problems 11.24 and 11.25

11.25 (a) For the system sketched, find the vibration frequencies $\lambda_1 = \omega_1^2$ and $\lambda_2 = \omega_2^2$ by extracting roots of the characteristic polynomial.

(b) Compute λ_1 by three cycles of inverse iteration, starting with $\{\bar{\mathbf{D}}\} = \{1\ 0\}$. Let $\lambda_0 = 0$.

(c) Compute λ_2 by three cycles of inverse iteration, starting with $\{\bar{\mathbf{D}}\} = \{1\ 0\}$. Let $\lambda_0 = 3$ and set $k = m = 1$.

11.26 Let $k = 2$ and $m = 1$ in Eq. 11.4.10. Verify the Sturm sequence property by using $\omega^2 = 3.9$, then $\omega^2 = 4.1$, in Eq. 11.4.10.

11.27 (a) Prove Eqs. 11.7.1. *Suggestion.* Write Eq. 11.4.2 for a mode i and then for a mode j, where $i \neq j$. Premultiply each by the other eigenvector, transpose one equation, and subtract.

(b) What computational procedure serves to normalize eigenvectors so that $\{\bar{\mathbf{D}}_i\}^T[\mathbf{M}]\{\bar{\mathbf{D}}_i\} = 1$?

(c) Use Eqs. 11.7.2 to derive Eq. 11.4.4, in which eigenvectors are not scaled.

11.28 Show that Eq. 11.7.1 is satisfied when applied to:

(a) The system of Fig. 11.4.1.

(b) The system in the sketch for Problem 11.24. Consider the case $k = m = 1$.

11.29 Obtain the lowest frequency for the system of Problem 11.24 by applying the transformations in Eqs. 11.7.3 to obtain a single uncoupled equation (Eq. 11.7.5). Let $k = m = 1$. (This is an exercise; it does not illustrate the practical purpose of writing Eq. 11.7.5.)

11.30 The uncoupled equations produced by a modal analysis (Eqs. 11.7.5) have a lower ω_{max} than ω_{max} of the original system. Would you therefore elect an explicit method to integrate the decoupled equations? Why or why not?

11.31 (a) Derive Eq. 11.8.1.

(b) Derive Eq. 11.8.2.

(c) Derive Eq. 11.8.3.

(d) Derive Eq. 11.8.4.

(e) Derive Eq. 11.8.7.

(f) Show that Eqs. 11.8.5 and 11.8.6 yield Eqs. 11.8.1 and 11.8.2 if $\gamma = 0.5$ and $\beta = 0$.

11.32 To take the initial time step in a direct integration method (Eq. 11.8.3 or Eq. 11.8.7), we must know $\{\ddot{\mathbf{D}}\}_0$. How might Eq. 11.2.4 be solved for $\{\ddot{\mathbf{D}}\}_0$ if $[\mathbf{M}]$ is singular?

11.33 Let $k = m = 1$ in the system of Problem 11.24. As initial conditions, use $u_1 = u_2 = \dot{u}_1 = 0$, $\dot{u}_2 = 1$.
   (a) Find the dynamic response by the modal method of Section 11.7. Include both modes in $[\boldsymbol{\phi}]$. Evaluate u_1 and u_2 at $t = 1, 2, 3, 4$, and 5.
   (b) Find the dynamic response by the central difference method (Eq. 11.8.3). Use seven steps of $\Delta t = 1.0$.
   (c) Repeat part (b) using four steps of $\Delta t = 0.5$.
   (d) Repeat part (b) trying three steps of $\Delta t = 2.0$. Why does this solution fail?
   (e) Find the dynamic response by Newmark's method (Eq. 11.8.7), with $\gamma = 0.5$ and $\beta = 0.25$. Use three steps of $\Delta t = 1.0$. At each time step, do accelerations computed from Eq. 11.8.5 agree with accelerations computed from Eq. 11.2.4?

11.34 A block of mass M is at rest on a horizontal platform. The coefficient of friction between block and platform is 0.2. At $t = 0$, the platform commences rectilinear horizontal oscillation with velocity $\dot{u} = A \sin 10t$, where A is a constant.
   (a) Program the central difference method to calculate the motion of the block (D, $\dot{D}$, and $\ddot{D}$). Compute the motion for $0 < t < 2$ seconds, using first $A = 0.015g$ and then $A = 0.025g$, where g is the acceleration of gravity.
   (b) Repeat part (a) using Newmark's method.
   (c) Is there an upper limit to the time step Δt in part (a)?

Several additional programming projects are possible in the areas of eigenproblem solution and dynamic response, where hand calculations are more tedious than enlightening.

CHAPTER 12

BUCKLING AND OTHER EFFECTS OF MEMBRANE FORCES

12.1 INTRODUCTION

Membrane forces in bars, plates and shells may produce buckling. A membrane force acts tangent to a member axis or midsurface (see, for example, Fig. 12.2.2*b*). Buckling occurs when a member or structure converts membrane strain energy into strain energy of bending. A thin-walled structure with low bending stiffness but high membrane stiffness may fail dramatically because sudden large bending deformations are needed to store the membrane energy released by buckling.

Usually, and quite properly, we associate buckling with displacements perpendicular to the surface of a plate or shell element. But occasionally buckling displacements lie in the plane of the element. A case in point is an I beam divided into flat elements. During buckling by rotation of cross sections about the web axis, elements in the flange move in their own plane. A buckling formulation that accounts only for surface-normal buckling displacements fails to solve this problem.

To study buckling, we need a matrix that accounts for the change in potential energy associated with rotation of volume elements under load. This matrix, designated $[\mathbf{k}_\sigma]$ for an element and $[\mathbf{K}_\sigma]$ for a structure, has been called the *initial stress stiffness matrix*, the *differential stiffness matrix*, the *geometric stiffness matrix*, and the *stability coefficient matrix*. Here we adopt the name *stress stiffness matrix*. Matrix $[\mathbf{k}_\sigma]$ is independent of elastic properties. It depends only on the element's geometry, displacement field, and state of stress. We can say that $[\mathbf{k}_\sigma]$ accounts for the effect of existing forces on bending stiffness. For example, it accounts for the fact that a laterally loaded beam deflects more when axial compression is added but less when axial tension is added.

Example

We illustrate concepts with the problem of Fig. 12.1.1. It has the single d.o.f. θ. The total potential is

$$\Pi_p = \tfrac{1}{2} k_\theta \theta^2 + \tfrac{1}{2} k_s [(L - L\cos\theta) + D]^2 - QL\sin\theta \qquad (12.1.1)$$

From $d\Pi_p/d\theta = 0$, the equilibrium value of θ is

$$k_\theta \theta + k_s [(L - L \cos \theta) + D] L \sin \theta = QL \cos \theta \tag{12.1.2}$$

If θ is restricted to the small angle $\delta\theta$, then $\cos \delta\theta \approx 1$ and $\sin \delta\theta \approx \delta\theta$, and Eq. 12.1.2 becomes

$$(k_\theta + k_s DL)\, \delta\theta = QL \tag{12.1.3}$$

or, since $k_s D = P$,

$$(k_\theta + PL)\, \delta\theta = QL \tag{12.1.4}$$

Here k_θ corresponds to the conventional stiffness $[\mathbf{k}]$ and PL to the stress stiffness $[\mathbf{k}_\sigma]$. The overall stiffness that resists load Q is the coefficient of $\delta\theta$. Thus the overall stiffness is greater than k for tensile P ($P > 0$) but less than k for compressive P ($P < 0$). If $Q = 0$, a nonzero $\delta\theta$ is possible if $k_\theta = -PL$, which defines the critical (buckling) load $P_{cr} = -k_\theta/L$.

Small-angle approximations show that during displacement $\delta\theta$, the membrane force P in Fig. 12.1.1 moves a distance $L(\delta\theta)^2/2$. The change in membrane energy is $PL(\delta\theta)^2/2$ and the change in bending energy is $k_\theta(\delta\theta)^2/2$. The sum of the two must vanish if $Q = 0$. Thus we again obtain the buckling load $P_{cr} = -k_\theta/L$.

With the exception of Section 12.5, where limitations and extensions of the theory are discussed, analyses in this chapter are subject to the following restrictions. Forces applied to a structure are fixed in magnitude, global direction, and point of application on the structure. Buckling displacements and rotations are small, forces and stresses remain essentially constant during a buckling displacement, and the problem is linear in the displacement variables (see Eq. 12.1.4). Each of these features is seen in the problem of Fig. 12.1.1.

The restriction to linear problems implies that stresses used to form $[\mathbf{k}_\sigma]$ can be found by a standard linear analysis, as discussed in Chapters 1 to 10. Then, in this chapter, $[\mathbf{k}_\sigma]$ is used in an *independent* analysis to find displacements under loads (such as Q in Fig. 12.1.1) or to define buckling. In a buckling problem we maintain constant *distribution* of forces and ask what their *intensity* must be to produce elastic instability. *Nonlinear* action is more often found in doubly curved shells than in flat plates and developable shells (Section 12.5).

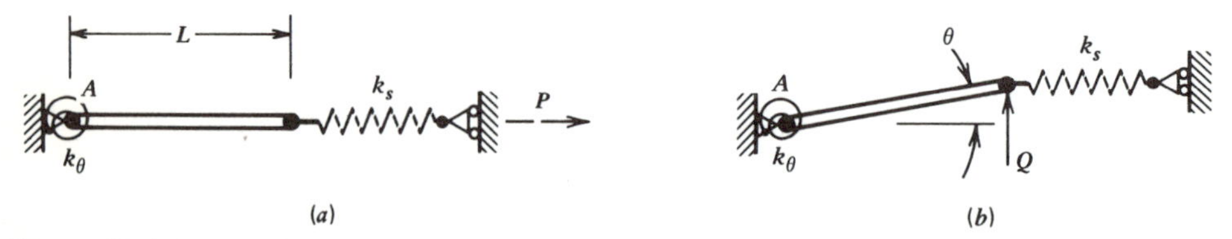

Figure 12.1.1. (*a*) Initial and (*b*) displaced configurations of a weightless, rigid bar of length L. Spring k_s is constrained to remain horizontal and is initially stretched an amount D to produce a force P. Load Q is externally applied. The rotational springs has stiffness k_θ.

A stress stiffness matrix is derived by adding higher-order terms to the strain-displacement relations. In linear problems only the simplest of the higher-order terms are used.

The structure matrix $[\mathbf{K}_\sigma]$ is built by summing overlapping terms of the component $[\mathbf{k}_\sigma]$ matrices, in the same way that $[\mathbf{K}]$ is built by summing terms of the conventional element stiffness matrices $[\mathbf{k}]$.

12.2 STRESS STIFFNESS MATRICES FOR BARS, BEAMS, AND PLATES

In what follows we derive $[\mathbf{k}_\sigma]$ matrices under the convention that tensile membrane forces are positive.

Consider first a bar or truss element (Fig. 12.2.1*a*). Asume that it can deform axially but is infinitely stiff in bending, so that it is straight in any displaced configuration. The axial strain for small displacements is

$$\epsilon_x = \epsilon_u + \epsilon_v, \qquad \text{where} \qquad \epsilon_u = \frac{u_2 - u_1}{L} \qquad \text{and} \qquad \epsilon_v = \frac{1}{2}\left(\frac{v_2 - v_1}{L}\right)^2 \tag{12.2.1}$$

Strain ϵ_v is a linearization based on the first two terms of the secant series. Specifically, $\epsilon_v = \Delta L/L$, where ΔL is the lengthening associated with rotation of the bar through a *small* angle θ without x-direction motion of any point (Fig. 12.2.1*b*).

$$\Delta L = \frac{L}{\cos\theta} - L = L(\sec\theta - 1) = L\frac{\theta^2}{2} = \frac{L}{2}\left(\frac{v_2 - v_1}{L}\right)^2 \tag{12.2.2}$$

Strain energy in the bar is $U = AEL\epsilon_x^2/2$. Thus

$$U = AEL\left(\frac{\epsilon_u^2}{2} + \epsilon_u\epsilon_v + \frac{\epsilon_v^2}{2}\right) = \frac{AEL}{2}(\epsilon_u^2 + \epsilon_v^2) + PL\epsilon_v \tag{12.2.3}$$

where $AE\epsilon_u$ has been identified as an axial force P, positive in tension. With displacements interpolated linearly from nodal values,

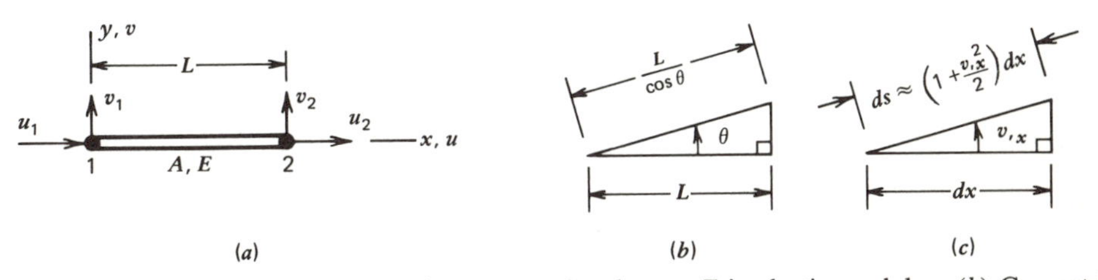

Figure 12.2.1. (*a*) Truss element. A is cross-sectional area, E is elastic modulus. (*b*) Geometric relation used in Eq. 12.2.2. (*c*) Corresponding geometric relation for an element of differential length dx.

$$\epsilon_u = \frac{1}{L}\lfloor -1 \quad 1 \rfloor \{u_1 \quad u_2\} \tag{12.2.4}$$

$$\epsilon_v = \frac{1}{2L^2}\begin{Bmatrix} v_1 \\ v_2 \end{Bmatrix}^T \lfloor -1 \quad 1 \rfloor^T \lfloor -1 \quad 1 \rfloor \begin{Bmatrix} v_1 \\ v_2 \end{Bmatrix} \tag{12.2.5}$$

The terms ϵ_u^2 and ϵ_v in Eq. 12.2.3 are quadratic in nodal d.o.f., but ϵ_v^2 is quartic, so it may be discarded as negligible in comparison with ϵ_u^2. Thus Eq. 12.2.3 yields, with $\{\mathbf{d}\} = \{u_1 \; v_1 \; u_2 \; v_2\}$,

$$U = \frac{1}{2}\{\mathbf{d}\}^T \left(\frac{AE}{L} \begin{bmatrix} 1 & 0 & -1 & 0 \\ 0 & 0 & 0 & 0 \\ -1 & 0 & 1 & 0 \\ 0 & 0 & 0 & 0 \end{bmatrix} + \frac{P}{L} \begin{bmatrix} 0 & 0 & 0 & 0 \\ 0 & 1 & 0 & -1 \\ 0 & 0 & 0 & 0 \\ 0 & -1 & 0 & 1 \end{bmatrix} \right) \{\mathbf{d}\} \tag{12.2.6}$$

We identify the first 4-by-4 matrix, with its coefficient AE/L, as the conventional stiffness matrix $[\mathbf{k}]$ for a plane truss element. We identify the second 4-by-4 matrix, with its coefficient P/L, as the stress stiffness matrix $[\mathbf{k}_\sigma]$ for a plane truss element. With it we could analyze *overall* buckling of a plane truss (buckling of individual bars is not accounted for). The physical meaning and computational use of $[\mathbf{k}_\sigma]$ matrices are considered in Section 12.4.

Consider next a beam element (Fig. 12.2.2*a*). A fiber at distance y above the neutral axis of bending has the strain

$$\epsilon_x = u_{,x} - y w_{,xx} + \frac{w_{,x}^2}{2} \tag{12.2.7}$$

where $u_{,x}$ represents axial stretching: $-zw_{,xx}$ is the strain produced by curvature $w_{,xx}$, negative if z and $w_{,xx}$ are of the same sign; and $w_{,x}^2/2$ is ϵ_v of Eq. 12.2.1, but written for a differential element of length dx (Fig. 12.2.1*c*). With $dV = dA\,dx$, the element strain energy is

$$U = \int_0^L \int_A \tfrac{1}{2} E \epsilon_x^2 \, dA \, dx \tag{12.2.8}$$

We substitute Eq. 12.2.7 into Eq. 12.2.8 and note that

$$\int dA = A, \qquad \int z \, dA = 0, \qquad \int z^2 \, dA = I, \qquad \int E u_{,x} \, dA = P \tag{12.2.9}$$

where P is again the axial force, positive in tension. A term $w_{,x}^4$, quartic in nodal d.o.f., is discarded. Thus

$$U = \int_0^L \frac{AE}{2} u_{,x}^2 \, dx + \int_0^L \frac{EI}{2} w_{,xx}^2 \, dx + \int_0^L \frac{P}{2} w_{,x}^2 \, dx \tag{12.2.10}$$

The first integral yields **[k]** for a truss element, associated with d.o.f. u_1 and u_2, and last seen in Eq. 12.2.6. The second integral yields **[k]** for a standard beam element (Eq. 4.4.5). The third integral sums the work done, and hence strain energy stored, when differential elements dx are stretched an amount $w_{,x}^2 \, dx/2$ by tensile force P. From this third integral we get $[\mathbf{k}_\sigma]$ for a beam element, first published in Ref. 12.1 and derived as follows.

Lateral displacement w of the beam, and its first derivative $w_{,x}$, are

$$w = \lfloor \mathbf{N} \rfloor \{\mathbf{d}\} \qquad w_{,x} = \lfloor \mathbf{G} \rfloor \{\mathbf{d}\} \tag{12.2.11}$$

where $\lfloor \mathbf{N} \rfloor = \lfloor N_1 \; N_2 \; N_3 \; N_4 \rfloor$ as in Eq. 4.2.5. The four coefficients in $\lfloor \mathbf{G} \rfloor$ are $G_i = N_{i,x}$. Also, $\{\mathbf{d}\} = \{w_1 \; \theta_1 \; w_2 \; \theta_2\}$. With $w_{,x}^2 = w_{,x}^T w_{,x}$ we have, from the third integral of Eq. 12.2.10,

$$\tfrac{1}{2} \{\mathbf{d}\}^T [\mathbf{k}_\sigma] \{\mathbf{d}\} = \tfrac{1}{2} \{\mathbf{d}\}^T \left[\int_0^L \lfloor \mathbf{G} \rfloor^T P \lfloor \mathbf{G} \rfloor \, dx \right] \{\mathbf{d}\} \tag{12.2.12}$$

Integration yields, for constant P,

$$[\mathbf{k}_\sigma] = \frac{P}{30L} \begin{bmatrix} 36 & 3L & -36 & 3L \\ 3L & 4L^2 & -3L & -L^2 \\ -36 & -3L & 36 & -3L \\ 3L & -L^2 & -3L & 4L^2 \end{bmatrix} \tag{12.2.13}$$

where P is positive in *tension*. Equation 12.2.13 reduces to the $[\mathbf{k}_\sigma]$ of Eq. 12.2.6 if bending is suppressed by setting $\theta_1 = \theta_2 = (w_2 - w_1)/L$.

Flat plate elements yield to the same approach used for beam elements [12.1, 12.2]. Let N_x, N_y, and N_{xy} be membrane forces per unit length (Fig. 12.2.2 *b*). Membrane strains produced by small rotations $w_{,x}$ and $w_{,y}$ of the midsurface are [9.1, 13.7]

$$\epsilon_x = \frac{w_{,x}^2}{2} \qquad \epsilon_y = \frac{w_{,y}^2}{2} \qquad \gamma_{xy} = w_{,x} w_{,y} \tag{12.2.14}$$

Strain energy associated with these strains is

$$\tfrac{1}{2} \{\mathbf{d}\}^T [\mathbf{k}_\sigma] \{\mathbf{d}\} = \tfrac{1}{2} \iint \begin{Bmatrix} w_{,x} \\ w_{,y} \end{Bmatrix}^T \begin{bmatrix} N_x & N_{xy} \\ N_{xy} & N_y \end{bmatrix} \begin{Bmatrix} w_{,x} \\ w_{,y} \end{Bmatrix} dx \, dy \tag{12.2.15}$$

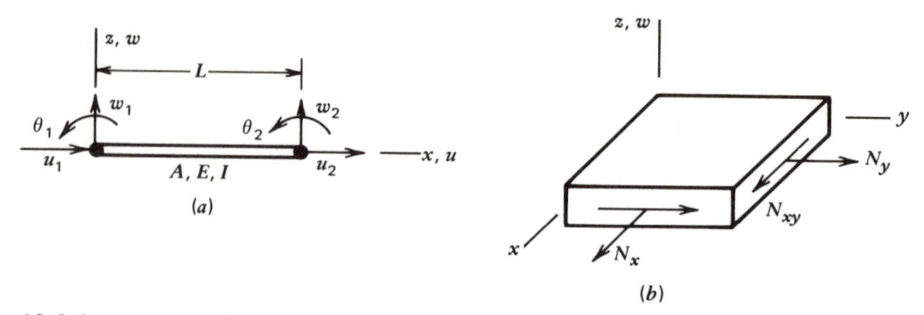

Figure 12.2.2. (*a*) Beam element. *A* is cross-sectional area, *E* is elastic modulus, *I* is area moment of inertia. Coordinate *z* originates at the centroid of a cross section. (*b*) Differential element of a flat plate, showing membrane forces (see Eqs. 9.1.15 for definitions of the *N*'s).

We select a displacement assumption $w = \lfloor \mathbf{N} \rfloor \{\mathbf{d}\}$ appropriate to the element shape and number of d.o.f., such as Eq. 9.2.2. From it we obtain the rotations

$$\begin{Bmatrix} w_{,x} \\ w_{,y} \end{Bmatrix} = \underset{2\times n}{[\mathbf{G}]} \underset{n\times 1}{\{\mathbf{d}\}} \tag{12.2.16}$$

where $[\mathbf{G}]$ is in general a function of x and y. Hence the stress stiffness matrix is

$$[\mathbf{k}_\sigma] = \iint [\mathbf{G}]^T \begin{bmatrix} N_x & N_{xy} \\ N_{xy} & N_y \end{bmatrix} [\mathbf{G}]\, dx\, dy \tag{12.2.17}$$

where integration extends over the element area. If the element is isoparametric, shape functions $\lfloor \mathbf{N} \rfloor$ are expressed in terms of ξ and η, and we must therefore invoke $[\mathbf{J}]$, the Jacobian matrix of Eqs. 5.3.6 and 5.3.7.

$$\begin{Bmatrix} w_{,\xi} \\ w_{,\eta} \end{Bmatrix} = [\mathbf{G}_I]\{\mathbf{d}\}, \qquad \begin{Bmatrix} w_{,x} \\ w_{,y} \end{Bmatrix} = [\mathbf{J}]^{-1} \begin{Bmatrix} w_{,\xi} \\ w_{,\eta} \end{Bmatrix} \tag{12.2.18}$$

where $[\mathbf{G}_I]$ contains derivatives of the N_{ij} with respect to ξ and η. Equation 12.2.15 yields

$$[\mathbf{k}_\sigma] = \int_{-1}^{1}\int_{-1}^{1} [\mathbf{G}_I]^T[\mathbf{J}]^{-T} \begin{bmatrix} N_x & N_{xy} \\ N_{xy} & N_y \end{bmatrix} [\mathbf{J}]^{-1}[\mathbf{G}_I]\, J\, d\xi\, d\eta \tag{12.2.19}$$

where J is the Jacobian determinant. Membrane forces may vary over an element. Equation 12.2.19 can use different membrane forces at the different sampling points.

A stress stiffness matrix is termed "consistent" if built from the same shape functions used to build the conventional stiffness matrix. If the structure geometry is well modeled and if elements are compatible and not softened by low-order integration rules, then such a formulation yields an upper bound to the magnitude of the true buckling load.

However, $[\mathbf{k}]$ and $[\mathbf{k}_\sigma]$ can be based on different displacement fields. We recall from the convergence requirements of Section 4.9 that if a strain energy expression involves displacement derivatives or order m, the displacement assumption must provide interelement continuity of displacement derivatives or order $m - 1$ as the mesh is refined. Preceding energy integrals that yield $[\mathbf{k}_\sigma]$ involve first derivatives of displacement, so continuity of displacement is all that is required. Thus, for example, we would expect to be able to find P_{cr} for a pin-ended column by using the conventional beam $[\mathbf{k}]$ but the $[\mathbf{k}_\sigma]$ of Eq. 12.2.6. This is indeed the case but, for a given accuracy, we must divide the column into more elements than when we use the consistent $[\mathbf{k}_\sigma]$. (However, in plate buckling with $N_x = -N_y$ in Eq. 12.2.17, the matrix analogous to $[\mathbf{k}_\sigma]$ in Eq. 12.2.6 fails miserably.)

It is recommended that $[\mathbf{k}_\sigma]$ for complicated elements be based on a simpler displacement field than used to construct the conventional stiffness matrix in order to increase computational efficiency with little if any loss in accuracy [11.28, 12.3]. The "best" $[\mathbf{k}_\sigma]$ is probably intermediate to the consistent $[\mathbf{k}_\sigma]$ and the simplest acceptable $[\mathbf{k}_\sigma]$. Numerical examples show that computed buckling loads are increased when $[\mathbf{k}_\sigma]$ is simplified. If $[\mathbf{k}_\sigma]$ is generated by numerical integration, a simplified displacement field is effectively employed by adopting a reduced order of quadrature. Unlike the conventional stiffness matrix, the stress stiffness matrix need not include constant strain (or constant curvature) in the assumed displacement field. However, a valid $[\mathbf{k}_\sigma]$ must not generate nodal forces during rigid-body translation.

An assumed displacement field is rarely competent enough to represent the actual structural deformations, whether the problem is static deflection or buckling. The error can be appreciable in the case of frames [12.4]. For good accuracy we may have to divide each frame member into two elements or adopt $[\mathbf{k}_\sigma]$ matrices that accurately reflect the deformed shape of a bar under the combined action of axial load and bending [12.5, 12.24]. (See also Appendix A.)

A cubic displacement field produces a satisfactory $[\mathbf{k}_\sigma]$ for a thin-plate element. For thick or sandwich plates, where transverse shear deformation is appreciable, we need a quadratic displacement field or reduced integration to produce a $[\mathbf{k}_\sigma]$ that provides convergence to correct results [12.6].

Stress stiffness matrices for plates and prismatic members are explicitly stated in various publications [12.6–12.11].

A matrix derived for an element that lies on a local x-axis can be transformed to another orientation by the standard transformation $[\mathbf{T}]^T[\mathbf{k}_\sigma]\,[\mathbf{T}]$ described in Chapter 6.

12.3 A MORE GENERAL FORMULATION

Section 12.2 approached each type of element as a special case. When applied to an element of complicated geometry, this method of derivation leaves us wondering if all important terms have been included. It is advantageous to have a general expression, analogous to that for a conventional stiffness matrix (Eq. 4.3.4), that may be specialized to specific cases and that wants only a specific displacement field to yield a matrix $[\mathbf{k}_\sigma]$. Such an expression is now derived.

Strains can be written

$$\{\boldsymbol{\epsilon}\} = \{\boldsymbol{\epsilon}_L\} + \{\boldsymbol{\epsilon}_{NL}\} \tag{12.3.1}$$

where $\{\boldsymbol{\epsilon}_L\}$ contains the usual definition of strain (Eq. 1.4.4), linear in the displacement derivatives. Nonlinear strains $\{\boldsymbol{\epsilon}_{NL}\}$ arise from higher-order terms. Work is done, and strain energy U is stored, as constant stresses $\{\boldsymbol{\sigma}_0\}$ act through strains $\{\boldsymbol{\epsilon}\}$ [3.4].

$$\begin{aligned} \int_V \{\boldsymbol{\epsilon}\}^T\{\boldsymbol{\sigma}_0\}\, dV &= \int_V \{\boldsymbol{\epsilon}_L\}^T\{\boldsymbol{\sigma}_0\}\, dV + \int_V \{\boldsymbol{\epsilon}_{NL}\}^T\{\boldsymbol{\sigma}_0\}\, dV \\ \text{or} \qquad U &= U_L + U_{NL} \end{aligned} \tag{12.3.2}$$

In section 12.2 we saw that $[\mathbf{k}_\sigma]$ is produced by stresses acting through displacements associated with higher-order contributions to strain. Accordingly, we extract $[\mathbf{k}_\sigma]$ from U_{NL} as follows.

Expressions for strain appear in elasticity texts. Typical terms, in conventional notation, are [3.4]

$$\begin{aligned} \epsilon_x &= u_{,x} + \tfrac{1}{2}(u_{,x}^2 + v_{,x}^2 + w_{,x}^2) \\ \gamma_{xy} &= u_{,y} + v_{,x} + (u_{,x}u_{,y} + v_{,x}v_{,y} + w_{,x}w_{,y}) \end{aligned} \tag{12.3.3}$$

where contributions to $\{\boldsymbol{\epsilon}_{NL}\}$ are set off by parentheses, and x, y, and z are global directions. To introduce matrix notation so that a formula for $[\mathbf{k}_\sigma]$ can be developed, we do the following. Let $\{\mathbf{d}\}$ be element nodal d.o.f., as usual. We define

$$\{\boldsymbol{\delta}\} = [\mathbf{G}]\{\mathbf{d}\}, \qquad \text{where} \qquad \{\boldsymbol{\delta}\} = \{u_{,x}\; u_{,y}\; u_{,z}\; v_{,x}\; v_{,y}\; v_{,z}\; w_{,x}\; w_{,y}\; w_{,z}\} \tag{12.3.4}$$

Coefficients in $[\mathbf{G}]$ are found by differentiating shape functions in $[\mathbf{N}]$, as in Eq. 12.2.11. We also define

$$[\mathbf{Q}] = \begin{bmatrix} u_{,x} & 0 & 0 & v_{,x} & 0 & 0 & w_{,x} & 0 & 0 \\ 0 & u_{,y} & 0 & 0 & v_{,y} & 0 & 0 & w_{,y} & 0 \\ 0 & 0 & u_{,z} & 0 & 0 & v_{,z} & 0 & 0 & w_{,z} \\ u_{,y} & u_{,x} & 0 & v_{,y} & v_{,x} & 0 & w_{,y} & w_{,x} & 0 \\ 0 & u_{,z} & u_{,y} & 0 & v_{,z} & v_{,y} & 0 & w_{,z} & w_{,y} \\ u_{,z} & 0 & u_{,x} & v_{,z} & 0 & v_{,x} & w_{,z} & 0 & w_{,x} \end{bmatrix} \tag{12.3.5}$$

With these definitions, strains $\{\boldsymbol{\epsilon}\}$ of Eqs. 12.3.1 [3.4] can be written in the form

$$\{\boldsymbol{\epsilon}\} = \{\boldsymbol{\epsilon}_L\} + \{\boldsymbol{\epsilon}_{NL}\} = \{\boldsymbol{\epsilon}_L\} + \tfrac{1}{2}[\mathbf{Q}][\mathbf{G}]\{\mathbf{d}\} \tag{12.3.6}$$

Displacements u, v, and w are functions of x, y, and z, where x, y, and z is the position of a point in the continuum in the unstrained configuration. With this definition, $\{\epsilon\}$ is called Green strain or Lagrangian strain, and $\{\epsilon\} = 0$ for any rigid-body motion, no matter how large.

The vector of existing stresses is

$$\{\boldsymbol{\sigma}_0\} = \{\sigma_{x0}\ \sigma_{y0}\ \sigma_{z0}\ \tau_{xy0}\ \tau_{yz0}\ \tau_{zx0}\} \tag{12.3.7}$$

Equations 12.3.2, 12.3.6, and 12.3.7 yield

$$U_{NL} = \tfrac{1}{2}\{\mathbf{d}\}^T \int_V [\mathbf{G}]^T[\mathbf{Q}]^T\{\boldsymbol{\sigma}_0\}\, dV = \tfrac{1}{2}\{\mathbf{d}\}^T \int_V [\mathbf{G}]^T \begin{bmatrix} \mathbf{s} & \mathbf{O} & \mathbf{O} \\ \mathbf{O} & \mathbf{s} & \mathbf{O} \\ \mathbf{O} & \mathbf{O} & \mathbf{s} \end{bmatrix} \{\boldsymbol{\delta}\}\, dV \tag{12.3.8}$$

The second integral is a change in form of the first, obtained by using the following identity.

$$\underset{9\times 6}{[\mathbf{Q}]^T}\ \underset{6\times 1}{\{\boldsymbol{\sigma}_0\}} = \underset{9\times 9}{\begin{bmatrix} \mathbf{s} & \mathbf{O} & \mathbf{O} \\ \mathbf{O} & \mathbf{s} & \mathbf{O} \\ \mathbf{O} & \mathbf{O} & \mathbf{s} \end{bmatrix}} \underset{9\times 1}{\{\boldsymbol{\delta}\}}, \quad \text{where} \quad \underset{3\times 3}{[\mathbf{s}]} = \begin{bmatrix} \sigma_{x0} & \tau_{xy0} & \tau_{zx0} \\ \tau_{xy0} & \sigma_{y0} & \tau_{yz0} \\ \tau_{zx0} & \tau_{yz0} & \sigma_{z0} \end{bmatrix} \tag{12.3.9}$$

After substitution of $\{\boldsymbol{\delta}\}$ from Eq. 12.3.4, we identify the stress stiffness matrix as

$$[\mathbf{k}_\sigma] = \int_V [\mathbf{G}]^T \begin{bmatrix} \mathbf{s} & \mathbf{O} & \mathbf{O} \\ \mathbf{O} & \mathbf{s} & \mathbf{O} \\ \mathbf{O} & \mathbf{O} & \mathbf{s} \end{bmatrix} [\mathbf{G}]\, dV \tag{12.3.10}$$

We see that $[\mathbf{k}_\sigma]$ is symmetric. As written here, $[\mathbf{G}]$ must be arranged so as to yield displacement derivatives in the order seen in Eq. 12.3.4.

Example

Matrix $[\mathbf{k}_\sigma]$ for a beam, derived in Section 12.2, is obtained from Eq. 12.3.10 as follows. The problem is plane and the beam lies on the x-axis, so v and all y and z derivatives vanish and σ_{x0} is the only membrane stress. Buckling involves significant rotation but no axial displacement of the midsurface. That is, during buckling, $w_{,x}$ is significant but $u_{,x}$ is not (imagine, for example, a column completely fixed at both ends and heated enough to make it buckle). Thus Eq. 12.3.10 reduces to

$$[\mathbf{k}_\sigma] = \int_V [\mathbf{G}]^T\, \sigma_{x0}\, [\mathbf{G}]\, dV = \int_0^L [\mathbf{G}]^T P\, [\mathbf{G}]\, dx \tag{12.3.11}$$

where $P = A\sigma_{x0}$ is the integral of σ_{x0} over $dy\,dz$, and $\lfloor \mathbf{G} \rfloor$ is the row vector of Eq. 12.2.11. If $u_{,x}$ had been retained, a $[\mathbf{k}_\sigma]$ matrix that operates on axial d.o.f. u_1 and u_2 would also appear. But it would be negligible in comparison with the conventional truss element stiffness matrix (which also operates on u_1 and u_2).

Example

A flat plate in the xy plane yields $[\mathbf{k}_\sigma]$ of Eq. 12.2.17 if we recognize σ_{x0}, σ_{y0}, and τ_{xy0} as the membrane stresses and recognize $w_{,x}$ and $w_{,y}$ as the rotations of significance. Integration with respect to z converts σ_{x0}, σ_{y0}, and τ_{xy0} to N_x, N_y, and N_{xy} of Fig. 12.2.2b.

Example

Consider next three-node triangular element. It lies in the xy plane but may have displacements u, v, and w. We adopt a linear field for each displacement. Let $\phi = u$, v, or w, with $\phi = \phi\,(x, y)$. We find, from equations in Section 4.5,

$$\phi = \lfloor 1 \quad x \quad y \rfloor\,[\mathbf{A}]^{-1}\,\{\phi_1 \quad \phi_2 \quad \phi_3\} \tag{12.3.12}$$

$$\{\phi_{,x} \quad \phi_{,y}\} = [\mathbf{G}_\phi]\,\{\phi_1 \quad \phi_2 \quad \phi_3\} \tag{12.3.13}$$

where

$$\underset{2\times 3}{[\mathbf{G}_\phi]} = \frac{1}{2A}\begin{bmatrix} y_2 - y_3 & y_3 - y_1 & y_1 - y_2 \\ x_3 - x_2 & x_1 - x_3 & x_2 - x_1 \end{bmatrix} \tag{12.3.14}$$

Membrane stresses are σ_{x0}, σ_{y0}, and τ_{xy0}. If these stresses and the thickness t are constant, then [11.13]

$$\underset{9\times 9}{[\mathbf{k}_\sigma]} = \begin{bmatrix} \mathbf{Q} & \mathbf{O} & \mathbf{O} \\ \mathbf{O} & \mathbf{Q} & \mathbf{O} \\ \mathbf{O} & \mathbf{O} & \mathbf{Q} \end{bmatrix} \tag{12.3.15}$$

where

$$\underset{3\times 3}{[\mathbf{Q}]} = At\,[\mathbf{G}_\phi]^T \begin{bmatrix} \sigma_{x0} & \tau_{xy0} \\ \tau_{xy0} & \sigma_{y0} \end{bmatrix} [\mathbf{G}_\phi] \tag{12.3.16}$$

and $[\mathbf{k}_\sigma]$ operates $\{u_1\ u_2\ u_3\ v_1\ \ldots\ w_3\}$. For buckling motion normal to the xy plane we would set $[\mathbf{k}_\sigma] = [\mathbf{Q}]$ and retain only w d.o.f. For buckling motion *in* the xy plane we would set $[\mathbf{k}_\sigma] = \lceil \mathbf{Q}\ \mathbf{Q} \rfloor$ and *suppress* all w d.o.f. In this way the beam flange mentioned in the second paragraph of Section 12.1 could be modeled.

If an element is isoparametric, transformations such as those of Eq. 12.2.18 are invoked. For the *shell* element of Fig. 10.4.2c, the displacement field of Eq. 10.5.5 is a poor choice for $[\mathbf{k}_\sigma]$ because midsurface displacements are then independent of nodal rotations.

Example

How might Eq. 12.3.10 be applied to a nonisoparametric thin shell? Consider, for example, the shell element of Fig. 10.8.1. Let z of Eq. 12.3.10 represent a local shell normal, so that x and y are always tangent to the midsurface. Then all stresses with z subscripts can be omitted from $[\mathbf{s}]$. Also, we can assume that $w_{,x}$ and $w_{,y}$ are the only displacement derivatives of importance. Then

$$[\mathbf{k}_\sigma] = \int_V [\mathbf{G}_w]^T \begin{bmatrix} \sigma_{x0} & \tau_{xy0} \\ \tau_{xy0} & \sigma_{y0} \end{bmatrix} [\mathbf{G}_w] \, dV \qquad (12.3.17)$$

where $\{w_{,x} \; w_{,y}\} = [\mathbf{G}_w]\{\mathbf{d}\}$. In Fig. 10.8.1, study of the geometry shows that

$$w_{,x} = w_{,s} + u\phi_{,s} \qquad\qquad w_{,y} = \frac{1}{r} w_{,\theta} - \frac{v \cos \phi}{r} \qquad (12.3.18)$$

where $w_{,x}$ and $w_{,y}$ can be identified as rotations of the midsurface. With Eq. 12.3.18 and shape functions $[\mathbf{N}]$ that give u, v, and w in terms of independent variables s and θ and nodal d.o.f. $\{\mathbf{d}\}$, matrix $[\mathbf{G}_w]$ can be established.

12.4 CRITICAL LOADS (AN EIGENVALUE PROBLEM)

If stresses are fixed in *distribution,* what must be their *intensity* so that the reference configuration and an infinitesimally close (buckled) configuration are both equilibrium configurations? The mathematics follows.

Let $[\mathbf{K}_\sigma]$ of a structure be based on an arbitrary reference intensity of membrane stresses. Then, if λ is arbitrary scalar multiplier, $\lambda[\mathbf{K}_\sigma]$ is the stress stiffness matrix for some other intensity. Our restriction to linearity, noted in Section 12.1, means that neither $[\mathbf{K}_\sigma]$ nor the conventional stiffness matrix $[\mathbf{K}]$ is a function of displacement. If displacements $\{\mathbf{D}\}$ of the reference configuration are augmented by virtual displacements $\{\bar{\mathbf{D}}\}$ while applied loads $\{\mathbf{R}\}$ remain constant, what must λ be so that states $\{\mathbf{D}\}$ and $\{\mathbf{D} + \bar{\mathbf{D}}\}$ are both equilibrium configurations? Thus

$$([\mathbf{K}] + \lambda[\mathbf{K}_\sigma])\{\mathbf{D}\} = ([\mathbf{K}] + \lambda[\mathbf{K}_\sigma])\{\mathbf{D} + \bar{\mathbf{D}}\} = \{\mathbf{R}\} \qquad (12.4.1)$$

Subtraction of the first equation from the second yields the eigenvalue problem

$$([\mathbf{K}] + \lambda[\mathbf{K}_\sigma])\{\bar{\mathbf{D}}\} = 0, \qquad \text{or} \qquad [\mathbf{K}_{\text{total}}]\{\bar{\mathbf{D}}\} = 0 \qquad (12.4.2)$$

where the critical (buckling) load is associated with λ_{cr}, the lowest magnitude eigenvalue of Eq. 12.4.2. The computed λ_{cr} will be negative if membrane forces are taken as positive in tension. Displacement $\{\bar{\mathbf{D}}\}$ identify the buckled shape, but not its magnitude. (The corresponding eigenvalue problem in vibration analysis is considered in Section 11.4.)

On occasion we want the lowest *several* eigenvalues and their characteristic modes in order to gain insight into ways of stiffening or supporting the structure to make buckling less likely.

Two physical interpretations of Eq. 12.4.2 are as follows.

1. Because the product $[\mathbf{K}_{\text{total}}]\{\bar{\mathbf{D}}\}$ yields zero force, stresses of critical intensity reduce the overall structure stiffness to zero with respect to buckling mode $\{\bar{\mathbf{D}}\}$.
2. Premultiplication by $\{\bar{\mathbf{D}}\}^T/2$ yields

$$\{\bar{\mathbf{D}}\}^T[\mathbf{K}]\{\bar{\mathbf{D}}\}/2 + \lambda\{\bar{\mathbf{D}}\}^T[\mathbf{K}_\sigma]\{\bar{\mathbf{D}}\}/2 = 0 \tag{12.4.3}$$

This equation states that the change in strain energy is equal in magnitude to work done by membrane forces as they act through displacements associated with $\{\bar{\mathbf{D}}\}$. The net energy change is zero, since external forces have done no work. Equation 12.4.3 reflects an exchange of membrane energy for bending energy.

In general, $\{\bar{\mathbf{D}}\}$ contains d.o.f. that govern bending displacements *and* d.o.f. that govern membrane displacements. The latter d.o.f. can be omitted from $\{\bar{\mathbf{D}}\}$ for flat plates, where the buckling mode is an out-of-plane displacement.

Example

Consider a flagpole of length L and constant cross section, fixed at end 1 but free at end 2, where a unit axial tensile force is placed. We take $[\mathbf{k}]$ from Eq. 4.4.5 and $[\mathbf{k}_\sigma]$ from Eq. 12.2.13, with the reference intensity $P = 1.0$. We omit axial d.o.f. u_1 and u_2 and suppress d.o.f. w_1 and θ_1 at the fixed end. Thus, with a single element, Eq. 12.4.2 becomes

$$\left(\frac{EI}{L^3}\begin{bmatrix} 12 & -6L \\ -6L & 4L^2 \end{bmatrix} + \frac{\lambda}{30L}\begin{bmatrix} 36 & -3L \\ -3L & 4L^2 \end{bmatrix}\right)\begin{Bmatrix} w_2 \\ \theta_2 \end{Bmatrix} = 0 \tag{12.4.4}$$

A solution other than $w_2 = \theta_2 = 0$ requires that the expression in parentheses have a zero determinant. The roots are

$$\lambda_1 = -32.1807\ EI/L^2, \qquad \lambda_2 = -2.4859\ EI/L^2 \tag{12.4.5}$$

We identify the critical load as $P_{\text{cr}} = \lambda_2(1.0) = -2.4859EI/L^2$, where the minus sign shows that the axial force must be compressive to produce buckling. The theoretical solution is $P_{\text{cr}} = -\pi^2 EI/4L^2 = -2.4674EI/L^2$, which shows that Eq. 12.4.4 produces the expected upper bound to the magnitude of P_{cr}. If we take $[\mathbf{k}_\sigma]$ from Eq. 12.2.6, we find $P_{\text{cr}} = -3EI/L^2$. Eigenmodes are computed in the usual way: for example, we can find $\{\bar{\mathbf{D}}\}$ by setting $\lambda = \lambda_2$ and $\theta_2 = 1$ in Eq. 12.4.4 and solving for v_2.

Table 12.4.1 gives another example. Reference 12.2 underestimates critical loads because the element (Eq. 9.2.1) is incompatible. Reference 12.6 bases $[\mathbf{k}]$ on assumed-stress hybrid elements and $[\mathbf{k}_\sigma]$ on an assumed displacement field (equation 6 of the

TABLE 12.4.1. PERCENT ERROR IN COMPUTED BUCKLING LOAD FOR A SIMPLY SUPPORTED THIN SQUARE PLATE UNDER UNIFORM COMPRESSION IN ONE DIRECTION. MESH n BY n REPRESENTS n^2 IDENTICAL ELEMENTS IN THE ENTIRE PLATE.

Source of $[\mathbf{k}_\sigma]$	2-by-2 mesh	4-by-4 mesh	8-by-8 mesh
Ref. 12.2	—	−5.75	−1.68
Ref. 12.6	11.85	2.54	0.59
Ref. 12.20	0.39	0.03	0.002

reference, with 2-by-2 Gauss quadrature). Reference 20 uses a fully compatible formulation whose 16-d.o.f. element includes $w_{,xy}$ as a nodal freedom.

Computational methods for extraction of λ_{cr} are numerous. For the most part, we can use the same methods used to compute natural frequencies (Chapter 11). Note, however, that $[\mathbf{k}_\sigma]$ and $[\mathbf{K}_\sigma]$ are generally indefinite: they cannot be inverted. We avoid trying to extract roots of the characteristic polynomial, except for small pencil-and-paper problems such as the preceding example. Determinant tracking may be an attractive method, but it has the usual scaling problem. Also, a plot of det$[\mathbf{K}_{\text{total}}]$ versus load level may display a sudden and nearly vertical plunge across the load axis, which may befuddle the algorithm.

Example

The condensation of Section 11.5 can be invoked. Consider eliminating θ_2 from Eq. 12.4.4. Equation 11.5.3 yields

$$[\mathbf{T}]^T = \left\lfloor 1 \quad -\left(\frac{L}{4EI}\right)\left(-\frac{6EI}{L^2}\right) \right\rfloor = \left\lfloor 1 \quad \frac{1.5}{L} \right\rfloor \tag{12.4.6}$$

The transformations of Eq. 11.5.5 convert Eq. 12.4.4 to

$$3EI/L^3 + 1.2\lambda/L = 0 \tag{12.4.7}$$

from which we find $\lambda_{\text{cr}} = -2.5EI/L^2$.

A shell of revolution usually has a nonaxisymmetric buckling mode even if the loading is axially symmetric. Many waves will probably appear in each hoop circle. It is commonly assumed that the buckling mode varies circumferentially as a single Fourier harmonic [12.12–12.14]. (If the load is not axially symmetric this assumption is false, yet it may lead to acceptable results [12.15].) The procedure is similar to that described in Section 8.4. The first step is to select a specific number n of circumferential waves, then compute $[\mathbf{K}]_n$ and $[\mathbf{K}_\sigma]_n$. Next we solve the eigenvalue problem to get λ_{cr} for n waves. The entire procedure is repeated for $n + 1$ waves, for $n + 2$ waves, and so on. The lowest of the sequence of λ_{cr} values is identified as the desired buckling parameter. It is not obvious which value of n will govern, and many analyses may be needed: Ref. 12.12 mentions a case where buckling is associated with $n = 39$. Despite simple geometry, as for a cylindrical shell, many elements may be needed to model buckling waves properly in the meridional direction.

The computational procedure just described can be used with structures discretized into finite strips (Section 9.2). Stiffened flat panels are effectively treated in this way [12.16, 12.17]. Alternatively, such a panel can be viewed as a small part of a shell of revolution that has a very large radius (see Ref. 12.18 and Section 10.10).

Buckling of a structure does not imply buckling of each element. That is, at the structure's critical load λ_{cr}, from Eq. 12.4.3,

$$\{\bar{\mathbf{d}}\}^T ([\mathbf{k}] + \lambda_{cr}[\mathbf{k}_\sigma]) \{\bar{\mathbf{d}}\} \geqq 0 \tag{12.4.8}$$

where $\{\bar{\mathbf{d}}\}$ are element d.o.f. that correspond to $\{\bar{\mathbf{D}}\}$ of Eq. 12.4.2. Equality prevails only in the highest buckling mode or for a one-element structure.

Dynamic analysis of undamped structures with membrane forces leads to the equations

$$[\mathbf{K} + \mathbf{K}_\sigma] \{\mathbf{D}\} + [\mathbf{M}] \{\ddot{\mathbf{D}}\} = \{\mathbf{R}\} \tag{12.4.9}$$

$$([\mathbf{K} + \mathbf{K}_\sigma] - \omega^2[\mathbf{M}]) \{\bar{\mathbf{D}}\} = 0 \tag{12.4.10}$$

Equation 12.4.9 governs the dynamic response problem and Eq. 12.4.10 governs the natural frequency problem. Tensile membrane forces increase the natural frequencies. Compressive membrane forces decrease them and produce a root $\omega = 0$ if $\lambda = \lambda_{cr}$.

Consider a linkage of pin-connected rigid bars, either hanging under its own weight or loaded by an axial force. If we set $[\mathbf{K}] = 0$ and take $[\mathbf{k}_\sigma]$ matrices from Eq. 12.2.6, we can use Eqs. 12.4.9 and 12.4.10 to study dynamic response and natural frequencies. Vibrating strings and membranes can be analyzed by using $[\mathbf{m}]$ and $[\mathbf{k}_\sigma]$ matrices based on cubic displacement functions. If $[\mathbf{K}] = [\mathbf{M}] = 0$, $[\mathbf{K}_\sigma]\{\mathbf{D}\} = \{\mathbf{R}\}$ describes a static problem, such as a hanging chain deflected by a lateral force.

12.5 BIFURCATION, IMPERFECTIONS, LIMIT POINTS, AND NONLINEARITY

A real structure contains imperfections and may collapse at a load quite different than the load predicted by linear theory. We must therefore remark on how reality corresponds to the restrictions made in Section 12.1. Throughout the present section we presume that the *material* is linearly elastic.

As load and deflection increase, a perfect structure follows its primary (or fundamental) path, OA in Fig. 12.5.1. The structure buckles at its *bifurcation point* (point A, where

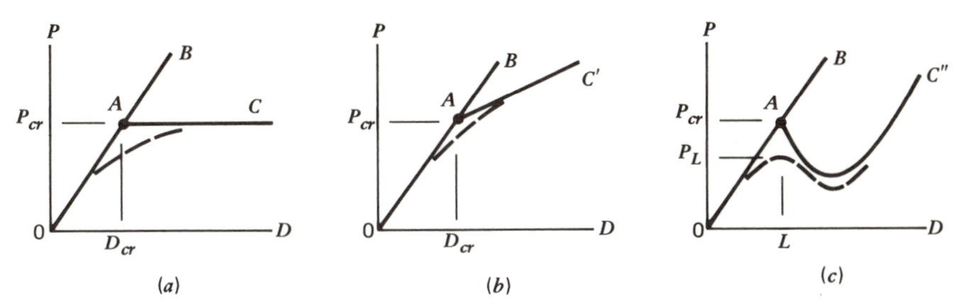

Figure 12.5.1. Plots of axial (membrane) force P versus its corresponding displacement D. These plots approximate the behavior of (*a*) a straight column, (*b*) a flat plate, and (*c*) an axially compressed cylindrical shell. Solid lines represent perfect structures. Dashed lines represent imperfect (real) structures.

primary path OA and secondary path AC intersect). Here more than one equilibrium configuration is possible, and $\det[\mathbf{K}_{\text{total}}] = 0$. In the postbuckling state, $D > D_{\text{cr}}$, path AB is never followed. It is the path of greatest resistance, and nature opts for a secondary path. The slope of the secondary path at the bifurcation point determines the nature of equilibrium at bifurcation. In the respective parts of Fig. 12.5.1, equilibrium is neutral (zero slope), stable (positive slope) and unstable (negative slope). A positive slope characterizes a structure with postbuckling strength: the structure can carry a load greater than P_{cr} without collapse. A negative slope characterizes a structure that may buckle precipitously by snap-through, as discussed shortly. In any of these cases, the structure is in its postbuckled state by the time buckling displacements are apparent to the unaided eye.

A real structure has geometric imperfections, such as dimples and out-of-roundness. It also has loading imperfections, such as nonuniformity of pressure and small, concentrated loads on a shell. (In analysis, imperfections can be introduced by elements that deviate slightly from a perfect geometry or by loading imperfections that can be deliberately introduced to mimic the effect of geometric imperfections.) Practical shells are imperfect, since they deliberately incorporate stiffeners, cutouts, flat spots, or bumps. Imperfections make the primary path curved and may prohibit bifurcation (Fig. 12.5.1). They may also introduce *limit points*, $D = L$ in Figs. 12.5.1*c* and 12.5.2*a*. A limit point is defined as a relative maximum on the P versus D curve, with no secondary path. Also, $[\mathbf{K}_{\text{total}}]$ becomes displacement dependent (see Eq. 12.5.1). At a limit point, $\det[\mathbf{K}_{\text{total}}]$

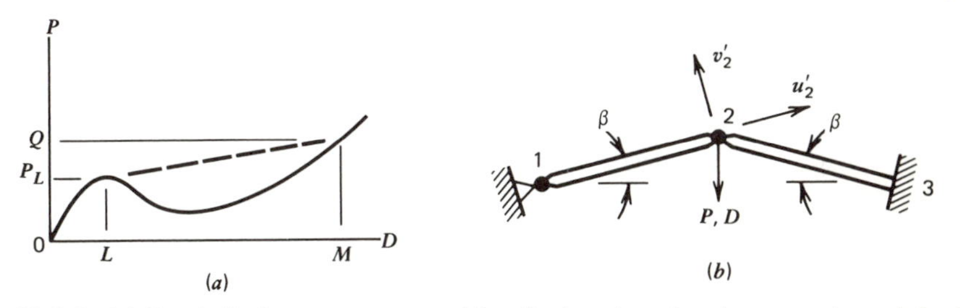

Figure 12.5.2. (*a*) Load-displacement curve with a limit point, showing snap-through buckling. (*b*) Frame that displays bifurcatiion if angle β is large. It may display snap-through if β is small.

= 0. Another way to find a limit point is to plot the P versus D curve by methods discussed in Chapter 13. These concepts are thoroughly discussed in Ref. 12.22.

Snap-through buckling is illustrated by Fig. 12.5.2*a*. When load P reaches P_L, the structure suddenly leaps from $D = L$ to $D = M$, where equilibrium is again possible.

A structure is *imperfection sensitive* if a small imperfection produces a large change in its load-carrying capacity. A "small" imperfection may be on the order of a shell thickness. If the secondary path has negative slope at bifurcation, the structure is imperfection sensitive (Fig. 12.5.1*c*). Such structures fail suddenly, usually at loads much less than predicted by linear bifurcation theory.

In preceding sections we assumed that membrane stresses are independent of bending. This assumption is false for the structure of Fig. 12.5.2*b*. Its primary path, *OA* in Fig. 12.5.1, is curved. We argue as follows. Axial force P_{12} in bar 1-2 depends on the moment M_3 at node 3, as a free-body diagram of bar 2-3 will show. In turn, M_3 depends on the bending stiffness of bar 2-3. In matrix notation,

$$([\mathbf{K}] + [\mathbf{K}_\sigma])\,\{u_2' \;\; v_2'\} = \{\mathbf{R}\} \tag{12.5.1}$$

where $[\mathbf{K}_\sigma]$ is *fully populated*. We can find P_{12} if we know u_2', but we cannot find u_2' because $[\mathbf{K}_\sigma]$ depends on P_{12}. In other words, the problem is nonlinear, and we must attack it with the tools of Chapter 13.

A primary path is linear only if the deformation state is rotation free. Such structures are rare: they include the perfect structures of Fig. 12.5.1 and complete spheres under external pressure. Subsequent loss of stability, at bifurcation or at a limit point, is *always* associated with rotation of structural elements.

Analyses of various shells show that limit points can appear at loads from one-quarter to 10 times the critical load predicted by linear bifurcation analysis [12.21, 12.22]. And, if we account for the change in direction of pressure loads caused by prebuckling rotations, the critical load may decrease by 50% [12.23].

We are now persuaded that linear theory is applicable to a limited class of structures. Have we then demolished the usefulness of this chapter? No. Such structures exist. And a *non*linear stability analysis must incorporate and extend linear theory.

PROBLEMS

12.1 In Chapter 11 we discussed the diagonal mass matrix $[\mathbf{m}]$. But why is a diagonal stress stiffness matrix unacceptable? Construct one that yields the exact critical load for a one-element, pin-ended column. Then try using it to find the critical load of a one-element, fixed-free column.

12.2 (a) Verify Eq. 12.2.6.
(b) Verify Eq. 12.2.10.
(c) Verify Eq. 12.2.13.

(d) Verify Eq. 12.3.9.

(e) Construct the $[\mathbf{k}_\sigma]$ produced by $u_{,x}$ and associated with u_1 and u_2, as suggested below Eq. 12.3.11. Show that $(k_\sigma)_{11}/(k)_{11} = \sigma_{x0}/E$, a negligible quantity.

(f) Verify Eqs. 12.3.14 through 12.3.16.

12.3 Derive the $[\mathbf{k}_\sigma]$ contained in Eq. 12.2.6 by imposing appropriate restrictions on the $[\mathbf{k}_\sigma]$ of Eq. 12.2.13.

12.4 (a) A bar element (Fig. 12.2.1) lies on the x-axis of local coordinates xyz, which are arbitrarily oriented with respect to global coordinates. What is the 6-by-6 $[\mathbf{k}_\sigma]$ in local coordinates? How is $[\mathbf{k}_\sigma]$ found in the global coordinates?

(b) Similarly, how would you establish $[\mathbf{k}_\sigma]$ in global coordinates for an arbitrarily oriented frame element?

12.5 Buckling of a tapered column is to be studied. Instead of using prismatic elements and stepwise thickness changes at nodes, elements themselves are to be tapered [11.4]. In which element matrices ($[\mathbf{k}]$ or $[\mathbf{k}_\sigma]$) does the effect of taper appear, and how is it to be included?

12.6 Construct a stress stiffness matrix for a beam element by using the displacement field of Problem 11.8 in Eq. 12.2.11.

12.7 Use the matrix derived in Problem 12.6 to solve the column buckling example problem in Section 12.4.

12.8 (a) Use $[\mathbf{k}_\sigma]$ from Eq. 12.2.6, if possible, to find the buckling loads of the one-element column sketched. Express P_{cr} as a function of E, I, and L.

(b) Repeat part (a) for the *two*-element column, but impose symmetry about the center node 2, so that the eigenvalue problem involves only two d.o.f. associated with element 1-2.

(c) Solve as stated in part (a). Let the linear spring have stiffness $k = 2EI/L^3$. The lower end is fully fixed.

(d) Solve as stated in part (a). Let the rotational spring have stiffness $k = EI/L$. The lower end is fully fixed.

(e) Repeat part (c) letting $k \to \infty$ (top free to rotate but cannot translate).

(f) Repeat part (d) letting $k \to \infty$ (top free to translate but cannot rotate).

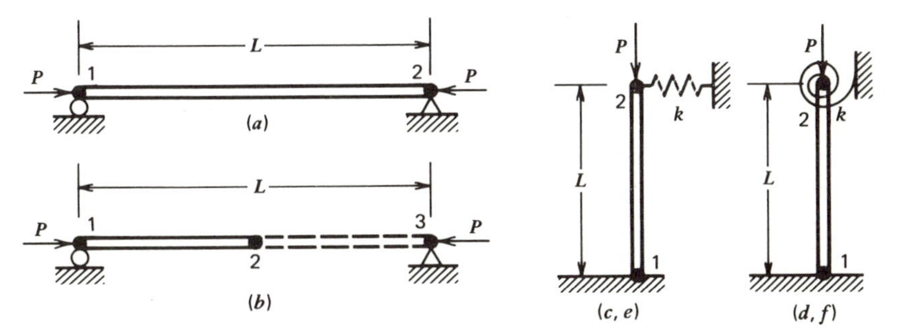

Problem 12.8

12.9 Repeat Problem 12.8 using $[\mathbf{k}_\sigma]$ from Eq. 12.2.13.

12.10 Repeat Problem 12.8 using the $[\mathbf{k}_\sigma]$ derived in Problem 12.6.

12.11 Solve for P_{cr} in Problem 12.8(c) if the bar is rigid and node 1 is made a hinge.

12.12 Consider a one-element cantilever beam of length $L = 3$ m and stiffness $EI = 110$ N·m^2. At node 2, the free end, it carries an axial force P and a transverse force $Q = 0.1$ N.

(a) Find the transverse displacement w_2 of node 2 if $P = 0$.
(b) Find w_2 if $P = 30.0$ N in compression. Take $[\mathbf{k}_\sigma]$ from Eq. 12.2.13.
(c) Find w_2 if $P = 30.0$ N in tension. Take $[\mathbf{k}_\sigma]$ from Eq. 12.2.13.

12.13 (a) Consider using the condensation technique described in Section 11.5 in the context of the present chapter. How will you decide which d.o.f. to elect as masters?
(b) Apply this technique to solve Problem 12.12(b).
(c) Apply this technique to solve Problem 12.12(c).
(d) If $[\mathbf{T}]$ in Eq. 11.5.3 is formed from coefficients in $[\mathbf{K} + \mathbf{K}_\sigma]$ instead of from coefficients in $[\mathbf{K}]$ alone, what standard operation is being pursued?

12.14 Use the condensation technique of Section 11.5 to remove d.o.f. θ_1 and solve for P_{cr} in (a) Problem 12.8(b), (b) Problem 12.9(b), and (c) Problem 12.10(b).

12.15 Consider a simply supported beam, modeled by a single element. Let $L = 1.0$ m, $A = 0.2(10)^{-3}$m^2, $EI = 300.0$ N·m^2, and $\rho = 2100.0$ kg/m^3. Use the stress stiffness matrix of Eq. 12.2.13. Impose symmetry (and reduce the problem to a single d.o.f.) by setting $\theta_2 = -\theta_1$.

(a) Find the fundamental frequency ω_1 if there is no axial force.
(b) Find the axial force that makes the frequency 347 rad/sec.
(c) Find the frequency if the axial force is 1200 N in compression.

12.16 A string of length $2L$ hangs from the ceiling. It carries a particle of weight W at its middle and another at its lower end. A lateral force P is applied at the lower end. Use the appropriate $[\mathbf{k}_\sigma]$ formulation and solve for the lateral deflections of the weights. Check your answers by elementary statics.

12.17 The weightless string is under tension T and carries a particle mass m. Use the appropriate $[\mathbf{m}]$ and $[\mathbf{k}_\sigma]$ matrices to find (*a*) the static deflection if gravity acts downward, and (*b*) the natural frequency of vibration.

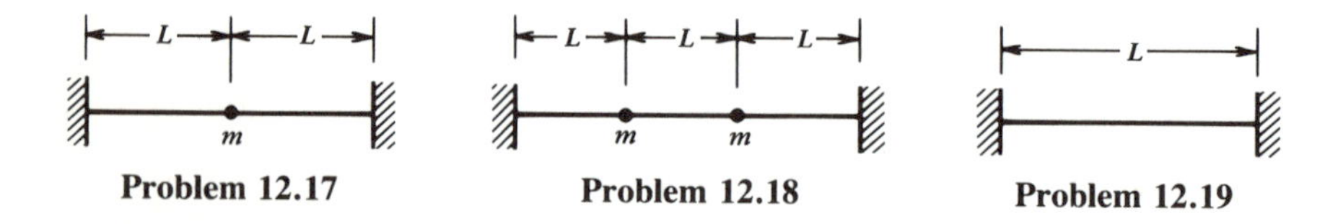

Problem 12.17 **Problem 12.18** **Problem 12.19**

12.18 Add a second mass and second d.o.f. to Problem 12.17 (see sketch). Find the natural frequencies and mode shapes.

12.19 The string is under tension T and has mass ρ per unit length. Model it by a single element. Use $[\mathbf{m}]$ and $[\mathbf{k}_\sigma]$ matrices associated with a cubic displacement field. Omit the conventional stiffness matrix. Solve for the natural frequencies and mode shapes. (The *exact* answer is $\omega_1{}^2 = \pi^2 T/\rho L^2$.)

12.20 Imagine a pin-ended column, built of several beam elements, in a buckled shape under compressive axial load. Write Eq. 12.4.2 in the form $[\mathbf{K}]\{\bar{\mathbf{D}}\} = -\lambda[\mathbf{K}_\sigma]\{\bar{\mathbf{D}}\}$ and assume that $[\mathbf{K}_\sigma]$ comes from Eq. 12.2.6. Give a physical interpretation of this equation, with particular reference to the "load" on the right-hand side.

12.21 Consider Fig. 4.4.2, but add an axial compressive force. Does this force make nodal loads computed from Eq. 4.3.5 less accurate? Explain.

12.22 The one-element column has length $L = 1.0$ m and bending stiffness $EI = 300$ N·m². The only d.o.f. permitted is θ_2.

(a) Solve for the critical load if the eccentricity e is zero. Take $[\mathbf{k}_\sigma]$ from Eq. 12.2.13.

(b) Plot P (ordinate) versus θ_2 (abscissa) for $e = 0$, $e = 0.005$ m, and $e = 0.010$ m. Consider loads P of 0, 3000 N, 6000 N, 8000 N, and 9000 N to generate the curves. (The results show that an imperfect structure has a nonlinear primary path.)

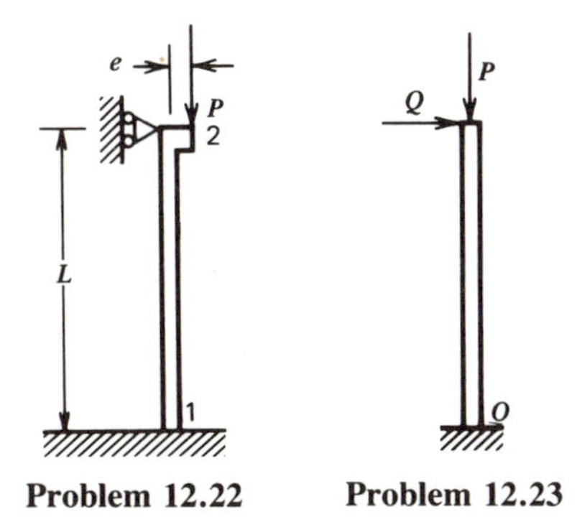

Problem 12.22 **Problem 12.23**

12.23 Force P is always vertical. As the column deflects under load Q, P generates a moment about O. Is this effect accounted for by the methods of this chapter? Why or why not?

12.24 Consider the eight-node linear solid isoparametric element. How many rows and columns are there in $[\mathbf{G}]$ in Eq. 12.3.4? Express the G_{ij} in terms of shape function derivatives and coefficients Γ_{ij} of the inverse Jacobian matrix.

12.25 (a) Use Eq. 12.3.10 to derive the 8-by-8 matrix $[\mathbf{k}_\sigma]$ for in-plane deformation of the four-node rectangle of Fig. 4.6.1.

(b) Write a computer program that uses this matrix to solve buckling problems. Use QM6 elements (Section 7.7) to generate the conventional stiffness matrix.

Solve for P_{cr} in the sketch for Problem 12.8(b) by using one layer of four elements to model the left half.

12.26 Consider the two-element structure of Fig. 12.5.2*b*, for which $[\mathbf{K}_{total}]$ is deformation dependent. Can you outline a solution procedure that will calculate nodal displacements produced by a given load?

12.27 Imagine that force P in Problem 12.12(c) is not always horizontal. Let P be applied through a cable of length c, which acquires a small slope of magnitude w_2/c as node 2 undergoes deflection w_2. How can w_2 be calculated by the methods of this chapter?

CHAPTER 13

INTRODUCTION TO NONLINEAR PROBLEMS

13.1 INTRODUCTION. GEOMETRIC NONLINEARITY

Various behaviors are called "nonlinear" (the term states only what a behavior is *not*). Stress-strain relations may be nonlinear in either a time-dependent or a time-independent way. Displacements may cause loads to alter their distribution or magnitude. Mating parts may stick or slip. Gaps may open or close. Nonlinearity may be mild or severe. The problem may be static or dynamic. Many solution schemes have been proposed, and it is no surprise that no one scheme is best for all problems.

Everyday problems are usually considered linear, and for good reasons. Materials and structures are commonly used in their linearly elastic, small-deflection range. Slight nonlinearity does not invalidate a linear design basis. Nonlinear analysis is harder to understand and more expensive to do. Computer runs that cost as much as a new car are not rare. Nevertheless, nonlinear analyses are becoming more common because of stringent design requirements and because finite elements and the computer have made nonlinear analysis a practical possibility.

To begin our study, it is enough to say that the analysis objective is to find the load versus displacement relation of a given structure, or to find either load or displacement when the other is prescribed, or to find the collapse load. Most analysis programs allow the user to choose among various kinds of nonlinearity and solution algorithms.

A user must understand the problem and the analytical tools well enough to make an intelligent choice. Even then, several analyses may be needed to get a satisfactory result. An incremental analysis gives *an* answer, but another analysis with a different step length is needed to estimate the quality of the answer. An iterative analysis may fail to converge because of a bug in the program or in the data, numerical error, greater nonlinearity than the algorithm can accommodate, or a prescribed load greater than the structure's collapse load [13.1].

Nonlinear problems are usually solved by taking a series of linear steps. The procedure resembles following a winding path in a dense fog. We step along the path, look at the part we stand on to check its direction, take a step in that direction, look again, and so on. In structural terms, this process is explained by writing equilibrium equations in the incremental form $[\mathbf{K}]\{\Delta\mathbf{D}\} = \{\Delta\mathbf{R}\}$. Here the stiffness matrix $[\mathbf{K}]$ is a function of displacements $\{\mathbf{D}\}$ because the problem is nonlinear. In turn, the current $\{\mathbf{D}\}$ is the sum of preceding $\{\Delta\mathbf{D}\}$'s. The current $[\mathbf{K}]$, called the *tangent stiffness,* is used to compute the next step, $\{\Delta\mathbf{D}\}$. Then we update $\{\mathbf{D}\}$, update $[\mathbf{K}]$, and are ready to take another step. In this way we approximate a load versus displacement curve by a series of straight line segments. Accordingly, we need make no wholesale revision of preceding chapters that were devoted to linear problems.

The first part of this chapter treats *geometric* nonlinearity. The latter part treats *material* nonlinearity. Similar solution techniques apply to both, so a computer program can allow both to be active at the same time. This chapter is introductory and oriented toward algorithm concepts, not generality and rigor. The mathematical highlands are left to others [13.2]. Sections 11.7 through 11.9 contain remarks about nonlinearity in dynamic problems.

In discussing geometric nonlinearity, we exclude material nonlinearity and time-dependent problems unless stated otherwise.

The essential feature of geometric nonlinearity is that equilibrium equations must be written with respect to the *deformed* geometry—which is not known in advance. Only if the nature of the problem is substantially unchanged by deformation do we call the problem "linear" and so presume that equilibrium equations can refer to the initial configuration.

A large-displacement problem can be analyzed in *Lagrangian coordinates* or in *Eulerian coordinates*. Both are credited to Euler [3.2].

The Lagrangian approach is also called "stationary Lagrangian" and "total Lagrangian." Its definition is that the original reference frame remains stationary, and everything is referred to it, regardless of how big the strains and rotations become: displacements, differentiations, and integrations are all with respect to the original frame. As displacements become larger and larger, more and more terms must be added to the strain-displacement relations in order to account for the nonlinearities. In a finite element context this means that the conventional stiffness matrix $[\mathbf{K}]$ is augmented by additional matrices derived from higher-order terms: first by $[\mathbf{K}_\sigma]$ to represent stiffness effects that depend linearly on displacements (Chapter 12), then by $[\mathbf{K}_2]$ to represent stiffness effects that depend on squares of displacements [13.3, 13.4]. The formulation can be expressed in terms of either displacement increments or the accumulated (total) displacements.

Strictly, the Eulerian approach involves convected coordinates: a reference frame that deforms with the structure so that the (convected) coordinates of a point never change [13.2]. As actually implemented, the Eulerian approach takes a form that is usually called the "updated Lagrangian" approach. We can describe it as follows (see Section 13.2 for details). A local coordinate system, called a *corotational* system, is attached to each element. The local system moves with the element and therefore shares its rigid-body motion. Differentiations and integrations are done with respect to local coordinates. The current deformed state is used as the reference state prior to the next incremental step of the solution. Then the local coordinates are updated to produce a new reference state. Local coordinates of points *do* change, so the method is not strictly Eulerian. But strains and rotations in the local system are usually small enough that $[\mathbf{K}_2]$, and sometimes even $[\mathbf{K}_\sigma]$, can be omitted. Essential nonlinearities are accounted for by tracking the orientations of the several local systems. The equations developed are expressed in terms of displacement *increments*.

Predictably, the Lagrangian and updated Lagrangian methods each have advocates who claim that *their* favorite method is easiest to formulate or most efficient in computation. Adding to the confusion is the use of the same term for different concepts (or different terms for the same concept), or no term at all when an author does not state a choice of method [13.4–13.6]. Also, definitions of stress and strain must suit the solution method, but they are often not stated clearly. Fortunately, much of what we have to say about solution algorithms is generally applicable.

13.2 UPDATED LAGRANGIAN FORMULATION

As the vehicle for this discussion, consider the beam-column of Fig. 13.2.1. The elements are plane frame elements, which resist both bending and axial strain. As deflection changes, so does the structure stiffness [**K**]. Clearly this is so, since α is not directly proportional to the load P that produces it. We first consider how to find [**K**] in any configuration.

Figure 13.2.2 shows a typical plane frame element (in Fig. 13.2.1, $\beta_0 = 0$ for all elements). Coordinates $x'y'$ rotate and translate with the element, such that $x' = y' = 0$ for node 1 and $y' = 0$ for node 2. Thus, *in the local system* $x'y'$, three nodal d.o.f. are always zero: $u_1 = v_1 = v_2 = 0$. All d.o.f. in *global* directions, D_1 to D_6, are generally nonzero. After D_1 to D_6 take place, the new global projections and orientation of the element are

$$x_L = x_0 + D_{41} \qquad y_L = y_0 + D_{52} \qquad \beta = \arctan(y_L/x_L)$$
$$\text{where} \qquad D_{41} = D_4 - D_1 \qquad \text{and} \qquad D_{52} = D_5 - D_2 \tag{13.2.1}$$

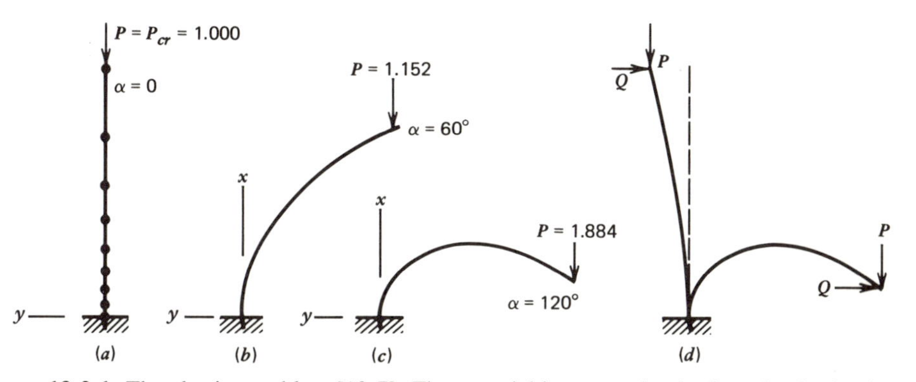

Figure 13.2.1. The elastica problem [13.7]. The material is assumed to be linearly elastic always. $P = 1.000$ for bifurcation buckling. α = rotation angle of the tip. Part (*a*) shows a possible division into elements. Part (*d*) shows two equilibrium positions for the same P; the lower is stable but the higher is not. Q is a very small constant lateral load that initiates lateral displacement.

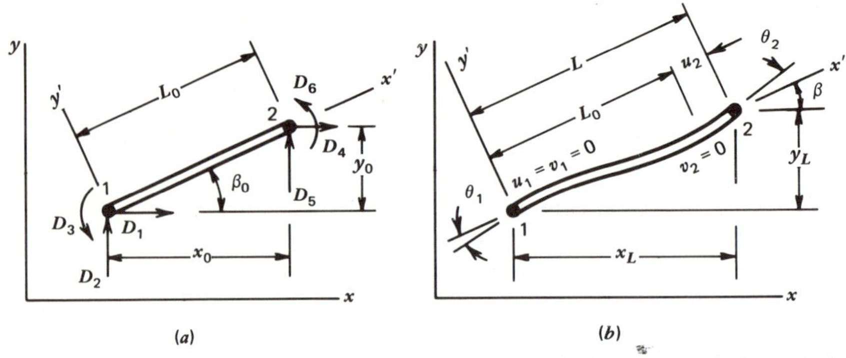

Figure 13.2.2. (*a*) Frame element, identifying its global d.o.f. D_1 to D_6, and shown before any rigid-body motion or strain; that is, when all $D_i = 0$. (*b*) The same element, after rigid-body motion and strain, identifying its local d.o.f. Axes $x'y'$ move with the element. Axes xy are global and fixed.

If $\beta_0 = 0$, and β is computed in Fortran as DATAN2(YL,XL), then β can reach 180° in either direction before it is ambiguously defined. The stiffness matrix of the structure in its current configuration is called the *instantaneous* or *tangent* stiffness matrix. To generate it by the updated Lagrangian method, we do as follows.

1. Establish local coordinates $x'y'$ for the element at hand.
2. Generate the element stiffness matrix $[\mathbf{k}]$ in local coordinates so that it operates on the local d.o.f.
3. Transform $[\mathbf{k}]$ to global coordinates (Eq. 6.3.7), so that it operates on global d.o.f. $D_1, D_2, \ldots, D_6$ and assemble it into the structure matrix.
4. Repeat Steps 1 through 3 until all elements have been treated. The tangent stiffness $[\mathbf{K}]$ is now complete.

In Section 13.3 we comment on whether $[\mathbf{k}]$ in Step 2 must include the stress stiffness $[\mathbf{k}_\sigma]$.

The transformation matrix $[\mathbf{T}]$ used in Step 3 (and in Step 6 below) is as stated by Eq. 6.3.3. It does not matter that global d.o.f. may be large or that some local d.o.f. are zero, because $[\mathbf{T}]$ is used only to relate *directions* of force and displacement between local and global systems.

13.3. A SOLUTION ALGORITHM (NEWTON-RAPHSON)

We continue the discussion begun in Section 13.2. The plane frame element is used again as an example. First, we must explain how to compute the nonzero local d.o.f.: θ_1, θ_2, and u_2. From Fig. 13.2.2,

$$\theta_1 = D_3 - (\beta - \beta_0), \qquad \theta_2 = D_6 - (\beta - \beta_0), \qquad u_2 = L - L_0$$
$$\text{where} \qquad L_0^2 = x_0^2 + y_0^2 \qquad \text{and} \qquad L^2 = x_L^2 + y_L^2 \tag{13.3.1}$$

If computed in this way, u_2 is the small difference between two large numbers. A less error-prone calculation is as follows [13.8]. Write $L^2 - L_0^2$, substitute for x_L^2 and y_L^2 from Eq. 13.2.1, and factor $L^2 - L_0^2$. Thus

$$(L - L_0)(L + L_0) = f(x_0, y_0, D_1, D_2, D_4, D_5) \tag{13.3.2}$$

from which

$$u_2 = L - L_0 = \frac{1}{L + L_0}\left[(2x_0 + D_{41})\,D_{41} + (2y_0 + D_{52})\,D_{52}\right] \tag{13.3.3}$$

where again $D_{41} = D_4 - D_1$ and $D_{52} = D_5 - D_2$.

In the denominator, $L + L_0 \approx 2L_0$, following our premise that the mesh is fine enough that strains in the local system remain small.

To solve a problem, such as finding the deflected shapes in Figs. 13.2.1*b* and 13.2.1*c*, we argue as follows. In an equilibrium configuration, each node is in equilibrium: loads externally applied, plus loads applied by elements, sum to zero (see Eq. 2.6.6). If the sum is $\{\Delta\mathbf{R}\}$ instead of zero, equilibrium does not prevail. Instead, the load imbalance $\{\Delta\mathbf{R}\}$ produces a displacement increment $\{\Delta\mathbf{D}\}$, computed by solving $[\mathbf{K}]\{\Delta\mathbf{D}\} = \{\Delta\mathbf{R}\}$ for $\{\Delta\mathbf{D}\}$. We deal in increments $\{\Delta\mathbf{D}\}$, not total displacements $\{\mathbf{D}\}$, because $[\mathbf{K}]$ depends on $\{\mathbf{D}\}$. (The dependence is introduced by the coordinate transformations.) We seek the $[\mathbf{K}]$ for which $\{\Delta\mathbf{R}\} = 0$ by the iterative process

$$[\mathbf{K}]_i\{\Delta\mathbf{D}\}_{i+1} = \{\mathbf{R}\} - \sum [\mathbf{k}]_i\{\mathbf{d}\}_i \tag{13.3.4}$$

$$\{\mathbf{D}\}_{i+1} = \{\mathbf{D}\}_i + \{\Delta\mathbf{D}\}_{i+1} \tag{13.3.5}$$

where $[\mathbf{k}]_i$, $\{\mathbf{d}\}_i$, and $[\mathbf{K}]_i$ are based on the current configuration i and are updated every cycle. Loads $\{\mathbf{R}\}$ come from external forces $\{\mathbf{P}\}$, body forces, element initial strain, and so on. In words, the algorithm is as follows.

1–4. Do the four steps cited below Eq. 13.2.1. Simultaneously do Steps 5 and 6 that follow.

5. Compute local d.o.f. $\{\mathbf{d}\} = \{0\ 0\ \theta_1\ u_2\ 0\ \theta_2\}$.

6. Compute forces $-[\mathbf{k}]\{\mathbf{d}\}$. With the negative sign, these forces are applied *to* nodes by distorted elements. Transform element stiffness and force arrays to global coordinates (Eqs. 6.3.5 and 6.3.7), and assemble with other elements.

7. Solve $[\mathbf{K}]\{\Delta\mathbf{D}\} = \{\mathbf{R}\} - \sum [\mathbf{k}]\{\mathbf{d}\}$ for $\{\Delta\mathbf{D}\}$. Add $\{\Delta\mathbf{D}\}$ to the vector $\{\mathbf{D}\}$ accumulated in preceding iterations.
8. Test for convergence. If not satisfied, return to Step 1.

Each cycle of this algorithm is a step tangent to the load versus displacement curve, as suggested in Section 13.1. We interpret the process graphically in Section 13.4 and there recognize it as the Newton-Raphson method.

Forces applied to nodes by a distorted element can also be found by integrating $[\mathbf{B}]^T\{\boldsymbol{\sigma}\}$ over the element volume. This is the more accurate method, especially with less competent elements such as constant-strain triangles (see Eq. 4.10.4).

After convergence, prescribed loads (or prescribed displacements) can be changed to another level and iteration can be begun again to find the new equilibrium configuration. Several monotonically increasing load levels may be needed on the way to the final load level. For example, if we want α for $P = 1.884$ in Fig. 13.2.1, we could prescribe (say) five load levels in the range $1.000 < P < 1.884$, with most of them near $P = 1.000$. Several cycles of iteration may be needed for convergence at each load level. Or, additional intermediate load levels could be prescribed, with fewer iterations needed in each. If the final load is applied all at once, the structure may jump to an unstable equilibrium configuration (Fig. 13.2.1*d*).

The tangent stiffness matrix $[\mathbf{K}]$ becomes ill-conditioned as we approach collapse ($D = L$ in Fig. 12.5.2*a* or $P = P_{cr}$ in Fig. 13.2.1). Low tangent stiffness results in large diagonal decay as $[\mathbf{K}]$ is processed in equation solving (Section 15.8). If any K_{ii} becomes negative, the structure is unstable. Incipient collapse can also be signalled by a $\{\Delta\mathbf{D}\}$ much larger than those previously calculated.

Stress stiffness matrices must be included in $[\mathbf{K}]$ when the configuration and loading place the structure in the neighborhood of collapse or buckling. In Fig. 13.2.1, for example, $[\mathbf{k}_\sigma]$ matrices are needed near $\alpha = 0$ but not near $\alpha = 120°$. If used, matrices $[\mathbf{k}_\sigma]$ are added to conventional matrices $[\mathbf{k}]$ in Steps 2 through 4 below Eq. 13.3.5. The combined $[\mathbf{k}]$ is used to compute element forces in Step 6.

Stresses needed to compute $[\mathbf{k}_\sigma]$ are not known a priori, and they change as the structure deforms. They must be calculated from strains $\{\boldsymbol{\epsilon}\} = [\mathbf{B}]\{\mathbf{d}\}$ in element local coordinates. Thus, if we begin iterative cycling with $\{\mathbf{D}\} = 0$, then $[\mathbf{k}_\sigma] = 0$ for all elements in the first cycle because all $\{\mathbf{d}\}$'s are zero. Only in the second and subsequent cycles are nonzero $[\mathbf{k}_\sigma]$'s produced.

The form of $[\mathbf{k}_\sigma]$ that gives convergence in the fewest cycles is problem dependent. Near $P = P_{cr}$ in Fig. 13.2.1, $[\mathbf{k}_\sigma]$ from Eq. 12.2.13 is better. At larger loads such as $1.152 < P < 1.884$, $[\mathbf{k}_\sigma]$ from Eq. 12.2.6 is better. Indeed, at the higher loads, fewer cycles are needed when $[\mathbf{k}_\sigma]$ is omitted altogether than when $[\mathbf{k}_\sigma]$ from Eq. 12.2.13 is used.

Convergence could be defined in terms of the relation between load imbalances and the total externally applied load. But a criterion based on displacements is preferable [13.9]. For each structural d.o.f. D_i we compute the ratio $e_i = |\Delta D_i/\bar{D}_i|$, where ΔD_i is the most recent increment and $\bar{D}_i$ is the largest total displacement of the same type. For example, if ΔD_i is a rotation increment, $\bar{D}_i$ is the largest nodal rotation in the structure.

Convergence is achieved when the largest e_i is less than c, where c is in the range $10^{-6} < c < 10^{-2}$, with the exact value of c dependent on the accuracy required.

The algorithm described in this section is generally applicable: elements need not be frame elements. In an application to plates, elements are allowed to rotate up to 90° with respect to their original positions [13.10]. If an element has more nodes than needed to define a local coordinate system uniquely, an ad hoc definition is needed. This happens, for example, with four-node quadrilaterals.

13.4 INTERPRETATION OF SOLUTION ALGORITHMS

Consider the load versus displacement function $R = f(D)$ for a single d.o.f. system. Imagine that we have found the displacement D_A that corresponds to load R_A. The load is now increased to R_B, and the corresponding displacement D_B is sought. A Taylor series expansion of $R = f(D)$ about D_A, truncated after the first derivative, is

$$f(D_A + \Delta D_1) = f(D_A) + \left(\frac{dR}{dD}\right)_A (\Delta D_1) \tag{13.4.1}$$

Now $f(D_A) = R_A$, and

$$\left(\frac{dR}{dD}\right)_A = K_A = \text{tangent stiffness at } A \tag{13.4.2}$$

We seek ΔD_1 such that $f(D_A + \Delta D_1) = R_B$, so

$$K_A(\Delta D_1) = R_B - R_A \tag{13.4.3}$$

where $R_B - R_A$ can be interpreted as the difference between the applied load and the resisting force of the structure. Thus Eq. 13.4.3 is the single d.o.f. case of Eq. 13.3.4. In the notation of Fig. 13.4.1, the next steps are:

- Update displacements: $D_1 = D_A + \Delta D_1$.
- Use D_1 to get stiffness K_1 and resistance R_1.
- Find the next increment ΔD_2 from $K_1(\Delta D_2) = R_B - R_1$.

Here $R_B - R_1$ is the current force unbalance. Eventually, $D_A + \Delta D_1 + \Delta D_2 + \cdots = D_B$ to a close approximation. Then the load can be increased from R_B to R_C, and the process can be started again. This solution algorithm is known as the Newton-Raphson method.

Variants of the Newton-Raphson method are numerous. We can write Eq. 13.3.4 in the form

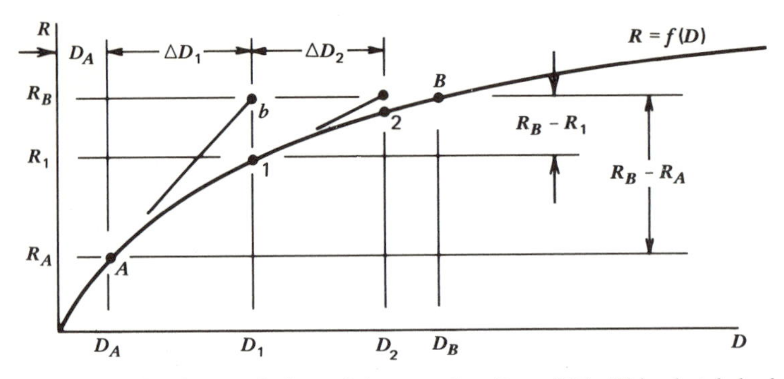

Figure 13.4.1. Newton-Raphson solution of the equation $R = f(D)$. This sketch looks more like the sketch usually seen in mathematics texts if we turn the book upside down and regard line $R = R_B$ as the abscissa.

$$[\mathbf{K}]_i\{\Delta\mathbf{D}\}_{i+1} = \{\Delta\mathbf{R}\}_{i+1} + \{\Delta\mathbf{R}_c\}_i \tag{13.4.4}$$

where $\{\Delta\mathbf{R}_c\}_i$ is the force unbalance, represented by the entire right side in Eq. 13.3.4, and $\{\Delta\mathbf{R}\}_{i+1}$ is an increment in the externally applied load ($\{\Delta\mathbf{R}\}_{i+1} = 0$ in Eq. 13.3.4). This algorithm is interpreted graphically in Fig. 13.4.2. It is sometimes called "incremental with one-step Newton-Raphson correction." The computed points (A, B, C) form our approximation to $R = f(D)$. The approximation clearly could be improved by iterating at each load level with $\{\Delta\mathbf{R}\}_{i+1} = 0$, as in Fig. 13.4.1.

Indeed, all gradations between Figs. 13.4.1 and 13.4.2 are possible: few load levels with many corrective iterations in each, or many load levels with few iterations in each.

A purely incremental scheme omits the corrective terms $\{\Delta\mathbf{R}_c\}_i$ from Eq. 13.4.4. It yields the dashed line AXY in Fig. 13.4.2. This is Euler's method for the numerical solution of a differential equation. Drift from the correct path $R = f(D)$ is evident.

The Newton-Raphson method requires that tangent stiffness matrix $[\mathbf{K}]$ be generated and reduced for equation solving in every iterative cycle. This is expensive. An alternative

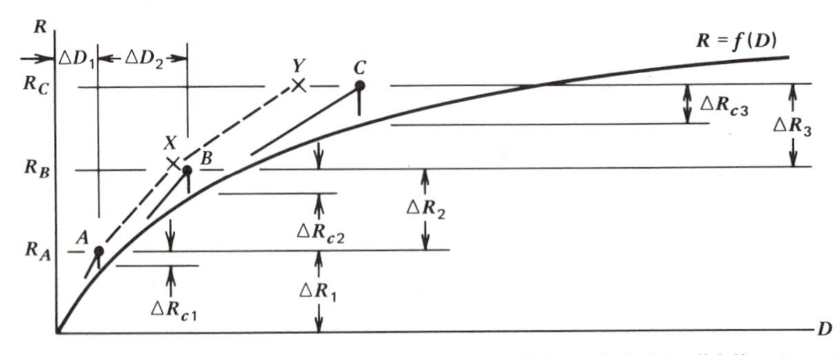

Figure 13.4.2. Single d.o.f. example of the algorithm of Eq. 13.4.4 (solid lines), and a purely incremental algorithm (dashed lines). The *effective* load increments are ΔR_1, $\Delta R_2 + \Delta R_{c1}$, and $\Delta R_3 + \Delta R_{c2}$.

is *modified* Newton-Raphson iteration, also called *constant stiffness iteration*. Here the same tangent stiffness matrix is used for several iterative cycles. It is updated only occasionally, perhaps at every load level or only when the convergence rate becomes poor. As compared with Newton-Raphson iteration, the modified method requires more cycles, but each is done more quickly because **[K]** is formed and reduced in only the first cycle. The process is sketched in Fig. 13.4.3*a*.

Thus far we have dealt with softening structures, whose tangent stiffness decreases with increasing displacement. A shell under compressive load may behave this way. But a membrane or a cable net may get stiffer as it displaces more. Usually the Newton-Raphson method still converges (Fig. 13.4.3*b*), but the modified method may converge slowly or diverge (Fig. 13.4.3*c*).

When applied to a hardening structure, algorithms may benefit from underrelaxation. Thus we use increments in Eq. 13.3.5 as $\beta\{\Delta\mathbf{D}\}_{i+1}$ rather than simply the $\{\Delta\mathbf{D}\}_{i+1}$ computed from Eq. 13.3.4. Here β is an underrelaxation factor in the range $0 < \beta < 1$. Underrelaxation is particularly helpful in Fig. 13.4.3*c*. If nonlinearity is so severe that underrelaxation fails, one may succeed by introducing artificial damping and viewing the problem as dynamic [13.63]. Thus the desired static solution is approached asymptotically as time increases.

Schemes intended to reduce drift and thus accelerate the convergence of incremental solutions have been proposed [13.11, 13.56, 13.62]. An *over*relaxation scheme can be explained with reference to Fig. 13.4.2. We would normally step off from point A using the slope at $D = D_A$. It would be more accurate to use a tangent that is roughly the *average* of the slope at $D = D_A$ and the slope at $D = D_B$. But D_B and the corresponding slope are not known in advance. So we assume that ΔD_2 will be little different that ΔD_1 and estimate the desired average slope from displacements $D_A + \beta(\Delta D_1)$, where β is (say) 0.5. This method is also called the *chord stiffness* scheme.

Solution methods and acceleration schemes have more trouble with hardening structures than with softening structures [13.15, 13.18, 13.56]. A method that is ideal in one situation may fail in another.

An algorithm that estimates error and adjusts step size accordingly has been proposed [13.12]. This is useful when the analyst has little idea what step size to use.

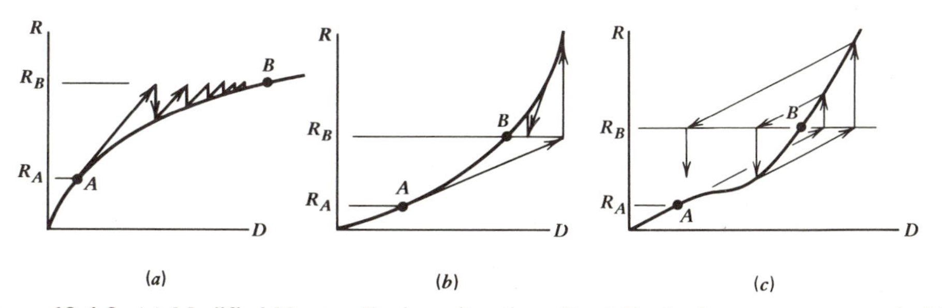

Figure 13.4.3. (*a*) Modified Newton-Raphson iteration. (*b, c*) Hardening structures, attacked by Newton-Raphson and modified Newton-Raphson methods, respectively.

13.5 TOTAL LAGRANGIAN FORMULATION

We consider a very restricted example intended to convey the flavor of the total Langrangian method but avoid the symbolism and lengthy manipulations that generality requires [13.3, 13.4, 13.13, 13.14].

In Fig. 13.5.1, axial strain ϵ_x is

$$\epsilon_x = \frac{u_1}{L} + \frac{\theta^2}{2}, \qquad \text{where} \qquad \theta = \frac{v_1}{L} \tag{13.5.1}$$

This approximation comes from Eq. 12.2.1. It restricts the analysis to small rotations. Strain energy U is

$$U = \frac{AE}{L}\int_0^L \epsilon_x^2\,dx = \frac{1}{2}\frac{AE}{L}\left(u_1^2 + \frac{u_1 v_1^2}{L} + \frac{v_1^4}{4L^2}\right) \tag{13.5.2}$$

For static equilibrium, as usual,

$$\left\{\frac{\partial \Pi_p}{\partial \mathbf{D}}\right\} = 0, \qquad \text{where} \qquad \Pi_p = U - \{\mathbf{D}\}^T\{\mathbf{R}\} \tag{13.5.3}$$

Here $\{\mathbf{D}\} = \{u_1\ v_1\}$ and $\{\mathbf{R}\}$ = applied loads that correspond to $\{\mathbf{D}\}$. Equations 13.5.2 and 13.5.3 yield

$$\frac{AE}{L}\left(\begin{bmatrix}1 & 0\\ 0 & 0\end{bmatrix} + \frac{1}{2}\begin{bmatrix}0 & v_1/L\\ v_1/L & u_1/L\end{bmatrix} + \frac{1}{3}\begin{bmatrix}0 & 0\\ 0 & 3v_1^2/2L^2\end{bmatrix}\right)\begin{Bmatrix}u_1\\ v_1\end{Bmatrix} = \begin{Bmatrix}R_1\\ R_2\end{Bmatrix} \tag{13.5.4}$$

The coefficient matrix is called a *secant stiffness matrix*. It operates on *total* displacements u_1 and v_1. To find an *incremental* form, we write $R_1 = f_1(u_1, v_1)$ and $R_2 = f_2(u_1, v_1)$, then differentiate to find ΔR_1 and ΔR_2. Thus Eq. 13.5.4 is transformed to

$$\frac{AE}{L}\left(\begin{bmatrix}1 & 0\\ 0 & 0\end{bmatrix} + \begin{bmatrix}0 & v_1/L\\ v_1/L & u_1 L\end{bmatrix} + \begin{bmatrix}0 & 0\\ 0 & 3v_1^2/2L^2\end{bmatrix}\right)\begin{Bmatrix}\Delta u_1\\ \Delta v_1\end{Bmatrix} = \begin{Bmatrix}\Delta R_1\\ \Delta R_2\end{Bmatrix} \tag{13.5.5}$$

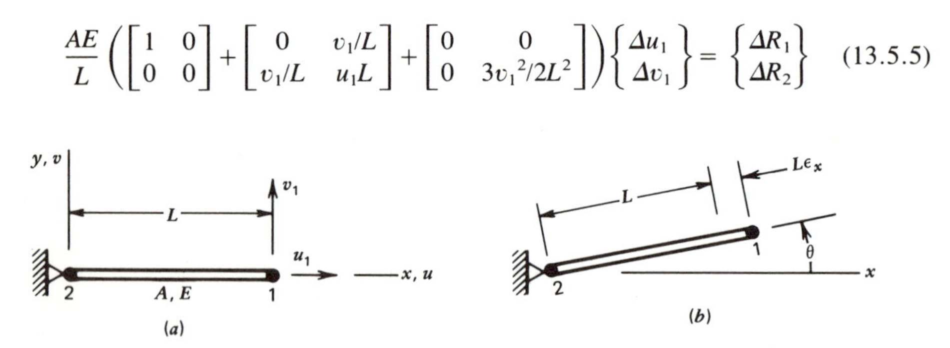

Figure 13.5.1. (*a*) A uniform plane truss element, with u_1 and v_1 the only nonzero nodal d.o.f. (*b*) The same element after stretching an amount $L\epsilon_x$ and rotating through angle θ.

The expression in parentheses, with the factor AE/L, is a tangent stiffness matrix.

We can relate the foregoing formulation to the updated Langrangian method. If we substitute $v/L = \theta$, use ϵ_x from Eq. 13.5.1, and express ϵ_x in terms of the axial force $P = AE\epsilon_x$, Eq. 13.5.5 assumes the form

$$\left(\frac{AE}{L}\begin{bmatrix} 1 & 0 \\ 0 & 0 \end{bmatrix} + \frac{P}{L}\begin{bmatrix} 0 & 0 \\ 0 & 1 \end{bmatrix} + \frac{AE}{L}\begin{bmatrix} 0 & \theta \\ \theta & \theta^2 \end{bmatrix}\right)\begin{Bmatrix} \Delta u_1 \\ \Delta v_1 \end{Bmatrix} = \begin{Bmatrix} \Delta R_1 \\ \Delta R_2 \end{Bmatrix} \tag{13.5.6}$$

For comparison, consider the conventional linear stiffness matrix $[\mathbf{k}]$ and the stress stiffness matrix $[\mathbf{k}_\sigma]$, both referred to an axis that rotates with the element, and the transformation matrix $[\mathbf{T}]$ that relates them to xy coordinates:

$$[\mathbf{k}] = \frac{AE}{L}\begin{bmatrix} 1 & 0 \\ 0 & 0 \end{bmatrix}, \qquad [\mathbf{k}_\sigma] = \frac{P}{L}\begin{bmatrix} 0 & 0 \\ 0 & 1 \end{bmatrix}, \qquad [\mathbf{T}] = \begin{bmatrix} \cos\theta & \sin\theta \\ -\sin\theta & \cos\theta \end{bmatrix} \tag{13.5.7}$$

With $\sin\theta \approx \theta$ and $\cos\theta \approx 1$, we find that the transformation $[\mathbf{T}]^T([\mathbf{k}] + [\mathbf{k}_\sigma])[\mathbf{T}]$ yields

$$\frac{AE}{L}\begin{bmatrix} 1 & 0 \\ 0 & 0 \end{bmatrix} + \frac{P}{L}\begin{bmatrix} \theta^2 & -\theta \\ -\theta & 1 \end{bmatrix} + \frac{AE}{L}\begin{bmatrix} 0 & \theta \\ \theta & \theta^2 \end{bmatrix} \tag{13.5.8}$$

This is the same tangent stiffness matrix as in Eq. 13.5.6, except that the middle matrix contains extra θ and θ^2 terms, but these can be discarded as negligible [13.4].

Now we can see that Eq. 13.5.5 incorporates the essential feature that is accomplished by coordinate transformation in Section 13.2: basing the tangent stiffness matrix on the deformed configuration.

The forms of Eqs. 13.5.4 and 13.5.5 suggest the following notation, in which the coefficients of $\{\mathbf{D}\}$ and $\{\Delta\mathbf{D}\}$ are respectively secant and tangent stiffness matrices.

$$[\mathbf{K} + \tfrac{1}{2}\mathbf{N}_1 + \tfrac{1}{3}\mathbf{N}_2]\,\{\mathbf{D}\} = \{\mathbf{R}\} \tag{13.5.9}$$

$$[\mathbf{K} + \mathbf{N}_1 + \mathbf{N}_2]\,\{\Delta\mathbf{D}\} = \{\Delta\mathbf{R}\} \tag{13.5.10}$$

where $[\mathbf{K}]$ represents here the conventional *linear* stiffness matrix. The second equation follows from the first because $[\mathbf{N}_1]$ and $[\mathbf{N}_2]$ are, respectively, linear and quadratic functions of the displacements. With some reservations, this symbolism is generally applicable [13.4]. Equation 13.5.10 is the purely incremental form of Eq. 13.4.4, where corrective terms $\{\Delta\mathbf{R}_c\}_i$ are absent. Appendix A gives a form of Eq. 13.5.10 applicable to a beam-column.

The formulation method, whether updated or total Lagrangian, does not dictate the solution algorithm: Eq. 13.5.10 can be solved by the Newton-Raphson method or by one of its variants.

Matrices should be symmetric so that efficient equation solvers can be used. Occasionally, even with proper derivation, a matrix such as $[\mathbf{N}_1]$ in Eq. 13.5.9 is without symmetry (Problem 13.11). If it is not clear how the secant stiffness matrix can then be cast in symmetric form, another option is to transfer some terms to the right side. A way to do this is as follows. Express $[\mathbf{N}_1]$ as the sum of a symmetric matrix $[\mathbf{S}]$ and an antisymmetric matrix $[\mathbf{A}]$. Their coefficients are

$$S_{ij} = \tfrac{1}{2}(N_{1ij} + N_{1ji}), \qquad A_{ij} = \tfrac{1}{2}(N_{1ij} - N_{1ji}) \tag{13.5.11}$$

The product $[\mathbf{A}]\{\mathbf{D}\}$ is a vector of deformation-dependent pseudo-loads that can be combined with loads $\{\mathbf{R}\}$. Matrix $[\mathbf{S}]$ is combined with the other stiffness matrices. Thus symmetry is preserved. Matrices in the *incremental* equation (Eq. 13.5.10) are always symmetric.

13.6 FURTHER SOLUTION ALGORITHMS. DIRECT ITERATION

The method of *direct iteration* is also called *functional iteration* and *successive substitution*. Unknowns are total displacements $\{\mathbf{D}\}$, not their increments $\{\Delta\mathbf{D}\}$. An iterative form of Eq. 13.5.9 is obtained by writing it as

$$[\mathbf{K} + \tfrac{1}{2}\mathbf{N}_1 + \tfrac{1}{3}\mathbf{N}_2]_i\{\mathbf{D}\}_{i+1} = \{\mathbf{R}\} \tag{13.6.1}$$

which implies that initially the coefficient matrix is based on initial displacements $\{\mathbf{D}\}_0$ and is used to compute $\{\mathbf{D}\}_1$. Typically, $\{\mathbf{D}\}_0 = 0$. Then $\{\mathbf{D}\}_1$ is used to update $[\mathbf{N}_1]$ and $[\mathbf{N}_2]$. Next we solve for $\{\mathbf{D}\}_2$, and so on. Figure 13.6.1*a* illustrates this method. It tends to be expensive, since the coefficient matrix must be formed and reduced in each iteration.

Another form of direct iteration is obtained by writing Eq. 13.6.1 as

$$[\mathbf{K}]\{\mathbf{D}\}_{i+1} = \{\mathbf{R}\} - [\tfrac{1}{2}\mathbf{N}_1 + \tfrac{1}{3}\mathbf{N}_2]_i\{\mathbf{D}\}_i \tag{13.6.2}$$

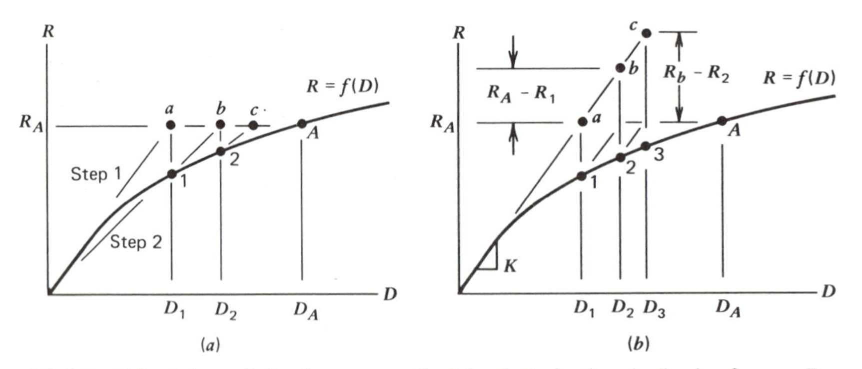

Figure 13.6.1. Calculation of D_A for a prescribed load R_A in the single d.o.f. case $R = f(D)$. (*a*) By Eq. 13.6.1. (*b*) By Eq. 13.6.2.

All nonlinearities appear on the right side, and [**K**] is formed and reduced only once. Figure 13.6.1*b* illustrates this method. The *effective* loads—the entire right side of Eq. 13.6.2—are R_A in the first step, $R_A + (R_a - R_1)$ in the second, $R_A + (R_b - R_2)$ in the third, and so on. This load sequence must converge if the displacement is to converge to D_A. Because Fig. 13.6.1 shows a softening structure, $[\mathbf{N}_1]/2 + [\mathbf{N}_2]/3$ is negative, and the effective loads exceed the applied load R_A.

Equation 13.6.2 is more attractive than Eq. 13.6.1 if more than one load must be analyzed (why?).

Direct iteration converges if nonlinearities are mild to moderate [13.15]. The method diverges if nonlinearity is severe. Divergence is more likely for hardening structures than for softening structures. Divergence is *less* likely if a final load is approached by intermediate load levels or if underrelaxation is used. For underrelaxation, we do not update displacements to their full values $\{\mathbf{D}\}_{i+1}$ after an iteration but use instead $\{\mathbf{D}\}_i + \beta\{\Delta\mathbf{D}\}_{i+1}$; that is,

$$\{\mathbf{D}\}_{i+1} = (1 - \beta)\,\{\mathbf{D}\}_i + \beta\{\mathbf{D}\}_{i+1} \tag{13.6.3}$$

where = means "is replaced by," as in Fortran, and β is a number in the range $0 < \beta < 1$.

At a limit point, $D = L$ in Fig. 12.5.2*a*, $\det[\mathbf{K}_{\text{total}}] = 0$. Also, according to the Sturm sequence property, one or more diagonals of $[\mathbf{K}_{\text{total}}]$ change from positive to negative during equation solving. An attempt to increase the load from P_L to Q in Fig. 12.5.2*a* results in failure of the solution algorithm or eventual convergence at $D = M$. The solution algorithm must be modified if the path between $D = L$ and $D = M$ is to be followed. One way is to use negative load increments when $D > L$. Another way is to augment the tangent stiffness [**K**] by a matrix $[\mathbf{K}_A]$ such that $[\mathbf{K}] + [\mathbf{K}_A]$ exhibits no limit point [13.16]. A drawback of this method is that the sparsity of $[\mathbf{K}_A]$ declines as the sparsity of {**R**} declines. A third method is to prescribe D and solve for the force P that produces it. If there is more than one load we cannot prescribe all nodal D's because we do not know in advance how they will be related. But we can prescribe one of the D's, alter loads on the other D's, and iteratively adjust the prescribed D until the computed and prescribed loads are in the desired ratio to one another. Other techniques are noted in Ref. 13.17.

13.7 JOINTS, GAPS, AND CONTACT PROBLEMS

Loading may cause parts of a structure to come in contact or to separate. Contact areas may change in size as loads change. Weak layers may crack and allow slip, or an adhesive layer may have nonlinear material properties. These problems include both geometric and material nonlinearity, but here we lump them together for a summary discussion.

Consider two structures, separated by a gap that contains an elastic bumper. The bumper does not fill the gap. After partial closure of the gap, contact is made, and the bumper adds its stiffness to the existing structure stiffness. We explain a way to handle this situation by means of the following single d.o.f. example.

In Fig. 13.7.1*a*, K is the structure stiffness and k is the bumper stiffness. The force F in spring k is

$$\begin{aligned} F &= 0 && \text{if} \quad u < g \\ F &= k(u - g) && \text{if} \quad u > g \end{aligned} \tag{13.7.1}$$

So the structure stiffness equation is

$$\begin{aligned} Ku &= P && \text{if} \quad u < g \\ Ku + k(u - g) &= P && \text{if} \quad u > g \end{aligned} \tag{13.7.2}$$

Let f represent the fraction of k that is active. Then we can write Eqs. 13.7.2 in the form [13.58]

$$(K + fk)u = P, \quad \text{where}$$
$$f = 0 \quad \text{if} \quad u < g \quad \text{or} \quad f = \frac{u - g}{u} \quad \text{if} \quad u > g \tag{13.7.3}$$

The response u is nonlinear because the stiffness $(K + fk)$ is a function of u.

An incremental solution is possible, where we prescribe ΔP and compute the resulting Δu. From Eqs. 13.7.2, the incremental equation with no unbalance correction is $K\,\Delta u = \Delta P$ if $u < g$ or $(K + k)\,\Delta u = \Delta P$ if $u > g$. At $u = g$ we have the problem of "rounding the corner" in Fig. 13.7.1*b*. This problem is discussed in connection with Fig. 13.11.1*a*.

A direct-iteration solution can be cast in either of the forms

$$(K + f_i k)u_{i+1} = P \tag{13.7.4}$$

or

$$Ku_{i+1} = P - f_i k u_i \tag{13.7.5}$$

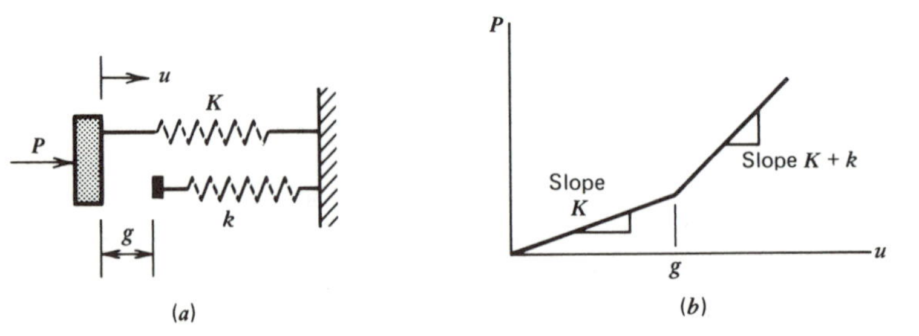

Figure 13.7.1. (*a*) Single d.o.f. model of a problem where an element becomes active only after a gap is closed. (*b*) Load versus displacement plot for this model.

where f_i is defined like f in Eq. 13.7.3. These forms correspond to Eqs. 13.6.1 and 13.6.2, respectively. Underrelaxation (Eq. 13.6.3) may be needed to make these equations converge, particularly if $k >> K$. In dynamic analysis, the sudden change in stiffness from K to $K + k$ requires a very small time step Δt near the instant of contact.

An *interface* element (Fig. 13.7.2*a*) can be used to model a thin layer of adhesive or a joint in a rock mass. To keep notation simple, we show the element parallel to the x-axis. In fact, it could be arbitrarily oriented. Its stiffness matrix would then be formulated in local coordinates and transformed to global coordinates.

Because h is small, the interface element has high y-direction stiffness and invites numerical trouble (Section 15.5). The likelihood of trouble is reduced by using *relative* displacements as nodal d.o.f. on one side of the element [13.19]. *Total* displacements, for example at node 4 in Fig. 13.7.2*b*, are $u_1 + u_{4r}$ and $v_1 + v_{4r}$. The quadrilateral element above element I must use the same displacements at nodes A and B that the interface element uses at its nodes 4 and 3. So the stiffness matrix of the quadrilateral element must be transformed from size 8 by 8 to size 12 by 12, since it involves d.o.f. at nodes C, D, and 1 to 4.

The strain-displacement relation for the interface element, after neglecting small quantities, is [13.19]

$$\{\epsilon_x \ \epsilon_y \ \gamma_{xy}\} = [\mathbf{B}] \ \{u_1 \ v_1 \ u_2 \ v_2 \ u_{3r} \ v_{3r} \ u_{4r} \ v_{4r}\} \tag{13.7.6}$$

where

$$[\mathbf{B}] = \frac{1}{h}\begin{bmatrix} -h/L & 0 & h/L & 0 & 0 & 0 & 0 & 0 \\ 0 & 0 & 0 & 0 & 0 & x/L & 0 & 1-(x/L) \\ 0 & -h/L & 0 & h/L & x/L & 0 & 1-(x/L) & 0 \end{bmatrix}$$

The element stiffness matrix is found in the usual way (Eq. 4.3.4). Three-dimensional and curved interface elements are clearly possible [13.20].

Material properties need not be constant. Fig. 13.7.3 represents properties of a joint that can carry compression but no tension, and that slips when friction is overcome. In

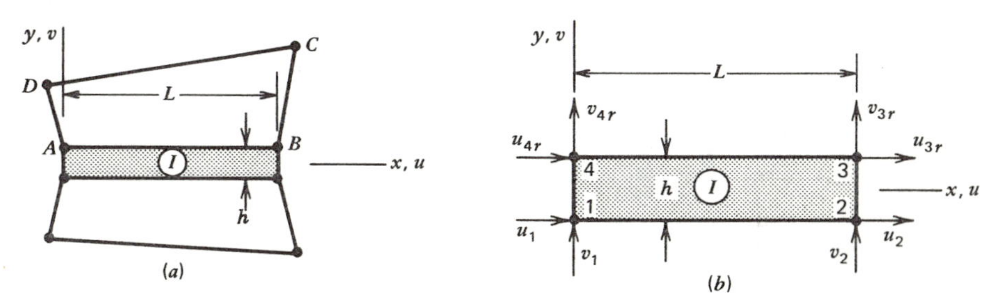

Figure 13.7.2. (*a*) Interface element I between two plane four-node linear elements. (*b*) Nodal displacements of the interface element, where u_{4r} and v_{4r} are displacements relative to u_1 and v_1, and u_{3r} and v_{3r} are displacements relative to u_2 and v_2.

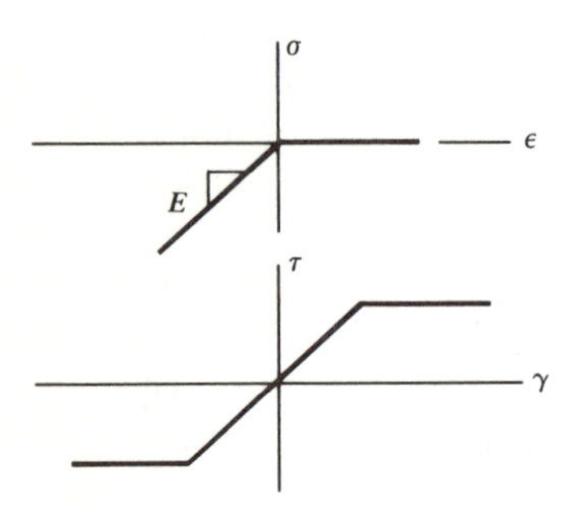

Figure 13.7.3. Stress-strain properties for an interface element that models a joint.

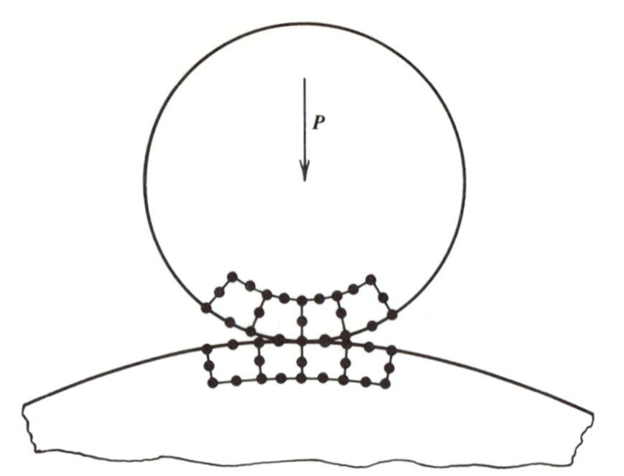

Figure 13.7.4. A very coarse mesh in the contact region between two cylinders.

practice, modulus E might be set to 10^5 times the modulus of the contacting bodies if $\epsilon < 0$, but to zero if $\epsilon > 0$ [13.20].

The contact-stress problem (Fig. 13.7.4) is nonlinear because the contact area grows as P increases. The same thing happens in a bolt and nut if there is a taper or a pitch mismatch [13.58]. Special interface elements, without "relative" d.o.f., have been used for the bolt-nut problem. But in contact-stress problems it is more common to detect when nodes on adjacent bodies have become coincident and constrain these nodes to have the same displacement normal to the contact surface. Relative displacement tangent to the contact surface is allowed if slip is permitted. Separation is possible if load decreases or if a sequence of loads is applied [13.21].

13.8 CABLES, MEMBRANES, AND SHELLS

A cable network usually deforms so much under load that its behavior must be considered nonlinear. Cable networks stiffen under load, as in Figs. 13.4.3*b* and 13.4.3*c*. If initially flat and unstressed, the network has *no* resistance to lateral load. To start a numerical solution, we require some lateral stiffness. It can be provided by adding a linear spring to all nodal d.o.f., thus adding a diagonal matrix $[\mathbf{K}_d]$ to the tangent stiffness matrix of the cable net. After the first solution step, the net is no longer flat. So it has lateral stiffness of its own, and $[\mathbf{K}_d]$ can be discarded. The same goal is served by using a stress stiffness matrix $[\mathbf{K}_\sigma]$ in place of $[\mathbf{K}_d]$. Arbitrary cable tensions can be assigned so that a $[\mathbf{K}_\sigma] > 0$ can be computed for the first solution step. After the first step, $[\mathbf{K}_\sigma]$ is not essential. But it should be included to reduce the tendency of the solution to oscillate and to promote the convergence rate. Besides oscillation, another computational problem is that in-plane stiffnesses are likely to be much bigger than lateral stiffnesses (Section 15.5). See Refs. 13.22 and 13.23.

A network that is not initially flat can usually benefit from either an estimated $[\mathbf{K}_\sigma]$ or underrelaxation on the first iteration, to avoid excessive displacements $\{\Delta\mathbf{D}\}_1$.

A long cable can be treated as a single element [13.64]. This may be far more efficient than dividing the cable into many simple elements.

Membranes present problems similar to those posed by cable networks. Also, there may be large strains, such as when a balloon is inflated. When elements stretch and rotate, pressure loads change both in magnitude and in direction [13.24].

Reference 13.25 uses membrane analysis to define the geometry of shells and arch dams. The authors note that a shell acts most effectively in a membrane mode. They therefore propose that shell geometry be the same as the deflected shape of an initially flat membrane, loaded by the same forces that the shell must support. As the object is only to find a suitable shape, not to analyze an actual membrane, the hypothetical membrane is assumed to have a linear stress-strain relation.

In a nonlinear analysis of a shell of revolution under a general load, nonlinear terms couple the Fourier harmonics. The coupling terms can be viewed as effective loads and combined with other loads $\{\mathbf{R}\}$ on the right side. Thus the various harmonics can still be analyzed separately. Convergence failure in any harmonic may indicate that a limit point has been reached.

13.9 MATERIAL NONLINEARITY. INCREMENTAL THEORY

In the next few sections we consider, in an introductory way, the formulation and solution of problems in which the stress-strain relation is nonlinear. Geometric nonlinearity is excluded, except in Section 13.14. However, we will see that solution algorithms are much the same, regardless of the source of nonlinearity.

If stress-strain relations are linear, or nonlinear but elastic, there is a unique relation between stress and strain. But if there are plastic strains, the stress-strain relation is path dependent, not unique: a given state of stress can be produced by many different straining procedures. In addition, different materials require different material theories.

The essential computational problem of material nonlinearity is that equilibrium equations must be written using material properties that depend on the strains, but the strains are not known in advance. Our discussion concentrates on plasticity. But the solution algorithms are not as restrictive: they apply to material nonlinearity, regardless of its origin.

Next, we summarize the equations of von Mises' plasticity theory. It is an *incremental* or *flow* theory: it relates *increments* of stress to *increments* of strain. (*Deformation* theory, which relates *total* stress to *total* strain, is considered in Section 13.10.) To a beginner, the definitions (Eqs. 13.9.1 and 13.9.2) and the flow rule (Eq. 13.9.9) may look cumbersome. Their justification is that they successfully predict the behavior of ductile isotropic metals.

The *effective plastic strain increment* $d\bar{\epsilon}_p$ is defined as a combination of the separate plastic strain increments.

$$d\bar{\epsilon}_p = \frac{\sqrt{2}}{3}\left[(d\epsilon_{xp} - d\epsilon_{yp})^2 + (d\epsilon_{yp} - d\epsilon_{zp})^2 + (d\epsilon_{zp} - d\epsilon_{xp})^2 + \frac{3}{2}(d\gamma_{xyp})^2 + \frac{3}{2}(d\gamma_{yzp})^2 + \frac{3}{2}(d\gamma_{zxp})^2\right]^{1/2} \tag{13.9.1}$$

The engineering definition of shear strain is used, for example, $\gamma_{xy} = u_{,y} + v_{,x}$. According to von Mises' theory, yielding begins under any state of stress when the *effective stress* $\bar{\sigma}$ exceeds a certain limit, where

$$\bar{\sigma} = \frac{\sqrt{2}}{2}\left[(\sigma_x - \sigma_y)^2 + (\sigma_y - \sigma_z)^2 + (\sigma_z - \sigma_x)^2 + 6(\tau_{xy}^2 + \tau_{yz}^2 + \tau_{zx}^2)\right]^{1/2} \tag{13.9.2}$$

For uniaxial stress σ_x, and in the plastic range where Poisson's ratio is 0.5, we find

$$\bar{\sigma} = \sigma_x \qquad \text{and} \qquad d\bar{\epsilon}_p = d\epsilon_{xp} \tag{13.9.3}$$

Thus in Fig. 13.9.1 we see the initiation of yield at $\sigma_x = \bar{\sigma}_0$. At $\sigma_x = \sigma_A$, the strain ϵ_x is $\epsilon_{xp} + \epsilon_{xe}$, of which only the elastic part ϵ_{xe} is recovered on release of load. As a result of plastic strain ϵ_{xp}, the material has hardened. Upon reloading, $\bar{\sigma}$ must exceed σ_A for yielding to resume. If we assume that $\bar{\sigma} > \sigma_A$ marks the resumption of yield, whether the state of stress is tensile, compressive, or multiaxial, we ignore the material anisotropy generated by plastic strain (the Bauschinger effect is a manifestation of this anisotropy). In other words, we assume the rule of *isotropic hardening,* which dictates the same shape of stress-strain curve whether the uniaxial stress is tensile or compressive. Other rules can be invoked, but they are not as easy to use.

The slope H' of the stress versus plastic strain plot (Fig. 13.9.1) is obtained by writing

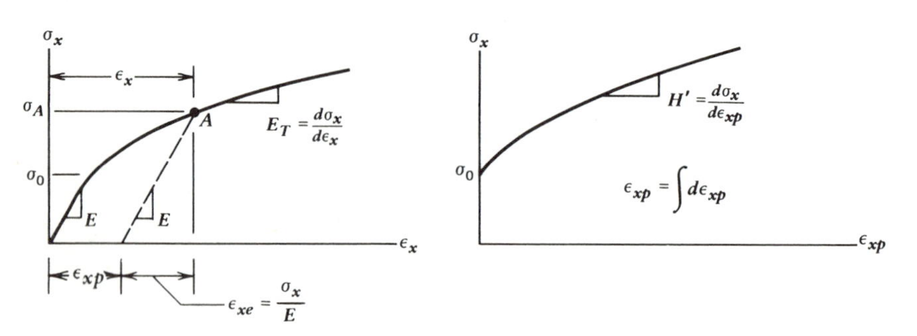

Figure 13.9.1. Stress-strain relations in uniaxial tension. E = elastic modulus, E_T = tangent modulus. The second plot is obtained from the first by removing the elastic strains.

$$d\sigma_x = E_T\, d\epsilon_x \qquad d\sigma_x = H'\, d\epsilon_{xp} \qquad d\epsilon_{xp} = d\epsilon_x - \frac{d\sigma_x}{E} \tag{13.9.4}$$

then substituting the first two equations into the third to yield

$$H' = \frac{EE_T}{E - E_T} \tag{13.9.5}$$

The following additional notation is needed. Increments of stress, strain, and plastic strain, all 6-by-1 vectors, are

$$\begin{aligned} \{d\boldsymbol{\sigma}\} &= \{d\sigma_x\, d\sigma_y \cdots d\tau_{zx}\}, \qquad \{d\boldsymbol{\epsilon}\} = \{d\epsilon_x\, d\epsilon_y \cdots d\gamma_{zx}\}, \\ \{d\boldsymbol{\epsilon}_p\} &= \{d\epsilon_{xp}\, d\epsilon_{yp} \cdots d\gamma_{zxp}\} \end{aligned} \tag{13.9.6}$$

Deviatoric stresses—σ'_x and so on—are defined as

$$\begin{aligned} \sigma'_x &= \sigma_x - \sigma_a \qquad \sigma'_y = \sigma_y - \sigma_a \qquad \sigma'_z = \sigma_z - \sigma_a \\ \tau'_{xy} &= \tau_{xy} \qquad \tau'_{yz} = \tau_{yz} \qquad \tau'_{zx} = \tau_{zx} \end{aligned} \tag{13.9.7}$$

where σ_a is the average stress, $\sigma_a = (\sigma_x + \sigma_y + \sigma_z)/3$. By differentiation of Eq. 13.9.2, we find

$$d\bar{\sigma} = \{\mathbf{Q}\}^T\{d\boldsymbol{\sigma}\}, \qquad \text{where} \qquad \{\mathbf{Q}\} = \frac{3}{\bar{\sigma}}\left\{\frac{\sigma'_x}{2}\ \frac{\sigma'_y}{2}\ \frac{\sigma'_z}{2}\ \tau'_{xy}\ \tau'_{yz}\ \tau'_{zx}\right\} \tag{13.9.8}$$

We note in passing that $\bar{\sigma}^2 = 9\tau_{\text{oct}}^2/2 = 3J_2$, where τ_{oct} = octahedral shear stress and J_2 = second invariant of deviatoric stress. The theory we are summarizing is sometimes called J_2 flow theory.

The flow rule of von Mises' theory is called the Prandtl-Reuss relation. It states that

$$\{d\boldsymbol{\epsilon}_p\} = \{\mathbf{Q}\}\, d\bar{\epsilon}_p \tag{13.9.9}$$

In words, Eq. 13.9.9 means that plastic strain increments $\{d\boldsymbol{\epsilon}_p\}$ result when the effective plastic strain increment $d\bar{\epsilon}_p$ occurs under the state of stress $\{\mathbf{Q}\}$. The corresponding stress increments $\{d\boldsymbol{\sigma}\}$ can be written in terms of elastic strain increments $\{d\boldsymbol{\epsilon}_e\}$ as follows.

$$\begin{aligned} &\{d\boldsymbol{\sigma}\} = [\mathbf{E}]\{d\boldsymbol{\epsilon}_e\} \qquad \text{or} \qquad \{d\boldsymbol{\sigma}\} = [\mathbf{E}]\,(\{d\boldsymbol{\epsilon}_{ep}\} - \{d\boldsymbol{\epsilon}_p\}), \\ &\text{where} \qquad \{d\boldsymbol{\epsilon}_e\} = \{d\boldsymbol{\epsilon}\} - \{d\boldsymbol{\epsilon}_p\} - \{d\boldsymbol{\epsilon}_T\} - \{d\boldsymbol{\epsilon}_c\} \\ &\text{and} \qquad \{d\boldsymbol{\epsilon}_{ep}\} = \{d\boldsymbol{\epsilon}\} - \{d\boldsymbol{\epsilon}_T\} - \{d\boldsymbol{\epsilon}_c\} \end{aligned} \tag{13.9.10}$$

Here [E] is the conventional matrix of *elastic* constants, and the five strain vectors in the middle equation are, respectively, the increments in elastic, total, plastic, thermal (or initial), and creep strains. If there is no temperature change and no creep while $\{d\boldsymbol{\epsilon}\}$ takes place, $\{d\boldsymbol{\epsilon}_T\} = \{d\boldsymbol{\epsilon}_c\} = 0$.

Finite element calculations yield $\{d\boldsymbol{\epsilon}\}$. So we can find $\{d\boldsymbol{\sigma}\}$ from Eqs. 13.9.9 and 13.9.10 if $d\bar{\epsilon}_p$ can be found. To find $d\bar{\epsilon}_p$, we substitute Eq. 13.9.9 into Eq. 13.9.10 and premultiply both sides by $\{\mathbf{Q}\}^T$. We also substitute $\{\mathbf{Q}\}^T\{d\boldsymbol{\sigma}\} = H'\,d\bar{\epsilon}_p$, which is obtained from Eqs. 13.9.3, 13.9.4 and 13.9.8. Thus

$$d\bar{\epsilon}_p = \lfloor \mathbf{W} \rfloor \{d\boldsymbol{\epsilon}_{ep}\}, \qquad \text{where} \qquad \lfloor \mathbf{W} \rfloor = \frac{\{\mathbf{Q}\}^T[\mathbf{E}]}{H' + \{\mathbf{Q}\}^T[\mathbf{E}]\{\mathbf{Q}\}} \tag{13.9.11}$$

An incremental stress-strain relation, analogous to $\{\boldsymbol{\sigma}\} = [\mathbf{E}]\{\boldsymbol{\epsilon}\}$ but valid into the elastic-plastic regime, results from substitution of Eq. 13.9.11 into Eq. 13.9.9 and the result into Eq. 13.9.10. It is

$$\begin{gathered} \{d\boldsymbol{\sigma}\} = [\mathbf{E}_{ep}]\,(\{d\boldsymbol{\epsilon}\} - \{d\boldsymbol{\epsilon}_T\} - \{d\boldsymbol{\epsilon}_c\}) = [\mathbf{E}_{ep}]\,\{d\boldsymbol{\epsilon}_{ep}\}, \\ \text{where} \qquad [\mathbf{E}_{ep}] = [\mathbf{E}] - [\mathbf{E}]\,\{\mathbf{Q}\}\,\lfloor \mathbf{W} \rfloor \end{gathered} \tag{13.9.12}$$

Matrix $[\mathbf{E}_{ep}]$ is symmetric. Also, it is valid for elastic-perfectly plastic materials, for which $E_T = H' = 0$.

In linear problems, stress and deformation depend on load but not on how a load state is reached: the load sequence makes no difference. In elastic-plastic problems, results *are* dependent on the loading sequence. Loading is called *proportional* if stresses at any point maintain the same ratio to one another throughout loading.

The assumption of isotropic hardening is good if loads are proportional. Otherwise, and also if loads reverse, isotropic hardening becomes a less accurate rule. Yet it may be accurate enough for engineering work. Other hardening rules produce other $[\mathbf{E}_{ep}]$ matrices. A physically motivated "overlay" scheme can accommodate various hardening rules and even strain softening [13.26, 13.41].

Plasticity is a time-independent phenomenon. Even so, equations are sometimes written in terms of rates to convey the flow nature of the phenomenon. Thus, for example, Eq. 13.9.9 becomes $\{\dot{\boldsymbol{\epsilon}}_p\} = \{\mathbf{Q}\}\,\dot{\bar{\epsilon}}_p$.

For more extensive arguments, and for yield and flow rules that apply to other materials, see the literature [3.2, 10.5, 13.26–13.34, 13.55; see also issues of *NucEDe* and the *SMiRT Conferences*].

13.10 DEFORMATION THEORY. DIRECT ITERATION SOLUTION

First, we try to make deformation theory plausible. The incremental theory of Section 13.9 and its corresponding deformation theory can be written in compact notation. If $\{\mathbf{S}\}$ is the vector of deviatoric stresses (Eq. 13.9.7), then [13.27]

$$\{d\epsilon_p\} = \frac{3}{2\bar{\sigma}}\{\mathbf{S}\}\, d\bar{\epsilon}_p \qquad \{\epsilon_p\} = \frac{3}{2\bar{\sigma}}\{\mathbf{S}\}\, \bar{\epsilon}_p \tag{13.10.1}$$

provided that shear strains are written in their tensor form, such as $\epsilon_{xy} = (u_{,y} + v_{,x})/2$ (thus $\gamma_{xy} = 2\epsilon_{xy}$). To get from the first equation to the second, we assume that loading is proportional. Then $\bar{\sigma} = c\bar{\sigma}_0$ and $\{\mathbf{S}\} = c\{\mathbf{S}\}_0$, where c is a constant and $\bar{\sigma}_0$ and $\{\mathbf{S}\}_0$ pertain to an arbitrary reference state of stress. Hence the first equation can be immediately integrated to yield the second, which relates *total* deviatoric stresses $\{\mathbf{S}\}$ to *total* plastic strains $\{\epsilon_p\}$. Loading is proportional in many practical situations. In many other situations it is a good approximation.

Next we consider the relation between total stresses and total strains $\{\epsilon\}$. The ratio $\bar{\sigma}/\bar{\epsilon}_p$ in Eq. 13.10.1 can be related to E and the secant modulus E_s determined from a tension test. From the first curve in Fig. 13.9.1,

$$\frac{\sigma_x}{\epsilon_{xp}} = \frac{\sigma_x}{\epsilon_x - (\sigma_x/E)} = \frac{E(\sigma_x/\epsilon_x)}{E - (\sigma_x/\epsilon_x)} = \frac{E\,E_s}{E - E_s} \tag{13.10.2}$$

Hence the second of Eqs. 13.10.1 yields equations such as

$$\epsilon_{xp} = \frac{E - E_s}{E\,E_s}\left(\sigma_x - \frac{\sigma_y}{2} - \frac{\sigma_z}{2}\right), \qquad \gamma_{xyp} = \frac{3(E - E_s)}{E\,E_s}\tau_{xy}, \text{ etc.} \tag{13.10.3}$$

These equations can be symbolized as $\{\epsilon_p\} = [\mathbf{F}_p]\{\sigma\}$. Elastic strains are $\{\epsilon_e\} = [\mathbf{E}]^{-1}\{\sigma\}$, and total strains are $\{\epsilon\} = \{\epsilon_p\} + \{\epsilon_e\}$. Thus the total stress versus total strain relation is

$$\{\sigma\} = [\mathbf{E}_D]\{\epsilon\}, \qquad \text{where} \qquad [\mathbf{E}_D] = \Big[[\mathbf{F}_p] + [\mathbf{E}]^{-1}\Big]^{-1} \tag{13.10.4}$$

Here $[\mathbf{E}_D]$ can be called the secant stiffness matrix of the isotropic material. It is a function of E, E_s, and Poisson's ratio ν. For a material point that has yet to yield, $[\mathbf{F}_p] = 0$; then $[\mathbf{E}_D] = [\mathbf{E}]$. If $E_s << E$, then $[\mathbf{F}_p] >> [\mathbf{E}]^{-1}$, therefore

$$[\mathbf{E}_D] \approx \frac{E_s}{E}[\mathbf{E}] \tag{13.10.5}$$

Direct iteration is a correct method for proportional loading without any unloading. It is also correct when $\{\sigma\} = [\mathbf{E}_D]\{\epsilon\}$ describes a nonlinear but elastic material, where loading and unloading follow the same stress-strain curve. Otherwise it is only approximate because path dependence is ignored. Theory demands that path-dependent problems be solved by incremental theory. Yet deformation theory may be surprisingly accurate [13.34].

To carry out a direct iteration solution, we must obtain, and store in numerical form, a relation between E_s and a *total* effective strain, $\bar{\epsilon}$. A suggested definition of $\bar{\epsilon}$ is one that makes it proportional to the second invariant of deviatoric strain [13.35].

$$\bar{\epsilon} = \frac{\sqrt{2}}{2(1 + \bar{\nu})} \left[(\epsilon_x - \epsilon_y)^2 + (\epsilon_y - \epsilon_z)^2 + (\epsilon_z - \epsilon_x)^2 + \frac{3}{2}(\gamma_{xy}^2 + \gamma_{yz}^2 + \gamma_{zx}^2) \right]^{1/2} \tag{13.10.6}$$

The $\bar{\sigma}$ versus $\bar{\epsilon}$ relation can be found by a tension test. In the linearly elastic range, $\bar{\nu} = \nu$. Then, with a uniaxial stress σ_x, Eqs. 13.9.2 and 13.10.6 yield $\bar{\sigma}/\bar{\epsilon} = \sigma_x/\epsilon_x = E$. (Also, if all strains are preceded by d to make them $d\bar{\epsilon}$, $d\epsilon_x$, and so on, Eqs. 13.9.1 and 13.10.6 have the same form in the plastic range where $\bar{\nu} = 0.5$.) To make the elastic-to-plastic transition, $\bar{\nu}$ can be expressed as a function of E, E_s, and the elastic value of ν (Problem 13.28). The substitution of $\bar{\nu}$ for ν can also be made in [**E**] of Eq. 13.10.5. The relation $E_s = \bar{\sigma}/\bar{\epsilon}$ implies isotropic hardening because the definitions of $\bar{\sigma}$ and $\bar{\epsilon}$ do not distinguish between tension and compression.

Two forms of direct iteration can be described: the *secant modulus* or *variable stiffness* method, in which the stiffness matrix is repeatedly revised, and the *initial-stress* method, in which all nonlinear effects are incorporated in the load vector on the right side. These procedures correspond to Eqs. 13.6.1 and 13.6.2, respectively. Other solution procedures, such as Newton-Raphson, could also be applied to deformation theory.

To begin a secant modulus solution, we pretend that all material is linearly elastic. We compute displacements and strains in the usual way. The effective strain $\bar{\epsilon}$ is computed at all sampling points. At points where $\bar{\epsilon}$ exceeds the proportional limit value, we compute E_s from the stored $\bar{\sigma}$ versus $\bar{\epsilon}$ relation. Hence, for each yielded element, we find $[\mathbf{E}_D]$, then a new $[\mathbf{k}]$ by integrating $[\mathbf{B}]^T[\mathbf{E}_D][\mathbf{B}]$ over the element. Assembly of $[\mathbf{k}]$'s gives an updated $[\mathbf{K}]$. The same load is applied again, new $\bar{\epsilon}$'s are found, $[\mathbf{K}]$ is again updated, and so on. Final stresses are given by Eq. 13.10.4. Clearly, this procedure is that of Fig. 13.6.1*a*. It is shown in the present context by Fig. 13.10.1*a*.

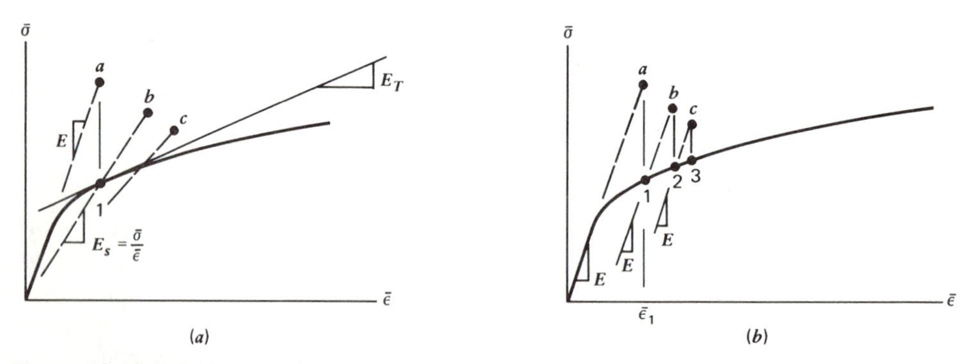

Figure 13.10.1. Direct iteration solutions in terms of $\bar{\sigma}$ versus $\bar{\epsilon}$ at a typical point. (*a*) Secant modulus. (*b*) Initial stress. Points *a*, *b*, and *c* are shown descending to suggest the transfer of load to other elements as yielding spreads. Thus, convergence is more rapid than in Fig. 13.6.1*a*.

The initial-stress method is most easily introduced by using $[\mathbf{E}_D]$ from Eq. 13.10.5. By using $[\mathbf{E}_D]$ instead of $[\mathbf{E}]$ in the standard expression for an element stiffness matrix, the secant element stiffness matrix $[\mathbf{k}_s]$ based on $[\mathbf{E}_D]$ can be written

$$[\mathbf{k}_s] = \frac{E_s}{E}[\mathbf{k}] = \left(1 - \frac{E - E_s}{E}\right)[\mathbf{k}] = [\mathbf{k}] - [\mathbf{k}_r] \tag{13.10.7}$$

where $[\mathbf{k}_r] = [\mathbf{k}]\,(E - E_s)/E$. Thus the usual assembly process yields the structural equations, written here in a form that implies an iterative solution like that in Eq. 13.6.2.

$$\begin{aligned} &[\mathbf{K}]\,\{\mathbf{D}\}_{i+1} = \{\mathbf{R}\} + \sum \{\mathbf{r}_c\}_i, \qquad \text{where} \\ &[\mathbf{K}] = \sum [\mathbf{k}] \qquad \text{and} \qquad \{\mathbf{r}_c\}_i = [\mathbf{k}_r]_i\,\{\mathbf{d}\}_i \end{aligned} \tag{13.10.8}$$

and $\{\mathbf{d}\}_i$ = element nodal displacements in the current configuration i. We see that $\{\mathbf{r}_c\}_i$ corresponds to a corrective load such as $R_a - R_1$ in Fig. 13.6.1b. In the present context, Fig. 13.10.1b shows that

$$\begin{aligned} &\{\mathbf{r}_c\}_i = \int_V [\mathbf{B}]^T\,(\{\boldsymbol{\sigma}\}_a - \{\boldsymbol{\sigma}\}_1)_i\,dV, \qquad \text{where} \\ &\{\boldsymbol{\sigma}\}_a = [\mathbf{E}]\{\boldsymbol{\epsilon}\}_i \qquad \text{and} \qquad \{\boldsymbol{\sigma}\}_1 = [\mathbf{E}_D]_i\,\{\boldsymbol{\epsilon}\}_i \end{aligned} \tag{13.10.9}$$

and $\{\boldsymbol{\epsilon}\}_i$ = current strains. Here $\{\mathbf{r}_c\}_i$ is an element nodal load vector, produced in the same way that initial stresses $\{\boldsymbol{\sigma}_0\}$ produce nodal loads in Eq. 4.3.5. The calculation of $\{\mathbf{r}_c\}_i$ in Eq. 13.10.9 is more efficient than the calculation in Eq. 13.10.8.

The integral of $[\mathbf{B}]^T\{\boldsymbol{\sigma}\}_a$ in Eq. 13.10.9 balances the applied load $\{\mathbf{R}\}$. Accordingly, one can show that the algorithm of Eq. 13.10.8 can be restated as

$$[\mathbf{K}]\{\Delta\mathbf{D}\}_{i+1} = \{\Delta\mathbf{R}_c\}_i \qquad \{\mathbf{D}\}_{i+1} = \{\mathbf{D}\}_i + \{\Delta\mathbf{D}\}_{i+1} \tag{13.10.10}$$

where, for $i = 1$, $\{\Delta\mathbf{R}_c\}_i$ is given by Eq. 13.11.7.

A trick that may speed convergence, especially when $E_s << E$, is to use a sequence of approximate $\bar{\sigma}$ versus $\bar{\epsilon}$ relations. Rough convergence is first sought for an almost linear $\bar{\sigma}$ versus $\bar{\epsilon}$ curve. Then the $\bar{\sigma}$ versus $\bar{\epsilon}$ curve is changed a bit toward its correct form, and rough convergence is again sought. The correct curve used last, when accurate convergence is desired.

Direct-iteration methods have the merit of simplicity. Their theoretical limitations and computational problems are noted in this section and in Section 13.6.

As plastic zones spread, a material becomes incompressible. As a result, $[\mathbf{F}_p]$ in Eq. 13.10.4 becomes singular and cannot be inverted except in the case of plane stress. As a rule, numerical trouble is likely whenever Poisson's ratio approaches 0.5 (see Section 16.6).

In any solution method, whether iterative or incremental, we ask, at what points in an element should we sample plastic strains and material properties? Properties may vary from point to point in Eq. 13.10.9, for example. In very simple elements, the centroid is a good choice. In a quadratic isoparametric element, the Gauss points of a 2-by-2 rule are good. The general guideline is that sampling points should be placed where strains are likely to be most accurate.

13.11 AN ALGORITHM FOR INCREMENTAL PLASTICITY

An incremental algorithm follows plastic action as it develops and so accounts for plasticity's path-dependent nature. Effectively, a problem is treated as a sequence of linearly elastic problems. Each step of the sequence is based on material properties appropriate to that step. See Ref. 13.28, 13.34, and 13.36 to 13.38.

For the sake of simplicity, we first describe a *purely* incremental algorithm like the one depicted by the dashed line *AXY* in Fig. 13.4.2. Corrective terms that eliminate the progressive drift are introduced later.

Imagine that under loads $\{\mathbf{R}\}_A$, the correct displacements $\{\mathbf{D}\}_A$ are known (see the single d.o.f. example in Fig. 13.4.1). The tangent stiffness at point A is $[\mathbf{K}]_A = \Sigma[\mathbf{k}]_A$, where each element stiffness matrix $[\mathbf{k}]_A$ is either

$$\int_V [\mathbf{B}]^T[\mathbf{E}][\mathbf{B}]\, dV \qquad \text{or} \qquad \int_V [\mathbf{B}]^T[\mathbf{E}_{ep}][\mathbf{B}]\, dV \tag{13.11.1}$$

depending on whether the element has yet to yield or has already yielded. Properties $[\mathbf{E}_{ep}]$ are given by Eq. 13.9.12. Loads are now increased to $\{\mathbf{R}\}_B$ and the corresponding displacements $\{\mathbf{D}\}_B$ are sought. We compute $\{\mathbf{D}\}_b$ as follows.

$$\begin{aligned} [\mathbf{K}]_A \{\Delta\mathbf{D}\}_1 &= \{\Delta\mathbf{R}\}_{AB} = \{\mathbf{R}\}_B - \{\mathbf{R}\}_A \\ \{\mathbf{D}\}_b &= \{\mathbf{D}\}_A + \{\Delta\mathbf{D}\}_1 \end{aligned} \tag{13.11.2}$$

and so arrive at point b in Fig. 13.4.1. Temporarily we ignore the drift and update the solution as follows. For a typical sampling point, we extract the element's $\{\Delta\mathbf{d}\}_1$ from $\{\Delta\mathbf{D}\}_1$ and compute strain increments $\{\Delta\epsilon\} = [\mathbf{B}]\{\Delta\mathbf{d}\}_1$. Next we compute the elastic-plastic strain increment $\{\Delta\epsilon_{ep}\}$ for elements that have already yielded.

$$\{\Delta\epsilon_{ep}\} = \{\Delta\epsilon\} - \{\Delta\epsilon_T\} - \{\Delta\epsilon_c\} \tag{13.11.3}$$

Thermal strains $\{\Delta\epsilon_T\}$ are known from the expansion coefficients and temperature change during the load increment [13.59]. Creep strains $\{\Delta\epsilon_c\}$ are known from stresses at the beginning of the increment and the time t associated with the load increment [13.34]. Now we apply the equations of Section 13.9 to each sampling point in each element.

$$\text{Eq. 13.9.11:} \quad \Delta\bar{\epsilon}_p = \int_A^b \lfloor \mathbf{W} \rfloor \{d\boldsymbol{\epsilon}_{ep}\} \approx \lfloor \mathbf{W} \rfloor_A \{\Delta\boldsymbol{\epsilon}_{ep}\}$$

$$\text{Eq. 13.9.9:} \quad \{\Delta\boldsymbol{\epsilon}_p\} = \int_A^b \{\mathbf{Q}\}\, d\bar{\epsilon}_p \approx \{\mathbf{Q}\}_A\, \Delta\bar{\epsilon}_p \tag{13.11.4}$$

$$\text{Eq. 13.9.10:} \quad \{\Delta\boldsymbol{\sigma}\} = [\mathbf{E}]\,(\{\Delta\boldsymbol{\epsilon}_{ep}\} - \{\Delta\boldsymbol{\epsilon}_p\})$$

Values of $\bar{\epsilon}_p$, $\{\boldsymbol{\epsilon}_p\}$ and $\{\boldsymbol{\sigma}\}$ are updated by adding the increments from Eqs. 13.11.4 to the previous values. The updated $\bar{\sigma}$ can be computed from the updated $\{\boldsymbol{\sigma}\}$. With the updated $\bar{\epsilon}_p$, Fig. 13.9.1 or its numerical equivalent yields the updated H'. Next we can compute updated matrices $\{\mathbf{Q}\}$, $\lfloor \mathbf{W} \rfloor$, and $[\mathbf{E}_{ep}]$ and apply Eqs. 13.11.1 to update the stiffness matrices. Now we are prepared to increase the load from $\{\mathbf{R}\}_B$ to $\{\mathbf{R}\}_C$ and take another step like that in Eq. 13.11.2. In this procedure we must keep records of $\bar{\epsilon}_p$, $\{\boldsymbol{\epsilon}_p\}$, $\bar{\sigma}$, and $\{\boldsymbol{\sigma}\}$ at each sampling point in each element and store the updated $\{\mathbf{D}\}$ if structure displacements are needed. Variants of the foregoing procedure are described in the literature.

When a sampling point makes the transition from elastic to plastic within a single load increment, we should switch from the first of Eqs. 13.11.1 to the second *within* the increment. This goal can be approximated as follows [13.28]. Before incrementing the load, estimate the next *elastic* increment $\Delta\bar{\sigma}$. A possible estimate is that $\Delta\bar{\sigma}$ will be equal to the preceding $\Delta\bar{\sigma}$. If $\Delta\bar{\sigma}$ pushes the sampling point past the proportional limit $\bar{\sigma}_{PL}$, proceed as follows. Compute $\Delta\bar{\epsilon} = \Delta\bar{\sigma}/E$. Thus we would move from (say) point A to point B in Fig. 13.11.1*a*. The fraction of $\Delta\bar{\epsilon}$ required to initiate yield is

$$m = (\bar{\epsilon}_{PL} - \bar{\epsilon}_A)/\Delta\bar{\epsilon} \tag{13.11.5}$$

Then we modify Eq. 13.9.12 to read

$$[\mathbf{E}_{ep}] = [\mathbf{E}] - (1 - m)\,[\mathbf{E}]\,\{\mathbf{Q}\}\,\lfloor \mathbf{W} \rfloor \tag{13.11.6}$$

where $\{\mathbf{Q}\}$ and $\lfloor \mathbf{W} \rfloor$ are based on stresses that correspond to point B. This $[\mathbf{E}_{ep}]$ is used to update the stiffness matrix. Only now is the load increment $\{\Delta\mathbf{R}\}_{AB}$ applied. The $\{\Delta\mathbf{D}\}_1$ that results may move us to point C in Fig. 13.11.1*a*. If desired, another iteration, using the *same* load increment and with m calculated from line AC, can be used to improve results. After convergence, Eqs. 13.11.3 and 13.11.4 are applied, but with $\{\Delta\boldsymbol{\epsilon}\}$ in Eq. 13.11.3 replaced by $(1 - m)\{\Delta\boldsymbol{\epsilon}\}$ and integration in Eq. 13.11.4 spanning the latter $(1 - m)$ fraction of the total step. Elastic stress increments are $\{\Delta\boldsymbol{\sigma}\} = m[\mathbf{E}]\{\Delta\boldsymbol{\epsilon}\}$. Use of factor m is analogous to the chord stiffness method (Section 13.4).

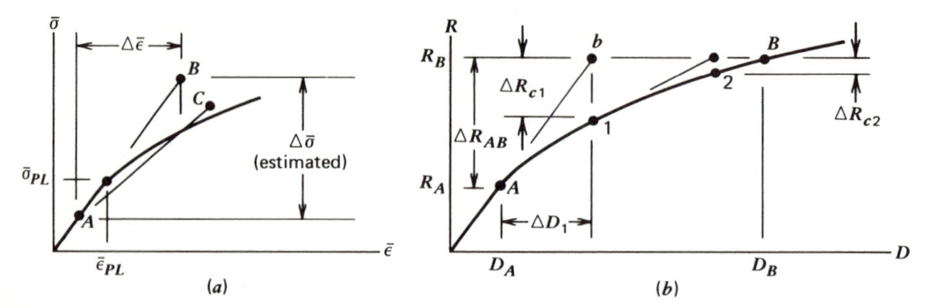

Figure 13.11.1. (*a*) Rounding the corner with a sampling point that makes the elastic-to-plastic transition. (*b*) Single d.o.f. representation of Newton-Raphson iteration with a load step, using corrective loads ΔR_{c1} and ΔR_{c2}.

If $\Delta\bar{\epsilon}_p < 0$, elastic unloading is indicated. Then $\Delta\bar{\epsilon}_p$ is set to zero in Eqs. 13.11.4, and stiffness matrices are updated using [**E**] instead of [$\mathbf{E}_{ep}$]. For greater accuracy we can, before incrementing the load, repeat the load step using [**E**] for the sampling points at which $\Delta\bar{\epsilon}_p < 0$. If $\Delta\bar{\epsilon}_p < 0$ at many sampling points, collapse may be indicated. According to the assumption of isotropic hardening, plastic action resumes only when $\bar{\sigma}$ exceeds its previous maximum magnitude.

Drift in an incremental solution is reduced if integrals in Eqs. 13.11.4 are evaluated more exactly by subincrements *within each load step*. To do this, we drop the subscripts p in Eq. 13.9.1 and use the just-computed *total* strain increments $\{\Delta\boldsymbol{\epsilon}\}$ to compute a number $\Delta\bar{\epsilon}$. Let Δq be a reference number, say $\Delta q = 0.0002$ for ductile metals. Compute $M = \Delta\bar{\epsilon}/\Delta q$, then replace M by the nearest integer greater than zero. Use M integrals instead of one to go from A to b in Eqs. 13.11.4, updating the integrands after each of the M subincrements. Note that $\lfloor \mathbf{W} \rfloor = 0$ within the elastic subincrements of a point that makes the elastic-plastic transition during the current load step. Load steps can be larger when subincrements are used than when they are omitted [13.34, 13.37, 13.38].

Drift can be further reduced by adopting the corrective loads ΔR_{ci} of Fig. 13.4.2. In this way we convert path AXY to path ABC [13.39]. The corrective loads are to be added to $\{\Delta\mathbf{R}\}_{AB}$ in Eq. 13.11.2. As shown in Fig. 13.11.1b, after load increment ΔR_{AB} has been applied, corrective load ΔR_{c1} permits the *next* load increment (or next iteration, explained shortly) to step off from point 1 instead of from point b. Here

$$\{\Delta\mathbf{R}_c\}_1 = \{\mathbf{R}\}_B - \sum \int_V [\mathbf{B}]^T \{\boldsymbol{\sigma}\}_1 \, dV \tag{13.11.7}$$

where the summation spans all elements of the structure and expresses the forces that elements apply to nodes because they sustain stresses $\{\boldsymbol{\sigma}\}_1$. Stresses $\{\boldsymbol{\sigma}\}_1$ are updated values, corresponding to point 1 in Fig. 13.11.1b. Stresses may have to be computed at more than one point in an element if their variation over the element is to be known well enough to get an accurate $\{\Delta\mathbf{R}_c\}$.

Instead of incrementing the load for every solution, as just suggested, we can do Newton-Raphson iterations *within* a load step to improve convergence. This is the method of Fig. 13.4.1. In Fig. 13.11.1*b* it corresponds to the steps tangent to points 1 and 2. Alternatively, the modified procedure of Fig. 13.4.3*a* can be used to avoid repeated updating of [**K**] within the load step. During these iterations there should be no test for unloading of the material [13.40]. The reason is seen in Figs. 13.4.3*b* and 13.4.3*c*: with a hardening structure, descending Newton-Raphson paths can imply unloading where, in fact, there is none.

We note in passing the resemblance of Eqs. 11.8.9 and 13.11.7 to Eq. 4.10.4. The similarity suggests that corrective load equations would be more effective if their stress vectors represent a continuous stress field.

It is possible to use modified Newton-Raphson iteration in such a way that the original *elastic* stiffness matrix [**K**] is used in all iterations and in all load steps [13.41]. This scheme is sometimes called the *initial-stress* method. We argue as follows. Subtract $\{\mathbf{R}\}_A$ from $\{\mathbf{R}\}_B$ in Eq. 13.11.7 but also add the equivalent of $\{\mathbf{R}\}_A$: the summed integrals of $[\mathbf{B}]\{\boldsymbol{\sigma}\}_A$. With Eq. 13.9.10, express the stress increment as an integral over the strain change from point A to point 1. Thus, with $\{\Delta\mathbf{R}\}_{AB} = \{\mathbf{R}\}_B - \{\mathbf{R}\}_A$,

$$\{\Delta\mathbf{R}_c\}_1 = \{\Delta\mathbf{R}\}_{AB} - \sum \int_V \left\{ \int_A^1 [\mathbf{B}]^T[\mathbf{E}] \left(\{d\boldsymbol{\epsilon}\} - \{d\boldsymbol{\epsilon}_p\} - \{d\boldsymbol{\epsilon}_T\} - \{d\boldsymbol{\epsilon}_c\}\right) \right\} dV \tag{13.11.8}$$

But if $\{\Delta\mathbf{D}\}_1$, and hence $\{d\boldsymbol{\epsilon}\}$, is calculated from a [**K**] based on *elastic* matrices [**E**], then $\{\Delta\mathbf{R}\}_{AB}$ is exactly balanced by the summed integrals of $[\mathbf{B}]^T[\mathbf{E}]\{d\boldsymbol{\epsilon}\}$. Therefore

$$\{\Delta\mathbf{R}_c\}_1 = \sum \int_V \left\{ \int_A^1 [\mathbf{B}]^T[\mathbf{E}] \left(\{d\boldsymbol{\epsilon}_p\} + \{d\boldsymbol{\epsilon}_T\} + \{d\boldsymbol{\epsilon}_c\}\right) \right\} dV \tag{13.11.9}$$

Thus, in each iteration, increments of plastic, thermal, and creep strains play the same role as initial strains $\{\boldsymbol{\epsilon}_0\}$ in Eq. 4.3.5. Iteration within each load step is mandatory, since the initial step under a new load increment may lead to a point far from the correct path. Matrix $[\mathbf{E}_{ep}]$ is never used to construct [**K**], so incremental plasticity relations enter the analysis only through $\{d\boldsymbol{\epsilon}_p\}$ in Eq. 13.11.9. Increments $\{d\boldsymbol{\epsilon}_p\}$, expressed as $\{\Delta\boldsymbol{\epsilon}_p\}$, are evaluated by Eqs. 13.11.4. Again, subincrements $M = \Delta\bar{\epsilon}/\Delta q$ can be used in evaluating the integrals.

The initial stress method is best suited to problems where nonlinearities are mild and not extensive.

By now we see that solution algorithms for geometric nonlinearity apply also to material nonlinearity. And again, the solution strategy can use few load increments or many, with (respectively) many or few iterations in each. Subincrements $M = \Delta\bar{\epsilon}/\Delta q$ can be used in each iteration, so we have what amounts to a three-level process: subincrements within iterations within load increments.

13.12 BENDING ACTION WITH MATERIAL NONLINEARITY

Integration through the thickness of beam, plate, and shell elements can be done explicitly if the material is linearly elastic. But a nonlinear material requires numerical integration and may generate coupling between bending moments and membrane forces (see the latter part of Section 9.4). A Gauss quadrature rule is inappropriate because every sampling point lies within the region: none is at either surface, where yielding begins. And it is surface layers that contribute most to bending stiffness and to corrective moment loads. So, for example, to evaluate the second of Eqs. 13.11.1, an n-by-n-by-m rule might be used for an isoparametric shell element, where an n-by-n Gauss rule is used in midsurface-tangent surfaces, and m sampling points of a trapezoidal rule or Simpson's rule are used to span the thickness (Fig. 13.12.1*a*). For a given m, Simpson's rule is more accurate than the trapezoidal rule [13.34, 13.35]. Depending on the problem, m may range from 7 to 21. To reduce run times and storage demands, low values of m and n are preferred.

In problems of pure bending of plates where material properties remain symmetric with respect to the midsurface, numerical integration through the thickness can be avoided [13.42]. Thickness integration can be done at the outset, once and for all, to produce an approximate moment-curvature relation $\bar{M}$ versus $\bar{\kappa}_p$, analogous to the σ_x versus ϵ_{xp} relation of Fig. 13.9.1. To find $\bar{M}$ and $\bar{\kappa}$, we pretend that the plate is of sandwich construction, with thin facings, but without transverse shear deformation. Then we set $\sigma_z = \tau_{yz} = \tau_{zx} = 0$ and

$$\sigma_x = c_1 M_x \qquad \sigma_y = c_1 M_y \qquad \tau_{xy} = c_1 M_{xy} \tag{13.12.1}$$

where c_1 is a constant that describes the geometry of layering. Then, with $\bar{\sigma}$ defined by $\bar{\sigma} = c_1 \bar{M}$, Eq. 13.9.2 yields

$$\bar{M} = (M_x^2 + M_y^2 - M_x M_y + 3M_{xy}^2)^{1/2} \tag{13.12.2}$$

Corresponding to $\sigma_z = \tau_{yz} = \tau_{zx} = 0$, we set $d\epsilon_{zp} = -(d\epsilon_{xp} + d\epsilon_{yp})$ and $d\gamma_{yzp} = d\gamma_{zxp} = 0$. Then, from Eqs. 9.1.2,

$$d\epsilon_{xp} = -z\, d\kappa_{xp} \qquad d\epsilon_{yp} = -z\, d\kappa_{yp} \qquad d\gamma_{xyp} = -2z\, d\kappa_{xyp} \tag{13.12.3}$$

Finally, with $d\bar{\kappa}_p$ defined by $d\bar{\epsilon}_p = -z\, d\bar{\kappa}_p$, Eq. 13.9.1 yields

$$d\bar{\kappa}_p = \frac{2\sqrt{3}}{3}\left(d\kappa_{xp}^2 + d\kappa_{yp}^2 + d\kappa_{xp}\, d\kappa_{yp} + d\kappa_{xyp}^2\right)^{1/2} \tag{13.12.4}$$

The $\bar{M}$ versus $\bar{\kappa}$ relation (Fig. 13.12.1*b*) can be obtained by a pure bending experiment. Equations analogous to those in Section 13.9 can be written. Element stiffness matrices have the form of Eq. 9.1.14, thus avoiding the thickness integration required in Eqs.

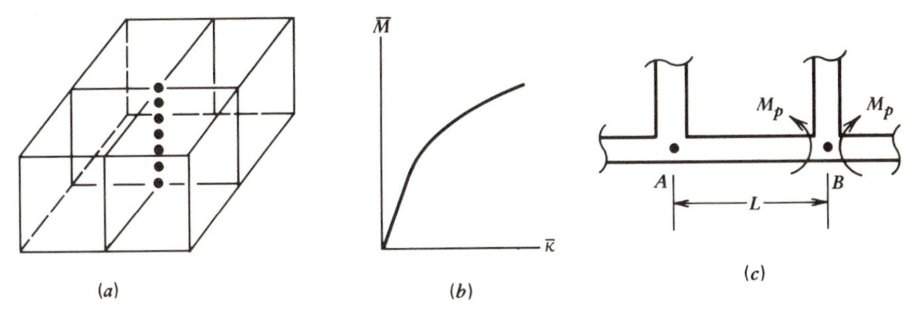

Figure 13.12.1. (*a*) A 1-by-1-by-7 integration rule for a plate bending element, showing seven Simpson sampling points in the thickness direction. (*b*) Typical plot of effective moment $\bar{M}$ versus effective curvature $\bar{\kappa}$. (*c*) Plastic hinge at *B* in frame element *AB*.

13.11.1. Equations for shells, analogous to Eqs. 13.12.1 through 13.12.4, are available [13.61].

The concept of a plastic hinge is a useful idealization for the limit analysis of frames. Consider a beam whose moment-curvature relation is linear until yield begins, whereupon the bending moment M stays constant at $M = M_p$ as curvature increases. Suppose that in a member *AB*, under concave-up bending, a hinge forms first at *B* (Fig. 13.12.1*c*). Now the rotation of the joint becomes independent of the rotation θ_B in member *AB*. Section 6.7 describes how to introduce the necessary hinge.

The load versus deflection plot of a structure that develops plastic hinges consists of straight line segments. Each segment has a smaller slope than the one before (unless a gap closes, which increases the stiffness). Each cusp on the plot marks the development of another plastic hinge and the consequent reduction of structure stiffness. A load increment $\{\Delta \mathbf{R}\}$ moves the solution along one line segment. Moments M_p of Fig. 13.12.1*c* should not be included in $\{\Delta \mathbf{R}\}$ because the deformations produced by M_p have already been accumulated by the end of the preceding step. Collapse is indicated by very large displacements or by finding negative pivots during reduction of the structure tangent stiffness matrix [13.43].

Plastic action and bifurcation buckling can occur in combination. Equation 12.4.2 still defines the problem, but now $[\mathbf{K}]$ and $[\mathbf{K}_\sigma]$ depend on $\{\mathbf{D}\}$. In other words, stiffnesses and stresses are functions of plastic deformation prior to buckling. Because $[\mathbf{K}]$ and $[\mathbf{K}_\sigma]$ are functions of the loads used in the prebuckling analysis, the correct loads $\{\mathbf{R}\}$ have been used only if the eigenvalue problem yields $\lambda = 1$. So a search is needed.

If $\lambda > 1$, $\{\mathbf{R}\}$ was too small; increase $\{\mathbf{R}\}$; reanalyze
If $\lambda < 1$, $\{\mathbf{R}\}$ was too large; decrease $\{\mathbf{R}\}$; reanalyze

Thus a series of eigenvalue problems must be solved to establish the buckling load $\{\mathbf{R}\}_{cr}$. Computations are analogous to those used to find the tangent-modulus buckling load of a column. The deformation theory of plasticity usually yields good results [13.44, 13.45].

13.13 OTHER INSTANCES OF MATERIAL NONLINEARITY

Creep causes strains to change with time at a rate that depends on stress (Fig. 13.13.1*a*). In contrast, plasticity is time independent. Nevertheless, the flow rules of plasticity are also successful in creep studies [13.27, 13.34, 13.46]. For example, in the first of Eqs. 13.10.1, we simply replace plastic strain increments by creep strain increments $\{d\epsilon_c\}$ and $d\bar{\epsilon}_c$. In *steady-state* or *secondary* creep, the effective stress versus effective strain relation can be written [13.34]

$$d\bar{\epsilon} = a\bar{\sigma}^b \, dt \qquad \text{or} \qquad \frac{d\bar{\epsilon}_c}{dt} = a\bar{\sigma}^b \tag{13.13.1}$$

where a and b are temperature-dependent material parameters [13.48]. More complicated relations can be invoked if it is thought to be worthwhile [3.2, 13.27, 13.48].

If we presume that a creeping material has an elastic but nonlinear stress-strain relation, a deformation-theory solution is possible [13.47]. What is needed is a $\bar{\sigma}$ versus $\bar{\epsilon}$ relation for each time at which a solution is desired. An "isochronous" curve (Fig. 13.13.1*b*) is extracted from Fig. 13.13.1*a* by plotting the $(\bar{\sigma}, \bar{\epsilon})$ pairs intercepted by a given time ordinate. Section 13.10 describes the solution algorithm, which has the same merits and drawbacks for creep that it has for elastic-plastic problems.

An *incremental* solution for a creep problem is now outlined [13.34, 13.46]. Assume that stresses are known and that an increment of time Δt passes. From Eq. 13.13.1, $\Delta\bar{\epsilon}_c$ is known. From the second of Eqs. 13.11.4, written for creep, we compute $\{\Delta\epsilon_c\}$ from $\Delta\bar{\epsilon}_c$. Structural equations are solved to yield $\{\Delta\mathbf{D}\}$, from which $\{\Delta\epsilon\}$ in Eq. 13.11.3 is calculated. Now the last of Eqs. 13.11.4 yields $\{\Delta\boldsymbol{\sigma}\}$. Division of these calculations into M subincrements, as described in Section 13.11, increases their accuracy. When stresses $\{\boldsymbol{\sigma}\}$ at the end of step Δt have been accumulated, Eq. 13.11.7 yields corrective loads $\{\Delta\mathbf{R}_c\}$ to be used in the next Δt increment. If external loads are constant, the structural equations become simply $[\mathbf{K}]\{\Delta\mathbf{D}\} = \{\Delta\mathbf{R}_c\}$ in each Δt.

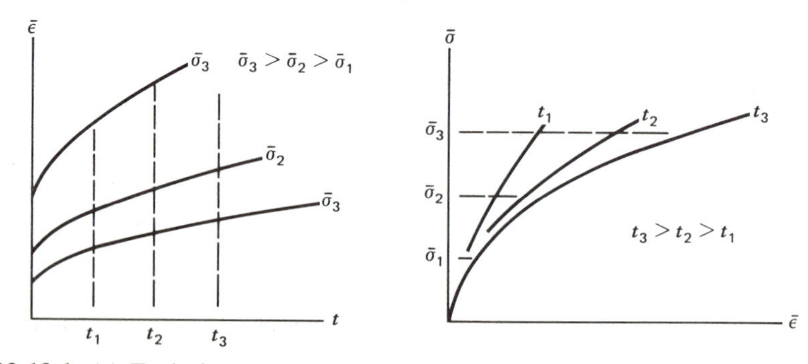

Figure 13.13.1. (*a*) Typical creep curves, where t = time. (*b*) Effective material properties at different times: the isochronous stress-strain relations.

Errors can accumulate if the stiffness matrix [**K**] changes during time steps Δt, perhaps because of temperature changes and temperature-dependent material properties. The error is countered by corrective loads [13.49].

The following three plastic problems are technically significant but, for lack of space, we do more than note them and cite references. *Slip zones* may develop, so that blocks of material move with respect to one another but with little deformation in each block [13.50]. Problems that involve *large plastic deformation* can be approximated as problems of viscous, nearly incompressible flow. These problems can be attacked with a conventional program for linear elasticity if we modify the material law and regard {**D**} as a vector of displacement velocities and $\{\boldsymbol{\epsilon}\}$ as a vector of strain rates [5.10, 13.51]. *Limit analysis* gives upper and lower bounds on collapse loads. The information obtained is confined to the collapse load and the collapse mechanism, but this is often the only information desired. In structures composed of elastic-perfectly plastic materials, limit analysis reduces to a linear programming problem [13.52].

Brittle materials such as rock and concrete will crack if the tensile stress is high enough. Von Mises' theory is inappropriate. Failure is better described by Mohr and Mohr-Coulomb envelopes. Theory agrees more closely with experiment if some shear stiffness parallel to a crack is retained. If crack locations are somehow known or are guessed a priori, the gap element techniques of Section 13.7 can be used. If the crack pattern is initially unknown, high stress at a Gauss point can signal that in subsequent calculations the sampling point should represent cracked material. Effectively, this approach models a crack by a cracked *zone* and calls on concepts of fracture mechanics [13.53]. Alternatively, cracks can be modeled by disconnecting nodes along interelement boundaries. Disconnection is accomplished by condensing of d.o.f. on one side of the crack so that these d.o.f. do not appear in {**D**}. This approach forces a crack to follow interelement boundaries. The usual array of solution algorithms can be invoked, ranging from direct iteration to pure incremental. An incremental form is more appropriate if the final crack pattern depends on loading history, that is, on the crack propagation sequence [13.53, 13.54].

13.14 CHOICE OF SOLUTION METHOD

No single algorithm is good for all problems. The choice of method is influenced by several factors [13.15].

1. Type of analysis (stability, geometric or material nonlinearity).
2. Formulation (such as incremental or deformation theory).
3. Ease of use and competence of the user.
4. Problem size, computation cost, storage space available.
5. Whether nonlinearity is mild or severe.
6. Desired accuracy.

A single computer program may allow the user to choose the algorithm and to adjust its behavior by assigning control parameters such as the number of steps, number of iteration cycles within a step, how often [**K**] is to be updated, convergence tolerance, and so on [13.18]. Reference 13.18 displays sample problems solved by various schemes.

The literature [13.8, 13.15, 13.17, 13.18, 13.34, 13.37, 13.38, 13.44, 13.56, 13.59] suggests the following choices of algorithms. For problems with geometric nonlinearity but linear elasticity, use the modified or unmodified Newton-Raphson method or its variant, the incremental scheme with load unbalance correction. In path-dependent problems the Newton-Raphson method may fail to converge, such as when there is unloading in an elastic-plastic problem. For elastic-plastic problems, the incremental method is favored. Further advice abounds: see the current literature for ongoing developments.

When well programmed, the total and updated Lagrangian methods should have about the same computational efficiency. With either method, if there are large strains and large rotations, care is needed to ensure that stresses and strains are consistently defined.

When [**K**] is to be updated because a few sampling points have yielded, we have the same problem that appears in design when a few elements are to be strengthened or weakened (Section 16.5).

Reference 13.57 proposes to reduce computational costs greatly by using a "reduced basis" for nonlinear problems. The idea is analogous to condensation (Section 11.5).

Nonlinear dynamic problems are briefly considered in Sections 11.7 to 11.10.

PROBLEMS

13.1 (a) Derive Eq. 13.3.3 from Eqs. 13.2.1 and 13.3.2.
(b) Check that Eq. 13.3.3 gives $u_2 = 0$ for a rigid-body rotation β about node 1, starting from $\beta_0 = 0$.

13.2 Consider the two upper nodes in Fig. 13.2.1. For $P = 1.152$, and for each of the two nodes, sketch the contributions to $\{\Delta\mathbf{R}\}$ in (*a*) global coordinates, and (*b*) local coordinates $x'y'$ appropriate to the upper two elements (see Eq. 13.3.4).

13.3 Devise a scheme to assign local coordinates analogous to $x'y'$ in Fig. 13.2.2 to a four-node plane quadrilateral that moves in its own plane.

13.4 As a programming project, code the large-deflection analysis of plane frames described in Section 13.3. It is left to you (or your instructor) to decide on the test cases and on the complexity and versatility required (see Section 13.14).

13.5 Points 1 and 2 in Fig. 13.4.1 are "on the curve." If R is the only load on a *multi*-d.o.f. structure and D its displacement, will this be so? That is, having computed D_1, can we say that the corresponding load is R_1?

13.6 Let loads P_1 and P_2 be functions of displacements D_1 and D_2: $P_1 = P_1(D_1, D_2)$ and $P_2 = P_2(D_1, D_2)$. Let D_A and D_B be exact displacements produced by loads P_A and P_B, let $\tilde{D}_A$ and $\tilde{D}_B$ be approximate displacements, and let $D_A = \tilde{D}_A + \Delta D_A$ and $D_B = \tilde{D}_B + \Delta D_B$. Derive the following equations, which are analogous to Eq. 13.4.3.

$$\begin{bmatrix} \partial P_1/\partial D_1 & \partial P_1/\partial D_2 \\ \partial P_2/\partial D_1 & \partial P_2/\partial D_2 \end{bmatrix}_{\tilde{D}_A,\tilde{D}_B} \begin{Bmatrix} \Delta D_A \\ \Delta D_B \end{Bmatrix} = \begin{Bmatrix} P_A - P_1(\tilde{D}_A, \tilde{D}_B) \\ P_B - P_2(\tilde{D}_A, \tilde{D}_B) \end{Bmatrix}$$

13.7 (a) Sketch a curve $R = f(D)$, concave down as in Fig. 13.4.1. Consider three load levels, R_A, R_B, and R_C. Sketch the progress of a Newton-Raphson solution as it cycles from A toward B and then from B toward C. Repeat for a curve $R = f(D)$ that is concave up.
(b) Repeat part (a) using the modified Newton-Raphson method, with K updated at A and at B.

13.8 Imagine that for a certain structure and a single load case, the Newton-Raphson and modified Newton-Raphson methods are equally efficient. Which of the two methods would you choose if there is more than one load case for the structure?

13.9 (a) A chord stiffness scheme is mentioned near the end of Section 13.4. Sketch this process on a concave-down P versus D curve. Consider two equal ΔP increments: use $\beta = 0$ for the first step and $\beta = 1.0$ for the second. Include the force unbalance correction of Eq. 13.4.4.
(b) Repeat part (a), but with a concave-up curve and $\beta = 0.5$ for the second step.

13.10 (a) Verify the right-hand expression in Eq. 13.5.2.
(b) Derive Eq. 13.5.4.
(c) Derive Eq. 13.5.5.
(d) Derive Eq. 13.5.6.
(e) Derive Eq. 13.5.8.

13.11 (a) The middle matrix in Eq. 13.5.4 can be made unsymmetric by writing coefficient (2,1) as $2v_1/L$ and making the diagonal null. Why is this form still valid?
(b) Generate $[\mathbf{S}]$ and $[\mathbf{A}]$ of Eq. 13.5.11, using $[\mathbf{N}_1]$ from part (a).
(c) If the unsymmetric matrix in part (a) is used, are the matrices in Eq. 13.5.5 still symmetric? Prove your answer.

13.12 From Fig. 13.6.1*b* and Eq. 13.6.2 we can write

$$KD_1 = R_A, \qquad KD_2 = R_A + R_a - R_1, \qquad KD_3 = R_A + R_b - R_2$$

By taking differences between these equations and using the geometry of Fig. 13.6.1*b*, show that we thus produce modified Newton-Raphson iteration, with K the initial stiffness at $D = 0$.

13.13 (a) The bar is of length L when unstressed, where $L^2 = a^2 + c^2$. When load P is zero, displacement D is also zero. The bar has axial stiffness AE/L, rolls without friction at B, and does not buckle as a column. Assume that the roller always contacts the wall. If $a >> c$, show that $\partial \Pi_p/\partial D = 0$ yields $P = (AE/2a^3)(2cD - D^2)(c - D)$. Also show that the tangent stiffness is $K = (AE/2a^3)(2c^2 - 6Dc + 3D^2)$ and that limit points are at $D = c(1 \pm \sqrt{3}/3)$.

(b) Let constants of this system be such that points A and F on the P versus D curve are at $P_A = 241$, $D_A = 0.211$, $P_F = 250$, and $D_F = 1.080$. After convergence at $P = 200$, P is increased to 250, and the following sequence of displacements D is generated by a Newton-Raphson algorithm: 0.173, 0.219, 0.071, 0.143, 0.190, 0.249, 0.199, 0.294, 0.235, 0.175, 0.222, 0.108, 0.166, 0.210, 1.178, 1.096, 1.080, 1.080. Explain this path to convergence by sketching it on the P versus D plot.

(c) Similarly, sketch the path that would be taken by a *modified* Newton-Raphson algorithm.

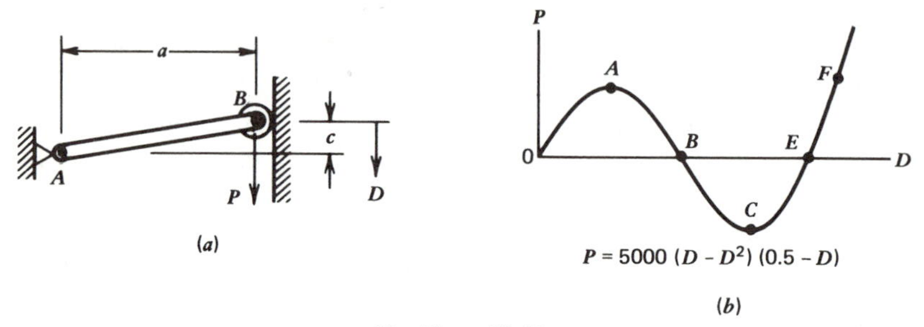

Problem 13.13

13.14 (a) Load P_A is prescribed and a Newton-Raphson solution is begun from point O. What happens? Show several iterative cycles on the sketch of P versus D.

(b) Repeat part (a), using *modified* Newton-Raphson iteration.

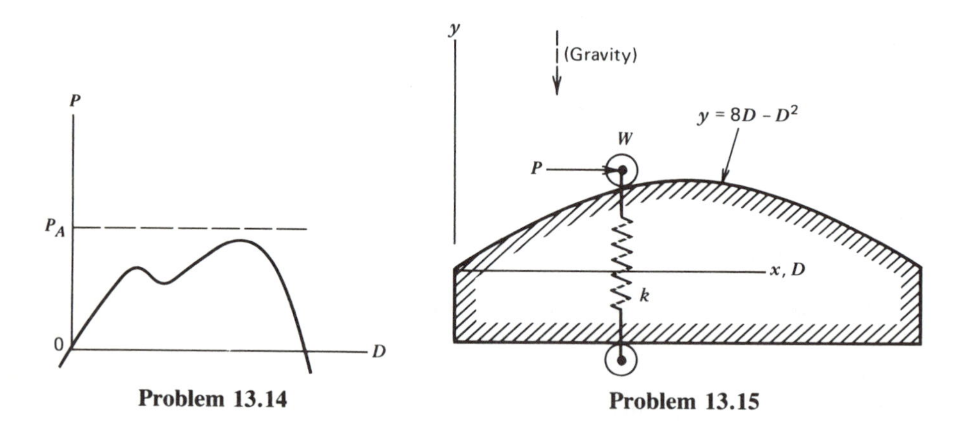

Problem 13.14

Problem 13.15

13.15 The linear spring k is unstressed when $D = 0$. The two rollers, each of weight W, are pushed along the frictionless block by a force P.

(a) Write the total potential Π_P in terms of k, P, D, and W.

(b) Write the equilibrium equation in the form (linear stiffness)$\cdot D$ = (actual load + pseudo load).

(c) Find K in the incremental equation $K \cdot \Delta D = \Delta P$. *Suggestion*. Review Section 13.6.

(d) Write the equation that defines the two values D_1 and D_2 for which the tangent stiffness becomes zero. On physical grounds, why should $0 < D_1 < 4$ and $4 < D_2 < 8$? *Suggestion*. Sketch the P versus D plot.

(e) Imagine that the given system is augmented by a horizontal spring that attaches W to the y-axis, so that the tangent stiffness remains positive for $0 < D < 8$. Describe a calculation method that can be applied to the augmented system to generate the P versus D plot of the given system.

13.16 Sketch the two direct iteration solution processes, as in Figs. 13.6.1*a* and 13.6.1*b*, but let the curves be concave up, as for a hardening structure.

13.17 The sketch shows a nonlinear load versus deflection curve. In the exercises of this problem we pretend that P and dP/dD can be found when D is known, but that an explicit expression for D in terms of P is not available. Solve for D, using the situations and methods indicated. Sketch the progress of each solution on a plot of P versus D.

(a) What D is predicted by five cycles of Newton-Raphson iteration if $P = 8$, starting from $P = D = 0$?

(b) What D is predicted by five cycles of modified Newton-Raphson iteration if $P = 8$, starting from $P = 7.5$, $D = 3$?

(c) Repeat part (b) using three cycles and update the tangent stiffness after the first cycle.

(d) What D is predicted by four purely incremental steps of $\Delta P = 2$, from $P = 1$ and going to $P = 9$? *Given*. $D = 0.11111$ at $P = 1$.

(e) Repeat part (d) but include a force unbalance correction at every step.

(f) Repeat part (e) but include the overrelaxation technique suggested near the end of Section 13.4, with $\beta = 0.5$. (The first incremental step is unchanged.)

(g) What D is predicted by five cycles of the direct iteration algorithm of Eq. 13.6.1 if $P = 8$, starting from $P = D = 0$?

(h) Show that direct iteration by the algorithm of Eq. 13.6.2 diverges if $D > 1$.

(i) Approximate the curve as $P = 4.5D - D^2/2$. What D is predicted by five cycles of the direct iteration algorithm of Eq. 13.6.2 if $P = 7$, starting from $P = D = 0$?

(j) For the approximate curve of part (i), which algorithm converges faster, Eq. 13.6.1 or Eq. 13.6.2?

13.18 The introductory remarks of Problem 13.17 again apply, but now to the hardening curve sketched.

(a) What D is predicted by five cycles of Newton-Raphson iteration of $P = 0.8$, starting from $P = D = 0$?

(b) What is the maximum value of P for which Newton-Raphson iteration will converge if iteration starts from $P = D = 0$?

(c) What D is predicted by five cycles of modified Newton-Raphson iteration if $P = 3$, starting from $P = 1.5$, $D = 6$?

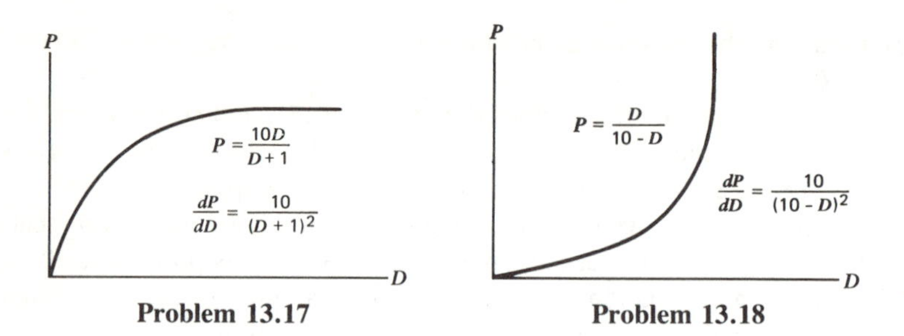

Problem 13.17 **Problem 13.18**

(d) Repeat part (c) but apply an underrelaxation factor $\beta = 0.6$ to the increments ΔD_i.
(e) What is the maximum value of P for which modified Newton-Raphson iteration will converge, if iteration starts from $P = D = 0$?
(f) Repeat part (c), using four cycles and updating the tangent stiffness after the first cycle.
(g) What D is predicted by three purely incremental steps of $\Delta P = 1$, starting from $P = 1.5$ and going to $P = 4.5$? *Given.* $D = 6$ at $P = 1.5$.
(h) Repeat part (g) but include a force unbalance correction at every step.
(i) Repeat part (h) but include the overrelaxation technique suggested near the end of Section 13.4, with $\beta = 0.5$. (The first incremental step is unchanged.)
(j) What D is predicted by four cycles of the direct iteration algorithm of Eq. 13.6.1 if $P = 4$, starting from $P = D = 0$?
(k) Repeat part (j) but include the underrelaxation of Eq. 13.6.3, with $\beta = 0.3$.
(l) Approximate the curve as $P = 0.1D + 0.025D^2$. What D is predicted by four cycles of the direct iteration algorithm of Eq. 13.6.2 if $P = 1.5$, starting from $P = D = 0$?
(m) Repeat part (l) but include the underrelaxation of Eq. 13.6.3, with $\beta = 0.3$.
(n) Repeat part (m) using the algorithm of Eq. 13.6.1.

13.19 In Fig. 13.7.1, let $K = k = 1$ and $g = 7$. Consider an incremental solution with $\Delta P = 2$, starting from $P = u = 6$. Compute, and show on a plot of P versus u, two solution steps (*a*) without unbalance correction, and (*b*) with unbalance correction.

13.20 Consider the problem of "rounding the corner" at $u = g$ in Fig. 13.7.1*b* during an incremental solution. What expression for tangent stiffness is appropriate if β is the anticipated fraction of the total step Δu for which $u > g$?

13.21 Use the data and methods prescribed here and find u_1, u_2, and u_3 in direct-iteration solutions for the problem of Fig. 13.7.1, starting with $f_0 = 0$ in Eqs. 13.7.4 and 13.7.5.
(a) $K = 2$, $k = 1$, $P = 3$, $g = 1$. Use Eq. 13.7.5.
(b) $K = 1$, $k = 2$, $P = 3$, $g = 1$. Use Eq. 13.7.5.
(c) $K = 1$, $k = 2$, $P = 3$, $g = 1$. Use Eq. 13.7.4.
(d) Repeat part (b) using underrelaxation: set $\beta = 0.4$ in Eq. 13.6.3.

13.22 (a) In Fig. 13.7.2, what [**T**] must operate on the 8-by-8 [**k**] of element *ABCD* so that it can be attached to element *I*?
(b) Does [**B**] of Eq. 13.7.6 yield zero strain for rigid-body motions?
(c) Derive [**B**] of Eq. 13.7.6 from a displacement field.

13.23 Data for a cable is given in the sketch. The sag stiffness dT/dL and the elasticity of the cable act as springs in series. Find the effective modulus E^* so that the cable stiffness k can be written $k = AE^*/L$. (This is a convenient way to approximate the tangent stiffness of a sagging cable [13.60].)

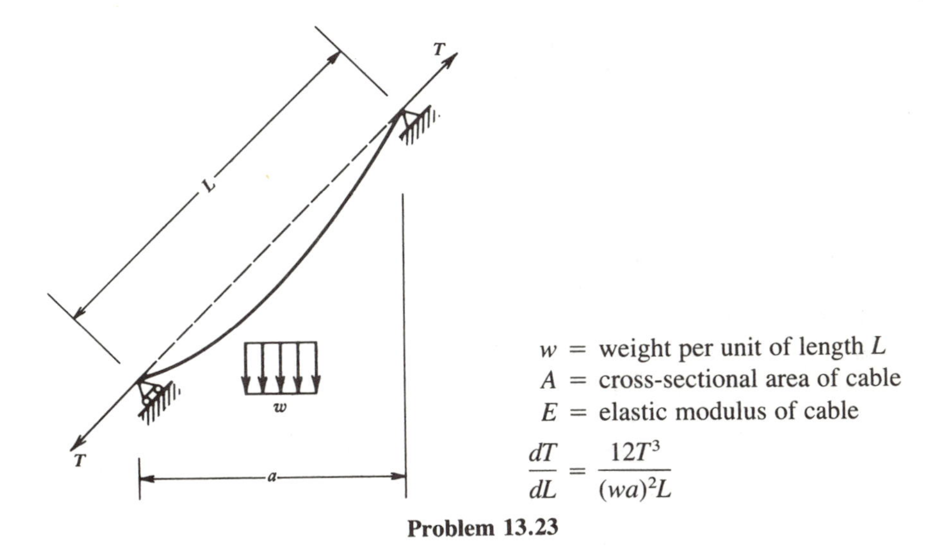

Problem 13.23

13.24 To illustrate the path-dependent nature of plasticity, consider a loading-unloading sequence in Fig. 13.9.1. Identify points that have the same stress but different strain and points that have the same strain but different stress.

13.25 (a) Show that Eqs. 13.9.3 follow from Eqs. 13.9.1 and 13.9.2.
(b) Verify that Eq. 13.9.5 follows from Eq. 13.9.4.
(c) Verify {**Q**} in Eq. 13.9.8.
(d) Verify ⌊**W**⌋ in Eq. 13.9.11.
(e) Verify [$\mathbf{E}_{ep}$] in Eq. 13.9.12.
(f) When all elements of an elastic-perfectly plastic material have yielded, why is {**Q**}⌊**W**⌋ almost a unit matrix?

13.26 (a) Show that Eqs. 13.9.7 and 13.9.9 yield the first of Eqs. 13.10.1.
(b) Show that the first of Eqs. 13.10.1 can be integrated to yield the second.

13.27 (a) Justify Eq. 13.10.5 as an acceptable approximation.
(b) Show that Eqs. 13.10.4 and 13.10.5 both yield $E_D = E_S$ for the case of uniaxial stress.

(c) Show that Eqs. 13.9.2 and 13.10.6 yield $\bar{\sigma}/\bar{\epsilon} = E$ for uniaxial stress σ_x in the linearly elastic range.

13.28 The bulk modulus B of an elastic isotropic material is $B = E/(3 - 6\nu)$. Assume that B remains constant despite extensive yielding (when E becomes the secant modulus E_s and ν becomes $\bar{\nu}$, where $\nu \leq \bar{\nu} < 0.5$). Find a formula for $\bar{\nu}$ of Eq. 13.10.6 in terms of E_s, E, and ν.

13.29 Show that the expression for $\{\mathbf{r}_c\}_i$ in Eq. 13.10.8 can be transformed to that in Eq. 13.10.9.

13.30 Someone proposes to incorporate incremental plasticity theory in the direct iteration method by solving the problem at each of several monotonically increasing load levels. Comment on this proposal.

13.31 Equations 13.10.7 and 13.10.8 may suggest the following procedure.

$$[\mathbf{k}_s]\{\mathbf{d}\} = \{\mathbf{r}\}, \qquad \frac{1}{E}[\mathbf{k}]\{\mathbf{d}\} = \frac{1}{E_s}\{\mathbf{r}\}, \qquad \left(\sum \frac{1}{E}[\mathbf{k}]\right)\{\mathbf{D}\} = \{\mathbf{P}\} + \sum \frac{1}{E_s}\{\mathbf{r}\}$$

where the latter equation implies element-to-structure assembly. But, if $\{\mathbf{r}\} = 0$, this is simply an elastic solution, except that $[\mathbf{k}]$'s are too small by the factor $1/E$. What is wrong?

13.32 The bars are fixed to rigid walls at their ends and are joined where load P is applied. The material behavior is shown in the sketch. If $P = 30$, solve for D and the bar stresses using the direct iteration algorithm of (*a*) Eq. 13.6.1, and (*b*) Eq. 13.6.2. Sketch the progress of the solution on a P versus D plot.

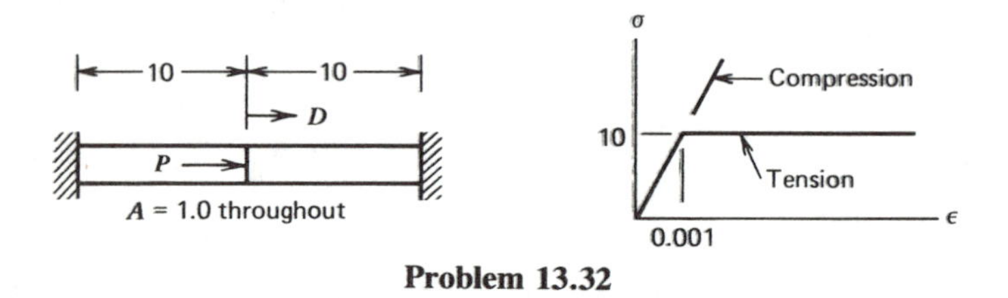

Problem 13.32

13.33 For the structure of Problem 13.32, $D = 0.008$ when $P = 16$. Start from this point and apply the initial stress algorithm (Eq. 13.10.8). Let $\Delta P = 8$ and carry out three cycles of iteration. Show the progress of the solution toward convergence at $P = 24$ on a P versus D plot.

13.34 A material has the following stress-strain behavior in uniaxial tension: $\sigma = 10\epsilon$ for $0 < \epsilon < 1.0$, and $\sigma = 10 + (\epsilon - 1)$ for $\epsilon > 1.0$. An axial load $P = 11.0$ is applied to one end of a bar 10 units long and fixed at the other end. The bar has cross section $A = 1.0$. Carry out two steps of a direct iteration solution and show the progress of the solution on a sketch of σ versus ϵ, if the solution algorithm:

(a) Keeps the load constant and changes the stiffness.

(b) Keeps the stiffness constant and changes the effective load.

13.35 Idealize the $\bar{\sigma}$ versus $\bar{\epsilon}$ curve of Fig. 13.11.1*a* as two straight lines. Also assume that there are no creep or thermal strains and that Eq. 13.9.12 pertains to uniaxial stress $\bar{\sigma}$. Thus, derive Eq. 13.11.6.

13.36 The horizontal bar is perfectly rigid and is constrained to remain horizontal as load P and displacement v increase. The three vertical bars are elastic-perfectly plastic with $A = E = 1$ and $L = 2$. The bars have yield point loads $F_1 = 2$, $F_2 = 4$, and $F_3 = 6$. Generate the P versus v plot. Use a three-step incremental solution. Modify the stiffness after each bar yields.

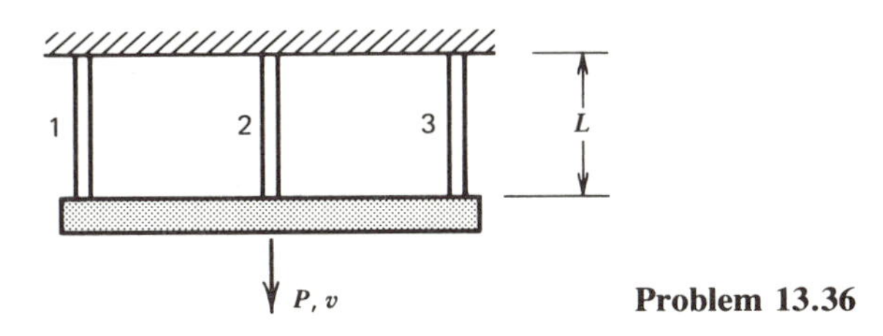

Problem 13.36

13.37 Assume that bars of a truss carry only uniaxial stress. In compression, bars buckle before they yield. Tensile members have a yield point, and subsequently may unload elastically. Loads may not be applied proportionally. Outline an incremental solution algorithm for this problem.

13.38 As a programming project, code the algorithm of Problem 13.37 for plane trusses.

13.39 Describe a procedure for incremental plasticity calculations such that a single sampling point is brought to the initiation of yield by each load increment [13.36].

13.40 Suppose that correction forces $\{\Delta\mathbf{R}_c\}$ are to be computed for a mesh of elements having internal d.o.f. Should internal d.o.f. carry forces that result from $\{\boldsymbol{\sigma}\}$, or should these forces be omitted from internal d.o.f.? If carried, should they be distributed to remaining d.o.f. (that is, condensed) by means of the *elastic* element stiffness equations? What bearing does your answer have on computational efficiency?

13.41 (a) Verify Eqs. 13.12.1 and 13.12.2. What is c_1 in terms of sandwich plate dimensions?

(b) Verify Eq. 13.12.4. What is the physical meaning of $d\epsilon_{zp} = -(d\epsilon_{xp} + d\epsilon_{yp})$?

13.42 The material of a beam is linearly elastic, but moduli in tension and compression are different. Describe an algorithm that calculates the flexural stresses, given the moduli and applied bending moment. Let the cross section be rectangular.

13.43 Section 13.12 discusses plastic hinges that do not strain harden. How might the effects of strain hardening be included?

13.44 Summarize the steps of an incremental algorithm that will trace the load versus deflection path of a frame that develops plastic hinges.

13.45 As a programming project, code the algorithm of Problem 13.44. The choice of test cases is left to you and your instructor.

13.46 For exercises such as Problems 13.37 and 13.44, why is an incremental solution to be preferred over a solution that keeps the elastic [**K**] and transfers all nonlinearities to the right side?

13.47 Cantilever beam *AB* is pin-connected to a tension member *AC* at *A*. A plastic hinge forms at *B* when the bending moment reaches $M_B = \pm 3$. The tension member has a yield point $\sigma_{yp} = 2$. Use the method of Problem 13.44 to generate a plot of *P* versus its displacement v_A, as *P* increases from zero to the fully plastic (limit) load.

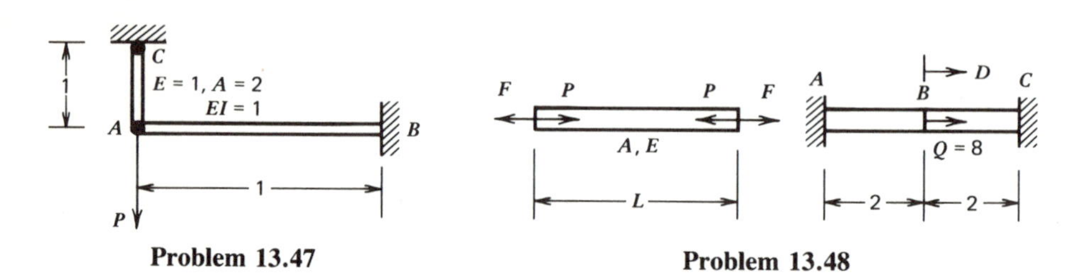

Problem 13.47 **Problem 13.48**

13.48 (a) An algorithm for the analysis of rock as a "no tension" material can be developed by physical argument [13.54]. Consider its uniaxial form. The sketch shows elastically developed forces *P* applied *to* nodes by the deformed element. If $\epsilon > 0$, forces *P* must be canceled by forces $F = \epsilon EA$, to be added to external loads in the next iteration. To which of the solution algorithms described in this chapter does the scheme correspond?

(b) Apply this scheme to bar *ABC*, if $E = 1000$, $A = 1$, and part *AB* can carry no tension. Sketch the solution process on a plot of load *Q* versus its deflection *D*.

13.49 Consider a plane structure modeled by constant strain triangles. The material is isotropic, but it cracks when the tensile stress in any direction exceeds a value σ^*.

(a) Outline a direct-iteration solution algorithm for this problem. Let [**K**] be repeatedly modified.

(b) Repeat part (a) letting the effective loads be repeatedly modified while [**K**] is kept unchanged.

(c) Outline an incremental solution algorithm in which each load increment causes cracking at a single sampling point.

13.50 Let $c = 0$ in the sketch for Problem 13.13. Again, the roller contacts the wall for any *D*. The bar has length $a = 20$, cross-section area $A = 1.0$, and material properties shown in the sketch for Problem 13.32. If $P = 0.6$ *upward*, solve for *D* by the method of Section 13.3. Include $[\mathbf{k}_\sigma]$ (a scalar in this problem). Assume an axial tension of eight units for the first iteration.

CHAPTER **14**

COPING WITH DATA, PROGRAMS, AND PROGRAMMING

14.1 INTRODUCTION

Conference proceedings, journal articles, and books about computer science fill many shelves. Engineers may overlook this information because their experience has not brought it to their attention. Those who program, work with programmers, or use programs can benefit by knowing about the many options available in software and hardware.

Computer programs reflect the separation between researchers and practitioners. Researchers concentrate on analysis tools. Their programs usually do not address the pre- and postprocessor functions so necessary in production. Practitioners tend to ignore the results of research because the old analysis tools are workable.

In this chapter we introduce considerations of practical value and point toward the wealth of literature available—some of it quite readable and even entertaining [14.1–14.9].

14.2 DOCUMENTATION

Our comments pertain to information needed to use, maintain, and modify an existing program. Development documentation, which includes plans, schedules, and progress reports, is excluded. We also distinguish between internal and external documentation.

Good internal documentation consists of straightforward code (Section 14.8) that includes adequate comment statements. Good comments describe—and describe *correctly*—the logic of small blocks of code and add information not otherwise obvious. They do not echo the code in English, such as "now replace N by N plus 1" to describe the statement N = N + 1. The *quantity* of comments is also important: if miserly, the listing is inscrutable; if lavish, the listing is ponderous and a labor to read.

External documentation includes the programmer's manual. It augments the listing and addresses matters useful in maintenance and modification of the code. These matters include descriptions (of files, overlays, data structure, and algorithms), advice about how to change the capacity of the program and how it might be modified or extended, estimates of accuracy, and the programmer's thoughts about possible trouble spots [14.1, 14.3]. A flow chart is desirable if it fits on a single page and if it describes the major logic structure, the interrelation of modules, or the flow of data. A detailed chart that meanders over several pages is too complicated to be useful.

A user's manual is also external documentation. It should be written from the user's viewpoint, with the assumption that the user knows nothing about the program. This assumption is often true. In any case, it is easy for a user to overlook what is obvious to the programmer. The prose should not bristle with acronyms. A good introductory section states what the program is for, gives an overview, and remarks on how to use the manual itself. An index is helpful. In addition to detailed instruction on data preparation and advice on output expected, the manual should comment on: the machines and operating systems on which the program will run, restrictions on problem type and input data, running time for typical cases, accuracy and internal checking, diagnostic messages, and sample input and output data [14.1, 14.3].

Ideally, a user's manual is written before the program. Thus a user-oriented manual guides the coding and is not a cryptic document pieced together as an afterthought. User confusion about any part of the manual suggests that a revision is needed.

Programs must be maintained. They must be debugged, improved, expanded, and adapted to new machines and new operating systems. Maintenance cost may exceed development cost over the life of a program. Good documentation is essential to maintenance. A program must be discarded if it falls into eccentric ways because of undocumented changes.

14.3 MESH GENERATION

"Mesh generation" refers to automatic generation of nodal coordinates and to automatic numbering of nodes and elements based on a minimal amount of user-supplied data. Automation reduces errors and spares the user a long, tedious chore. Solution accuracy may increase because a computer-generated mesh is more regular than one supplied by the user.

Many approaches to mesh generation have been advocated. Here we introduce some simple but useful concepts and cite some of the many papers available. Sections 7.13 and 14.4 also bear upon mesh generation.

Consider Fig. 14.3.1*a*. Let us specify the locations of nodes 9 and 90, the location of the center of the arc, and that the arc is to be divided into nine intervals. A simple program can then place nodes at equal intervals along the arc and assign to them the numbers 18, 27, . . ., 81. The same program can interpolate nodes along a straight line from data about the end points. Thus data for many boundary nodes of a mesh can be generated by breaking the boundary into line segments and supplying data for nodes at the ends of the segments.

When all boundary nodes have been located, individually or by generation, interior nodes can be generated automatically by placing each one at the average of its neighbor's positions.

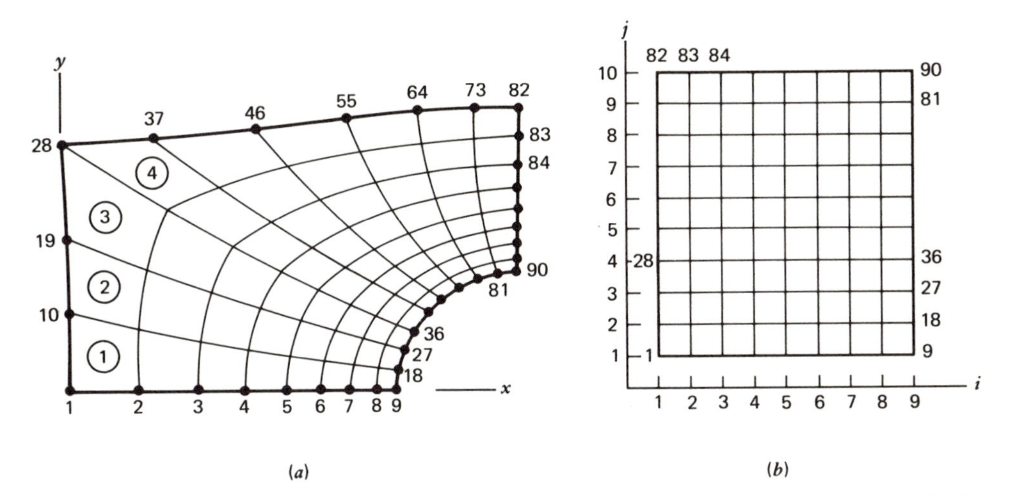

Figure 14.3.1. (*a*) Laplacian grid, from Eq. 14.3.1. (*b*) Rectangular mesh, of which the Laplacian grid is a distorted version.

$$x_{i,j} = (x_{i-1,j} + x_{i+1,j} + x_{i,j-1} + x_{i,j+1})/4 \tag{14.3.1}$$

Here a comma does *not* denote partial differentiation. A similar equation is written for $y_{i,j}$. In this calculation, nodes in Fig. 14.3.1*a* are identified by the i and j subscripts in Fig. 14.3.1*b*. So, in Eq. 14.3.1, i runs from 2 to 8 and j from 2 to 9 to span all interior nodes. The equations can be solved iteratively: the most recent values of nodal x's and y's are used in the right side of Eq. 14.3.1 to define a new $x_{i,j}$, then we increment i or j and repeat. An overrelaxation factor will speed convergence. The grid is called Laplacian because Eq. 14.3.1 is the finite difference form of Laplace's equation [14.11].

A Laplacian grid has drawbacks. Elements may become undesirably distorted. Interior nodes may pile up along curved edges as in Fig. 14.3.1*a*, sometimes to the extent that they fall *outside* the edge. These problems are avoided by a generalization of Eq. 14.3.1 (and its corresponding y equation). The x coordinate of an interior node k is given by [14.12]

$$x_k = \frac{1}{N_k(2-\beta)} \sum_{n=1}^{N_k} (x_{a1} + x_{a2} - \beta x_0) \tag{14.3.2}$$

where N_k is the number of elements that share node k, subscripts a and 0 refer to adjacent and diagonally opposite nodes in element n, and β is an arbitrary factor between 0 and 1 but usually near 0.9. The Laplacian scheme results if $\beta = 0$. For example, consider node 12 of Fig. 14.3.1*a*, for which $N_k = 4$ and coordinates in Fig. 14.3.1*b* are $i = 3$ and $j = 2$.

$$x_{3,2} = \frac{1}{4(2-\beta)}(x_{2,2} + x_{3,1} - \beta x_{2,1} + x_{3,1} + x_{4,2} - \beta x_{4,1} + x_{4,2} + x_{3,3} - \beta x_{4,3} + x_{3,3} + x_{2,2} - \beta x_{2,3}) \tag{14.3.3}$$

Other attacks on the same problems, some invoking areas of elements, are reported in Ref. 14.13.

Element generation can also be automated. In Fig. 14.3.1*a*, we see that node numbers of element 2 result when 9 is added to node numbers of element 1. Thus an entire row of elements can be generated from data about the first.

Electromechanical devices are also useful in mesh generation [14.11]. A mesh is sketched on paper or on a three-dimensional object or model. Then a probe is manually positioned at a node and, at the touch of a button, a digitizer records the nodal coordinates. Similarly, a structure can be "noded" from its image on a cathode-ray display. In *element* generation the digitizer can be programmed to seek out the three or four nodes closest to the probe, thus identifying a triangular or quadrilateral element.

Miscellaneous techniques include automatic deletion of elements and changes in nodal connectivity [14.13] and definition of element boundaries by the ξ = constant and η = constant lines in a single isoparametric element that spans most or all of a structure [14.14, 14.42]. Triangular elements [14.15] and three-dimensional structures [14.16–14.18] have also received attention.

To achieve an optimum mesh layout with a fixed number of nodes, nodal d.o.f. *and* nodal coordinates must be treated as variables. The total potential Π_p becomes an objective function and nodal coordinates become optimization variables in a nonlinear programming problem that searches for minimum Π_p. Results show that optimum node placement requires that element boundaries follow lines of constant strain energy [14.19]. Regrettably, optimality requires a different mesh for each different load case and a changing mesh for nonlinear problems.

Mesh refinement or mesh optimization can be restricted to small portions of the mesh. For example, we can compute strain-energy densities U_B and U_C, where B represents an arbitrary point in an element and C its centroid, or where B and C are in different elements but touch one another across the interelement boundary. A large difference $U_B - U_C$ indicates large departure from constant strain conditions and suggests the need for mesh refinement near B [14.19, 14.20]. Large residuals (the right side of Eq. 4.10.4) may also suggest a need for refinement.

14.4 COMPUTER GRAPHICS

Practical (nonacademic) structures are oddly shaped, with bosses, holes, protrusions, and stiffeners. A great amount of input data is needed to describe them, so errors in data preparation are likely. Also, an analysis can produce an overwhelming amount of output data. Money spent for computer time may be less than one-tenth the money spent to prepare data and interpret output. When these data-handing tasks are done by interactive graphics instead of manually, the total time and money spent for an analysis may be cut in half, despite somewhat greater computer expense.

''Computer graphics'' refers to computer-generated pictures displayed on paper or CRT (cathode-ray tube). The picture may show structure geometry, mesh layout, deflected shapes, and so on. The necessary data is often stored on a disc file. A CRT terminal responds quickly and has more display options than paper output. The user may peruse all data on a CRT terminal, then select portions of it to be printed or plotted on paper. Often the paper output can be incorporated directly in a report or used as shop drawings. Also, data about geometry might be used to prepare tapes for numerically controlled machine tools.

Input data should be checked by study of a machine-generated plot of the mesh, since errors in node location and connectivity are much easier to see in a picture than in a list of numbers produced by a line printer. Display options include perspective views, views from different vantage points, magnified views, sectioning, removal of hidden lines, and shading of a surface to identify its orientation. Different parts or different materials can be represented by different colors. If all edges common to only one element of a plane mesh are made to blink or glow brightly, the highlighted lines should identify the structure boundary. A ''shrink plot'' is a mesh plot in which all elements are shrunk by (say) 20%. Thus all element edges become visible, and it is easy to see if there is no element where one was intended. An error can be corrected as soon as it is discovered.

Displays of output data include the following. The deformed structure can be displayed, either superposed on its deformed state or oscillating back and forth between deformed and undeformed positions. Vibration modes can be similarly displayed. Stress trajectories and stress contours can be plotted [14.21]. Zones of high stress can be colored red, shading to blue in zones of low stress.

Of the many engineering publications we cite only Refs. 14.21 to 14.26. Much more information is available in computer science publications.

14.5 BLUNDERS AND ERROR TRAPS

Here we discuss how to test a program to see if it is working properly, and how a program can itself test for errors in input data. Errors of the numerical kind are considered in Chapter 15.

When testing a new program—or the coding of a new element—*start simply*. Try statics before dynamics and linear problems before nonlinear. Begin with problems such as a simple patch test and a one-element beam. Before a program is applied to a practical problem, it should yield good results in a variety of test cases for which answers are known. Bad results may arise from bad data as well as from a faulty program. Proper data preparation requires the user to understand the program's capabilities and limitations and to know something about the analysis method.

A program that survives the foregoing tests still cannot be fully trusted. The test cases and mesh geometries were, no doubt, not exhaustive. Those used may not have exercised all possible computational paths. If a different computer is used, trouble is likely. Trouble is possible on the *same* computer because of changes in its operating system. Accordingly, test cases should be run periodically. Some should exercise the program's main functions. Some should be barely legitimate and others barely illegitimate to test the diagnostics [14.1]. Unfortunately, tests can detect errors but cannot prove their absence.

In production use, the first run should be a data-check run, without analysis. A three-dimensional analysis, because of its expense, can be preceded by an analogous two-dimensional analysis to get a feeling for the mesh required and the probable results. The mesh, as understood by the program, should be plotted and checked (Section 14.4). Input data should be kept on file with output data in case of subsequent questions about the analysis. Output data should be compared with available standards: simple approximate models, experiments, previous similar analyses, and engineering judgment (also known as the do-the-results-look-strange method). Some analysts distrust the results until they are confirmed by an independent program, preferably having a different analytical basis [14.26].

Some tests for input data errors can be coded into a program. All tests should be applied before execution is terminated by failure of the data to pass any single test. Tests include the following. If an element with more than three nodes must be planar, each set of three noncolinear nodes must define the same unit normal. Nodal coordinates, moduli, cross-sectional areas, and so on can be scanned to see if they fit within user-defined limits. Elements may have limits on aspect ratio, internal angles, and out-of-planeness. Excessive distortion of a quadrilateral (Fig. 14.5.1) can be detected by drawing both diagonals so as to form the four triangles 123, 234, 341, and 412: at least one triangle in a "bad" quadrilateral will have zero or negative area according to Eq. 4.5.4. Analogous tests for isoparametric elements are noted in Section 5.12. In two dimensions, if an interelement boundary is shared by more than two elements, an element may have been defined twice [14.43]. Nodes used to couple substructures must agree in number and location. Elements that share a node must define the same number of d.o.f. for the node. Repetition of a node or element number, nodes not referenced by any element, and different nodes with identical coordinates may signal an error. Node and element numbers must be positive, as must element thickness and the number of load cases. Program capacity must not be exceeded. Adequate boundary conditions must be prescribed.

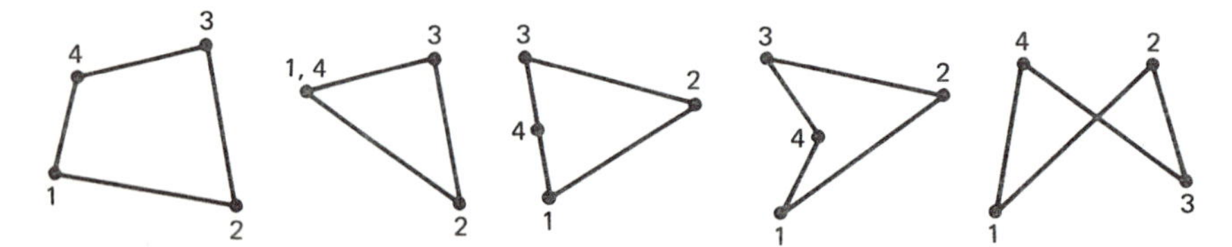

Figure 14.5.1. Quadrilateral element and four excessive distortions. The last is a "bow tie" element.

14.6 REMARKS ABOUT LARGE AND SMALL PROGRAMS

A "large" program is large in both size and capability and is usually general purpose. Its capacity, or number of d.o.f. allowed, is governed by the available auxiliary storage available, not by the size of primary storage. A general-purpose program has a library of several element types and can use them in combination. It permits various analysis options such as static, vibration, or buckling and permits various options and formats for input and output. Restart capability is provided: intermediate results are saved, so that after scheduled or unscheduled stops, an analysis can be resumed instead of redone. Table 14.6.1 summarizes program characteristics [14.28–14.33].

Large programs are apt to rely on *preprocessors* and *postprocessors* [14.34, 14.35]. A preprocessor is a program used to prepare, check, and alter data. A postprocessor is a program used by the analyst to interpret and display results. These functions are peripheral to the analysis phase yet may represent 60 to 80% of the total program, and a like amount of the cost to develop, maintain, and use the program [14.22]. Some tasks suited to pre- and postprocessors are noted in Sections 14.3, 14.4, and 14.5. A preprocessor may also reorder node or element numbering to reduce solution time, estimate the time and cost of solution, allocate data to storage files, and compute nodal loads produced by pressure on a curved surface. A postprocessor may also scan for displacements higher than allowed, output (say) the dozen largest stresses and their locations, and test a trial design to see if it meets the requirements of a standard design code.

Modular design is a feature of large general-purpose programs. A module is a set of statements that comprises one logical subset of the program. (*NASTRAN*'s 400,000 statements are grouped into 175 modules and 1675 subroutines.) Modules and subroutines can be separately tested and optimized and separately documented. Program structure should be hierarchical: a module on one level may call modules on the next lower level

TABLE 14.6.1. CHARACTERISTICS OF STRUCTURAL MECHANICS SOFTWARE [14.29]. K = MULTIPLIER OF 1000.

Type of Program	single element static	multiple element static	large general purpose
Number of statements	500–2000	2K–10K	100K–400K
Development cost[a]	\$5K–\$50K	\$25K–\$200K	\$2000K–\$10 000K
Pages of documentation	20–100	50–500	2000–7000
Machine words needed	10K–30K	20K–50K	50K–150K
Diagnostics	few–moderate	few–moderate	extensive
Cost per run[a]	\$1–\$100	\$1–\$500	\$10–\$10K
Examples	piping program, plane stress	shell program, 3-D solid	*NASTRAN*, *ASKA*

[a]When this table was written, a graduating engineer started work at roughly \$1000 per month.

but may not call a module on the same level. Modular programs are (comparatively) easy to design, maintain, and modify.

Modules of a program must share a data base. A data base is a collection of files or records [14.35, 14.41]. Ideally, it has little waste space, can be accessed quickly either sequentially or directly, and is flexible to allow for growth and change. Data transferred between modules should be minimal. The data base clearly influences the selection and design of modules. The data base of a general-purpose program overflows primary storage into mass storage devices. Special-purpose programs are usually not equipped to manage such a data base.

What are the relative merits of special-purpose and general-purpose programs? The small or special-purpose program can be developed, maintained, and used in-house, usually at modest cost if it is repeatedly used. But the user often finds that it lacks enough capacity or a feature needed for a current problem. The large or general-purpose program must be purchased, usually at great cost. Engineers may have to take a training course in order to use it. The extent of an analysis is limited by the user's purse, not by program capacity. It probably has the resources to do whatever the user requires [14.29]. However, straightforward problems may be more easily and cheaply solved by a special-purpose program.

Except for students, few will find it worthwhile to develop software unless it is their business to sell it. Large programs can be purchased or their use rented at a service bureau [14.37]. One must ask if the supplier will answer users' questions and correct bugs. Small programs are sometimes confidential unless considered obsolete by their developers, but then they may be obtained cheaply.

A major software development project and a major analysis project are similarly managed [14.1–14.6, 14.38, 14.39]. A project is divided into parts, and each part is assigned to a separate group. The more interaction between groups, the slower the progress, and adding manpower may *delay* completion. It has been said that a programming project goes through six stages [14.4]: enthusiasm, confusion, disillusionment, search for the culprits, punishment of the innocent, and rewarding of the nonparticipants!

14.7 RECOMMENDATIONS FOR PROGRAMMING

Some attributes of a good program have been noted elsewhere in this chapter. Here we note a few of many additional rules for Fortran programs that enhance portability, simplify maintenance, and increase efficiency. See Refs. 14.7, 14.8, and 14.34.

A portable program, to the extent possible, uses standard code that runs on all machines. It avoids mixed-mode arithmetic and does not mix integer and real (or double precision) variables in COMMON statements. Arrays are cleared to zero: we do not assume the computer will do it without being asked. Each DO loop ends on a separate CONTINUE statement. Character I/O is done under a field accepted by almost all computers, such as A4. Variables designate I/O units, for example

```
IN = 5
READ (IN, 1010) List
```

Thus the change of the single statement IN = 5 adapts all READ statements to a computer that uses a different unit number for input.

For ease of maintenance, we value clarity and simplicity more than cleverness and efficiency. Variables should have good mnemonic names. Parentheses in arithmetic expressions can avoid ambiguity. Prettyprinting is recommended by some (Fig. 5.5.2 is prettyprinted). Statement numbers and card numbers *increase monotonically* and do so by increments of 10 or more to allow room for later additions. GO TO statements are used sparingly and never to leap all over the place in an attempt to repair ramshackle organization (instead, start over). Updates are tested on a duplicate program before insertion into the production code. Security is needed to prevent modification of the code by casual users.

Efficiency of calculation may be overemphasized in academic circles. In practice, user convenience and manpower cost are of greater concern. In any case, make the code correct before making it faster: it is pointless to turn out wrong answers at great speed. For efficiency, set constants by DATA statements if possible and compute other constants and subscripts outside loops. Arrays should have single, not multiple, subscripts. Do data I/O in large blocks. Element matrices need be generated only once for a series of identical elements. On some machines it is very helpful to code the inner-most loop of a nest as an assembly language subroutine. Few such subroutines are needed, since inner loops are similar; a common form is one variable plus the product of two others. The original Fortran is retained as comment cards to explain the procedure.

The engineer should seek expert advice about organization of data [14.35, 14.36], I/O strategies and buffering, and storage of arrays in packed form. Printed output is less confusing if F formats prevail and if each sheet identifies the problem, load case, program, date, and time.

14.8 DYNAMIC STORAGE ALLOCATION

We begin with an example of *fixed* dimensioning in Fortran (Fig. 14.8.1). This is a simple, hypothetical example, with only enough detail for illustration. Subroutine INPUT reads data for all NUMNP nodes. Subroutine BUILD constructs a stiffness matrix S and load vector R. All arrays ever to be active in the program are dimensioned at the outset, each to the maximum of anticipated use.[1] Thus, in this example, NUMNP $\leqq$ 300, NEQ $\leqq$ 600, and MBAND $\leqq$ 30. Note that dimension statements in the subroutines need not be "fullsize."

[1]Do you think Fig. 14.8.1 is for plane structures, for solids, or what? Can you suggest a more reasonable DIMENSION statement?

```
C --- FIRST MAIN ROUTINE, BASED ON FIXED DIMENSIONS.
      DIMENSION X(300),Y(300),ID(6,300),S(600,30),R(600)
C ---------(STATEMENTS NOT ESSENTIAL TO THIS EXAMPLE OMITTED HERE)
      CALL INPUT (X,Y,ID,NUMNP,IN)
C ---------(STATEMENTS NOT ESSENTIAL TO THIS EXAMPLE OMITTED HERE)
      CALL BUILD (S,R,NEQ,MBAND)
C ---------(STATEMENTS NOT ESSENTIAL TO THIS EXAMPLE OMITTED HERE)
      END

      SUBROUTINE INPUT (X,Y,ID,NUMNP,IN)
      DIMENSION X(1),Y(1),ID(6,1)
      DO 20 K=1,NUMNP
   20 READ (IN,1000)  X(K),Y(K),(ID(J,K),J=1,6)
 1000 FORMAT (2F10.4,4X,6I1)
      RETURN
      END

      SUBROUTINE BUILD (S,R,NEQ,MBAND)
C --- ASSEMBLY OF NEQ BY MBAND MATRIX S AND NEQ BY 1 VECTOR R.
      DIMENSION S(NEQ,1),R(1)
C ---------(STATEMENTS NOT ESSENTIAL TO THIS EXAMPLE OMITTED HERE)
      RETURN
      END
```

Figure 14.8.1. A main routine and two subroutines, in outline form.

Fixed dimensions waste storage space. The following strategy uses it more effectively. (1) Store node data in primary storage. (2) Read element data one element at a time. Each time combine node and element data, generate element matrices, and write them on a mass storage device. (3) Destroy node data and use primary storage space to build the structure matrices by recalling element matrices from mass storage.

Dynamic dimensioning makes it easy to adopt the preceding strategy. By use of Fig. 14.8.2, each array in a subroutine is automatically given the space necessary. We need not guess in advance about number of nodes, number of equations, or bandwidth. All

```
C --- ALTERNATE MAIN ROUTINE. USES DYNAMIC DIMENSIONING.
      COMMON A(15000)
      LIM = 15000
C ---------(STATEMENTS NOT ESSENTIAL TO THIS EXAMPLE OMITTED HERE)
      N1 = 1
      N2 = N1 + NUMNP
      N3 = N2 + NUMNP
      N4 = N3 + NUMNP*6 - 1
      IF (N4 .GT. LIM)  CALL ERROR
      CALL INPUT (A(N1),A(N2),A(N3),NUMNP,IN)
C ---------(STATEMENTS NOT ESSENTIAL TO THIS EXAMPLE OMITTED HERE)
      N1 = 1
      N2 = N1 + NEQ*MBAND
      N3 = N2 + NEQ - 1
      IF (N3 .GT. LIM)  CALL ERROR
      CALL BUILD (A(N1),A(N2),NEQ,MBAND)
C ---------(STATEMENTS NOT ESSENTIAL TO THIS EXAMPLE OMITTED HERE)
      END
```

Figure 14.8.2. Alternate main routine that uses dynamic dimensioning.

data resides in a single array A. An error routine is called if the problem calls for more than the available space. The same space in array A is used first for node data and then for structure matrices. Each pointer (N1, N2, and so on) identifies the locations in A where another array (such as ID) is to begin. Subroutines in Fig. 14.8.1 need not be changed at all if the alternate main routine is adopted. Calls from subroutines are straightforward. For example, we could invoke an equation solver from BUILD by writing CALL SOLVE(S,R,NEQ,MBAND) and including in SOLVE the same dimension statement used in BUILD. A program of any length and any number of subroutines is changed in capacity merely by changing the dimension of A in the single COMMON statement.

The scheme of Fig. 14.8.2 wastes some space, since the dimension of array A is fixed instead of adapting to the problem at hand. Also, if program modification demands new arrays, the programmer must redefine the pointers, with attendant chance for error. These drawbacks can be overcome [14.34, 14.35].

Virtual-memory machines handle data transfer to and from mass storage without instruction from the programmer, so that primary storage seems unlimited to the user. However, a poor data structure produces *thrashing*. This is a condition in which productive computation all but ceases as the machine searches through large blocks of data in mass storage to select a few numbers from each block.

14.9 COSTS

Problems vary so much that no blanket statement about computer cost is possible. A very small problem can be run on a hand computer. At the other extreme, structural analyses for a nuclear power plant design may consume 180 engineering man-months and require 30 different mathematical models [14.40]. An interesting viewpoint is that if a program is used about four hours per week, an increase of 10% in its efficiency will pay an engineer's annual salary.

The effort of writing a program varies as roughly the three-halves power of the number of instructions [14.1]. The coding cost per instruction, exclusive of comment cards, is estimated at one man-hour per mathematical statement and two man-hours per logical statement [14.3]. These estimates refer to the cost of programming, debugging, and documenting a program and do not include the cost of working out the theory on which the program is based. The productivity of programmers may vary by a factor of 10 [14.1]. In one instance, each of several programmers with 2 to 11 years of experience coded a logic problem. The ratio of worst to best was 25/1 for coding time, 5/1 for code size, and 13/1 for running time. There was no correlation between productivity and experience [14.10]. Given this uncertainty, we resort to the circular estimate (multiply the programmer's cost estimate by π).

It seems safe to say that manpower cost far exceeds computer cost and that maintenance cost exceeds development cost over the life of a large program. It takes practice to learn to use a large program effectively, and this requires time, money, and management support for the engineer. A readable user's manual, good diagnostics, and good I/O formatting may save more money than an efficient program.[2]

Data is often stored just in case it may be needed in the future. If it is stored on high-speed discs, the cost of storing data may exceed the cost of computer processing.

In running a finite element program, more than half the "number-crunching" cost of a static analysis may result from the generation of element matrices. Equation solving may account for 25 to 75% of the cost, depending on whether the problem is static or nonlinear and/or dynamic.

[2]My experience is that all manuals are confusing or fail to answer the user's questions. For this reason, in part, students fail to get good results. Most often, answers are clearly wrong or have aspects that are at odds with physical reality or with the way the program purports to work.

CHAPTER 15

DETECTING AND AVOIDING NUMERICAL DIFFICULTIES

15.1 INTRODUCTION

There are many reasons why computed results are rarely exact. We divide a structure into elements whose displacement fields exclude many of the possible modes. The type, number, size, shape, and grading of elements may each be chosen within a broad range of possibilities. Node and element numbering schemes affect the sequence of calculations. The computer represents numbers by only a finite number of digits.

In this chapter we describe and categorize sources of error. We also discuss how to set up a problem to reduce the chances for error, how to test for error during computation, and how to improve results known to certain error.

Difficulties discussed in this chapter may arise while using a program strictly for its intended purpose. We exclude outright blunders in using the program. These are discussed in Section 14.5.

Equation solving is a major component of analysis, not only in linear static problems, but also in nonlinear problems, eigenvalue extraction, and transient response calculations. Accordingly, we emphasize errors associated with equation solving. Information about errors in eigenvalue problems appears in Refs. 15.1 to 15.5.

There is no single definitive test for solution accuracy short of knowing what the result should be. A calculation that survives one error test may fail another. Rosanoff et al. [15.6] give an example of how easy it is to be misled. They began with a single-precision representation of a tenth-order Hilbert matrix $[\mathbf{H}]$, which is notoriously ill-conditioned. They computed $[\mathbf{H}]^{-1}$, then the inverse of $[\mathbf{H}]^{-1}$. The final result agreed with $[\mathbf{H}]$ to some seven digits, which implies the correctness of $[\mathbf{H}]^{-1}$. In fact, $[\mathbf{H}]^{-1}$ was in error by *three orders of magnitude*.

15.2 EIGENVALUE TEST OF ELEMENTS

The eigenvalue test can detect zero-energy modes, lack of invariance, and other defects. It is only one of several tests of element quality (see also Section 15.3).

We describe the calculations and then comment on how to interpret results. Let nodal loads $\{\bar{\mathbf{r}}\}$ on an element be proportional to nodal displacements $\{\mathbf{d}\}$ through a factor λ.

$$[\mathbf{k}]\{\mathbf{d}\} = \{\bar{\mathbf{r}}\} = \lambda\{\mathbf{d}\}, \qquad \text{or} \qquad ([\mathbf{k}] - \lambda[\mathbf{I}])\{\mathbf{d}\} = 0 \tag{15.2.1}$$

This is an eigenproblem. If each eigenvector $\{\mathbf{d}_i\}$ is normalized so that $\{\mathbf{d}_i\}^T\{\mathbf{d}_i\} = 1$, Eq. 15.2.1 yields

$$\lambda_i = \{\mathbf{d}_i\}^T[\mathbf{k}]\{\mathbf{d}_i\} = 2U_i \tag{15.2.2}$$

Thus each eigenvalue λ_i of $[\mathbf{k}]$ is twice the element strain energy U_i when (normalized) nodal displacements $\{\mathbf{d}_i\}$ are imposed [15.7]. Usually an element is unrestrained for the eigenvalue test, so $[\mathbf{k}]$ is the complete element stiffness matrix.

We find $\lambda_i = 0$ when the corresponding $\{\mathbf{d}_i\}$ represents rigid-body motion—translation or rotation, singly or in combination. There are three linearly independent rigid-body modes for a plane element and six for a solid element. There is only one for axially symmetric behavior of a solid or shell of revolution. Zero-energy deformation modes (Section 5.10) also yield zero eigenvalues. The associated eigenvector may represent a combination of the zero-energy mode and a rigid-body mode.

In testing an element, one first checks that $[\mathbf{k}]$ has the proper number of zero eigenvalues. Too few suggests that the displacement field excludes a desired rigid-body motion capability (see Table 10.3.1). Too many suggests the presence of zero-energy deformation modes.

Nonzero eigenvalues must be real and positive if $[\mathbf{k}]$ is symmetric and positive semidefinite. Eigenvalues should not change if the element is differently oriented in global coordinates: if they change, the element is not invariant. Similar modes, such as the flexure modes of Fig. 15.2.1, should have the same eigenvalue.

Eigenvalues of $[\mathbf{k}]$ can sometimes be used to assess the relative merit of competing element formulations. Valid elements of equal size, shape, and modulus are equally stiff in their constant strain modes. Stiffness differs in the higher modes. For elements based on assumed displacement fields, the higher λ_i are upper bounds on the correct stiffness. Therefore, among differently formulated elements with the same number of d.o.f., the

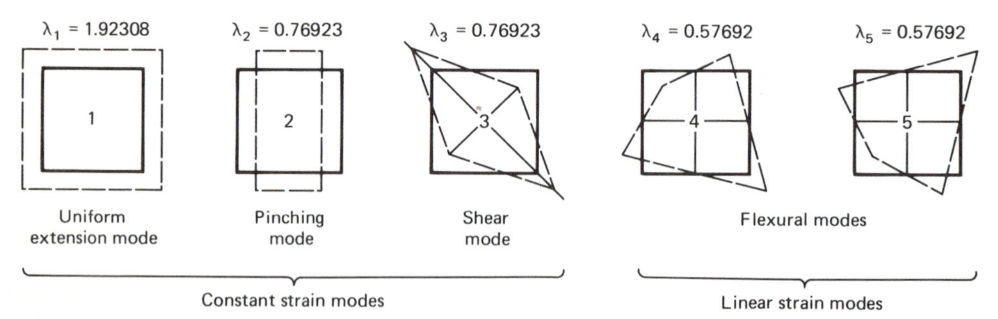

Figure 15.2.1. Nonzero eigenvalues and deformation modes of a linear isoparametric element in plane strain [7.9]. $E = 1.0$, $\nu = 0.3$, and each side length $= 1.0$.

least stiff element is best. It is this element that has the lowest eigenvalues for the higher modes and therefore the smallest trace (tr[**k**] equals the sum of the eigenvalues of [**k**]). This argument fails if the higher modes do not yield upper bonds. Accordingly, we must exclude elements softened by low-order integration rules, elements with zero-energy deformation modes, and incompatible elements. Also, an element judged "best" by this test may have drawbacks that prevent it from being best in practical use.

15.3 TESTS OF ELEMENT QUALITY

An element is "good" if it yields acceptable results in a coarse mesh and yields rapid convergence toward correct results as the mesh is refined. Desirable properties of elements include ability to pass a patch test, invariance, no zero-energy deformation modes in the assembled structure, and ability to mate with elements of different type. The theory should be simple and its computer implementation economical.

We cannot expect to discover this peerless element. Always we must accept a mix of properties, some desirable and others not. The evidence that seems most forceful is performance in test cases whose theoretical solution is known.

Robinson [15.8] suggests a *single-element test* in which response is examined as element aspect ratio is changed. Consider, for example, Fig. 15.3.1*a*. Imagine that there are two competing element formulations, A and B, and that they work equally well when $L = H$. As L/H becomes large, A may work much better than B. Therefore, A has proved superior in this particular test.

But several tests should be used—patch, eigenvalue, single element and multielement. Multielement problems can have few elements and yet test the effects of aspect ratio, element distortion, different loading and support conditions, and type of mesh layout (Fig. 15.3.1*b*). The quality ranking of competing elements is likely to be different in different test cases and is likely to depend on whether stress or displacement is the indicator of quality.

Large aspect ratio and large shape distortion can precipitate numerical troubles discussed next, so that bad answers do not necessarily imply bad elements.

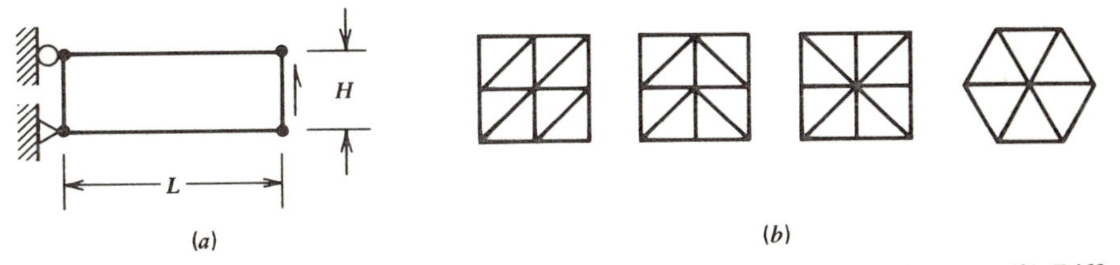

Figure 15.3.1. (*a*) Cantilever beam modeled by a single four-node plane element. (*b*) Different arrangements of triangular elements in a rectangular mesh.

15.4 IDEALIZATION ERROR. CONVERGENCE RATE

To model a problem, we decide on the number, type, and shape of elements and the grading of the mesh. We also choose elastic constants, allocate loads to nodes, and select displacement boundary conditions. The model has only a finite number of d.o.f. Each of these factors may contribute to the misrepresentation of reality called *idealization error* or *discretization error*.

Bad idealization can lead to numerical difficulties. However, our present concern is with the error that the computer inherits because our model differs from reality.

Volume misrepresentation is possible if elements with straight edges have nodes on a curved surface. But this error is easy to reduce (Fig. 15.4.1). Usually, input data such as nodal coordinates need not be of high precision. Slight changes in node placement or elastic moduli cause only slight changes in the results because element stiffness matrices remain well behaved in the patch test sense. This is not to say that subsequent *manipulation* of the input data requires only limited precision (see Section 15.5).

The *Babuska paradox* describes an error associated with modeling a curved boundary by straight-sided elements [15.30]. Imagine that a mesh whose boundary is a regular polygon serves to model a circular region. Mesh refinement causes the polygon to have more sides and to converge to a circle. We would then expect stresses to converge to values that are exact for the circular region. But strains ϵ_n and stresses σ_n normal to the boundary converge to wrong values. Errors in ϵ_n and σ_n can be of the order of the tangential strains and stresses. To avoid this error, elements with curved edges should be used.

Idealization error, and convergence rate as a mesh is refined, can be found by an order of error analysis. Consider, for example, the problem of Fig. 15.4.2*a* [15.9]. From Fig. 15.4.2*b*, the governing equation of the problem is $A\sigma_{,x} + q = 0$ or, with $\sigma = Eu_{,x}$,

$$u_{,xx} + \frac{q}{AE} = 0 \tag{15.4.1}$$

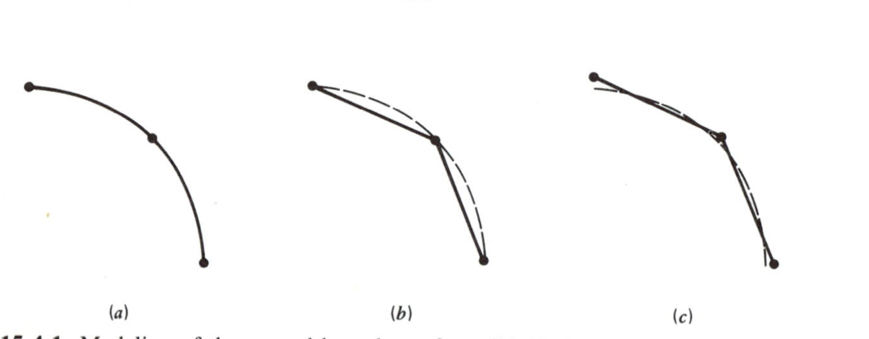

Figure 15.4.1. Modeling of the curved boundary of a solid (dashed line). (*a*) By elements with curved edges. (*b*) By elements with straight edges and nodes on the boundary. (*c*) By elements with straight edges but good placement of nodes.

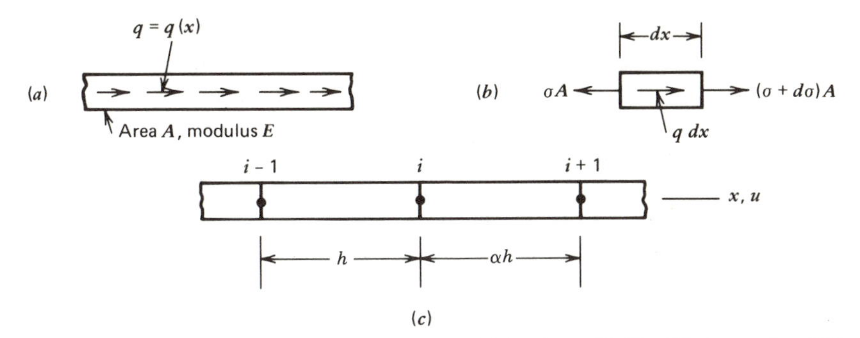

Figure 15.4.2. (*a*) Straight bar with distributed axial load q. (*b*) Equilibrium of a differential element. (*c*) Typical nodes of a finite element model. Elements have dimensions h and αh.

The finite element model (Fig. 15.4.2*c*) consists of simple bar elements. Load on an element is distributed equally to its nodes. Thus the equilibrium equation at node i is

$$\frac{AE}{h}(-u_{i-1} + u_i) + \frac{AE}{\alpha h}(u_i - u_{i+1}) = \frac{q_i h(1 + \alpha)}{2} \tag{15.4.2}$$

Displacements u_{i-1} and u_{i+1} can be expanded in Taylor series about node i.

$$\begin{aligned} u_{i-1} &= u_i - hu_{i,x} + \frac{h^2}{2}u_{i,xx} - \frac{h^3}{6}u_{i,xxx} + \cdots \\ u_{i+1} &= u_i + \alpha h u_{i,x} + \frac{(\alpha h)^2}{2}u_{i,xx} + \frac{(\alpha h)^3}{6}u_{i,xxx} + \cdots \end{aligned} \tag{15.4.3}$$

Substitution of Eqs. 15.4.3 into Eq. 15.4.2 yields

$$u_{i,xx} - \frac{h}{3}(1 - \alpha)\,u_{i,xxx} + \frac{h^2}{12}\left(\frac{1 + \alpha^3}{1 + \alpha}\right)u_{i,xxxx} + \cdots + \frac{q_i}{AE} = 0 \tag{15.4.4}$$

Because Eq. 15.4.4 reduces to Eq. 15.4.1 as h approaches zero, the finite element results converge to exact answers as the mesh is refined. (Imagine here that the entire structure has unit length; then, typically, $h = 0.1$ or less.)

Equation 15.4.4 shows a displacement discretization error of $O(h)$ if $\alpha \neq 1$ and $O(h^2)$ if $\alpha = 1$, where "O" means "order." If a finite element model is to produce valid equations, the decretization error must be $O(h^p)$ with $p > 0$ (because if $p = 0$, then $h^p = 1$, which does not vanish as h approaches zero).

The standard four d.o.f. straight *beam* element has a displacement discretization error of $O(h^4)$ if elements are of equal length. Therefore, if all elements are halved in length, the error in displacement is divided by approximately 16.

An order of error analysis is tedious. It is not even possible if closed-form equations for stiffness coefficients are lacking, as is the case for numerically integrated elements. In a given idealization, the displacement, stress, and strain energy have different rates of convergence (see the following). An order of error analysis is prone to misinterpretation and has trouble with concentrated loads [15.10]. The analysis is too crude to say which is the better of two elements that have equal orders of error. Its main value is in predicting the convergence rate produced by mesh refinement.

Additional symbols are needed if convergence rates are to be discussed. We define:

h = "diameter", for example, an element's length or largest span
$q - 1$ = degree of highest *complete* polynomial in the displacement field
$2m$ = order of highest derivative in the governing equilibrium equation expressed in terms of displacements

For the standard four d.o.f. beam element, $q - 1 = 3$ and $2m = 4$. For a plane quadratic isoparametric element, $q - 1 = 2$ and $2m = 2$.

The following convergence rate argument can be made [15.11]. An element can fit exactly a displacement field of degree $q - 1$ and therefore has an error of $O(h^q)$ in modeling a displacement field of higher degree. Consequently, the error in the mth displacement derivative is of $O(h^{q-m})$. In plane problems stresses are given by first derivatives of displacement, so the convergence rate for stress is $O(h^{q-1})$. Bending moments in beams are given by second derivatives, which therefore have a convergence rate of $O(h^{q-2})$, or $O(h^2)$ for the standard four d.o.f. beam element. A strain energy expression contains squares of the mth displacement derivatives. Therefore the strain energy error is of $O(h^{2q-2m})$. These conclusions apply to continua modeled a uniform mesh of undistorted elements not softened by low-order integration rules (these restrictions preserve the definition of $q - 1$).

If convergence is monotonic, two approximate results can be extrapolated to a better result [15.12]. Let ϕ represent the quantity of interest (displacement, stress, or other) and suppose that its convergence rate is known to be of $O(h^n)$. Let ϕ_1 and ϕ_2 represent values of ϕ produced by meshes with elements of size h_1 and h_2, respectively. Then the value of ϕ predicted as h approaches zero is

$$\phi_{\text{extrapolated}} = \frac{\phi_1 h_2{}^n - \phi_2 h_1{}^n}{h_2{}^n - h_1{}^n} \tag{15.4.5}$$

If convergence is not monotonic, Eq. 15.4.5 is not applicable, and near-agreement between two analyses that use different meshes does not necessarily mean that results are nearly exact.

Elements should not be severely distorted. For fastest convergence, triangular elements should be equilateral, four-sided elements should be square, and node patterns should be regular. Equation 15.4.4 shows the effect of uneven node placement: if $\alpha = 1$, convergence is quadratic in h, but if $\alpha \neq 1$, convergence is only linear in h. As an element assumes curved edges, opposite edges of unequal length, obtuse or acute corner angles, or unevenly spaced side nodes, it has decreased ability to assume deformation patterns contained in its assumed field. Theoretical studies and numerical trials [15.13, 15.14] show that a quadratic or cubic isoparametric element may converge no faster than a linear element if its sides are too highly curved or if side nodes are placed too close to corner nodes, as explained in Section 5.12.

The preceding remarks do not mean that all meshes must be uniform. When using a *given number* of elements, it is best to have a finer mesh where stress gradients are higher. Thus, for example, Fig. 7.13.3*b* may give more accurate results than filling the same space with an even larger number of equal-size elements: convergence rate and accuracy are different things. A need for mesh refinement is indicated if plots of stress contours look erratic or if different meshes yield quite different results.

Convergence may be slower with concentrated loads than with distributed loads, if both loadings are used on the same series of mesh layouts.

Error is subsequent numerical processing is more likely when the condition number $C(\mathbf{K})$ is large. Sections 15.5 and 15.6 discuss conditioning in more detail. Here we note that $C(\mathbf{K})$ is increased by distorted elements, nonuniform meshes, or an increased number of d.o.f. Reference 15.15 finds that $C(\mathbf{K})$ is proportional to

$$(h_{max}/h_{min})^{2m-1}\, N^{2m/n} \tag{15.4.6}$$

where h_{max} and h_{min} are, respectively, the greatest and least node spacings in the mesh, $2m$ is again the differential equation order, N is the number of elements, and n is the dimension ($n = 2$ for plane problems, $n = 3$ in three dimensions). Thus, in a beam problem, if we mix elements of 10 to 1 length ratio, $C(\mathbf{K})$ increases by a factor of 1000. If we also double the number of elements, $C(\mathbf{K})$ increases by a factor of 16,000.

The influence of N in Eq. 15.4.6 can be restated: $C(\mathbf{K})$ is proportional to h^{-2m}. However, the condition number of the mass matrix, $C(\mathbf{M})$, is independent of h [15.16].

Improving a result by using a finer mesh of the same kind of elements is called h-convergence. Alternatively, we can keep the same mesh but improve the displacement field in each element. This is called p-convergence. It is faster than h-convergence and is under active development [7.2].

15.5 ILL-CONDITIONING. TRUNCATION AND ROUNDING ERROR

Equations $[\mathbf{K}]\{\mathbf{D}\} = \{\mathbf{R}\}$ are termed "ill-conditioned" if small changes in $[\mathbf{K}]$ or $\{\mathbf{R}\}$ *may* lead to large changes of coefficients in the solution vector $\{\mathbf{D}\}$. We emphasize "may" because $\{\mathbf{D}\}$ may be sensitive to some types of change in $[\mathbf{K}]$ or $\{\mathbf{R}\}$ but not to others. Ill-conditioning may reflect the physical reality of a structure with low tangent stiffness because it is near buckling or collapse. For example, if $P = 1.001$ in Fig. 13.2.1*a*, a small change in $[\mathbf{K} + \mathbf{K}_\sigma]$ or in P may greatly change the computed α. But ill-conditioning may also arise even when the *physical* problem is stable because of the way the computer manipulates numbers. Included in this manipulation error are *truncation error* and *rounding error* [15.17–15.21].

Truncation error is the more important. Consider a computer that uses p bits per word. Only the leading p bits of a stiffness coefficient K_{ij} can be stored. It is possible, as shown in the following example, that information essential to an accurate solution resides in *trailing* bits of K_{ij}. If the number is truncated, this information is discarded. The information content of $[\mathbf{K}]$ is therefore inadequate and cannot be restored by subsequent processing of $[\mathbf{K}]$, however accurately done.

Rounding error refers to the adjustment of the last bit during computation (for example, during equation solving). Experience in finite element analysis indicates that rounding error is less important than truncation error. However, gratuitous rounding is to be avoided: constants, such as π and Gauss point coordinates and weights, should be written with as many significant digits as the computer word can accommodate.

Example

The problem of Fig. 15.5.1 yields insight into truncation error [15.18]. The structure stiffness matrix and its inverse are

$$[\mathbf{K}] = \begin{bmatrix} k_A & -k_A \\ -k_A & k_A + k_B \end{bmatrix}, \qquad [\mathbf{K}]^{-1} = \begin{bmatrix} \dfrac{1}{k_A} + \dfrac{1}{k_B} & \dfrac{1}{k_B} \\ \dfrac{1}{k_B} & \dfrac{1}{k_B} \end{bmatrix} \tag{15.5.1}$$

If $k_A >> k_B$, k_A dominates $[\mathbf{K}]$, but k_B dominates $[\mathbf{K}]^{-1}$ and hence the compute displacements. Numerical computation of $[\mathbf{K}]^{-1}$ is of satisfactory accuracy only if the coefficient $k_A + k_B$ in $[\mathbf{K}]$ is represented accurately enough that k_B is not lost in comparison with k_A. If, for example, $k_A = 40$ and $k_B = 0.0014$, the computer must carry at least six digits and k_A must be represented as 40.0000 if the last digit of k_B is to be retained in the coefficient $k_A + k_B$, despite the fact that k_A is not really known to six significant figures. If only four digits were carried, $k_A + k_B$ would be represented as 40.00, and $[\mathbf{K}]$ becomes singular. This problem is ill-conditioned because its solution is sensitive to changes in the sixth digit of the coefficient $k_A + k_B$.

Consider also a Gauss elimination solution for displacements in this example. Elimination of u_1 changes the last diagonal coefficient to $(k_A + k_B) - k_A$. We see that information needed for an accurate solution may be lost if $k_A >> k_B$ but not if $k_B >> k_A$.

Reformulation reduces ill-conditioning (but in general does not eliminate it). If in Fig. 15.5.1 displacements $\{u_1 \; u_2\}$ are replaced by $\{\Delta \; u_2\}$, where $\Delta = u_1 - u_2$, the structure equations become $[k_A \; k_B]\{\Delta \; u_2\} = \{P \; P\}$. These equations clearly give no trouble even if k_A is huge (why?). Note that the transformation to new displacements must be made on individual elements before assembly, since no purpose is served by transforming a structure matrix that is already defective. These considerations are applied in Section 13.7.

Figure 15.5.1. Two d.o.f. structure with springs of stiffness k_A and k_B.

The example of Fig. 15.5.1 is an instance of a general rule: *the major cause of ill-conditioning is an element or region of high stiffness supported by an element or region of low stiffness*. This circumstance shifts essential information to the latter bits of computer words. These bits may be so few in number that the solution is unacceptable. The limiting case is the unsupported structure, whose stiffness matrix is so ill-conditioned that it is singular.

As a mesh is refined, the stiffness of the path between an interior element and a support does not change much, but the stiffness of each element may increase without limit (see Problem 4.44). Accordingly, great mesh refinement may lead to great truncation error.

In summary, large stiffnesses supported by comparatively small stiffnesses can be troublesome because essential information is lost by truncation when computer words carry too few bits. The consequences may appear later: in a large condition number, in decay of diagonal coefficienets, or in unreasonable displacements and stresses. If matrix coefficients are generated in single precision, it is of marginal value at best to use double precision in any portion of the equation solver because doing so can reduce only the rounding error, whose effect is usually small. Such a use of partial double precision may even give worse answers than use of single precision throughout, apparently because truncation and rounding errors tend to balance one another [15.19]. Changing to a finer mesh reduces discretization error but promotes truncation and rounding errors, so that a profusion of elements will eventually make answers worse (Fig. 15.5.2).

15.6 THE CONDITION NUMBER

A numerical measure of ill-conditioning in a coefficient matrix [**K**] is the *condition number,* denoted here by $C(\mathbf{K})$. Its definition and calculation are described shortly. Its significance is that if coefficients K_{ij} are presented with d digits in the computer, computed results are accurate to s digits, where

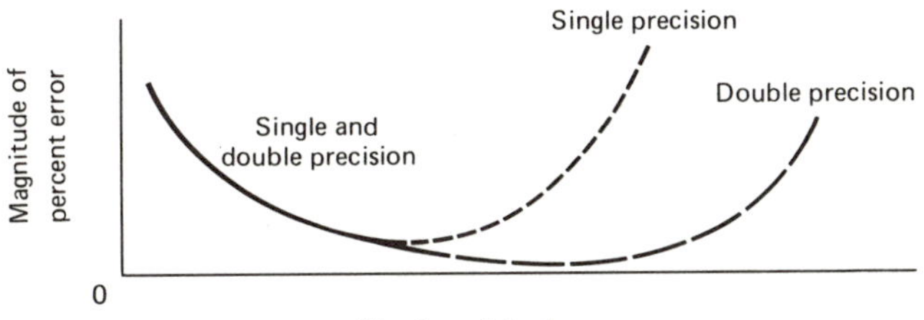

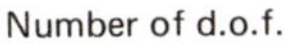

Figure 15.5.2. Solution accuracy under competing effects of discretization error and truncation error. Not to scale: only qualitative effects are shown.

$$s \geqq d - \log_{10} C(\mathbf{K}) \tag{15.6.1}$$

Thus, if $d = 7$ and $C(\mathbf{K}) = 1000$, then $s \geqq 4$. Rounding error is not included: s represents only the effect of truncation. Note that that $\log_{10} C(\mathbf{K})$ is a worst-case estimate of the number of digits lost in solving the equations $[\mathbf{K}]\{\mathbf{D}\} = \{\mathbf{R}\}$. The actual loss is often less.

Equation 15.6.1 may be pessimistic because it ignores the right side $\{\mathbf{R}\}$. If $\{\mathbf{R}\}$ is almost orthogonal to the lowest eigenvector of $[\mathbf{K}]$, subsequent arguments indicate that the solution vector $\{\mathbf{D}\}$ does not mobilize the most severely truncated information in $[\mathbf{K}]$. In this case a better estimate of s is found by using the next-to-last eigenvalue $\lambda_{min\text{-}1}$ rather than λ_{min} in Eq. 15.6.2 [15.19].

The condition number of $[\mathbf{K}]$, also known as the spectral condition number, is defined as

$$C(\mathbf{K}) = \lambda_{max}/\lambda_{min} \tag{15.6.2}$$

where λ_{max} and λ_{min} are the maximum and minimum eigenvalues of $[\mathbf{K}]$ *after scaling*. Scaling is accomplished by transforming $[\mathbf{K}]$ with a diagonal matrix $\lceil\mathbf{S}\rfloor$ that is built from diagonal coefficients in $[\mathbf{K}]$. The scaled matrix $[\mathbf{K}_s]$ is

$$[\mathbf{K}_s] = \lceil\mathbf{S}\rfloor [\mathbf{K}] \lceil\mathbf{S}\rfloor, \qquad S_{ii} = \frac{1}{\sqrt{K_{ii}}} \tag{15.6.3}$$

Diagonal coefficients of $[\mathbf{K}_s]$ are unity. Eigenvalues of $[\mathbf{K}_s]$ are used in Eq. 15.6.2.

The purpose of scaling is to avoid an artificially high condition number. In Fig. 15.5.1, for example, the condition number of the *unscaled* $[\mathbf{K}]$ becomes large if $k_A >> k_B$ *and* if $k_B >> k_A$. However, $C(\mathbf{K})$ of the *scaled* coefficient matrix becomes large only if $k_A >> k_B$, and it is only in this case that several digits might be lost during solution. The case $k_B >> k_A$ is an example of artificial ill-conditioning.

Scaling is done only to calculate $C(\mathbf{K})$. Scaling does not change the accuracy of a direct solution algorithm, provided that the choice of pivots and sequence of eliminations are unchanged [15.22]. Scaling improves the convergence rate of certain iterative solution methods [2.34].

A fine mesh may yield a $C(\mathbf{K})$ in excess of 10^6. A very thin shell may yield $C(\mathbf{K}) > 10^{12}$ because its membrane stiffness is much greater than its bending stiffness. I found that increasing the aspect ratio of a rectangular plate element from 1/1 to 10/1 raised $C(\mathbf{k})$ of the element stiffness matrix from 14 to 31,000. (The element was supported against rigid-body motion, so that $\lambda_{min} > 0$.)

The theory that connects Eqs. 15.6.1 and 15.6.2 is outlined as follows [2.1, 15.17, 15.18, 15.19]. Let $[\mathbf{K}]$ have eigenvalues λ_i and eigenvectors $\{\mathbf{V}_i\}$, normalized so that $\{\mathbf{V}_i\}^T\{\mathbf{V}_i\} = 1$. Then we can write, for an *n-by-n* matrix $[\mathbf{K}]$ that is symmetric and nonsingular,

$$[\mathbf{K}] = \sum_{i=1}^{n} \lambda_i \{\mathbf{V}_i\}\{\mathbf{V}_i\}^T, \qquad [\mathbf{K}]^{-1} = \sum_{i=1}^{n} \frac{1}{\lambda_i} \{\mathbf{V}_i\}\{\mathbf{V}_i\}^T \tag{15.6.4}$$

The argument resembles that used with Eqs. 15.5.1. Matrix [**K**] is dominated by λ_{max} and $[\mathbf{K}]^{-1}$ by λ_{min}. Information needed to compute the lowest mode is buried in the rightmost bits of coefficients K_{ij}, so this mode is computed with less accuracy than any other. For each power of 10 in the ratio $\lambda_{max}/\lambda_{min}$, the λ_{min} information in K_{ij} is shifted right one digit with respect to the λ_{max} information in K_{ij}. Therefore one digit of the λ_{min} information is dropped off the right end of the computer word.

The eigenvalue problem that yields λ_{max} and λ_{min} of the scaled matrix $[\mathbf{K}_s]$ is

$$([\mathbf{K}_s] - \lambda [\mathbf{I}]) \{\mathbf{D}\} = 0 \tag{15.6.5}$$

which may be viewed as a vibration problem with a unit mass matrix and natural frequencies $\omega_i^2 = \lambda_i$. An alternative form of Eq. 15.6.5 is possible because new coordinates leave eigenvalues unchanged. Accordingly, we can introduce the transformation $\{\mathbf{D}\} = [\mathbf{S}]^{-1} \{\mathbf{D}_1\}$, where $[\mathbf{S}]$ is the diagonal scaling matrix of Eq. 15.6.3 [15.23].

$$([\mathbf{S}]^{-1} [\mathbf{K}_s] [\mathbf{S}]^{-1} - \lambda [\mathbf{S}]^{-1} [\mathbf{S}]^{-1}) \{\mathbf{D}_1\} = 0 \tag{15.6.6}$$

$$([\mathbf{K}] - \lambda [K_{11} K_{22} \cdots K_{nn}]) \{\mathbf{D}_1\} = 0 \tag{15.6.7}$$

Thus λ_{max} and λ_{min} of the *scaled* matrix $[\mathbf{K}_s]$ can be computed from Eq. 15.6.7, using the unscaled [**K**] and a diagonal "mass" matrix consisting of the diagonal coefficients of [**K**]. This is a manipulation, not an approximation. Equation 15.6.7 shows why isolated stiff regions raise the condition number: large "masses" K_{ii} placed far from supports reduce the lowest frequencies but have little effect on the highest.

An approximation for $C(\mathbf{K})$ is adequate, so λ_{max} and λ_{min} may be approximated. A close upper bound on λ_{max} of $[\mathbf{K}_s]$ results from summing magnitudes in each row i of the *scaled* matrix [15.18, 15.19]:

$$\lambda_{max} \approx \max Q_i, \qquad \text{where} \qquad Q_i = \sum_{j=1}^{n} |K_{sij}| \tag{15.6.8}$$

The column index is j, and n is the order of $[\mathbf{K}_s]$. Reference 15.23 finds $1.5 \leqq \lambda_{max} \leqq 3.2$ in a series of test cases. If this approximation is accepted, only λ_{min} need be calculated. As even a crude estimate suffices, one might adopt the interpretation of Eq. 15.6.7 and find the "fundamental vibration frequency" λ_{min} by hand calculation with an assumed mode shape (Rayleigh's method). Or, to automate the process, two or three cycles of inverse iteration will approximate λ_{min}. The latter procedure should not be expensive because [**K**] must be triangularized anyway to solve equations. However, the error estimate is then unavailable before procesing. An *a priori* estimate of $C(\mathbf{K})$ by automatic calculation may cost as much as the solution itself, and few users are willing to pay such a price.

15.7 EQUATION SEQUENCING

The accuracy of an elimination solution depends on the choice of pivots [15.24]. When diagonal coefficients are always used as pivots, good choice of pivots means good sequencing of equations.

Example

We illustrate the point with the series of linear springs in Fig. 15.7.1, with $k = 100$. Let a Gauss elimination solution treat equations in the order that nodes are numbered. After 99 eliminations in Fig. 15.7.1a, the stiffness of the condensed (eliminated) part of the structure is reduced to unity, leaving the following two equations yet to be processed.

$$\begin{bmatrix} 101 & -100 \\ -100 & 100 \end{bmatrix} \begin{Bmatrix} u_{100} \\ u_{101} \end{Bmatrix} = \begin{Bmatrix} 0 \\ P \end{Bmatrix} \tag{15.7.1}$$

After 99 eliminations in Fig. 15.7.1b, we have instead

$$\begin{bmatrix} 100 & -100 \\ -100 & 200 \end{bmatrix} \begin{Bmatrix} u_{100} \\ u_{101} \end{Bmatrix} = \begin{Bmatrix} P \\ 0 \end{Bmatrix} \tag{15.7.2}$$

where elimination has merely transferred load P from node to node, and the condensed part of the structure has zero stiffness because node 1 is free. Equation 15.7.1, but not Eq. 15.7.2, is the ill-conditioned and error-prone situation of a stiff spring supported by a flexible spring (see Fig. 15.5.1 and Eqs. 15.5.1). In Fig. 15.7.1b the estimate of $\log_{10} C(\mathbf{K})$ digits lost is too pessimistic. Note that the condition number of the original 101-by-101 stiffness matrix does not depend on how nodes are numbered.

In general, operations in an equation solver should proceed from the more flexible part of a structure toward the stiffer part. Thus, in Fig. 15.7.1,the tip-to-root numbering is preferred whether or not all springs have the same stiffness. An analysis of a cantilever beam shows that 370 elements with root-to-tip sequencing have about the same accuracy

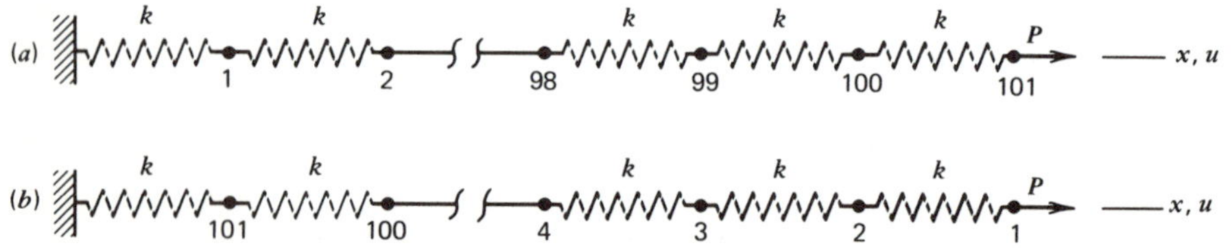

Figure 15.7.1. Two node numbering schemes for springs of stiffness k in series.

as $134(10)^6$ elements with tip-to-root sequencing [15.20, 15.21]. However, different-solution algorithms may have different sensitivity to equation sequencing [15.25].

15.8 DECAY OF DIAGONAL COEFFICIENTS

We describe a simple and inexpensive test that operates during equation solving. It can be used to terminate a run as soon as serious trouble is detected [15.20, 15.21, 15.26].

As each equation is processed by Gauss elimination (or by Choleski decomposition), a subtraction operation reduces the magnitude of diagonal coefficients in **[K]** that correspond to d.o.f. not yet eliminated. Let K_{ii} be a diagonal coefficient in the original stiffness matrix, and let D_{ii} be its reduced value at the time it acts as a pivot in processing the ith equation. If, for example, $D_{ii}/K_{ii} = 10^{-6}$, then six leading digits of K_{ii} have disappeared because of subtraction. This is clearly unacceptable if numbers are represented to only seven digits. The reader should check that elimination of u_{100} in Eq. 15.7.1 yields $D_{ii}/K_{ii} = 0.01$, where $i = 101$. In this particular case, $D_{ii}/K_{ii} > 0.5$ for i from 1 through 100, and trouble is detected in only the final elimination.

The test, then, consists of duplicating the diagonal coefficients K_{ii} of the original coefficient matrix **[K]** in a separate array and computing D_{ii}/K_{ii} just before eliminating the ith equation. A warning message can be printed if D_{ii}/K_{ii} is uncomfortably small, and execution can be terminated if D_{ii}/K_{ii} is unacceptably small. In a computer with a p-bit mantissa, a ratio as low as $1000/2^p$ might be acceptable for engineering purposes [15.20]. If $D_{ii} \leqq 0$, the structure is unstable.

The causes of low D_{ii}/K_{ii} are the causes of ill-conditioning: too few supports, isolated large stiffnesses, and so on. Accordingly, the diagonal decay test can be invoked prior to the actual solution phase. We might replace the off-diagonal K_{ij}'s of lower magnitude by zeros and solve this abbreviated system in single-precision arithmetic. Such a solution is crude but cheap; if it detects large diagonal decay we are warned of possible disaster when solving the complete system.

It is the *decay* of diagonals that is significant, not their smallness. Small pivots (and large multipliers) *per se* do not provoke large error [15.29]. Even a low value of D_{ii}/K_{ii} does not *guarantee* large error. A case in point is a stiff structure on soft supports but loaded by self-equilibrating forces (see the second paragraph of Section 15.6).

15.9 RESIDUALS AND ITERATIVE IMPROVEMENT

After solving equations $[\mathbf{K}]\{\mathbf{D}\} = \{\mathbf{R}\}$ for $\{\mathbf{D}\}$, we can compute *residuals* $\{\Delta\mathbf{R}\}$.

$$\{\Delta\mathbf{R}\} = \{\mathbf{R}\} - [\mathbf{K}]\{\mathbf{D}\} \tag{15.9.1}$$

If the equation solver introduces no rounding error, then $\{\Delta\mathbf{R}\} = 0$. But if we regard the solution as an approximation $\{\mathbf{D}\}_1$ and assume that the exact solution is $\{\mathbf{D}\}_1 + \{\Delta\mathbf{D}\}_1$, we are led to the following iterative improvement scheme [15.1, 15.24, 15.28].

$$\begin{aligned}
\{\Delta\mathbf{R}\}_1 &= \{\mathbf{R}\} - [\mathbf{K}]\{\mathbf{D}\}_1 \\
[\mathbf{K}]\{\Delta\mathbf{D}\}_1 &= \{\Delta\mathbf{R}\}_1, \qquad \text{solve for } \{\Delta\mathbf{D}\}_1 \\
\{\mathbf{D}\}_2 &= \{\mathbf{D}\}_1 + \{\Delta\mathbf{D}\}_1 \\
[\mathbf{K}]\{\Delta\mathbf{D}\}_2 &= \{\Delta\mathbf{R}\}_2 = \{\mathbf{R}\} - [\mathbf{K}]\{\mathbf{D}\}_2, \qquad \text{solve for } \{\Delta\mathbf{D}\}_2 \\
\{\mathbf{D}\}_3 &= \{\mathbf{D}\}_2 + \{\Delta\mathbf{D}\}_2, \qquad \text{and so on}
\end{aligned} \tag{15.9.2}$$

(This scheme resembles constant-stiffness iterative methods used in nonlinear problems.) Ill-conditioned equations converge slowly. Failure to converge indicates that $\{\mathbf{D}\}_1$ is worthless.

A scalar measure of error is

$$e = \frac{\{\mathbf{D}\}_1^T\{\Delta\mathbf{R}\}_1}{\{\mathbf{D}\}_1^T\{\mathbf{R}\}} \tag{15.9.3}$$

Physically, e represents the ratio of work done by residual forces to work done by applied forces. If $|e|$ is (say) 10^{-8} or less, the solution is probably satisfactory. The algorithm of Eq. 15.9.2 diverges if $|e|$ exceeds unity [15.27].

Truncation error is not reduced unless residuals are calculated to greater precision than used in computing the successive approximations $\{\mathbf{D}\}_i$. For example, in Eqs. 15.9.2 we could accumulate each $\{\Delta\mathbf{R}\}$ in double precision but do all other processing in single precision. Thus we assume that $[\mathbf{K}]$ is sufficiently accurate in single precision and that the rounding error of equation solving is to be reduced. Truncation error is *not* reduced. Accordingly, successive $\{\mathbf{D}\}_i$ are improved solutions of the finite element model if rounding error is both significant and not partially compensated by truncation error in the original system of equations.

The following iterative improvement scheme reduces both rounding error *and* truncation error [15.19]. Generate $[\mathbf{K}]$ in double precision and duplicate it in a mass storage file. Truncate $[\mathbf{K}]$ and solve $[\mathbf{K}]\{\mathbf{D}\} = \{\mathbf{R}\}$ in single precision. Recall the double precision $[\mathbf{K}]$ from mass storage and use it to compute residuals $\{\Delta\mathbf{R}\}$ in double precision. Truncate $\{\Delta\mathbf{R}\}$ to single precision and use it with the previously reduced single precision $[\mathbf{K}]$ to compute the next $\{\Delta\mathbf{D}\}$ in single precision.

A small residual does not guarantee accuracy. Of two approximate solutions, it is possible for the more exact to have the larger residual [15.24]. Accordingly, great faith should not be placed in the *equilibrium check*, in which we ask if each node is in static equilibrium under loads applied to it. These loads include external forces and forces computed from strains in elements that share the node. In another form of equilibrium check, computed stresses are integrated to see if forces and moments they apply to a cross section are in equilibrium with applied loads. *Near-satisfaction of equilibrium is no guarantee that stresses and displacements are accurate*. But failure to satisfy equilibrium warns that something is wrong. That is, small residuals are a necessary but not a sufficient condition for accuracy.

Truncation error, rounding error, and residuals can all be zero and yet results can be worthless because of bad idealization. In Fig. 15.9.1 the mesh is so coarse that computed

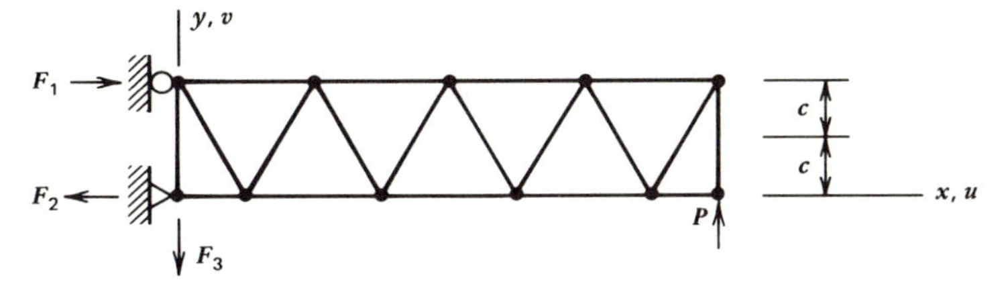

Figure 15.9.1. Cantilever beam modeled by constant strain triangles.

displacements {**D**} are far below exact values. Yet, because [**K**] is too high, the product [**K**] {**D**} is correct: nodes are in equilibrium and reactions F_i agree with values predicted by elementary statics.

An iterative improvement scheme that reduces discretization error is discussed in connection with Eq. 4.10.4.

15.10 CONCLUDING REMARKS

In this chapter we have discussed the testing of a new element. We have also discussed how poor modeling can produce manipulation error in subsequent processing. However, poor discretization does not always produce numerical difficulty. Nor does the absence of numerical difficulty certify that the results mirror physical reality. Similarly, of the tests discussed (condition number, size of residual, and so on), none is by itself fully reliable in predicting trouble, and none can show that results are correct. Tests can detect errors but cannot prove their absence.

Again, physical situations that make error make likely include elements with great distortion or large aspect ratio, elements with much higher membrane or shear stiffness than bending stiffness, stiff elements used to force nodes to move together, and a very fine mesh. Each is an instance of a stiff region supported by a flexible region.

The cantilever beam is a common structure that is particularly susceptible to error. Loads are apt to produce large rigid-body motion but small strain in elements near the tip.

The test for diagonal decay is recommended as easy to program and inexpensive in storage space and computer time. Partial double precision in equation solving and full double-precision solution of equations generated in single precision are discouraged as profitless in most cases. The Hilbert matrix is a good test case for an equation solver.

It is risky to generalize from a single example. Yet it is interesting that practical problems with $1.5(10)^6$ d.o.f. have been solved with good engineering accuracy.

If a set of equations is badly ill-conditioned it is better to reformulate the problem than to make heroic attempts to extract a good solution: "If a thing is not worth doing, it is not worth doing well" [15.24]. Ill-conditioning can be reduced by changing element stiffnesses and aspect ratios, resequencing equations, and so on.

The following sections in other chapters also relate to errors, accuracy, and blunders: 1.5, 1.8, 2.9, 4.3, 4.6, 4.8–4.10, 5.8–5.12, 7.6, 7.8, 7.12, 7.13, 10.2, 10.4, 11.3, 11.9, 12.5, 13.1, 14.2, 14.3, 14.5, 14.7, 16.6.

PROBLEMS

15.1 Verify Eqs. 15.4.1 through 15.4.4.

15.2 Rates of convergence for displacement, stress, and strain energy are noted in Section 15.4. They have the form $O(h^p)$, where p depends on q and m. Consider the elements described by:
(a) Section 4.5.
(b) Figure 5.2.1*b*.
(c) Figure 5.7.1*a*.
(d) Figure 7.7.1.
In each case, is the error measure valid? If so, what is p for the respective cases of displacement, stress, and strain energy?

15.3 Let n^2 be the number of elements in a plane mesh. For meshes $n^2 = 1$, 4, and 16, the respective computed values of deflection at a certain point are 10.4, 11.6, and 11.9 units. Use Eq. 15.4.5 as best you can to predict the exact deflection.

15.4 Why, in Eq. 15.4.6, can $N^{2m/n}$ be replaced by h^{-2m}?

15.5 In the problem of Fig. 13.2.1, what changes in $[\mathbf{K} + \mathbf{K}_\sigma]$ and $\{\mathbf{R}\}$ do you think would be *un*likely to produce great change in the computed $\{\mathbf{D}\}$?

15.6 Let $P = 1$, $k_A = 1.3$, and $k_B = 0.06$ in Fig. 15.5.1. Carry out a Gauss elimination solution for the displacement of P. Assume that the computer stores only two digits and rounds answers after every operation. Compare exact and computed results.

15.7 Repeat Problem 15.6 using $k_A = 0.06$ and $k_B = 1.3$.

15.8 (a) By inspection, what is $C(\mathbf{K})$ if $[\mathbf{K}]$ is diagonal?
(b) How small can $C(\mathbf{K})$ be?
(c) Physically, what is implied if $C(\mathbf{K})$ is infinite?

15.9 (a) Let $k_A = 10$ and $k_B = 1$ in Fig. 15.5.1 and then let $k_A = 1$ and $k_B = 10$. In each case, calculate the condition number of $[\mathbf{K}]$ using the *unscaled* stiffness matrix.
(b) Repeat part (a), but now use the *scaled* stiffness matrix for the two situations.

15.10 In Fig. 15.5.1, let $k_A = 2$ and $k_B = 3$. Show that Eqs. 15.5.1 and 15.6.4 yield the same values of the stiffness matrix and its inverse.

15.11 (a) Use a single element to model a cantilever beam of length L. Calculate $C(\mathbf{K})$.
(b) Use Eq. 15.6.8 and the Rayleigh approximation for λ_{min} suggested below it to estimate $C(\mathbf{K})$ for the problem of part (a).
(c) Repeat part (b), but with reference to the problem of Fig. 15.7.1.

15.12 For each numbering in Fig. 15.7.1, write the first four equations in [**K**]{**D**} = {**R**}. In each case carry out three steps of Gauss elimination. Observe what happens to the diagonal coefficients in [**K**].

15.13 A structure has two d.o.f. and two springs, of stiffness $k = 1$ and $k = 6$. The sketch shows four different arrangements of spring placement and node numbering. Rank them by the order of solution accuracy you expect, going from most accurate to least. State your reason in each case.

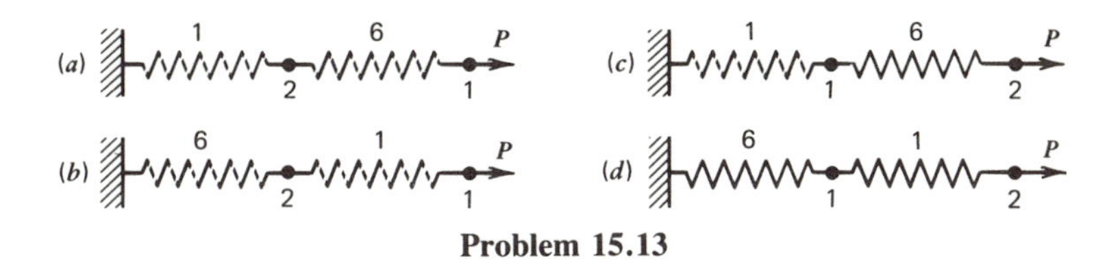

Problem 15.13

15.14 Verify your prediction in Problem 15.13 by solving for displacements $\{u_1 \; u_2\}$ in each of the four cases. Use Gauss elimination, process equations in node-number order, and assume that the machine rounds results to a single digit after each operation. Let $P = 6$.

15.15 Repeat Problem 15.14 using Choleski decomposition. However, round to *two* digits after each operation.

15.16 Suppose that in parts (a) and (b) of this problem the analyst has forgotten to specify any displacement boundary conditions, so the structure is not connected to earth.
(a) A plane frame with three d.o.f. per node.
(b) A solid or revolution with two d.o.f. per node.
In each case what, in theory, is the condition number? And in what equation will a test for diagonal decay detect trouble?

15.17 What Fortran statements should be added, and where, if a test for diagonal decay is to be incorporated in the equation solver of Fig. 2.11.2?

15.18 What convergence criterion would you recommend for the iterative improvement algorithm of Eqs. 15.9.2?

15.19 In Fig. 15.5.1 let $k_A = \frac{1}{3}$, $k_B = \frac{1}{4}$, and $P = 1$. Let the stiffness matrix and its inverse be approximated as follows.

$$[\mathbf{K}] \approx \begin{bmatrix} 0.3 & -0.3 \\ -0.3 & 0.5 \end{bmatrix}, \qquad [\mathbf{K}]^{-1} \approx \begin{bmatrix} 8.0 & 5.0 \\ 5.0 & 5.0 \end{bmatrix}$$

(a) Use the preceding [**K**] and $[\mathbf{K}]^{-1}$ and apply three cycles of iterative improvement (Eqs. 15.9.2). Do displacements converge toward the expected values?
(b) Use the preceding $[\mathbf{K}]^{-1}$ but the *correct* [**K**] in three cycles of the alternate iterative improvement scheme that also reduces truncation error. Do displacements converge toward expected values?

15.20 With mesh refinement, answers may improve, get worse, or not change. For each of these three possibilities, give an example of a problem (or situation, or mesh) that would behave this way and state your reason why.

15.21 Recently one of your co-workers (who has now left the company) wrote a computer program based on an element of her own devising. Her test case gave good results. You try the program with your test case, and the results are terrible. List several possible reasons for the discrepancy.

CHAPTER 16

MISCELLANEOUS TOPICS IN STRUCTURAL MECHANICS

16.1 EQUILIBRIUM, MIXED, AND HYBRID METHODS

Most structural elements in common use are based on an assumed displacement field and the potential energy functional Π_p of Eq. 3.4.9. Other bases and other functionals are possible [16.1]. Some yield elements that are better than displacement elements.

Equilibrium elements are based on an assumed stress field and the complementary energy functional. The stress field must satisfy equilibrium (Eqs. 1.3.3), so it is convenient to begin with a stress function, such as the Airy function ϕ of elasticity theory [16.2]. Generalized *force* d.o.f., not displacement d.o.f., are the primary unknowns. Stresses are continuous across element boundaries. Approximate solutions err by being too flexible, so an exact solution can be bracketed by displacement and equilibrium solutions. A disadvantage of an equilibrium model is that displacements must be obtained by integration of the stress field. Accordingly, the commonplace zero-displacement boundary condition is very difficult to handle. Also, a stress function for a multiply-connected region must be constrained to make the function single valued.

Equilibrium and displacement models have a dual relationship. That is, equations and principles in equilibrium analysis have analogues in displacement analysis. An example is the slab analogy: the displacement field w that solves a plate bending problem differs only by a constant from the stress function ϕ that solves a plane elasticity problem, because w and ϕ both satisfy the biharmonic equation. From the duality viewpoint, the library of equilibrium elements is also a library of displacement elements for plane stress, plate, and shell problems [16.3, 16.4].

Mixed elements have as primary unknowns both force *and* displacement d.o.f. Since neither is given preferential treatment, they may be computed with comparable accuracy. No bound is given: the finite element model may be too stiff or too flexible. Like equilibrium elements, mixed elements cannot be attached to displacement elements. The coefficient matrix of a mixed element is indefinite, so the equation solver must be chosen with care. Also, a mixed model may require more d.o.f. than a displacement model to achieve a given accuracy. For an example of a mixed element, imagine that Fig. 7.9.2*a* represents a plate bending element: the mixed element has six d.o.f.—lateral displace ment at the corners and bending moment at the midsides [16.5]. A mixed beam element is discussed in Section 18.8 and in Problems 17.20, 18.11, 18.13, and 18.14. For mixed elements, see Refs. 16.5 to 16.8.

A displacement formulation that uses Lagrange multipliers (Section 6.9) can be called a mixed formulation. Like side moments in a mixed plate element, the multipliers are interelement force quantities that constrain elements to be compatible.

Hybrid elements can be of various types. The most successful is the assumed-stress hybrid, which is based on an assumed stress field within the element and assumed displacement patterns on its boundaries. The two assumptions are independent. Nodal d.o.f. are displacement quantities. Therefore hybrid elements can be attached to displacement elements.

A plane hybrid element, for example, can be based on the assumed stress field

$$\begin{aligned} \sigma_x &= \beta_1 + \beta_4 x + \beta_7 y + \beta_{10} x^2 + \beta_{13} xy + \cdots \\ \sigma_y &= \beta_2 + \beta_5 x + \beta_8 y + \beta_{11} x^2 + \beta_{14} xy + \cdots \\ \tau_{xy} &= \beta_3 + \beta_6 x + \beta_9 y + \beta_{12} x^2 + \beta_{15} xy + \cdots \end{aligned} \tag{16.1.1}$$

where the β_i are generalized coordinates. A finite (and usually small) number of the β_i are used in formulating an element. The β_i are not all independent because stresses must satisfy equilibrium (Eqs. 1.3.3). None of the β_i appear in the final element stiffness equation $[\mathbf{k}]\{\mathbf{d}\} = \{\mathbf{r}\}$. Displacement along an edge is governed by d.o.f. on only that edge. For example, along the lower edge of the element in Fig. 4.6.1*a*, we can write

$$u = \frac{b-x}{2b} u_1 + \frac{b+x}{2b} u_2 \qquad v = \frac{b-x}{2b} v_1 + \frac{b+x}{2b} v_2 \tag{16.1.2}$$

Three additional pairs of equations account for the three additional edges.[1]

Assumed-stress hybrid elements become stiffer as more β_i are used but more flexible as edges are permitted more complicated displacement patterns. Thus a hybrid element structure may be either too stiff or too flexible. Hybrid elements have been particularly effective for plate bending problems. It is easy to include transverse shear deformation capability. Hybrid elements deserve to be more widely used. See Ref. 16.1 and 16.9 to 16.16.

New mass matrices $[\mathbf{m}]$ and stress stiffness matrices $[\mathbf{k}_\sigma]$ must be derived if consistent formulations for equilibrium, mixed, or hybrid elements are demanded. But apart from mathematical esthetics, there is no reason not to use displacement-based $[\mathbf{m}]$ and $[\mathbf{k}_\sigma]$ matrices with hybrid elements.

[1]A good element is based on Eqs. 16.1.2 and $\sigma_x = \beta_1 + \beta_7 y$, $\sigma_y = \beta_2 + \beta_5 x$, $\tau_{xy} = \beta_3$, with all other β_i zero. The element is identical to element QM6 of Fig. 7.7.3 if rectangular and almost as good as QM6 if generally shaped [16.13, 16.63].

16.2 STRAIGHT BEAMS AND CURVED BEAMS

Beams and girders serve as structures in their own right. Beams also serve as ribs and edge stiffeners for plates and shells. So much can be done with these essentially one-dimensional elements that a book could be devoted to them. Unfortunately, their methodology involves lengthy expressions and tedious manipulations that are necessary but provide little insight. Here we offer only a synopsis and references for beams and beamlike elements.

Three methods of deriving element matrices for straight or curved beams have been frequently used. In one method, governing differential equations of the problem are integrated. A particular solution for a specific load or displacement condition yields a coefficient in the element stiffness or flexibility matrix. In a second method, Castigliano's complementary energy theorem is used with equilibrium equations to find displacements produced by a specific load. Thus flexibility coefficients are produced. The flexibility matrix is inverted to yield a stiffness matrix. The third method is the familiar one of assumed displacement fields. Element properties are often improved if nodeless d.o.f. are added (Table 10.3.1). But the added d.o.f. may damage compatibility between the beam and an adjacent plate or shell element. The compatibility question may be of little consequence—indeed, it may not even come to mind if one uses the first or second method of element formulation.

Equation 4.4.5 gives the stiffness matrix of a straight, uniform beam element. If modified to account for transverse shear deformation, this matrix becomes [16.17, 16.18]

$$[\mathbf{k}] = \frac{EI}{L^3 + 12Lg}\begin{bmatrix} 12 & 6L & -12 & 6L \\ 6L & 4L^2 + 12g & -6L & 2L^2 - 12g \\ -12 & -6L & 12 & -6L \\ 6L & 2L^2 - 12g & -6L & 4L^2 + 12g \end{bmatrix} \tag{16.2.1}$$

where $g = nEI/AG$, A = cross-sectional area, G = shear modulus, and n = shape factor ($n = 6/5$ for a rectangular cross section). Nodal d.o.f. θ_1 and θ_2 must now be regarded as rotations of lines initially normal to the beam axis, not as slopes $w_{,x1}$ and $w_{,x2}$. As the beam becomes more slender, g approaches zero, Eq. 16.2.1 reduces to the conventional $[\mathbf{k}]$, and $\theta_1 = w_{,x1}$ and $\theta_2 = w_{,x2}$.

Beam elements considered in Figs. 9.5.1 and 9.5.2 also include transverse shear deformation. The former element has two internal d.o.f. that must be condensed to yield a 4-by-4 matrix $[\mathbf{k}]$. The latter element cannot model a linearly varying moment.

In dynamic problems, transverse shear deformation and rotary inertia can significantly affect the higher modes. The Timoshenko beam includes these effects [16.19].

Finite elements are available for many beam geometries. Published papers often include explicit expressions for element matrices $[\mathbf{k}]$, $[\mathbf{m}]$, and even $[\mathbf{k}_\sigma]$. The element menu includes beams that are straight, tapered, vertically curved, horizontally curved, solid, thin walled, or pretwisted. Sometimes these properties appear in combination. Static, dynamic, and stability problems have been considered. References 9.12, 9.17, 10.3, 10.4, 10.7, 10.32, 10.36, 11.4, 11.46, 11.48, 12.4, 12.7 to 12.9, and 16.17 to 16.33 are sampling of the literature.

Torque caused by restraint of warping can be significant in thin-walled and open sections. An extra d.o.f. is needed to account for this effect. Thus there are more than the six d.o.f. per node that are usual in a three-dimensional problem.

A curved beam can serve as a preliminary design model for a horizontally curved box-girder bridge. More accurate analysis requires that the geometry be better modeled by assemblages of horizontally curved elements [16.30, 16.31].

Sometimes a beam element must be connected to an element that has no rotational d.o.f. The connection can be made by the methods of Chapter 6 or by a transition element [16.33].

16.3 ELASTIC FOUNDATIONS

A nonrigid foundation, such as soil, influences structural behavior. Often we want to model the effect of a foundation on the structure but are not concerned with details of stress or deformation in the foundation itself. Toward this end, various foundation models have been proposed [16.34–16.39].

The simplest elastic foundation is made of discrete springs that connect structure nodes to a rigid base. A spring can resist axial and/or angular displacement. Each spring can be modeled as a two-node bar that resists axial strain and/or twisting. The base node is fixed; thus active d.o.f. of the element are only those at the structure end. In this sense the element has only one node (although the base node can be retained to define element orientation). Such an element is called a *scalar element*. During the element-assembly process, spring stiffness associated with the ith d.o.f. is added to the diagonal coefficient K_{ii} of the spring-supported structure.

The next more complicated model is the Winkler foundation (Fig. 16.3.1a). It "smears" discrete springs to form a continuous base. If a differential area $dx\,dy$ of the foundation surface is given deflection w, it resists with a force $dF = \beta w\,dx\,dy$. Here β is the foundation modulus. It has units of pressure per unit displacement. Let displacement w over an area A of the foundation surface be interpolated from nodes on A according to $w = \lfloor \mathbf{N} \rfloor \{\mathbf{d}\}$. By analogy with the strain energy $U = (kd)d/2$ in a linear spring, strain energy in the foundation is

$$U = \frac{1}{2}\iint w(\beta w\,dx\,dy) = \frac{1}{2}\{\mathbf{d}\}^T[\mathbf{k}_f]\{\mathbf{d}\} \tag{16.3.1}$$

where $[\mathbf{k}_f]$ is the stiffness matrix of the Winkler foundation.

$$[\mathbf{k}_f] = \iint \beta \lfloor \mathbf{N} \rfloor^T \lfloor \mathbf{N} \rfloor\,dx\,dy \tag{16.3.2}$$

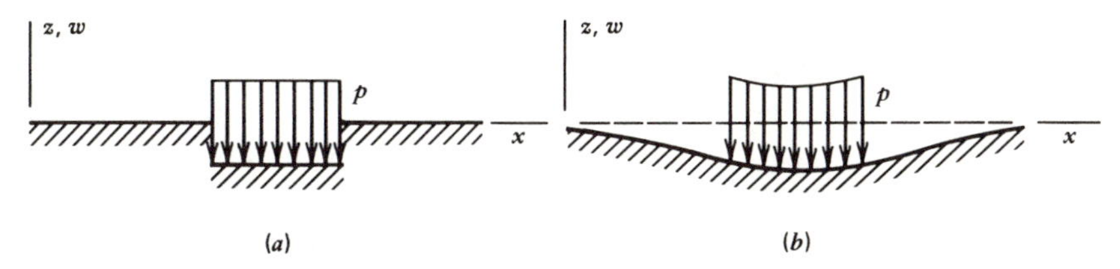

Figure 16.3.1. Deflections of elastic foundations under uniform pressure p. (a) Winkler foundation. (b) Elastic solid.

Matrix $[\mathbf{k}_f]$ can be added in the usual way to the matrix $[\mathbf{K}]$ of the rest of the structure. If only foundation matrices are assembled, to yield $[\mathbf{K}_f] = \Sigma[\mathbf{k}_f]$, we find that for a Winkler foundation $[\mathbf{K}_f]$ is sparse (and banded if the mesh is suitably numbered).

Note that $[\mathbf{k}_f]$ and the mass matrix $[\mathbf{m}]$ of Eq. 11.2.3 have the same form. As with an element mass matrix, any of a variety of shape functions can be used to construct $[\mathbf{k}_f]$. A "consistent" $[\mathbf{k}_f]$ results when $w = \lfloor \mathbf{N} \rfloor \{\mathbf{d}\}$ matches the w field of the structure element that contacts the foundation. The lumped mass matrix has its analogy in assigning a fraction of βA to each w d.o.f. of an element of area A. This reduces the foundation model to a set of linear springs and makes $[\mathbf{k}_f]$ a diagonal matrix.

An elastic half-space foundation (Fig. 16.3.1b) yields a *full* matrix $[\mathbf{k}_f]$. This happens because a displacement pattern in which only one node on the foundation surface has nonzero displacement w requires that forces be applied to *all* nodes. A banded $[\mathbf{k}_f]$ can be achieved by adopting the approximation that nodes separated by more than (say) a one-element span do not interact with one another [16.38].

There is evidence that neither foundation model in Fig. 16.3.1 is adequate for soil. Additional models are available. These include the Pasternak models, which use two or three coefficients to describe foundation properties [16.35–16.37].

Vogt applied distributed loads to a rectangular area on the surface of an elastic half-space foundation and computed the resulting average displacements and rotations of the area. Thus he derived force-deflection influence coefficients. These coefficients can be used to generate Vogt foundation elements useful in the analysis of arch dams [16.39].

On occasion, a linear analysis demands a deflected shape such that the foundation must pull on the structure instead of push against it. If negative foundation pressure cannot, in fact, be sustained, the problem becomes nonlinear because the foundation contact area must be determined by an iterative calculation sequence.

16.4 INCREMENTAL CONSTRUCTION

In concept, a linear stress analysis proceeds as if the structure were assembled in a stress-free condition, followed by the loads being "switched on." Actually, as a large structure is built, stresses accumulate during construction: each piece adds load to a structure already loaded by preceding parts [16.40].

Incremental construction can be analyzed as follows. Consider, the construction of an earthen embankment. Model the first layer by finite elements and analyze it under its own weight. Add a second layer and analyze the two-layer structure under the weight of the second layer only. Superpose the stress and displacement results of the two analyses. Add a third layer and analyze the three-layer structure under the weight of the third layer only, and so on. In general, in a structure of N layers, N analyses are superposed to find the final state of layer 1, $N - 1$ analyses are superposed to find the final state of layer 2, and so on. The number of layers used in analysis can be much smaller than the number used in actual construction [16.40].

Because each layer adds to the existing structure but does not otherwise change its stiffness (unless material properties are nonlinear), an added layer does not demand complete reanalysis (see Section 16.5).

In comparison with conventional gravity-switch-on analysis, incremental analysis may display similar stresses but different displacements. However, it is possible for conventional analysis to give negative backfill pressures. This is clearly an unsatisfactory result [16.41].

If material properties are linear, the final state of stress around an excavation does not depend on the excavation sequence [16.42].

16.5 REANALYSIS AFTER STRUCTURAL MODIFICATION

We presume that a trial design, or perhaps the first analysis in an iterative solution of a nonlinear problem, has produced the usual stiffness equations

$$[\mathbf{K}]\{\mathbf{D}\} = \{\mathbf{R}\} \tag{16.5.1}$$

Now, after $\{\mathbf{D}\}$ is computed, the structure is changed. Stiffnesses and hence displacements are altered. Equation 16.5.1 becomes

$$[\mathbf{K}^*]\{\mathbf{D}^*\} = \{\mathbf{R}\}, \quad \text{where} \quad [\mathbf{K}^*] = [\mathbf{K}] + [\Delta\mathbf{K}] \quad \text{and} \quad \{\mathbf{D}^*\} = \{\mathbf{D}\} + \{\Delta\mathbf{D}\} \tag{16.5.2}$$

We want to find the new displacements $\{\mathbf{D}^*\}$, if possible with less effort than required for a *complete* reanalysis, which requires building all of $[\mathbf{K}^*]$ and solving Eq. 16.5.2 for $\{\mathbf{D}^*\}$. In vibration problems, the analogous task is to compute modified frequencies without completely redoing the eigenvalue extraction. We will call this the *partial reanalysis* problem. Many methods have been proposed—89 references are listed in a survey paper [16.43]. In this section we introduce the subject in the context of static analysis.

Partial reanalysis can yield exact results. For example, we can operate on the Choleski factor of $[\mathbf{K}]$ and produce the exact Choleski factor of $[\mathbf{K}^*]$. If N is the number of rows (or columns) in $[\mathbf{K}]$ that are changed, the break-even point in cost between partial and complete reanalysis is about $N = 0.35B$, where B is the semibandwidth of [**K**]. So the method is a good one only if there are few changes [16.43–16.45].

Another method of partial reanalysis yields an approximate $\{\mathbf{D}^*\}$ that can be iteratively improved. From Eq. 16.5.2,

$$[\mathbf{K}]\{\mathbf{D}^*\}_{i+1} = \{\mathbf{R}\} - [\Delta\mathbf{K}]\{\mathbf{D}^*\}_i \tag{16.5.3}$$

where, for the first iteration, $\{\mathbf{D}^*\}_0 = \{\mathbf{D}\}$. Matrix $[\mathbf{K}^*]$ is neither formed nor reduced. It is efficient to assemble $[\Delta\mathbf{K}]\{\mathbf{D}^*\}_i$ on the element level instead of on the structural level.

$$[\Delta\mathbf{K}]\{\mathbf{D}^*\}_i = \sum[\Delta\mathbf{k}]\{\mathbf{d}^*\}_i \tag{16.5.4}$$

where the sum extends over only the altered elements. The similarity of these calculations to Eq. 13.10.8 is evident. Iteration is usually efficient but, if changes are large or extensive, the process may converge slowly or diverge [16.43, 16.46].

Structural changes may be confined to a small region of the structure. Then partial reanalysis can be done more efficiently than otherwise. Reduction of the right side of Eq. 16.5.3 can begin near the end of the vector if changes are associated with only the highest-numbered d.o.f. With substructuring, if changes are confined to one substructure, condensed matrices of all other substructures need not be changed. If $[\mathbf{K}^*]$ is formed, by revising only the lower part of [**K**], the reduced [**K**] and the reduced $[\mathbf{K}^*]$ are identical in their upper parts. Therefore only the lower part of $[\mathbf{K}^*]$ need be reduced [16.47]. (Forming and reducing *all* of $[\mathbf{K}^*]$ is a complete reanalysis. Reference 16.48 suggests that this is usually more efficient than revising an existing solution.)

In redesign, we can borrow a procedure from photoelasticity. To design (say) a fillet of optimum shape, we cut away material that carries low stress until the boundary of the fillet coincides with an isochromatic fringe. The analogous analytical procedure is to reshape lightly stressed boundaries until the boundary coincides with a principal stress contour.

16.6 INCOMPRESSIBLE MEDIA

As Poisson's ratio ν approaches 0.5, a material becomes incompressible. Values of ν near 0.5 occur in rubberlike materials and in materials that flow, such as fluids and plastic solids. Unless the problem is one of plane stress, $\nu = 0.5$ is forbidden because denominators in Eqs. 1.5.6, 1.5.8, 1.6.3, 1.6.5, and 8.2.3 become zero. It is tempting to approximate incompressibility by using (say) $\nu = 0.49$. But, near $\nu = 0.5$, stresses are strongly dependent on ν—stresses may double as ν goes from 0.48 to 0.50 [16.49]. Also, structural equations become ill-conditioned as ν approaches 0.5, for reasons explained next.

The shear modulus G and bulk modulus B are

$$G = \frac{E}{2(1 + \nu)} \qquad B = \frac{E}{3(1 - 2\nu)} \tag{16.6.1}$$

In terms of G and B, $[\mathbf{E}]$ of Eq. 1.5.6 is

$$[\mathbf{E}] = G\begin{bmatrix} 2 & 0 & 0 & 0 & 0 & 0 \\ 0 & 2 & 0 & 0 & 0 & 0 \\ 0 & 0 & 2 & 0 & 0 & 0 \\ 0 & 0 & 0 & 1 & 0 & 0 \\ 0 & 0 & 0 & 0 & 1 & 0 \\ 0 & 0 & 0 & 0 & 0 & 1 \end{bmatrix} + \left(B - \frac{2G}{3}\right)\begin{bmatrix} 1 & 1 & 1 & 0 & 0 & 0 \\ 1 & 1 & 1 & 0 & 0 & 0 \\ 1 & 1 & 1 & 0 & 0 & 0 \\ 0 & 0 & 0 & 0 & 0 & 0 \\ 0 & 0 & 0 & 0 & 0 & 0 \\ 0 & 0 & 0 & 0 & 0 & 0 \end{bmatrix} \tag{16.6.2}$$

or, abbreviated, $[\mathbf{E}] = G[\mathbf{E}_G] + \alpha[\mathbf{E}_B]$, where $\alpha = (B - 2G/3)$. The element stiffness matrix (Eq. 4.3.4) becomes

$$[\mathbf{k}] = G\int_V [\mathbf{B}]^T[\mathbf{E}_G][\mathbf{B}]\, dV + \alpha\int_V [\mathbf{B}]^T[\mathbf{E}_B][\mathbf{B}]\, dV \tag{16.6.3}$$

Therefore structural equations have the form

$$(G[\mathbf{K}_G] + \alpha[\mathbf{K}_B])\{\mathbf{D}\} = \{\mathbf{R}\} \tag{16.6.4}$$

This is the same form as Eq. 6.10.3. Because α approaches infinity as ν approaches 0.5, we see that the denominator $(1 - 2\nu)$ in Eqs. 1.5.6, 1.5.8, and 8.2.3 acts as a penalty number (Section 6.10). As α grows, it enforces the constraint of incompressibility. Accordingly, as ν approaches 0.5, numerical trouble becomes more likely, and finally the mesh "locks"—*unless* $[\mathbf{K}_B]$ *is singular*.

Singularity of $[\mathbf{K}_B]$ can be achieved by underintegration. The rationale, procedure, and possible pitfalls are discussed, in another context, in Section 9.5. To recapitulate, each element added to the structure should bring with it more d.o.f. than constraints, and this is achieved by a low-order integration rule for $[\mathbf{K}_B]$. Despite such integration, values of ν extremely close to 0.5 may have to be artificially reduced, just as shear stiffness is artificially reduced in Section 9.5. The eight-node plane isoparametric element of Fig. 5.6.1*a* is workable, but the Lagrange element noted in Section 7.3 is better. Both should be integrated with a 2-by-2 Gauss rule. The analogous 20-and 21-node solid elements should be integrated with a 2-by-2-by-2 Gauss rule. Limited tests suggest that the QM6 elements of Section 7.7 also work as nearly incompressible elements. Values of ν up to 0.499999 or more are acceptable. Further information appears in Refs. 5.10, 11.53, and 16.50 to 16.52.

Another approach to incompressible media *includes* $\nu = 0.5$ as an acceptable value. It is based on a mixed variational principle, in which mean pressure at nodes joins nodal displacements in the list of primary d.o.f. See Refs. 16.52 to 16.54.

The methods of Sections 6.8 and 6.9 could be used to enforce a zero-volume-change constraint on each element. These approaches are less attractive than those just noted because either there is substantial added processing, or many Lagrange multipliers are added to the list of unknowns.

16.7 FLUID-STRUCTURE INTERACTION

This topic includes offshore structures and dam-reservoir systems loaded by earthquake motion. Problems of flow and shock loading are excluded. This section introduces a voluminous literature [11.53, 16.55–16.60, 16.62].

A simple approach is to model the structure *and* the fluid by finite elements, as in Fig. 16.7.1*a*. Fluid elements are formulated as if they were solid elements, but with special material properties. We take [**E**] from Eq. 16.6.2 but set $G = 0$. Thus, for the plane elements of Fig. 16.7.1*a*,

$$[\mathbf{E}] = B\begin{bmatrix} 1 & 1 & 0 \\ 1 & 1 & 0 \\ 0 & 0 & 0 \end{bmatrix} \tag{16.7.1}$$

which describes a material that resists the volumetric strain $\epsilon_x + \epsilon_y$ but has no resistance to other distortion. The element stiffness matrix is 8 by 8. But it has rank one (and is invariant) if integrated with a one-point Gauss rule. With a one-point rule, a mode associated with zero volume change has zero frequency. This is desirable for direct integration in time, where a proliferation of large eigenvalues is to be avoided.

These "mock fluid" elements of low rank cannot support arbitrary nodal loads. When assembled, their structure stiffness matrix is singular. The fluid mesh has zero-energy deformation modes unless boundary conditions are adequate (they *are* adequate in Fig.

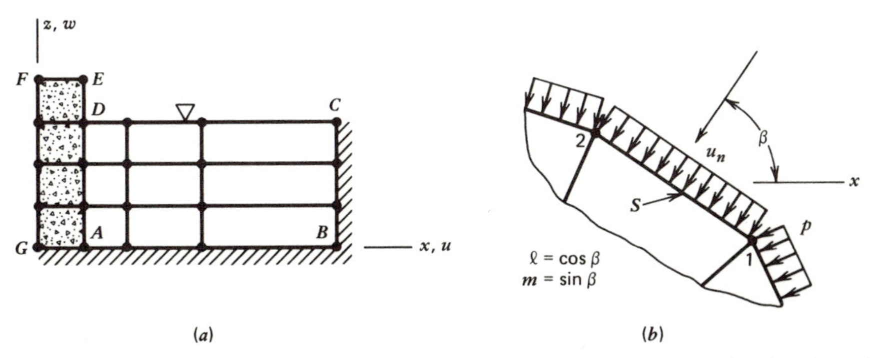

Figure 16.7.1. (*a*) Plane strain idealization of a fluid region *ABCD*, contained by rigid boundaries and a finite element structure *AEFG*. (*b*) Fluid pressure *p* against a *plane* structure. Displacement u_n is perpendicular to the structure surface *S*.

16.7.1*a*). Or, stability can be provided by retaining a small positive G in Eq. 16.6.2; and the "hourglassing" of Fig. 5.10.1*a* that is still possible with a one-point rule can be controlled by adding viscous damping. If zero-energy modes are under control, elements that are statically unstable can still be used for dynamics because the mass matrix keeps the numerical process from coming unstuck. As with solid elements, the mock fluid mass matrix can be consistent, lumped, or otherwise formulated. But if mass condensation is used (Section 11.5), fluid nodes should be slave nodes. If they are masters, low-frequency modes of the system may be fluid-circulation modes, not structure vibration modes.

If fluid *surrounds* a structure instead of being contained *within* it as in Fig. 16.7.1*a*, we have the infinite-medium problem. Suitable techniques are discussed in Sections 4.13 and 7.14.

Another approach to fluid-structure interaction uses fluid mechanics concepts to model the fluid. It is more elegant but more difficult to use than mock fluid elements. In its simplest form, explained next, the method adds a portion of the fluid mass to the structure mass. The added mass is called "virtual" or "hydrodynamic" mass.

Let pressure p act on the structure-fluid interface S. Positive pressure is *compressive* on S and is the same in all directions. If ℓ, m, n, are direction cosines of a normal to S, the surface traction vector of Eq. 1.3.4 becomes

$$\{\mathbf{\Phi}\} = -\, p\{\ell \; m \; n\} \tag{16.7.2}$$

Consider now the contribution of $\{\mathbf{\Phi}\}$ to element nodal forces $\{\mathbf{r}\}$ (Eq. 4.3.5). The contribution of $\{\mathbf{\Phi}\}$ to $\{\mathbf{r}\}$ is

$$\{\mathbf{r}_p\} = -\int_S [\mathbf{N}]^T \{\ell \; m \; n\}\, p \, dS = \int_S \lfloor \bar{\mathbf{N}} \rfloor^T p \, dS \tag{16.7.3}$$

where $\lfloor \bar{\mathbf{N}} \rfloor = -\lfloor \ell \; m \; n \rfloor [\mathbf{N}]$ is a row vector associated with surface-normal displacement. That is, $u_n = \lfloor \bar{\mathbf{N}} \rfloor \{\mathbf{d}\}$ is displacement normal to S, positive *toward* the structure as in Fig. 16.7.1*b*. Next we borrow from Eq. 17.9.7 and add element contributions $\{\mathbf{r}_p\}$ to get the structure force vector $\{\mathbf{R}_p\}$.

$$\{\mathbf{R}_p\} = \sum_{m=1}^{\text{numel}} \int_S \lfloor \bar{\mathbf{N}} \rfloor^T \lfloor \mathbf{N} \rfloor \, dS \, \{\mathbf{p}\} = [\mathbf{F}]\{\mathbf{P}\} \tag{16.7.4}$$

where $\{\mathbf{P}\}$ is the structural vector of nodal pressures and $[\mathbf{F}]$ represents the evaluation and assembly of element integrals. Now, from Eq. 17.9.8,

$$\{\mathbf{B}\} = \sum_{m=1}^{\text{numel}} \int_S \lfloor \mathbf{N} \rfloor^T \rho \lfloor \bar{\mathbf{N}} \rfloor \, dS \, \{\ddot{\mathbf{d}}\} = [\mathbf{Y}]\{\ddot{\mathbf{D}}\} \tag{16.7.5}$$

where $[\boldsymbol{Y}]$ represents the collection of element matrices and $\{\ddot{\mathbf{D}}\}$ is the vector of structure node accelerations. Comparison of Eqs. 16.7.4 and 16.7.5 shows that

$$[\boldsymbol{Y}] = \rho[\mathbf{F}]^T \tag{16.7.6}$$

Also, if surface waves are small and if the fluid is incompressible, Eq. 17.9.10 yields

$$\{\mathbf{P}\} = -[\mathbf{H}]^{-1}\{\mathbf{B}\} \tag{16.7.7}$$

Therefore Eq. 16.7.4 becomes

$$\{\mathbf{R}_p\} = -[\mathbf{F}][\mathbf{H}]^{-1}\{\mathbf{B}\} = -[\mathbf{F}][\mathbf{H}]^{-1}[\mathbf{F}]^T\rho\{\ddot{\mathbf{D}}\} \tag{16.7.8}$$

Equation 11.2.4 is the dynamic equation of the structure alone. We now interpret loads $\{\mathbf{R}\}$ on its right side as representing all forces except forces $\{\mathbf{R}_p\}$ from fluid pressure. So we augment $\{\mathbf{R}\}$ by $\{\mathbf{R}_p\}$ from Eq. 16.7.8 and, after rearrangement, obtain

$$[\mathbf{K}]\{\mathbf{D}\} + [\mathbf{C}]\{\dot{\mathbf{D}}\} + ([\mathbf{M}] + \rho[\mathbf{F}][\mathbf{H}]^{-1}[\mathbf{F}]^T)\{\ddot{\mathbf{D}}\} = \{\mathbf{R}\} \tag{16.7.9}$$

where the (symmetric) matrix that augments $[\mathbf{M}]$ is the virtual mass matrix of the fluid. Every element boundary on the fluid-structure interface contributes to this matrix.

As compared with mock fluid elements, the latter method ignores compressibility but has smaller matrices. If compressibility is included, the governing equation (analogous to Eq. 16.7.9) couples displacement and pressure d.o.f. and has unsymmetric matrices with large bandwidth.

PROBLEMS

16.1 (a) If β_1, β_2, and β_3 are the only nonzero β_i in Eqs. 16.1.1, what defect would you expect to see in the stiffness matrix of a plane quadrilateral element?
(b) If the three stresses in Eqs. 16.1.1 are each to contain constant terms and terms linear in x and y, which of the β_i are needed, and how must they be related?

16.2 A uniform cantilever beam carries force P and moment M at its free end. Find the tip displacement and rotation. Use this *flexibility* information to obtain the lower right 2-by-2 portion of the *stiffness* matrix (Eq. 16.2.1). (Section 7.1 explains how to expand this kernel to obtain the complete $[\mathbf{k}]$.)

16.3 Isoparametric beam elements are noted in the last paragraph of Section 10.5. If used as stiffeners with rigid offsets at their nodes, will the error noted below Eq. 6.7.2 be introduced? Consider the element of Fig. 9.5.1*b*.

16.4 The sketch shows two five-element simply supported beams and two sets of linear springs. The upper set joins nodes of the two beams. The lower set joins nodes of the lower beam to ground. Explain specifically how you would add the stiffness of these springs to a structure [**K**] that operates on d.o.f. of the beams.

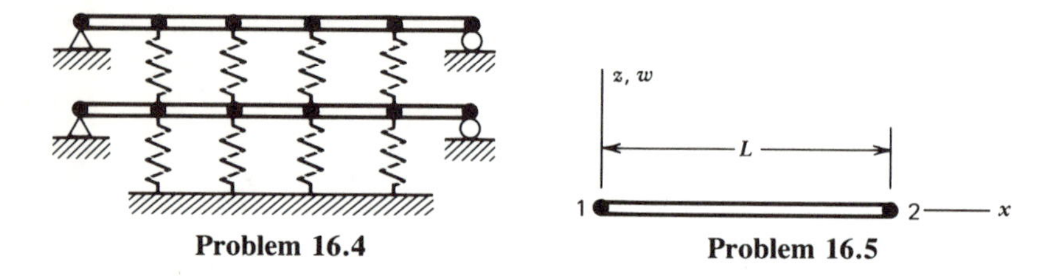

Problem 16.4 **Problem 16.5**

16.5 The bar shown has y-direction width b, is perfectly rigid, and rests on a Winkler elastic foundation. Nodal d.o.f. are w_1 and w_2. Find the 2-by-2 matrix $[\mathbf{k}_f]$.

16.6 A standard four d.o.f. beam element rests on a Winkler elastic foundation. For what end conditions (at the two beam nodes) would it be unacceptable to approximate the foundation as discrete linear springs? For these end conditions, what $\lfloor \mathbf{N} \rfloor$ do you recommend be used in Eq. 16.3.2? Why not use some other $\lfloor \mathbf{N} \rfloor$?

16.7 Consider a triangular area A on the surface of a Winkler elastic foundation. Let the surface-normal displacement w be linearly interpolated over A from corner values w_1, w_2, and w_3. Find the element foundation stiffness matrix $[\mathbf{k}_f]$. Can this $[\mathbf{k}_f]$ be used with a nine d.o.f. plate bending element?

16.8 Imagine that a Winkler elastic foundation has rotational stiffness α (units of force divided by length) as well as translational stiffness β. What formula for $[\mathbf{k}_f]$ replaces Eq. 16.3.2?

16.9 (a) A beam rests on a Winkler elastic foundation and has no other support. Consider a very stiff beam and then a very flexible beam. In each case, apply a concentrated center load downward at midspan. In each case, apply a concentrated center load downward at midspan. Without calculation, sketch your expectation of the deformed configurations of beam and foundation and the distribution of foundation pressure on the beam.

(b) Repeat part (a) for an elastic solid foundation.

16.10 A procedure that generates a diagonal mass matrix is discussed in connection with Eqs. 11.3.4 and 11.3.5.

(a) Can this procedure be used to generate a diagonal $[\mathbf{k}_f]$ for an element of a Winkler elastic foundation? How?

(b) Repeat part (a) but consider an elastic-solid foundation.

16.11 The sketch represents a beam of flexural stiffness EI on a three-parameter Pasternak foundation: two Winkler layers of stiffness β_1 and β_2 are separated by a beam of stiffness S that deforms only in shear. The shear force in this beam is $Sw_{,x}$. Let L be a one-element span and let b be the thickness normal to the xz plane. What is the stiffness matrix of this construction? A symbolic answer, with terms clearly defined, is acceptable. Let $\{\mathbf{d}\} = \{w_1\ \theta_1\ w_2\ \theta_2\ w_3\ w_4\}$.

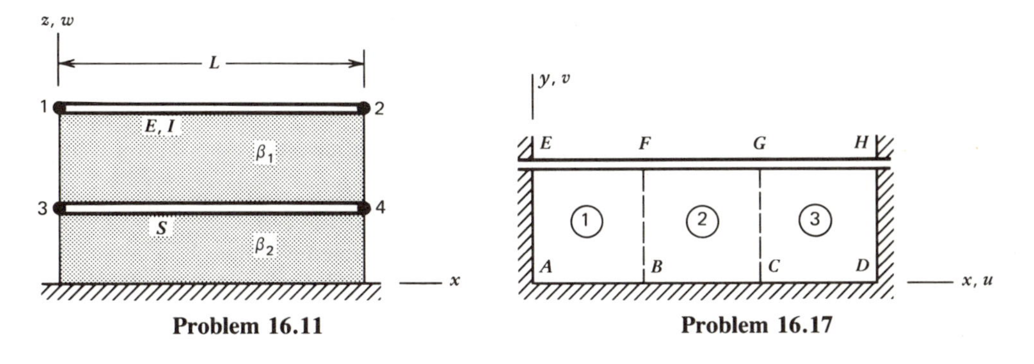

Problem 16.11 **Problem 16.17**

16.12 The foundation valley of an arch dam might be idealized as a Winkler model or as a Vogt model. For which model will displacements of the dam be larger? Why?

16.13 A structure rests on an elastic foundation that sustains only positive contact pressure. In other words, the structure can separate from the foundation. Outline two algorithms for the solution of such a problem. One algorithm should not require repeated reduction of the coefficient matrix.

16.14 An elastic, simply supported beam of length L, width b, and height $6c$ is incrementally constructed in three layers, each of thickness $2c$. The material is concrete, of mass density ρ. The first layer is poured, it sets, and the forms are removed. The second layer is poured and supported by the first while it sets. Then the third is poured and supported by the first two while it sets. Layers bond together as they set. Find the stresses and displacements after each increment of construction. Compare with the results predicted by a "gravity-switch-on" analysis. Use mechanics of materials methods, not finite elements.

16.15 (a) Show that Eq. 16.5.3 can be recast in the form $[\mathbf{K}]\{\Delta\mathbf{D}\}_{i+1} = -[\Delta\mathbf{K}]\{\mathbf{D}^*\}_i$, where $\{\mathbf{D}^*\}_{i+1} = \{\mathbf{D}\} + \{\Delta\mathbf{D}\}_{i+1}$.

(b) Consider a linear spring of stiffness $k = 0.5$ units, loaded by a two-unit static force. Now imagine that the spring stiffness is increased to 0.8. Use the method of part (a) to solve for D^* in this single d.o.f. problem.

(c) Solve the problem of part (b) by Eq. 16.5.3.

(d) How much can the spring stiffness be increased or decreased before convergence fails?

16.16 (a) Show that Eq. 16.6.2 implies incompressibility as ν approaches 0.5.

(b) Under what circumstances will Eq. 16.6.3 approach the constraint condition of Problem 6.33?

16.17 The sketch represents three four-node fluid elements, 1, 2, and 3, like those in region *ABCD* of Fig. 16.7.1*a*. These fluid elements are confined by rigid walls and the three-element, fixed-fixed beam *EFGH*. Motion is confined to the *xy* plane. Answer qualitatively the following.

(a) If the fluid is ignored, sketch the first two vibration modes of the beam. Which represents the higher natural frequency of vibration?

(b) Now replace the fluid and consider the two modes of part (a). What effects act to raise or lower the natural frequencies? Which mode has the higher frequency, and why?

(c) What d.o.f. are active in part (b)? Assume that no condensation to a 2-by-2 eigenvalue problem has been made.

(d) What incompatibility exists? Is convergence to correct answers precluded?

16.18 If a differential element dA of a fluid surface is elevated w units, the mass center of the fluid volume $w\ dA$ is elevated $w/2$ units. The increase in potential is $\int(\rho g w^2/2)\ dA$, where ρg represents weight density. This integral yields a "slosh stiffness matrix" $[\mathbf{k}_s]$, analogous to $[\mathbf{k}_f]$ of Eq. 16.3.2 [16.55, 16.62].

(a) Find $[\mathbf{k}_s]$ for a typical linear edge in Fig. 16.7.1*a*. Element dimensions are b (width), h (height), and t (thickness). There are four nonzero terms in $[\mathbf{k}_s]$.

(b) Use a single rectangular element to compute the slosh frequency of an incompressible fluid of depth h in a rectangular tank of width b. Lump fluid mass at the nodes. *Suggestion*. There are only two nonzero d.o.f., which are related by the incompressibility constraint.

(c) The answer in part (b) is poor mainly because of mass lumping. But what can be done to improve $[\mathbf{k}_s]$?

(d) Solve part (b) again using a cubic surface displacement having zero slope at the walls and defined by two displacement d.o.f. This defines a new $[\mathbf{k}_s]$. Also use a consistent mass matrix, based on the assumption that w within the fluid is directly proportional to z.

16.19 Is fluid viscosity accounted for in the elements of Fig. 16.7.1*a* if a small positive value of G is retained in Eq. 16.6.2?

16.20 (a) Consider the mock fluid elements in Section 16.7 ($G = 0$ in Eq. 16.6.2). Show that the element stiffness matrix integrand can be written in the form $\lfloor \mathbf{B}^+ \rfloor^T B \lfloor \mathbf{B}^+ \rfloor\ dV$, where $\lfloor \mathbf{B}^+ \rfloor$ is a row vector.

(b) For the four-node rectangular elements of Fig. 16.7.1*a*, what is the maximum possible rank of $[\mathbf{k}]$, if $G = 0$ in Eq. 16.6.2?

16.21 If nodal d.o.f. at nodes 1 and 2 in Fig. 16.7.1*b* are $\{u_1\ v_1\ u_2\ v_1\}$, what row vector $\lfloor \bar{\mathbf{N}} \rfloor$ operates on these d.o.f. to yield u_n? Let η be a dimensionless edge-tangent coordinate, such that $\eta = 0$ at node 1 and $\eta = 1$ at node 2.

16.22 A shell roof that carries snow load is supported at discrete points on its edges. How might the influence surface for support reaction be generated? Or the influence surface for bending moment at an internal location [16.61]?

16.23 The sketch depicts a cast iron manhole cover on a supporting ring. Radial ribs are 45° apart, and buttresses B occur every 60°.

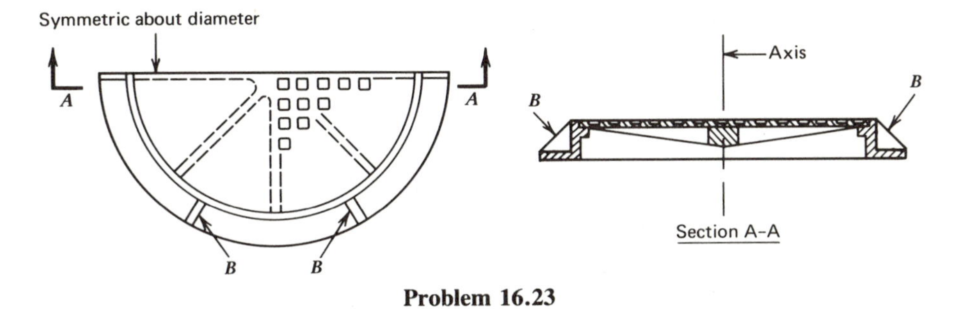

Problem 16.23

(a) Consider making a preliminary analysis with uniform pressure against the cover. Describe a suitable mesh, choice of element types, exploitation of near-symmetry, treatment of supports, and treatment of surface dimples.

(b) How would you change your answer to part (a) if the pressure is not uniform, or if buttresses *B* are massive, or if the cover and ring are not perfectly flat where they meet?

(c) How would you refine the analysis of part (a)?

(d) If the cover and ring come directly from foundry molds, what important contribution to stress has been overlooked?

CHAPTER 17

INTRODUCTION TO CONDUCTION HEAT TRANSFER AND OTHER NONSTRUCTURAL PROBLEMS

17.1 FORMULATIONS FOR HEAT TRANSFER AND OTHER PROBLEMS

This chapter emphasizes the heat conduction problem. However, heat conduction is only one example of many problems described by the *quasiharmonic equation* (Section 17.7). Accordingly, the terminology and procedures of heat transfer are widely applicable.

One reason to use finite elements for heat transfer analysis is that node point temperatures are needed for thermal-stress analysis. Fortunately, a single mesh layout can be used for both problems: a program can read a single data file, compute nodal temperatures, then use them to compute stresses. A finite *difference* analysis for nodal temperatures cannot cope as easily with irregular boundaries and the irregular pattern of node points common in finite element stress analysis.

In preceding chapters we assumed a displacement field and found the finite element formulation from the stationary value of the total potential functional (Section 4.3). A similar approach is effective in nonstructural problems. An assumed field, for temperature or another quantity, is defined element by element in terms of nodal values. The stationary value of the appropriate functional yields matrices analogous to the element stiffness matrix and the load vector. This approach is used, for example, in Section 17.3.

There are problems for which no functional exists. A finite element formulation is still possible by weighted residual methods (Chapter 18).

The following symbols are used in the discussion of heat transfer. The unit of heat is J = 1 joule = 1 N·m.

c = specific heat, J/kg·°C
h = convection heat transfer coefficient (also called film coefficient), J/m^2·s·°C
k = thermal conductivity, J/m·s·°C
q = heat flux per unit area, J/m^2·s
ρ = mass density, kg/m^3
T = temperature, °C
T_∞ = fluid temperature, °C

t = time, s
$\dot{T} = \partial T/\partial t$

Unless stated otherwise, we assume that c, h, k, and ρ are independent of temperature.

17.2 EQUATIONS FOR HEAT CONDUCTION IN A PLANE

The Fourier heat conduction equation is

$$q = -k\frac{\partial T}{\partial s} \tag{17.2.1}$$

It states that heat flux q in direction s is proportional to the gradient of temperature in direction s. The negative sign indicates that heat flow is opposite to the direction of temperature increase.

If the material is thermally orthotropic (Fig. 17.2.1), Eq. 17.2.1 yields

$$\{q_{x'} \ \ q_{y'}\} = -\lceil k_{x'} \ \ k_{y'} \rfloor \{T_{,x'} \ \ T_{,y'}\} \tag{17.2.2}$$

where $\lceil k_{x'} \ k_{y'} \rfloor$ is a diagonal matrix of principal conductivities and $q_{x'}$ and $q_{y'}$ are heat fluxes in the principal directions. From the chain rule of differentiation,

$$\{T_{,x'} \ \ T_{,y'}\} = [\boldsymbol{\Lambda}] \{T_{,x} \ \ T_{,y}\} \tag{17.2.3}$$

where $\Lambda_{11} = \partial x/\partial x' = \cos\beta$, $\Lambda_{12} = \partial y/\partial x' = \sin\beta$, $\Lambda_{21} = \partial x/\partial y' = -\sin\beta$, and $\Lambda_{22} = \partial y/\partial y' = \cos\beta$. Heat flux q is a vector and transforms like displacement (Section 6.3).

$$\{q_x \ \ q_y\} = [\boldsymbol{\Lambda}]^T \{q_{x'} \ \ q_{y'}\} \tag{17.2.4}$$

Equations 17.2.2 through 17.2.4 yield the relation between x- and y-direction heat fluxes and the corresponding temperature gradients.

$$\{q_x \ \ q_y\} = -[\mathbf{k}] \{T_{,x} \ \ T_{,y}\} \tag{17.2.5}$$

where

$$\underset{2\times 2}{[\mathbf{k}]} = \begin{bmatrix} k_x & k_{xy} \\ k_{xy} & k_y \end{bmatrix} = [\boldsymbol{\Lambda}]^T \begin{bmatrix} k_{x'} & 0 \\ 0 & k_{y'} \end{bmatrix} [\boldsymbol{\Lambda}] \tag{17.2.6}$$

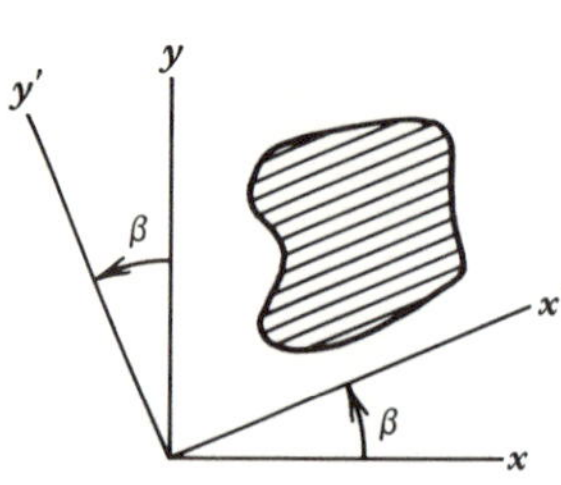

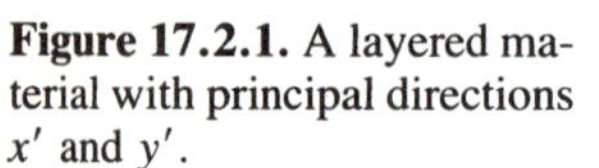
Figure 17.2.1. A layered material with principal directions x' and y'.

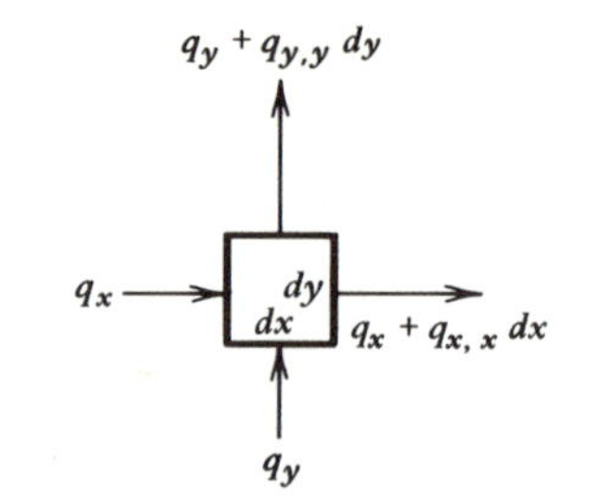

Figure 17.2.2. Heat flow associated with a differential element of volume.

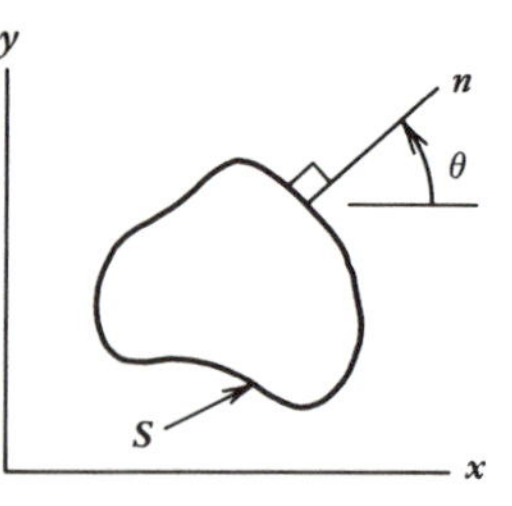

Figure 17.2.3. Plane region A with outward normal n on its boundary S.

The net flow into a region is (flow in) − (flow out). The contribution to net flow from flux across element boundaries (Fig. 17.2.2) is

$$[(q_x\,dy + q_y\,dx)\tau] - [(q_x + q_{x,x}\,dx)\,dy\,\tau] - [(q_y + q_{y,y}\,dy)\,dx\,\tau] \\ = (-q_{x,x} - q_{y,y})\,dx\,dy\,\tau \qquad (17.2.7)$$

where τ is the *constant* thickness of the material. Let Q represent internally generated heat flow, J/s/m³. Sources of Q include dialectic heating and flow into the interior of a plane region across its lateral surfaces. With Eq. 17.2.7, the net inward flow is

$$(Q - q_{x,x} - q_{y,y})\,dx\,dy\,\tau \qquad \text{(J/s)} \qquad (17.2.8)$$

This flow produces a time rate of change of stored energy

$$c\,\rho\,dx\,dy\,\tau\,\dot{T} \qquad \text{(J/s)} \qquad (17.2.9)$$

Expressions 17.2.8 and 17.2.9 are equated, and Eq. 17.2.5 is substituted. Thus

$$\frac{\partial}{\partial x}(k_xT_{,x} + k_{xy}T_{,y}) + \frac{\partial}{\partial y}(k_{xy}T_{,x} + k_yT_{,y}) + Q = c\rho\dot{T} \qquad (17.2.10)$$

If the medium is homogeneous and isotropic, Eq, 17.2.10 becomes

$$k(T_{,xx} + T_{,yy}) + Q = c\rho\dot{T} \qquad (17.2.11)$$

where $k_{xy} = 0$ and $k_x = k_y = k$, a constant. If, in addition, $Q = 0$ and a steady state prevails, we obtain Laplace's equation, $T_{,xx} + T_{,yy} = 0$.

The plane heat conduction problem is to solve Eq. 17.2.10 for $T = T(x, y, t)$ in area A, subject to boundary conditions on S and initial conditions at time $t = 0$ (Fig. 17.2.3). Boundary conditions take various forms, as follows.

1. Prescribed temperature $T = T^*$ on portion S_D of S. (This is a *first* or *Dirichlet condition:* the dependent variable has prescribed values on the boundary.)
2. Prescribed flux $q = q^*$ on portion S_N of S. (This is a *second* or *Neumann condition:* the normal derivative of the dependent variable has prescribed values on the boundary.) Here q^* is normal to the boundary and positive outward. The temperature gradient on S_N in an isotropic material adjusts itself to satisfy the Fourier equation $q^* = -kT_{,n}$. From Eqs. 17.2.4 and 17.2.5 and Fig. 17.2.3, the flux equation for an orthotropic material in terms of temperature gradients is

$$\begin{aligned} q^* = & -(k_x T_{,x} + k_{xy} T_{,y}) \cos\theta \\ & -(k_{xy} T_{,x} + k_y T_{,y}) \sin\theta \end{aligned} \tag{17.2.12}$$

   On an insulated boundary, $q^* = 0$.
3. Prescribed convection conditions on portion S_C of S. (This is a *third* or *mixed* or *Cauchy condition:* the dependent variable *and* its derivative have a prescribed relation on the boundary.) Equations written for Condition 2 again apply if q^* is replaced by $h(T - T_\infty.)$

Boundary S may consist entirely of S_D, S_N, S_C, or any two or three of these in any sequence.

A functional Π_T for the problem is

$$\begin{aligned} \Pi_T = & \frac{1}{2} \iint \left(\begin{Bmatrix} T_{,x} \\ T_{,y} \end{Bmatrix}^T [\mathbf{k}] \begin{Bmatrix} T_{,x} \\ T_{,y} \end{Bmatrix} - 2QT + 2\,\rho c T\dot{T} \right) dx\,dy \\ & + \int_{S_N} q^* T\,dS + \int_{S_C} \frac{h}{2} (T^2 - 2TT_\infty)\,dS \end{aligned} \tag{17.2.13}$$

where dS is an increment of length on S. At every instant of time, the temperature field must adjust itself so that $\delta\Pi_T = 0$. Quantities T, $T_{,x}$, and $T_{,y}$ are subject to variation, but $\dot{T}$ is not. Hence calculus of variations produces Eqs. 17.2.10 and nonessential boundary conditions (preceding conditions 2 and 3) from the stationary condition $\delta\Pi_T = 0$. Readers familiar with the mathematics can verify this statement.

17.3 A FINITE ELEMENT FORMULATION

The formulation is analogous to that in Section 4.3. We make no restriction as to element shape or number of nodes.

The spatial field $T = T(x, y)$ for a single element is

$$T = \lfloor \mathbf{N}_T \rfloor \{\mathbf{T}_e\} \tag{17.3.1}$$

where $\lfloor \mathbf{N}_T \rfloor$ is a row vector of interpolation functions (shape functions) and $\{\mathbf{T}_e\}$ are nodal d.o.f. These d.o.f. include nodal temperatures and may also include spatial derivatives of temperature. Temperature gradients are

$$\{T_{,x} \quad T_{,y}\} = [\mathbf{B}_T]\{\mathbf{T}_e\} \tag{17.3.2}$$

where

$$[\mathbf{B}_T] = \{\partial/\partial x \quad \partial/\partial y\} \lfloor \mathbf{N}_T \rfloor \tag{17.3.3}$$

Also, let Q be defined in terms of nodal values $\{\mathbf{Q}_e\}$.

$$Q = \lfloor \mathbf{N}_Q \rfloor \{\mathbf{Q}_e\} \tag{17.3.4}$$

It is convenient, but not necessary, to have $\lfloor \mathbf{N}_Q \rfloor = \lfloor \mathbf{N}_T \rfloor$.

We define element matrices as follows. Let $dV = dx\,dy$, for unit thickness.

$$\begin{aligned}
[\mathbf{k}_T] &= \int_V [\mathbf{B}_T]^T [\mathbf{k}][\mathbf{B}_T]\, dV \\
[\mathbf{c}_T] &= \int_V \lfloor \mathbf{N}_T \rfloor^T \lfloor \mathbf{N}_T \rfloor \rho c\, dV \\
[\mathbf{h}_T] &= \int_{S_C} \lfloor \mathbf{N}_T \rfloor^T \lfloor \mathbf{N}_T \rfloor h\, dS \\
\{\mathbf{r}_Q\} &= \int_V \lfloor \mathbf{N}_T \rfloor^T \lfloor \mathbf{N}_Q \rfloor\, dV\, \{\mathbf{Q}_e\} \\
\{\mathbf{r}_q\} &= \int_{S_N} \lfloor \mathbf{N}_T \rfloor^T q^*\, dS \\
\{\mathbf{r}_\infty\} &= \int_{S_C} \lfloor \mathbf{N}_T \rfloor^T hT_\infty\, dS
\end{aligned} \tag{17.3.5}$$

If $k_{xy} = 0$, the expression for $[\mathbf{k}_T]$ can be written more simply as

$$[\mathbf{k}_T] = \iint \left(\lfloor \mathbf{N}_{T,x} \rfloor^T k_x \lfloor \mathbf{N}_{T,x} \rfloor + \lfloor \mathbf{N}_{T,y} \rfloor^T k_y \lfloor \mathbf{N}_{T,y} \rfloor \right) dx\, dy \tag{17.3.6}$$

Usually, $\{\mathbf{r}_q\}$ and $\{\mathbf{r}_\infty\}$ are nonzero for only a few boundary edges of a few boundary elements. Similarly, $[\mathbf{h}_T]$ comes from edges and is nonzero only for edges where $h \neq 0$ and temperature T_∞ is prescribed.

With the notation of Eq. 17.3.5, substitution of Eqs. 17.3.1 through 17.3.4 into Eq. 17.2.13 yields the functional for an element.

$$\Pi_{Te} = \tfrac{1}{2}\{\mathbf{T}_e\}^T([\mathbf{k}_T] + [\mathbf{h}_T])\{\mathbf{T}_e\} + \{\mathbf{T}_e\}^T([\mathbf{c}_T]\{\dot{\mathbf{T}}_e\} - \{\mathbf{r}_Q\} - \{\mathbf{r}_\infty\} + \{\mathbf{r}_q\}) \qquad (17.3.7)$$

By adding the Π_{Te} contributions of the elements, we obtain Π_T for the assembled structure. Assembly implies the usual conceptual expansion of element arrays to "structure size," so that the nodal temperature array $\{\mathbf{T}\}$ of the entire structure replaces $\{\mathbf{T}_e\}$, and $[\mathbf{K}_T] = \Sigma[\mathbf{k}_T]$, and so on. Hence equations that make Π_T stationary are $\{\partial \Pi_T / \partial \mathbf{T}\} = 0$ and, following differentiation rules of Eqs. 3.3.9 through 3.3.13,

$$([\mathbf{K}_T] + [\mathbf{H}_T])\{\mathbf{T}\} + [\mathbf{C}_T]\{\dot{\mathbf{T}}\} = \{\mathbf{R}_Q\} + \{\mathbf{R}_\infty\} - \{\mathbf{R}_q\} \qquad (17.3.8)$$

The simplest special form of Eq. 17.3.8 is $[\mathbf{K}_T]\{\mathbf{T}\} = 0$, which represents steady-state conditions without internal heat sources or sinks, certain of the nodal d.o.f. in $\{\mathbf{T}\}$ set to zero, and no heat flow across the boundary other than that dictated by prescribed nodal d.o.f. Prescribed d.o.f. are set in a standard fashion (Section 17.6).

We note the following *analogies of form*.

1. $[\mathbf{k}_T]$—conventional stiffness matrix (Eq. 4.3.4).
2. $[\mathbf{h}_T]$—elastic foundation stiffness matrix (Eq. 16.3.2).
3. $[\mathbf{c}_T]$—mass matrix (Eq. 11.2.3).
4. $\{\mathbf{r}_Q\}$—nodal loads from body force (Eq. 4.3.5).
5. $\{\mathbf{r}_q\}$, $\{\mathbf{r}_\infty\}$—nodal loads from surface traction (Eq. 4.3.5).

Matrix $[\mathbf{c}_T]$ is analogous to mass matrix $[\mathbf{m}]$ in that both multiply time derivatives of nodal variables and both provide resistance to time rates of change. But $[\mathbf{c}_T]$ multiplies *first* derivatives and $[\mathbf{m}]$ multiplies *second* derivatives.

Like $[\mathbf{m}]$, $[\mathbf{c}_T]$ may be consistent, as in Eq. 17.3.5, or it may be lumped. For an element with n nodal temperatures in $\{\mathbf{T}_e\}$, a lumped form of $[\mathbf{c}_T]$ results from multiplying ρc by the element area A and assigning $\rho cA/n$ to each element node. Thus $\rho cA/n$ appears n times in a diagonal matrix $\lceil \mathbf{c}_T \rfloor$. Similar ad hoc lumping can be used to simplify calculation of the nodal "load" vector $\{\mathbf{r}_Q\}$.

17.4 GENERAL SOLIDS AND SOLIDS OF REVOLUTION

Extension of the preceding treatment to three dimensions is straightforward. Changes required in Eq. 17.2.13 are: include $T_{,z}$ in the temperature gradient vector, include k_z, k_{yz}, and k_{zx} in $[\mathbf{k}]$, do volume integration over $dV = dx\,dy\,dz$, and regard dS in the surface integrals as an element of area.

A functional for the solid of revolution follows directly from the functional for the three-dimensional solid. We substitute as follows: dr and $r\,d\theta$ replace dx and dy, and $T_{,r}$ and $T_{,\theta}/r$ replace $T_{,x}$ and $T_{,y}$. Hence

$$\Pi_T = \frac{1}{2}\int_{-\pi}^{\pi}\iint\left(\begin{Bmatrix} T_{,r} \\ T_{,\theta}/r \\ T_{,z}\end{Bmatrix}^T \begin{bmatrix} k_r & k_{r\theta} & k_{zr} \\ k_{r\theta} & k_\theta & k_{\theta z} \\ k_{zr} & k_{\theta z} & k_z \end{bmatrix} \begin{Bmatrix} T_{,r} \\ T_{,\theta}/r \\ T_{,z}\end{Bmatrix} - 2QT + 2\,\rho c T\dot{T}\right) r\,dr\,dz\,d\theta \qquad (17.4.1)$$

$$+ \int_{S_N} q^* T\,dS + \int_{S_C} \frac{h}{2}(T^2 - 2TT_\infty)\,dS$$

Here dS is the increment of boundary surface area $dS = r\,d\theta\,d\ell$, where $d\ell$ is an increment of boundary length in an rz plane.

If the temperature field is axially symmetric, derivatives with respect to θ in Eq. 17.4.1 disappear. A two-dimensional interpolation of temperature, like that in Eq. 17.3.1, is then appropriate.

If the temperature field is *not* axially symmetric, a Fourier series treatment like that in Sections 8.4 through 8.6 is possible. A three-dimensional problem is then replaced by a series of two-dimensional problems. The simplest case is that where $k_{r\theta} = k_{\theta z} = 0$ and T is symmetric with respect to a plane $\theta = 0$. Then

$$T = \sum_0^\infty \bar{T}_n \cos n\theta \qquad (17.4.2)$$

where $\bar{T}_n$ is a function of r and z but is independent of θ. Equations 8.3.2 are used for θ-direction integration. Boundary conditions are written in terms of their Fourier series components, and $\bar{T}_n$ is found for $n = 0$, $n = 1$, $n = 2$, and so on. Superposition of these solutions yields the resultant temperature field.

If $k_{r\theta}$ or $k_{\theta z}$ is nonzero, the general treatment of Section 8.6 may be used.

17.5 THE TRANSIENT THERMAL PROBLEM

Two useful solution methods for transient problems are mode superposition and direct temporal integration. The choice is influenced by the same considerations that apply to structural mechanics (Section 11.9): is the solution dominated by a few low eigenmodes instead of by a sharp transient that excites many modes, is the solution needed for many time steps, and so on. (An affirmative answer to both questions favors mode superposition.)

Next we develop a method for direct integration with respect to time. Equation 17.3.8 has the form

$$[\mathbf{K}]\{\mathbf{T}\} + [\mathbf{C}_T]\{\dot{\mathbf{T}}\} = \{\mathbf{R}\} \qquad (17.5.1)$$

where $\{\mathbf{R}\}$ is a known function of time and $[\mathbf{K}]$ and $[\mathbf{C}_T]$ are, for the present, assumed independent of T. An integration scheme is based on the assumption that temperatures $\{\mathbf{T}\}_t$ at time t and temperatures $\{\mathbf{T}\}_{t+\Delta t}$ at time $t + \Delta t$ have the relation [15.27]

$$\{\mathbf{T}\}_{t+\Delta t} = \{\mathbf{T}\}_t + \left\{(1 - \beta)\{\dot{\mathbf{T}}\}_t + \beta\{\dot{\mathbf{T}}\}_{t+\Delta t}\right\}(\Delta t) \tag{17.5.2}$$

Like Newmark's method for second-order equations (Section 11.8), Eq. 17.5.2 contains a factor β that the analyst may select. If $\beta = 0$, we have *Euler's method*. If $\beta = 0.5$, we have the *central difference formula* (also known as the *balanced Euler method* and the *trapezoidal rule*).

An implicit method is developed from Eqs. 17.5.1 and 17.5.2 as follows. We write Eq. 17.5.1 for time t and then for time $t + \Delta t$. We then multiply the first of these two equations by $1 - \beta$ and the second by β.

$$\begin{aligned} (1 - \beta)\left([\mathbf{K}]\{\mathbf{T}\}_t + [\mathbf{C}_T]\{\dot{\mathbf{T}}\}_t\right) &= (1 - \beta)\{\mathbf{R}\}_t, \\ \beta\left([\mathbf{K}]\{\mathbf{T}\}_{t+\Delta t} + [\mathbf{C}_T]\{\dot{\mathbf{T}}\}_{t+\Delta t}\right) &= \beta\{\mathbf{R}\}_{t+\Delta t} \end{aligned} \tag{17.5.3}$$

Equations 17.5.3 are added, and Eq. 17.5.2 is used to eliminate time derivatives of temperature. The result is

$$\begin{aligned} \left(\frac{1}{\Delta t}[\mathbf{C}_T] + \beta[\mathbf{K}]\right)\{\mathbf{T}\}_{t+\Delta t} &= \left(\frac{1}{\Delta t}[\mathbf{C}_T] - (1 - \beta)[\mathbf{K}]\right)\{\mathbf{T}\}_t \\ &\quad + (1 - \beta)\{\mathbf{R}\}_t + \beta\{\mathbf{R}\}_{t+\Delta t} \end{aligned} \tag{17.5.4}$$

From a known $\{\mathbf{T}\}_0$ at $t = 0$, Eq. 17.5.4 yields $\{\mathbf{T}\}_{\Delta t}$. Then, using $\{\mathbf{T}\}_{\Delta t}$, we find $\{\mathbf{T}\}_{2(\Delta t)}$, and so on. In this way the history of temperature is generated. According to definitions in Section 11.8, the method becomes *explicit* if $\beta = 0$.

If Δt is not changed, the coefficient of $\{\mathbf{T}\}_{t+\Delta t}$ in Eq. 17.5.4 need be reduced only once. The equations are then repeatedly solved for a sequence of right-hand sides.

In linear problems, where neither $[\mathbf{K}]$ nor $[\mathbf{C}_T]$ depends on T, the stability limit of Eq. 17.5.4 is [15.27]

$$\Delta t_{\mathrm{cr}} = \frac{2}{(1 - 2\beta)\lambda_{\max}} \tag{17.5.5}$$

where $\lambda_{\max}$ is the largest eigenvalue of the system

$$\left([\mathbf{K}] - \lambda[\mathbf{C}_T]\right)\{\bar{\mathbf{T}}\} = 0 \tag{17.5.6}$$

Thus, if $\beta = 0$, Δt must not exceed $2/\lambda_{\max}$. If β is 0.5 or greater, the method is unconditionally stable. When $\beta = 0.5$, and when applied to heat conduction problems, the method is called the *Crank-Nicolson method*. When Δt exceeds $2/\lambda_{\max}$ in the Crank-

Nicolson method, there are undesirable oscillations in response to a step change in the forcing function. The oscillations can be reduced by using a smaller value of Δt, or they can be damped by using a larger value of β, say $0.67 < \beta < 0.88$.

Additional remarks appear in section 17.6 in connection with nonlinear problems.

17.6 MISCELLANEOUS REMARKS ON HEAT TRANSFER

Prescribed nodal temperatures—and their derivatives, if used as nodal d.o.f. in $\{\mathbf{T}\}$—can be treated like prescribed nodal displacements in structural mechanics (Section 2.9). Zero displacements correspond to zero temperature, where zero temperature means a base level that is arbitrarily placed (except when there is radiation heat transfer; see the following). Accordingly, we can discard equations that correspond to nodal temperatures of zero. A prescribed nonzero nodal temperature T^* can be imposed by adding a large conductivity K_D to the appropriate diagonal coefficient of $[\mathbf{K}]$ in Eq. 17.5.1 and also replacing the corresponding coefficient in $\{\mathbf{R}\}$ by the large heat flow $K_D T^*$. This treatment [17.2] greatly increases the maximum eigenvalues of $[\mathbf{K}]$ and therefore may cause trouble in transient analyses.

Various effects produce a nonlinear heat transfer problem. The easiest to solve, with the background of Chapter 13, is a steady-state problem where thermal conductivity is temperature dependent. To carry out a direct iteration solution, we compute an estimated $\{\mathbf{T}\}$, use these temperatures to recompute $[\mathbf{K}]$, use the new $[\mathbf{K}]$ to compute a new $\{\mathbf{T}\}$, and repeat until $\{\mathbf{T}\}$ converges. Alternatively, direct iteration can proceed with all nonlinearities transferred to the right side. Refinements are possible [17.3].

Other sources of nonlinearity include radiation and convection boundary conditions. Heat transferred by radiation is proportional to fourth powers of the *absolute* temperatures T_a of the radiating and absorbing surfaces; that is

$$q = f_r\,(T_a^{\,4} - T_{a\infty}^{\,4}) \tag{17.6.1}$$

where T_a and $T_{a\infty}$ are in degrees Kelvin and f_r is a constant. Factorization of Eq. 17.6.1 yields

$$q = [f_r\,(T_a^{\,2} + T_{a\infty}^{\,2})\,(T_a + T_{a\infty})]\,(T_a - T_{a\infty}) \tag{17.6.2}$$

which is analogous to the linear convection transfer condition

$$q = h(T - T_\infty) \tag{17.6.3}$$

Accordingly, we can write a "radiation" matrix, analogous to $[\mathbf{h}_T]$ in Eq. 17.3.5, but a function of T_a, $T_a^{\,2}$, and $T_a^{\,3}$. Thus $[\mathbf{h}_T]$ contributes nonlinear effects because h depends on temperature. To add further complexity, the fluid may move, its motion and temperature may depend on heat transfer effects, and h may depend on fluid velocity.

In a nonlinear *transient* problem we can keep [**K**] and [$\mathbf{C}_T$] in Eq. 17.5.1 unchanged by transferring all nonlinearities to the right side. So {**R**} is augmented by a temperature-dependent vector {**N**}. Reference 15.27 treats this vector by adding the terms

$$(1 + \beta)\{\mathbf{N}\}_t - \beta\{\mathbf{N}\}_{t-\Delta t} \tag{17.6.4}$$

to the right side of Eq. 17.5.4. For a good combination of accuracy, efficiency, and stability in nonlinear problems, β should slightly exceed 0.5. The default value used by [15.27] is $\beta = 0.55$.

A short list of references is given. Many more are available.

17.7 THE QUASI-HARMONIC EQUATION

Preceding sections in this chapter show that the finite element method is not limited to structural problems. Indeed, the heat transfer problem is only one interpretation of the *quasi-harmonic equation*, written here in its two-dimensional form.

$$\frac{\partial}{\partial x}(k_x \phi_{,x}) + \frac{\partial}{\partial y}(k_y \phi_{,y}) + Q = 0 \tag{17.7.1}$$

where k_x and k_y are properties of an orthotropic medium, and $\phi = \phi(x, y)$. Coordinates x and y are in the principal material directions (angle $\beta = 0$ in Fig. 17.2.1). For simplicity here we omit both the generalization to $\beta \neq 0$ and the transient conditions represented by eq. 17.2.10.

Table 17.7.1 lists problems described by eq. 17.7.1. Accordingly, the formulations of Sections 17.3 through 17.6 are applicable to these problems by suitable definition of symbols and material constants. *Caution*. It is risky to apply a solution technique without a sound grasp of the physical problem.

The remaining ingredients needed for a finite element formulation are as follows. The governing functional is

$$\begin{aligned}\Pi_Q = \frac{1}{2}\iint (k_x \phi^2_{,x} + k_y \phi^2_{,y} - 2Q\phi)\, dx\, dy \\ + \int \phi\left(q + \frac{1}{2}\alpha\phi - \alpha\phi_0\right) dS\end{aligned} \tag{17.7.2}$$

The stationary condition $\delta\Pi_Q = 0$ yields Eq. 17.7.1 and the nonessential boundary condition

$$\ell k_x\phi_{,x} + mk_y\phi_{,y} + q + \alpha(\phi - \phi_0) = 0 \tag{17.7.3}$$

TABLE 17.7.1. A SAMPLING OF PROBLEMS DESCRIBED BY EQ. 17.7.1 [17.3].

Field Problem	Unknown	k_x, k_y (or $k_x = k_y = k$)	Q
Heat conduction	Temperature	Thermal conductivity	Internal heat generated
Seepage flow	Hydraulic head	Permeability	None
Incompressible flow	Stream function	Unity	Twice the vorticity
Soap film	Deflection	Surface tension	Pressure
Elastic torsion	Stress function	(Shear modulus)$^{-1}$	Twice the rate of twist
Elastic torsion	Warping function	Unity	None
Electric conduction	Voltage	Electric conductivity	Internal current source
Magnetostatic	Magnetic potential	Reluctivity	Current density

where ℓ and m are direction cosines of an outward normal n to the boundary, and q, α, and ϕ_0 are prescribed functions of position on the boundary. If $k_x = k_y = k$, Eq. 17.7.3 takes the form

$$k\phi_{,n} + q + \alpha(\phi - \phi_0) = 0 \tag{17.7.4}$$

On portions of the boundary where Eq. 17.7.4 does not apply, the essential boundary condition of prescribed values of ϕ is imposed.

A finite element formulation follows immediately from the substitution $\phi = [\mathbf{N}]\{\boldsymbol{\phi}_e\}$, which gives ϕ within an element by interpolation from nodal d.o.f. $\{\boldsymbol{\phi}_e\}$. The result is essentially that of Section 17.3 (see Eq. 18.9.6).

How are variables and boundary conditions identified in specific applications? Consider, for example, the seepage flow problem. Element equations have the form $[\mathbf{k}]\{\mathbf{h}\} = \{\mathbf{v}\}$, where $[\mathbf{k}]$ is a matrix defined by permeability coefficients, element geometry and shape functions; $\{\mathbf{h}\}$ contains nodal values of hydraulic head; and $\{\mathbf{v}\}$ contains integrals of normal flow velocities over the length of an edge. One can specify either pressure or flow on any portion of the structure boundary. Thus, in the ith equation of the system, we can specify h_i or v_i but not both. Equations are solved for nodal h_i. Flow velocity is proportional to pressure gradient and permeability. Accordingly, the velocity field can be computed when the h_i are known by operations similar to those used in finding stresses from displacements.

The finite element analysis of seepage flow clearly has its counterpart in the analysis of stress and heat conduction, but familiarity with the physical problem is needed if we are to feel comfortable with it. Another application of the quasi-harmonic equation is given in Section 17.8.

A careless choice of the interpolation field $\phi = \lfloor \mathbf{N} \rfloor \{\boldsymbol{\phi}_e\}$ invites worthless results. A good choice must observe the usual rules. If derivatives of ϕ of order m appear in the functional, the assumed field must:

A. Be complete through order m and be able to display constant states of ϕ and its derivatives through order m.

B. Provide interelement continuity for ϕ and its derivatives through order $m - 1$, at least as the mesh is refined.

The same requirements are stated in Section 4.9. *The requirements apply in general*; they are not limited to the quasi-harmonic equation. Convergence toward exact results occurs if these requirements are satisfied in the limit of mesh refinement. In the quasi-harmonic equation, $m = 1$, so a linear interpolation of ϕ from nodal values is adequate, and only ϕ (not its derivatives) need be continuous across interelement boundaries.

17.8 AN APPLICATION IN TWO-DIMENSIONAL FLUID FLOW

A cylindrical obstacle of diameter D is centered in a channel of width $2H$ units (Fig. 17.8.1). An incompressible, ideal (nonviscous) fluid enters well upstream of the obstacle with uniform velocity u_0. We ask for the velocities of flow near the cylinder.

The necessary theory can be stated briefly. Flow velocities u and v are

$$u = \psi_{,y} \qquad v = -\psi_{,x} \tag{17.8.1}$$

where ψ is the stream function, m^3/s per meter in the z direction. The stream function satisfies Laplace's equation

$$\psi_{,xx} + \psi_{,yy} = 0 \tag{17.8.2}$$

Also, $\psi =$ constant on a streamline, and the rate of flow Q through a region bounded by two streamlines A and B is $Q = \psi_B - \psi_A$.

A finite element solution [17.12] is straightforward. Equation 17.8.2 is a special case of Eq. 17.7.1, with $k_x = k_y = 1$, $Q = 0$, and ϕ replaced by ψ. Because of symmetry,

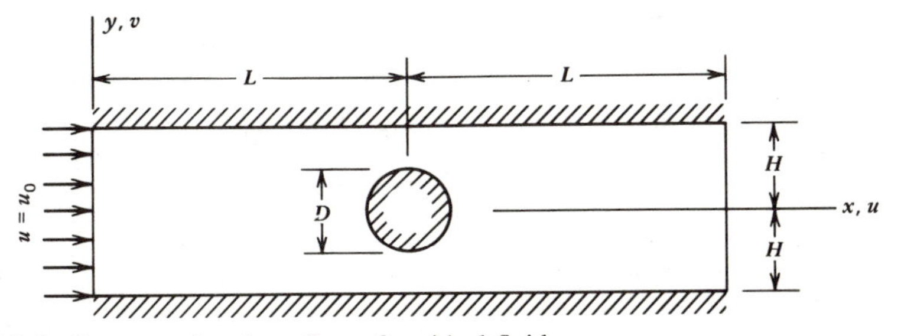

Figure 17.8.1. Geometry for plane flow of an ideal fluid.

we need model only one-quarter of Fig. 17.8.1. Figure 17.8.2 shows a possible mesh layout in one quadrant. Boundary conditions on ψ, depicted in Fig. 17.8.2, are chosen to satisfy Eq. 17.8.1: except for edge $x = L$, the flow velocity normal to every boundary is either zero or the known value u_0. The value $\psi = 0$ on $y = 0$ is chosen arbitrarily. Another constant, say $\psi = c$, would merely add c to all nodal values of ψ and would leave computed flow velocities unchanged.

With the usual interpolation from element nodal d.o.f. $\{\boldsymbol{\psi}_e\}$,

$$\psi = \lfloor \mathbf{N} \rfloor \{\boldsymbol{\psi}_e\} \tag{17.8.3}$$

we generate the element "stiffness" matrix $[\mathbf{k}]$.

$$[\mathbf{k}] = \iint \left(\lfloor \mathbf{N}_{,x} \rfloor^T \lfloor \mathbf{N}_{,x} \rfloor + \lfloor \mathbf{N}_{,y} \rfloor^T \lfloor \mathbf{N}_{,y} \rfloor \right) dx\, dy \tag{17.8.4}$$

Global equations are

$$[\mathbf{K}]\{\boldsymbol{\psi}\} = 0 \qquad \text{or} \qquad (\Sigma[\mathbf{k}])\{\boldsymbol{\psi}\} = 0 \tag{17.8.5}$$

Boundary conditions prescribe certain entries in global vector $\{\boldsymbol{\psi}\}$. Thus Eq. 17.8.5 is converted to a form in which nonzero constants appear on the right-hand side and $\{\boldsymbol{\psi}\}$ contains only the remaining unknowns. These unknowns include those at interior nodes and those at nodes on $x = L$. After solution for $\{\boldsymbol{\psi}\}$, Eqs. 17.8.1 and 17.8.3 yield fluid velocities.

If $\{\boldsymbol{\psi}_e\}$ contains only nodal values of ψ, the foregoing solution does not yield the exact result $v = -\psi_{,x} = 0$ on $x = L$. This situation is analogous to a structural problem in which stresses normal to a free edge, computed from the normal displacement gradient, are not precisely zero. In both situations mesh refinement improves the results. If derivatives $\psi_{,x}$ and $\psi_{,y}$ also appear in $\{\boldsymbol{\psi}_e\}$, velocities in Eq. 17.8.1 are given directly, so the condition $v = 0$ on $x = L$ can be explicitly enforced by setting nodal values of $\psi_{,x}$ to zero on $x = L$.

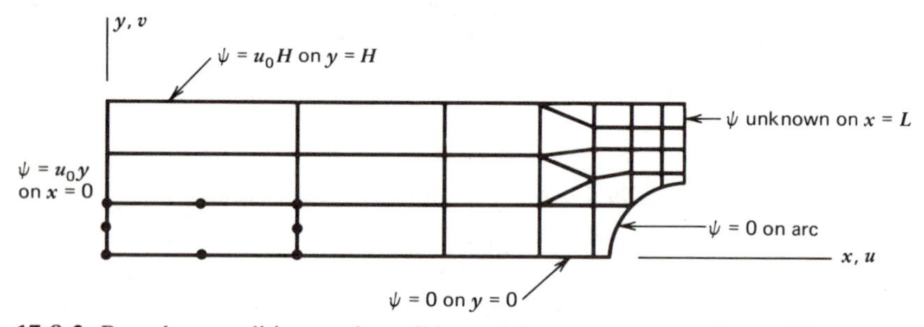

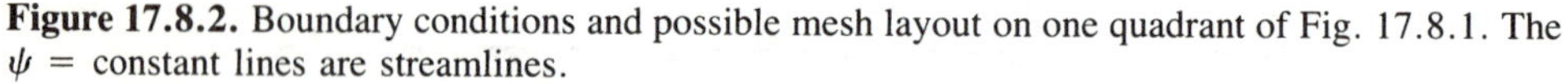

Figure 17.8.2. Boundary conditions and possible mesh layout on one quadrant of Fig. 17.8.1. The ψ = constant lines are streamlines.

17.9 THE WAVE EQUATION. ACOUSTIC MODES IN CAVITIES

The wave equation governs phenomena in which energy is propagated by waves. Here we consider a finite element formulation applicable to fluid-structure interaction problems (Section 16.7) and to finding the frequencies and mode shapes of acoustic cavities [17.13].

Consider an inviscid fluid (no shear stress; see Fig. 17.9.1*a*). Some of the total pressure is hydrostatic. Let p represent the remaining part of the total pressure. In general p is a function of the coordinates. Newton's law $\mathbf{F} = m\mathbf{a}$ is applied to a differential fluid element (Fig. 17.9.1*b*). From Fig. 17.9.1*b* and similar analyses in the y and z directions, we find

$$p_{,x} = -\rho\ddot{u} \qquad p_{,y} = -\rho\ddot{v} \qquad p_{,z} = -\rho\ddot{w} \tag{17.9.1}$$

where ρ = mass density, assumed constant. We differentiate each of Eqs. 17.9.1 with respect to its own spatial coordinate and add.

$$\begin{aligned} p_{,xx} + p_{,yy} + p_{,zz} &= -\rho(\ddot{u}_{,x} + \ddot{v}_{,y} + \ddot{w}_{,z}) \\ \text{or} \qquad \nabla^2 p &= -\rho\frac{\partial^2}{\partial t^2}(\epsilon_x + \epsilon_y + \epsilon_z) \end{aligned} \tag{17.9.2}$$

where $\epsilon_x = u_{,x}$, and so on. We next introduce bulk modulus B, defined as B = (pressure)/(resulting unit volume change).

$$B = -\frac{p}{dV/V} = -\frac{p}{\epsilon_x + \epsilon_y + \epsilon_z} \tag{17.9.3}$$

A negative sign appears because volume decreases under increasing pressure. Equation 17.9.2 becomes

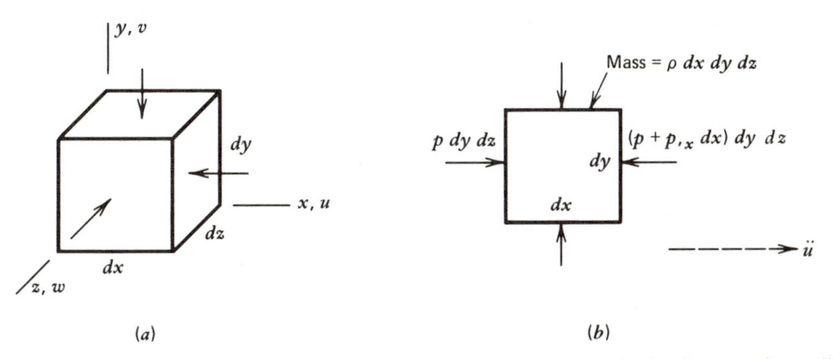

Figure 17.9.1. (*a*) Differential element under fluid pressure. (*b*) Free-body diagram for x-direction equation of motion.

$$\nabla^2 p = \frac{\rho}{B}\ddot{p} \qquad \text{or} \qquad \nabla^2 p = \frac{1}{c^2}\ddot{p} \tag{17.9.4}$$

where c is the speed of sound in the medium, $c = \sqrt{B/\rho}$. Equation 17.9.4 is the *wave equation*.

Equation 17.9.4 must be solved in region V, subject to boundary conditions on its surface S. The essential boundary condition is $p = 0$, which prevails on a free surface if waves are negligibly small. The nonessential boundary condition

$$\frac{\partial p}{\partial n} = -\rho \ddot{u}_n \tag{17.9.5}$$

prevails on an accelerating surface, where n = outward normal to S and $\ddot{u}_n$ = prescribed acceleration along the normal.

A functional for this problem is

$$\Pi_w = \frac{1}{2}\int_V \left(p_{,x}^2 + p_{,y}^2 + p_{,z}^2 + 2\frac{\ddot{p}}{c^2}p\right) dV + \int_S \rho \ddot{u}_n p \, dS \tag{17.9.6}$$

With p (but not $\ddot{p}$) subject to variation, the stationary condition $\delta\Pi_w = 0$ yields Eqs. 17.9.4 and 17.9.5.

A finite element formulation follows the now familiar pattern. We assume that pressure within an element may be interpolated from element nodal d.o.f. $\{\mathbf{p}\}$, where $\{\mathbf{p}\}$ may contain only nodal pressures or may also contain spatial derivatives of nodal pressures. Thus

$$p = \lfloor \mathbf{N} \rfloor \{\mathbf{p}\}, \qquad \ddot{p} = \lfloor \mathbf{N} \rfloor \{\ddot{\mathbf{p}}\} \tag{17.9.7}$$

As usual, elements may be two- or three-dimensional, have straight or curved sides, many nodes or few, and so on. The choice is reflected in $\lfloor \mathbf{N} \rfloor$. We define the following global matrices, where summation signs indicate assembly of element matrices.

$$\begin{aligned}
[\mathbf{H}] &= \sum \int_V \left(\lfloor \mathbf{N}_{,x} \rfloor^T \lfloor \mathbf{N}_{,x} \rfloor + \lfloor \mathbf{N}_{,y} \rfloor^T \lfloor \mathbf{N}_{,y} \rfloor + \lfloor \mathbf{N}_{,z} \rfloor^T \lfloor \mathbf{N}_{,z} \rfloor \right) dV \\
[\mathbf{Q}] &= \sum \int_V \frac{1}{c^2} \lfloor \mathbf{N} \rfloor^T \lfloor \mathbf{N} \rfloor \, dV \\
\{\mathbf{B}\} &= \sum \int_S \lfloor \mathbf{N} \rfloor^T \rho \ddot{u}_n \, dS
\end{aligned} \tag{17.9.8}$$

Equation 17.9.6 becomes

$$\Pi_w = \{\mathbf{P}\}^T\left(\tfrac{1}{2}[\mathbf{H}]\{\mathbf{P}\} + [\mathbf{Q}]\{\ddot{\mathbf{P}}\} + \{\mathbf{B}\}\right) \tag{17.9.9}$$

where $\{\mathbf{P}\}$ is the *global* array of nodal d.o.f. The stationary condition $\{\partial\Pi_w/\partial\mathbf{P}\} = 0$ yields the finite element formulation

$$[\mathbf{H}]\{\mathbf{P}\} + [\mathbf{Q}]\{\ddot{\mathbf{P}}\} = -\{\mathbf{B}\} \tag{17.9.10}$$

This formulation is applied to fluid-structure interaction problems in Section 16.7.

Acoustic Modes. In a cavity with rigid walls, the boundary condition on S is $\ddot{u} = 0$. In any of the natural modes, all particles move in phase with one another. Therefore $p = \bar{p} \sin \omega t$, where $\bar{p}$ is a function of spatial coordinates but is independent of time. Equation 17.9.4 becomes

$$\nabla^2\bar{p} + \frac{\omega^2}{c^2}\bar{p} = 0 \tag{17.9.11}$$

Equation 17.9.11 is the *Helmholtz equation*. In addition to acoustic modes, it describes vibration of a membrane, electromagnetic modes in a wave guide, and seiche motion (oscillation of water in a lake or harbor). A functional for Eq. 17.9.11 is [17.1]

$$\Pi_a = \int_V \left(\bar{p}_{,x}^2 + \bar{p}_{,y}^2 + \bar{p}_{,z}^2 - \frac{\omega^2}{c^2}\bar{p}^2\right) dV \tag{17.9.12}$$

The finite element formulation for acoustic modes follows from Eq. 17.9.12 or by specialization of Eq. 17.9.10. Adopting the latter course, we set $\{\mathbf{B}\} = 0$ and let $\{\mathbf{P}\} = \{\bar{\mathbf{P}}\} \sin \omega t$. Thus

$$\left([\mathbf{H}] - \omega^2 [\mathbf{Q}]\right)\{\bar{\mathbf{P}}\} = 0 \tag{17.9.13}$$

Equation 17.9.13 is an eigenvalue problem. It may be solved by methods noted in Section 11.6. To each frequency ω_i there corresponds a mode shape $\{\bar{\mathbf{P}}\}_i$. Numerical examples appear in Ref. 17.13.

PROBLEMS

17.1 Write equations analogous to those in Sections 17.2 and 17.3 for an isotropic bar that may be tapered. Heat flows only along the bar and enters or exits only at either end.

17.2 Let thickness τ in Section 17.2 be a function of x and y. Repeat the derivation to obtain a result analogous to Eq. 17.2.10.

17.3 Verify Eq. 17.2.12.

17.4 Consider an isotropic solid of revolution in cylindrical coordinates. Show that the equation analogous to Eq. 17.2.10 is

$$k\left(T_{,rr} + \frac{1}{r}T_{,r} + \frac{1}{r^2}T_{,\theta\theta} + T_{,zz}\right) + Q = \rho c\dot{T}$$

17.5 (For readers who understand calculus of variations.)
   (a) As suggested below Eq. 17.2.13, verify that $\delta\Pi_T = 0$ yields the proper equations.
   (b) Do likewise for the functional Π_T of Eq. 17.4.1. For simplicity, assume that homogeneity and isotropy prevail.
   (c) Augment each integral of Eq. 17.2.13 by τ to account for the variable thickness of Problem 17.2. Verify that $\delta\Pi_T = 0$ yields the proper equations.

17.6 For each of the following cases, formulate the finite element matrices. That is, working from Eqs. 17.3.5 or analogous general forms, find specific coefficients in element matrices in terms of material properties and element geometry. Assume isotropy and constant material poperties.
   (a) A bar of length L, constant cross-sectional area A, lying on the x-axis, and having a node at each end. Let heat flow across the lateral surface of the bar as well as across end cross-sections.
   (b) Same as part (a), but let A vary linearly from A_1 at node 1 to A_2 at node 2. Ignore heat flow across the lateral surface.
   (c) A plane triangular element that has three nodes and constant thickness. Evaluate surface integrals on edge 1-2 only.
   (d) A four-node plane isoparametric element. In this case do only the setup for computation, as in Section 5.3.

17.7 In the manner of Section 17.3, obtain a finite element formulation from Eq. 17.4.1. Use definitions such as those in Eq. 17.3.5.

17.8 The restriction $k_{r\theta} = k_{\theta z} = 0$ is made above Eq. 17.4.2. Why?

17.9 Follow the instructions of Problem 17.6, but with reference to an isotropic solid of revolution and an axially symmetric temperature field. Consider the following elements.
   (a) A two-node, washerlike element (see sketch). Surfaces z = constant are perfectly insulated, so heat flows only radially.
   (b) A three-node element of triangular cross section.
   (c) Same as Problem 17.6(d).

17.10 Repeat Problem 17.9(a), but let T depend on r *and* on θ. Apply the series of Eq. 17.4.2 (This problem is suitable for a programming project.)

17.11 Temperature distribution in the structure of Fig. 6.12.1*a* is to be found by analysis of Fig. 6.12.1*b* and the rules of cyclic symmetry. The quadrilateral elements have 12 d.o.f. and four nodes. In type and number of d.o.f., what plate elements do

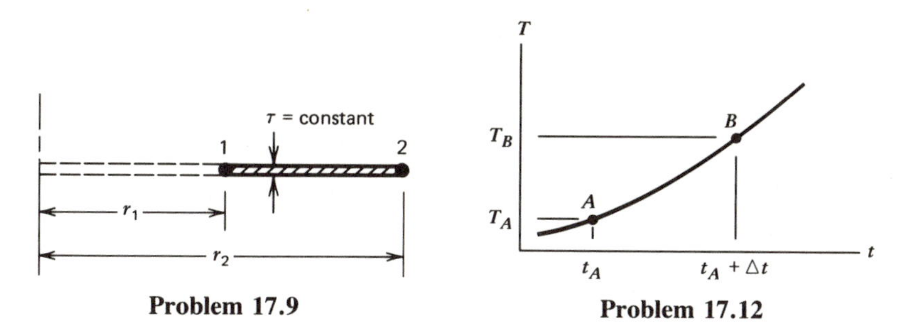

Problem 17.9 **Problem 17.12**

the triangles resemble? What are d.o.f. at nodes along *A-A* and *B-B*? How must they be related?

17.12 The sketch shows the actual variation of temperature with time at a cetain node. Imagine that we start at point *A* and use Eq. 17.5.2 to predict the temperature at $t = t_A + \Delta t$. For each of the following schemes make a sketch that shows the predicted temperature. Comment on the advantages and disadvantages of each scheme.

(a) Let $\beta = 0$ in Eq. 17.5.2.
(b) Let $\beta = 1.0$ in Eq. 17.5.2.
(c) Let $\beta = 0.5$ in Eq. 17.5.2.

17.13 A bar of cross-sectional area $A = 0.5$, perfectly insulated along its length, extends from $x = 0$ to $x = 2$. At time $t = 0$, the constant heat input $q_x^* = 3$ is imposed on the left end. At $x = 2$, $T = 0$ for all t. Let $k = 24$ and $\rho c = 6$. In parts (a) to (c) carry out five steps of direct integration to find $T = T(t)$ at the leftend, starting from $t = 0$.

(a) Use Crank-Nicolson, with $\Delta t = 1.0$.
(b) Use Crank-Nicolson, with $\Delta t = 0.1$.
(c) Use Euler's method with $\Delta t = 0.1$.
(d) In parts (a) to (c), what T is predicted at infinite time?
(e) What is the greatest Δt permitted in Euler's method? Verify your answer by attempting integration with two Δt's, one slightly greater and one slightly smaller than the limit.

17.14 Write a computer subroutine that accepts $[\mathbf{K}_T]$, $[\mathbf{C}_T]$, $\{\mathbf{T}_0\}$, β, and Δt as input and carries out the integration of Eq. 17.5.4. Assume that subroutines are available to solve equations or invert a matrix, to compute $\{\mathbf{R}\}$ at any time t, to do input and output, and to multiply matrices (Fig. 11.6.1). Make whatever arrangements you think suitable to terminate the procedure.

17.15 Interpret the patch test and other convergence requirements of Sections 4.8 and 4.9 with reference to a finite element solution of the quasi-harmonic equation (Secton 17.7).

17.16 Show that setting ψ to zero on the circular arc in Fig. 17.8.2 corresponds to a zero flow velocity normal to the arc.

17.17 Show that $dV/V = \epsilon_x + \epsilon_y + \epsilon_z$ (see Eq. 17.9.3).

17.18 Show that $\delta \Pi_w = 0$ yields the wave equation and its nonessential boundary condition, as suggested below Eq. 17.9.6. (Calculus of variations must be used.)

17.19 Derive Eq. 17.9.13 from Eq. 17.9.12 (rather than by specializing Eq. 17.9.10).

17.20 Section 18.8 discusses a "mixed" formulation for the beam problem. A functional for the problem is

$$\Pi_m = \int_0^L \left(\frac{F}{2} w_{,x}^2 + \frac{B}{2} w^2 - qw - \frac{M^2}{2EI} - M_{,x} w_{,x} \right) dx$$
$$- V_1 w_1 + V_2 w_2 - M_1 \theta_1 + M_2 \theta_2$$

where F and B are defined in Problem 18.6.

(a) Show that $\delta \Pi_m = 0$ yields the expected governing equations and nonessential boundary conditions when variations are given to w and M.

(b) Use the preceding functional to generate the finite element formulation of Section 18.8. There will be two additional matrices, one associated with F and another with B. *Suggestion*. The derivative of Π_m with respect to the nodal displacement vector must vanish, as must the derivative of Π_m with respect to the nodal moment vector.

CHAPTER 18

INTRODUCTION TO WEIGHTED RESIDUAL METHODS

18.1 REASONS FOR USE OF WEIGHTED RESIDUAL METHODS

Thus far we have presented the finite element method as an approximation technique that develops from a variational principle. A variational principle is an integral expression (a *functional*) that yields the governing differential equations and nonessential boundary conditions of a problem when given the standard treatment of calculus of variations. The principle of minimum potential energy is only one of many variational principles. In areas other than mechanics of solids, it is more likely that a variational principle may not be known or may not exist. This happens in fluid mechanics where, for some types of flow, all that is available are differential equations and boundary conditions. The finite element method can still be applied, as we describe in this chapter. The description presumes familiarity with stress analysis, so that it is easily accessible to students of structural mechanics. However, the methods are more likely to find application in nonstructural areas.

We present an introductory treatment. References listed for this chapter contain much more information. Consult them before making serious use of the methods. The problem area and the numerical method must both be understood: lack of understanding may lead to wrong equations merely because they contain the right symbols.

18.2 SOME WEIGHTED RESIDUAL METHODS

We adopt the following notation.

u = dependent variables (for example, displacements of a point)
x = independent variables (for example, coordinates of a point)
f, g = functions of x, or constants, or zero
L, B = differential operators

The governing differential equation and nonessential boundary condition of the problem are

$$\begin{aligned} Lu &= f \qquad \text{in region } V \\ Bu &= g \qquad \text{on boundary } S \text{ of } V \end{aligned} \tag{18.2.1}$$

See Eqs. 18.3.1 to 18.3.3 for an example of this notation.

The exact solution $u = u(x)$ is unknown. We seek an *approximate* solution, $\tilde{u}$. It may be a polynomial that satisfies the essential boundary conditions and contains undetermined coefficients $a_1, a_2, \ldots, a_n$. Thus $\tilde{u} = \tilde{u}(a, x)$. We must find values a_i such that u and $\tilde{u}$ are "close" in some sense.

If $\tilde{u}$ is substituted into Eq. 18.2.1, we obtain *residuals* R_L and R_B because $\tilde{u}$ is not exact. Residuals are functions of both x and the a_i.

$$\begin{aligned} R_L &= R_L(a, x) = L\tilde{u} - f \qquad \text{(interior residual)} \\ R_B &= R_B(a, x) = B\tilde{u} - g \qquad \text{(boundary residual)} \end{aligned} \tag{18.2.2}$$

Residuals vanish only for the exact solution, $\tilde{u} = u$. We presume that $\tilde{u}$ is a good approximation if the residuals are made small. This can be done by various schemes. Each intends to produce algebraic equations whose solution yields the n coefficients a_i. Some popular schemes are as follows.

1. **Collocation.** For n different values of x, the residuals are set to zero.

$$\begin{aligned} R_L(a, x_i) &= 0 \qquad \text{for} \qquad i = 1, 2, \ldots, j-1 \\ R_B(a, x_i) &= 0 \qquad \text{for} \qquad i = j, j+1, \ldots, n \end{aligned} \tag{18.2.3}$$

2. **Least Squares.** The a_i are chosen to minimize the integral of the square of the residual. A weight factor $\bar{W}$ is applied to R_B. It can be chosen arbitrarily and may be viewed as a penalty number, like α in Section 6.10. A large $\bar{W}$ makes R_B more important than R_L.

$$I = \int_V [R_L(a, x)]^2 \, dV + \bar{W}^2 \int_S [R_B(a, x)]^2 \, dS \tag{18.2.4}$$

$$\frac{\partial I}{\partial a_i} = 0 \qquad i = 1, 2, \ldots, n \tag{18.2.5}$$

The method is also called *continuous* least squares.

3. **Least Squares Collocation.** The integration of Eq. 18.2.4 is omitted. Instead, the residuals are evaluated and squared at each of i points, where i runs from 1 to m and $m \geqq n$. Thus

$$I = \sum_{i=1}^{k-1} [R_L(a, x_i)]^2 + \bar{W}^2 \sum_{i=k}^{m} [R_B(a, x_i)]^2 \tag{18.2.6}$$

Equation 18.2.5 is now applied. Although $m > n$, the result is n equations for the n values a_i (see the example in Section 18.3). The method is also called *point least squares* and *overdetermined collocation* [18.10, 18.11]. If $m = n$, the method becomes simple collocation.

4. **Galerkin.** We select "weight functions" $W_i = W_i(x)$ and set the weighted averages of the residual to zero. For $i = 1, 2, \ldots, n$,

$$R_i = \int_V W_i(x) R_L(a, x)\, dV + \int_S W_i(x) R_B(a, x)\, dS = 0 \qquad (18.2.7)$$

Weight functions W_i are, *by definition*, coefficients of the generalized coordinates. Thus $W_i = \partial \tilde{u}/\partial a_i$. In structural mechanics the residuals are proportional to force or moment, and the W_i can be regarded as virtual displacement or rotation. Accordingly, each integral represents virtual work, which should vanish at an equilibrium confrguration. Therefore both integrals should contribute to positive work (or both to negative work), so in specific applications the plus sign in Eq. 18.2.7 may have to be changed to a minus sign.

The unifying concept is that all weighted residual methods can be symbolized as

$$0 = \int_V W_i R\, dV + \int_S W_i R\, dS \qquad (18.2.8)$$

The methods differ in how weight W is defined [18.1]. In Galerkin's method, $W_i = \partial \tilde{u}/\partial a_i$; in least squares, $W_i = \partial R/\partial a_i$; in collocation, W_i is a unit delta function that is nonzero only at collocation points. As with the Rayleigh-Ritz method, the general procedure is to establish a trial family of solutions and apply a criterion to select the best member of the family. Our criterion is now Eq. 18.2.8 rather than the criterion $\{\partial \Pi_p/\partial \mathbf{D}\} = 0$ (Eq. 3.3.2).

Galerkin's method yields a symmetric coefficient matrix if the system of differential equations and boundary conditions is self-adjoint [18.1]. If differential equations *and* a variational principle are both available, the Galerkin method and the stationary-functional method yield identical solutions when both use the same approximating function $\tilde{u}$. The latter point is demonstrated by subsequent examples.

Least squares methods *always* produce a symmetric coefficient matrix. Other advantages include the "tuning" permitted by $\bar{W}$ in Eqs. 18.2.4 and 18.2.6 and the avoidance of integration in least squares collocation. A disadvantage is that the coefficient matrix tends to be ill-conditioned, so that special treatment may be needed to get an accurate solution. Also, despite the tuning factor $\bar{W}$ in Eq. 18.2.4, the continuous least squares method may be too strongly influenced by unimportant residuals and too little influenced by important ones. Further disadvantages are noted at the end of Section 18.7.

Given a choice, most readers will find that a variational principle yields a finite element formulation more easily than does a weighted residual approach—if a variational principle is already available and need not be derived.

18.3 NUMERICAL EXAMPLES

We consider a uniform bar of cross-sectional area A, elastic modulus E, and mass density ρ that rotates about the end $x = 0$ (Fig. 18.3.1). A stress σ_L is applied at $x = L$.

The governing equation and nonessential boundary condition are[1]

$$\begin{aligned} Eu_{,xx} + \rho\omega^2 x &= 0 \quad \text{for} \quad 0 < x < L \\ Eu_{,x} - \sigma_L &= 0 \quad \text{at} \quad x = L \end{aligned} \tag{18.3.1}$$

The essential boundary condition is $u = 0$ at $x = 0$. The exact solution is given in Problem 18.1. We seek approximate solutions by the methods of Section 18.2. Each solution is to have two undetermined coefficients. Numerical comparisons are made in Table 18.3.1.

Equations used by all four methods are as follows. Let the trial function be

$$\tilde{u} = a_1 x + a_2 x^2 \tag{18.3.2}$$

Note that $\tilde{u}$ satisfies the essential boundary condition. In the notation of Eqs. 18.2.1 and 18.2.2,

$$\begin{aligned} L &= E\frac{d^2}{dx^2} & B &= E\frac{d}{dx} \\ f &= -\rho\omega^2 x & g &= \sigma_L \\ R_L &= E\tilde{u}_{,xx} + \rho\omega^2 x & R_B &= E\tilde{u}_{,x} - \sigma_L \\ R_L &= 2Ea_2 + \rho\omega^2 x & R_B &= E(a_1 + 2a_2 L) - \sigma_L \end{aligned} \tag{18.3.3}$$

The nonessential boundary condition appears only at $x = L$, so R_B is evaluated only at $x = L$.

Collocation Solution. We arbitrarily elect to evaluate R_L at $x = L/3$. Of necessity, R_B is evaluated at $x = L$. We find from Eqs. 18.2.3

$$a_1 = \frac{\sigma_L}{E} + \frac{\rho\omega^2 L^2}{3E} \qquad a_2 = -\frac{\rho\omega^2 L}{6E} \tag{18.3.4}$$

Least Squares Solution. Let $\bar{W}^2 = 1/L$ in Eq. 18.2.4, so that the two integrals have the same units. Thus

[1]If derivatives of degree $2m$ appear in the differential equation, nonessential boundary conditions involve derivatives of degree m or greater, and essential boundary conditions involve derivatives of degree $m - 1$ or less.

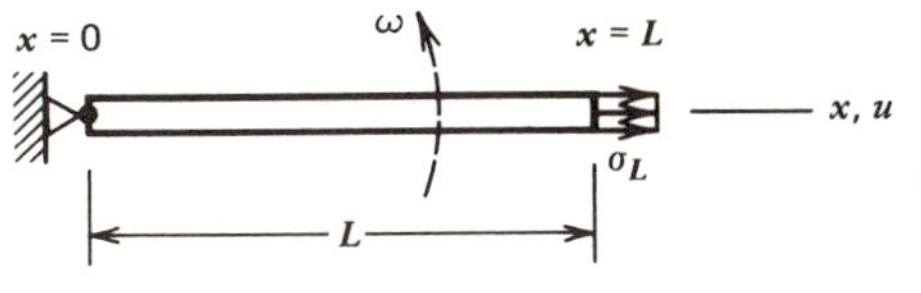

Figure 18.3.1. Bar with end load, rotating about $x = 0$ with constant angular velocity ω.

$$I = \int_0^L R_L{}^2 A\, dx + \frac{1}{L} R_B{}^2 A \tag{18.3.5}$$

Equation 18.2.5 then yields

$$a_1 = \frac{\sigma_L}{E} + \frac{\rho\omega^2 L^2}{2E} \qquad a_2 = -\frac{\rho\omega^2 L}{4E} \tag{18.3.6}$$

Least Squares Collocation Solution. Let $\bar{W} = 1/L$ in Eq. 18.2.6 to achieve dimensional homogeneity. We elect to evaluate R_L at $x = L/3$ and at $x = L$. R_B is evaluated at $x = L$. The three residuals can be written in the form

$$\begin{Bmatrix} R_{L1} \\ R_{L2} \\ R_B/L \end{Bmatrix} = \begin{bmatrix} 0 & 2E \\ 0 & 2E \\ E/L & 2E \end{bmatrix} \begin{Bmatrix} a_1 \\ a_2 \end{Bmatrix} - \begin{Bmatrix} -\rho\omega^2 L/3 \\ -\rho\omega^2 L \\ \sigma_L/L \end{Bmatrix} \tag{18.3.7}$$

Setting the residuals to zero would yield an overdetermined system: three equations in two unknowns. We introduce obvious notation for Eq. 18.3.7 and find the least squares solution for a_1 and a_2 as follows.

$$\begin{aligned} \{\mathbf{R}\} &= [\mathbf{Q}]\{\mathbf{a}\} - \{\mathbf{b}\} \\ I &= R_{L1}^2 + R_{L2}^2 + R_B{}^2/L^2 = \{\mathbf{R}\}^T\{\mathbf{R}\} \\ I &= \{\mathbf{a}\}^T[\mathbf{Q}]^T[\mathbf{Q}]\{\mathbf{a}\} - 2\{\mathbf{a}\}^T[\mathbf{Q}]^T\{\mathbf{b}\} + \{\mathbf{b}\}^T\{\mathbf{b}\} \\ \left\{\frac{\partial I}{\partial \mathbf{a}}\right\} &= 0 \qquad \text{yields} \qquad [\mathbf{Q}]^T[\mathbf{Q}]\{\mathbf{a}\} = [\mathbf{Q}]^T\{\mathbf{b}\} \end{aligned} \tag{18.3.8}$$

TABLE 18.3.1. RESULTS FOR THE PROBLEM OF FIG. 18.3.1, FOR THE SPECIAL CASE $\sigma_L = 0$, $\rho\omega^2 = L = E = 1$.

	Exact	Collo-cation	Least squares	Least squares collocation	Galerkin
u at $x = L/2$	0.2292	0.1250	0.1875	0.2500	0.2292
u at $x = L$	0.3333	0.1667	0.2500	0.3333	0.3333
σ_x at $x = 0$	0.5000	0.3333	0.5000	0.6667	0.5833
σ_x at $x = L/2$	0.3750	0.1667	0.2500	0.3333	0.3333
σ_x at $x = L$	0	0	0	0	0.0833

In the preceding equations we have noted that $\{\mathbf{b}\}^T[\mathbf{Q}]\{\mathbf{a}\} = \{\mathbf{a}\}^T[\mathbf{Q}]^T\{\mathbf{b}\}$ because each triple product is a scalar. Differentiation rules in Section 3.3 have been used. In the final result, the coefficient matrix $[\mathbf{Q}]^T[\mathbf{Q}]$ is symmetric and of the same order as $\{\mathbf{a}\}$. Hence we find

$$a_1 = \frac{\sigma_L}{E} + \frac{2\rho\omega^2L^2}{3E} \qquad a_2 = -\frac{\rho\omega^2L}{3E} \tag{18.3.9}$$

Galerkin Solution. The ith residual is

$$R_i = \int_0^L W_i(E\tilde{u}_{,xx} + \rho\omega^2 x)A\,dx - \Big[W_i(E\tilde{u}_{,x} - \sigma_L)A\Big]_{x=L} \tag{18.3.10}$$

The negative sign in Eq. 18.3.10 is chosen so that forces $\rho\omega^2 xA\,dx$ and $\sigma_L A$ both do positive work during a displacement W_i.

It is standard practice in the Galerkin method to begin with an integration by parts [18.9, 18.13]. A motivation is to reduce the order of differentiation in the integrals. If derivatives of order m appear, the integral is defined if the integrand has continuous derivatives through order $m - 1$. In the present example, $m = 2$. In a finite element context this means that if the bar is divided into elements, Eq. 18.3.10 demands that the element displacement field $\tilde{u}$ provide interelement continuity of both $\tilde{u}$ and $\tilde{u}_{,x}$. An integration by parts reduces m to unity and so permits interelement continuity of $\tilde{u}$ alone.

In one dimension, the formula for integration by parts is written in conventional notation as $\int u\,dv = uv - \int v\,du$ where u and v represent arbitrary functions of x. Here $u = W_i$ and $dv = \tilde{u}_{,xx}\,dx = d(\tilde{u}_{,x})$. Thus

$$EA\int_0^L W_i\tilde{u}_{,xx}\,dx = EA\Big[W_i\tilde{u}_{,x}\Big]_0^L - EA\int_0^L W_{i,x}\tilde{u}_{,x}\,dx \tag{18.3.11}$$

Substitution of Eq. 18.3.11 into Eq. 18.3.10 yields

$$R_i = -EA\Big[W_i\tilde{u}_{,x}\Big]_{x=0} + \int_0^L (W_i\rho\omega^2 x - EW_{i,x}\tilde{u}_{,x})A\,dx + \Big[AW_i\sigma_L\Big]_{x=L} \tag{18.3.12}$$

The W_i are $W_1 = \partial\tilde{u}/\partial a_1 = x$ and $W_2 = \partial\tilde{u}/\partial a_2 = x^2$. Accordingly, the first term on the right side of Eq. 18.3.12 vanishes. The remaining terms yield, from $R_1 = R_2 = 0$,

$$a_1 = \frac{\sigma_L}{E} + \frac{7\rho\omega^2L^2}{12E} \qquad a_2 = -\frac{\rho\omega^2L}{4E} \tag{18.3.13}$$

To see that Eq. 18.3.12 yields the same result given by the variational principle based on an assumed displacement $\tilde{u}$, consider the total potential.

$$\Pi_p = \int_0^L \frac{E}{2} \tilde{u}_{,x}^2 A \, dx - \int_0^L \rho\omega^2 x\tilde{u}A \, dx - \Big[A\sigma_L\tilde{u}\Big]_{x=L} \tag{18.3.14}$$

Equations that make Π_p stationary are $\partial\Pi_p/\partial a_i = 0$, so

$$0 = \int_0^L E\tilde{u}_{,x} \frac{\partial \tilde{u}_{,x}}{\partial a_i} A \, dx - \int_0^L \rho\omega^2 x \frac{\partial \tilde{u}}{\partial a_i} A \, dx - \left[A\sigma_L \frac{\partial \tilde{u}}{\partial a_i}\right]_{x=L} \tag{18.3.15}$$

Because $\partial\tilde{u}/\partial a_i = W_i$ and $\partial\tilde{u}_{,x}/\partial a_i = \partial(\partial\tilde{u}/\partial a_i)/\partial x = W_{i,x}$, it is clear that Eqs. 18.3.12 and 18.3.15 yield the same solution.

Summary of Results. Table 18.3.1 lists results for the problem of Fig. 18.3.1. Stress is calculated from the equation $\sigma_x = E\tilde{u}_{,x}$. The Galerkin value of $\tilde{u}$ is exact at only three points: $x = 0$, $x = L/2$, and $x = L$. In collocation methods, *different points yield different results*. Also, a large *number* of collocation points is usually beneficial in least squares collocation. Solutions summarized in Table 18.3.1 are not the best possible.

18.4 GALERKIN FINITE ELEMENT METHOD

We use the beam problem to illustrate the procedure. As in classical beam theory, we omit deformations produced by transverse shear strain. The notation is as follows.

EI = bending stiffness (force times length squared), assumed constant
w = lateral displacement
M, V = bending moment, transverse shear force
q = distributed load (force per unit length)

Equilibrium of forces and moments in Fig. 18.4.1*a* requires that

$$M_{,x} = V \quad \text{and} \quad V_{,x} = q \tag{18.4.1}$$

Hence, using the moment-curvature relation $EIw_{,xx} = M$, we find the governing differential equation

$$EIw_{,xxxx} - q = 0 \quad \text{in each element} \tag{18.4.2}$$

The nonessential boundary conditions are

$$\begin{aligned} EIw_{,xx} - M^* &= 0 \\ EIw_{,xxx} - V^* &= 0 \end{aligned} \qquad \text{at element boundaries} \tag{18.4.3}$$

where M^* and V^* are prescribed values of M and V. Essential boundary conditions are prescribed values of w and its first derivative. They are usually prescribed at few nodes, such as those at the beam ends.

Let the beam be divided into *numel* elements of length L_j, where j runs from 1 to *numel*. Each element has the assumed displacement field

$$\tilde{w} = \lfloor \mathbf{N} \rfloor \{\mathbf{d}\} = \lfloor N_1 \ N_2 \ N_3 \ N_4 \rfloor \{\mathbf{d}\} \tag{18.4.4}$$

where the N_i are given by Eq. 4.2.5 and $\{\mathbf{d}\} = \{w_1 \ \theta_1 \ w_2 \ \theta_2\}$. The d.o.f. in $\{\mathbf{d}\}$ are the displacements w and rotations θ at the two ends, where $\theta = w_{,x}$. These d.o.f. play the same role as the a_i in Section 18.3. Since weights W_i are $W_i = \partial\tilde{w}/\partial d_i = N_i$, the residual equation (Eq. 18.2.7) takes the form

$$\sum_{j=1}^{\text{numel}} \int_0^{L_j} N_i\,(EI\tilde{w}_{,xxxx} - q)\,dx + \sum_{j=1}^{\text{numel}} \Big[N_{i,x}(EI\tilde{w}_{,xx} - M^*) \Big]_0^{L_j} - \sum_{j=1}^{\text{numel}} \Big[N_i(EI\tilde{w}_{,xxx} - V^*) \Big]_0^{L_j} \tag{18.4.5}$$

where index i ranges over the four shape functions; $i = 1, 2, 3, 4$. All three sums in Eq. 18.4.5 contribute to negative work: consider, for example, forces $-q\,dx$ and $V^*_{L_j}$, which act through positive displacements $\tilde{w}$, and moment $-M^*_{L_j}$, which acts through the positive rotation $\tilde{w}_{,x}$ (see also the argument that follows Eq. 18.5.7). Two integrations by parts in the first integral yield

$$EI \int_0^{L_j} N_i\tilde{w}_{,xxxx}\,dx = EI \int_0^{L_j} N_{i,xx}\tilde{w}_{,xx}\,dx + EI \Big[N_i\tilde{w}_{,xxx} - N_{i,x}\tilde{w}_{,xx} \Big]_0^{L_j} \tag{18.4.6}$$

Substitution of Eq. 18.4.6 into Eq. 18.4.5 yields

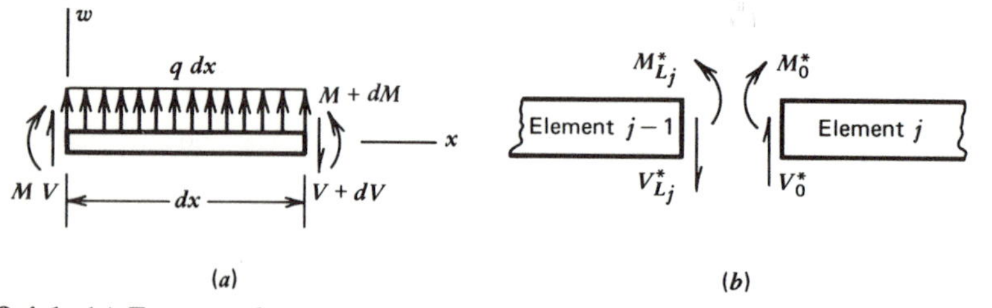

Figure 18.4.1. (*a*) Forces and moments on a differential beam element. (*b*) Forces and moments that act where adjacent elements meet at a node.

$$\sum_{j=1}^{\text{numel}} \int_0^{L_j} (N_{i,xx}EI\tilde{w}_{,xx} - N_i q)\, dx + \sum_{j=1}^{\text{numel}} \Big[N_i V^* - N_{i,x} M^* \Big]_0^{L_j} = 0 \quad (18.4.7)$$

We adopt the notation

$$\begin{aligned} \tilde{w}_{,x} &= \lfloor \boldsymbol{\theta} \rfloor \{\mathbf{d}\}, \qquad \text{where } \lfloor \boldsymbol{\theta} \rfloor = \frac{d}{dx} \lfloor \mathbf{N} \rfloor \\ \tilde{w}_{,xx} &= \lfloor \mathbf{B} \rfloor \{\mathbf{d}\}, \qquad \text{where } \lfloor \mathbf{B} \rfloor = \frac{d^2}{dx^2} \lfloor \mathbf{N} \rfloor \end{aligned} \quad (18.4.8)$$

Thus Eq. 18.4.7 takes the form

$$\begin{aligned} \sum_{j=1}^{\text{numel}} \int_0^{L_j} \lfloor \mathbf{B} \rfloor^T EI \lfloor \mathbf{B} \rfloor \, dx \, \{\mathbf{d}\} &= \sum_{j=1}^{\text{numel}} \int_0^{L_j} \lfloor \mathbf{N} \rfloor^T q \, dx \\ &+ \sum_{j=1}^{\text{numel}} \Big[\lfloor \boldsymbol{\theta} \rfloor^T M^* - \lfloor \mathbf{N} \rfloor^T V^* \Big]_0^{L_j} \end{aligned} \quad (18.4.9)$$

Except for the third summation, Eq. 18.4.9 is clearly the standard formulation, that is, Eq. 4.3.7.

$$\left(\sum_{j=1}^{\text{numel}} [\mathbf{k}] \right) \{\mathbf{D}\} = \sum_{j=1}^{\text{numel}} \{\mathbf{r}\} + \{\mathbf{P}\} \quad (18.4.10)$$

where $\{\mathbf{D}\}$ replaces $\{\mathbf{d}\}$ because of the usual expansion of element matrices to "structure size."

We must explain how the third summation in Eq. 18.4.9 can be viewed as $\{\mathbf{P}\}$, the vector of externally applied forces. At ends of a typical element,

$$\begin{aligned} \lfloor \mathbf{N} \rfloor_0 &= \lfloor 1 \quad 0 \quad 0 \quad 0 \rfloor, \qquad \lfloor \boldsymbol{\theta} \rfloor_0 = \lfloor 0 \quad 1 \quad 0 \quad 0 \rfloor \\ \lfloor \mathbf{N} \rfloor_{L_j} &= \lfloor 0 \quad 0 \quad 1 \quad 0 \rfloor, \qquad \lfloor \boldsymbol{\theta} \rfloor_{L_j} = \lfloor 0 \quad 0 \quad 0 \quad 1 \rfloor \end{aligned} \quad (18.4.11)$$

Therefore, at a typical node (Fig. 18.4.1(b), the third summation in Eq. 18.4.9 produces the terms

$$\begin{Bmatrix} 0 \\ 0 \\ 0 \\ 1 \end{Bmatrix} M_{L_j}^* - \begin{Bmatrix} 0 \\ 1 \\ 0 \\ 0 \end{Bmatrix} M_0^* - \begin{Bmatrix} 0 \\ 0 \\ 1 \\ 0 \end{Bmatrix} V_{L_j}^* + \begin{Bmatrix} 1 \\ 0 \\ 0 \\ 0 \end{Bmatrix} V_0^* \quad (18.4.12)$$

When elements are assembled, two moments from adjacent elements are associated with a single rotational d.o.f. and two shear forces from adjacent elements are associated with a single translational d.o.f. Thus we generate the concentrated moment $M_{L_j}^* - M_0^*$ and the concentrated force $V_0^* - V_{L_j}^*$ in the positive θ and w directions, respectively. These quantities may be zero at any node. If they are greater than zero, they imply the presence of concentrated nodal loads in the positive directions of nodal d.o.f. (Expression 18.4.12 is not used in actual computation; it serves only to explain the transition from Eq. 18.4.9 to Eq. 18.4.10.)

18.5 APPLICATION OF GALERKIN'S METHOD IN TWO DIMENSIONS

Preceding examples used a single independent variable x. What happens in the plane, where independent variables x and y are both present? We answer with an example. Consider the plane stress problem, with unit thickness, but without body forces or initial stresses and strains.

Necessary equations, repeated here, are explained in Sections 1.3 to 1.5.

$$\text{Equilibrium} \qquad \begin{aligned} \sigma_{x,x} + \tau_{xy,y} &= 0 \\ \tau_{xy,x} + \sigma_{y,y} &= 0 \end{aligned} \tag{18.5.1}$$

$$\text{Stress-strain relation} \qquad \underset{3\times 1}{\{\boldsymbol{\sigma}\}} = \underset{3\times 3}{[\mathbf{E}]}\ \underset{3\times 1}{\{\boldsymbol{\epsilon}\}} \tag{18.5.2}$$

$$\text{Strain-displacement relation} \qquad \{\boldsymbol{\epsilon}\} = \begin{bmatrix} \partial/\partial x & 0 \\ 0 & \partial/\partial y \\ \partial/\partial y & \partial/\partial x \end{bmatrix} \begin{Bmatrix} u \\ v \end{Bmatrix} \tag{18.5.3}$$

$$\text{Nonessential boundary condition} \qquad \begin{aligned} \ell\sigma_x + m\tau_{xy} &= \Phi_x \\ \ell\tau_{xy} + m\sigma_y &= \Phi_y \end{aligned} \tag{18.5.4}$$

Tractions Φ_x and Φ_y are usually prescribed on part on the boundary curve S. On the remainder of S, displacements u and v are prescribed.

We seek a finite element formulation based on assumed fields for u and v. For convenience of notation, we write the assumption in the form

$$\begin{Bmatrix} \tilde{u} \\ \tilde{v} \end{Bmatrix} = [\bar{\mathbf{N}}]\{\mathbf{d}\}, \qquad \text{where} \qquad [\bar{\mathbf{N}}] = \begin{bmatrix} \lfloor \mathbf{N} \rfloor & \lfloor \mathbf{O} \rfloor \\ \lfloor \mathbf{O} \rfloor & \lfloor \mathbf{N} \rfloor \end{bmatrix} \tag{18.5.5}$$

where the N_i are functions of the coordinates x and y. For the particular example of a linear isoparametric element (Section 5.3), Eq. 18.5.5 implies the following ordering of terms.

$$\begin{aligned} \lfloor \mathbf{N} \rfloor &= \lfloor N_1 \; N_2 \;\; N_3 \;\; N_4 \rfloor \\ \{\mathbf{d}\} &= \{u_1 \;\; u_2 \;\; u_3 \;\; u_4 \;\; v_1 \;\; v_2 \;\; v_3 \;\; v_4\} \end{aligned} \tag{18.5.6}$$

Equations 18.5.1 could be expressed in terms of $\tilde{u}$ and $\tilde{v}$ at the outset, but there is much less writing to do if this substitution is postponed. Let $\tilde{\sigma}_x$, $\tilde{\sigma}_y$, and $\tilde{\tau}_{xy}$ represent approximate stresses, produced by substitution of Eq. 18.5.5 into Eq. 18.5.3 and the result into Eq. 18.5.2.

In what follows we concentrate on a single element. We omit the description of how to assemble elements, symbolized by summation signs in the preceding section.

A Galerkin formulation is based on the residual equations $R_x = 0$ and $R_y = 0$. Here it is understood that R_x and R_y each represent four residuals.

$$\begin{aligned} R_x &= \iint \lfloor \mathbf{N} \rfloor^T (\tilde{\sigma}_{x,x} + \tilde{\tau}_{xy,y})\, dx\, dy - \int \lfloor \mathbf{N} \rfloor^T (\ell \tilde{\sigma}_x + m \tilde{\tau}_{xy} - \Phi_x)\, dS = 0 \\ R_y &= \iint \lfloor \mathbf{N} \rfloor^T (\tilde{\tau}_{xy,x} + \tilde{\sigma}_{y,y})\, dx\, dy - \int \lfloor \mathbf{N} \rfloor^T (\ell \tilde{\tau}_{xy} + m \tilde{\sigma}_y - \Phi_y)\, dS = 0 \end{aligned} \tag{18.5.7}$$

We could write Eqs. 18.5.7 in the form $\int [\bar{\mathbf{N}}]^T \{I_1 \; I_2\} = 0$, where $[\bar{\mathbf{N}}]$ is given by Eq. 18.5.5, and I_1 and I_2 represent the integrands of R_x and R_y, exclusive of the $\lfloor \mathbf{N} \rfloor^T$ terms. If we premultiply by $\{\delta \mathbf{d}\}^T$ and successively let $\{\delta \mathbf{d}\} = \{\delta u_1, 0, 0, \ldots, 0\}$, $\{\delta \mathbf{d}\} = \{0, \delta u_2, 0, \ldots, 0\}$, and so on, we see that $\{\delta \mathbf{d}\}^T [\bar{\mathbf{N}}]^T$ represents the virtual displacement fields produced by virtual displacements of the nodal d.o.f. in turn. Thus the equation $\int [\bar{\mathbf{N}}]^T \{I_1 \, I_2\} = 0$ corresponds to stating that virtual work vanishes for virtual displacement fields $\delta\tilde{u}$ and $\delta\tilde{v}$. Since $\tilde{u}$, $\tilde{v}$, Φ_x, Φ_y, and the stress derivatives are positive in the same sense in their respective equations, integrals in Eqs. 18.5.7 all contribute to positive work.

Formulas for integration by parts are summarized in Section 18.6. When applied to terms in the first integral in Eq. 18.5.7, the formulas yield

$$\begin{aligned} \iint \lfloor \mathbf{N} \rfloor^T \tilde{\sigma}_{x,x}\, dx\, dy &= -\iint \lfloor \mathbf{N}_{,x} \rfloor^T \tilde{\sigma}_x\, dx\, dy + \int \lfloor \mathbf{N} \rfloor^T \tilde{\sigma}_x\, \ell\, dS \\ \iint \lfloor \mathbf{N} \rfloor^T \tilde{\tau}_{xy,y}\, dx\, dy &= -\iint \lfloor \mathbf{N}_{,y} \rfloor^T \tilde{\tau}_{xy}\, dx\, dy + \int \lfloor \mathbf{N} \rfloor^T \tilde{\tau}_{xy}\, m\, dS \end{aligned} \tag{18.5.8}$$

The second double integral in Eq. 18.5.7 is similarly treated. Accordingly, Eq. 18.5.7 becomes

$$\begin{aligned} &-\iint \begin{bmatrix} \lfloor \mathbf{N}_{,x} \rfloor^T & \lfloor \mathbf{O} \rfloor^T & \lfloor \mathbf{N}_{,y} \rfloor^T \\ \lfloor \mathbf{O} \rfloor^T & \lfloor \mathbf{N}_{,y} \rfloor^T & \lfloor \mathbf{N}_{,x} \rfloor^T \end{bmatrix} \begin{Bmatrix} \tilde{\sigma}_x \\ \tilde{\sigma}_y \\ \tilde{\tau}_{xy} \end{Bmatrix} dx\, dy \\ &\qquad + \int \begin{bmatrix} \lfloor \mathbf{N} \rfloor & \lfloor \mathbf{O} \rfloor \\ \lfloor \mathbf{O} \rfloor & \lfloor \mathbf{N} \rfloor \end{bmatrix}^T \begin{Bmatrix} \Phi_x \\ \Phi_y \end{Bmatrix} dS = 0 \end{aligned} \tag{18.5.9}$$

The surface integral vanishes except on element boundaries where tractions are prescribed.
Now $\{\tilde{\sigma}_x \; \tilde{\sigma}_y \; \tilde{\tau}_{xy}\} = [\mathbf{E}]\{\tilde{\epsilon}_x \; \tilde{\epsilon}_y \; \tilde{\gamma}_{xy}\}$ and, from Eqs. 18.5.3 and 18.5.5,

$$\begin{Bmatrix} \tilde{\epsilon}_x \\ \tilde{\epsilon}_y \\ \tilde{\gamma}_{xy} \end{Bmatrix} = \begin{bmatrix} \lfloor \mathbf{N}_{,x} \rfloor & \lfloor \mathbf{O} \rfloor \\ \lfloor \mathbf{O} \rfloor & \lfloor \mathbf{N}_{,y} \rfloor \\ \lfloor \mathbf{N}_{,y} \rfloor & \lfloor \mathbf{N}_{,x} \rfloor \end{bmatrix} \{\mathbf{d}\} = [\mathbf{B}]\{\mathbf{d}\} \tag{18.5.10}$$

Therefore Eq. 18.5.9 yields

$$\iint [\mathbf{B}]^T[\mathbf{E}][\mathbf{B}] \, dx \, dy \, \{\mathbf{d}\} = \int [\mathbf{N}]^T\{\boldsymbol{\Phi}\} \, dS \tag{18.5.11}$$

which we recognize as the same formula that results from making the total potential Π_p stationary.

18.6 INTEGRATION BY PARTS

It is now clear that Galerkin's method relies heavily on integration by parts. Some useful formulas are developed next.

Let **i, j,** and **k** be unit vectors in the coordinate directions. We define a function **F** in volume V as

$$\mathbf{F} = F_1 \, \mathbf{i} + F_2 \, \mathbf{j} + F_3 \, \mathbf{k} \tag{18.6.1}$$

and a unit outward normal to the surface S of V as

$$\boldsymbol{\nu} = \ell \mathbf{i} + m\mathbf{j} + n\mathbf{k} \tag{18.6.2}$$

where ℓ, m, and n are direction cosines. The divergence theorem states that

$$\iiint \operatorname{div} \mathbf{F} \, dV = \iint \mathbf{F} \cdot \boldsymbol{\nu} \, dS \tag{18.6.3}$$

The theorem is true if **F** and its partial derivatives are continuous in V and on S and if we integrate over all bounderies, interior as well as exterior. In application to finite elements we assume "well-behaved" fields and integrate around the boundary of each element.

In rectangular and cylindrical coordinates, respectively, the divergence is

$$\begin{aligned} \operatorname{div} \mathbf{F} &= \frac{\partial F_1}{\partial x} + \frac{\partial F_2}{\partial y} + \frac{\partial F_3}{\partial z} \\ \operatorname{div} \mathbf{F} &= \frac{1}{r}\frac{\partial}{\partial r}(rF_1) + \frac{1}{r}\frac{\partial F_2}{\partial \theta} + \frac{\partial F_3}{\partial z} \end{aligned} \tag{18.6.4}$$

where F_1, F_2, and F_3 are independent functions of the coordinates.

Let N and P be functions of the coordinates. Then, for example,

$$\frac{\partial}{\partial z}(NP) = N,_z P + NP,_z \tag{18.6.5}$$

By rearrangement and integration, Eq. 18.6.5 yields

$$\iiint NP,_z \, dV = -\iiint N,_z P \, dV + \iiint (NP),_z \, dV \tag{18.6.6}$$

Now regard NP as F_3 in Eq. 18.6.4, and let $F_1 = F_2 = 0$. Equation 18.6.3 allows us to replace the last integral in Eq. 18.6.6 by a surface integral. Thus Eq. 18.6.6 becomes a formula for integration by parts.

$$\iiint NP,_z \, dV = -\iiint N,_z P \, dV + \iint NPn \, dS \tag{18.6.7}$$

Analogous formulas for the x and y derivatives in Eq. 18.6.4 are easy to derive.

Consider next the r derivative in Eq. 18.6.4. Corresponding to Eqs. 18.6.5 and 18.6.6, we now have

$$\frac{1}{r}\frac{\partial}{\partial r}(rNP) = N,_r P + N\frac{1}{r}\frac{\partial}{\partial r}(rP) \tag{18.6.8}$$

$$\iiint N\frac{1}{r}(rP),_r \, dV = -\iiint N,_r P \, dV + \iiint \frac{1}{r}(rNP),_r \, dV \tag{18.6.9}$$

Now regard NP as F_1 in Eq. 18.6.4, and let $F_2 = F_3 = 0$. Then, by Eq. 18.6.3, the last integral in Eq. 18.6.9 becomes a surface integral, and we have a formula for integration by parts.

$$\iiint N\frac{1}{r}(rP),_r \, dV = -\iiint N,_r P \, dV + \iint NP\ell \, dS \tag{18.6.10}$$

Similarly, we derive a formula for the θ derivative in Eq. 18.6.4.

$$\iiint \frac{1}{r}NP,_\theta \, dV = -\iiint \frac{1}{r}N,_\theta P \, dV + \iint NPm \, dS \tag{18.6.11}$$

Formulas for integration by parts in two dimensions can be obtained from the preceding equations by setting F_3 to zero in Eq. 18.6.4 and presuming that integation with respect to z has already been done across a unit thickness.

18.7 LEAST SQUARES COLLOCATION METHOD

We consider as our example the plane stress problem of Section 18.5. By combining Eqs. 18.5.2, 18.5.3, and 18.5.5, we may symbolize the stresses produced by the assumed displacement field as

$$\underset{3\times 1}{\{\tilde{\sigma}_x \ \tilde{\sigma}_y \ \tilde{\tau}_{xy}\}} = \underset{3\times n}{[\mathbf{H}]} \underset{n\times 1}{\{\mathbf{d}\}} \tag{18.7.1}$$

where

$$[\mathbf{H}] = [\mathbf{E}] \begin{bmatrix} \partial/\partial x & 0 \\ 0 & \partial/\partial y \\ \partial/\partial y & \partial/\partial x \end{bmatrix} [\mathbf{N}] \tag{18.7.2}$$

The interior residuals, from Eq. 18.5.1, are

$$\begin{aligned} R_{L1} &= \tilde{\sigma}_{x,x} + \tilde{\tau}_{xy,y} = \lfloor \mathbf{L}_1 \rfloor \{\mathbf{d}\} \\ R_{L2} &= \tilde{\tau}_{xy,x} + \tilde{\sigma}_{y,y} = \lfloor \mathbf{L}_2 \rfloor \{\mathbf{d}\} \end{aligned} \tag{18.7.3}$$

where

$$\begin{aligned} \lfloor \mathbf{L}_1 \rfloor &= \lfloor \partial/\partial x \quad 0 \quad \partial/\partial y \rfloor [\mathbf{H}] \\ \lfloor \mathbf{L}_2 \rfloor &= \lfloor\ 0 \quad \partial/\partial y \quad \partial/\partial x \rfloor [\mathbf{H}] \end{aligned} \tag{18.7.4}$$

The boundary residuals, from Eq. 18.5.4, are

$$\begin{aligned} R_{B1} &= \ell\tilde{\sigma}_x + m\tilde{\tau}_{xy} - \Phi_x = \lfloor \mathbf{B}_1 \rfloor \{\mathbf{d}\} - \Phi_x \\ R_{B2} &= \ell\tilde{\tau}_{xy} + m\tilde{\sigma}_y - \Phi_y = \lfloor \mathbf{B}_2 \rfloor \{\mathbf{d}\} - \Phi_y \end{aligned} \tag{18.7.5}$$

where

$$\begin{aligned} \lfloor \mathbf{B}_1 \rfloor &= \lfloor \ell \quad 0 \quad m \rfloor [\mathbf{H}] \\ \lfloor \mathbf{B}_2 \rfloor &= \lfloor 0 \quad m \quad \ell \rfloor [\mathbf{H}] \end{aligned} \tag{18.7.6}$$

Each of the four residuals may be evaluated at several points, for a total of m points, where $m > n$. Accordingly, the residual equations can be written

$$\underset{m\times 1}{\begin{Bmatrix} \mathbf{R}_{L1} \\ \mathbf{R}_{L2} \\ \bar{W}\mathbf{R}_{B1} \\ \bar{W}\mathbf{R}_{B2} \end{Bmatrix}} = \begin{Bmatrix} [\mathbf{L}_1]\{\mathbf{d}\} \\ [\mathbf{L}_2]\{\mathbf{d}\} \\ \bar{W}([\mathbf{B}_1]\{\mathbf{d}\} - \boldsymbol{\Phi}_x) \\ \bar{W}([\mathbf{B}_2]\{\mathbf{d}\} - \boldsymbol{\Phi}_y) \end{Bmatrix} = \underset{m\times n}{\begin{bmatrix} [\mathbf{L}_1] \\ [\mathbf{L}_2] \\ \bar{W}[\mathbf{B}_1] \\ \bar{W}[\mathbf{B}_2] \end{bmatrix}} \underset{n\times 1}{\{\mathbf{d}\}} - \begin{Bmatrix} \mathbf{0} \\ \mathbf{0} \\ \bar{W}\boldsymbol{\Phi}_x \\ \bar{W}\boldsymbol{\Phi}_y \end{Bmatrix} \tag{18.7.7}$$

where $[\mathbf{L}_1]$, $[\mathbf{L}_2]$, $[\mathbf{B}_1]$, and $[\mathbf{B}_2]$ must now be viewed as rectangular matrices that have several rows each. If $\bar{W}$ has units of $(\text{length})^{-1}$, the residuals agree dimensionally. Equation 18.7.7 can be written in the form of Eq. 18.3.7: $\{\mathbf{R}\} = [\mathbf{Q}]\{\mathbf{d}\} - \{\mathbf{b}\}$. The solution shown in Eq. 18.3.7 defines the stiffness equation $[\mathbf{k}]\{\mathbf{d}\} = \{\mathbf{r}\}$, where

$$[\mathbf{k}] = [\mathbf{Q}]^T[\mathbf{Q}], \qquad \{\mathbf{r}\} = [\mathbf{Q}]^T\{\mathbf{b}\} \tag{18.7.8}$$

Aside from conventional notation, this development makes no reference to finite elements. However, the element stiffness matrix is as defined by Eq. 18.7.8, *but with boundary residuals omitted from Eq. 18.7.7* [18.11]. Thus the method implicitly generates a nonzero $\{\mathbf{r}\}$ only if $f \neq 0$ in Eq. 18.2.2.

Both the continuous and collocation forms of least squares have a disadvantage that should be noted. Interelement continuity is required to a degree of one less than that appearing in the governing differential equation [18.11]. Unlike Galerkin's method, least squares does not benefit from integration by parts: weights in least squares are $W_i = \partial R/\partial a_i$, which contain derivatives of the same order as those in the residual R itself.

When Eq. 18.5.1 is stated in terms of displacements, second derivatives of u and v appear. Accordingly, interelement continuity of u and v is insufficient. The assumed field and nodal d.o.f. must provide continuity of u and v *and* their first derivatives. This means that a linear element such as the constant strain triangle cannot be constructed by a least squares method. To reduce the necessary degree of continuity, we could reformulate the problem in terms of differential equations of lower order. In so doing a "mixed" formulation is produced (Section 16.1). Reference 18.7 formulates a beam element in this way; it has eight nodal d.o.f. instead of the four needed in a pure displacement formulation. Mixed formulations have not been popular in structural mechanics, but these are useful elsewhere.

18.8 GALERKIN FINITE ELEMENT METHOD, MIXED FORMULATION

The beam problem of Section 18.4 will be cast in a finite element formulation that has translation and moment as nodal d.o.f. rather than translation and rotation. In the following formulation nodal d.o.f. are moments M_j and displacements w_j at the two ends of an element, $j = 1$ and $j = 2$. The formulation can also be derived from a variational principle (Problem 17.20).

Residual equations within the beam are

$$R_{L1} = EI\tilde{w}_{,xx} - \tilde{M}, \qquad R_{L2} = \tilde{M}_{,xx} - q \tag{18.8.1}$$

where $\tilde{w}$ and $\tilde{M}$ represent the approximating fields. R_{L1} represents the moment-curvature relation, and R_{L2} represents the equilibrium equation. Both equations contain derivatives of degree $2m = 2$, so nonessential boundary conditions are associated with derivatives of degree $m = 1$ or greater. The boundary residuals are therefore

$$R_{B1} = \left[\tilde{w}_{,x} - \theta_j \right]_0^L, \qquad R_{B2} = \left[\tilde{M}_{,x} - V_j \right]_0^L \tag{18.8.2}$$

where $\theta = w_{,x}$. Quantities in Eq. 18.8.2 are depicted in Fig. 18.8.1*b*. Note that $M_j\theta_j$ and w_jV_j both have units of work. Essential boundary conditions are prescribed values of w_j and M_j. At a typical node j, we can prescribe w_j or V_j (but not both), and M_j or θ_j (but not both).

Approximating fields are taken as linear polynomials.

$$\begin{aligned} \tilde{w} &= \lfloor \mathbf{N}_w \rfloor \{\mathbf{w}\} = \left\lfloor 1 - \frac{x}{L} \quad \frac{x}{L} \right\rfloor \{w_1 \ \ w_2\} \\ \tilde{M} &= \lfloor \mathbf{N}_m \rfloor \{\mathbf{M}\} = \left\lfloor 1 - \frac{x}{L} \quad \frac{x}{L} \right\rfloor \{M_1 \ \ M_2\} \end{aligned} \tag{18.8.3}$$

where $\{\mathbf{w}\}$ and $\{\mathbf{M}\}$ are the nodal d.o.f. Here $\lfloor \mathbf{N}_w \rfloor = \lfloor \mathbf{N}_m \rfloor$, but this need not be true in general.

A Galerkin formulation is based on the residual equations

$$\begin{aligned} &\int_0^L \lfloor \mathbf{N}_m \rfloor^T \left(\tilde{w}_{xx} - \frac{\tilde{M}}{EI} \right) dx - \left[\lfloor \mathbf{N}_m \rfloor^T (\tilde{w}_x - \theta_j) \right]_0^L = 0 \\ &\int_0^L \lfloor \mathbf{N}_w \rfloor^T (\tilde{M}_{,xx} - q)\, dx - \left[\lfloor \mathbf{N}_w \rfloor^T (\tilde{M}_{,x} - V_j) \right]_0^L = 0 \end{aligned} \tag{18.8.4}$$

We use $\lfloor \mathbf{N}_m \rfloor$ with R_{L1} and $\lfloor \mathbf{N}_w \rfloor$ with R_{L2} in Eq. 18.8.4 because virtual work is associated with $\delta\tilde{M}$ times the first equation and with $\delta\tilde{w}$ times the second equation. R_{L1} of Eq. 18.8.1 has been divided by EI for the sake of dimensional homogeneity. After integration by parts, Eq. 18.8.4 becomes

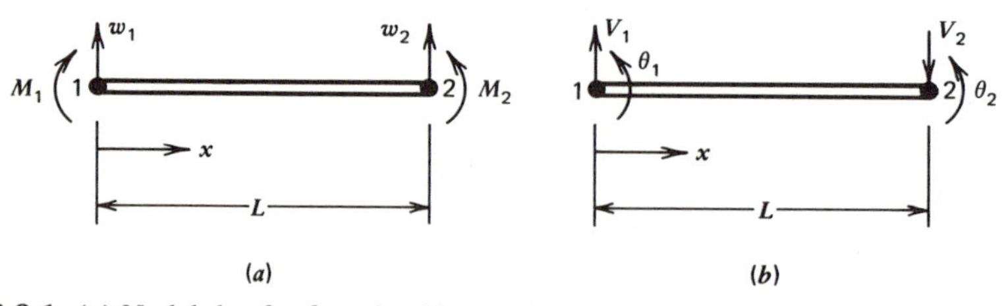

Figure 18.8.1. (*a*) Nodal d.o.f. of a mixed beam element. (*b*) Corresponding terms in the "load" vector of the element equations.

$$\begin{aligned}\int_0^L \lfloor \mathbf{N}_m \rfloor^T \frac{\tilde{M}}{EI}\, dx + \int_0^L \lfloor \mathbf{N}_{m,x} \rfloor^T \tilde{w}_{,x}\, dx &= \Big[\lfloor \mathbf{N}_m \rfloor^T \theta_j \Big]_0^L \\ \int_0^L \lfloor \mathbf{N}_{w,x} \rfloor^T \tilde{M}_{,x}\, dx &= - \int_0^L \lfloor \mathbf{N}_w \rfloor^T q\, dx + \Big[\lfloor \mathbf{N}_w \rfloor^T V_j \Big]_0^L\end{aligned} \tag{18.8.5}$$

From Eqs. 18.8.3 and the integrals in Eq. 18.8.5, we define

$$\begin{aligned}[\mathbf{H}_{11}] &= \int_0^L \frac{\lfloor \mathbf{N}_m \rfloor^T \lfloor \mathbf{N}_m \rfloor}{EI}\, dx = \frac{L}{6EI}\begin{bmatrix} 2 & 1 \\ 1 & 2 \end{bmatrix} \\ [\mathbf{H}_{12}] &= \int_0^L \lfloor \mathbf{N}_{m,x} \rfloor^T \lfloor \mathbf{N}_{w,x} \rfloor\, dx = \frac{1}{L}\begin{bmatrix} 1 & -1 \\ -1 & 1 \end{bmatrix} \\ [\mathbf{H}_{21}] &= \int_0^L \lfloor \mathbf{N}_{w,x} \rfloor^T \lfloor \mathbf{N}_{m,x} \rfloor\, dx = [\mathbf{H}_{12}]^T \\ \begin{Bmatrix} q_1 \\ q_2 \end{Bmatrix} &= \int_0^L \lfloor \mathbf{N}_w \rfloor^T q\, dx\end{aligned} \tag{18.8.6}$$

Terms in Eq. 18.8.5 associated with nonessential boundary conditions are

$$\begin{aligned}\Big[\lfloor \mathbf{N}_m \rfloor^T \theta_j \Big]_0^L &= \lfloor 0 \quad 1 \rfloor^T \theta_2 - \lfloor 1 \quad 0 \rfloor^T \theta_1 = \{ -\theta_1 \quad \theta_2 \} \\ \Big[\lfloor \mathbf{N}_w \rfloor^T V_j \Big]_0^L &= \lfloor 0 \quad 1 \rfloor^T V_2 - \lfloor 1 \quad 0 \rfloor^T V_1 = \{ -V_1 \quad V_2 \}\end{aligned} \tag{18.8.7}$$

Hence Eqs. 18.8.5 are

$$\begin{bmatrix} [\mathbf{H}_{11}] & [\mathbf{H}_{12}] \\ [\mathbf{H}_{12}]^T & [\mathbf{O}] \end{bmatrix} \begin{Bmatrix} M_1 \\ M_2 \\ w_1 \\ w_2 \end{Bmatrix} = \begin{Bmatrix} -\theta_1 \\ \theta_2 \\ -V_1 - q_1 \\ V_2 - q_2 \end{Bmatrix} \tag{18.8.8}$$

The preceding mixed formulation does not give preferential treatment to displacement: moments are calculated directly and not as derivatives of displacement. More generally, mixed formulations may predict stresses more accurately than displacement formulations. For example, Eq. 18.8.8 yields the exact moment at the root of a uniformly loaded cantilever beam. The standard beam element, with only displacement and rotation d.o.f., requires the calculation $M = EIw_{,xx}$ and yields an approximate root moment [16.5–16.8].

Although compatible displacement formulations are guaranteed not to be too flexible, a mixed formulation may be either too stiff or too flexible. Also, we need a special library of mixed elements if we want to join elements of different type (such as plate elements and edge beams).

18.9 GALERKIN TREATMENT OF THE QUASI-HARMONIC EQUATION

In Section 17.7 we discussed the quasi-harmonic equation and the generation of its finite element formulation from a variational principle. Here we generate the formulation by the Galerkin method.

In the notation of Section 17.7 the residual in region V is

$$R_L = \frac{\partial}{\partial x}(k_x\tilde{\phi}_{,x}) + \frac{\partial}{\partial y}(k_y\tilde{\phi}_{,y}) + Q \tag{18.9.1}$$

where $\tilde{\phi}$ is the approximating field. The residual on the boundary S of V is

$$R_B = \ell k_x\tilde{\phi}_{,x} + mk_y\tilde{\phi}_{,y} + q + \alpha(\tilde{\phi} - \phi_0) \tag{18.9.2}$$

As in Section 17.2, the boundary can be split into portions S_N and S_C where Eq. 18.9.2 applies and portion S_D where the essential boundary condition $\phi = \phi^*$ is prescribed. The appoximating field is written

$$\tilde{\phi} = \lfloor \mathbf{N} \rfloor \{\boldsymbol{\phi}_e\} = \lfloor N_1 \; N_2 \cdots N_n \rfloor \{\boldsymbol{\phi}_e\} \tag{18.9.3}$$

where $\{\boldsymbol{\phi}_e\}$ are element nodal d.o.f. Galerkin's method reqires that

$$\iint \lfloor \mathbf{N} \rfloor^T R_L \, dx \, dy - \int \lfloor \mathbf{N} \rfloor^T R_B \, dS = 0 \tag{18.9.4}$$

From Section 18.6, integration by parts yields

$$\begin{aligned} \iint \lfloor \mathbf{N} \rfloor^T \frac{\partial}{\partial x}(k_x\tilde{\phi}_{,x})\, dx\, dy &= -\iint \lfloor \mathbf{N}_{,x} \rfloor^T k_x\tilde{\phi}_{,x}\, dx\, dy + \int \lfloor \mathbf{N} \rfloor^T k_x\tilde{\phi}_{,x}\ell \, dS \\ \iint \lfloor \mathbf{N} \rfloor^T \frac{\partial}{\partial y}(k_y\tilde{\phi}_{,y})\, dx\, dy &= -\iint \lfloor \mathbf{N}_{,y} \rfloor^T k_y\tilde{\phi}_{,y}\, dx\, dy + \int \lfloor \mathbf{N} \rfloor^T k_y\tilde{\phi}_{,y} m \, dS \end{aligned} \tag{18.9.5}$$

Substitution of Eqs. 18.9.1 to 18.9.3 and 18.9.5 into Eq. 18.9.4 yields

$$\begin{aligned} \Bigg[\iint \left(k_x \lfloor \mathbf{N}_{,x} \rfloor^T \lfloor \mathbf{N}_{,x} \rfloor + k_y \lfloor \mathbf{N}_{,y} \rfloor^T \lfloor \mathbf{N}_{,y} \rfloor\right) dx\, dy \\ + \int \lfloor \mathbf{N} \rfloor^T \lfloor \mathbf{N} \rfloor \alpha \, dS \Bigg] \{\boldsymbol{\phi}_e\} = \iint \lfloor \mathbf{N} \rfloor^T Q \, dx\, dy - \int \lfloor \mathbf{N} \rfloor^T (q - \alpha\phi_0)\, dS \end{aligned} \tag{18.9.6}$$

We recognize that this set of element equations is the same as the set in Section 17.3.

PROBLEMS

18.1 (a) Derive the first of Eqs. 18.3.1.

(b) Show that the exact solution for the problem of Fig. 18.3.1 is

$$u = \frac{\rho\omega^2}{6E}(3L^2x - x^3) + \frac{\sigma_L x}{E}, \qquad \sigma_x = \frac{\rho\omega^2}{2}(L^2 - x^2) + \sigma_L$$

(c) Show that despite collocation at $x = L/3$, Eqs. 18.3.2 and 18.3.4 do not yield $u = \tilde{u}$ at $x = L/3$. Why is there this disagreement?

18.2 Solve the problem of Fig. 18.3.1 in the following ways. Compare your answers with those in Table 18.3.1.

(a) Use collocation, with the sampling point at $x = L/2$.

(b) Use least squares, with $\bar{W}^2 = 1/2L$.

(c) Use least squares collocation. Evaluate R_L at $x = 0$, at $x = L/2$, and at $x = L$. Use $\bar{W} = 1/L$.

(d) Omit the residual R_{L2} in Eq. 18.3.7.

(e) Modify Eq. 18.3.7 so that $\bar{W} = 1/2L$.

(f) Use Galerkin's method, but use Eq. 18.2.7 "as is," without an initial integration by parts.

(g) Explain why, in parts (b) and (e) but not in general, a change in $\bar{W}$ does not change the results.

18.3 Let $\sigma_L = \omega = 0$ in Fig. 18.3.1 and let the loading be a uniformly distributed traction q_0 (force/length) on $L/2 < x < L$ only. Find two-parameter approximate solutions by the following methods. Compare approximate and exact results for the case $q_0 = L = AE = 1$. (a) Collocation. (b) Least squares. (c) Least squares collocation. (d) Galerkin's method. Collocate at $x = 2L/3$ in part (a); otherwise, use the choice of points and weights used in Section 18.3.

18.4 The sketch represents a rigid bar embedded in a linearly elastic isotropic material that fills the space between r_1 and r_2. The loads shown are a force P and a torque T on the bar. Assume that $L >> r_2$.

(a) For the case $P \neq 0$, $T = 0$, axial displacement w is the dependent variable. Formulate expresssions for residuals R_L and R_B.

(b) Write Eq. 18.2.7 for the present problem; then do an integration by parts. Is the results dimensionally consistent?

(c) Write a two-parameter polynomial $\tilde{w} = \tilde{w}(r)$ that is appropriate for approximate solutions. What then are Galerkin weights W_1 and W_2?

18.5 Repeat Problem 18.4 for the case $P = 0$, $T \neq 0$. Circumferential displacement v is the dependent variable.

18.6 Let F = axial force, positive in tension, and B = elastic foundation constant, whose dimensions are force per unit length per unit of deflection w. The terms $Bw - (Fw_{,x})_{,x}$ are added to the left side of Eq. 18.4.2. Asume that neither F nor B

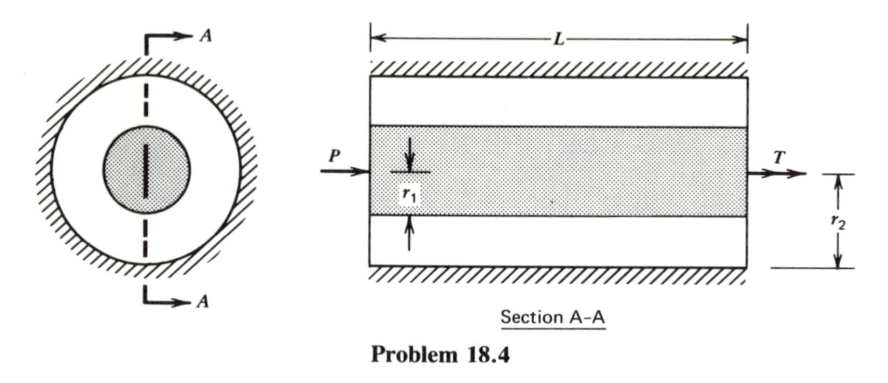

Problem 18.4

is a function of x. Repeat the derivation of Section 18.4 to find the contribution of these added efffects to the finite element formulation.

18.7 Repeat the derivation of Section 18.5 but include body force $\{\mathbf{F}\}$ and initial stresses $\{\boldsymbol{\sigma}_0\}$.

18.8 Repeat the derivation of Section 18.5 for a solid of revolution under axially symmetric loads.

18.9 The problem of Fig. 18.3.1 can be stated in terms of two first-order differential equations.

$$\sigma_{x,x} + \rho\omega^2 x = 0, \qquad \sigma_x - Eu_{,x} = 0$$

Esssential boundary conditions are $\sigma_x = 0$ at $x = L$ and $u = 0$ at $x = 0$. Assume $\tilde{\sigma}_x = a_1(L - x)$ and $\tilde{u} = a_2 x$. Find a_1 and a_2 by the following methods and compare your results with those in Table 18.3.1.

(a) Collocation with the sampling point at $x = L/2$.
(b) Least squares.
(c) Least squares collocation with sampling points at $x = 0$ and $x = L$.
(d) Least squares collocation with a sampling pont at $x = L/2$.

18.10 (a) Consider a uniform straight bar of cross-sectional area A, elastic modulus E, and length L. Load it by a constant axial traction q, units of force per unit length. Make the problem one-dimensional by allowing only axial displacement. Generate an element stiffness matrix and load vector by Galerkin's method.

(b) If, in the problem of part (a), generation of $[\mathbf{k}]$ by least squares collocation is contemplated, what shape functions are appropriate? And how many collocation points are needed?

18.11 An eight d.o.f. mixed beam element is mentioned at the end of Section 18.7. First-order differential equations of the problem are

$$M_{,x} - V = V_{,x} - q = w_{,x} - \theta = \theta_{,x} - M/EI = 0$$

Nodal d.o.f. are $\{\mathbf{d}\} = \{w_1\ \theta_1\ M_1\ V_1\ w_2\ \theta_2\ M_2\ V_2\}$.

(a) Use least squares to derive an expression for the element "stiffness" matrix. A symbolic result, analogous to that in Section 18.5, is desired. *Suggestion.* Set up a 4-by-1 residual vector $\{\mathbf{R}\} = [\mathbf{M}]\{\mathbf{d}\}$, where $[\mathbf{M}]$ is roughly analogous to $[\mathbf{H}]$ in Eq. 18.7.2.

(b) If a specific matrix is desired, what should the shape functions be?

18.12 Consider the cylindrical-coordinate form of the quasi-harmonic equation for an isotropic solid of revolution. Generate a finite element formulation by Galerkin's method, analogous to that for the plane case in Section 18.9. *Suggestion.* See Section 17.4.

18.13 Use the finite element formulation of Section 18.8 (and, where necessary, that of Problem 17.20) to solve the following beam problems. A prescribed load or deflection is imposed in each case. Each element (1–2, 2–3) is of length L. An axial load indicates a buckling problem. Compare computed nodal M's and w's and critical loads with exact values.

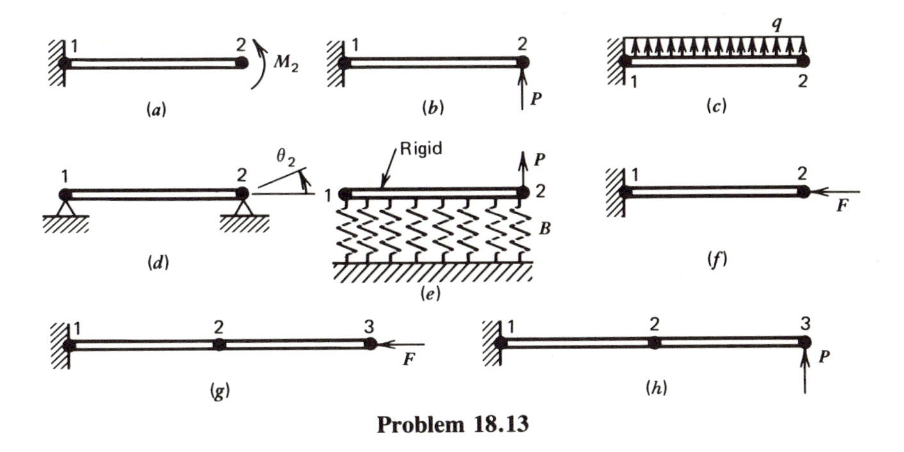

Problem 18.13

18.14 Aside from questions of accuracy, the formulation of Eq. 18.8.8 has a substantial awkwardness. What is it? *Suggestion.* Consider a tee or cross intersection, where three or four elements terminate at a single node.

18.15 Use Galerkin's method to derive Eq. 17.9.13 from Eq. 17.9.11.

18.16 If ρ_L represents the mass per unit length of a beam, an accceleration $\ddot{w}$ produces a downward "inertia force" $\rho_L \ddot{w}\, dx$ in Fig. 18.4.1*a*. Follow the Galerkin finite element method of Section 18.4 to discover what matrices are produced by this added term. A symbolic result is desired, not all elements in the new matrix. *Note.* Time derivatives are not to be integrated by parts; the time-varying term is treated much like a distributed load q during formulation.

18.17 The differential equation of a freely vibrating string is $Fv_{,xx} = \rho\ddot{v}$, where F = constant axial force, v = lateral displacement, ρ= mass per unit length, x = axial coordinate, and $\ddot{v}$ = acceleration.

(a) Assume a displacement field $\tilde{v}$ and one d.o.f. per node and generate finite element matrices by the Galerkin method (see the note in Problem 18.16).

(b) Solve for the fundamental frequency ω_0 of a simply supported string of length $2L$. Use two elements, each of length L.

18.18 The differential equation for wind-driven circulation in a shallow lake is

$$\psi_{,xx} + \psi_{,yy} + A\psi_{,x} + B\psi_{,y} + C = 0$$

where ψ = stream function, A, B, and C = functions of x and y, and x and y = coordinates tangent to the lake surface. The nonessential boundary condition is $\psi_{,n} = 0$ on the shoreline, where n is a direction normal to the shoreline. Derive a finite element formulation by the Galerkin method [18.12]. A symbolic result is desired, analogous to that in Section 18.9, not the details of a particular element.

18.19 A beam element of length L rests on an elastic foundation of modulus B (units F/L^2) and carries a distributed load q (units F/L). The beam has shear stiffness AG but no bending stiffness. Equations that describe the problem are

$$AGw_{,xx} + q - Bw = 0 \qquad \text{for} \qquad 0 < x < L$$
$$AGw_{,x} + V = 0 \qquad \text{at} \qquad x = 0, L$$

Symbols and signs are defined in Fig. 18.4.1.

(a) Verify the foregoing equations.

(b) Asume $w = \lfloor \mathbf{N} \rfloor \{w_1 \; w_2\}$ and derive a finite element formulation by the Galerkin method. A symbolic result is desired, analogous to that in Section 18.9.

(c) Write a functional Π_w that yields the foregoing equations.

(d) Derive the finite element formulation of part (b) from the functional Π_w.

18.20 For the beam of Problem 18.19, consider the least squares collocation method. Describe what sort of shape functions $\lfloor \mathbf{N} \rfloor$ are appropriate. How many collocation points are needed in each element?

APPENDIX A

STIFFNESS MATRIX OF A PLANE BEAM-COLUMN

The notation is that of Fig. 4.4.1*b*. The method of derivation is elementary beam theory, retaining an axial force P. The element tangent stiffness matrix $[\mathbf{k}]$ that results is symmetric, a function of axial force P, and operates on d.o.f. $\{\mathbf{d}\} = \{w_1 \ \theta_1 \ w_2 \ \theta_2\}$. Coefficients in its upper triangle are

$$\begin{aligned} k_{11} &= k_{33} = -k_{13} = 12EI\phi_1/L^3 \\ k_{12} &= k_{14} = -k_{23} = -k_{34} = 6EI\phi_2/L^2 \\ k_{22} &= k_{44} = 4EI\phi_3/L \\ k_{24} &= 2EI\phi_4/L \end{aligned} \tag{A.1}$$

The ϕ_i are as follows [12.5, A.1]. Let F be the magnitude of P and let $\beta = L\sqrt{F/EI}$. Then

Compressive Force P	*Tensile Force P*
$\phi_1 = \dfrac{\beta^3 \sin\beta}{12\phi_-}$	$\phi_1 = \dfrac{\beta^3 \sinh\beta}{12\phi_+}$
$\phi_2 = \dfrac{\beta^2(1 - \cos\beta)}{6\phi_-}$	$\phi_2 = \dfrac{\beta^2(\cosh\beta - 1)}{6\phi_+}$
$\phi_3 = \dfrac{\beta(\sin\beta - \beta\cos\beta)}{4\phi_-}$	$\phi_3 = \dfrac{\beta(\beta\cosh\beta - \sinh\beta)}{4\phi_+}$
$\phi_4 = \dfrac{\beta(\beta - \sin\beta)}{2\phi_-}$	$\phi_4 = \dfrac{\beta(\sinh\beta - \beta)}{2\phi_+}$
$\phi_- = 2 - 2\cos\beta - \beta\sin\beta$	$\phi_+ = 2 - 2\cosh\beta + \beta\sinh\beta$

(A.2)

For $P = 0$, $\phi_1 = \phi_2 = \phi_3 = \phi_4 = 1.0$. As β approaches zero, Eqs. A.2 approach the indeterminate form 0/0. For low values of β, say $\beta < 0.4$, we can use the following expressions, in which $\beta^2 < 0$ for compressive P and $\beta^2 > 0$ for tensile P.

$$\begin{aligned} \phi_1 &= 1 + \frac{\beta^2}{10} - \frac{\beta^4}{8400} + \cdots & \phi_3 &= 1 + \frac{\beta^2}{30} - \frac{11\,\beta^4}{25{,}200} + \cdots \\ \phi_2 &= 1 + \frac{\beta^2}{60} - \frac{\beta^4}{8400} + \cdots & \phi_4 &= 1 - \frac{\beta^2}{60} + \frac{13\,\beta^4}{25{,}200} + \cdots \end{aligned} \tag{A.3}$$

Expressions for fixed-end forces and moments—the {**r**} of Eq. 4.3.5, but with *P* active—are available [12.5, 12.24, A.1].

PROBLEMS

A.1 What general expression relates ϕ_4 to the other three ϕ_i?
A.2 Show that if only the β^2 terms are retained in Eqs. A.3, Eqs. A.1 yield Eqs. 4.4.5 and 12.2.13.

REFERENCE

A.1 E. Chwalla, "Second Order Theory," in *Handbook of Engineering Mechanics*, W. Flugge, ed., McGraw-Hill, New York, 1962.

APPENDIX B

EXAMPLE FINITE ELEMENT MESHES

These meshes were used in Ref. 2.11. Photographs and data were provided by Gordon C. Everstine of the David W. Taylor Naval Ship Research and Development Center. His help is gratefully acknowledged.

Element notation is as follows.

A = two-force member
B = beam element
C = triangular membrane
D = quadrilateral membrane
E = triangular plate
F = quadrilateral plate
G = triangular shell
H = quadrilateral shell
I = solid tetrahedron
J = solid wedge
K = solid hexahedron

The "RMS wavefront" is defined in Ref. 2.11. A good equation solver has a solution time proportional to the number of equations times the square of the RMS wavefront of [**K**]. Data on bandwidth, profile, and RMS wavefront pertain to the mesh as renumbered by the GPS algorithm cited in Ref. 2.11 and are constructed on the assumption that there is a single d.o.f. per node.

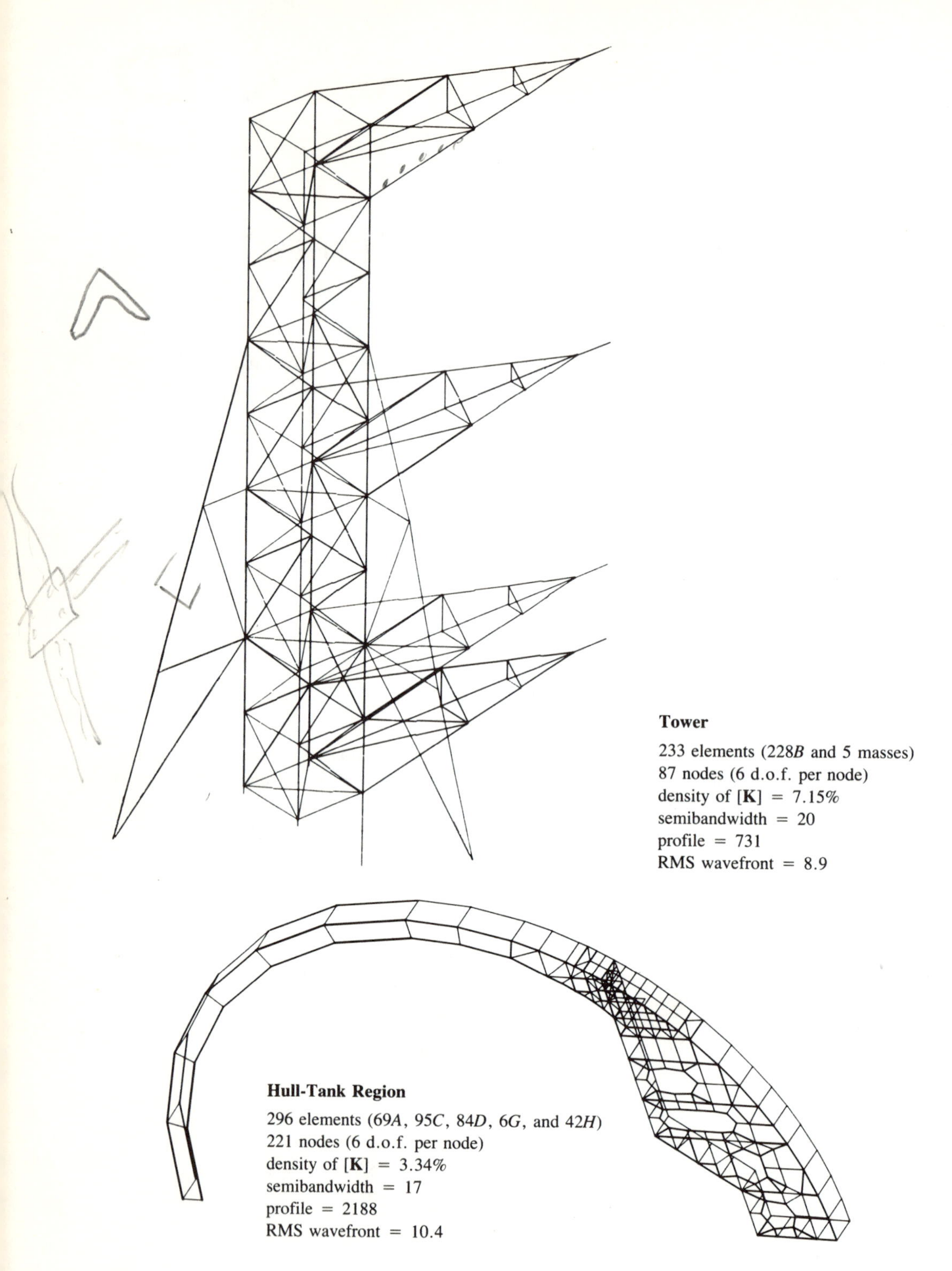
Tower
233 elements (228B and 5 masses)
87 nodes (6 d.o.f. per node)
density of [K] = 7.15%
semibandwidth = 20
profile = 731
RMS wavefront = 8.9
Hull-Tank Region
296 elements (69A, 95C, 84D, 6G, and 42H)
221 nodes (6 d.o.f. per node)
density of [K] = 3.34%
semibandwidth = 17
profile = 2188
RMS wavefront = 10.4

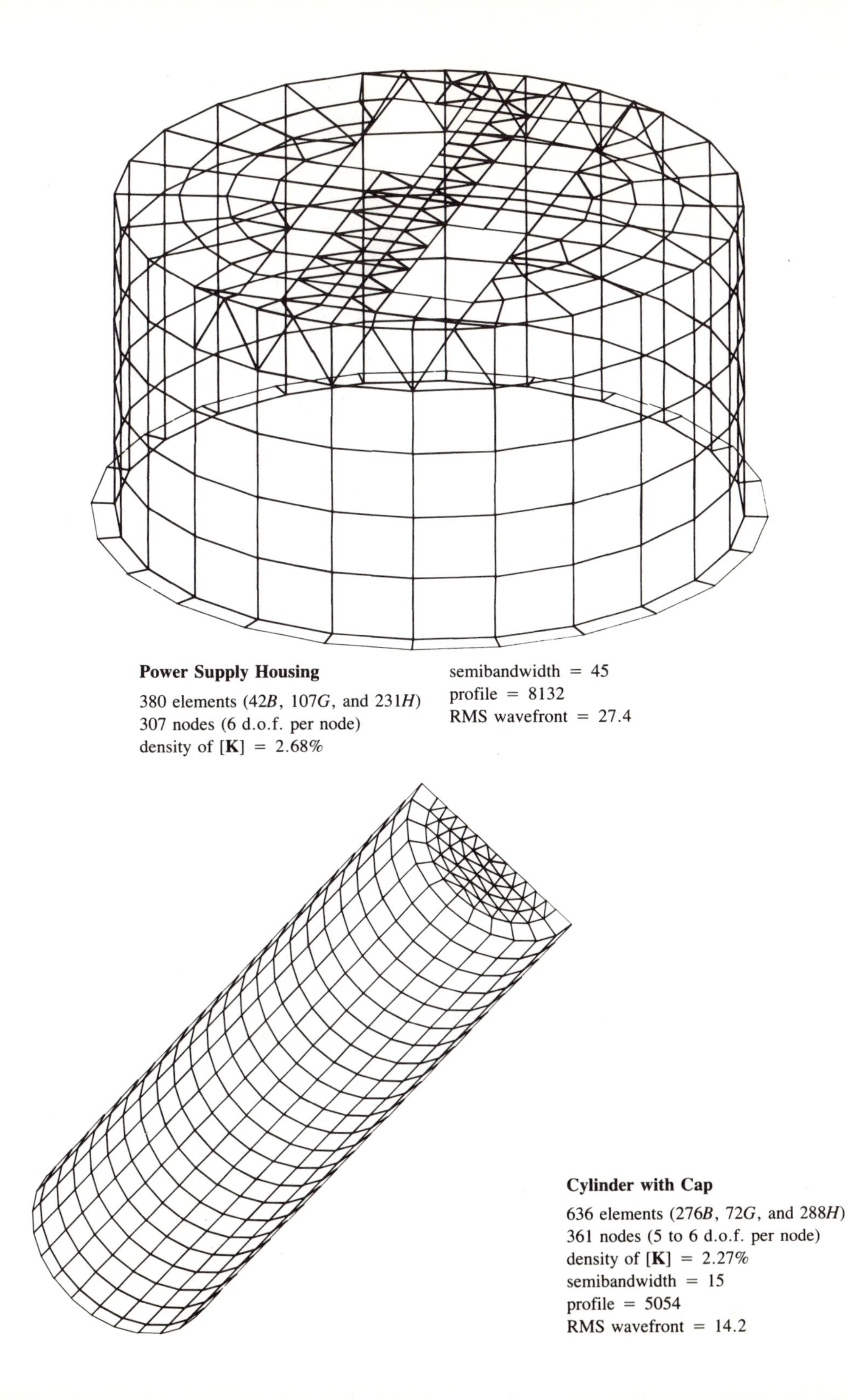

Power Supply Housing

380 elements (42*B*, 107*G*, and 231*H*)
307 nodes (6 d.o.f. per node)
density of [**K**] = 2.68%
semibandwidth = 45
profile = 8132
RMS wavefront = 27.4

Cylinder with Cap

636 elements (276*B*, 72*G*, and 288*H*)
361 nodes (5 to 6 d.o.f. per node)
density of [**K**] = 2.27%
semibandwidth = 15
profile = 5054
RMS wavefront = 14.2

Baseplate

440 elements (18*F*, 34*G*, 200*H*, 45*I*, 71*J*, and 72*K*)
503 nodes (3 to 6 d.o.f. per node)
density of [**K**] = 2.38%
semibandwidth = 55
profile = 16,096
RMS wavefront = 34.6

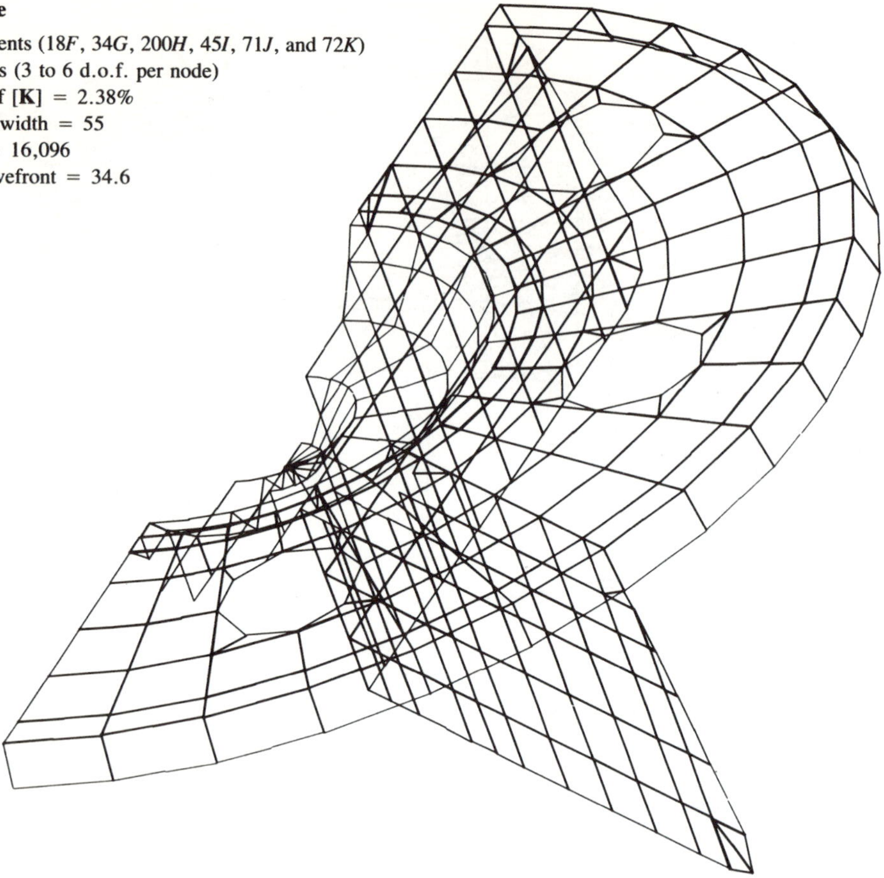

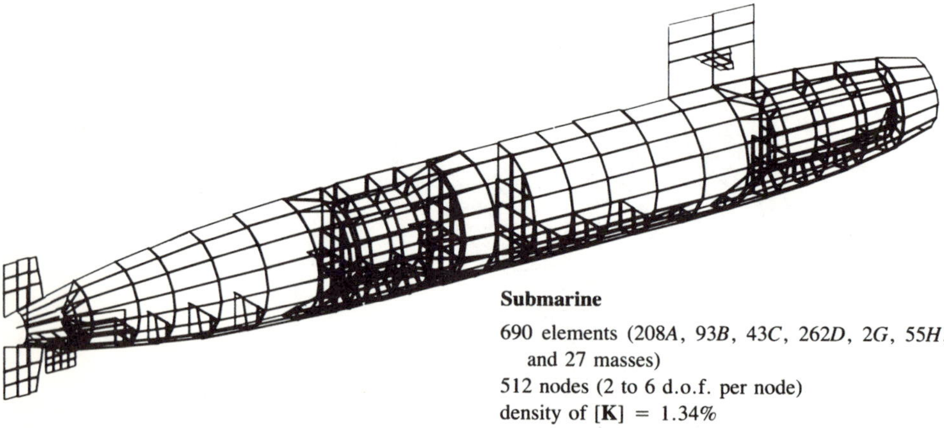

Submarine

690 elements (208*A*, 93*B*, 43*C*, 262*D*, 2*G*, 55*H*, and 27 masses)
512 nodes (2 to 6 d.o.f. per node)
density of [**K**] = 1.34%
semibandwidth = 30
profile = 5171
RMS wavefront = 12.5

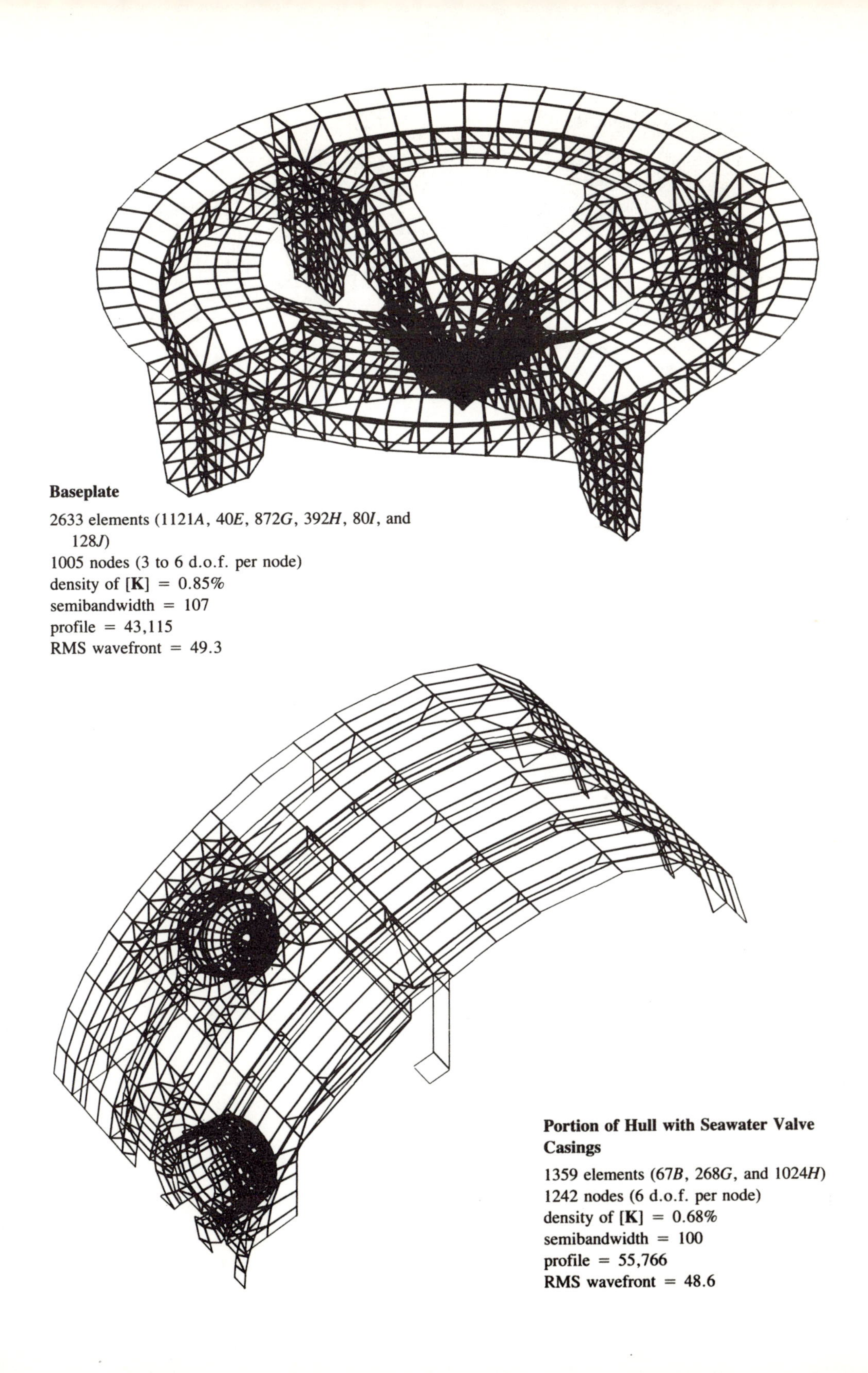

Baseplate

2633 elements (1121*A*, 40*E*, 872*G*, 392*H*, 80*I*, and 128*J*)
1005 nodes (3 to 6 d.o.f. per node)
density of [**K**] = 0.85%
semibandwidth = 107
profile = 43,115
RMS wavefront = 49.3

Portion of Hull with Seawater Valve Casings

1359 elements (67*B*, 268*G*, and 1024*H*)
1242 nodes (6 d.o.f. per node)
density of [**K**] = 0.68%
semibandwidth = 100
profile = 55,766
RMS wavefront = 48.6

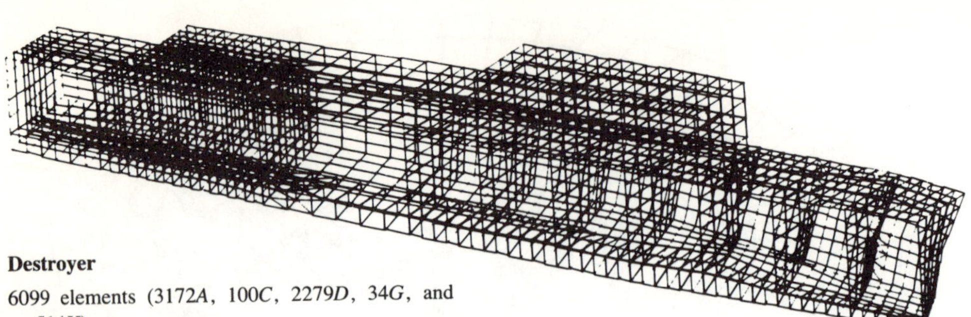

Destroyer

6099 elements (3172*A*, 100*C*, 2279*D*, 34*G*, and 514*H*)
2680 nodes (2 to 6 d.o.f. per node)
density of [**K**] = 0.35%
semibandwidth = 69
profile = 104,252
RMS wavefront = 39.9

REFERENCES

NOTATION FOR REFERENCES

The following abbreviations are used in the citations. This list is also a guide to the more useful publications about numerical methods in structural mechanics.

AdEnSo *Advances in Engineering Software.*
AeroJ *Aeronautical Journal* of the Royal Aeronautical Society.
AIAAJ Journal of the American Institute of Aeronautics and Astronautics.
CanJCE *Canadian Journal of Civil Engineering.*
CMAME *Computer Methods in Applied Mechanics and Engineering.*
CoMaAp *Computers and Mathematics with Applications.*
CompAD *Computer Aided Design.*
CompAE Sympos. *Computer Aided Engineering,* G. M. L. Gladwell, ed., University of Waterloo Press, Waterloo, Ontario, 1971.
CompSh Conf. *Computer Oriented Analysis of Shell Structures* (Conference Proceedings), August 1971 (AFFDL-TR-71-79, June 1972; AD-740-547, N.T.I.S.).
CompGr *Computers and Graphics,* An International Journal.
CompSt *Computers and Structures,* An International Journal.
EEStDy *International Journal of Earthquake Engineering and Structural Dynamics.*
EnFraM *Engineering Fracture Mechanics.*
FElCEl Sympos. *Application of Finite Element Methods in Civil Engineering,* W. H. Rowan, Jr., and R. M. Hackett, eds., American Society of Civil Engineers, New York, 1969.
FETSCM *Finite Elements for Thin Shells and Curved Members,* D. G. Ashwell and R. H. Gallagher, eds., John Wiley & Sons, New York, 1976.
IJFrac *International Journal of Fracture* (was *International Journal of Fracture Mechanics*).
IJMSci *International Journal of Mechanical Sciences.*
IJNAMG *International Journal for Numerical and Analytical Methods in Geomechanics.*
IJNME *International Journal for Numerical Methods in Engineering.*
IJSoSt *International Journal of Solids and Structures.*
IUTAM Sympos. *High Speed Computing of Elastic Structures,* B. Fraeijs de Veubeke, ed., University of Leige Press, Liege, Belgium, 1971.
JAeroS *Journal of the Aeronautical* (or *Aerospace*) *Sciences.*
JApplM *Journal of Applied Mechanics,* Transactions of ASME.
JEInd *Journal of Engineering for Industry,* Transactions of ASME.
JEMDiv *Journal of the Engineering Mechanics Division,* Proceedings of ASCE.
JFrIns *Journal of the Franklin Institute.*

JHeTra *Journal of Heat Transfer,* Transactions of ASME.

JMESci *Journal of Mechanical Engineering Science.*

JPrVeT *Journal of Pressure Vessel Technology,* Transactions of ASME.

JSMDiv *Journal of the Soil Mechanics and Foundations Division,* Proceedings of ASCE (now *Journal of the Geotechnical Division*).

JSouVi *Journal of Sound and Vibration.*

JStDiv *Journal of the Structural Division,* Proceedings of ASCE.

JStrAn *Journal of Strain Analysis.*

MTAC *Mathematical Tables and Other Aids to Computation.*

NASTRAN U.C. NASTRAN User's Colloquium. Sponsored by NASA and held annually. Proceedings indexed and abstracted in STAR [1.15].

NHT *Numerical Heat Transfer.*

NucEDe *Nuclear Engineering and Design.*

ONR Sympos. *Numerical and Computer Methods in Structural Mechanics,* S. J. Fenves, N. Perrone, A. R. Robinson, and W. C. Schnobrich, eds., Academic Press, New York, 1973.

PInsCE *Proceedings of the Institution of Civil Engineers.*

ShViDi *The Shock and Vibration Digest* [1.17].

SMCP Sympos. *Structural Mechanics Computer Programs: Surveys, Assessments, and Availability,* W. Pilkey, K. Saczalski, and H. Schaeffer, eds., University Press of Virginia, Charlottesville, 1974.

SMiRTI Conf. "Proceedings of the *I*th International Conference on Structural Mechanics in Reactor Technology."

Sparse2 Conf. *Sparse Matrices and Their Applications,* D. J. Rose and R. A. Willoughby, eds., Plenum Press, New York, 1972.

SSDM Conf. "AIAA/ASME/SAE Structures, Structural Dynamics and Materials Conference" (held annually).

USJap1 Conf. *Recent Advances in Matrix Methods of Structural Analysis and Design,* R. H. Gallagher, Y. Yamada, and J. T. Oden, eds., University of Alabama Press, Huntsville, 1971.

USJap2 Conf. *Advances in Computational Methods in Structural Mechanics and Design,* J. T. Oden, R. W. Clough, and Y. Yamamoto, eds., University of Alabama Press, Huntsville, 1972.

USJap3 Conf. *Theory and Practice in Finite Element Structural Analysis,* Y. Yamada and R. H. Gallagher, eds., University of Tokyo Press, Tokyo, 1973.

WPAFB1 Conf. "Proceedings of the [First] Conference on Matrix Methods in Structural Mechanics," Wright-Patterson Air Force Base, Ohio, October 1965 (AFFDL-TR-66-80, November 1966; AD-646-300, N.T.I.S.).

WPAFB2 Conf. "Proceedings of the Second Conference on Matrix Methods in Structural Mechanics," Wright-Patterson Air Force Base, Ohio, October 1968 (AFFDL-TR-68-150, December 1969; AD-703-685, N.T.I.S.).

WPAFB3 Conf. "Proceedings of the Third Conference on Matrix Methods in Structural Mechanics," Wright-Patterson Air Force Base, Ohio, October 1971.

CHAPTER 1

1.1 J. Ergatoudis, B. M. Irons, and O. C. Zienkiewicz, "Three-Dimensional Analysis of Arch Dams and Their Foundations," *Research Report No. C/R/74/67,* University of Wales, Swansea, 1968.

1.2 K. Wieghardt, "Über einen Grenzübergang der Elastizitätslehre und seine Anwendung auf die Statik hochgradig statisch unbestimmter Fachwerke," *Verhandlungen des Vereins z. Beförderung des Gewerbefleisses. Abhandlungen,* Vol. 85, 1906, pp. 139-176.

1.3 W. Riedel, "Beiträge zur Lösung des ebenen Problems eines elastischen Körpers mittels der Airyschen Spannungsfunktion," *Zeitschrift für Angewandte Mathematik und Mechanik,* Vol. 7, No. 3, 1927, pp. 169-188.

1.4 A. L. Yettram and H. M. Husain, "Plane Framework Models for Plates in Extension, *JEMDiv,* Vol. 92, No. EM1, 1966, pp. 157-168.

1.5 R. Courant, "Variational Methods for the Solution of Problems of Equilibrium and Vibrations," *Bulletin of the American Mathematical Society,* Vol. 49, 1943, pp. 1-23.

1.6 S. Levy, "Structural Analysis and Influence Coefficients for Delta Wings," *JAeroS,* Vol. 20, No. 7, 1953, pp. 449-454.

1.7 M. J. Turner, R. W. Clough, H. C. Martin, and L. J. Topp, "Stiffness and Deflection Analysis of Complex Structures," *JAeroS,* Vol. 23, No. 9, 1956, pp. 805-823.

1.8 J. H. Argyris and S. Kelsey, *Energy Theorems and Structural Analysis,* Butterworths, London, 1960 (collection of papers published in *Aircraft Engineering* in 1954 and 1955).

1.9 R. W. Clough, "The Finite Element Method in Plane Stress Analysis," *Proceedings of the Second Conference on Electronic Computation,* American Society of Civil Engineers, New York, 1960, pp. 345-377.

1.10 S. W. Key and R. D. Krieg, "Comparison of Finite-Element and Finite Difference Methods," *ONR Sympos.,* pp. 337-352.

1.11 F. Birch, "The Effect of Pressure upon the Elastic Parameters of Isotropic Solids, According to Murnaghan's Theory of Finite Strain," *Journal of Applied Physics,* Vol. 9, 1938, pp. 279-288.

1.12 P. W. Bridgman, "The Effect of Tension on the Thermal and Electrical Conductivity of Metals," *Proceedings of the Academy of Arts and Sciences,* Vol. 59, 1923, pp. 119-137.

1.13 A. R. Rosenfield and B. L. Averbach, "Effect of Stress on the Expansion Coefficient," *Journal of Applied Physics,* Vol. 27, 1956, pp. 154-156.

1.14 *Government Reports Announcements and Index,* N.T.I.S. (National Technical Information Service), U.S. Department of Commerce, Springfield, Va. (published every two weeks).

1.15 *STAR (Scientific and Technical Aerospace Reports),* prepared by NASA, available from U.S. Government Printing Office, Washington, D.C. (published every two weeks).

1.16 *International Aerospace Abstracts,* AIAA, New York (abstracts and indexes periodicals, books, conference proceedings, and translations; published every two weeks).

1.17 *The Shock and Vibration Digest,* The Shock and Vibration Information Center, Naval Research Laboratory, Washington, D.C. (published monthly).

1.18 J. R. Whiteman, *A Bibliography for Finite Elements,* Academic Press, London, 1975.

1.19 D. Norrie and G. de Vries, *Finite Element Bibliography,* IFI/Plenum, New York, 1976.

1.20 S. P. Timoshenko and J. N. Goodier, *Theory of Elasticity,* 3rd. Ed., McGraw-Hill, New York, 1970.

CHAPTER 2

2.1 K. J. Bathe and E. L. Wilson, *Numerical Methods in Finite Element Analysis,* Prentice-Hall, Englewood Cliffs, N.J., 1976.

2.2 A. Jennings, *Matrix Computation for Engineers and Scientists,* John Wiley & Sons, London, 1977.

2.3 I. S. Duff, "A Survey of Sparse Matrix Research," *Proceedings of the IEEE,* Vol. 65, No. 4. 1977, pp. 500-535 (cites 604 references).

2.4 E. H. Cuthill, "Several Strategies for Reducing the Bandwidth of Matrices," *Sparse2 Conf.,* pp. 157-166.

2.5 N. E. Gibbs, W. G. Poole, Jr., and P. K. Stockmeyer, "An Algorithm for Reducing the Bandwidth and Profile of a Sparse Matrix," *SIAM Journal on Numerical Analysis,* Vol. 13, No. 2, 1976, pp. 236-250.

2.6 K. J. Bathe, E. L. Wilson, and F. E. Peterson, "SAP IV: A Structural Analysis Program for Static and Dynamic Response of Linear Systems," *Report EERC 73-11,* University of California, Berkeley, 1973.

2.7 E. L. Wilson, "A Computer Program for the Dynamic Analysis of Underground Structures," *Report 68-1,* Civil Engineering Department, University of California, Berkeley, January 1968 (AD-832-681, N.T.I.S.).

2.8 W. Weaver, *Computer Programs for Structural Analysis,* D. Van Nostrand, New York, 1967.

2.9 J. H. Wilkinson and C. Reinsch, eds., *Handbook of Automatic Computation, Vol. II—Linear Algebra,* Springer-Verlag, Berlin, 1971, p. 10.

2.10 P. Tong and J. N. Rossettos, *Finite-Element Method: Basic Technique and Application,* MIT Press, Cambridge, Mass., 1977.

2.11 G. C. Everstine, "A Comparison of Three Resequencing Algorithms for the Reduction of Matrix Profile and Wavefront," *IJNME,* Vol. 14, No. 6, 1979, pp. 837-853.

2.12 J. H. Argyris and O. E. Brönlund, "The Natural Factor Formulation of the Stiffness for the Matrix Displacement Method," *CMAME,* Vol. 5, No. 1, 1975, pp. 97-119.

2.13 C. Meyer, "Solution of Linear Equations—State-of-the-Art," *JStDiv,* Vol. 99, No. ST7, 1973, pp. 1507-1526 (discussion: Vol. 100, No. ST4, pp. 828-829; Vol. 100, No. ST8, pp. 1724-1728).

2.14 C. Meyer, "Special Problems Related to Linear Equation Solvers," *JStDiv,* Vol. 101, No. ST4, 1975, pp. 869-890.

2.15 D. J. Harman, "Structural Analysis Using a Basic Frontal Method," *CanJCE,* Vol. 4, No. 3, 1977, pp. 392-403.

2.16 H. G. Jensen and G. A. Parks, "Efficient Solutions for Linear Matrix Equations," *JStDiv,* Vol. 96, No. ST1, 1970, pp. 49-64.

2.17 W. F. Tinney and J. W. Walker, "Direct Solutions of Sparse Network Equations by Optimally Ordered Triangular Factorization," *Proceedings of the IEEE,* Vol. 55, No. 11, 1967, pp. 1801-1809.

2.18 E. L. Wilson, K. J. Bathe, and W. P. Doherty, "Direct Solution of Large Systems of Linear Equations," *CompSt,* Vol. 4, No. 2, 1974, pp. 363-372.

2.19 D. P. Mondkar and G. H. Powell, "Towards Optimal In-Core Equation Solving," *CompSt,* Vol. 4, No. 3, 1974, pp. 531-548.

2.20 E. L. Wilson and H. H. Dovey, "Solution or Reduction of Equilibrium Equations for Large Complex Structural Systems," *AdEnSo,* Vol. 1, No. 1, 1978, pp. 19-25.

2.21 C. A. Felippa, "Solution of Linear Equations with Skyline-Stored Symmetric Matrix," *CompSt,* Vol. 5, No. 1, 1975, pp. 13-29.

2.22 S. K. Gupta and K. K. Tanji, "Computer Program for Solution of Large, Sparse, Unsymmetric Systems of Linear Equations," *IJNME,* Vol. 11, No. 8, 1977, pp. 1251-1259.

2.23 C. P. Vendhan, M. P. Kapoor, and Y. C. Das, "An Integrated Sequential Solver for Large Matrix Equations," *IJNME,* Vol. 8, No. 2, 1974, pp. 227-248.

2.24 P. Hood, "Frontal Solution Program for Unsymmetric Matrices," *IJNME,* Vol. 10, No. 2, 1976, pp. 379-399 (comment: Vol. 11, No. 3, p. 604).

2.25 B. M. Irons, "A Frontal Solution Program for Finite Element Analysis," *IJNME,* Vol. 2, No. 1, 1970, pp. 5-32 (comment: Vol. 3, No. 1, p. 149).
2.26 C. J. Apelt and L. T. Isaacs, "On the Estimation of the Optimum Accelerator for SOR Applied to Finite Element Methods," *CMAME,* Vol. 12, No. 3, 1977, pp. 383-391.
2.27 Y. R. Rashid, "Three Dimensional Analysis of Elastic Solids," *IJSoSt,* "I—Analysis Procedure," Vol. 5, No. 12, 1969, pp. 1311-1331; "II—The Computational Problem," Vol. 6, No. 1, 1970, pp. 195-207.
2.28 I. Fried, "A Gradient Computational Procedure for the Solution of Large Problems Arising from the Finite Element Discretization Method," *IJNME,* Vol. 2, No. 4, 1970, pp. 477-494.
2.29 R. H. Mallett and L. A. Schmit, "Nonlinear Structural Analysis by Energy Search," *JStDiv,* Vol. 93, No. ST3, 1967, pp. 221-234.
2.30 O. Axelsson, "A Class of Iterative Methods for Finite Element Equations," *CMAME,* Vol. 9, No. 2, 1976, pp. 123-137.
2.31 E. Hinton and D. R. J. Owen, *Finite Element Programming,* Academic Press, London, 1977.
2.32 H. A. Kamel and M. W. McCabe, "Direct Numerical Solution of Large Sets of Simultaneous Equations," *CompSt,* Vol. 9, No. 2, 1978, pp. 113-123.
2.33 A. K. Noor and J. J. Lambiotte, Jr., "Finite Element Dynamic Analysis on CDC STAR-100 Computer," *CompSt,* Vol. 10, Nos. 1/2, 1979, pp. 7-20.
2.34 R. L. Fox and E. L. Stanton, "Developments in Structural Analysis by Direct Energy Minimization," *AIAAJ,* Vol. 6, No. 6, 1968, pp. 1036-1042.

CHAPTER 3

3.1 R. W. McLay, "On Natural Boundary Conditions in the Finite Element Method, *JApplM,* Vol. 39, No. 4, 1972, pp. 1149-1150.
3.2 Y. C. Fung, *Foundations of Solid Mechanics,* Prentice-Hall, Englewood Cliffs, N.J., 1965.
3.3 R. E. Jones, "A Generalization of the Direct-Stiffness Method of Structural Analysis," *AIAAJ,* Vol. 2, No. 5, 1964, pp. 821-826.
3.4 H. L. Langhaar, *Energy Methods in Applied Mechanics,* John Wiley & Sons, New York, 1962.
3.5 I. S. Sokolnikoff, *Mathematical Theory of Elasticity,* McGraw-Hill, New York, 1956.
3.6 K. Washizu, *Variational Methods in Elasticity and Plasticity,* Pergamon Press, Oxford, England, 1968.
3.7 L. R. Herrmann, "Interpretation of Finite Element Procedure as Stress Error Minimization Procedure," *JEMDiv,* Vol. 98, No. EM5, 1972, pp. 1330-1336.
3.8 E. Hinton and J. S. Campbell, "Local and Global Smoothing of Discontinuous Finite Element Functions Using a Least Squares Method," *IJNME,* Vol. 8, No. 3, 1974, pp. 461-480.

CHAPTER 4

4.1 R. J. Melosh, "Basis for Derivation of Matrices for the Direct Stiffness Method," *AIAAJ,* Vol. 1, No. 7, 1963, pp. 1631-1637.
4.2 A. K. Gupta and B. Mohraz, "A Method of Computing Numerically Integrated Stiffness Matrices," *IJNME,* Vol. 5, No. 1, 1972, pp. 83-89.

4.3 R. H. Gallagher and C. H. Lee, "Matrix Dynamic and Stability Analyses With Nonuniform Elements," *IJNME,* Vol. 2, No. 2, 1970, pp. 265-275.

4.4 G. P. Bazeley, Y. K. Cheung, B. M. Irons, and O. C. Zienkiewicz, "Triangular Elements in Plate Bending—Conforming and Non-Conforming Solutions, *WPAFB1 Conf.*, pp. 547-576.

4.5 G. Strang and G. J. Fix, *An Analysis of the Finite Element Method,* Prentice-Hall, Englewood Cliffs, N.J., 1973.

4.6 R. Narayanaswami, "Dependence of Plate-Bending Finite Element Deflections and Eigenvalues on Poisson's Ratio," *AIAAJ,* Vol. 12, No. 10, 1974, pp. 1420-1421.

4.7 R. J. Melosh and D. W. Lobitz, "On a Numerical Sufficiency Test for Monotonic Convergence of Finite Element Models," *AIAAJ,* Vol. 13, No. 5, 1975, pp. 675-678.

4.8 J. T. Oden and H. J. Brauchli, "On the Calculation of Consistent Stress Distributions in Finite Element Approximations," *IJNME,* Vol. 3, No. 3, 1971, pp. 317-325.

4.9 E. Stein and R. Ahmad, "An Equilibrium Method for Stress Calculation Using Finite Element Displacement Models," *CMAME,* Vol. 10, No. 2, 1977, pp. 175-198.

4.10 L. R. Herrmann, "Improved Stress Calculation for Simple Quadrilateral Elements," *CompSt,* Vol. 6, No. 2, 1976, pp. 141-148.

4.11 G. Loubignac, G. Cantin, and G. Touzot, "Continuous Stress Fields in Finite Element Analysis," *AIAAJ,* Vol. 15, No. 11, 1977, pp. 1645-1647.

4.12 I. U. Ojalvo, "Improved Thermal Stress Determination by Finite Element Methods," *AIAAJ,* Vol. 12, No. 8, 1974, pp. 1131-1132.

4.13 D. Bushnell, "Finite Difference Energy Models versus Finite Element Models: Two Variational Approaches in One Computer Program," *ONR Sympos.*, pp. 291-336.

4.14 "Finite Elements versus Finite Differences," panel discussion at *CompSh Conf.*, pp. 798-824.

4.15 B. O. Almroth and C. A. Felippa, "Structural Stability," *SMCP Sympos.*, pp. 498-539.

4.16 T. A. Cruse and F. J. Rizzo, eds., *Boundary-Integral Equation Method: Computational Applications in Applied Mechanics,* Applied Mechanics Division, Vol. 11, American Society of Mechanical Engineers, New York, 1975.

4.17 O. C. Zienkiewicz, D. W. Kelly, and P. Bettess, "The Coupling of the Finite Element Method and Boundary Solution Procedure," *IJNME,* Vol. 11, No. 2, 1977, pp. 355-375.

CHAPTER 5

5.1 B. M. Irons, "Engineering Applications of Numerical Integration in Stiffness Methods," *AIAAJ,* Vol. 4, No. 11, 1966, pp. 2035-2037.

5.2 J. Robinson, *Integrated Theory of Finite Element Methods,* John Wiley & Sons, London, 1973.

5.3 Z. Kopal, *Numerical Analysis,* John Wiley & Sons, New York, 1955.

5.4 A. H. Stroud and D. Secrest, *Gaussian Quadrature Formulas,* Prentice-Hall, Englewood Cliffs, N.J., 1966.

5.5 K. J. Bathe and E. L. Wilson, *Numerical Methods in Finite Element Analysis,* Prentice-Hall, Englewood Cliffs, N.J., 1976.

5.6 B. M. Irons, "Economical Computer Techniques for Numerically Integrated Finite Elements," *IJNME,* Vol. 1, No. 2, 1969, pp. 201-203.

5.7 I. Ergatoudis, B. M. Irons, and O. C. Zienkiewicz, "Curved Isoparametric, 'Quadrilateral' Elements for Finite Element Analysis," *IJSoSt,* Vol. 4, No. 1, 1968, pp. 31-42.

5.8 R. L. Taylor, "On Completeness of Shape Functions for Finite Element Analysis," *IJNME,* Vol. 4, No. 1, 1972, pp. 17-22.
5.9 O. C. Zienkiewicz, "Isoparametric and Allied Numerically Integrated Elements—A Review," *ONR Sympos.,* pp. 13-41.
5.10 O. C. Zienkiewicz, *The Finite Element Method,* 3rd Ed., McGraw-Hill, London, 1977.
5.11 B. M. Irons, "Quadrature Rules for Brick Based Finite Elements," *IJNME,* Vol. 3, No. 2, 1971, pp. 293-294.
5.12 T. K. Hellen, "Effective Quadrature Rules for Quadratic Solid Isoparametric Finite Elements," *IJNME,* Vol. 4, No. 4, 1972, pp. 597-599.
5.13 P. C. Hammer and A. H. Stroud, "Numerical Evaluation of Multiple Integrals II," *MTAC,* Vol. 12, No. 64, 1958, pp. 272-280.
5.14 E. Hinton and J. S. Campbell, "Local and Global Smoothing of Discontinuous Finite Element Functions Using a Least Squares Method," *IJNME,* Vol. 8, No. 3, 1974, pp. 461-480.
5.15 J. H. Argyris and K. J. Willam, "Some Considerations for the Evaluation of Finite Element Models," *NucEDe,* Vol. 28, No. 1, 1974, pp. 76-96.
5.16 E. Hinton, F. C. Scott, and R. E. Ricketts, "Local Least Squares Smoothing for Parabolic Isoparametric Elements," *IJNME,* Vol. 9, No. 1, 1975, pp. 235-238.
5.17 J. Barlow, "Optimal Stress Locations in Finite Element Models," *IJNME,* Vol. 10, No. 2, 1976, pp. 243-251 (discussion: Vol. 11, No. 3, p. 604).
5.18 R. D. Cook, "Advoidance of Parasitic Shear in Plane Element," *JStDiv,* Vol. 101, No. ST6, 1975, pp. 1239-1253.
5.19 J. A. Stricklin, W. S. Ho, E. Q. Richardson, and W. E. Haisler, "On Isoparametric vs. Linear Strain Triangular Elements," *IJNME,* Vol. 11, No. 6, 1977, pp. 1041-1043.
5.20 J. Backlund, "On Isoparametric Elements," *IJNME,* Vol. 12, No. 4, 1978, pp. 731-732.
5.21 V. T. Nicholas and E. Citipitioglu, "A General Isoparametric Finite Element Program SDRC SUPERB," *CompSt,* Vol. 7, No. 2, 1977, pp. 303-313.
5.22 L. N. Gifford, "More on Distorted Isoparametric Elements," *IJNME,* Vol. 14, No. 2, 1979, pp. 290-291.
5.23 J. S. Holt and P. S. Hope, "Displacement Oscillation in Plane Quadratic Isoparametric Elements in Orthotropic Situations," *IJNME,* Vol. 14, No. 6, 1979, pp. 913-920.
5.24 N. Bicanic and E. Hinton, "Spurious Modes in Two-Dimensional Isoparametric Elements," *IJNME,* Vol. 14, No. 10, 1979, pp. 1545-1557.

CHAPTER 6

6.1 S. J. Brown, Jr., "A Finite Plate Method to Solve Cylinder-to-Cylinder Structures Subjected to Internal Pressure," *JPrVeT,* Vol. 99, No. 3, 1977, pp. 404-412.
6.2 C. A. Felippa and T. L. Geers, "Constraint Techniques for Coupling of Discrete Axisymmetric and General Structures," *SMiRT3 Conf.,* 1975, Vol. 5, Part M, paper M5/1.
6.3 R. E. Miller, "Reduction of the Error in Eccentric Beam Modelling," *IJNME,* Vol. 15, No. 4, 1980, pp. 575-582.
6.4 B. E. Greene, R. E. Jones, R. W. McLay, and D. R. Strome, "Generalized Variational Principles in the Finite Element Method," *AIAAJ,* Vol. 7, No. 7, 1969, pp. 1254-1260.
6.5 R. W. McLay, "A Special Variational Principle for the Finite Element Method," *AIAAJ,* Vol. 7, No. 3, 1969, pp. 533-534.

6.6 C. A. Felippa, "Iterative Procedures for Improving Penalty Function Solutions of Algebraic Systems," *IJNME,* Vol. 12, No. 5, 1978, pp. 821-836.
6.7 P. G. Glockner, "Symmetry in Structural Mechanics," *JStDiv,* Vol. 99, No. ST1, 1973, pp. 71-89.
6.8 A. K. Noor and R. A. Camin, "Symmetry Considerations for Anisotropic Shells," *CMAME,* Vol. 9, No. 3, 1976, pp. 317-335.
6.9 P. D. Mangalgiri, B. Dattaguru, and T. S. Ramamurthy, "Specification of Skew Conditions in Finite Element Formulation," *IJNME,* Vol. 12, No. 6, 1978, pp. 1037-1041.
6.10 O. C. Zienkiewicz and F. C. Scott, "On the Principle of Repeatability and Its Application in Analysis of Turbine and Pump Impellers," *IJNME,* Vol. 4, No. 3, 1972, pp. 445-448.
6.11 D. L. Thomas, "Dynamics of Rotationally Periodic Structures," *IJNME,* Vol. 14, No. 1, 1979, pp. 81-102.
6.12 J. S. Przemieniecki, "Matrix Structural Analysis of Substructures," *AIAAJ,* Vol. 1, No. 1, 1963, pp. 138-147.
6.13 B. E. Kluttz, Jr., and S. Utku, "Best Partitions of a Structure in a Given Computing Environment," *CompSt,* Vol. 7, No. 1, 1977, pp. 35-45.
6.14 F. W. Williams, "Comparison Between Sparse Stiffness Matrix and Sub-Structure Methods," *IJNME,* Vol. 5, No. 3, 1973, pp. 383-394.
6.15 T. Furike, "Computerized Multiple Level Substructuring Analysis," *CompSt,* Vol. 2, Nos. 5/6, 1972, pp. 1063-1073.
6.16 H. Petersson and E. P. Popov, "Substructuring and Equation System Solutions in Finite Element Analysis," *CompSt,* Vol. 7, No. 2, 1977, pp. 197-206.
6.17 W. C. Hurty, "Dynamic Analysis of Structural Systems Using Component Modes," *AIAAJ,* Vol. 3, No. 4, 1965, pp. 678-685.
6.18 R. R. Craig, Jr., and M. C. C. Bampton, "Coupling of Substructures for Dynamic Analyses," *AIAAJ,* Vol. 6, No. 7, 1968, pp. 1313-1319.
6.19 M. R. Trubert, "A Practical Approach to Spacecraft Structural Dynamics Problems," *Journal of Spacecraft and Rockets,* Vol. 9, No. 11, 1972, pp. 818-824.
6.20 G. H. Holze and A. P. Boresi, "Free Vibration Analysis Using Substructuring," *JStDiv,* Vol. 101, No. ST12, 1975, pp. 2627-2639.
6.21 R. R. Craig, Jr., and C. J. Chang, "Free-Interface Methods of Substructure Coupling for Dynamic Analysis," *AIAAJ,* Vol. 14, No. 11, 1976, pp. 1633-1635.
6.22 R. R. Craig, Jr., "Methods of Component Mode Synthesis," *ShViDi,* Vol. 9, No. 11, 1977, pp. 3-10.
6.23 R. H. Gallagher and R. H. Mallett, "Efficient Solution Processes for Finite Element Analysis of Transient Heat Conduction," *JHeTra,* Vol. 93, No. 3, 1971, pp. 257-263.
6.24 J. F. Abel and M. S. Shephard, "An Algorithm for Multipoint Constraints in Finite Element Analysis," *IJNME,* Vol. 14, No. 3, 1979, pp. 464-467.

CHAPTER 7

7.1 R. W. Clough, "Comparison of Three Dimensional Finite Elements," *FElCE1 Sympos.,* pp. 1-26.
7.2 I. N. Katz, A. G. Peano, and M. P. Rossow, "Nodal Variables for Complete Conforming Finite Elements of Arbitrary Polynomial Order," *CoMaAp,* Vol. 4, No. 2, 1978, pp. 85-112.

7.3 C. D. Mote, Jr., "Global-Local Finite Element," *IJNME,* Vol. 3, No. 4, 1971, pp. 565-574.

7.4 R. D. Cook and V. N. Shah, "A Cost Comparison of Two Static Condensation-Stress Recovery Algorithms," *IJNME,* Vol. 12, No. 4, 1978, pp. 581-588.

7.5 E. L. Wilson, "The Static Condensation Algorithm," *IJNME,* Vol. 8, No. 1, 1974, pp. 198-203.

7.6 R. D. Cook, "Ways to Improve the Bending Response of Finite Elements," *IJNME,* Vol. 11, No. 6, 1977, pp. 1029-1039.

7.7 R. D. Cook, "More About 'Artificial' Softening of Finite Elements," *IJNME,* Vol. 11, No. 8, 1977, pp. 1334-1339.

7.8 R. D. Cook, "Avoidance of Parasitic Shear in Plane Element," *JStDiv,* Vol. 101, No. ST6, 1975, pp. 1239-1253.

7.9 W. P. Doherty, E. L. Wilson, and R. L. Taylor, "Stress Analysis of Axisymmetric Solids Utilizing Higher-Order Quadrilateral Finite Elements," *Report UC-SESM-69-3,* Civil Engineering Department, University of California, Berkeley, 1969 (PB-190-321, N.T.I.S.).

7.10 D. Kosloff and G. A. Frazier, "Treatment of Hourglass Patterns in Low Order Finite Element Codes," *IJNAMG,* Vol. 2, No. 1, 1978, pp. 57-72.

7.11 J. L. Bretl and R. D. Cook, "A New Eight-Node Solid Element," *IJNME,* Vol. 14, No. 4, 1979, pp. 593-615.

7.12 E. L. Wilson, R. L. Taylor, W. P. Doherty, and J. Ghaboussi, "Incompatible Displacement Models," *ONR Sympos.,* pp. 43-57.

7.13 R. L. Taylor, P. J. Beresford, and E. L. Wilson, "A Non-Conforming Element for Stress Analysis," *IJNME,* Vol. 10, No. 6, 1976, pp. 1211-1219.

7.14 R. T. Haftka and J. C. Robinson, "Effect of Out-of-Planeness of Membrane Quadrilateral Finite Elements," *AIAAJ,* Vol. 11, No. 5, 1973, pp. 742-744.

7.15 N. M. Ferrers, *An Elementary Treatise on Trilinear Coordinates, the Method of Reciprocal Polars, and the Theory of Projections,* Macmillan, London, 1861.

7.16 J. B. Mertie, "Transformation of Trilinear and Quadriplanar Coordinates to and from Cartesian Coordinates," *The American Mineralogist,* Vol. 49, Nos. 7/8, 1964, pp. 926-936.

7.17 M. A. Eisenberg and L. E. Malvern, "On Finite Element Integration in Natural Coordinates," *IJNME,* Vol. 7, No. 4, 1973, pp. 574-575.

7.18 I. Fried, "Some Aspects of the Natural Coordinate System in the Finite-Element Method," *AIAAJ,* Vol. 7, No. 7, 1969, pp. 1366-1368.

7.19 R. H. Gallagher, *Finite Element Analysis: Fundamentals,* Prentice-Hall, Englewood Cliffs, N.J., 1975.

7.20 P. C. Hammer, O. J. Marlow, and A. H. Stroud, "Numerical Integration Over Simplexes and Cones," *MTAC,* Vol. 10, No. 3, 1956, pp. 130-137.

7.21 J. Radon, "Zur Mechanischen Kubatur," *Monatshefte für Mathematik,* Vol. 52, No. 4, 1948, pp. 286-300.

7.22 M. E. Laursen and M. Gellert, "Some Criteria for Numerically Integrated Matrices and Quadrature Formulas for Triangles," *IJNME,* Vol. 12, No. 1, 1978, pp. 67-76.

7.23 A. H. Stroud, *Approximate Calculation of Multiple Integrals,* Prentice-Hall, Englewood Cliffs, N.J., 1971.

7.24 G. R. Cowper, E. Kosko, G. M. Lindberg, and M. D. Olson. "A High Precision Triangular Plate Bending Element," *Aeronautical Report LR-514,* National Research Council of Canada, Ottawa, December 1968 (AD-685-576, N.T.I.S.).

7.25 R. A. Tinawi, "Anisotropic Tapered Elements Using Displacement Models," *IJNME,* Vol. 4, No. 4, 1972, pp. 475-489.

7.26 K. Bell, "A Refined Triangular Plate Bending Finite Element," *IJNME,* Vol. 1, No. 1, 1969, pp. 101-122 (discussion: Vol. 1, No. 4, p. 395; Vol. 2, No. 1, pp. 146-147).

7.27 C. Caramanlian, K. A. Selby, and G. T. Will, "A Quintic Conforming Plate Bending Triangle," *IJNME,* Vol. 12, No. 7, 1978, pp. 1109-1130.

7.28 P. Pedersen, "On Computer-Aided Analytic Element Analysis and the Similarities of Tetrahedron Elements," *IJNME,* Vol. 11, No. 4, 1977, pp. 611-622.

7.29 J. Jensen and F. Niordson, "Symbolic and Algebraic Manipulation Languages and their Applications in Mechanics," *Structural Mechanics Software Series,* Vol. I, N. Perrone and W. Pilkey, eds., University Press of Virginia, Charlottesville, 1977, pp. 541-576.

7.30 A. K. Noor and C. M. Andersen, "Computerized Symbolic Manipulation in Structural Mechanics—Progress and Potential," *CompSt,* Vol. 10, Nos. 1/2, 1979, pp. 95-118.

7.31 B. M. Irons, "A Technique for Degenerating Brick-Type Isoparametric Elements Using Hierarchical Midside Nodes," *IJNME,* Vol. 8, No. 1, 1974, pp. 203-209.

7.32 R. E. Newton, "Degeneration of Brick-Type Isoparametric Elements," *IJNME,* Vol. 7, No. 4, 1973, pp. 579-581.

7.33 G. E. Ramey, "Some Effects of System Idealizations, Singularities and Mesh Patterns on Finite Element Solutions," *CompSt,* Vol. 4, No. 6, 1974, pp. 1173-1184.

7.34 P. F. Walsh, "Intensive Finite Element Grading for Stress Concentrations," *EnFraM,* Vol. 10, No. 2, 1978, pp. 211-213.

7.35 A. K. Gupta, "A Finite Element for Transition from a Fine to a Coarse Grid," *IJNME,* Vol. 12, No. 1, 1978, pp. 35-45.

7.36 O. Orringer and G. Stalk, "A Hybrid Finite Element of Stress Analysis of Fastener Details," *EnFraM,* Vol. 8, No. 4, 1976, pp. 719-729.

7.37 H. A. Kamel, D. Liu, W. McCabe, and V. Philippopoulos, "Some Developments in the Analysis of Complex Ship Structures," *USJap2 Conf.,* pp. 703-726.

7.38 I. C. Taig, "Modelling and Interpretation of Results on Finite Element Structural Analysis," *(First) World Congress on Finite Element Methods in Structural Mechanics,* J. Robinson, ed., Bournemouth, Dorset, England, October 1975, pp. B1-B36.

7.39 A. M. Ebner, ed., *Guidelines for Finite Element Idealization,* Preprint 2504, American Society of Civil Engineers National Structural Engineering Convention, New Orleans, 1975.

7.40 E. Stein and R. Ahmad, "Upon Sector Elements for Two-Dimensional Problems," *CompSt,* Vol. 5, Nos. 5/6, 1975, pp. 275-278.

7.41 F. Sawko and P. A. Merriman, "An Annular Segment Finite Element for Plate Bending," *IJNME,* Vol. 3, No. 1, 1971, pp. 119-129 (discussion and closure: Vol. 3, No. 4, pp. 589-591).

7.42 S. Singh and G. S. Ramaswamy, "A Sector Element for Thin Plate Flexure," *IJNME,* Vol. 4, No. 1, 1972, pp. 133-142.

7.43 P. V. T. Babu and D. V. Reddy, "Frequency Analysis of Skew Orthotropic Plates by the Finite Strip Method," *JSouVi,* Vol. 18, No. 4, 1971, pp. 465-474.

7.44 G. R. Monforton and M. G. Michail, "Finite Element Analysis of Skew Sandwich Plates," *JEMDiv,* Vol. 98, No. EM3, 1972, pp. 763-769.

7.45 P. Bettess and O. C. Zienkiewicz, "Diffraction and Refraction of Surface Waves Using Finite and Infinite Elements," *IJNME,* Vol. 11, No. 8, 1977, pp. 1271-1290.

7.46 P. Bettess, "Infinite Elements," *IJNME,* Vol. 11, No. 1, 1977, pp. 53-64.

7.47 T. K. Hellen, "On the Method of Virtual Crack Extensions," *IJNME,* Vol. 9, No. 1, 1975, pp. 187-207.

7.48 O. Orringer et al., "K-Solutions with Assumed-Stress Elements," *JStDiv,* Vol. 103, No. ST2, 1977, pp. 321-334.
7.49 L. N. Gifford, Jr., and P. D. Hilton, "Stress Intensity Factors by Enriched Finite Elements," *EnFraM,* Vol. 10, No. 3, 1978, pp. 485-496.
7.50 J. E. Akin, "The Generation of Elements with Singularities," *IJNME,* Vol. 10, No. 6, 1976, pp. 1249-1259.
7.51 R. D. Henshell and K. G. Shaw, "Crack Tip Finite Elements are Unnecessary," *IJNME,* Vol. 9, No. 3, 1975, pp. 495-507.
7.52 Y. Yamada, Y. Ezawa, and I. Nishiguchi, "Reconsiderations on Singularity or Crack-Tip Elements," *IJNME,* Vol. 14, No. 10, 1979, pp. 1525-1544.
7.53 C. F. Shih, H. G. deLorenzi, and M. D. German, "Crack Extension Modeling with Singular Quadratic Isoparametric Elements," *IJFrac,* Vol. 12, No. 4, 1976, pp. 647-651.
7.54 R. S. Barsoum, "Triangular Quarter-Point Elements as Elastic and Perfectly-Plastic Crack Tip Elements," *IJNME,* Vol. 11, No. 1, 1977, pp. 85-98.
7.55 C. E. Freese and D. M. Tracey, "The Natural Isoparametric Triangle versus Collapsed Quadrilateral for Elastic Crack Analysis," *IJFrac,* Vol. 12, No. 5, 1976, pp. 767-770.
7.56 J. M. Bloom, "Determination of Stress Intensity Factors for Axisymmetric Crack Problems," *IJFrac,* Vol. 12, No. 5, 1976, pp. 771-774.
7.57 R. S. Barsoum, "A Degenerate Solid Element for Linear Fracture Analysis of Plate Bending and General Shells," *IJNME,* Vol. 10, No. 3, 1976, pp. 551-564.
7.58 S. L. Pu, M. A. Hussain, and W. E. Lorensen, "The Collapsed Cubic Isoparametric Element as a Singular Element for Crack Problems," *IJNME,* Vol. 12, No. 11, 1978, pp. 1727-1742.
7.59 H. D. Hibbit, "Some Properties of Singular Isoparametric Elements," *IJNME,* Vol. 11, No. 1, 1977, pp. 180-184.
7.60 P. P. Lynn and A. R. Ingraffea, "Transition Elements to be Used with Quarter-Point Crack-Tip Elements," *IJNME,* Vol. 12, No. 6, 1978, pp. 1031-1036.
7.61 R. D. Cook, "Strain Resultants in Certain Finite Elements," *AIAAJ,* Vol. 7, No. 3, 1969, p. 535.
7.62 J. W. Harvey and W. C. Clark, "Smooth Interfacing of Finite Element Models," *IJNME,* Vol. 14, No. 7, 1979, pp. 1073-1078.

CHAPTER 8

8.1 E. L. Wilson, "Structural Analysis of Axisymmetric Solids," *AIAAJ,* Vol. 3, No. 12, 1965, pp. 2269-2274.
8.2 J. G. Crose and R. M. Jones, *SAAS III: Finite Element Analysis of Axisymmetric and Plane Solids with Different Orthotropic, Temperature-Dependent Material Properties in Tension and Compression,* Aerospace Corp., San Bernardino, Calif., 1971 (AD-729-188, N.T.I.S.).
8.3 R. D. Cook, "A Note on Certain Incompatible Elements," *IJNME,* Vol. 6, No. 1, 1973, pp. 146-147.
8.4 O. C. Zienkiewicz and Y. K. Cheung, "Stresses in Shafts," *The Engineer* (London), Vol. 224, No. 5835, 1967, pp. 696-697.
8.5 T. Belytschko, "Finite Elements for Axisymmetric Solids Under Arbitrary Loadings with Nodes on Origin," *AIAAJ,* Vol. 10, No. 11, 1972, pp. 1532-1533 (discussion and closure: Vol. 11, No. 9, pp. 1357-1358).

8.6 C. P. Stavrinidis, "Elimination of Singularities in Harmonic Elements," *CMAME,* Vol. 10, No. 3, 1977, pp. 355-357.

8.7 J. Padovan, "Quasi-Analytical Finite Element Procedures for Axisymmetric Anisotropic Shells and Solids," *CompSt,* Vol. 4, No. 3, 1974, pp. 467-483.

8.8 J. Frater, J. Lestingi, and J. Padovan, "Complex Stiffness Formulation for the Finite Element Analysis of Anisotropic Axisymmetric Solids Subjected to Nonsymmetric Loads," *SMiRT4 Conf.,* Vol. M, 1977.

8.9 J. G. Crose, "Bandwidth Minimization of Stiffness Matrices," *JEMDiv,* Vol. 97, No. EM1, 1971, pp. 163-167.

8.10 J. G. Crose, "Stress Analysis of Axisymmetric Solids with Asymmetric Properties," *AIAAJ,* Vol. 10, No. 7, 1972, pp. 866-871.

8.11 G. C. Pardoen, "Axisymmetric Stress Analysis of Axisymmetric Solids with Anisotropic Material Properties," *AIAAJ,* Vol. 15, No. 10, 1977, pp. 1498-1500.

8.12 E. L. Wilson and P. C. Pretorius, "A Computer Program for the Analysis of Prismatic Solids," *Report UC-SESM-70-21,* Civil Engineering Department, University of California, Berkeley, 1970 (PB-196-462, N.T.I.S.).

8.13 O. C. Zienkiewicz and J. M. Too, "The Finite Prism in Analysis of Thick Simply Supported Bridge Boxes," *PInsCE,* Vol. 53, Part 2, 1972, pp. 147-172.

8.14 G. A. Greenbaum, L. D. Hofmeister, and D. A. Evenson, "Pure Moment Loading of Axisymmetric Finite Element Models," *IJNME,* Vol. 5, No. 4, 1973, pp. 459-463.

8.15 P. Pedersen and M. M. Megahed, "Axisymmetric Element Analysis Using Analytical Computing," *CompSt,* Vol. 5, No. 4, 1975, pp. 241-247.

8.16 K. J. Bathe and C. A. Almeida, "A Simple and Effective Pipe Elbow Element—Linear Analysis," *JApplM,* Vol. 47, No. 1, 1980, pp. 93-100.

CHAPTER 9

9.1 S. Timoshenko and S. Woinowsky-Krieger, *Theory of Plates and Shells,* 2nd Ed., McGraw-Hill, New York, 1959.

9.2 Comments on Ref. 4.1 in *AIAAJ:* Vol. 2, No. 2, 1964, p. 403; Vol. 2, No. 6, 1964, p. 1161; Vol. 3, No. 6, 1965, pp. 1215-1216.

9.3 R. W. Clough and J. L. Tocher, "Finite Element Stiffness Matrices for Analysis of Plate Bending," *WPAFB1 Conf.,* pp. 515-545.

9.4 B. M. Irons and K. J. Draper, "Inadequacy of Nodal Connections in a Stiffness Solution for Plate Bending," *AIAAJ,* Vol. 3, No. 5, 1965, p. 961.

9.5 G. P. Bazeley, Y. K. Cheung, B. M. Irons, and O. C. Zienkiewicz, "Triangular Elements in Plate Bending—Conforming and Non-Conforming Solutions," *WPAFB1 Conf.,* pp. 547-576.

9.6 J. A. Stricklin, W. E. Haisler, P. R. Tisdale, and R. Gunderson, "A Rapidly Converging Triangular Plate Element," *AIAAJ,* Vol. 7, No. 1, 1969, pp. 180-181.

9.7 J. F. Abel and C. S. Desai, "Comparison of Finite Elements for Plate Bending," *JStDiv,* Vol. 98, No. ST9, 1972, pp. 2143-2148.

9.8 M. P. Rossow and K. C. Chen, "Computational Efficiency of Plate Elements," *JStDiv,* Vol. 103, No. ST2, 1977, pp. 447-451.

9.9 R. D. Henshell, D. Walters, and G. B. Warburton, "A New Family of Curvilinear Plate Bending Elements for Vibration and Stability," *JSouVi*, Vol. 20, No. 3, 1972, pp. 381-397 (discussion and closure: Vol. 23, No. 4, pp. 507-513).
9.10 E. D. L. Pugh, E. Hinton, and O. C. Zienkiewicz, "A Study of Quadrilateral Plate Bending Elements with 'Reduced' Integration," *IJNME*, Vol. 12, No. 7, 1978, pp. 1059-1079.
9.11 T. P. Khatua and Y. K. Cheung, "Triangular Element for Multilayer Sandwich Plates," *JEMDiv*, Vol. 98, No. EM5, 1972, pp. 1225-1238.
9.12 Y. K. Cheung, *Finite Strip Method in Structural Analysis*, Pergamon Press, Oxford, England, 1976.
9.13 S. Ahmad, B. M. Irons, and O. C. Zienkiewicz, "Analysis of Thick and Thin Shell Structures by Curved Finite Elements," *IJNME*, Vol. 2, No. 3, 1970, pp. 419-451.
9.14 O. C. Zienkiewicz, R. L. Taylor, and J. M. Too, "Reduced Integration Technique in General Analysis of Plates and Shells," *IJNME*, Vol. 3, No. 2, 1971, pp. 275-290.
9.15 F. J. Plantema, *Sandwich Construction*, John Wiley & Sons, New York, 1966.
9.16 J. M. Whitney and A. W. Leissa, "Analysis of Heterogeneous Anisotropic Plates," *JApplM*, Vol. 36, No. 2, 1969, pp. 261-266.
9.17 T. J. R. Hughes, R. L. Taylor, and W. Kanoknukulchai, "A Simple and Efficient Finite Element for Plate Bending," *IJNME*, Vol. 11, No. 10, 1977, pp. 1529-1543.
9.18 T. J. R. Hughes and M. Cohen, "The 'Heterosis' Finite Element for Plate Bending," *CompSt*, Vol. 9, No. 5, 1978, pp. 445-450.
9.19 T. J. R. Hughes, M. Cohen, and M. Haroun, "Reduced and Selective Integration Techniques in the Finite Element Analysis of Plates," *NucEDe*, Vol. 46, No. 1, 1978, pp. 203-222.
9.20 R. D. Cook, "More on Reduced Integration and Isoparametric Elements," *IJNME*, Vol. 5, No. 1, 1972, pp. 141-142.
9.21 P. G. Bergan, "Finite Elements Based on Energy Orthogonal Functions," *IJNME*, Vol. 15, No. 10, 1980, pp. 1541-1555 (comment: Vol. 17, No. 1, pp. 154-155).

CHAPTER 10

10.1 B. M. Irons, "The SemiLoof Shell Element," *FETSCM*, pp. 197-222.
10.2 N. C. Knowles, A. Razzaque, and J. B. Spooner, "Experience of Finite Element Analysis of Shell Structures," *FETSCM*, pp. 245-262.
10.3 D. G. Ashwell and A. B. Sabir, "On the Finite Element Calculation of Stress Distributions in Arches," *IJMSci*, Vol. 16, No. 1, 1974, pp. 21-29.
10.4 D. J. Dawe, "Curved Finite Elements for the Analysis of Shallow and Deep Arches," *CompSt*, Vol. 4, No. 3, 1974, pp. 559-580.
10.5 O. C. Zienkiewicz, *The Finite Element Method*, 3rd Ed., McGraw-Hill, London, 1977.
10.6 W. E. Haisler and J. A. Stricklin, "Rigid-Body Displacements of Curved Elements in the Analysis of Shells by the Matrix Displacement Method," *AIAAJ*, Vol. 5, No. 8, 1967, pp. 1525-1527.
10.7 P. M. Mebane and J. A. Stricklin, "Implicit Rigid Body Motion in Curved Finite Elements," *AIAAJ*, Vol. 9, No. 2, 1971, pp. 344-345.
10.8 G. Cantin, "Rigid Body Motions and Equilibrium in Finite Elements," *FETSCM*, pp. 55-61.

10.9 G. A. Fonder, "Studies in Doubly-Curved Elements of Shells of Revolution," *FETSCM*, pp. 113-129.

10.10 A. J. Morris, "A Summary of Appropriate Governing Equations and Functionals in the Finite Element Analysis of Thin Shells," *FETSCM*, pp. 15-39.

10.11 D. G. Ashwell and A. B. Sabir, "A New Cylindrical Shell Finite Element Based on Simple Independent Strain Functions," *IJMSci*, Vol. 14, No. 3, 1972, pp. 171-183.

10.12 R. W. Clough and E. L. Wilson, "Dynamic Finite Element Analysis of Arbitrary Thin Shells," *CompSt*, Vol. 1, Nos. 1/2, 1971, pp. 33-56.

10.13 G. Cantin, "Strain Displacement Relationships for Cylindrical Shells," *AIAAJ*, Vol. 6, No. 9, 1968, pp. 1787-1788.

10.14 H. Takemoto and R. D. Cook, "Some Modifications of an Isoparametric Shell Element," *IJNME*, Vol. 7, No. 3, 1973, pp. 401-405.

10.15 S. F. Pawsey and R. W. Clough, "Improved Numerical Integration of Thick Shell Finite Elements," *IJNME*, Vol. 3, No. 4, 1971, pp. 575-586.

10.16 R. H. Gallagher, "Problems and Progress in Thin Shell Finite Element Analysis," *FETSCM*, pp. 1-14.

10.17 L.S.D. Morley and B. C. Merrifield, "Polynomial Comparison Solutions in the Sanders-Koiter Theory of Circular Cylindrical Shells," *FETSCM*, pp. 41-53.

10.18 G. R. Cowper, "Stress Analysis at Corners of Box Structures," *CANCAM 75*, University of New Brunswick, Canada, May 26-30, 1975.

10.19 G. Cantin and R. W. Clough, "A Curved, Cylindrical Shell, Finite Element," *AIAAJ*, Vol. 6, No. 6, 1968, pp. 1057-1062.

10.20 S. Nor, *Ph.D. dissertation*, Université de Technologie, Compiegne, France, 1978.

10.21 E. A. Witmer and J. J. Kotanchik, "Progress Report on Discrete-Element Elastic and Elastic-Plastic Analyses of Shells of Revolution Subjected to Axisymmetric and Asymmetric Loading," *WPAFB2 Conf.*, pp. 1341-1453.

10.22 J. H. Percy, T. H. H. Pian, S. Klein, and D. R. Navaratna, "Application of Matrix Displacement Method to Linear Elastic Analysis of Shells of Revolution," *AIAAJ*, Vol. 3, No. 11, 1965, pp. 2138-2145.

10.23 J. A. Stricklin, D. R. Navaratna, and T. H. H. Pian, "Improvements on the Analysis of Shells by the Matrix Displacement Method," *AIAAJ*, Vol. 4, No. 11, 1966, pp. 2069-2071.

10.24 R. E. Ball, "A Program for the Static and Dynamic Analysis of Arbitrarily Loaded Shells of Revolution," *CompSh Conf.*, pp. 590-617.

10.25 J. A. Stricklin, W. E. Haisler, and W. A. Von Riesemann, "Large Deflection Elastic-Plastic Dynamic Response of Stiffened Shells of Revolution," *JPrVeT*, Vol. 96, No. 2, 1974, pp. 87-95.

10.26 L. J. Brombolich and P. L. Gould, "A High-Precision Curved Shell Finite Element," *12th SSDM Conf.*, April 1971.

10.27 D. Bushnell, "Thin Shells," *SMCP Sympos.*, pp. 277-358.

10.28 D. Bushnell, "Stress, Buckling and Vibration of Prismatic Shells," *AIAAJ*, Vol. 9, No. 10, 1971, pp. 2004-2013.

10.29 D. Bushnell, "Stress, Stability, and Vibration of Complex, Branched Shells of Revolution," *CompSt*, Vol. 4, No. 2, 1974, pp. 399-435.

10.30 S. Ahmad, B. M. Irons, and O. C. Zienkiewicz, "Curved Thick Shell and Membrane Elements with Particular Reference to Axisymmetric Problems," *WPAFB2 Conf.*, pp. 539-572.

10.31 O. C. Zienkiewicz, J. Bauer, K. Morgan, and E. Onate, "A Simple and Efficient Element for Axisymmetric Shells,'' *IJNME,* Vol. 12, No. 10, 1977, pp. 1545-1548 (discussion: Vol. 12, No. 7, pp. 1196-1197).

10.32 D. N. Buragohain, S. B. Agrawal, and R. S. Ayyar, ''A Matching Superparametric Beam Element for Shell Beam Problems,'' *CompSt,* Vol. 9, No. 2, 1978, pp. 175-182.

10.33 B. M. Irons and A. Razzaque, ''A Further Modification to Ahmad's Shell Element,'' *IJNME,* Vol. 5, No. 4, 1973, pp. 588-589.

10.34 M. D. Olson and T. W. Bearden, ''A Simple Flat Triangular Shell Element Revisited,'' *IJNME,* Vol. 14, No. 1, 1979, pp. 51-68.

10.35 W. Kanok-Nukulchai, ''A Simple and Efficient Finite Element for General Shell Analysis,'' *IJNME,* Vol. 14, No. 2, 1979, pp. 179-200.

10.36 G. H. Ferguson and R. D. Clark, ''A Variable Thickness, Curved Beam and Shell Stiffening Element with Shear Deformations,'' *IJNME,* Vol. 14, No. 4, 1979, pp. 581-592.

10.37 H. R. Meck, ''An Accurate Polynomial Displacement Function for Finite Ring Elements,'' *CompSt,* Vol. 11, No. 4, 1980, pp. 265-269.

CHAPTER 11

11.1 R. W. Clough and J. Penzien, *Dynamics of Structures,* McGraw-Hill, New York, 1975.

11.2 J. W. S. Rayleigh, *Theory of Sound,* 2nd Ed., Vols. I and II, Dover Publications, New York, 1945 (originally published in 1894).

11.3 J. S. Archer, ''Consistent Matrix Formulations for Structural Analysis Using Finite-Element Techniques,'' *AIAAJ,* Vol. 3, No. 10, 1965, pp. 1910-1918.

11.4 R. H. Gallagher and C. H. Lee, ''Matrix Dynamic and Instability Analyses with Non-Uniform Elements,'' *IJNME,* Vol. 2, No. 2, 1970, pp. 265-275.

11.5 S. H. Crandall, *Engineering Analysis,* McGraw-Hill, New York, 1956.

11.6 E. Dokumaci, ''A Critical Examination of Discrete Models in Vibration Problems of Continuous Systems,'' *JSouVi,* Vol. 53, No. 2, 1977, pp. 153-164.

11.7 T. Belytschko, ''A Survey of Numerical Methods and Computer Programs for Dynamic Structural Analysis,'' *NucEDe,* Vol. 37, No. 1, 1976, pp. 23-34.

11.8 Z. Bazant, J. L. Glazik, and J. D. Achenbach, ''Finite Element Analysis of Wave Diffraction by a Crack,'' *JEMDiv,* Vol. 102, No. EM3, 1976, pp. 479-496 (closure to discussion: Vol. 103, No. EM6, 1977, pp. 1181-1185).

11.9 E. Hinton, T. Rock, and O. C. Zienkiewicz, ''A Note on Mass Lumping and Related Processes in the Finite Element Method,'' *EEStDy,* Vol. 4, No. 3, 1976, pp. 245-249.

11.10 K. S. Surana, ''Lumped Mass Matrices with Non-Zero Inertia for General Shell and Axisymmetric Shell Elements,'' *IJNME,* Vol. 12, No. 11, 1978, pp. 1635-1650.

11.11 W. C. Hurty and M. F. Rubinstein, *Dynamics of Structures,* Prentice-Hall, Englewood Cliffs, N.J., 1964.

11.12 R. Craig, Jr., and M. C. C. Bampton, ''On the Iterative Solution of Semidefinite Eigenvalue Problems,'' *AeroJ,* Vol. 75, No. 724, 1971, pp. 287-290.

11.13 C. A. Felippa, ''Refined Finite Element Analysis of Linear and Nonlinear Two-Dimensional Structures,'' Ph.D. dissertation, University of California, Berkeley, 1966 (also available as PB-178-418 and PB-178-419, N.T.I.S.; computer programs in the latter).

11.14 R. J. Guyan, ''Reduction of Stiffness and Mass Matrices,'' *AIAAJ,* Vol. 3, No. 2, 1965, p. 380.

11.15 M. Geradin, "Error Bounds for Eigenvalue Analysis by Elimination of Variables," *JSouVi,* Vol. 19, No. 2, 1971, pp. 111-132.

11.16 R. G. Anderson, B. M. Irons, and O. C. Zienkiewicz, "Vibration and Stability of Plates Using Finite Elements," *IJSoSt,* Vol. 4, No. 10, 1968, pp. 1031-1055.

11.17 R. L. Kidder, "Reduction of Structural Frequency Equations," *AIAAJ,* Vol. 11, No. 6, 1973, p. 892 (discussion and closure: Vol. 13, No. 5, 1975, pp. 701-703).

11.18 J. D. Sowers, "Condensation of Free Body Mass Matrices Using Flexibility Coefficients," *AIAAJ,* Vol. 16, No. 3, 1978, pp. 272-273.

11.19 R. D. Henshell and J. H. Ong, "Automatic Masters for Eigenvalue Economization," *EEStDy,* Vol. 3, No. 4, 1975, pp. 375-383.

11.20 V. B. Watwood, T. Y. Chow, Z. Zudans, and W. H. Miller, "Combined Analysis and Evaluation of Piping Systems Using the Computer," *NucEDe,* Vol. 27, No. 3, 1974, pp. 334-342.

11.21 J. H. Wilkinson, *The Algebraic Eigenvalue Problem,* Clarendon Press, Oxford, England, 1965.

11.22 L. Fox, *An Introduction to Numerical Linear Algebra,* Oxford University Press, New York, 1965.

11.23 O. C. Zienkiewicz, R. W. Lewis, and K. G. Stagg, *Numerical Methods in Offshore Engineering,* John Wiley & Sons, Chichester, England, 1978.

11.24 P. C. Chowdhury, "The Truncated Lanczos Algorithm for Partial Solution of the Symmetric Eigenproblem," *CompSt,* Vol. 6, No. 6, 1976, pp. 439-446.

11.25 I. U. Ojalvo, "ALARM—A Highly Efficient Eigenvalue Extraction Routine for Very Large Matrices," *ShViDi,* Vol. 7, No. 12, 1975, pp. 3-9.

11.26 V. N. Shah, G. J. Bohm, and A. N. Nahavandi, "Modal Superposition Method for Computationally Economical Nonlinear Structural Analysis," *JPrVeT,* Vol. 101, No. 2, 1979, pp. 134-141.

11.27 J. A. Stricklin and W. E. Haisler, "Formulations and Solution Procedures for Nonlinear Structural Analysis," *CompSt,* Vol. 7, No. 1, 1977, pp. 125-136.

11.28 R. W. Clough and E. L. Wilson, "Dynamic Finite Element Analysis of Arbitrary Thin Shells," *CompSt,* Vol. 1, Nos. 1/2, 1971, pp. 33-56.

11.29 G. L. Goudreau and R. L. Taylor, "Evaluation of Numerical Integration Methods in Elastodynamics," *CMAME,* Vol. 2, No. 1, 1973, pp. 69-97.

11.30 S. Nagarajan and E. P. Popov, "Elastic-Plastic Dynamic Analysis of Axisymmetric Solids," *CompSt,* Vol. 4, No. 6, 1974, pp. 1117-1134.

11.31 E. M. Buturla and R. W. McLay, "On Sparse Matrices in Finite Element Wave Propagation," *JEInd,* Vol. 96, No. 3, 1974, pp. 1048-1053.

11.32 L. Lapidus and J. H. Seinfeld, *Numerical Solution of Ordinary Differential Equations,* Academic Press, New York, 1971.

11.33 K. C. Park, "An Improved Stiffly Stable Method for Direct Integration of Nonlinear Structural Dynamic Equations," *JApplM,* Vol. 42, No. 2, 1975, pp. 464-470.

11.34 J. R. Tillerson, "Selecting Solution Procedures for Nonlinear Structural Dynamics," *ShViDi,* Vol. 7, No. 4, 1975, pp. 2-13.

11.35 T. Belytschko, "Efficient Large Scale Nonlinear Transient Analysis by Finite Elements," *IJNME,* Vol. 10, No. 3, 1976, pp. 579-596.

11.36 D. Shantaram, D. R. J. Owen, and O. C. Zienkiewicz, "Dynamic Transient Behavior of Two- and Three-Dimensional Structures Including Plasticity, Large Deformation Effects and Fluid Interaction," *EEStDy,* Vol. 4, No. 6, 1976, pp. 561-578.

11.37 H. M. Hilber, T. J. R. Hughes, and R. L. Taylor, "Improved Numerical Dissipation for

Time Integration Algorithms in Structural Dynamics,'' *EEStDy*, Vol. 5, No. 3, 1977, pp. 283-292.

11.38 J. C. Houbolt, ''A Recurrence Matrix Solution for the Dynamic Response of Elastic Aircraft,'' *JAeroS*, Vol. 17, No. 9, 1950, pp. 540-550.

11.39 K. C. Park, ''Practical Aspects of Numerical Time Integration,'' *CompSt*, Vol. 7, No. 3, 1977, pp. 343-353.

11.40 D. M. Trujillo, ''An Unconditionally Stable Explicit Algorithm for Structural Dynamics,'' *IJNME*, Vol. 11, No. 10, 1977, pp. 1579-1592.

11.41 N. M. Newmark, ''A Method of Computation for Structural Dynamics,'' *JEMDiv*; Vol. 85, No. EM3, 1959, pp. 67-94.

11.42 S. W. Key, ''Concepts Underlying Finite Element Methods for Structural Analysis,'' *NucEDe*, Vol. 48, No. 1, 1978, pp. 259-268.

11.43 T. Belytschko, ''Transient Analysis,'' *SMCP Sympos.*, pp. 256-276.

11.44 T. Belytschko, J. R. Osias, and P. V. Marcal, eds., *Finite Element Analysis of Transient Nonlinear Structural Behavior*, Applied Mechanics Division, Vol. 14, American Society of Mechanical Engineers, New York, 1975.

11.45 K. K. Gupta, ''Free Vibration Analysis of Spinning Structural Systems,'' *IJNME*, Vol. 5, No. 3, 1973, pp. 395-418.

11.46 S. Putter and H. Manor, ''Natural Frequencies of Radial Rotating Beams,'' *JSouVi*, Vol. 56, No. 2, 1978, pp. 175-185.

11.47 F. V. Filho, ''Finite Element Analysis of Structures Under Moving Loads,'' *ShViDi*, Vol. 10, No. 8, 1978, pp. 27-35.

11.48 C. Mei, ''Finite Element Displacement Method for Large Amplitude Free Flexural Vibrations of Beams and Plates,'' *CompSt*, Vol. 3, No. 1, 1973, pp. 163-174.

11.49 G. V. Rao, I. S. Raju, and K. K. Raju, ''A Finite Element Formulation for Large Amplitude Flexural Vibrations of Thin Rectangular Plates,'' *CompSt*, Vol. 6, No. 3, 1976, pp. 163-167.

11.50 T. Y. Yang, ''Flutter of Flat Finite Element Panels in a Supersonic Potential Flow,'' *AIAAJ*, Vol. 13, No. 11, 1975, pp. 1502-1507.

11.51 C. D. Mote and G. Y. Matsumoto, ''Coupled, Nonconservative Stability-Finite Element,'' *JEMDiv*, Vol. 98, No. EM3, 1972, pp. 595-608.

11.52 S. Levy and J. P. D. Wilkinson, *The Component Element Method in Dynamics*, McGraw-Hill, New York, 1976.

11.53 T. J. R. Hughes, ''Recent Developments in Computer Methods for Structural Analysis,'' *NucEDe*, Vol. 57, No. 2, 1980, pp. 427-439.

11.54 O. E. Hansteen and K. Bell, ''On the Accuracy of Mode Superposition Analysis in Structural Dynamics,'' *EEStDy*, Vol. 7, No. 5, 1979, pp. 405-411.

11.55 A. Jennings, ''Eigenvalue Methods for Vibration Analysis,'' *ShViDi*, Vol. 12, No. 2, 1980, pp. 3-16 (146 references).

11.56 T. J. R. Hughes, K. S. Pister, and R. L. Taylor, ''Implicit-Explicit Finite Elements in Nonlinear Transient Analysis,'' *CMAME*, Vol. 17/18, Part I, 1979, pp. 159-182.

CHAPTER 12

12.1 R. H. Gallagher and J. Padlog, ''Discrete Element Approach to Structural Stability Analysis,'' *AIAAJ*, Vol. 1, No. 6, 1963, pp. 1437-1439.

12.2 K. K. Kapur and B. J. Hartz, "Stability of Plates Using the Finite Element Method," *JEMDiv,* Vol. 92, No. EM2, 1966, pp. 177-195.

12.3 R. W. Clough and C. A. Felippa, "A Refined Quadrilateral Element for Analysis of Plate Bending," *WPAFB2 Conf.,* pp. 399-440.

12.4 D. E. Beskos, "Framework Stability by Finite Element Method," *JStDiv,* Vol. 103, No. ST11, 1977, pp. 2273-2276.

12.5 C. K. Wang, *Computer Methods in Advanced Structural Analysis,* Intext Press, New York, 1973.

12.6 R. D. Cook, "Finite Element Buckling Analysis of Homogeneous and Sandwich Plates," *IJNME,* Vol. 9, No. 1, 1975, pp. 39-50.

12.7 R. S. Barsoum and R. H. Gallagher, "Finite Element Analysis of Torsional and Torsional-Flexural Stability Problems," *IJNME,* Vol. 2, No. 3, 1970, pp. 335-352.

12.8 R. S. Barsoum, "Finite Element Method Applied to the Problem of Stability of a Non-Conservative System," *IJNME,* Vol. 3, No. 1, 1971, pp. 63-87.

12.9 N. Tebedge and L. Tall, "Linear Stability Analysis of Beam Columns," *JStDiv,* Vol. 99, No. ST12, 1973, pp. 2439-2457.

12.10 J. S. Przemieniecki, "Discrete-Element Methods for Stability Analysis," *AeroJ,* Vol. 72, No. 12, 1968, pp. 1077-1086.

12.11 R. A. Tinawi, "Anisotropic Tapered Elements Using Displacement Models," *IJNME,* Vol. 4, No. 4, 1972, pp. 475-489.

12.12 D. R. Navaratna, T. H. H. Pian, and E. A. Witmer, "Analysis of Elastic Stability of Shells of Revolution by the Finite Element Method," *AIAAJ,* Vol. 6, No. 2, 1968, pp. 355-361.

12.13 D. Bushnell, "Analysis of Ring-Stiffened Shells of Revolution Under Combined Thermal and Mechanical Loading," *AIAAJ,* Vol. 9, No. 3, 1971, pp. 401-410.

12.14 P. Tong and T. H. H. Pian, "Postbuckling Analysis of Shells of Revolution by the Finite Element Method," in *Thin-Shell Structures: Theory, Experiment and Design,* Y. C. Fung and E. E. Sechler, eds., Prentice-Hall, Englewood Cliffs, N.J., 1974, pp. 435-452.

12.15 P. P. Cole, J. F. Abel, and D. P. Billington, "Buckling of Cooling-Tower Shells: Bifurcation Results," *JStDiv,* Vol. 101, No. ST6, 1975, pp. 1205-1222.

12.16 J. S. Przemieniecki, "Matrix Analysis of Local Instability in Plates, Stiffened Panels and Columns," *IJNME,* Vol. 5, No. 2, 1972, pp. 209-216.

12.17 J. S. Przemieniecki, "Finite Element Structural Analysis of Local Instability," *AIAAJ,* Vol. 11, No. 1, 1973, pp. 33-39.

12.18 D. Bushnell, "Stress, Stability and Vibration of Complex, Branched Shells of Revolution," *14th SSDM Conf.,* March 1973.

12.19 J. M. T. Thompson, comment on "An Engineering Approach to Interactive Buckling," *IJMSci,* Vol. 16, No. 5, 1974, pp. 335-336.

12.20 W. G. Carson and R. E. Newton, "Plate Buckling Analysis Using a Fully Compatible Finite Element," *AIAAJ,* Vol. 7, No. 3, 1969, pp. 527-529.

12.21 A. D. Kerr and M. T. Soifer, "The Linearization of the Prebuckling State and Its Effect on the Determined Stability Loads," *JApplM,* Vol. 36, No. 4, 1969, pp. 775-783.

12.22 D. Bushnell, "Buckling of Shells—Pitfall for Designers," *21st SSDM Conf.,* May 1980, pp. 1-56.

12.23 R. F. Jones, M. G. Costello, and T. E. Reynolds, "Buckling of Pressure Loaded Rings and Shells by the Finite Element Method," *CompSt,* Vol. 7, No. 2, 1977, pp. 267-274.

12.24 R. J. Roark and W. C. Young, *Formulas for Stress and Strain,* 5th Ed., McGraw-Hill, New York, 1975.

CHAPTER 13

13.1 D. Bushnell, "A Computerized Information Retrieval System," *SMCP Sympos.*, pp. 735-804.

13.2 J. T. Oden, *Finite Elements of Nonlinear Continua*, McGraw-Hill, New York, 1972.

13.3 R. H. Mallett and P. V. Marcal, "Finite-Element Analysis of Nonlinear Structures," *JStDiv*, Vol. 94, No. ST9, 1968, pp. 2081-2105.

13.4 S. Rajasekaran and D. W. Murray, "Incremental Finite Element Matrices," *JStDiv*, Vol. 99, No. ST12, 1973, pp. 2423-2438.

13.5 R. D. Wood and B. Schrefler, "Geometrically Nonlinear Analysis—A Correlation of Finite Element Notations," *IJNME*, Vol. 12, No. 4, 1978, pp. 635-642.

13.6 P. L. Boland and T. H. H. Pian, "Large Deflection Analysis of Thin Elastic Structures by the Assumed Stress Hybrid Finite Element Method," *CompSt*, Vol. 7, No. 1, 1977, pp. 1-12.

13.7 S. P. Timoshenko and J. M. Gere, *Theory of Elastic Stability*, 2nd Ed., McGraw-Hill, New York, 1961.

13.8 T. Belytschko and B. J. Hsieh, "Non-Linear Transient Finite Element Analysis with Convected Coordinates," *IJNME*, Vol. 7, No. 3, 1973, pp. 255-271.

13.9 P. G. Bergan and R. W. Clough, "Convergence Criteria for Iterative Processes," *AIAAJ*, Vol. 10, No. 8, 1972, pp. 1107-1108.

13.10 D. W. Murray and E. L. Wilson, "Finite-Element Large Deflection Analysis of Plates," *JEMDiv*, Vol. 95, No. EM1, 1969, pp. 143-165.

13.11 O. C. Zienkiewicz and B. M. Irons, "Matrix Iteration and Acceleration Processes in Finite Element Problems of Structural Mechanics," in *Methods for Nonlinear Algebraic Equations*, P. Rabinowitz, ed., Gordon & Breach, London, 1970, pp. 183-194.

13.12 G. R. Thomas, "A Variable Step Incremental Procedure," *IJNME*, Vol. 7, No. 4, 1973, pp. 563-566.

13.13 J. A. Stricklin, W. E. Haisler, and W. A. Von Riesemann, "Geometrically Nonlinear Analysis by Stiffness Method," *JStDiv*, Vol. 97, No. ST9, 1971, pp. 2299-2314.

13.14 P. G. Bergan and R. W. Clough, "Large Deflection Analysis of Plates and Shallow Shells Using the Finite Element Method," *IJNME*, Vol. 5, No. 4, 1973, pp. 543-556.

13.15 J. R. Tillerson, J. A. Stricklin, and W. E. Haisler, "Numerical Methods for the Solution of Nonlinear Problems in Structural Analysis," in *Numerical Solution of Nonlinear Structural Problems*, R. F. Hartung, ed., Applied Mechanics Division, Vol. 6, American Society of Mechanical Engineers, New York, 1973, pp. 67-101.

13.16 P. Sharifi and E. P. Popov, "Nonlinear Buckling Analysis of Sandwich Arches," *JEMDiv*, Vol. 97, No. EM5, 1971, pp. 1397-1412.

13.17 J. L. Batoz and G. Dhatt, "Incremental Displacement Algorithms for Nonlinear Problems," *IJNME*, Vol. 14, No. 8, 1979, pp. 1262-1267.

13.18 D. P. Mondkar and G. H. Powell, "Evaluation of Solution Schemes for Nonlinear Structures," *CompSt*, Vol. 9, No. 3, 1978, pp. 223-236.

13.19 E. L. Wilson, "Finite Elements for Foundations, Joints, and Fluids," in *Finite Elements in Geomechanics*, G. Gudehus, ed., John Wiley & Sons, London, 1977, pp. 319-350.

13.20 D. N. Buragohain and V. L. Shah, "Curved Isoparametric Interface Surface Element," *JStDiv*, Vol. 104, No. ST1, 1978, pp. 205-209.

13.21 D. Gaertner, "Investigation of Plane Elastic Contact Region Allowing for Friction," *CompSt*, Vol. 7, No. 1, 1977, pp. 59-63.

13.22 F. Baron and M. S. Venkatesan, "Nonlinear Analysis of Cable and Truss Structures," *JStDiv,* Vol. 97, No. ST2, 1971, pp. 679-710.
13.23 A. H. Peyrot and A. M. Goulois, "Analysis of Flexible Transmission Lines," *JStDiv,* Vol. 104, No. ST5, 1978, pp. 763-779.
13.24 J. T. Oden, "Note on an Approximate Method for Computing Nonconservative Generalized Forces on Finitely Deformed Finite Elements," *AIAAJ,* Vol. 8, No. 11, 1970, pp. 2088-2090.
13.25 P. G. Smith and E. L. Wilson, "Automatic Design of Shell Structures," *JStDiv,* Vol. 97, No. ST1, 1971, pp. 191-201.
13.26 D. R. J. Owen, A. Prakash, and O. C. Zienkiewicz, "Finite Element Analysis of Nonlinear Composite Materials by Use of Overlay Systems," *CompSt,* Vol. 4, No. 6, 1974, pp. 1251-1267.
13.27 A. Mendelson, *Plasticity: Theory and Application,* Macmillan, New York, 1968.
13.28 P. V. Marcal, "Large Deflection Analysis of Elastic-Plastic Shells of Revolution," *AIAAJ,* Vol. 8, No. 9, 1970, pp. 1627-1633.
13.29 C. S. Desai, ed., *Applications of the Finite Element Method in Geotechnical Engineering,* Proceedings of the Symposium held at Vicksburg, Miss., May 1972, U.S. Army Engineer Waterways Experiment Station, Corps of Engineers, Vicksburg, Miss., 1972.
13.30 R. S. Barsoum, "A Convergent Method for Cyclic Plasticity Analysis with Application to Nuclear Components," *IJNME,* Vol. 6, No. 2, 1973, pp. 227-236.
13.31 G. Gudehus, *Finite Elements in Geomechanics,* John Wiley & Sons, London, 1977.
13.32 S. Valliappan, P. Boonlaulohr, and I. K. Lee, "Nonlinear Analysis for Anisotropic Materials," *IJNME,* Vol. 10, No. 3, 1976, pp. 597-606.
13.33 G. Strang, "Some Recent Contributions to Plasticity Theory," *JFrlns,* Vol. 302, Nos. 5/6, 1976, pp. 429-442.
13.34 D. Bushnell, "Large Deflection Elastic-Plastic Creep Analysis of Axisymmetric Shells," in *Numerical Solution of Nonlinear Structural Problems,* R. F. Hartung, ed., Applied Mechanics Division, Vol. 6, American Society of Mechanical Engineers, New York, 1973, pp. 103-138.
13.35 D. W. Murray and E. L. Wilson, "An Approximate Nonlinear Analysis of Plates," *WPAFB2 Conf.,* pp. 1207-1230.
13.36 Y. Yamada, N. Yoshimura, and T. Sakurai, "Plastic Stress-Strain Matrix and Its Application for the Solution of Elastic-Plastic Problems by the Finite Element Method," *IJMSci,* Vol. 10, No. 5, 1968, pp. 343-354.
13.37 J. A. Stricklin, W. E. Haisler, and W. A. Von Riesemann, "Computation and Solution Procedures for Nonlinear Analysis by Combined Finite Element-Finite Difference Methods," *CompSt,* Vol. 2, Nos. 5/6, 1972, pp. 955-974.
13.38 J. A. Stricklin, W. E. Haisler, and W. A. Von Riesemann, "Formulation, Computation and Solution Procedures for Material and/or Geometric Non-linear Structural Analysis by the Finite Element Method," *Report SC-CR-72-3102,* Sandia Laboratories, Albuquerque, N.M., July 1972.
13.39 L. D. Hofmeister, G. A. Greenbaum, and D. A. Evenson, "Large Strain, Elasto-Plastic Finite Element Analysis," *AIAAJ,* Vol. 9, No. 7, 1971, pp. 1248-1254.
13.40 D. Bushnell, "BOSOR5—Program for Buckling of Elastic-Plastic Complex Shells of Revolution Including Large Deflections and Creep," *CompSt,* Vol. 6, No. 3, 1976, pp. 221-239.

13.41 G. C. Nayak and O. C. Zienkiewicz, "Elasto-Plastic Stress Analysis. A Generalization for Various Constitutive Relations Including Strain Softening," *IJNME,* Vol. 5, No. 1, 1972, pp. 113-135.
13.42 M. A. Crisfield, "Elastic-Plastic Finite Element Analysis of Plates," *PInsCE,* Vol. 59, Part 2, 1975, pp. 219-221.
13.43 K. I. Majid, *Non-Linear Structures,* Butterworths, London, 1972.
13.44 A. Pifko and G. Isakson, "A Finite Element Method for the Plastic Buckling Analysis of Plates," *AIAAJ,* Vol. 7, No. 10, 1969, pp. 1950-1957.
13.45 D. Bushnell, "Bifurcation Buckling of Shells of Revolution Including Large Deflections, Plasticity and Creep," *IJSoSt,* Vol. 10, No. 11, 1974, pp. 1287-1305.
13.46 G. A. Greenbaum and M. F. Rubinstein, "Creep Analysis of Axisymmetric Bodies Using Finite Elements," *NucEDe,* Vol. 7, No. 4, 1968, pp. 379-397.
13.47 J. O. Smith and O. M. Sidebottom, *Inelastic Behavior of Load-Carrying Members,* John Wiley & Sons, New York, 1965.
13.48 A. P. Boresi and O. M. Sidebottom, "Creep of Metals Under Multiaxial States of Stress," *NucEDe,* Vol. 18, No. 3, 1972, pp. 415-456.
13.49 N. A. Cyr and R. D. Teter, "Finite Element Elastic-Plastic-Creep Analysis of Two-Dimensional Continuum with Temperature Dependent Material Properties," *CompSt,* Vol. 3, No. 4, 1973, pp. 849-863.
13.50 P. G. Hodge, Jr., and H. M. van Rij, "A Finite-Element Model for Plane-Strain Plasticity," *JApplM,* Vol. 46, No. 3, 1979, pp. 536-542.
13.51 O. C. Zienkiewicz and P. N. Godbole, "A Penalty Function Approach to Problems of Plastic Flow of Metals with Large Surface Deformations," *JStrAn,* Vol. 10, No. 3, 1975, pp. 180-183.
13.52 R. H. Mallett, ed., *Limit Analysis Using Finite Elements,* papers presented at Winter Annual Meeting, American Society of Mechanical Engineers, New York, 1976.
13.53 Z. P. Bazant and L. Cedolin, "Blunt Crack Band Propagation in Finite Element Analysis," *JEMDiv,* Vol. 105, No. EM2, 1979, pp. 297-315.
13.54 O. C. Zienkiewicz, S. Valliappan, and I. P. King, "Stress Analysis of Rock as a 'No Tension' Material," *Geotechnique,* Vol. 18, No. 1, 1968, pp. 56-66.
13.55 B. Hunsaker, Jr., D. K. Vaughn, and J. A. Stricklin, "A Comparison of the Capability of Four Hardening Rules to Predict a Material's Plastic Behavior," *JPrVeT,* Vol. 98, No. 1, 1976, pp. 66-74.
13.56 D. P. Mondkar and G. H. Powell, "Finite Element Analysis of Nonlinear Static and Dynamic Response," *IJNME,* Vol. 11, No. 3, 1977, pp. 499-520.
13.57 A. K. Noor, C. M. Anderson, and J. M. Peters, "Global-Local Approach for Nonlinear Shell Analysis," *Seventh Conference on Electronic Computation,* American Society of Civil Engineers, New York, 1979, pp. 634-657.
13.58 J. L. Bretl and R. D. Cook, "Modelling the Load Transfer in Threaded Connections by the Finite Element Method," *IJNME,* Vol. 14, No. 9, 1979, pp. 1359-1377.
13.59 A. Grill and K. Sorimachi, "The Thermal Loads in the Finite Element Analysis of Elasto-Plastic Stresses," *IJNME,* Vol. 14, No. 4, 1979, pp. 499-505.
13.60 H. J. Ernst, "Der E-Modul von Seilen unter Berücksichtigung des Durchhanges," *Der Bauingenieur,* Vol. 40, No. 2, 1965, pp. 52-55.
13.61 A. A. Ilyushin, *Plasticité,* Editions Eyrolles, Paris, 1956.
13.62 M. A. Crisfield, "A Faster Modified Newton-Raphson Iteration," *CMAME,* Vol. 20, No.

3, 1979, pp. 267-278.

13.63 R. L. Webster, "On the Static Analysis of Structures with Strong Geometric Nonlinearity," *CompSt,* Vol. 11, Nos. 1/2, 1980, pp. 137-145.

13.64 A. M. Peyrot and A. M. Goulois, "Analysis of Flexible Transmission Lines," *JStDiv,* Vol. 104, No. ST5, 1978, pp. 763-779.

CHAPTER 14

14.1 F. B. Brooks, Jr., *The Mythical Man-Month,* Addison-Wesley, Reading, Mass., 1975.

14.2 P. W. Metzger, *Managing a Programming Project,* Prentice-Hall, Englewood Cliffs, N.J., 1973.

14.3 R. M. Jones, "Effective Development, Documentation, and Distribution of Computer Programs," *CompSt,* Vol. 2, Nos. 5/6, 1972, pp. 1089-1095.

14.4 J. L. Tocher, "Specifications for Successful Programming Projects," in *Methods of Structural Analysis,* W. E. Saul and A. H. Peyrot, eds., American Society of Civil Engineers, New York, 1976, pp. 837-850.

14.5 G. M. Weinberg, *The Psychology of Computer Programming,* Van Nostrand Reinhold, New York, 1971.

14.6 A. Ralston and C. L. Meek, *Encyclopedia of Computer Science,* Petrocelli-Charter, New York, 1976.

14.7 B. W. Kernighan and P. J. Plauger, *The Elements of Programming Style,* 2nd Ed., McGraw-Hill, New York, 1978.

14.8 H. F. Ledgard, *Programming Proverbs for Fortran Programmers,* Hayden Book Co., Rochelle Park, N.J., 1975.

14.9 J. M. McCormick, "Programming for Effective Interchange," *SMCP Sympos.,* pp. 651-667.

14.10 P. Naur, B. Randell, and J. N. Buxton, *Software Engineering: Concepts and Techniques,* Petrocelli-Charter, New York, 1976.

14.11 W. R. Buell and B. A. Bush, "Mesh Generation—A Survey," *JEInd,* Vol. 95, No. 1, 1973, pp. 332-338.

14.12 L. R. Herrmann, "Laplacian-Isoparametric Grid Generation Scheme," *JEMDiv,* Vol. 102, No. EM5, 1976, pp. 749-756.

14.13 R. E. Jones, "A Self-Organizing Mesh Generation Program," *JPrVeT,* Vol. 96, No. 3, 1974, pp. 193-199.

14.14 G. Steinmueller, "Restrictions in the Application of Automatic Mesh Generation Schemes by 'Isoparametric' Coordinates," *IJNME,* Vol. 8, No. 2, 1974, pp. 289-294.

14.15 A. Bykat, "Automatic Generation of Triangular Grid: I—Subdivision of a General Polygon into Convex Subregions: II—Triangulation of Convex Polygons," *IJNME,* Vol. 10, No. 6, 1976, pp. 1329-1342.

14.16 K. Sagawa, "Automatic Mesh Generation for Three Dimensional Structures Based on Their Three Views," *USJap3 Conf.,* pp. 687-703.

14.17 W. A. Cook, "Body Oriented (Natural) Coordinates for Generating Three-Dimensional Meshes," *IJNME,* Vol. 8, No. 1, 1974, pp. 27-43.

14.18 W. J. Gordon and C. A. Hall, "Construction of Curvilinear Coordinate Systems and Applications to Mesh Generation," *IJNME,* Vol. 7, No. 4, 1973, pp. 461-477.

14.19 D. J. Turcke and G. M. McNeice, "Guidelines for Selecting Finite Element Grids Based on an Optimization Study," *CompSt,* Vol. 4, No. 3, 1974, pp. 499-519.
14.20 R. J. Melosh and P. V. Marcal, "An Energy Basis for Mesh Refinement of Structural Continua," *IJNME,* Vol. 11, No. 7, 1977, pp. 1083-1091.
14.21 W. H. Gray and J. E. Akin, "An Improved Method for Contouring on Isoparametric Surfaces," *IJNME,* Vol. 14, No. 3, 1979, pp. 451-458.
14.22 J. L. Tocher and C. A. Felippa, "Computer Graphics Applied to Production Structural Analysis," *IUTAM Sympos.,* pp. 522-545.
14.23 R. Douty and S. Shore, "Technique for Interactive Computer Graphics in Design," *JStDiv,* Vol. 97, No. ST1, 1971, pp. 273-288.
14.24 C. A. Felippa, "An Alphanumeric Finite Element Mesh Plotter," *IJNME,* Vol. 5, No. 2, 1972, pp. 217-236.
14.25 R. D. Bousquet and O. N. Yates, "A Low Cost Interactive Graphics System for Large Scale Finite Element Analysis," *CompSt,* Vol. 3, No. 6, 1973, pp. 1321-1330.
14.26 H. N. Christiansen, "Applications of Continuous Tone Computer-Generated Images in Structural Mechanics," *SMCP Sympos.,* pp. 1003-1015.
14.27 C. F. Beck, "Responsibilities of Computer-Aided Design," *Civil Engineering,* June 1972, pp. 58-60.
14.28 Panel discussion, "The Large General Purpose Code," *CompSh Conf.,* pp. 1063-1106.
14.29 J. A. Swanson, "The Development of General Purpose Software, or What is a Software Supplier?", *SMCP Sympos.,* pp. 687-702.
14.30 G. Ruoff and E. Stein, "The Development of General Purpose Programs and Systems," *SMCP Sympos.,* pp. 703-719.
14.31 C. W. McCormick, "Shell Analysis with Large General Purpose Computer Programs," *CompSt,* Vol. 1, Nos. 1/2, 1971, pp. 323-332.
14.32 R. K. Henrywood, "The Design, Development, Documentation and Support of a Major Finite Element System," *CompAD,* Vol. 5, No. 3, 1973, pp. 160-165.
14.33 E. Schrem, "Computer Implementation of the Finite Element Procedure," *ONR Sympos.,* pp. 79-121.
14.34 See several papers in *SMCP Sympos.,* pp. 805-1015.
14.35 J. M. McCormick, *Data Handling Techniques in Civil Engineering,* Meeting Preprint 2503, American Society of Civil Engineers, New York, 1975.
14.36 J. A. Bettess, "A Data Structure for Finite Element Analysis," *IJNME,* Vol. 11, No. 12, 1977, pp. 1779-1799.
14.37 E. C. E. Moone, "Software Coordination Efforts," *CompSt,* Vol. 6, No. 6, 1976, pp. 489-496.
14.38 S. D. Hansen, G. L. Anderton, N. E. Connacher, and C. S. Dougherty, "Analysis of the 747 Aircraft Wing-Body Intersection," *WPAFB2 Conf.,* pp. 743-788.
14.39 H. S. Iyengar, N. Amin, and L. Carpenter, "Computerized Design of World's Tallest Building," *CompSt,* Vol. 2, Nos. 5/6, 1972, pp. 771-783.
14.40 I. Kadar, "Three-Dimensional Structural Programs in Everyday Usage," *CompSt,* Vol. 6, No. 6, 1976, pp. 481-487.
14.41 C. A. Felippa, "Database Management in Scientific Computing—I. General Description," *CompSt,* Vol. 10, Nos. 1/2, 1979, pp. 53-62.
14.42 L. A. Durocher, "A Versatile Two-Dimensional Mesh Generator with Automatic Bandwidth Reduction," *CompSt,* Vol. 10, No. 4, 1979, pp. 561-575.

14.43 K. Preiss, "Checking the Topological Consistency of a Finite Element Mesh," *IJNME,* Vol. 14, No. 12, 1979, pp. 1805-1812.

CHAPTER 15

15.1 L. Fox, *An Introduction to Numerical Linear Algebra,* Oxford University Press, New York, 1965.

15.2 M. Geradin, "Error Bounds for Eigenvalue Analysis by Elimination of Variables," *JSouVi,* Vol. 19, No. 2, 1971, pp. 111-132.

15.3 I. Fried, "Condensation of Finite Element Eigenproblems," *AIAAJ,* Vol. 10, No. 11, 1972, pp. 1529-1530.

15.4 W. C. Hurty, "Truncation Errors in Natural Frequencies as Computed by the Method of Component Mode Synthesis," *WPAFB1 Conf.,* pp. 803-821.

15.5 G. Strang and G. J. Fix, *An Analysis of the Finite Element Method,* Prentice-Hall, Englewood Cliffs, N.J., 1973.

15.6 R. A. Rosanoff, J. F. Gloudemann, and S. Levy, "Numerical Conditioning of Stiffness Matrix Formulations for Frame Structures," *WPAFB2 Conf.,* pp. 1029-1060.

15.7 G. L. Rigby and G. M. McNeice, "A Strain Energy Basis for Studies of Element Stiffness Matrices," *AIAAJ,* Vol. 10, No. 11, 1972, pp. 1490-1493.

15.8 J. Robinson, "A Single Element Test," *CMAME,* Vol. 7, No. 2, 1976, pp. 191-200.

15.9 J. E. Walz, R. E. Fulton, and N. J. Cyrus, "Accuracy and Convergence of Finite Element Approximations," *WPAFB2 Conf.,* pp. 995-1027.

15.10 J. T. Oden and H. J. Brauchli, "A Note on Accuracy and Convergence of Finite Element Approximations," *IJNME,* Vol. 3, No. 2, 1971, pp. 291-292.

15.11 I. Fried, "Accuracy and Condition of Curved (Isoparametric) Finite Elements," *JSouVi,* Vol. 31, No. 3, 1973, pp. 345-355.

15.12 S. H. Crandall, *Engineering Analysis,* McGraw-Hill, New York, 1956, pp. 171-173.

15.13 I. Fried, "Accuracy of Complex Finite Elements," *AIAAJ,* Vol. 10, No. 3, 1972, pp. 347-349.

15.14 R. D. Henshell, D. Walters, and G. B. Warburton, "A New Family of Curvilinear Plate Bending Elements for Vibration and Stability," *JSouVi,* Vol. 20, No. 3, 1972, pp. 381-397 (discussion and closure: Vol. 23, No. 4, pp. 507-513).

15.15 I. Fried, "Condition of Finite Element Matrices Generated from Nonuniform Meshes," *AIAAJ,* Vol. 10, No. 2, 1972, pp. 219-221.

15.16 I. Fried, "Bounds on the Extremal Eigenvalues of the Finite Element Stiffness and Mass Matrices and Their Spectral Condition Number," *JSouVi,* Vol. 22, No. 4, 1972, pp. 407-418.

15.17 R. A. Rosanoff and T. A. Ginsburg, "Matrix Error Analysis for Engineers," *WPAFB1 Conf.,* pp. 887-910.

15.18 R. A. Rosanoff, J. F. Gloudemann, and S. Levy, "Numerical Conditioning of Stiffness Matrix Formulations for Frame Structures," *WPAFB2 Conf.,* pp. 1029-1060.

15.19 J. R. Roy, "Numerical Error in Structural Solutions," *JStDiv,* Vol. 97, No. ST4, 1971, pp. 1039-1054 (author's closure to discussion: Vol. 98, No. ST7, pp. 1663-1666).

15.20 R. J. Melosh, "Manipulation Errors in Finite Element Analysis," *USJap1 Conf.,* pp. 857-877.

15.21 R. J. Melosh and E. L. Palacol, "Manipulation Errors in Finite Element Analysis of Structures," *Report NASA CR-1385,* Philco-Ford Corp., August 1969 (N69-34360, N.T.I.S.).
15.22 G. Forsythe and C. B. Moler, *Computer Solution of Linear Algebraic Systems,* Prentice-Hall, Englewood Cliffs, N.J., 1967.
15.23 S. Kelsey, K. N. Lee, and C. K. K. Mak, "The Condition of Some Finite Element Coefficient Matrices," *CompAE Sympos.,* pp. 267-283.
15.24 B. Noble, *Applied Linear Algebra,* Prentice-Hall, Englewood Cliffs, N.J., 1969.
15.25 Discussion and closure on "Efficient Solution of Load-Deflection Equations," *JStDiv,* Vol. 96, No. ST2, 1970, pp. 421-426; Vol. 97, No. ST2, 1971, pp. 713-717.
15.26 B. M. Irons, "Roundoff Criteria in Direct Stiffness Solutions," *AIAAJ,* Vol. 6, No. 7, 1968, pp. 1308-1312.
15.27 R. H. MacNeal, ed., *The NASTRAN Theoretical Manual (Level 16.0),* NASA-SP-221(03), March 1976 (N79-27531, N.T.I.S.).
15.28 I. U. Ojalvo, F. Austin, and A. Levy, "Iterative Analysis Method for Structural Components with Diverse Stiffnesses," *AIAAJ,* Vol. 14, No. 9, 1976, pp. 1219-1224.
15.29 B. Parlett, "Roundoff Error in the Solution of Finite Element Equations," in *Formulations and Computational Algorithms in Finite Element Analysis,* K. J. Bathe, J. T. Oden, and W. Wunderlich, eds., M.I.T. Press, Cambridge, Mass., 1977, pp. 632-654.
15.30 T. Krathammer, "Accuracy of the Finite Element Method Near a Curved Boundary," *CompSt,* Vol. 10, No. 6, 1979, pp. 921-929.

CHAPTER 16

16.1 T. H. H. Pian and P. Tong, "Basis of Finite Element Methods for Solid Continua," *IJNME,* Vol. 1, No. 1, pp. 3-28.
16.2 B. Tabarrok and A. Simpson, "An Equilibrium Finite Element Model for Buckling Analysis of Plates," *IJNME,* Vol. 11, No. 11, 1977, pp. 1733-1751.
16.3 Z. M. Elias, "Duality in Finite Element Methods," *JEMDiv,* Vol. 94, No. EM4, 1968, pp. 931-946.
16.4 B. Fraejis de Veubeke and O. C. Zienkiewicz, "Strain Energy Bounds in Finite Element Analysis by Slab Analogy," *JStrAn,* Vol. 2, No. 4, 1967, pp. 265-271.
16.5 L. R. Herrmann, "Finite-Element Bending Analysis for Plates," *JEMDiv,* Vol. 93, No. EM5, 1967, pp. 13-26.
16.6 Y. Yoshida, "Equivalent Finite Elements of Different Bases," *USJap2 Conf.,* pp. 133-149.
16.7 L. R. Herrmann and W. E. Mason, "Mixed Formulations for Finite Element Shell Analysis," *CompSh Conf.,* pp. 290-336.
16.8 A. K. Noor, W. B. Stephens, and R. E. Fulton, "An Improved Numerical Process for Solution of Solid Mechanics Problems," *CompSt,* Vol. 3, No. 6, 1973, pp. 1397-1437.
16.9 T. H. H. Pian, "Derivation of Element Stiffness Matrices by Assumed Stress Functions," *AIAAJ,* Vol. 2, No. 7, 1964, pp. 1333-1336.
16.10 T. H. H. Pian and P. Tong, "Rationalization in Deriving Element Stiffness Matrix by Assumed Stress Approach," *WPAFB2 Conf.,* pp. 441-469.
16.11 P. Tong and T. H. H. Pian, "A Variational Principle and the Convergence of a Finite-Element Method Based on Assumed Stress Distribution," *IJSoSt,* Vol. 5, No. 5, 1969, pp. 463-472.

16.12 R. D. Cook, "Two Hybrid Elements for Analysis of Thick, Thin and Sandwich Plates," *IJNME,* Vol. 5, No. 2, 1972, pp. 277-288.
16.13 R. D. Cook, "Improved Two-Dimensional Finite Element," *JStDiv,* Vol. 100, No. ST9, 1974, pp. 1851-1863.
16.14 J. P. Wolf, "Alternate Hybrid Stress Finite Element Models," *IJNME,* Vol. 9, No. 3, 1975, pp. 601-615.
16.15 R. D. Cook, "Further Improvement of an Effective Plate Bending Element," *CompSt,* Vol. 6, No. 2, 1976, pp. 93-97.
16.16 R. D. Cook and S. G. Ladkany, "Observations Regarding Assumed-Stress Hybrid Plate Elements," *IJNME,* Vol. 8, No. 3, 1974, pp. 513-519.
16.17 R. T. Severn, "Inclusion of Shear Deflection in the Stiffness Matrix for a Beam Element," *JStrAn,* Vol. 5, No. 4, 1970, pp. 239-241.
16.18 R. Narayanaswami and H. M. Adelman, "Inclusion of Transverse Shear Deformation in Finite Element Displacement Formulations," *AIAAJ,* Vol. 12, No. 11, 1974, pp. 1613-1614 (discussion: Vol. 13, No. 9, pp. 1253-1254; application to plates: Vol. 12, No. 12, pp. 1761-1763).
16.19 D. J. Dawe, "A Finite Element for the Vibration Analysis of Timoshenko Beams," *JSouVi,* Vol. 60, No. 1, 1978, pp. 11-20.
16.20 H. P. Lee, "Generalized Stiffness Matrix of a Curved-Beam Element," *AIAAJ,* Vol. 7, No. 10, 1969, pp. 2043-2045.
16.21 L. R. L. Wang, "Parametric Matrices of Some Structural Members," *JStDiv,* Vol. 96, No. ST8, 1970, pp. 1735-1759.
16.22 R. Davis, R. D. Henshell, and G. B. Warburton, "Curved Beam Finite Elements for Coupled Bending and Vibration," *EEStDy,* Vol. 1, No. 2, 1972, pp. 165-175.
16.23 F. M. El-Amin and D. M. Brotton, "Horizontally Curved Beam Finite Element Including Warping," *IJNME,* Vol. 10, No. 6, 1976, pp. 1397-1403 (discussion: Vol. 14, No. 4, pp. 621-623).
16.24 W. A. Thornton and B. G. Master, "Direct Stiffness Formulation for Horizontally Curved Beams," *JStDiv,* Vol. 103, No. ST1, 1977, pp. 284-289.
16.25 S. K. Chauduri and S. Shore, "Thin-Walled Curved Beam Finite Element," *JEMDiv,* Vol. 103, No. EM5, 1977, pp. 921-937.
16.26 A. K. Chugh, "Stiffness Matrix for a Beam Element Including Transverse Shear and Axial Force Effects," *IJNME,* Vol. 11, No. 11, 1977, pp. 1681-1697.
16.27 L. L. Durocher and J. Kane, "Stiffness Matrix for Eccentric Beam Segment," *JStDiv,* Vol. 105, No. ST2, 1979, pp. 447-450.
16.28 K. L. Lawrence, "Twisted Beam Element Matrices for Bending," *AIAAJ,* Vol. 8, No. 6, 1970, pp. 1160-1162.
16.29 E. Dokumaci, J. Thomas, and W. Carnegie, "Matrix Displacement Analysis of Coupled Bending-Bending Vibrations of Pretwisted Blading," *JMESci,* Vol. 9, No. 4, 1967, pp. 247-254.
16.30 M. A. Crisfield, "Finite Element Methods for the Analysis of Multicellular Structures," *PInsCe,* Vol. 48, 1971, pp. 413-437.
16.31 K. J. Willam and A. C. Scordelis, "Cellular Structures of Arbitrary Plan Geometry," *JStDiv,* Vol. 98, No. ST7, 1972, pp. 1377-1394.
16.32 T. Kawai, "The Application of Finite Element Methods to Ship Structures," *CompSt,* Vol. 3, No. 5, 1973, pp. 1175-1194.
16.33 K. S. Surana, "Isoparametric Elements for Cross-Sectional Properties and Stress Analysis of Beams," *IJNME,* Vol. 14, No. 4, 1979, pp. 475-497.

16.34 Y. K. Cheung and O. C. Zienkiewicz, "Plates and Tanks on Elastic Foundations—An Application of Finite Element Method," *IJSoSt,* Vol. 1, No. 4, 1965, pp. 451-461.

16.35 A. D. Kerr, "Elastic and Viscoelastic Foundation Models," *JApplM,* Vol. 31, No. 3, 1964, pp. 491-498.

16.36 D. Q. Fletcher and L. R. Herrmann, "Elastic Foundation Representation of Continuum," *JEMDiv,* Vol. 97, No. EM1, 1971, pp. 95-107.

16.37 T. Y. Yang, "A Finite Element Analysis of Plates on a Two Parameter Foundation Model," *CompSt,* Vol. 2, No. 4, 1972, pp. 593-614.

16.38 M. S. Cheung, "A Simplified Finite Element Solution for the Plates on Elastic Foundation," *CompSt,* Vol. 8, No. 1, 1978, pp. 139-145.

16.39 G. A. Hartley, G. M. McNeice, and W. Stensch, "Vogt Boundary for Finite Element Arch Dam Analysis," *JStDiv,* Vol. 100, No. ST1, 1974, pp. 51-62.

16.40 R. W. Clough and R. J. Woodward, III, "Analysis of Embankment Stresses and Deformations," *JSMDiv,* Vol. 93, No. SM4, 1967, pp. 529-549.

16.41 J. M. Duncan and G. W. Clough, "Finite Element Analysis of Port Allen Lock," *JSMDiv,* Vol. 97, No. SM8, 1971, pp. 1053-1068.

16.42 A. K. Chugh, "On the Boundary Conditions in Analysis of Underground Openings," *CompSt,* Vol. 10, No. 4, 1979, pp. 637-645.

16.43 J. S. Arora, "Survey of Reanalysis Techniques," *JStDiv,* Vol. 102, No. ST4, 1976, pp. 783-802.

16.44 M. Kleiber and A. Lutoborski, "Modified Triangular Factors in the Incremental Finite Element Analysis with Nonsymmetric Stiffness Changes," *CompSt,* Vol. 9, No. 6, 1978, pp. 599-602.

16.45 J. H. Argyris and J. R. Roy, "General Treatment of Structural Modifications," *JStDiv,* Vol. 98, No. ST2, 1972, pp. 465-492 (errata: Vol. 99, No. ST2, pp. 280-281).

16.46 A. I. Raibstein, I. Kalev, and A. Pipano, "Efficient Reanalysis of Structures by a Direct Modification Method," *Fifth NASTRAN U.C.,* 1976.

16.47 D. G. Row, G. H. Powell, and D. P. Mondkar, "Solution of Progressively Changing Equilibrium Equations for Nonlinear Structures," *CompSt,* Vol. 7, No. 5, 1977, pp. 659-665.

16.48 D. Kavlie and G. H. Powell, "Efficient Reanalysis of Modified Structures," *JStDiv,* Vol. 97, No. ST1, 1971, pp. 377-392 (discussion: Vol. 97, No. ST10, pp. 2612-2619).

16.49 M. E. Fourney, "A Pseudo Two-Dimensional Photoelastic Method of Testing Axisymmetric Geometries," *Experimental Mechanics,* Vol. 11, No. 1, 1971, pp. 19-25.

16.50 D. S. Malkus, "A Finite Element Displacement Model Valid for any Value of the Compressibility," *IJSoSt,* Vol. 12, Nos. 9/10, 1976, pp. 731-738.

16.51 D. S. Malkus and T. J. R. Hughes, "Mixed Finite Element—Reduced and Selective Integration Techniques: A Unification of Concepts," *CMAME,* Vol. 15, No. 1, 1978, pp. 63-81.

16.52 T. J. R. Hughes, "Equivalence of Finite Elements for Nearly Incompressible Elasticity," *JApplM,* Vol. 44, No. 1, 1977, pp. 181-183.

16.53 L. R. Herrmann, "Elasticity Equations for Incompressible and Nearly Incompressible Materials by a Variational Theorem," *AIAAJ,* Vol. 3, No. 10, 1965, pp. 1896-1900.

16.54 P. Tong, "An Assumed Stress Hybrid Finite Element Method for an Incompressible and Near-Incompressible Material," *IJSoSt,* Vol. 5, No. 5, 1969, pp. 455-461.

16.55 R. D. Cook, "Comment on 'Discrete Element Idealization of an Incompressible Liquid for Vibration Analysis' and 'Discrete Element Structural Theory of Fluids,'" *AIAAJ,* Vol. 11, No. 5, 1973, pp. 766-767.

16.56 *Fourth* and *Fifth NASTRAN U.C.*, 1975 and 1976.

16.57 C. Y. Liaw and A. K. Chopra, "Earthquake Analysis of Axisymmetric Towers Partially Submerged in Water," *EEStDy,* Vol. 3, No. 2, 1975, pp. 233-248.

16.58 T. Belytschko and T. L. Geers, eds., *Computational Methods for Fluid-Structure Interaction Problems,* Applied Mechanics Division, Vol. 26, American Society of Mechanical Engineers, New York, 1977.

16.59 O. C. Zienkiewicz, R. W. Lewis, and K. G. Stagg, eds., *Numerical Methods in Offshore Engineering,* John Wiley & Sons, Chichester, England, 1978.

16.60 *IJNME,* Vol. 13, No. 1, 1978 (entire issue devoted to fluid-structure interaction).

16.61 H. C. Fu, "Indirect Structural Analysis by Finite Element Method," *JStDiv,* Vol. 99, No. ST1, 1973, pp. 91-111.

16.62 L. Kiefling and G. C. Feng, "Fluid-Structure Finite Element Vibrational Analysis," *AIAAJ,* Vol. 14, No. 2, 1976, pp. 199-203.

16.63 M. Fröier, L. Nilsson, and A. Samuelsson, "The Rectangular Plane Stress Element by Turner, Pian and Wilson," *IJNME,* Vol. 8, No. 2, 1974, pp. 433-437.

CHAPTER 17

17.1 D. H. Norrie and G. deVries, "Non-Structural Applications of the Finite Element Method," *USJap3 Conf.*, pp. 511-539.

17.2 E. L. Wilson, K. J. Bathe, and F. E. Peterson, "Finite Element Analysis of Linear and Nonlinear Heat Transfer," *NucEDe,* Vol. 29, No. 1, 1974, pp. 110-124.

17.3 J. F. Lyness, D. R. J. Owen, and O. C. Zienkiewicz, "The Finite Element Analysis of Engineering Systems Governed by a Non-Linear Quasi-Harmonic Equation," *CompSt,* Vol. 5, No. 1, 1975, pp. 65-79.

17.4 G. Myers, *Analytical Methods in Conduction Heat Transfer,* McGraw-Hill, New York, 1971.

17.5 O. C. Zienkiewicz, C. J. Parekh, and A. J. Wills, "The Application of Finite Elements to Heat Conduction Problems Involving Latent Heat," *Rock Mechanics,* Vol. 5, No. 2, 1973, pp. 65-76.

17.6 J. Donea and S. Giuliani, "Finite Element Analysis of Steady-State Nonlinear Heat Transfer Problems," *NucEDe,* Vol. 30, No. 2, 1974, pp. 205-213.

17.7 A. F. Emery and W. W. Carson, "An Evaluation of the Use of the Finite-Element Method in the Computation of Temperature," *JHeTra,* Vol. 93, No. 2, 1971, pp. 136-145.

17.8 J. Donea, "On the Accuracy of Finite Element Solutions to the Transient Heat-Conduction Equation," *IJNME,* Vol. 8, No. 1, 1974, pp. 103-110.

17.9 J. Padovan, "Semi-Analytical Finite Element Procedure for Conduction in Anisotropic Axisymmetric Solids," *IJNME,* Vol. 8, No. 2, 1974, pp. 295-310.

17.10 R. E. Beckett and S.-C. Chu, "Finite Element Method Applied to Heat Conduction in Solids With Nonlinear Boundary Conditions," *JHeTra,* Vol. 95, No. 1, 1973, pp. 126-128.

17.11 G. Comini, S. Del Guidice, R. W. Lewis, and O. C. Zienkiewicz, "Finite Element Solution of Nonlinear Heat Conduction Problems with Special Reference to Phase Change," *IJNME,* Vol. 8, No. 3, 1974, pp. 613-624.

17.12 H. C. Martin, "Finite Element Analysis of Fluid Flows," *WPAFB2 Conf.*, pp. 517-535.

17.13 M. Petyt, J. Lea, and G. H. Koopmann, "A Finite Element Method for Determining the Acoustic Modes of Irregular Shaped Cavities," *JSouVi,* Vol. 45, No. 4, 1976, pp. 495-502.

CHAPTER 18

18.1 S. H. Crandall, *Engineering Analysis,* McGraw-Hill, New York, 1956.

18.2 B. A. Finlayson and L. E. Scriven, "The Method of Weighted Residuals—A Review," *Applied Mechanics Reviews,* Vol. 19, No. 9, 1966, pp. 735-748.

18.3 O. C. Zienkiewicz and C. J. Parekh, "Transient Field Problems: Two-Dimensional and Three-Dimensional Analysis by Isoparametric Finite Elements," *IJNME,* Vol. 2, No. 1, 1970, pp. 61-71.

18.4 J. W. Leonard and T. T. Bramlette, "Finite Element Solutions to Differential Equations," *JEMDiv,* Vol. 96, No. EM6, 1970, pp. 1277-1283.

18.5 S. G. Hutton and D. L. Anderson, "Finite Element Method: A Galerkin Approach," *JEMDiv,* Vol. 97, No. EM5, 1971, pp. 1503-1520.

18.6 P. P. Lynn and S. K. Arya, "Finite Elements Formulated by the Weighted Discrete Least Squares Method," *IJNME,* Vol. 8, No. 1, 1974, pp. 71-90.

18.7 O. C. Zienkiewicz, D. R. J. Owen, and K. N. Lee, "Least Square Finite Element for Elasto-Static Problems. Use of 'Reduced' Integration," *IJNME,* Vol. 8, No. 2, 1974, pp. 341-358.

18.8 J. F. Lyness, D. R. J. Owen, and O. C. Zienkiewicz, "The Finite Element Analysis of Engineering Systems Governed by a Nonlinear Quasi-Harmonic Equation," *CompSt,* Vol. 5, No. 1, 1975, pp. 65-79.

18.9 J. J. Connor and C. A. Brebbia, *Finite Element Techniques for Fluid Flow,* Newnes-Butterworths, London, 1976.

18.10 E. D. Eason, "A Review of Least-Squares Methods for Solving Partial Differential Equations," *IJNME,* Vol. 10, No. 5, 1976, pp. 1021-1046.

18.11 W. L. Kwok, Y. K. Cheung, and C. Delcourt, "Application of Least Squares Collocation Technique in Finite Element and Finite Strip Formulation," *IJNME,* Vol. 11, No. 9, 1977, pp. 1391-1404.

18.12 R. H. Gallagher, J. A. Liggett, and S. T. K. Chan, "Finite Element Shallow Lake Circulation Analysis," *Journal of the Hydraulics Division, ASCE,* Vol. 99, No. HY7, 1973, pp. 1083-1096.

18.13 G. Strang and G. J. Fix, *An Analysis of the Finite Element Method,* Prentice-Hall, Englewood Cliffs, N.J., 1973.

18.14 B. A. Finlayson, *The Method of Weighted Residuals and Variational Principles,* Academic Press, New York, 1972.

18.15 P. C. M. Lau and C. A. Brebbia, "The Cell Collocation Method in Continuum Mechanics," *IJMSci,* Vol. 20, No. 2, 1978, pp. 83-95.

18.16 C. A. J. Fletcher, "An Improved Finite Element Formulation Derived from the Method of Weighted Residuals," *CMAME,* Vol. 15, No. 2, 1978, pp. 207-222.

ANSWERS TO PROBLEMS

CHAPTER 1

1.1 (a) $\begin{bmatrix} + & + & - & + \\ + & + & - & + \\ - & - & + & - \\ + & + & - & + \end{bmatrix}$ 1.2 (a, b) $\begin{bmatrix} + & + & - & - & - & - \\ + & + & + & - & - & + \\ - & + & + & + & - & + \\ - & - & + & + & + & + \\ - & - & - & + & + & - \\ - & + & + & + & - & + \end{bmatrix}$

1.2 (c) See Section 4.4.

1.3 $u_3 = -P/k_3$, $u_2 = -P/k_2 + u_3$, $u_1 = -P/k_1 + u_2$.

1.5 (b) $\sigma_{r,r} + \dfrac{\tau_{r\theta,\theta}}{r} + \tau_{rz,z} + \dfrac{\sigma_r - \sigma_\theta}{r} + F_r = 0$

$$\tau_{r\theta,r} + \frac{\sigma_{\theta,\theta}}{r} + \tau_{\theta z,z} + \frac{2\tau_{r\theta}}{r} + F_\theta = 0$$

$$\tau_{rz,r} + \frac{\tau_{\theta z,\theta}}{r} + \sigma_{z,z} + \frac{\tau_{rz}}{r} + F_z = 0$$

1.6 (c) None but to replace x by r and y by θ.

1.7 (d) $a_2 + a_9 = 0$, $a_6 + a_8 = 0$.

1.8 $\epsilon_{x,yy} + \epsilon_{y,xx} = \gamma_{xy,xy}$.

1.10 The companion equation is $v_{,xx} + v_{,yy} = (1 + \nu)(v_{,xx} - u_{,xy})/2$.

1.11 (a) Satisfied. (b) $a_5 = a_{11} = 0$.

(c) $a_4 = -\dfrac{1 + \nu}{4} a_{11}$, $a_{12} = -\dfrac{1 + \nu}{4} a_5$.

(d) $\{a_6 \; a_{10}\} = -\dfrac{A\nu + G}{2G}\{a_{11} \; a_5\}$, where $A = \dfrac{E}{1 - \nu^2}$.

(f) All are suitable: equilibrium need not be satisfied identically.

1.13 (b) In the first equation, σ is not known in advance, so neither is α. We must assume α, compute σ, revise α, again compute σ, until convergence. In the second equation we simply use the E and $\partial E/\partial T$ appropriate to the temperature at hand.

CHAPTER 2

2.2 With c = constant, $\{\mathbf{d}\}_a = c\{-3\ 4\ -3\ 4\ -3\ 4\}$, $\{\mathbf{d}\}_b = c\{0\ 0\ 0\ 3\ 4\ 3\}$, $\{\mathbf{d}\}_c = c\{-4\ 0\ -4\ 3\ 0\ 3\}$. These vectors are linearly independent.

2.3 The derivation of $[\mathbf{K}]$ presumes *small* rotations.

2.6 (a) Can be zero (consider $\beta = 0$ in Fig. 2.4.1) but not negative.
(b) $\{\mathbf{d}\} = \{1\ 1\ 1\ 1\}$ is not rigid-body motion for a beam.
2.7 (a) Row 1 of $[\mathbf{k}]$ is $AE/L \lfloor c^2\ cs\ -c\ 0 \rfloor$.
(b) $\{\mathbf{d}\} = \{1\ 0\ c\ -s\}$, $\{0\ 1\ s\ c\}$, and $\{0\ 0\ 0\ Ld\theta\}$.
2.9 Row 7 is $\lfloor a_4\ 0\ (a_6 + b_3)\ b_2\ 0\ 0\ (a_5 + b_1) \rfloor$.
2.13 KK(4) = 2*NOD(2,N), KK(3) = 2*NOD(1,N), KK(2) = KK(4) − 1, KK(1) = KK(3) − 1.
2.15 (a) One possibility that gives $B = 10$ is $L = 1$, $K = 2$, $M = 3$, $J = 4$, $N = 5$, $H = 6$, $G = 7$, $F = 8$, $I = 9$, $E = 10$, $D = 11$, $A = 12$, $C = 13$, and $B = 14$.
(b) To $K_{23,23}$: FA, AC, AB. To $K_{24,24}$: AD, AC.
(c) Activate d.o.f. in turn and examine the resulting forces at A (node 12).
Row 23: 24(−), 25(−), 26(+), 27(−). Row 24: 25(+), 26(−).
2.17 (a) $B = 6$. (c) Interchange node numbers 3 and 7, or nodes 2 and 7.
(d) $P = 56$.
2.18 Use information in array KK (Fig. 2.9.3) and assume that each $[\mathbf{k}]$ is full (thus some zeros in Fig. 2.7.2 are not detected).
2.21 (c)

$$\begin{bmatrix} 0 & 0 & 0 & 0 & 1 & 0 & 1 \\ 1 & 0 & 0 & 0 & 1 & 0 & 0 \\ 1 & 1 & 1 & 1 & 1 & 1 & 1 \\ 1 & 1 & 1 & 1 & 1 & 1 & 1 \\ 1 & 1 & 1 & 1 & 1 & 1 & 1 \\ 0 & 0 & 0 & 0 & 0 & 0 & 0 \end{bmatrix}, \quad \begin{bmatrix} 1 & 3 & 6 & 9 & 0 & 13 & 0 \\ 0 & 4 & 7 & 10 & 0 & 14 & 16 \\ 0 & 0 & 0 & 0 & 0 & 0 & 0 \\ 0 & 0 & 0 & 0 & 0 & 0 & 0 \\ 0 & 0 & 0 & 0 & 0 & 0 & 0 \\ 2 & 5 & 8 & 11 & 12 & 15 & 17 \end{bmatrix}$$

2.22 Consider Fig. 1.2.2. If all v_i are suppressed, $B = 2$, not 4.
2.24 Place large axial forces on a stiff bar. Attach nodes of this bar to nodes 3 and 4 of the truss. *Caution*. See Section 15.5.
2.28 (a) Only those that are below the skyline.
(b) Overflow or attempt to divide by zero. The matrix is singular if the truss has no supports.
2.33 The determinant is 1.0.
2.35, 2.36 Exact results are $D_1 = 4\ D_2 = 8$, $D_3 = 12$.
2.37, 2.38 Exact results are $D_1 = 1$, $D_2 = 2$, $D_3 = 3$.

CHAPTER 3

3.1 $\Pi_p = k(D_1^2 + \bar{D}_2^2 + \bar{D}_3^2)/2 - PD_1 - P(D_1 + \bar{D}_2) - P(D_1 + \bar{D}_2 + \bar{D}_3)$, where the $\bar{D}$'s are relative displacements.
3.2 k_4 is added to the second diagonal term and k_5 to the third.
3.3 Add to Π_p in Problem 3.1: $k_4(D_1 + \bar{D}_2)^2/2 + k_5(D_1 + \bar{D}_2 + \bar{D}_3)^2/2$.
3.7 $a = \kappa_0/2 - m_0/2EI + M/2EI$. Also, $\sigma_x = -M_L y/I$.
3.8 $w_{,x} \neq 0$ at $x = 0$. Acceptable: $w = cx^2/L^2 + \Sigma\, a_i(1 - \cos 2i\pi x/L)$.
3.10 $\sigma_{x,x} + q/A = 0$ satisfied by only the exact solution.
3.11 (a) $0.010417\ qL^4/EI$ if $w = a_1 x(x - L)$.
(b) $0.013071\ qL^4 EI$ if $w = a_1 \sin(\pi x/L)$.
(c) $M = 0$ at $x = 0$ and at $x = L$ from (b), but not from (a).
3.12 One term: $u_L = 3PL/4AE$. Two terms: $u_L = 8PL/9AE$.
3.13 (a) $D = 2\alpha LT/3$. (b) $D = \alpha LT/3$. (Exact $D = \alpha LT/2$.)
3.14–3.17 Collected results, for part (b):

$$a_1 = \frac{11PL}{8EI} - \frac{M_0}{2EI} + \frac{5q_0L^2}{6EI} + \frac{\alpha T}{t}, \qquad a_2 = -\frac{P}{4EI} - \frac{q_0L}{6EI}$$

3.14 (c) Infinite. 3.15 (c) One. 3.16 (c) Three. 3.17 (c) One.

3.18 $U_1 = c^2\ell^5/18AE$, $U_2 = 19c^2\ell^5/288AE$, $U_{\text{exact}} = c^2\ell^5/15AE$. The second integral is twice the magnitude of the first.

3.19 From Eq. 3.7.13: $U = 95c^2L^5/6AE$. Exact: $U = 81c^2L^5/5AE$.

3.20 At right end, in all cases, $u = c\ell^3/6AE$. Exact at nodes.

3.21 The common factor is (a) $k = 2A_0E/\ell$, (b) $k = 1.820A_0E/\ell$.

3.22 $\{u_2 \; u_3 \; u_4\} = (c\ell^3/6AE)\,\{0.495 \;\; 1.375 \;\; 2.000\}$.

3.23 Results are exact according to beam theory.

3.24 $w_2 = qL^4/8EI$, $\theta_2 = qL^3/6EI$. Both are exact.

CHAPTER 4

4.1
$$[A]^{-1} = \begin{bmatrix} 1 & 0 & 0 & 0 \\ 0 & 1 & 0 & 0 \\ -3/L^2 & -2/L & 3/L^2 & -1/L \\ 2/L^3 & 1/L^2 & -2/L^3 & 1/L^2 \end{bmatrix}$$

4.2 For example, at $x = 0$, $\lfloor \mathbf{N} \rfloor = \lfloor 1 \;\, 0 \;\, 0 \;\, 0 \rfloor$ and $\lfloor \mathbf{N}_{,x} \rfloor = \lfloor 0 \;\, 1 \;\, 0 \;\, 0 \rfloor$.

4.8 (a) It is easiest to use $\lfloor \mathbf{B} \rfloor = \lfloor 0 \;\, 0 \;\, 2 \;\, 6x \rfloor \, [\mathbf{A}]^{-1}$.
(b) $\{\mathbf{d}\} = \{0 \;\; \theta \;\; L\theta \;\; \theta\}$.

4.10 (c) The load does work in going through displacements produced *by activating the nodal d.o.f.*, not those produced by the load itself.

4.11 No. Such a model does not allow joints to rotate.

4.12 At the left end, $r_1 = -[2L/b + 3]\, A\rho\omega^2 bL/6$.

4.13 $[E\alpha T(A_1 + A_2)/2]\,\{-1 \;\; 1\}$. Exact: $[E\alpha T(A_2 - A_1)/\ln(A_2/A_1)]\,\{-1 \;\; 1\}$.

4.14 $[EA\alpha(T_1 + T_2)/2]\,\{-1 \;\; 1\}$.

4.15 Use the m_0 term in Eq. 3.4.16; get $(2EI\alpha T_0/t)\,\{0 \;\; -1 \;\; 0 \;\; 1\}$.

4.16 $\{\mathbf{r}\} = (q_1L/60)\,\{21 \;\; 3L \;\; 9 \;\; -2L\} + (q_2L/60)\,\{9 \;\; 2L \;\; 21 \;\; -3L\}$.

4.17 Changes signs in rows and columns 3 and 4 (but k_{33} and k_{44} remain positive). Makes assembly awkward.

4.19 (c) A partial answer is given under Problem 7.5(b).
(d) $u_3 = 0$, $v_3 = 2Pc/Ebt$.

4.22 (a) 37, 38, 59, 60, 61, 62, 39, 40.
(b) (1) 0. (2) 0. (3) $Et/2$. (4) $Et/2$.

4.26 $$[\mathbf{k}] = \frac{4AE}{3L}\begin{bmatrix} 1 & -1 \\ -1 & 1 \end{bmatrix}$$

4.27 $$[\mathbf{k}] = \frac{AE}{3L}\begin{bmatrix} 7 & 1 \\ 1 & 7 \end{bmatrix}$$

4.29 Yes. Nodal moments go to zero faster than nodal forces.

4.30 Rigid-body motion and constant curvature states are correct, but $w_{,x}$ does not agree with the nodal θ's.

4.31 The major defect is lack of slope continuity between elements.

4.32 Displacement along an edge is not completely defined by d.o.f. on that edge.

4.33 (b) (1) xyz. (2) x^3, y^3, z^3. (3) x^3, y^3, z^3, xyz. (4) x^2y, x^2z, xy^2, xz^2, yz^2, y^2z. (6) All but xyz.

4.34 (a) No. It is incompatible.
(b) It is still a complete quadratic.

4.36 (a) -16.7% error at the root. (b) -66.7% error at the root.
4.38 A square wave pattern.
4.39 One must ask if the strain and temperature fields are of equal competence.
4.40 (a) No. (b) For σ_x but not for σ_y.
4.41 (a) $u_2 = \alpha cL/2$, $\sigma_x = E\alpha c(L - 2x)/2L$.
(b) $u_2 = \alpha cL/2$, $\sigma_x = 0$.
(c) Part (a) is incorrect. Part (b) is correct if node 2 is free.
(d) Use $T = cx/L$. Then, since $u = 0$, $\sigma_x = -E\alpha cx/L$.
(e) If u_2 is the computed d.o.f. and $\bar{u}_2$ its value for unrestrained expansion, use $T = (c/2)(u_2/\bar{u}_2) + (cx/L)[1 - (u_2/\bar{u}_2)]$ in an otherwise standard stress calculation.
4.42 (a) $NB^2 \approx 8i^3k^4$. (b) $NB^2 \approx 36i^3k^4$. (c) $NB^2 \approx 343i^3k^7$.

CHAPTER 5

5.1 (d) $\dfrac{AE}{3L}\begin{bmatrix} 7 & 1 & -8 \\ 1 & 7 & -8 \\ -8 & -8 & 16 \end{bmatrix}$ 5.2 (b) $\dfrac{1}{4}\begin{bmatrix} 1 & 1 & 1 & 1 \\ -1 & 1 & 1 & -1 \\ -1 & -1 & 1 & 1 \\ 1 & -1 & 1 & -1 \end{bmatrix}$

5.3 $d\xi = dy/2$, $d\eta = -dx/2$.
5.4 (a) $I_1 = 2.000$, $I_2 = 1.232$, $I_3 = 1.274$, exact $I = 1.273$.
(b) $I_1 = 0.000$, $I_2 = 0.667$, $I_3 = 0.667$, exact $I = 0.667$.
(c) $I_1 = 1.500$, $I_2 = 1.846$, $I_3 = 1.925$, exact $I = 1.946$.
5.7 (b) Place it in the DO 20 loop, to calculate $[\mathbf{B}]^T\, dV$. Then remove DV from statements 30, 30 − 1, and 60 to save 36 multiplications per Gauss point.
5.9 Find I at Gauss points by shape-function interpolation from I_1 and I_2.
5.13 (a) ϵ_x contains up to x and y^2, ϵ_y contains up to x^2 and y, and γ_{xy} contains up to x^2 and y^2. Exact for tip M and tip P but approximate for distributed load.
5.15 Consider the expense of element generation and equation solving, ability to match structure boundaries, and competence of stress representation.
5.16 Answers will be locally bad: this assembly will fail a patch test.
5.17 Note that integrals of first powers vanish when the integration limits are -1 to $+1$.
5.18 On (say) $\zeta = 1$, the N_i are those of Table 5.6.1.
5.19 $\{\mathbf{r}\} = (Lt/6)\ \{\sigma_A\ \ 2(\sigma_A + \sigma_B)\ \ \sigma_B\}$.
5.20 On a straight x-parallel edge with a midside node, $\tau = J_{12} = 0$, $\sigma = \sigma_y$, and $J_{11} = L/2$.
5.27 (a) 2. (b) An infinite number.
5.29 (a) 2 by 2. (b) 3 by 3.
5.30 (a) 2 by 2. (b) 3 by 3. (c) 2 by 2 by 2. (d) 3 by 3 by 3.
5.32 (a) There are two.
(b) Attempt to divide by zero, or quotient overflow.
(d) A mode that makes $x = L/2$ an inflection point.
5.33 No change. Both rules integrate the k_{ij} exactly.
5.34 This mode is the one shown in Fig. 5.10.1*b*.
5.35 As in Fig. 5.10.1*a*, a *mesh* can have the mode.
5.36 Imagine that all points lie on the axis of a beam in pure bending.
5.37 (a) Row 1 of $[\mathbf{Q}]$ is, with a and b as in Eq. 5.11.4, $\lfloor a^3\ a^2b\ ab^2\ a^2b\ a^2b\ ab^2\ b^3\ ab^2 \rfloor/8$.
5.38 With a general shape, $\{\boldsymbol{\sigma}\}$ does not vary linearly with ξ and η.

5.39 Yes for the linear element. No for the quadratic element.
5.40 Node 3 can be moved to either quarter point.

CHAPTER 6

6.1 The matrix is 4 by 4 and expresses a transformation in only an *rz* plane.
6.2 $[\mathbf{E}]_i = [EV_i \quad 0 \quad 0]$, where V_i is the volume fraction of the *i*th fibers ($i = 1, 2$).
6.3 (c) $[\mathbf{T}]^T[\mathbf{T}]$ involves scalar products of mutually orthogonal unit vectors.
  (e) $[\mathbf{T}]$ of Eq. 6.2.5 is orthogonal, but $[\mathbf{Q}]$ is not.
  (h) $\{\boldsymbol{\sigma}\} = [\mathbf{T}_\epsilon]^T \{\boldsymbol{\sigma}'\} = [\mathbf{T}_\epsilon]^T [\mathbf{E}'] \{\boldsymbol{\epsilon}'\} = [\mathbf{T}_\epsilon]^T[\mathbf{E}'] [\mathbf{T}_\epsilon]\{\boldsymbol{\epsilon}\}$.
6.6 (a, b) Row 1 of $[\mathbf{k}]$ is $(AE/L)[c^2 \; cs \; -c \; 0]$. (c) See Fig. 4.4.3.
6.10 A 14-statement algorithm works nicely.
6.12 $T_{ij} = 0$ except $T_{11} = T_{22} = T_{37} = T_{48} = T_{54} = T_{65} = 1$.
6.13 Use $[\mathbf{N}]$ of the quadrilateral to define u_5 and v_5 in terms of $\{\mathbf{d}\}$ of the quadrilateral. Similarly, define $\theta_5 = (v_{,x} - u_{,y})/2$ and use $[\mathbf{N}]$ to express it in terms of $\{\mathbf{d}\}$.
6.14 (a) Nonzero T_{ij} are $T_{11} = T_{22} = c/L_2$, $T_{33} = T_{44} = a/L_1$, $T_{17} = T_{28} = d/L_2$, $T_{35} = T_{46} = b/L_1$.
  (b) If the quadrilateral is warped, see Section 7.8. Replace the 12 d.o.f. of the quadrilateral by the 24 d.o.f. of the solid. $[\mathbf{E}]$ depends on E and A of each bar, the number of bars, and the orientation of the bars in local coordinates.
6.15 (a) $[\mathbf{T}_1]$ contains 36 zeros, 8 ones, and 4 $(\pm H/2)$'s.
  (b) $[\mathbf{T}_2]$ contains 36 zeros, 8 (1/2)'s, and 4 $(\pm 1/H)$'s.
  (d) $[\mathbf{T}_1]\,[\mathbf{T}_2] \neq [\mathbf{I}]$. $[\mathbf{T}_2]\,[\mathbf{T}_1] = [\mathbf{I}]$.
6.16 $[\boldsymbol{\Lambda}_1] = [\mathbf{I}]$ except $(\Lambda_1)_{15} = -(\Lambda_1)_{24} = -(z_1 - z_i)$, $(\Lambda_1)_{16} = -(\Lambda_1)_{34} = (y_1 - y_i)$, $(\Lambda_1)_{26} = -(\Lambda_1)_{35} = -(x_1 - x_i)$.
6.17 *Ad hoc* procedures: set $m_{ij} = 0$ for $i \neq j$, or appy the method that yields Eq. 11.3.5.
6.18 (b) Division by zero. If $y_1 \approx y_2$, define $\theta = (v_2 - v_1)/(x_2 - x_1)$.
6.19 $[\mathbf{K}]$ is singular if θ_i remains in $\{\mathbf{D}\}$. Rigidly joined bars: fix u_i and v_i (but not θ_i) by a standard boundary condition treatment.
6.20 $[\mathbf{T}] = [\mathbf{I}]$ except 1/4 appears in columns 1, 3, 5, and 7 of row 9 and in columns 2, 4, 6, and 8 of row 10.
6.21 (a) Write $u_5 = u_{5\text{rel}} + (u_1 + u_2)/2$, and so on.
  (b) Use a standard boundary condition treatment to set the relative displacements to zero.
6.22 $\bar{N}_i = N_i + N_9/2$ for $i = 1, 2$; $\bar{N}_i = N_i + N_{11}/2$ for $i = 3, 4$; $\bar{N}_i = N_i + N_{13}/2$ for $i = 5, 6$; $\bar{N}_i = N_i + N_{15}/2$ for $i = 7, 8$. $\bar{N}_i = N_i$ for $i = 10, 12, 14, 16$ to 20. $\bar{N}_i = 0$ for $i = 9, 11, 13, 15$.
6.23 $[\mathbf{K}] = [\mathbf{C}_{er}^T \mathbf{K}_{ee} \mathbf{C}_{er} + \mathbf{C}_{er}^T \mathbf{K}_{er} + \mathbf{K}_{er}^T \mathbf{C}_{er} + \mathbf{K}_{rr}]$.
  $\{\mathbf{R}_r\} = \{\mathbf{C}_{er}^T \mathbf{R}_e + \mathbf{R}_r - (\mathbf{C}_{er}^T \mathbf{K}_{ee} + \mathbf{K}_{er}^T)\, \mathbf{C}_e^{-1}\, \mathbf{Q}\}$.
6.24 $[-\mathbf{T} \quad \mathbf{I}]\,\{\mathbf{D}_r \quad \mathbf{D}_e\} = 0$.
6.25 See Problem 6.24. Three equations for the triangle; five for the quadrilateral.
6.26 (a) In the answer to Probem 6.24, $T_{ij} = 0$ except
  $T_{11} = T_{12} = T_{22} = T_{23} = T_{33} = T_{34} = T_{44} = T_{41} = -1/2$.
  (b) Add a null column for each added d.o.f.
6.27 (a) $\begin{bmatrix} 1 & 0 & -1 & 0 & \cdots & \text{null} & \cdots \\ 0 & 1 & 0 & -1 & \cdots & \text{null} & \cdots \end{bmatrix} \{u_A \;\; v_A \;\; u_B \;\; v_B \cdots\} = 0.$

(b) $[\mathbf{T}] = [\mathbf{I}]$ except $T_{31} = T_{42} = 1$.

6.28 $\begin{bmatrix} x_2/x_1 & -1 & 0 \\ x_3/x_1 & 0 & -1 \end{bmatrix} \{u_1 \quad u_2 \quad u_3\} = 0.$

6.29 Invoke Eq. 6.3.7 where, for example, for element 2, $[\mathbf{T}] = [\mathbf{I}]$ except for $T_{11} = T_{22} = A/L$ and $T_{17} = T_{28} = B/L$, where A, B, and L are lengths as in Problem 6.14.

6.30 (a) $[\mathbf{T}]$ is null except for $T_{11} = T_{22} = T_{33} = T_{44} = T_{53} = T_{64} = T_{75} = T_{86} = 1$.

(b) $\{\mathbf{R}\} = [\mathbf{T}]^T \{\mathbf{r}_1 \quad \mathbf{r}_2\}$.

6.31 (b) Nodes 1 and 2 have the same w_i and the same θ_i (not a rigid-body constraint). The element contributes $12EI/L$ to resist θ_2.

6.33 $\lfloor -c \ -b \ c \ -b \ c \ b \ -c \ b \rfloor \{u_1 \ v_1 \ u_1 \ \cdots \ v_4\} = 0$.

6.35 (b) $u_1 = 2 + 3/\alpha, \quad u_2 = 5 + 3/\alpha$.

6.37 $[\mathbf{C}] = \lfloor 0\ 0\ 1\ 0\ 0\ 0\ -1\ 0\ 0\ 0\ 0\ 0\ 0\ 0 \rfloor$,

$\lceil \boldsymbol{\alpha} \rfloor = AE/L, \quad \{\mathbf{Q}\} = L\bar{\epsilon}$.

6.38 (a)

$$[\mathbf{T}] = \begin{bmatrix} 0.866 & -0.500 & 0 & 0 \\ 0.500 & 0.866 & 0 & 0 \\ 0 & 0 & 0 & -1 \\ 0 & 0 & 1 & 0 \end{bmatrix}$$

(b) Set the θ d.o.f. to zero. Make the radial d.o.f. equal.

(c) Radial displacement $= 2PL/AE$ (this is exact).

6.39 $u = 0$ on $x = 0$, $v = 0$ on $y = 0$, and $w = 0$ on $z = 0$.

6.40 See the example problem in Section 11.5.

6.41 $\theta_2 = M_0L/12EI$.

6.42 (b) If $t \perp r$, $w_{,t}(r) = -w_{,t}(-r)$, $M_t(r) = M_t(-r)$, $Q_t(r) = -Q_t(-r)$. Change the initial signs on the right sides for skew antisymmetry.

6.43 Case 1: load $q/2$ over the entire span.

Case 2: load $q/2$ up on one half and down on the other.

Economical if the entire structure has too many d.o.f.

CHAPTER 7

7.4 $[\mathbf{k}_a] = AEL$. See Eq. 2.4.5 for $[\mathbf{k}]$.

7.5 (a) $u = a_1x + a_2y$, $v = a_4 \, [y - (y_2/x_2)\, x]$. Fails if $x_2 = 0$.

(b)

$$[\mathbf{T}]^T[\mathbf{k}_{EE}][\mathbf{T}] = \frac{Et}{4bc} \begin{bmatrix} 2c^2 + b^2 & cb & -cb \\ cb & c^2 + 2b^2 & -c^2 \\ -cb & -c^2 & c^2 \end{bmatrix} \begin{matrix} u_1 \\ v_1 \\ v_2 \end{matrix}$$

7.6

$$[\mathbf{k}] = C \begin{bmatrix} 1/r_1{}^2 & -1/r_1r_2 \\ -1/r_1r_2 & 1/r_2{}^2 \end{bmatrix}$$

7.10 The condensed matrix is null.

7.11 (a) Expand $[\mathbf{k}]$ to 8 by 8 by adding β_1 and β_2 to $\{\mathbf{d}\}$. Add two matrices like that of Eq. 2.4.5 to $[\mathbf{k}]$, then condense θ_1 and θ_2.

(b) Successively, apply unit values of u_i, v_i, and β_i; the required forces and moments are columns of $[\mathbf{k}]$. Note that $\beta_i = 0$ does not imply $\theta_i = 0$.

7.12 $w_A = (P/2)L^3/3EI, \quad \theta_A = (P/2)L^2/2EI$.

7.13 Apply $P = 1$, solve for θ_B. Actual P = (prescribed θ_B)/(computed θ_B).

7.14 (a) $k_{11} = k_{22} = -k_{12} = -k_{21} = AE/L$, $k_{33} = AEL^3/3$.
(b) Row 1 of [**k**] is $(AE/L)\lfloor 7/3 \quad 1/3 \quad -8/3 \rfloor$.
(c) $\lfloor \mathbf{N} \rfloor = \lfloor 1 - (3x/L) + (2x^2/L^2),\ -(x/L) + (2x^2/L^2),\ -4x(x - L)/L^2 \rfloor$.
(d) In both cases, [**k**] is as seen in Eq. 2.4.5.

7.15 (a) Integrate $(1 - \xi^2)(1 - \eta^2)$ from -1 to $+1$ in both directions.
(b) They do not: edge displacements are not affected.

7.16 (a) $[\mathbf{T}] = [\mathbf{I}]$ except for $T_{51} = T_{52} = T_{53} = T_{54} = -1/4$.
(b) $[\mathbf{T}]^{-1} = [\mathbf{I}]$ except for $T_{51} = T_{52} = T_{53} = T_{54} = 1/4$.

7.17 Yes, but there is negligible benefit (or detriment). Polynomials are needed for N_1 through N_4 to satisfy rigid-body and constant strain modes. The new "center" force is 1.62.

7.18 Standard: only when there is a distributed load. Tapered: almost always. To provide $w = w_{,x} = 0$ at the ends, write $x^2(x - L)^2a$, where a is a nodeless d.o.f.

7.19 $N_1 = \frac{1}{4}\xi\eta(\xi - 1)(\eta - 1)$,
$N_2 = \frac{1}{4}\xi\eta(\xi + 1)(\eta - 1)$,
$N_3 = \frac{1}{4}\xi\eta(\xi + 1)(\eta + 1)$,
$N_4 = \frac{1}{4}\xi\eta(\xi - 1)(\eta + 1)$,
$N_5 = \frac{1}{2}\eta(\eta - 1)(1 - \xi^2)$,
$N_6 = \frac{1}{2}\xi(\xi + 1)(1 - \eta^2)$,
$N_7 = \frac{1}{2}\eta(\eta + 1)(1 - \xi^2)$,
$N_8 = \frac{1}{2}\xi(\xi - 1)(1 - \eta^2)$,
$N_9 = (1 - \xi^2)(1 - \eta^2)$

7.20 Relative $u_9 = -0.9$. Absolute $u_9 = -0.1$.

7.23 Not if they are displacements *relative* to the boundary d.o.f.

7.25 Partial answer: $\sigma_x = -p$ and $\tau_{xy} = 0$ yield equations of constraint along AB; $u = v = u_{,x} = v_{,x} = 0$ along BC.

7.26 (b) It is correct: it is the Poisson effect.
(c) $R = \dfrac{1}{1 - \nu^2} + \dfrac{GL^2}{EH^2}$

7.29 Top and bottom: these edges must curve to avoid parasitic shear.

7.30 (a) Connection at node 5 is lost. The top and bottom triangles deform, but the side triangles have only rigid-body motion.
(b) Separate the top and bottom layers. View each as loaded by bending plus tension (or compression).

7.32 (a) Relative.
(b) There will be a spurious τ_{xy} except at $\xi = \eta = 0$. Unless $\nu = 0$, σ_x and σ_y will be a bit off except at $\xi = \eta = 0$, (because $\epsilon_y = 0$ is computed, but really $\epsilon_y \neq 0$).

7.33 (b) For x moments, $\lfloor \mathbf{Q} \rfloor = \lfloor 0, H, y_1, 0, -H, y_2, 0, H, y_3, 0, -H, y_4 \rfloor$.

7.34 No change in substance, only in notation.

7.36 $6Vk!\,\ell!\,m!\,n!/(3 + k + \ell + m + n)!$ The surface integral is Eq. 7.9.8.

7.37 (b) $N_1 = \alpha$, $N_2 = \beta$, $N_3 = \gamma$. Others unchanged.

7.38 (f) For any Gauss point i, $\alpha_i + \beta_i + \gamma_i = 1$.

7.39 No. For example, the y^2 term in ϵ_x cannot be interpolated using only linear shape functions.

7.40 (a) $N_1 = \alpha(1 + \zeta)/2$, $N_2 = \beta(1 + \zeta)/2$, $N_3 = \gamma(1 + \zeta)/2$,
$N_4 = \alpha(1 - \zeta)/2$, $N_5 = \beta(1 - \zeta)/2$, $N_6 = \gamma(1 - \zeta)/2$.

7.41 (a) $\alpha = 1 - s - t$, $\beta = s$, $\gamma = t$.

7.42, 7.43 The "expected values" include second moments of plane areas.

7.45 Corners at $(-1, -1, 0)$, $(0, 1, -1)$, $(1, -1, 0)$, and $(1, 1, 1)$. $(N_1)_{new} = (N_1 + N_2 + N_4)_{old}$, and so on.
7.46 (b) On the lower edge, $\{\mathbf{r}\} = -2\sigma_y tL\ \{1/2\ \ 2/3\ \ 1/2\}$.
7.48 Constant: if linear, the number of nodes must be at least double the number of elements.
7.49 $L_1 = (1 - \eta)/2$, $L_2 = (1 + \eta)/2$, $M_1 = \exp(-s/L)$, $M_2 = 1 - M_1$, and no change in Eqs. 7.14.6.
7.51 (b) $(\text{force})(\text{length})^{-3/2}$

CHAPTER 8

8.1 (a) Rigid: 1. Zero-energy deformation: 3, 7, 8.
(b) Rigid: 1. Zero-energy deformation: none.
(c) Rigid: 1, 2, 3. Zero-energy deformation: 7, 8.
(d) Rigid: 1, 2, 3. Zero-energy deformation: none.
8.2 Radius r replaces thickness t, otherwise no change.
8.3 Mohr's strain circle must be a point.
8.4 $(p/3)\ \{r_3L\quad r_3^2 - r_4^2\quad r_4L\}$ on a one-radian segment.
8.6 The w field is unreasonable: it has a cusp at $r = 0$.
8.7 (d)

$$\text{Pressure} = \frac{P}{\pi tR} + \sum_{n\ \text{even}} \frac{2P}{\pi tR} \cos\frac{n\pi}{2} \cos n\theta$$

8.8 (a) One term: $qL^2/7.75$. Two terms: $qL^2/8.05$.
(a) One term: $0.203PL$. Two terms: $0.225PL$.
8.10 For 36 linear elements around, ratio ≈ 2200.
8.12 The following constants in the equations of Problem 8.11 are nonzero. (a) c_1. (b) b_1. (c) a_1. (d) c_2. (e) a_2. (f) b_2.
8.16 Set $n = v = 0$ in the answers to Problem 8.17.
8.17

$$\underset{3\times 4}{[\mathbf{B}]} = \frac{1}{r(r_2 - r_1)} \begin{bmatrix} -r\cos n\theta & 0 & \cdots \\ (r_2 - r)\cos n\theta & (r_2 - r)n\cos n\theta & \cdots \\ -(r_2 - r)n\sin n\theta & -r_2\sin n\theta & \cdots \end{bmatrix}$$

where $\{\boldsymbol{\epsilon}\} = \{u_{,r}\quad (v_{,\theta} + u)/r\quad \gamma_{\theta r}\}$ and $\{\mathbf{d}\} = \{u_1\ v_1\ u_2\ v_2\}$.
8.18 By Mohr's circle, $\sigma_r = (\sigma_x/2)(1 + \cos 2\theta)$ and $\tau_{r\theta} = -(\sigma_x/2)\sin 2\theta$.
For $n = 0, 2$, respectively, $\{\mathbf{r}\}$ is $\begin{Bmatrix} t\pi r_2\sigma_x \\ 0 \end{Bmatrix}$, $(t\pi r_2\sigma_x/2)\begin{Bmatrix} 1 \\ -1 \end{Bmatrix}$.

8.19 (a) $k_{12} = -\dfrac{Etr_1r_2}{(r_2 - r_1)^2}\ell n\dfrac{r_2}{r_1}$, $\quad k_{22} = \dfrac{Et}{(r_2 - r_1)^2}\left[r_1^2\,\ell n\dfrac{r_2}{r_1} + (r_2 - r_1)^2\right]$.
(b) We need only k_{22}, which reduces to simply Et.

CHAPTER 9

9.3 (a) $w = a_1 + a_2 r\cos\theta + a_3 r\sin\theta$.
9.4 (c) From part (b), $w_{,x}$ depends on all d.o.f. in element 1 but $w_{,x1}$ and $w_{,x4}$. So even if $\{\mathbf{d}\} = 0$ in element 2, $w_{,x}$ appears along the common edge if w_1, w_4, $w_{,y1}$, or $w_{,y4}$ is nonzero.

(e) $w,_y$ is quadratic and should be fully defined by w_2, w_3, $w,_{y2}$, and $w,_{y3}$. Interelement compatible.

9.5 See the discussion in Section 7.5.

9.6 (a) Zero, first and second derivatives of w at nodes 1 to 3; normal slopes $w,_n$ at nodes 4 to 6. At least one other arrangement is possible.

9.7 Eliminate eight: w and tangential slopes at the side nodes (retain only $w,_n$ at the side nodes).

9.9 The second is worse because its lower left element is inactive.

9.10 Thickness t may vary, but ϵ_z is ignored.

9.14 No: transverse shears γ_{yz} and γ_{zx} couple nodal w_i to nodal θ_i.

9.16 (c) When [**E**] is constant, the integrands are functions of z^1, which integrates to zero when the limits are $-t/2$ to $t/2$.

(e) No. Only nodal moments appear.

9.17 (b) $Q_y = \mathfrak{D}_{44} \Sigma (b_i w_i - N_i \theta_{yi})$.

9.18 $N_y = act\epsilon_y - (bct^3/12)\theta_{y,y}$, $M_y = (ct^3/12)(b\epsilon_y - a\theta_{y,y})$.

9.19 M is linear and is Q_x is quadratic. But beam theory says $Q_x = dM/dx$. Two-point quadrature implies linear Q_x.

9.21 QM6 of Section 7.7: ϵ_x is independent of x, γ_{zx} is correct.

9.22 Row 2 of each matrix is as follows, for a beam of unit width.

(a) $(Et^3/12L) \lfloor 0 \quad 1 \quad 0 \quad -1 \rfloor$.

(b) $(Gt/1.2L) \lfloor L/2 \quad L^2/4 \quad -L/2 \quad L^2/4 \rfloor$.

(c) $(Gt/1.2L) \lfloor L/2 \quad L^2/3 \quad -L/2 \quad L^2/6 \rfloor$.

9.23 In the thin-beam limit, only bending deformation is present.

9.24 In the thin-beam limit, only shear deformation is present.

9.25 (a) We can eliminate w_M and θ_M.

(b) $0 = (w_2 - w_1)/L - (\theta_1 + \theta_2)/2$. Agrees with Problem 9.23.

9.26 (a) Constant $w,_{xx}$ superposed on any rigid-body motion.

9.28 Mode $w = 0$, $\theta_x = 0$, $\theta_y = xy$. Also $w = 0$, $\theta_x = xy$, $\theta_y = 0$.

9.29 Since nodal θ_i do not enter the assumed w field, we get nodal forces but not moments.

(a) Equal forces at corners. (b) See Figure. 5.8.1*c*.

(c) See Fig. 7.3.1*a* for the "center" force.

9.30 (a) The absence of nodal moments means that this single-element plate will not respond.

(b) Nodal moments appear as a result of condensation.

(c) No: there is no center *force* to be manipulated.

9.31 (b) Absolute center w is c, so (relative) $w_9 = -3c$.

9.32 (a) Only w at the center is active. Deformation is caused by shear only.

(b) Element 1: 1, 8, 40, 176. Element 2: 5, 28, 128, 544.
Element 3: 7, 36, 160, 672. Element 4: 6, 32, 144, 608.

CHAPTER 10

10.1 Compatibility is approached as the mesh is refined.

10.2 Row 1 of the matrix is $\alpha EV \lfloor 1 \quad -1/3 \quad -1/3 \quad -1/3 \rfloor$.

10.3 (b) $u = H \cos [(L - s)\theta/L]$, $u = H[\cos \theta + (1 - \cos \theta)s/L]$.

10.4 Nodal moment loads will not change the conclusion.

10.5 No. Bubble functions do not add the needed constant and linear terms.

10.6 Find a_5 and a_6 to fit the end points, then make a least squares fit.

10.7 (b) $\{\mathbf{d}\} = \left\{u_1 \quad w_1 \quad 0 \quad \left(u_1 - \dfrac{8a}{\ell}w_1\right) \quad w_1 \quad 0\right\}$

10.8 (a) Follow (for example) the procedure of Eqs. 4.2.1 to 4.2.5.
(b) The $w,_{xi}$ d.o.f. do, the β_i d.o.f. do not.

10.10 See Section 9.2.

10.11 Retain nodal θ_i about the fold line. Better: use finite strips.

10.12 No (cubic w, linear u, v). Yes (quadratic u, v, w).

10.13 $u = \Sigma N_i u_{ip} + \Sigma N_i (1 + \zeta) u_{ipq}/2$.

10.14 Thickness may vary, but normal strain is ignored.

10.15 (a) The vectors must only be nonparallel.

(b) $$\begin{Bmatrix} u \\ v \\ w \end{Bmatrix} = \sum N_i \begin{Bmatrix} u_i \\ v_i \\ w_i \end{Bmatrix} + \sum N_i \frac{\zeta}{2} t_i \begin{bmatrix} 0 & n_{3i} & -m_{3i} \\ -n_{3i} & 0 & \ell_{3i} \\ m_{3i} & -\ell_{3i} & 0 \end{bmatrix} \begin{Bmatrix} \alpha_{1i} \\ \alpha_{2i} \\ \alpha_{3i} \end{Bmatrix}$$

where ℓ_{3i}, m_{3i}, and n_{3i} are direction cosines of z' at a node.

(c) Use the cross-product of vectors along adjacent sides.

10.16 Three rotational d.o.f. per node should be used. Or, use transition elements that have nodes at $\zeta = \pm 1$ along the fold line but midsurface nodes elsewhere.

10.22 A 3-by-3-by-2 rule, or a 2-by-2-by-2 rule if the thickness is constant. The limit, with mesh refinement, is a single point.

10.23 If $\sigma_0 = E\alpha T = E\alpha\zeta$, then $M_h = 2\sigma_0 bh^2/3$, $M_s = \sigma_0 bh(c + h)$.

10.24 Omit thickness, rotational d.o.f., vectors $\mathbf{C}_i$, γ_{yz}, and γ_{zx}.

10.25 Evaluate the γ_{BC} and γ_{CA} rows of $[\mathbf{B}]$ a total of eight times to set these γ's to zero at the Gauss points.

10.26 (a) Note the coefficients of x^0, x^1, and x^2 in the expression for dy/dx.
(b) Nodes 1 and 2 have the same elevation above the chord (zero in this case).

10.28 (a-d) 1.000152, 1.000609, 1.002434, 1.033788.

10.29 See Refs. 1.20 and 9.1, but note the confusion in $\kappa_{s\theta}$.

10.32 $w = -H + \left(\dfrac{3H}{L^2} - \dfrac{H}{LR}\right)s^2 + \left(\dfrac{H}{L^2R} - \dfrac{2H}{L^3}\right)s^3$

10.35 The β_2 row is: $\lfloor 0 \quad 1 \quad 2L \quad 3L^2 \quad -1/R_{s2} \quad -L/R_{s2} \quad 0 \quad 0 \rfloor$

10.36 Condensation must take place *after* $\{\mathbf{d}\}$ replaces $\{\mathbf{a}\}$.

10.38 (a) Element *geometry* changes, not just the directions of its nodal d.o.f.
(b) Make u' and w' linear in s, for example,

$$u' = \left(1 - \frac{s}{L}\right)(u_1 \cos\beta + w_1 \sin\beta) + \frac{s}{L}(\cdot\cdot\cdot)$$

10.39 Curvature $w,_{\theta\theta}/r^2$ is used in Section 10.9 but not in Section 10.11.

10.40 At $\eta = \sqrt{3}/3$ and at $\eta = -\sqrt{3}/3$.

10.42 Rigid-body axial motion will produce strain.

10.44 A thin-plate disc model could be connected to a solid-of-revolution shaft model by transformation equations.

10.45, 10.46 See Section 6.6.

CHAPTER 11

11.1 $\omega_2/\omega_1 = 0.316$.

11.3 A damper might connect nonadjacent nodes.

11.4 (a) No. (b) Consistent: otherwise, effectively $[\mathbf{m}] = 0$.
(c) Consider a one-element plate.

11.5 (d) Only that in Eq. 11.2.3 dV becomes $A\,dx$, where A depends on x.
(f) Note that kinetic energy is independent of our reference frame.

11.6 (a) Row 1 is $(\rho tA/12)\,[2\ 1\ 1\ 0\ 0\ 0]$.
(b) There are only the three rigid-body motions w, $w_{,x}$, and $w_{,y}$.

11.7 $m_{22} = m_{44} = -mL^2/12$, where m = total element mass. Fails for a simply supported beam. Moral: keep m_{ii}'s positive.

11.8 (b)
$$[\mathbf{m}] = \frac{\rho AL}{120}\begin{bmatrix} 40 & 5L & 20 & -5L \\ 5L & L^2 & 5L & -L^2 \\ 20 & 5L & 40 & -5L \\ -5L & -L^2 & -5L & L^2 \end{bmatrix}$$

11.9 (b) $[\mathbf{m}] = (\rho tA/3)\,[\mathbf{I}]$. (c) $[\mathbf{m}] = (\rho tA/4)\,[\mathbf{I}]$.
(d) $[\mathbf{m}] = (m/80)\,[40\ L^2\ 40\ L^2]$, where $m = \rho AL$.
(e) When there are rotational d.o.f. or when not all translational m_{ii} are equal.

11.10 Eq. 11.4.4 yields $\lambda_1 = 1.023$ and $\lambda_2 = 5.971$.

11.11 $\omega = 1$ for each d.o.f. so treated. To find eigenvectors, set one nonfixed d.o.f. to unity and all fixed d.o.f. to zero.

11.12 (c) $\omega_1{}^2 = k/2m$, $\omega_2{}^2 = \infty$.

11.14 (a) $\omega_1{}^2 = 3AE/mL$ and $\omega_1{}^2 = 2.597AE/mL$.
(b) $\omega_1{}^2 = 2AE/mL$ and $\omega_1{}^2 = 2.343AE/mL$.
(c) $\omega_1{}^2 = 2.4AE/mL$ and $\omega_1{}^2 = 2.463AE/mL$.

11.15 In (a) and (b), the highest complete polynomial is of degree one, so error should be quartered.

11.16 (a) Impossible: $[\mathbf{m}]$ becomes null.
(b) $\omega_1 = 2.739\sqrt{EI/\rho AL^4}$, $\omega_2 = 12.55\sqrt{EI/\rho AL^4}$.
(c) Impossible: $[\mathbf{m}]$ becomes null.
(d) $\omega_1 = 3.122\sqrt{EI/\rho AL^4}$, $\omega_2 = 5.408\sqrt{EI/\rho AL^4}$.
(e) $\omega_1 = 2.739\sqrt{EI/\rho AL^4}$, ω_2 cannot be found.

11.17 Symmetry excludes the second mode (ω_2).
(a) $\omega_1 = 3\sqrt{EI/\rho AL^4}$, ω_3 cannot be found.
(b) $\omega_1 = 2.477\sqrt{EI/\rho AL^4}$, $\omega_3 = 27.53\sqrt{EI/\rho AL^4}$.
(c) $\omega_1 = 2.449\sqrt{EI/\rho AL^4}$, ω_3 cannot be found.
(d) $\omega_1 = 2.381\sqrt{EI/\rho AL^4}$, $\omega_3 = 18.18\sqrt{EI/\rho AL^4}$.
(e) $\omega_1 = 2.496\sqrt{EI/\rho AL^4}$, $\omega_3 = 43.01\sqrt{EI/\rho AL^4}$.

11.18 $[\mathbf{T}]$ then depends on ω. An iterative solution is required.

11.19 (b) Partition $[\mathbf{K}]$ and $[\mathbf{F}]$ and work with $[\mathbf{K}][\mathbf{F}] = [\mathbf{I}]$.
(c) Column j of $[\mathbf{F}_{mm}]$ gives the deflected shape under a unit load on the jth master d.o.f.
(d) Usually $m << s$, so $[\mathbf{F}_{mm}]$ is smaller than $[\mathbf{K}_{ss}]$.

11.20 (a) See Eq. 7.2.3.
(b) $\mathbf{M}_r = \mathbf{M}_{mm} - \mathbf{M}_{ms}\,\mathbf{K}_{ss}^{-1}\mathbf{K}_{ms}^T - \mathbf{K}_{ms}\,\mathbf{K}_{ss}^{-1}\,\mathbf{M}_{ms}^T + \mathbf{K}_{ms}\,\mathbf{K}_{ss}^{-1}\,\mathbf{M}_{ss}\,\mathbf{K}_{ss}^{-1}\,\mathbf{K}_{ms}^T$.

11.21 (a) Yes: $M_{11}/K_{11} = 0.031mL^3/EI$, $M_{22}/K_{22} = 0.0024mL^3/EI$.
(b) $\omega^2 = 7.50EI/mL^3$.
11.22 (c) $\omega_1 = 0.2911\sqrt{k/m}$ in (a), $\omega_1 = 0.2675\sqrt{k/m}$ in (b).
11.23 (a) No change.
(b) $\omega^2 = 6.1362EI/mL^3$.
(c) $\omega^2 = 6.2050EI/mL^3$.
11.24 $\omega^2 = 0.200$. Improved $\omega^2 = 0.191011$. Exact $\omega^2 = 0.190983$.
11.25 (a) $\lambda_1 = 0.382\ (k/m)$, $\lambda_2 = 2.618\ (k/m)$
(b) $\lambda_1 = 0.375\ (k/m)$. (c) $\lambda_2 = 2.615$.
11.26 One negative diagonal, then two.
11.27 (b) Find $P_i = \{\bar{\mathbf{D}}_i\}^T [\mathbf{M}] \{\bar{\mathbf{D}}_i\}$, then form $\{\bar{\mathbf{D}}_i\}/\sqrt{P_i}$.
11.28 (b) The modes are $\{\bar{\mathbf{d}}_1\} = \{1.000 \quad 1.618\}$ and $\{\bar{\mathbf{d}}_2\} = \{1.000 \quad -0.618\}$.
11.29 $\omega^2 = 0.382$.
11.30 Probably not. *Decoupled* equations are not expensive to integrate by an implicit method.
11.33 At $t = 1$ and $t = 2$, respectively, u_1 is:
(a) 0.143, 0.709. (b) 0, 1. (c) 0.125, 0.750.
(e) 0.138, 0.571. The accelerations agree.

CHAPTER 12

12.1 Force-free rigid-body motion is not possible.
12.2 (e) $(k_\sigma)_{11} = (k_\sigma)_{22} = -(k_\sigma)_{12} = -(k_\sigma)_{21} = P/L$.
12.3 Write $[\mathbf{T}]^T[\mathbf{k}_\sigma][\mathbf{T}]$, where $[\mathbf{T}]$ expresses $\theta_1 = \theta_2 = (w_2 - w_1)/L$.
12.4 (a) Nonzero coefficients are in positions (2, 2), (3, 3), (5, 5), (6, 6)—positive P/L; and (2, 5), (3, 6), (5, 2), (6, 3)—negative P/L.
12.5 In Eq. 4.4.5, EI becomes a function of x. There is no change in Eq. 12.2.12.
12.6

$$[\mathbf{k}_\sigma] = \frac{P}{12L}\begin{bmatrix} 12 & 0 & -12 & 0 \\ 0 & L^2 & 0 & -L^2 \\ -12 & 0 & 12 & 0 \\ 0 & -L^2 & 0 & L^2 \end{bmatrix}$$

12.7 $P_{cr} = -2.505EI/L^2$. Exact $P_{cr} = -2.467EI/L^2$.
12.8 (a) No solution. (b) $P_{cr} = -12EI/L^2$.
(c) $P_{cr} = -5EI/L^2$. (d) $P_{cr} = -4.8EI/L^2$.
(e) No solution. (f) $P_{cr} = -12EI/L^2$.
12.9 (a) $P_{cr} = -12EI/L^2$. (b) $P_{cr} = -9.944EI/L^2$.
(c) $P_{cr} = -4.126EI/L^2$. (d) $P_{cr} = -4.155EI/L^2$.
(e) $P_{cr} = -30EI/L^2$. (f) $P_{cr} = -10EI/L^2$.
12.10 (a) $P_{cr} = -12EI/L^2$. (b) $P_{cr} = -10.018EI/L^2$.
(c) $P_{cr} = -4.149EI/L^2$. (d) $P_{cr} = -4.251EI/L^2$.
(e) $P_{cr} = -48EI/L^2$. (f) $P_{cr} = -12EI/L^2$.
12.11 $P_{cr} = -2EI/L^2$. This value is exact.
12.12 (a) $8.182(10)^{-3}$ m. (b) $643.5(10)^{-3}$ m. (c) $4.138(10)^{-3}$ m.
12.13 (a) Those with a large magnitude of $K_{\sigma ii}/K_{ii}$.
(b) $w_2 = 450.0(10)^{-3}$ m. (c) $w_2 = 4.128(10)^{-3}$ m.
(d) A Gauss elimination solution (exact, not approximate).

12.14 (a) $-12EI/L^2$. (b) $-10EI/L^2$. (c) $-10.105EI/L^2$.
12.15 (a) 293 rad/s. (b) +1457 N. (c) 239 rad/s.
12.16 For small deflections, $w_L = PL/2W$ and $w_{2L} = 3PL/2W$.
12.17 (a) $w = mgL/2T$. (b) $\omega^2 = 2T/mL$.
12.18 $\omega_1{}^2 = T/mL$, $\omega_2{}^2 = 3T/mL$.
12.19 $\omega^2 = 10T/\rho L^2$.
12.20 Pseudoloads are nodal forces normal to the column.
12.21 Accuracy declines. An analysis based on the differential equation of a beam-column is appropriate.
12.22 (a) $P_{cr} = 9000\ N$ (compression).
(b) $\theta_2 = eP/(1200 - 0.1333P)$.
12.23 No. We need the displacement-dependent matrices of Chapter 13.
12.24 [**G**] is 9 by 24. It involves, for example, $N_{1,x} = \Gamma_{11}N_{1,\xi} + \Gamma_{12}N_{1,\eta} + \Gamma_{13}N_{1,\zeta}$.
12.25 Follow the procedure of Eqs. 12.3.12 to 12.3.16.

CHAPTER 13

13.3 See Ref. 7.8 or 16.13.
13.5 No. We are assuming that the deformed shape is real for *some* load, then computing that load. But the shape is not reliable, since it is found by use of an approximate [**K**].
13.8 Modified: the same [**K**] applies to all load vectors.
13.9 Sketches show slightly improved accuracy.
13.11 (a) Yes: multiply out. (c) Yes.
13.13 (b) The solution dances about point *A* until it happens to strike off to the right with a small enough slope that it intersects the $P = 250$ line to the right of point *C*.
13.15 (a) $\Pi_p = 32kD^2 - 8kD^3 + (kD^4/2) + 8WD - WD^2 - PD$.
(b) $(64k - 2W)D = P + (-8W + 24kD^2 - 2kD^3)$.
(c) $[(64k - 2W) + (-48kD + 6kD^2)]\ \Delta D = \Delta P$.
(d) $D = 4 \pm \sqrt{(16k + W)/3k}$.
(e) See Ref. 13.16.
13.17 (a) 3.996. (b) 3.997. (c) 3.999. (d) 2.403.
(e) 3.557. (f) 4.832. (g) 2.689. (i) 1.986.
(j) Eq. 13.6.1.
13.18 (a) 4.448. (b) 1.000. (c) 8.400. (d) 7.495. (e) 0.500.
(f) 7.596. (g) 8.509. (h) 8.265. (i) 8.294.
(j) −2040. (k) 7.500. (l) −42000. (m) 6.005.
(n) 5.810.
13.19 (a) $D_1 = 8, D_2 = 9$. (b) $D_1 = 8, D_2 = 8.5$.
13.20 $K_T = (1 - \beta)K + \beta(K + k) = K + \beta k$.
13.21 (a) Third iteration: $u_3 = 1.375$ (exact $u = 1.333$).
(b) $u_1 = 3$, $u_2 = -1$, $u_3 = 3$ (exact $u = 1.667$).
(c) Third iteration: $u_3 = 2.076$ (exact $u = 1.667$).
(d) Third iteration: $u_3 = 1.648$ (exact $u = 1.667$).
13.22 (a) [**T**] expresses $u_A = u_1 + u_{4r}$, $u_B = u_2 + u_{3r}$, $u_C = u_C$, and so on.
(b) Yes.
(c) $u = u_1 + (u_2 - u_1)x/L + y(1 - x/L)u_{4r}/h + xyu_{3r}/hL$.

13.23 $E^* = E\left[1 + \dfrac{w^2a^2AE}{12T^3}\right]^{-1}$

13.25 (f) Combine Eqs. 13.9.9 and 13.11 and note that $\{d\epsilon_e\}$ becomes negligible.

13.28 $\bar{\nu} = \dfrac{1}{2} - \dfrac{E_s}{E}\left(\dfrac{1}{2} - \nu\right)$

13.30 No good: each load level constitutes a problem *independent* of the others.
13.31 E_s differs between sampling points. It should not be factored out.
13.32 (a) Third iteration: $D = 0.0193$ (exact $D = 0.020$).
(b) Third iteration: $D = 0.01875$ (exact $D = 0.020$).
13.33 Third iteration: $D = 0.01350$ (exact $D = 0.014$).
13.34 (a) $D_2 = 11.98$ (exact $D = 20$).
(b) $D_2 = 11.90$ (exact $D = 20$).
13.36 $v_1 = 4, P_1 = 6.$ $v_2 = 8, P_2 = 10.$ $v_3 = 12, P_3 = 12.$
13.41 (b) The equation states that the material is incompressible.
13.46 There are only as many steps as plastic hinges.
13.47 Hinges are at $P = 5$, $v_A = -1$ and at $P = 7$, $v_A = -2$.
13.48 (a) Methods where [**K**] is not changed and nonlinearities appear on the right side (for example, Eqs. 13.6.2, 13.7.5, 13.10.8, and 13.11.9).
(b) $D_1 = 0.008$, $D_2 = 0.012$, $D_3 = 0.014$ (exact $D = 0.016$).
13.50 $D_1 = -1.500$, $D_2 = -1.202$ (exact $D = -1.202$).

CHAPTER 15

15.2 Displacement, stress, and strain energy errors have the following orders: (a, c) h^2, h^1, h^2. (b, d) h^3, h^2, h^4.
15.3 The curve seems to be a parabola. Exact value ≈ 12.0.
15.4 If L = overall dimension, note that $(L/h)^n = N$.
15.5 Partial answer: sensitivity decreases as α increases.
15.6 $u_1 = 11P$, $u_2 = 10P$ (exact: $u_1 = 17.436P$, $u_2 = 16.667P$).
15.7 $u_1 = 18P$, $u_2 = 0.77P$ (exact: $u_1 = 17.436P$, $u_2 = 0.769P$).
15.8 (a, b) 1.0. (c) There is a rigid-body or other zero-energy mode.
15.9 (a) 42.1, 12.3. (b) 41.9, 1.86.
15.11 (a) 13.93. (c) $C(\mathbf{K}) \approx 2.21/0.00015 = 14{,}700$.
15.13 Consider sequencing and conditioning.
15.14 (a) $D_1 = 7, D_2 = 6$. (c) $D_1 = 5, D_2 = 6$.
(b) $D_1 = 7, D_2 = 1$. (d) $D_1 = 0.6, D_2 = 6$.
15.15 (a) $D_1 = 10, D_2 = 8.9$. (c) $D_1 = 7.7, D_2 = 8.5$.
(b) $D_1 = 7.2, D_2 = 1.2$. (d) $D_1 = 1.0, D_2 = 7.0$.
15.16 (a) $C(\mathbf{K}) = \infty$. If $\{\mathbf{D}\} = \{\ldots u_n\ v_n\ \theta_n\}$, trouble appears in the u_n equation.
(b) $C(\mathbf{K}) = \infty$. If $\{\mathbf{D}\} = \{\ldots u_n\ w_n\}$, trouble appears in the last (w_n) equation.
15.19 Exact: $u_1 = 7$, $u_2 = 4$. After the third cycle:
(a) $u_1 = 8.333$, $u_2 = 5.000$ (converges to wrong answer).
(b) $u_1 = 6.984$, $u_2 = 3.984$ (converges to correct answer).

15.20 Respectively: reduction of discretization error, development of numerical trouble, original answers exact.

15.21 Bad data. Zero-energy modes in your case. A different computer or compiler. An unrecorded program modification. Element not suited to your test case. Numerical troubles. Orientation-dependent element. Program error. Misunderstood or erroneous user's manual.

CHAPTER 16

16.1 (a) There is no resistance to bending.
(b) $\sigma_x = \beta_1 + \beta_4 x + \beta_7 y, \quad \sigma_y = \beta_2 + \beta_5 x + \beta_8 y, \quad \tau_{xy} = \beta_3 - \beta_8 x - \beta_4 y$.

16.3 No: here *all* displacements are quadratic.

16.4 Both sets contribute to the K_{ii}. The upper set also contributes to the off-diagonals (K_{ij}).

16.5 $(k_f)_{11} = (k_f)_{22} = bL\beta/3$, $(k_f)_{12} = (k_f)_{21} = bL\beta/6$.

16.6 The w field must include the effect of θ_1 and θ_2.

16.7 $(k_f)_{ii} = \beta A/6$, $(k_f)_{ij} = \beta A/12$, where $i, j = 1, 2, 3$.
Yes: converges with mesh refinement.

16.8 If $\{w \ w_{,x} \ w_{,y}\} = [\mathbf{Q}]\{\mathbf{d}\}$, the integrand of Eq. 16.3.2 becomes $[\mathbf{Q}]^T \lceil \beta \ \alpha \ \alpha \rfloor [\mathbf{Q}] \, dx \, dy$.

16.10 (a) Yes: scale the k_{fii} to give the proper force under constant w.
(b) No: displacements of elements cannot be independent.

16.12 The Winkler model.

16.14 Compared with gravity switch-on, displacements increase by the factor 19.1 and stress σ by the factor 6.75.

16.15 (b, c) The converged (and exact) result is $D^* = 2.50$. (d) By $\pm$ 100%.

16.16 (a) Volumetric strain ($\epsilon_v = \sigma_{ave}/B$) approaches zero.
(b) Use a single Gauss point to integrate Eq. 16.6.3.

16.17 (b) A mode that makes the fluid slosh has lower frequency than a mode that requires compressing the fluid.
(c) $u_B, u_C, u_F, u_G, v_F, v_G, \theta_F, \theta_G$.
(d) The cubic beam v does not match the linear edge v of the quadrilaterals. Convergence with mesh refinement.

16.18 (a) For d.o.f. w_1 and w_2, $k_{s11} = k_{s22} = 2k_{s12} = 2k_{s21} = \rho gtb/3$.
(b) $\omega^2 = 2g/3h$ (exact $\omega^2 = 3.13g/h$ if $h = b$).
(c) Add θ d.o.f. and represent w as cubic in x.
(d) $\omega^2 = 3g/h$ (exact $\omega^2 = 3.13g/h$ if $h = b$).

16.19 No: shear stress depends on the *rate* of strain.

16.20 (a) Use $(\epsilon_x + \epsilon_y + \epsilon_z) = \lfloor \mathbf{B}^+ \rfloor \{\mathbf{d}\}$.
(b) Rank $= 8 - 5 = 3$. The rigid-body, pinching, and shear modes (Fig. 15.2.1) are zero energy.

16.21 $\lfloor \bar{\mathbf{N}} \rfloor = \lfloor -(1-\eta)\ell \quad -(1-\eta)m \quad -\ell\eta \quad -m\eta \rfloor$.

16.22 Impose $w = 1$ where the reaction is desired. The deflections define the influence surface.

16.23 (d) Residual stresses.

CHAPTER 17

17.2 $(k_x T_{,x}\tau + k_{xy} T_{,y}\tau)_{,x} + (k_{xy} T_{,x}\tau + k_y T_{,y}\tau)_{,y} + Q\tau = c\rho T_{,t}\tau$.

17.5 The Euler equation is given by Problem 17.2.

17.8 Unless $k_{r\theta} = k_{\theta z} = 0$, the $\theta = 0$ plane will not be a plane of symmetry for both T and the "loads" that produce it, so decoupling of modes becomes impossible.

17.11 Element BCIZ of Section 9.2, with T in place of w. Nodal d.o.f. are T and its first two derivatives with respect to local directions along and normal to substructure boundaries.

17.12 T_B is (a) underestimated, (b) overestimated, and (c) well approximated. See remarks on implicit versus explicit methods in Sections 11.8 and 11.9.

17.13 (a) $T_5 = 0.250$. (b) $T_5 = 0.195$. (c) $T_5 = 0.208$.
(d) $T_\infty = 0.250$. (e) $\Delta t = 0.667$.

CHAPTER 18

18.1 (c) Collocation sets the *residual* to zero at certain points. It does not set $u = \tilde{u}$. There is no reason to expect agreement.

18.2 At $x = L/2$ in parts (a) through (e), with $\sigma_L = 0$ and $\rho\omega^2 = L = E = 1$, $\tilde{u}$ is 0.1875, 0.1875, 0.1875, 0.1250, 0.2500, and $\tilde{\sigma}_x$ is 0.2500, 0.2500, 0.2500, 0.1667, 0.3333. Part (f) is the same as Eq. 18.3.13. Part (g): $\bar{W}$ cancels out because a_1 does not appear in R_L.

18.3 At $x = L/2$ in parts (a) to (d), $\tilde{u}$ is 0.375, 0.183, 0.183, 0.250 (exact = 0.250), and $\tilde{\sigma}_x$ is 0.500, 0.250, 0.250, 0.375 (exact = 0.500).

18.4 (a) $R_L = G\left(\tilde{w},_{rr} + \dfrac{\tilde{w},_r}{r}\right)$

$$R_B = (\tilde{w},_r)_{r1} + \frac{P}{2\pi G L r_1}$$

(c) $\tilde{w} = a_1(r_2 - r) + a_2 r\,(r_2 - r)$

18.5 (a) $R_L = G\left(\tilde{v},_{rr} + \dfrac{\tilde{v},_r}{r} - \dfrac{\tilde{v}}{r^2}\right)$

$$R_B = G\left(\tilde{v},_r - \frac{\tilde{v}}{r}\right)_{r1} - \frac{T}{2\pi L r_1{}^2}$$

(c) $\tilde{v} = a_1(r_2 - r) + a_2 r(r_2 - r)$

18.7 Results agree with those in Section 4.3.

18.8 The two differential equations of equilibrium are

$$\frac{1}{r}(r\sigma_r),_r + \tau_{rz,z} - \frac{\sigma_\theta}{r} = \frac{1}{r}(r\tau_{rz}),_r + \sigma_{z,z} = 0.$$

Results agree with those in Section 8.2.

18.9 At $x = L/2$ in parts (a) through (d), $\tilde{u}$ is 0.125, 0.115, 0.100, 0.125, and $\tilde{\sigma}_x$ is 0.250, 0.231, 0.250, 0.250.

18.10 (a) See Eq. 2.4.5.
(b) Cubic, as in Eq. 4.2.5. There are four d.o.f., so at least four points are needed. If four, it is collocation; if more than four, it is least squares collocation.

18.13 (a) $M_1 = M_2$, $w_2 = M_2L^2/2EI$ (both are exact).
(b) $M_1 = PL$, $w_2 = PL^3/3EI$ (both are exact).
(c) $M_1 = qL^2/2$ (exact), $w_2 = qL^4/6EI$.
(d) $M_2 = 3EI\theta_2/L$ (exact).
(e) $w_1 = -2P/BL$, $w_2 = 4P/BL$ (both are exact).

(f) $F_{cr} = 3EI/L^2$ (exact $F_{cr} = \pi^2 EI/4L^2$).
(g) $F_{cr} = 0.648EI/L^2$ (exact $F_{cr} = 0.616EI/L^2$).
(h) $M_1 = 2M_2 = 2PL,\ w_2 = 5w_3/16 = 5PL^3/6EI$ (all are exact).

18.14 Element moments are in general unequal at the node. Equilibrium of moments at the node yields a constraint equation that relates them.

18.17 (a) $$[\mathbf{k}] = \frac{F}{L}\begin{bmatrix} 1 & -1 \\ -1 & 1 \end{bmatrix}, \qquad [\mathbf{k}_\sigma] = \frac{\rho L}{6}\begin{bmatrix} 2 & 1 \\ 1 & 2 \end{bmatrix}$$
(b) $\omega_0^2 = 3F/\rho L^2$ (exact $\omega_0^2 = \pi^2 F/4\rho L^2$).

18.18 $$[\mathbf{k}] = \iint (-\mathbf{N}^T_{,x}\mathbf{N}_{,x} - \mathbf{N}^T_{,y}\mathbf{N}_{,y} + A\mathbf{N}^T\mathbf{N}_{,x} + B\mathbf{N}^T\mathbf{N}_{,y})\, dx\, dy$$

18.19 (b) $$[\mathbf{k}] = \frac{AG}{L}\begin{bmatrix} 1 & -1 \\ -1 & 1 \end{bmatrix}, \qquad [\mathbf{k}_f] = BL\begin{bmatrix} 2 & 1 \\ 1 & 2 \end{bmatrix}$$
(c) $\Pi_w = \frac{1}{2}\int (AG\gamma^2 + Bw^2 - 2qw)\, dx - V_1 w_1 + V_2 w_2$.

18.20 Same answer as for Problem 18.10(b).

APPENDIX A

A.1 $\phi_4 = 3\phi_2 - 2\phi_3$.

Index

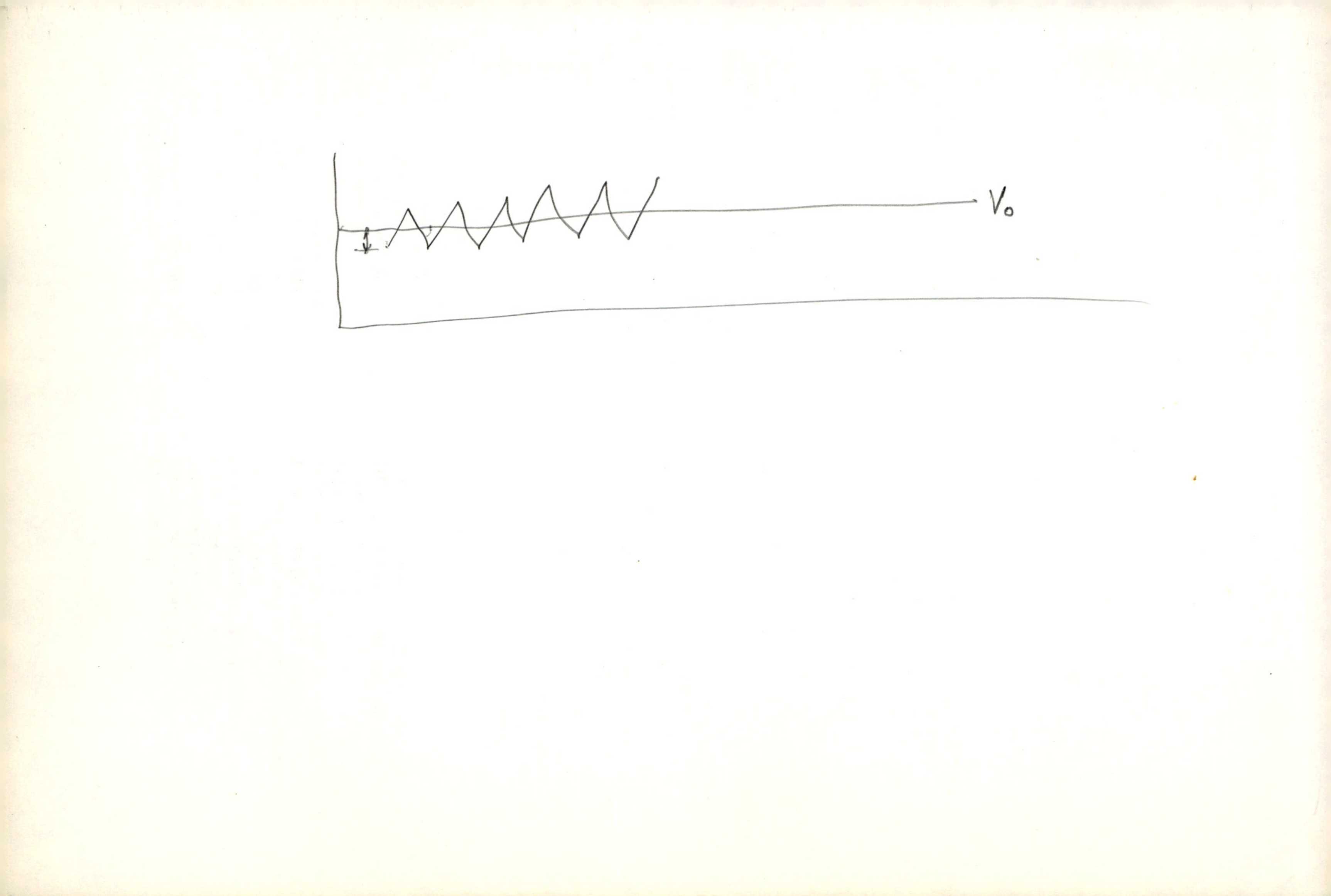

V_0

PROPERTIES OF LINES AND AREAS

Figure	Centroid	Area Moments of Inertia
Arc Segment	$\bar{r} = \dfrac{r \sin \alpha}{\alpha}$	—
Quarter and Semicircular Arcs	$\bar{y} = \dfrac{2r}{\pi}$	—
Triangular Area	$\bar{x} = \dfrac{a + b}{3}$ $\bar{y} = \dfrac{h}{3}$	$I_x = \dfrac{bh^3}{12}$ $\bar{I}_x = \dfrac{bh^3}{36}$ $I_{x_1} = \dfrac{bh^3}{4}$
Rectangular Area	—	$I_x = \dfrac{bh^3}{3}$ $\bar{I}_x = \dfrac{bh^3}{12}$ $\bar{J} = \dfrac{bh}{12}(b^2 + h^2)$
Area of Circular Sector	$\bar{x} = \dfrac{2}{3}\dfrac{r \sin \alpha}{\alpha}$	$I_x = \dfrac{r^4}{4}(\alpha - \tfrac{1}{2}\sin 2\alpha)$ $I_y = \dfrac{r^4}{4}(\alpha + \tfrac{1}{2}\sin 2\alpha)$ $J = \tfrac{1}{2}r^4\alpha$
Quarter Circular Area	$\bar{x} = \bar{y} = \dfrac{4r}{3\pi}$	$I_x = I_y = \dfrac{\pi r^4}{16}$ $J = \dfrac{\pi r^4}{8}$
Area of Elliptical Quadrant Area $A = \dfrac{\pi ab}{4}$	$\bar{x} = \dfrac{4a}{3\pi}$ $\bar{y} = \dfrac{4b}{3\pi}$	$I_x = \dfrac{\pi ab^3}{16}$ $I_y = \dfrac{\pi a^3 b}{16}$ $J = \dfrac{\pi ab}{16}(a^2 + b^2)$

Reprinted, by permission, from J. L. Meriam, *Statics,* 2nd ed., Table C4, page 369, Wiley, New York, 1966